Springer Proceedings in Mathematics & Statistics

Volume 34

For further volumes:
http://www.springer.com/series/10533

Springer Proceedings in Mathematics & Statistics

This book series features volumes composed of select contributions from workshops and conferences in all areas of current research in mathematics and statistics, including OR and optimization. In addition to an overall evaluation of the interest, scientific quality, and timeliness of each proposal at the hands of the publisher, individual contributions are all refereed to the high quality standards of leading journals in the field. Thus, this series provides the research community with well-edited, authoritative reports on developments in the most exciting areas of mathematical and statistical research today.

Frederi Viens • Jin Feng • Yaozhong Hu
Eulalia Nualart
Editors

Malliavin Calculus and Stochastic Analysis

A Festschrift in Honor of David Nualart

Springer

Editors

Frederi Viens
Department of Statistics
Purdue University
West Lafayette, IN, USA

Jin Feng
Department of Mathematics
University of Kansas
Lawrence, KS, USA

Yaozhong Hu
Department of Mathematics
University of Kansas
Lawrence, KS, USA

Eulalia Nualart
Department of Economics and Business
University Pompeu Fabra
Barcelona, Spain

ISSN 2194-1009 ISSN 2194-1017 (electronic)
ISBN 978-1-4899-9657-2 ISBN 978-1-4614-5906-4 (eBook)
DOI 10.1007/978-1-4614-5906-4
Springer New York Heidelberg Dordrecht London

Mathematics Subject Classification (2010): 60H07, 60H10, 60H15, 60G22, 60G15, 60H30, 62F12, 91G80

Printed on acid-free paper

Springer is part of Springer Science+Business Media (www.springer.com)

Preface

David Nualart was born in Barcelona on March 21, 1951. After high school he studied mathematics at the University of Barcelona, from which he obtained an undergraduate degree in 1972 and a PhD in 1975. He was a full professor at the University of Barcelona from 1984 to 2005. He moved to the University of Kansas in 2005, as a Professor in the Department of Mathematics, and was appointed Black-Babcock Distinguished Professor there in 2012.

David Nualart is among the world's most prolific authors in probability theory, with more than 200 research papers, many of which are considered pathbreaking, and several influential monographs and lecture notes. His most famous book is undoubtedly *Malliavin Calculus and Related Topics* (cited more than 530 times on MathSciNet), which has been serving as an ultimate reference on the topic since its publication. Its most recent edition contains two chapters which have become standard references in their own right, on state-of-the-art applications of the Malliavin calculus to quantitative finance and to fractional Brownian motion.

David Nualart has long influenced the general theory of stochastic analysis, including martingale theory, stochastic calculus of variations, stochastic equations, limit theorems, and mathematical finance. In the first part of his scientific life, he contributed to the development of a stochastic calculus for two-parameter martingales, setting the basis of stochastic integration in this context. Subsequently, one of his major achievements in probability theory has been his ability to develop and apply Malliavin calculus techniques to a wide range of concrete, interesting, and intricate situations. For instance, he is at the inception and is recognized as the leader in anticipating stochastic calculus, a genuine extension of the classical Itô calculus to non-adapted integrands. His other contributions to stochastic analysis include results related to integration-by-parts formulas, divergence and pathwise integrals, regularity of the laws of random variables through Malliavin calculus, and the study of various types of stochastic (partial) differential equations.

In the last decades, his research focused largely on the stochastic calculus with respect to Gaussian processes, especially fractional Brownian motion, to which he has become the main contributor. David Nualart's most recent work also includes important results on limit theorems in terms of Malliavin calculus.

David Nualart's prominent role in the stochastic analysis community and the larger mathematics profession is obvious by many other metrics, including membership in the Royal Academy of Exact Physical and Natural Sciences of Madrid since 2003, an invited lecture at the 2006 International Congress of Mathematicians, continuous and vigorous service as editor or associate editor for all the main journals in probability theory, and above all, the great number of Ph.D. students, postdoctoral scholars, and collaborators he has trained and worked with around the world. By being an open-minded, kind, generous, and enthusiastic colleague, mentor, and person, he has fostered a good atmosphere in stochastic analysis. All those working in this area have cause to be grateful and to celebrate the career of David Nualart.

In this context, the book you hold in your hands presents 25 research articles on various topics in stochastic analysis and Malliavin calculus in which David Nualart's influence is evident, as a tribute to his lasting impact in these fields of mathematics. Each article went through a rigorous peer-review process, led by this volume's four editors Jin Feng (Kansas), Yaozhong Hu (Kansas), Eulàlia Nualart (Pompeu Fabra, Barcelona), and Frederi Viens (Purdue) and six associate members of this volume's Editorial Board, Laure Coutin (Toulouse), Ivan Nourdin (Nancy), Giovanni Peccati (Luxembourg), Lluís Quer-Sardanyons (Autònoma, Barcelona), Samy Tindel (Nancy), and Ciprian Tudor (Lille), with the invaluable assistance of many anonymous referees.

The articles' authors represent some of the top researchers in these fields, all of whom are recognized internationally for their contributions to date; many of them were also able to participate in a conference in honor of David Nualart held at the University of Kansas on March 19–21, 2011, on Malliavin calculus and stochastic analysis, with major support from the US National Science Foundation, with additional support from the Department of Mathematics and the College of Liberal Arts and Sciences at the University of Kansas, the Department of Mathematics and the Department of Statistics at Purdue University, and the French National Agency for Research.

As the title of this volume indicates and the topics of many of the articles within emphasize, this Festschrift also serves as a tribute to the memory of Paul Malliavin and his extraordinary influence on probability and stochastic analysis, through the inception and subsequent constant development of the stochastic calculus of variations, known today as the Malliavin calculus. Professor Malliavin passed away in June 2010. He is dearly missed by many as a mathematician, colleague, mentor, and friend. Dan Stroock initially coined the term "Malliavin calculus" around 1980 to describe the stochastic calculus of variations developed by Paul Malliavin, which employs the Malliavin derivative operator. The term has been broadened to describe any mathematical activity using this derivative and related operators on standard or abstract Wiener space as well as, to some extent, calculus based on Wiener chaos expansions. We consider the Malliavin calculus in this broadest sense.

The term "stochastic analysis" originated in its use as the title of the 1978 conference volume edited by Avner Friedman and Mark Pinsky. It described results on finite- and infinite-dimensional stochastic processes that employ probabilistic tools as well as tools from classical and functional analysis. We understand stochastic analysis as being broadly rooted and applied this way in probability theory and stochastic processes, rather than a term to describe solely analysis results with a probabilistic flavor or origin.

The topics in this volume are divided by theme into five parts, presented from the more theoretical to the more applied. While these divisions are not fundamental in nature and can be interpreted loosely, they crystallize some of the most active areas in stochastic analysis today and should be helpful for readers to grasp the motivations of some of the top researchers in the field.

- Part I covers Malliavin calculus and Wiener space theory, with topics which advance the basic understanding of these tools and structures; these topics are then used as tools throughout the rest of the volume.
- Part II develops the analysis of stochastic differential systems.
- Part III furthers this development by focusing on stochastic partial differential equations and some of their fine properties.
- Part IV also deals largely with stochastic equations and now puts the emphasis on noise terms with long-range dependence, particularly using fractional Brownian motion as a building block.
- Part V closes the volume with articles whose motivations are solving specific applied problems using tools of Malliavin calculus and stochastic analysis.

A number of stochastic analysis methods cut across all of the five parts listed above. Some of these tools include:

- Analysis on Wiener space
- Regularity and estimation of probability laws
- Malliavin calculus in connection to Stein's method
- Variations and limit theorems
- Statistical estimators
- Financial mathematics

As the readers will find out by perusing this volume, stochastic analysis can be interpreted within several distinct fields of mathematics and has found many applications, some reaching far beyond the core mathematical discipline. Many researchers working in probability, often using tools of functional analysis, are still heavily involved in discovering and developing new ways of using the Malliavin calculus, making it one of the most active areas of stochastic analysis today and for some time to come. We hope this Festschrift will serve to encourage researchers to consider the Malliavin calculus and stochastic analysis as sources of new techniques that can advance their research.

The four editors of this Festschrift are indebted to the members of the Editorial Board, Laure Coutin, Ivan Nourdin, Giovanni Peccati, Lluís Quer-Sardanyons, Samy Tindel, and Ciprian Tudor, for their tireless work in selecting and editing the articles herein, to the many anonymous referees for volunteering their time to discern and help enforce the highest quality standards, and above all to David Nualart, for inspiring all of us to develop our work in stochastic analysis and the Malliavin calculus.

Thank you, David.

Lawrence, Kansas, USA — Jin Feng and Yaozhong Hu
Barcelona, Spain — Eulalia Nualart
West Lafayette, Indiana, USA — Frederi Viens

Contents

Part I
Malliavin Calculus and Wiener Space Theory

Chapter 1
An Application of Gaussian Measures to Functional Analysis

Daniel W. Stroock

Abstract In a variety of settings, it is shown that all Borel measurable, linear maps from one locally convex topological vector space to another must be continuous. When the image space is Polish, this gives a proof of L. Schwartz's Borel graph theorem. The proof is based on a simple probabilistic argument and, except for the application to Schwartz's theorem, avoids the descriptive set theory used in previous treatments of such results.

Received 6/14/2011; Accepted 11/22/2011; Final 11/22/2011

1 Introduction

This article is an expanded version of my note [6] dealing with Laurent Schwartz's Borel graph theorem. Schwartz's theorem [5] shows that, under appropriate technical conditions, the classical closed graph theorem can be improved to the statement that a linear map between topological vector spaces is continuous if its graph is Borel measurable. That is, the condition in the classical statement that the graph be closed can be replaced by the condition that it be Borel measurable. Schwartz's proof, as well as A. Martineu's (cf. [3] and the appendix in [9]) simplification of the original argument, is a *tour de force* in the use of descriptive set theory. In this paper, it is shown (cf. Corollaries 2.1 and 3.3) that, in a wide variety of circumstances, the same conclusion can be reached as an application of relatively elementary probabilistic ideas.

D.W. Stroock (✉)
M.I.T, 2-272, Cambridge, MA 02139, USA
e-mail: dws@math.mit.edu

F. Viens et al. (eds.), *Malliavin Calculus and Stochastic Analysis: A Festschrift in Honor of David Nualart*, Springer Proceedings in Mathematics & Statistics 34,
DOI 10.1007/978-1-4614-5906-4_1,

My formulation is somewhat different from Schwartz's. Instead of assuming that its graph is Borel measurable, I assume that the map itself is Borel measurable. When the spaces are Polish (i.e., complete, separable metric spaces), there is no difference between these two hypotheses. Indeed, if E and F are any pair of Polish spaces and $\Phi : E \longrightarrow F$, then the graph $G(\Phi)$ is Borel measurable if and only if Φ is Borel measurable. To check this, recall the fact (cf. [2] or [4]) that a one-to-one, Borel measurable map from a Borel measurable subset of one Polish space into a second Polish space takes Borel measurable sets to Borel measurable sets. Applying this to the map $x \in E \longmapsto \big(x, \Phi(x)\big) \in E \times F$, one sees that $G(\Phi)$ is a Borel measurable subset of $E \times F$ if Φ is Borel measurable. Conversely, if $G(\Phi)$ is Borel measurable and π_E and π_F are the natural projection maps from $E \times F$ onto E and F, respectively, then, because π_E and π_F are continuous and $\pi_E \restriction G(\Phi)$ is one-to-one, $\Phi = \pi_F \circ \big(\pi_E \restriction G(\Phi)\big)^{-1}$ is Borel measurable. Further, as shown in Corollaries 2.1 and 3.3, under reasonable conditions, it is possible to reduce some non-separable situations to separable ones.

In order to simplify the presentation of the basic ideas, in Sect. 2 I restrict my attention to Banach spaces, where my proof provides an independent proof of the classical closed graph theorem in the case when the image space is separable. In Sect. 3, I extend the result to more general settings, although my proof there relies on the closed graph theorem.

2 Banach Space Setting

Throughout, for a topological space S, $\mathcal{B}_S$ will denote the Borel σ-algebra over S, and, in this section, E and F will be Banach spaces over $\mathbb{R}$ with norms $\|\cdot\|_E$ and $\|\cdot\|_F$.

Set $\Omega = \mathbb{R}^{\mathbb{Z}^+}$, give Ω the product topology, and set $\mathbb{P} = \gamma_{0,1}^{\mathbb{Z}^+}$ on $(\Omega, \mathcal{B}_\Omega)$, where $\gamma_{0,1}(d\xi) = (2\pi)^{-\frac{1}{2}} e^{-\frac{\xi^2}{2}}\, d\xi$ is the standard Gauss distribution on $\mathbb{R}$. Given a sequence $\{x_n : n \geq 1\} \subseteq E$ set

$$S_n(\omega) = \sum_{m=1}^{n} \omega_m x_m \quad \text{for } n \in \mathbb{Z}^+ \text{ and } \omega \in \Omega,$$

$$A \equiv \Big\{\omega : \lim_{n\to\infty} S_n(\omega) \text{ exists in } E\Big\}$$

and

$$S(\omega) = \begin{cases} \lim_{n\to\infty} S_n(\omega) & \text{if } \omega \in A \\ 0 & \text{if } \omega \notin A. \end{cases}$$

Since

$$\mathbb{E}^{\mathbb{P}}\left[\sum_{m=1}^{\infty} |\omega_m| \|x_m\|_E\right] = \sqrt{\frac{2}{\pi}} \sum_{m=1}^{\infty} \|x_m\|_E,$$

$\mathbb{P}(A) = 1$ if $\sum_{m=1}^{\infty} \|x_m\|_E < \infty$.

The following is a minor variation on the renowned theorem of X. Fernique (cf. [1] or Theorem 8.2.1 in [7]).

Theorem 2.1. *Let $\{x_n : n \geq 1\} \subseteq E$, assume that $\sum_{n=1}^{\infty} \|x_n\|_E < \infty$, and define $\omega \rightsquigarrow S(\omega)$ accordingly. If $\Phi : E \longrightarrow F$ is a Borel measurable, linear map, then*

$$\|\Phi(x_m)\|_F^2 \leq \mathbb{E}^{\mathbb{P}}\big[(\Phi \circ S)^2\big] < \infty \quad \textit{for all } m \in \mathbb{Z}^+.$$

Proof. Since, for every $\omega \in \Omega$, $S(\omega)$ is an element of the closed linear span of $\{x_n : n \geq 1\}$, I will, without loss in generality, assume that E is separable and therefore that $\mathcal{B}_{E^2} = \mathcal{B}_E \times \mathcal{B}_E$. In particular, this means that the maps $(x, y) \in E^2 \longmapsto \frac{x \pm y}{\sqrt{2}} \in E$ are $\mathcal{B}_E \times \mathcal{B}_E$-measurable.

Next note that

$$\frac{\Phi \circ S(\omega^1) \pm \Phi \circ S(\omega^2)}{\sqrt{2}} = \Phi \circ S\left(\frac{\omega^1 \pm \omega^2}{\sqrt{2}}\right) \quad \text{for } (\omega^1, \omega^2) \in A^2.$$

Thus, since

$$\left(\frac{\omega^1 + \omega^2}{\sqrt{2}}, \frac{\omega^1 - \omega^2}{\sqrt{2}}\right)$$

has the same distribution under $\mathbb{P}^2$ as (ω^1, ω^2), we know that

$$\left(\frac{\Phi \circ S(\omega^1) + \Phi \circ S(\omega^2)}{\sqrt{2}}, \frac{\Phi \circ S(\omega^1) - \Phi \circ S(\omega^2)}{\sqrt{2}}\right)$$

has the same $\mathbb{P}^2$-distribution as $\big(\Phi \circ S(\omega^1), \Phi \circ S(\omega^2)\big)$.

Starting from the preceding, precisely the same argument as the one introduced by Fernique shows that $\mathbb{E}^{\mathbb{P}}\big[e^{\alpha\|\Phi \circ S\|_F^2}\big] < \infty$ for some $\alpha > 0$, which certainly means that $\mathbb{E}^{\mathbb{P}}\big[\|\Phi \circ S\|_F^2\big] < \infty$.

To complete the proof, let $m \in \mathbb{Z}^+$ be given and define $\omega \rightsquigarrow S^{(m)}(\omega)$ relative to the sequence $\{(1 - \delta_{m,n})x_n : n \geq 1\}$. Then $S^{(m)}(\omega)$ is $\mathbb{P}$-independent of ω_m, and $S(\omega) = \omega_m x_m + S^{(m)}(\omega)$ for $\omega \in A$. Hence, if $y^* \in F^*$ with $\|y^*\|_{F^*} \leq 1$, then

$$\mathbb{E}^{\mathbb{P}}\big[\|\Phi \circ S\|_F^2\big] \geq \mathbb{E}^{\mathbb{P}}\big[\langle \Phi \circ S, y^* \rangle^2\big] = \langle x_m, y^* \rangle^2 + \mathbb{E}^{\mathbb{P}}\big[\langle \Phi \circ S^{(m)}, y^* \rangle^2\big] \geq \langle x_m, y^* \rangle^2,$$

and so $\|\Phi(x_m)\|_F^2 \leq \mathbb{E}^P\big[\|\Phi \circ S\|_F^2\big]$. □

Corollary 2.1. *Let $\Phi : E \longrightarrow F$ be a linear map. If Φ is Borel measurable or if F is separable and the graph $G(\Phi)$ of Φ is Borel measurable, then Φ is continuous.*

Proof. Assume that Φ is Borel measurable and suppose that Φ were not continuous. Then we could find $\{x_n : n \geq 1\} \subseteq E$ such that $\|x_n\|_E \leq n^{-2}$ and $\|\Phi(x_n)\|_F \geq n$. But then, if $S(\omega)$ is defined relative to $\{x_n : n \geq 1\}$, we would have the contradiction that

$$m^2 \leq \|\Phi(x_m)\|_F^2 \leq \mathbb{E}^{\mathbb{P}}\big[\|\Phi \circ S\|_F^2\big] < \infty \quad \text{for all } m \in \mathbb{Z}^+.$$

Now assume that F is separable and that $G(\Phi)$ is Borel measurable. If E is separable as well, then, by the comments in the introduction, Φ is Borel measurable and therefore, by the preceding, continuous. To handle general E's, suppose that $\{x_n : n \geq 1\} \subseteq E$ and that $x_n \longrightarrow x$ in E. To see that $\Phi(x_n) \longrightarrow \Phi(x)$ in F, take E' to be the closed linear span of $\{x_n : n \geq 1\}$ in E. Then E' is separable and if $\Phi' = \Phi \restriction E'$, then $G(\Phi')$ is a Borel measurable subset of $E' \times F$. Hence, Φ' is Borel measurable and therefore continuous. In particular, this means that $\Phi(x_n) \longrightarrow \Phi(x)$. □

3 Some Generalizations

In this section, E will be a Fréchet space over $\mathbb{R}$ with a complete metric ρ having the property that $\rho(\lambda x, 0) \leq C(1 + |\lambda|)\rho(x, 0)$ for some $C < \infty$ and all $\lambda \in \mathbb{R}$ and $x \in E$. In particular, this will be the case if there exists a sequence of seminorms $\{p_k : k \geq 1\}$ for which $\rho(x, 0) = \sum_{k=1}^{\infty} 2^{-k} \frac{p_k(x)}{1+p_k(x)}$.

Given a sequence $\{x_n : n \geq 1\} \subseteq E$ with $\sum_{n=1}^{\infty} \rho(x_n, 0) < \infty$, define the random variables S_n as in Sect. 2. Then, without any substantive change in the argument given earlier, one can show that $\mathbb{P}(A) = 1$ when A is the set of $\omega \in \Omega$ for which $\lim_{n\to\infty} S_n(\omega)$ exists in E. Finally, define $S(\omega)$ as before. Then, by the same argument as was used to prove Theorem 2.1, we have the following.[1]

Lemma 3.1. *If* $\{x_n : n \geq 1\} \subseteq E$ *satisfies* $\sum_{n=1}^{\infty} \rho(x_n, 0) < \infty$ *and* $\varphi : E \longrightarrow \mathbb{R}$ *is a Borel measurable linear function, then* $\mathbb{E}^{\mathbb{P}}\big[(\varphi \circ S)^2\big] < \infty$.

Theorem 3.1. *If* $\varphi : E \longrightarrow \mathbb{R}$ *is a Borel measurable linear function, then* φ *is continuous.*

Proof. The proof is essentially the same as that for the first part of Corollary 2.1. Namely, if φ were not continuous, then we could find a sequence $\{x'_n : n \geq 1\} \subseteq E$ such that $\rho(x'_n, 0) \leq n^{-3}$ and $\varphi(x'_n) \geq \epsilon$ for some $\epsilon > 0$. Now set $x_n = nx'_n$ and define S relative to $\{x_n : n \geq 1\}$ and, for each $m \in \mathbb{Z}^+$, $S^{(m)}$ relative to $\{(1 - \delta_{m,n})x_n : n \geq 1\}$. Then, just as before, we get the contradiction $\epsilon m^2 \leq \mathbb{E}^{\mathbb{P}}\big[(\varphi \circ S)^2\big] < \infty$ for all $m \in \mathbb{Z}^+$. □

Given a locally convex, Hausdorff topological space F, say that a map $\Phi : E \longrightarrow F$ is w-Borel measurable if $\langle \Phi(\cdot), y^* \rangle$ is Borel measurable from E to $\mathbb{R}$ for each $y^* \in F^*$. Equivalently, Φ is w-Borel measurable if it is Borel measurable as a map from E into F with the weak topology. Obviously, every Borel measurable map is w-Borel measurable. Conversely, if F^* is separable in the weak* topology, then every w-Borel measurable map is Borel measurable.

[1]It should be observed that, because it deals with $\mathbb{R}$-valued maps, the proof of Lemma 3.1 does not require Fernique's theorem. All that one needs is the fact that centered Gaussian measures are the only probability measures μ on $\mathbb{R}$ with the property that μ is the distribution of $(\xi_1, \xi_2) \in \mathbb{R}^2 \longrightarrow \frac{\xi_1+\xi_2}{\sqrt{2}} \in \mathbb{R}$ under μ^2. See, for example, Exercise 2.3.21 in [7].

Corollary 3.1. *Suppose that F is a locally convex, Hausdorff topological vector space and that $\Phi : E \longrightarrow F$ is a w-Borel measurable linear map. Then Φ is continuous from E into the weak topology on F. In particular, $G(\Phi)$ is closed.*

Proof. To see that Φ is continuous into the weak topology on F, let $y^* \in E^*$ be given and define $\varphi : E \longrightarrow \mathbb{R}$ by $\varphi(x) = \langle \Phi(x), y^* \rangle$. Then, φ is Borel measurable and linear, and therefore, by Theorem 3.1, it is continuous.

To show that $G(\Phi)$ is closed, suppose that $\{x_\alpha\}$ is a net such that $(x_\alpha, \Phi(x_\alpha)) \longrightarrow (x, y)$ in $E \times F$. Then, by the first part, $\langle y, y^* \rangle = \langle \Phi(x), y^* \rangle$ for every $y^* \in F^*$, and so $y = \Phi(x)$. □

Given a topological vector space F, say that the pair (E, F) has the *closed graph property* if a linear map $\Phi : E \longrightarrow F$ is continuous whenever $G(\Phi)$ is closed. Since E is a Fréchet space, (E, F) has the closed graph property for every Fréchet space F (cf. Theorem 1 in Sect. 6 of Chap. II of [10]).

Corollary 3.2. *Assume that F is a locally convex, Hausdorff topological space for which (E, F) has the closed graph property. Then every w-Borel measurable linear map $\Phi : E \longrightarrow F$ is continuous. Furthermore, if in addition, (F, E) has the closed graph property and Φ is one-to-one and onto, then Φ^{-1} is continuous.*

Proof. The first assertion is an immediate consequence of Corollary 3.1. As for the second, observe that, by the first assertion, $G(\Phi)$ and therefore $G(\Phi^{-1})$ are closed. Hence, by the closed graph property for (F, E), Φ^{-1} is continuous. □

Corollary 3.3. *Assume that F is a separable Fréchet space. If $\Phi : E \longrightarrow F$ is linear and $G(\Phi)$ is Borel measurable, then Φ is continuous. Moreover, if, in addition, Φ is one-to-one and onto, then Φ^{-1} is continuous.*

Proof. The argument here is essentially the same as the one given in the proof of Corollary 2.1. Namely, when E is separable and $G(\Phi)$ is Borel measurable, then Φ is Borel measurable and therefore continuous, and, in general, one can reduce to the separable case by the same reasoning as was used in the proof of Corollary 2.1. As for the case when Φ is one-to-one and onto, note that, because $G(\Phi)$ is closed, so is $G(\Phi^{-1})$. Hence, the continuity of Φ^{-1} follows from the closed graph property for (F, E). □

Finally, it may be of some interest to observe that the linearity assumption in Corollaries 3.1, 3.2, and 3.3 can be replaced by additivity. Namely, assume that F is a locally convex, Hausdorff topological vector space. A map $\Phi : E \longrightarrow F$ is additive if it satisfies $\Phi(x_1 + x_2) = \Phi(x_1) + \Phi(x_2)$ for all $(x_1, x_2) \in E^2$. Now assume that Φ is a w-Borel measurable, additive map. Then, for each $x \in E$ and $y^* \in F^*$, $t \in \mathbb{R} \longrightarrow \langle \Phi(tx), y^* \rangle \in \mathbb{R}$ is a Borel measurable, additive function. Since every $\mathbb{R}$-valued, Borel measurable function on $\mathbb{R}$ is linear (cf. Exercise 2.2.36 in [8]), it follows that Φ is linear.

References

1. Fernique, X.: Régularité des trajectoires des fonctions aléatoires gaussiennes, Ecole d'Eté de Probabilités de Saint-Flour IV-1974. In: Hennequin, P.L. (ed.) Lecture Notes in Mathematics, vol. 480, pp. 1–97. Springer, Berlin (1975)
2. Kuratowski, K.: Topologie I. Academic, New York (1966)
3. Martineau, A.: Sur des théorèmes de S. Banach et L. Schwartz concernant le graphe fermé. Studia Mathematica **XXX**, 43–51 (1968)
4. Parthasarathy, K.P.: Probability Measures on Metric Spaces, vol. 276. AMS Chelsea Series, Providence (1967)
5. Schwartz, L.: Sur le théorème du graphe. C. R. Acad. Sci. Paris **263 série A**, 602–605 (1966)
6. Stroock, D.: On a Theorem of L. Schwartz. C. R. Acad. Sci. Paris Ser. I **349**(1–2), 5–6 (2010)
7. Stroock, D.: Probability Theory, an Analytic View, 2nd edn. Cambridge University Press, Cambridge (2011)
8. Stroock, D.: Essentials of integration theory for analysis. In: GTM series. Springer Heidelberg (2011)
9. Treves, F.: Topological Vector Spaces, Distributions and Kernels. Academic, New York (1967)
10. Yoshida, K.: Functional Analysis. In: Grundlehren Series # 123. Springer, Berlin (1965)

Chapter 2
Stochastic Taylor Formulas and Riemannian Geometry

Mark A. Pinsky

To David Nualart, with admiration and respect.

Received 11/22/2011; Accepted 2/12/2012; Final 2/21/2012

1 Introduction

Let (X_t, P_x) be the standard Brownian motion on a complete Riemannian manifold. We investigate the asymptotic behavior of the moments of the exit time from a geodesic ball when the radius tends to zero. This is combined with a "stochastic Taylor formula" to obtain a new expansion for the mean value of a function on the boundary of a geodesic ball.

Several authors [3–5] have considered mean value formulas on a Riemannian manifold, using the exponential mapping and integration over the unit sphere of the tangent space, at each point. In general the stochastic mean value is not equal to the exponential mean value. If these coincide, the manifold must be Einsteinian (constant Ricci curvature). Our expansions are used to answer some *inverse questions* in stochastic Riemannian geometry. If the mean exit time from every small ball is the same as for a flat manifold then the manifold is flat, in case $d = 2$ or $d = 3$. Our method of proof begins with an expansion of the Laplacian in a system of Riemannian normal coordinates [6]. The successive correction terms in the expansion of the moments is obtained by a perturbation expansion, the rigorous validity of which is established by systematic application of the stochastic Taylor formula.

M.A. Pinsky (✉)
Department of Mathematics, Northwestern University, 2033 Sheridan Road,
Evanston, IL 60208-2730, USA
e-mail: markpin@math.northwestern.edu

F. Viens et al. (eds.), *Malliavin Calculus and Stochastic Analysis: A Festschrift in Honor of David Nualart*, Springer Proceedings in Mathematics & Statistics 34,
DOI 10.1007/978-1-4614-5906-4_2,

We use the summation convention throughout. The notation $|x| < \epsilon \to 0$ means that the indicated asymptotic estimate holds uniformly in the ball of radius ϵ, when $\epsilon \to 0$.

2 Stochastic Taylor Formula

For completeness we include the proof of this formula, which was previously treated by several authors (Athreya-Kurtz, Airault-Follmer, and Van der Bei). Let (X_t, P_x) be a diffusion process on a locally compact and separable space V. Let f be a real-valued function such that

$$f \in \mathcal{D}(A^k) \qquad\qquad k = 1, 2, \ldots, N+1,$$

where A is the infinitesimal generator defined by

$$Af(x) = \lim_{t \to 0} t^{-1} E_x[f(X_t) - f(x)].$$

Let T be a stopping time with

$$E_x(T^{N+1}) < \infty.$$

Proposition 2.1. *Under the above conditions we have for $N \geq 1$*

$$E_x f(X_T) = f(x) + \sum_{k=1}^{N-1} \frac{(-1)^{k+1}}{k!} E_x\left(T^k A^k f(X_T)\right), + (-1)^{N-1} R_{N-1}, \tag{2.1}$$

where

$$R = R_{N-1} = \frac{1}{(N-1)!} E_x\left(\int_0^T u^{N-1} A^N f(X_u)\, \mathrm{d}u\right). \tag{2.2}$$

The empty sum is defined to be zero.

Proof. For $N = 1$ we have an empty sum and Dynkin's identity: $E_x f(X_T) = f(x) + R_1$. In general for $N \geq 1$ we have

$$A^k f(X_t) - A^k f(X_0) = \int_0^t A^{k+1} f(X_u)\, \mathrm{d}u + M_t, \tag{2.3}$$

where M_t is a local martingale. From the stochastic product rule we have

$$d\left(t^k A^k f(X_t)\right) = t^k\, \mathrm{d}M_t + \left(t^k A^{k+1} f(X_t) + k t^{k-1} A^k f(X_t)\right) \mathrm{d}t. \tag{2.4}$$

Integrating and taking the expectation, we have

$$E_x\left[T^k A^k f(X_T)\right] = E_x \int_0^T t^k A^{k+1} f(X_t)\, \mathrm{d}t + k E_x \int_0^T t^{k-1} A^k f(X_t) \mathrm{d}t. \tag{2.5}$$

Dividing by $k!$, we have

$$\frac{1}{k!}E_x\left[T^k A^k f(X_T)\right] = R_{k+1} + R_k. \tag{2.6}$$

Multiplying by $(-1)^{k+1}$ and summing for $k = 1, \ldots, N$ gives the stated result. □

2.1 Some Special Cases

The stochastic Taylor formula allows one to pass freely between classical solutions of equations and moments of various functionals.

Example 2.1a. Let B be the unit ball of R^n centered at the origin and T be the first exit time from B. If $f = 0$ on $S = \partial B$ and solves $Af = -1$ in B, then we have

$$0 = E_x f(X_T) = f(x) + E_x \int_0^T Af(X_u)\,\mathrm{d}u = f(x) - E_x(T).$$

Thus the mean exit time can be retrieved from the value at the starting point. Explicit computation gives $f(x) = (1 - |x|^2)/2n$.

The stochastic Taylor formula can be used to give a probabilistic representation of the solution of some higher-order elliptic boundary-value problems.

Example 2.1b. Let B be a bounded region of Euclidean space with boundary hypersurface $S = \partial B$. Let u be the solution of the boundary-value problem

$$u|_S = 0, \ldots, A^{N-1}u|_S = 0, A^N u|_B = g. \tag{2.7}$$

We first note that the left side of Eq. (2.1) is $0 - u(x)$ whereas all of the terms on the right are zero, save the last term, which $= 1/(N-1)!$ times

$$E_x\left(\int_0^T s^{N-1} A^N f(X_s)\,\mathrm{d}s\right) = E_x\left(\int_0^T s^{N-1} g(X_s)\mathrm{d}s\right),$$

so that we have

$$u(x) = \frac{1}{(N-1)!}\left(E_x \int_0^T s^{N-1} g(X_s))\,\mathrm{d}s\right),$$

which is the desired probabilistic representation. In the special case $g = 1$ we can perform the integration to obtain

$$u(x) = \frac{E_x(T^N)}{N!}.$$

Stated otherwise, the higher moments of the exit time are obtained by solving a higher-order boundary-value problem.

One can also make conclusions based on *approximate solutions.*

Example 2.1c. If $f = 0$ on $S = \partial B$ and solves[1] $Af = -1 + \epsilon$, then we have

$$0 = E_x f(X_T) = f(x) + E_x \int_0^t Af(X_u)\, du = f(x) - E_x(T)(1 \pm \epsilon);$$

hence $E_x(T) = f(x) \times (1 \pm \epsilon)$. Thus the mean exit time can be retrieved with ϵ accuracy from the value at the starting point.

3 Expansion of the Laplacian

Now consider the case of an n-dimensional Riemannian manifold V. Let $O \in V$ and consider the exponential mapping

$$\exp_O V_O \to V.$$

A choice of an orthonormal basis in V_O gives rise to normal coordinates $(x_1, \ldots, x_n)$.

Let Δ be the Laplacian in normal coordinates:

$$\Delta f = \frac{1}{\sqrt{g}} \partial_i \left(\sqrt{g} g^{ij} \partial_j f \right) \quad g := \det(g_{ij}). \tag{2.8}$$

From Eq. (2.8) it follows that Δ is self-adjoint with respect to the weight function $x \to \sqrt{g}$.

Let τ_ϵ be the dilation operator, defined by

$$(\tau_\epsilon f)(x) = f\left(\frac{x}{\epsilon}\right). \tag{2.9}$$

Definition. *A differential operator A is is said to be homogeneous of degree j if and only if for every homogeneous polynomial Q of degree k, AQ is homogeneous of degree $k + j$.*

Proposition 3.1. *For every integer $N \geq 0$, there exists a finite set of second-order differential operators $\Delta_0, \ldots, \Delta_N$ such that Δ_j is homogeneous of degree j and we have the asymptotic expansion*

$$\tau_\epsilon^{-1} \circ \Delta \circ \tau_\epsilon f = \epsilon^{-2} \Delta_{-2} f + \sum_{j=0}^{N} \epsilon^j \Delta_j f + O(\epsilon^{N+1}), \tag{2.10}$$

[1]This means that $\sup_{x \in B} |Af(x) + 1| \leq \epsilon$.

where

$$\Delta_{-2} f = \sum_{i=1}^{n} (\partial_i \partial_i) f$$

for any twice differentiable function f.

Proof. In normal coordinates we have the asymptotic expansion [5]

$$g_{ij} = \delta_{ij} + a_{ijkl} x^k x^l + \text{terms of order 3 and higher} \tag{2.11}$$

with similar expressions for $\sqrt{g}, g^{ij}$. Substitution into Eq. (2.8) and collecting terms give the formula

$$\Delta f - \Delta_{-2} f = P_{ij} \partial_i \partial_j f + Q_i \partial_i f, \tag{2.12}$$

where P_{ij} begins with the quadratic terms and Q_i begins with linear terms. Now replace f by $\tau_\epsilon f$ in Eq. (2.12). Clearly $\tau_\epsilon^{-1} \circ (\Delta_{-2}) \circ \tau_\epsilon f = \epsilon^{-2} \Delta_{-2} f$. On the other hand when we apply the right side of Eq. (2.12) to $\tau_\epsilon f$, we obtain an asymptotic series in ϵ. The coefficient of ϵ^0 is the quadratic term in P_{ij} plus the linear terms in Q_i. Proceeding to the next stage, the coefficient of ϵ consists of the cubic term in P_{ij} and the quadratic terms in $Q_j f$. Continuing this to higher powers of ϵ we can compute the coefficient operators $\Delta_0, \Delta_1, \ldots$; hence we have completed the proof of Proposition 3.1. □

3.1 Computation of Δ_0

To compute $\Delta_0 f$ we begin with Cartan's formula

$$g_{ij} = \delta_{ij} + \frac{1}{3} R_{ijkl} x^k x^l + O(|x|^3), (x \to 0), \tag{2.13}$$

$$g^{\alpha\beta} = \delta^{\alpha\beta} - \frac{1}{3} R_{\alpha\beta kl} x^k x^l + O(|x|^3), (x \to 0). \tag{2.14}$$

The Christoffel coefficients are computed from Eq. (2.13). We use the covariant form of the Laplacian:

$$\Delta f = g^{ij} \left(\partial_i \partial_j f - \Gamma_{ij}^k \partial_k f \right). \tag{2.15}$$

When we substitute these expansions and collect terms, there results

$$\Delta f = \partial_i \partial_i f + \left(-\frac{1}{3} R_{ijkl} x^k x^l + O(|x|^3) \right) \partial_i \partial_j f$$
$$+ \left(-\frac{2}{3} R_{ikk'i} x^{k'} \partial_i f + O(|x|^2) \right).$$

Hence

$$\Delta_0 f = -\frac{1}{3} R_{ijkl} x^k x^l \partial_i \partial_j f - \frac{2}{3} R_{ikk'i} x^{k'} \partial_k f. \tag{2.16}$$

Of particular interest is the case $f = \phi(r)$, where $\phi \in C^2[0, \infty)$. This yields

$$\Delta_0(\phi \circ r) = -\frac{1}{3} \rho_{kl} \frac{x^k x^l}{r} \phi'(r),$$

where ρ_{kl} is the *Ricci tensor*, defined by $\rho_{kl} = R_{ikil}$ (sum on i).

3.2 Computation of $\Delta_k, k \geq 1$

To compute the higher corrections $\Delta_1, \Delta_2 \ldots$, we introduce polar coordinates in V_O; letting $x = r\theta$, we have

$$\Delta_k = b_k(\theta) r^{1+k} \frac{\partial}{\partial r} + r^k \tilde{\Delta}_k, \tag{2.17}$$

where $\tilde{\Delta}_k$ are second-order differential operators in $\theta_1, \ldots, \theta_n$ and b_k is a homogeneous function of degree zero. By Gauss' lemma, the terms with $\partial^2/\partial r^2$ or $\partial^2/\partial r \partial\theta_i$ are not present. To compute $b_k(\theta)$, we recall the expansion of Gray–Vanhecke [4] for the determinant of the exponential map of a Riemannian manifold:

$$\omega = 1 - \frac{1}{6}\beta_2 r^2 - \frac{1}{12}\beta_3 r^3 + \frac{1}{24}\beta_4 r^4 + O(r^5), (r \to 0), \tag{2.18}$$

where

$$\beta_2 = \rho_{ij}\theta_i\theta_j, \tag{2.19}$$

$$\beta_3 = \nabla_i \rho_{jk} \theta_i \theta_j \theta_k, \tag{2.20}$$

$$\beta_4 = \left(1 - \frac{3}{5}\nabla^2_{ij} + \frac{1}{3}\rho_{ij}\rho_{kl} - \frac{2}{15} R_{iajk} R_{kalb}\right) \theta_i \theta_j \theta_k \theta_l. \tag{2.21}$$

Performing the long division we have

$$\frac{\omega'}{\omega} = b_0 r + b_1 r^2 + b_2 r^3 + O(r^4),$$

$$b_0 = -\frac{1}{3}\rho_i \rho_j \theta_i \theta_j,$$

$$b_1 = -\frac{1}{4}\rho_j \rho_k \partial_i \partial_i \partial_j \theta_j \theta_k,$$

$$b_2 = -\frac{1}{9}\nabla^2_{ij}\rho_{kl} - \frac{1}{45} R_{iajb} R_{kalb} \theta_i \theta_j \theta_k \theta_l.$$

4 Estimate of the Moments $E_x(T_r^k)$, $r \to 0$

Let V be an N-dimensional complete Riemannian manifold and (X_t, P_x) be the Brownian motion process on V, a diffusion process with infinitesimal generator Δ. The *exit time* from a ball of radius R centered at $O \in V$ is defined by

$$T_R := \inf\{t : t > 0, d(X_t, O) = R\}. \tag{2.22}$$

We emphasize that the computations are made as a function of the unspecified starting point $p = X_0$ where we set $p = O$ at the end. We will prove the following.

Proposition 4.1. *For each* $k \geq 1, E_O(T_R^k) \sim c_{nk} R^{2k}, R \to 0$, *where* c_{nk} *are positive constants. In case* $V = R^n$ *we have for each* $k \geq 1, E_O(T_R^k) = c_{nk} R^{2k}$ *for all* $R > 0$.

Proof. Let $B = \{y : d(y, O) \leq R\}$ and let $f_0^k(r)$ be the solution of

$$\Delta_{-2} f_0^k = -f_0^{k-1}, \quad f_0^k|_{\partial B} = 0, \quad f_0^0 \equiv 1 \qquad (k \geq 1). \tag{2.23}$$

In case $V = R^n$ $f_0^k(r)$ is a polynomial of degree $2k$. The substitution $r \to r/\epsilon$ with subsequent normalization allows one to study the case $R = 1$. In detail, we have $f_0^k(r) \to \epsilon^{2k} f_0^k(r/\epsilon)$.

By explicit computation, we have

$$-f_0^1(r) = \frac{r^2 - 1}{2n}, \quad f_0^1(0) = \frac{1}{2n}$$

$$-f_0^2(r) = \frac{r^4}{8n(n+2)} - \frac{r^2}{4n^2} + \frac{n+4}{8n^2(n+2)}, \quad f_0^2(0) = \frac{n+4}{8n^2(n+2)},$$

$$-f_0^3(r) = \frac{r^6}{48n(n+2)(n+4)} - \frac{r^4}{16n^2(n+2)} + \frac{r^2(n+4)}{16n^3(n+2)} + \frac{n^2+12n+48}{48n^3(n+2)(n+4)}.$$

In general f_0^k is a polynomial of the form

$$-f_0^k(r) = \frac{r^{2k}}{2^k} k!n(n+2)\cdots(n+2k-2) + \sum_{j=0}^{k-1} (-1)^j c_{knj} r^{2k-2j}.$$

□

In order to confirm the polynomial character of $f_0^k(r)$, one can either use mathematical induction or proceed directly, as follows: given functions $g(x), h(x)$, it is required to solve the differential equation $f''(x) + (g'/g) f'(x) = -h$ on the interval $0 \leq x < R$, with the boundary conditions that $f(R) = 0$ and f remains bounded when $x \to 0$. To do this, multiply the differential equation by g to obtain $(gf')' = -gh$, $gf' = C - \int_x^1 gh$. Solving for f' and doing the integral, we obtain the formula

$$f(x) = \int_x^1 \frac{1}{g(t)} \left(\int_0^t h(u)g(u)\mathrm{d}u \right) \mathrm{d}t, \tag{2.24}$$

which can be directly checked as follows:

$$f'' + (g'/g)f' = (1/g)(gf')' = (1/g)(-gh) = -h.$$

The boundary condition $f(1) = 0$ follows from Eq. (2.24). The integral defining $f(0)$ is absolutely convergent since f is displayed in terms of the ratio of two terms each of which tends to zero. l'Hospital's rule applies to the ratio of the derivatives, namely

$$\lim_{t\to 0} \frac{\int_0^t h(u)g(u)\mathrm{d}u}{g(t)} = \lim_{t\to 0} \frac{h(t)g(t)}{g'(t)} = h(0) \lim_{t\to 0} \frac{g(t)}{g'(t)}.$$

This shows that the integral in Eq. (2.24) is well defined whenever g/g' is integrable; in particular this is the case where $g(x) = x^{n-1}$ and $g/g' = x/(n-1)$.

In particular if $g(x)$ is a monomial—$g(x) = x^n$ and $h(x)$ is a polynomial, then $f(x)$ is also a polynomial.

From the stochastic Taylor formula we have

$$E_x[f(X_T)] = f(x) + \sum_{k=1}^{N-1} \frac{(-1)^{k+1}}{k!} E_x\left(T^k A^k f(X_T)\right) + (-1)^{N-1} R_{N-1}, \tag{2.25}$$

where

$$R = R_{N-1} = \frac{1}{(N-1)!} E_x \left(\int_0^T u^{N-1} A^N f(X_u)\,\mathrm{d}u \right). \tag{2.26}$$

By construction, f_0^k is identically zero on the sphere S when $k < N$. We can write $0 - f(x) = E_x(T^N)/N!$ which proves that $E_O(T^N) = c_N R^{2N}$. This proves the proposition in case of Euclidean space.

In the general case, we can still use the stochastic Taylor formula with $A = \Delta, T = T_B, f = f_0^k$. Then

$$f(X_T) = O(\epsilon), \ldots, A^{N-1} f(X_T) = O(\epsilon), A^N f(X_T) = -1 + O(\epsilon).$$

From the decomposition of the Laplacian of V we have $\Delta f - \Delta_{-2} f = O(\epsilon)$

$$f_0^N(r) = \frac{1}{N!} E_x(T^N)\left(1 + O(\epsilon^2)\right).$$

Taking $r = 0$, $R = \epsilon \to 0$, we see that

$$\limsup_{R \to 0} \frac{E_O(T_R)}{R^N} = 1.$$

To compute an explicit form, we specialize to Euclidean space, $V = R^N$, $f(x) = e^{<\alpha,x>} \alpha \in R^N$. The stochastic Taylor formula gives

$$E_0 e^{<\alpha, X_{T_r}>} = \sum_{k=0}^{\infty} \frac{|\alpha|^{2k}}{k!} E_0(T_r^k).$$

But the P_O hitting measure of X_T is uniform on the sphere $|x| = r$. Hence

$$\begin{aligned} E_0 e^{<\alpha, X_{T_r}>} &= \int_{S^{n-1}} \exp(|\alpha| r \cos\theta) d\theta \\ &= \sum_{m=0}^{\infty} \frac{|\alpha|^m}{m!}. \end{aligned}$$

Comparing the two expansions, we have

$$\frac{E(T_r^k)}{r^{2k}} = \frac{k!}{(2k)!} \int_{S^{n-1}} (\cos\theta)^{2k} = \frac{\Gamma(\frac{n}{2})}{\Gamma(k + \frac{n}{2})} 2^{-2k}.$$

Hence $c_{nk} = \frac{\Gamma(\frac{n}{2})}{\Gamma(k+\frac{n}{2})}$.

5 Refined Estimate of $E(T_r)$, $r \downarrow 0$

The mean exit time of Brownian motion is affected by the curvature σ of the manifold. When σ is zero, we have the Euclidean space, which has been well studied. If $\sigma > 0$ a geodesic disk has smaller volume than the Euclidean counterpart and the mean exit time is greater. Finally in the case of negative curvature the volume is strictly smaller and the mean exit time is strictly larger than the corresponding disk in Euclidean space. Here is a precise result.

Proposition 5.1.

$$E_0(T_r) = \frac{r^2}{2n} + \sigma \frac{r^4}{12n^2(n+2)} + O(r^5) \qquad (r \downarrow 0). \tag{2.27}$$

Proof. For this purpose, let

$$F = f_0^1 + \epsilon^2 f_2^1 + \epsilon^3 f_3^1 + \epsilon^4 f_4^1, \tag{2.28}$$

where

$$f_0^1 = (1 - r^2)/2n, \tag{2.29}$$

$$\Delta_{-2} f_2^1 + \Delta_0 f_0^1 = 0, \quad \text{in } B, \quad f_2|_{\partial B} = 0, \tag{2.30}$$

$$\Delta_{-2} f_3^1 + \Delta_0 f_1^1 + \Delta_1 f_0^1 = 0 \quad \text{in } B, \quad f_3|_{\partial B} = 0, \tag{2.31}$$

$$\Delta_{-2} f_4^1 + \Delta_0 f_2^1 + \Delta_2 f_0^1 = 0 \quad \text{in } B, \quad f_4|_{\partial B} = 0. \tag{2.32}$$

Now we set $f = \epsilon^2 \tau_\epsilon F$. From Eq. (2.10) and Proposition 2.1 we have

$$\Delta f = -1 + O(\epsilon^5), \quad \Delta^2 f = O(\epsilon^3), \quad f|\partial B_\epsilon = 0.$$

Therefore, from the stochastic Taylor formula with $N = 2$

$$0 = f(x) + E_x(T_\epsilon)(-1 + O(\epsilon^5)),$$

$$E_x(T_\epsilon) = \epsilon^2 \left[f_0^1(x) + \epsilon^2 f_2^1(x) + \epsilon^3 f_3(x) + \epsilon^4 f_4(x) \right] + O(\epsilon^7), \quad |x| < \epsilon \to 0.$$

In the previous section it was proved that $f_0^1(0) = 1/2n$. To compute $f_2^1(0)$ we first note that

$$\Delta_{-2} f_2^1 = -\rho_{ij} \theta_i \theta_j r^2 / 2n = \frac{r^2}{2n} b(\theta).$$

To compute $f_2^1(0), f_4^1(0)$ we invoke the following lemma. Integration over $S = S^{n-1}$ is taken with respect to the uniform probability measure $d\theta$. □

Lemma 5.1. *Let $j(\theta), g(\theta)$ be polynomials which satisfy the relation $\Delta_{-2} j = r^k g(\theta)$, $g|_{\partial S} = 0$ where k is an integer. Then*

$$j(0) = \frac{1}{(n+k)(k+2)} \int_S g(\theta).$$

Proof. We can write j in the form $j = r^{k+2} c(\theta) + h$ where c is a homogeneous polynomial and h is a harmonic function, solution of $\Delta_{-2} h = 0$. Then

$$[(k+2)(n+1) + (n-1)(k+2)]c(\theta)]\Delta_{-2} c(\theta) = -g(\theta).$$

Integrating over S, we have $(k+2)(n+k) \int_S c(\theta) = -\int g(\theta)$. But $j(0) = h(0) = -\int_S c(\theta) d\theta = 1/(n+k)(k+2) \int_S g(\theta)$.
Hence

$$f_2^1(0) = \frac{1}{4n(n+2)} \int_S b_0(\theta) = \frac{\sigma}{12n^2(n+2)}. \tag{2.33}$$

To compute $f_4^1(0)$, we have $\Delta_{-2} f_4^1 = -\Delta_0 f_2^1 - \Delta_2 f_0^1$, $f_4|_S = 0$. We treat the two terms separately.

Recalling that $f_0^1(r) = (1-r^2)/2n, \Delta_2 = r^3 b_2(\theta)\frac{\partial}{\partial r}$, it follows that $\Delta_2 f_0^1(r) = -r^4 b_2(\theta)\mathrm{d}\theta$. By Lemma 5.1, this term contributes to $f_4(0)$ in the amount

$$\frac{1}{6(n+4)} \int_S b_2(\theta)\mathrm{d}\theta.$$

To handle the second term, we use the fact that $\Delta_0 + \beta_2\Delta_{-2}$ is self-adjoint with respect to Lebesgue measure on B. Thus

$$\begin{aligned}\int_S \Delta_0 f_2 &= -\int_S \beta_2\Delta_{-2} f_2\\ &= \int_S \beta_2 f_0\\ &= 3\int_S b_0^2(\theta).\end{aligned}$$

Thus the second term contributes to $f_4^1(0)$ in the amount $1/2(n+4)\int_S b_0(\theta)^2$. Combining this with the previous computations, we have

$$f_4^1(0) = \frac{1}{6(n+4)}\int_S b_2(\theta) + \frac{1}{2(n+4)}\int_S b_0(\theta)^2.$$

Finally we note that $f_3^1(0) = 0$, since $\Delta_{-2} f_3^1 = -\Delta_1 f_0^1 = r^3 b_1(\theta)/n$ which is a cubic polynomial; hence the integral is zero. □

6 Mean Value Formulas for General Manifolds

On Euclidean space one may define the mean value of a function on a sphere. This can be effectively computed in terms of the radius of the sphere and the values of $\Delta f, \Delta^2 f, \ldots$ at the center of the sphere. The resulting series expansion is known as Pizetti's theorem in classical differential geometry.

When we pass to manifolds of variable curvature, the corresponding series expansion is much more complicated than for the Euclidean case, leading one to ask for a simpler mean value. In the following sections we will explore the properties of the *stochastic mean value formula*.

6.1 Stochastic Mean Value

The stochastic mean value of a continuous function f on a sphere is defined as the following linear functional on continuous functions:

$$\Phi_O^\epsilon f = E_O[f(X_{T_\epsilon})] = \int_S f(y)\mathrm{d}S^\epsilon(y),$$

where the existence of the probability measure dS_ϵ is guaranteed by the Riesz representation theorem. We propose to find a three-term asymptotic expansion of the measure dS^ϵ when $\epsilon \to 0$. Equivalently we can find a three-term expansion of the linear functional Φ^ϵ_O when $\epsilon \to 0$. The key to success is through Dynkin's formula, which is transformed into the identity

$$\Phi^\epsilon_O f = E_O[f(X_{T_\epsilon})] = E_O f + E_O\left(\int_0^{T_\epsilon} \Delta f(X_s),\mathrm{d}s\right), \quad \epsilon > 0, f \in C(V). \tag{2.34}$$

For any $f \in C^5(V)$, we can write for $|x| < \epsilon \to 0$

$$\Delta f = d_0 + \sum_{i=1}^n d_i x_i + \frac{1}{2}\sum_{i,j=1}^n d_{ij} x_i x_j + O(|x|^3).$$

Equivalently

$$d_0 = \Delta f(O), \quad d_i = \partial_i \Delta f(O), \quad d_{ij} = \partial_i \partial_j \Delta f(O).$$

Therefore it is sufficient to consider test functions f which are either constant, linear, or purely quadratic. From Dynkin's formula, to study Φ, it is equivalent to study the integral (2.34). In the next section we study the equation

$$E_p \int_0^{T_\epsilon} \Delta f(X_s)\,\mathrm{d}s = d_0 u^\epsilon_0(x) + \sum_{i=1}^n d_i u^\epsilon_i(x) + \sum_{i,j=1}^n d^\epsilon_{ij} u^\epsilon_{ij}(x) + O(\epsilon^5). \tag{2.35}$$

6.2 Solution of Poisson's Equation in a Geodesic Ball

In order to proceed further, we develop the properties of the mean value of three functionals as follows: the result of these computations is that the Euclidean mean value differs from the non-Euclidean mean value to within $O(\epsilon^5)$ when $\epsilon \to 0$.

Lemma 6.1. *For each $\epsilon > 0$, let $B = B_\epsilon$ denote the geodesic ball of radius ϵ in the tangent space. Functions u_0, u_i, and u_{ij} are defined as the solutions of the following boundary-value problems and respective stochastic representations:*

$$\Delta u^\epsilon_0|_B = -1, \quad u^\epsilon_0|_{\partial B} = 0 \qquad \Rightarrow u_0 = E_p(T_\epsilon),$$

$$\Delta u^\epsilon_i|_B = -x_i, \quad u^\epsilon_i|_{\partial B} = 0, 1 \le i \le n, \qquad \Rightarrow u_i(p) = E_p\left(\int_0^{T_\epsilon} X_i(s)\mathrm{d}s\right),$$

$$\Delta u^\epsilon_{ij}|_B = -x_i x_j,\ u^\epsilon_{ij}|_{\partial B} = 0, 1 \le i, j \le n \Rightarrow u_{ij}(p) = E_p\left(\int_0^{T_\epsilon} X_i(s) X_j(s)\right)\mathrm{d}s.$$

To see this, apply Dynkin's formula successively to $f = u_0^\epsilon$, $f = u_i^\epsilon$, $f = u_{ij}^\epsilon$. *We have the asymptotic formulas when* $\epsilon \to 0$*:*

$$u_0^\epsilon(x) = \epsilon^2 U_0(x/\epsilon) + \epsilon^4 V_0(x/\epsilon) + O(\epsilon^5), \tag{2.36}$$

$$u_i^\epsilon(x) = \epsilon^3 U_i(x/\epsilon) + O(\epsilon^5), \tag{2.37}$$

$$u_{ij}^\epsilon(x) = \epsilon^4 U_{ij}(x/\epsilon) + +O(\epsilon^5), \tag{2.38}$$

where

$$U_0(x) = \frac{1-|x|^2}{2n}, \quad U_i(x) = \frac{x_i(1-|x|^2)}{2n+4}, \quad U_{ij}(x) = \frac{x_i x_j(1-|x|^2)}{2n+8},$$

$$(n+4)V_0(x) = \frac{(1-|x|^2)}{6n}\left(\sum_{i=1}^n \lambda_i x_i^2\right) + \sigma\frac{(1-|x|^2)}{6n^2} - \sigma\frac{(1-|x|^4}{12n(n+2)},$$

where (λ_i) *are the eigenvalues of the Ricci tensor.*

Proof. The stochastic representations follow immediately from Dynkin's formula, applied successively to u_0, u_i, and u_{ij}.

In terms of the Euclidean Laplacian Δ_{-2}, we verify that U_0, U_i, and U_{ij} satisfy the following identities:

$$\Delta_{-2}U_0 = -1,$$

$$\Delta_{-2}(x_i x_j) = 2\delta_{ij},$$

$$\Delta_{-2}(x_i^2|x|^2) = 2|x|^2 + (2n+8)x_i^2,$$

$$\Delta_{-2}|x|^4 = (4n+8)|x|^2,$$

$$\Delta_{-2}(x_i x_j|x|^2) = (2n+8)x_i x_j,$$

where we use the identities

$$\Delta_{-2}(|x|^2) = 2n, \text{ etc, etc.}$$

We combine this with Dynkin's formula—the stochastic Taylor formula with $N = 1$

$$E_p f(X_{T_\epsilon}) = f(p) + E_p\left(\int_0^{T_\epsilon} \Delta f(X_s)\,ds\right), \qquad p \in B, f \in C^2(V). \tag{2.39}$$

To prove Eq. (2.36) apply Dynkin's formula with the choice $f = \epsilon^2 U_0(x/\epsilon) + \epsilon^4 V_0(x/\epsilon)$ for which $f = 0$ on S. Then we obtain

$$\Delta f = -1 + O(\epsilon^3) \Rightarrow 0 = f(p) + \int_0^T \left(-1 + O(\epsilon^3)\right) ds,$$

which proves Eq. (2.36).
To prove Eq. (2.37), apply Dynkin's formula with $f = \epsilon^2 U_i(x/\epsilon)$. Then $\Delta f = -x_i + O(\epsilon^3)$.
For Eq. (2.38), apply Dynkin to the choice $f = \epsilon^4 U_{ij}$, for which $\Delta f = -x_i x_j + O(\epsilon^4) \in B, f = 0$ on S. □

7 Comparison with the Gray–Willmore Expansion

When we combine the results of Sects. 1–6, we find the following asymptotic formula:

$$E_O f(X_{T_\epsilon}) - f(O) = \Delta f(0)\left[\epsilon^2 U_0(y) + \epsilon^4 V_0(y)\right] + \epsilon^3 \textstyle\sum_{i=1}^n \partial_i(\Delta f) + O(\epsilon^5),$$

$$E_O f(X_{T_r}) = f(O) + \tfrac{r^2}{2n}\Delta f(O) + r^4\left[\tfrac{\Delta^2 f(O)}{8n(n+2)} - \tfrac{\sigma\Delta f(O)}{12n^2(n+2)}\right] + O(r^5), r \to 0. \quad (2.40)$$

It is interesting to compare this with the Gray–Willmore expansion as defined by integration in the tangent space:

$$M(r,f) = f(O) + \frac{r^2}{2n}\Delta f(O) + \frac{r^4}{24n(n+2)}(3\Delta^2 f(O) - 2 < \nabla^2 f, \rho >$$
$$-3 < \nabla f, \nabla\sigma > + \frac{4}{n}\sigma\Delta f(O)) + O(r^5), \quad r \to 0.$$

Proposition 7.1. *Suppose that for all $f \in C^2(V)$, $M(r,f) = E_O f(X_{T_r})$ for all $r > 0$. Then V is an Einstein manifold.*

Proof. From the above formulas we have

$$\lim_{r\to 0} \frac{[E_O f(X_{T_r}) - M(r,f)]}{r^4} = \frac{< \nabla^2 f, \rho >}{12n(n+2)} + \frac{< \nabla f, \nabla\sigma >}{8n(n+2)} - \frac{< \sigma, \Delta f >}{12n^2(n+2)}. \quad (2.41)$$

If the right side is zero for all $f \in C^2(V)$, we first take f to be linear in the normal coordinates so that ∇f is perpendicular to $\nabla\sigma$, from which we find $\nabla\sigma = 0$. The remaining term of Eq. (2.41) is

$$\frac{1}{12n(n+2)}\left[< \nabla^2 f, \rho > - \frac{\sigma}{n}\Delta f\right].$$

Take normal coordinates so that the Ricci tensor is diagonal at O; thus $\rho_{ij}0 = 0$ for $i \neq j$. Now choose f so that $\partial_i \partial_j = 0$ for $i \neq j$ from which we conclude that $\rho_{jj} = \frac{1}{n}(\rho_{11} + \cdots + \rho_{nn})$. Hence the Ricci tensor is a multiple of the identity, i.e., V is an Einstein manifold. □

References

1. Athreya, K.B., Kurtz, T.G.: A generalization of Dynkin's identity and some applications. Ann. Probab. **1**, 520–529 (1973)
2. Airault, H., Follmer,H.: Relative densities of semimartingales. Invent. Math. **27**, 299–327 (1974)
3. Friedman, A.: Function-theoretic characterization of Einstein spaces and harmonic functions. Trans. Am. Math. Soc. **101**, 240–258 (1961)
4. Gray, A., Willmore T.J.: Mean value theorems on Riemannian manifolds. Proc. Roy. Soc. Edinburgh Sec A. **92**, 334–364 (1982)
5. Gray, A., vanHecke, L.: Riemannian geometry as determined by the volume of small geodesic balls. Acta Math. **142**, 157–198 (1979)
6. Levy-Bruhl, A.: Courbure Riemannienne et developpements infinitesimaux du laplacien. CRAS **279**, 197–200 (1974)
7. Van der Bei, R.: Ph.D. dissertation, Cornell University (1981)

Chapter 3
Local Invertibility of Adapted Shifts on Wiener Space and Related Topics

Rémi Lassalle and A.S. Üstünel

Abstract In this article we show that the invertibility of an adapted shift on the Brownian sheet is a local property in the usual sense of stochastic calculus. Thanks to this result we give a short proof of the invertibility for some processes which occur in free euclidean quantum mechanics and we relate this result to optimal transport. We also investigate some applications to information theory of a recent criterion which relates the invertibility of a shift to an equality between the energy of the signal and the relative entropy of the measure it induces. In particular, thanks to a change of measure, we interpret Shannon's inequality as a consequence of information loss in Gaussian channels and we extend it to any abstract Wiener space. Finally, we extend the criterion of invertibility to the case of some stochastic differential equations with dispersion.

Keywords Strong solutions • Entropy • Invertibility • Shannon's inequality

Received 12/5/2011; Accepted 6/12/2012; Final 7/10/2012

1 Introduction

The invertibility of adapted perturbations of the identity on Wiener space may be seen as an alternative approach to investigate the existence of a unique strong solution to stochastic differential equations of the form

$$dX_t = dB_t - \dot{v}_t(X)dt;\ X_0 = 0. \tag{3.1}$$

R. Lassalle (✉) • A.S. Üstünel
LTCI CNRS Dépt. Infres, Institut Telecom, Telecom ParisTech,
46, rue Barault, 75013, Paris, France
e-mail: lassalle@telecom-paristech.fr

F. Viens et al. (eds.), *Malliavin Calculus and Stochastic Analysis: A Festschrift in Honor of David Nualart*, Springer Proceedings in Mathematics & Statistics 34,
DOI 10.1007/978-1-4614-5906-4_3, © Springer Science+Business Media New York 2013

It has been developed in the last years (for instance see [17, 29–31, 35]) and offers many advantages with respect to other approaches. In particular it involves compact notations and it introduces a framework which enables to take a full advantage of stochastic analysis. It also provides naturally some new connections between many problems from filtering theory to optimal transport. In this paper we continue the study of invertibility and we provide new applications. To be precise, let us introduce some notations. In this introduction we note (W, H, μ) the classical Wiener space, where $W = C([0,1], \mathbb{R})$, where H is the associated Cameron–Martin space, and where μ is the Wiener measure which is such that the coordinate process $t \to W_t(\omega) := \omega(t)$ is a Wiener process on (W, μ). We also note $L_a^0(\mu, H)$ the set of the equivalence classes with respect to μ of measurable mappings $u : \omega \in W \to \int_0^{\cdot} \dot{u}_s(\omega) ds \in H$ such that $t \to \dot{u}_t$ is adapted to the filtration generated by the coordinate process. The notion of invertibility is related to a $v : W \to H$ which satisfies the two following conditions:

$$v \in L_a^0(\mu, H) \tag{3.2}$$

and

$$E_\mu\left[\exp\left(-\int_0^1 \dot{v}_s dW_s - \frac{1}{2}\int_0^1 \dot{v}_s^2 ds\right)\right] = 1. \tag{3.3}$$

Such a v defines a perturbation of the identity V which is given by

$$V := I_W + v,$$

where $I_W : \omega \in W \to \omega \in W$ is the identity on W. Under Eqs. (3.2) and (3.3) a notion of stochastic invertibility on Wiener space for V was introduced in [35] (see also [30] and [17]). To define it, it is sufficient to note that for any $U : W \to W$ such that $U\mu << \mu$ (i.e., absolutely continuous) the pullbacks $V \circ U$ and $U \circ V$ are well defined μ- almost surely [16]. In particular if $U := I_W + u$ where $u \in L_a^0(\mu, H)$, it is well known that $U\mu << \mu$ so that both pullbacks are well defined. Hence, under these hypothesis a notion of stochastic inverse for $V = I_W + v$ can be naturally defined in the following way: V is said to be invertible if there is a mapping $u \in L_a^0(\mu, H)$ such that $U := I_W + u$ satisfies $\mu - a.s.$

$$V \circ U = I_W$$

and

$$U \circ V = I_W.$$

As far as v satisfies the two hypothesis Eqs. (3.2) and (3.3), the relevance of this notion comes from the equivalence of the invertibility of $V = I_W + v$ with the existence of a unique strong solution to Eq. (3.1). In particular, the invertibility is an original and useful tool to prove the pathwise uniqueness. Many sufficient conditions of invertibility where given in [35] by means of Malliavin calculus, but a necessary and sufficient condition was still to find. This task was achieved in

[30] where a criterion of invertibility for adapted shifts was shown. This criterion relates the invertibility of a shift to an equality between the entropy of the measure it induces and the energy of the associated drift. More accurately, let $V := I_W + v$ where v satisfies the conditions (3.2) and (3.3). It was shown that for any u which satisfies the same conditions and which is such that the density of $U := I_W + u$ is related to v by

$$\frac{dU\mu}{d\mu} = \exp\left(-\int_0^1 \dot{v}_s dW_s - \frac{1}{2}\int_0^1 \dot{v}_s^2 ds\right)$$

we have

$$H(U\mu|\mu) \leq E_\mu\left[\frac{|u|_H^2}{2}\right]$$

with equality if and only if V is invertible with inverse U. We recall that the relative entropy $H(U\mu|\mu)$ is defined by

$$H(U\mu|\mu) = E_\mu\left[\frac{dU\mu}{d\mu} \ln \frac{dU\mu}{d\mu}\right]$$

In [17] we provided a general and simple proof of this criterion and we dropped the hypothesis that u had to satisfy Eq. (3.3). The persistence of invertibility under stopping was shown in [32] and an explicit formula for the stopped inverse was given in [17]. The notion of local invertibility which will be recalled accurately below was introduced in [17], and thanks to this last result, it was proved to be equivalent to invertibility under a finite entropy condition. In this paper we first extend this result to prove that the equivalence between invertibility and local invertibility always holds for adapted shifts. In other words, we show that invertibility is a local property in the usual sense of stochastic calculus. This result provides local versions of results related to the pathwise uniqueness. Under mild conditions, we illustrate the use of these local properties by showing that the pathwise uniqueness holds for the stochastic description of the free euclidean quantum mechanics [39]. To motivate this kind of results, we recall here that from the origin, stochastic mechanics provides a stochastic representation of physical phenomenons by means of weak solutions of stochastic differential equations. However, the existence of a unique strong solution may be seen as the stochastic counterpart of the classical picture of the underlying determinism. Hence, it seemed relevant to investigate the pathwise uniqueness for the related equations in great generality. In this paper, we only investigated the case of euclidean quantum mechanics which is well known to describe the continuous limit of thermodynamic systems. In that case the Brownian motion represents the effects of the thermal energy, i.e., the shocks of many small particles, and it is physically clear that pathwise uniqueness should hold. By using the local properties of invertibility, we give a straightforward proof of this result for the free particle, both in finite and infinite dimensions, and we also relate this result to optimal transport. The rest of the paper is devoted to applications and extensions of the criterion of [30]. We give some applications of this criterion to information theory and we give an extension of this criterion to some stochastic differential equations with dispersion. In particular, thanks to a change of measure, we give an abstract Wiener space version of the famous inequality of Shannon (see [1] and references therein) with a proof which receives a nice interpretation in terms of information loss in Gaussian channels. The structure of the paper is the following. In Sect. 2 we state the main notations which will be used in the whole paper. We also recall briefly the main tools of Malliavin calculus. In Sect. 3 we define the Girsanov shift in the same way as in [17] and we give some of its main properties. We also give some new results which rely on the main properties of the Girsanov shift. In Sect. 4 we recall the notion of invertibility and we give some results of [17] which will be used in the sequel. In Sect. 5 we recall the notion of local invertibility for adapted shifts and we show that it is equivalent to invertibility. In Sects. 6 and 7 we apply these notions to prove the invertibility of some Markovian shifts. Specifically in Sect. 6 we consider the classical Wiener space $W = C_0([0,1], \mathbb{R}^d)$ and we show a new sufficient condition for the invertibility of Markovian shifts on that space. In Sect. 7 we use this result to show the invertibility for shifts associated with the free euclidean quantum particle. Under mild conditions we also generalize this result to the Brownian sheet by means of Malliavin calculus. In Sect. 8 we investigate some applications of the criterion of [30] to information theory on Wiener space. In particular, thanks to a change of measure, we show that this criterion may be written in terms of variance. We then get easily a formulation of Shannon's inequality on abstract Wiener space as the consequence of the properties of variance and of information loss in Gaussian channel. In Sect. 9 we give some other results related to the invertibility of the processes we considered in Sect. 7: specifically we investigate the connection with optimal transport. Finally in Sect. 10 we generalize the criterion of [30] to some stochastic differential equations with dispersion.

2 Notations

Let (S, H_S, i_S) be an abstract Wiener space [15] where S is a separable Banach space, H_S the associated Cameron–Martin space, and i_S the injection of H_S into S which is dense and continuous. In this paper S will be the state space. Indeed we will work on the space $W := C_0([0,1], S)$ of the continuous paths vanishing at 0 with states in S. We recall that W is also a separable Banach space with an associated Cameron–Martin space H defined by

$$H = \left\{ \eta : [0,1] \to H_S, \eta_t = \int_0^t \dot{\eta}_s ds, \int_0^1 |\dot{\eta}_s|_{H_S}^2 ds < \infty \right\}.$$

Moreover the scalar product on H is given by

$$< h, k >_H = \int_0^1 < \dot{h}_s, \dot{k}_s >_{H_S} ds$$

for any $h, k \in H$. The injection i of H into W is also dense and continuous and (W, H, i) is also an abstract Wiener space. Let μ be the Wiener measure on $(W, \mathcal{B}(W))$ and let $\mathcal{F}$ be the completion of the Borelian sigma-field $\mathcal{B}(W)$ with respect to μ. We still note μ the unique extension of the Wiener measure on $(W, \mathcal{F})$ and we still call it the Wiener measure. We also note $I_W : \omega \in W \to \omega \in W$ the identity map on W. In the sequel I_W will be seen as an equivalence class of $M_\mu((W,\mathcal{F}),(W,\mathcal{F}))$, where $M_\mu((W,\mathcal{F}),(W,\mathcal{F}))$ is the set of the μ-equivalence classes of mappings from W into itself, which are $\mathcal{F}/\mathcal{F}$ measurable. To cope with adapted processes we need to introduce not only the filtration $(\mathcal{F}_t^0)$ generated by the coordinate process $t \to W_t$ but also the filtration $(\mathcal{F}_t)$ which is the usual augmentation [3] of $(\mathcal{F}_t^0)$ with respect to the Wiener measure μ. We note $L^0(\mu, H)$ (resp. for a probability law ν equivalent to μ, $L^2(\nu, H)$) the set of the equivalence classes with respect to μ of the measurable H-valued mappings $u : W \to H$ (resp. the subset of the $u \in L^0(\mu, H)$ such that $E_\nu[|u|_H^2] < \infty$). We also set $L_a^0(\mu, H)$ (resp. $L_a^2(\nu, H)$) the subset of the $u \in L^0(\mu, H)$ (resp. of the $u \in L^2(\nu, H)$) such that $t \to \dot{u}_t$ is adapted to $(\mathcal{F}_t)$. Let ν be a probability equivalent to μ, and $t \to B_t$ be a $(\mathcal{F}_t)$-Wiener process on $(W, \mathcal{F}, \nu)$. The abstract stochastic integral [34] of a $a \in L_a^0(\mu, H)$ with respect to B will be noted $\delta^B a$. In this context ($W = C_0([0,1], S)$) it can also be written

$$\delta^B a = \int_0^1 \dot{a}_s \mathrm{d}B_s$$

For a shift $U := B + u$ where $u \in L_a^0(\mu, H)$ we set

$$\delta^U a := \delta^B a + < u, a >_H .$$

In particular for a $u \in L_a^0(\mu, H)$, $\delta^W u = \int_0^1 \dot{u}_t \mathrm{d}W_t$. We recall that in the case where W is the classical Wiener space, the abstract stochastic integral is nothing but the usual stochastic integral. For convenience of notations, for any optional time τ with respect to $(\mathcal{F}_t)$ we note $(\pi_\tau a)_. = \int_0^. 1_{[0,\tau]}(s)\dot{a}_s \mathrm{d}s$ and $a^\tau := \pi_\tau a$. In particular for any $t \in [0,1]$ $(\pi_t a)_. = \int_0^. 1_{[0,t]}(s)\dot{a}_s \mathrm{d}s$. For a $U \in M_\mu((W,\mathcal{F}),(W,\mathcal{F}))$ and a probability ν equivalent to μ, the image measure of ν by U will be denoted by $U\nu$. Moreover for any random variable L on $(W, \mathcal{F})$ such that $E_\mu[L] = 1$ and $\mu - a.s.$ $L \geq 0$, we will note $L.\mu$ the probability on $(W, \mathcal{F})$ whose density with respect to the Wiener measure is L. To be consistent with [30] we set

$$\rho(-\delta^W u) := \exp\left(-\delta^W u - \frac{|u|_H^2}{2}\right). \tag{3.4}$$

For the sections below Sect. 5 we need to consider the Wiener measure on S: we note it $\widehat{\mu}$ and we recall that we then have $W_1\mu = \widehat{\mu}$. We will also consider some integrals over Borelian measures of S-valued or H_S-valued elements: all these integrals are Bochner integrals. Finally in the whole paper we adopt the convention $\inf(\emptyset) = 1$. We now give the notations of Malliavin calculus and then we give a brief reminder of it (see [19, 26, 28, 31], or [13] for more detail). We note $\{\delta h, h \in H\}$

(resp. $\{<.,h>, h \in H_S\}$) the isonormal Gaussian field which is the closure of $W^\star$ (resp. of $S^\star$) in $L^p(\mu)$ (resp. in and $L^p(\widehat{\mu})$) and ∇ (resp. D) will denote the Malliavin derivative on W (resp. on S). We also note $D_s\phi$ the density of $\nabla\phi$, i.e., $\nabla_h\phi = \int_0^1 < D_s\phi, \dot{h}_s >_{H_S} \mathrm{d}s$ for any $h \in H$. We now recall the construction of the derivative on W, but the construction on S is exactly the same (we don't use the time structure at this point). Let $(k_i)_{i\in\mathbb{N}} \subset H$ be an orthonormal basis of H, let E be a separable Hilbert space, and let $(e_i)_{i\in\mathbb{N}} \subset E$ be an orthonormal basis of E. For every $F \in \cap_{p>1} L^p(\mu, E)$ we say that F is a cylindrical function and we note $F \in \mathcal{S}_\mu(E) \subset \cap_{p>1} L^p(\mu, E)$ if there exist a $n \in \mathbb{N}$, $(l_1, \ldots, l_n) \in (\mathbb{N}^*)^n$, $(k_{l_1}, \ldots, k_{l_n}) \subset (k_i)_{i\in\mathbb{N}}$ and an f in the Schwartz space of the smooth rapidly decreasing functions $\mathcal{S}(\mathbb{R}^n)$ such that $\mu - a.s.$

$$F = \sum_{i=1}^{m} f^i(\delta k_{l_1}, \ldots, \delta k_{l_n}) e_i.$$

If we set

$$\nabla_h F = \frac{\mathrm{d}}{\mathrm{d}\lambda} F \circ \tau_{\lambda h}|_{\lambda=0},$$

where for any $h \in H$

$$\tau_h : \omega \in W \to \tau_h(\omega) := \omega + h \in W,$$

we then have

$$\nabla_h F = \sum_{i=1}^{m} \sum_{j=1}^{n} \partial_j f^i(\delta k_{l_1}, \ldots, \delta k_{l_n}) < h, k_{l_j} >_H e_i.$$

By construction, up to a negligible set, for every $\omega \in W$ the mapping defined by $(\nabla F)(\omega) : h \in H \to (\nabla_h F)(\omega) \in E$ is linear and continuous and even Hilbert–Schmidt with the property that $\nabla_h F(\omega) = (\nabla F)(\omega)(h)$. Therefore, by using Hilbert–Schmidt tensor products, we have the explicit formula:

$$\nabla F = \sum_{i=1}^{m} \sum_{j=1}^{n} \partial_j f^i(\delta k_{l_1}, \ldots, \delta k_{l_n}) k_{l_j} \otimes e_i$$

and we have defined a linear operator $\nabla : \mathcal{S}_\mu(E) \subset L^p(\mu, E) \to L^p(\mu, H \otimes E)$ which is such that $\mu - a.s.$

$$\frac{\mathrm{d}F(\omega + \lambda h)}{\mathrm{d}\lambda}|_{\lambda=0} = \nabla_h F.$$

Thanks to the Cameron–Martin theorem, it is easy to see that although ∇ is not a closed operator, it is however closable. We still denote by $\nabla : \mathrm{Dom}_p(\nabla, E) \subset$

$L^p(\mu, E) \to L^p(\mu, H \otimes E)$ the closure of $\nabla : \mathcal{S}_\mu(E) \subset L^p(\mu, E) \to L^p(\mu, H \otimes E)$ which can be built explicitly in the following way. Let $\mathrm{Dom}_p(\nabla, E)$ be the set of the $F \in L^p(\mu, E)$ for which there is a sequence of cylindrical random variables $(F_n)_{n\in\mathbb{N}} \subset \mathcal{S}_\mu(E)$ with the property that $\lim_{n\to\infty} F_n = F$ in $L^p(\mu, E)$ and ∇F_n is Cauchy in $L^p(\mu, H \otimes E)$. Then for any $F \in \mathrm{Dom}_p(\nabla, E)$ we can define $\nabla F = \lim_{n\to\infty} \nabla F_n$ which is unique since ∇ is closable. By construction $\mathrm{Dom}_p(\nabla, E)$ is the completion of $\mathcal{S}_\mu(E)$ with respect to the norm of the graph associated with ∇ which is defined by $||F||_{p,1;E} = ||F||_{L^p(\mu,E)} + ||\nabla F||_{L^p(\mu,H\otimes E)}$. We note $\mathbb{D}_{p,1}(E)$ the Banach space $\mathrm{Dom}_p(\nabla, E)$ endowed with the norm $||F||_{p,1;E}$. Of course ∇ is nothing but the infinite-dimensional version of the Sobolev derivative with respect to the Gaussian measure, and $\mathbb{D}_{p,1}(E)$ is the Sobolev space associated with the weak Gross–Sobolev derivative ∇. We define the higher-order derivatives and the associated Sobolev spaces by iterating the same procedure. Thus, if $\nabla^{k-1} F \in \mathbb{D}_{p,1}(E \otimes H^{\otimes(k-1)})$, we can define $\nabla^k F := \nabla(\nabla^{k-1} F)$ and the associated Sobolev space $\mathbb{D}_{p,k}(E)$ as being the set of such F equipped with the norm $||F||_{p,k;E} = \sum_{i=0}^{k} ||\nabla^i F||_{L^p(\mu,E\otimes H^{\otimes i})}$. In the sequel we will often deal with the case where $E = \mathbb{R}$. Note that in that case, because of the Riesz representation theorem, $H \otimes \mathbb{R} \simeq H$ so that we can identify (with fixed ω) $\nabla F(\omega)$ with a vector of H and we will write $\nabla_h F =< h, \nabla F >_H$. Still in that case we note $\mathbb{D}_{p,1}$ instead of $\mathbb{D}_{p,1}(\mathbb{R})$. Finally we define the so-called divergence operator. By the monotone class theorem and from the martingale convergence theorem it is easy to see that $\mathcal{S}_\mu(E)$ is dense in every $L^p(\mu, E)$, $p \geq 1$. Since $\mathcal{S}_\mu(E) \subset \mathbb{D}_{p,1}(E)$, the operator $\nabla : \mathrm{Dom}_p(\nabla, E) \subset L^p(\mu, E) \to L^p(\mu, H \otimes E)$ has a dense support. Therefore there is an operator δ which is the adjoint of ∇. The domain $\mathrm{Dom}_p(\delta, E)$ is defined classically as being the set of the random variables $\xi \in L^p(\mu, H \otimes E)$ such that for any $\phi \in \mathbb{D}_{q,1}(E)$ (where $\frac{1}{p} + \frac{1}{q} = 1$) $E_\mu[< \nabla\phi, \xi >_{H\otimes E}] \leq c_{p,q}(|\phi|_{L^q(\mu,E)})$. For any $\xi \in \mathrm{Dom}_p(\delta, E)$ $\delta\xi$ is characterized by the relation $E_\mu[< \phi, \delta\xi >_E] = E_\mu[< \nabla\phi, \xi >_{H\otimes E}]$ which holds for any $\phi \in \mathrm{Dom}_p(\nabla, E)$. Of course this relation is the infinite-dimensional counterpart of the integration by part with respect to the Gaussian measure. Note that the set of the constant H-valued random variables is a subset of all the $\mathrm{Dom}_p(\delta) := \mathrm{Dom}_p(\delta, \mathbb{R})$ and that the Cameron–Martin theorem implies $E_\mu[\phi\delta h] = E[< \nabla\phi, h >_H]$ for any $h \in H$. Hence it is clear that one may think to this operator as an extension of $\delta : H \to L^p(\mu, H)$, which justifies the notations. The divergence of a u on S can be defined likewise and we note it $< u, . >$.

3 The Girsanov Shift

Let ν be a probability which is equivalent to the Wiener measure μ. Then it is well known (see, for instance, Sect. 2.6 of [34] or [16]) that there is a unique $v \in L_a^0(\mu, H)$ such that $t \to W_t + v_t$ is a $(\mathcal{F}_t)$-abstract Wiener process on $(W, \mathcal{F}, \nu)$ and $\mu - a.s.$

$$\frac{d\nu}{d\mu} = \rho(-\delta^W v).$$

We say that v (resp. $V := I_W + v$) is the Girsanov drift (resp. shift) associated with ν. We also recall that when $\mu - a.s.$ $\frac{d\nu}{d\mu} = 1 + \delta\alpha$ for an $\alpha \in L^0_a(\mu, H)$ we have $dt \times d\mu - a.s.$ $\dot{v}_t = -\frac{\dot{\alpha}_t}{E_\mu[\frac{d\nu}{d\mu}|\mathcal{F}_t]}$. If we further assume that $\frac{d\nu}{d\mu} \in \mathbb{D}_{2,1}$ an easy application of the Clark–Ocone formula [31] yields

$$\dot{v}_s = -E_\nu\left[D_s \ln \frac{d\nu}{d\mu}|\mathcal{F}_s\right].$$

An old result of Föllmer [7, 8] relates the integrability of the Girsanov drift to the relative entropy. In the case of probabilities equivalent to μ, a generalization of this result is the Proposition 3.1 [to recover Föllmer result take $\widetilde{\nu} = \mu$ in Eq. (3.5)] which shows that the relative entropy of two probabilities equivalent to the Wiener measure is related to a distance between their Girsanov drifts. In particular it shows that the variance of the Girsanov drift may also be seen as an entropy. We recall that the relative entropy of a probability ν absolutely continuous with respect to a probability $\widehat{\nu}$ is defined by

$$H(\nu|\widehat{\nu}) = E_\nu\left[\ln \frac{d\nu}{d\widehat{\nu}}\right].$$

Proposition 3.1. *Let ν and $\widetilde{\nu}$ be two probabilities equivalent to the Wiener measure μ. Then we have*

$$2H(\nu|\widetilde{\nu}) = E_\nu\left[|v - \widetilde{v}|^2_H\right], \tag{3.5}$$

where v (resp. $\widetilde{v}$) is the Girsanov drift associated with ν (resp. with $\widetilde{\nu}$). In particular for any measure ν equivalent to μ with a Girsanov drift v, if we set

$$\frac{d\mu_\nu}{d\mu} := \rho(-\delta^W E_\nu[v]) \tag{3.6}$$

[see Eq. (3.4) for the definition of the right-hand term] we then have

$$2H(\nu|\mu_\nu) = \mathcal{V}ar_\nu(v),$$

where $\mathcal{V}ar_\nu(v) := E_\nu\left[|v - E_\nu[v]|^2_H\right]$.

Proof. By definition we have

$$\begin{aligned} E_\nu\left[\ln \frac{d\nu}{d\widetilde{\nu}}\right] &= E_\nu\left[\ln \frac{d\nu}{d\mu}\right] - E_\nu\left[\ln \frac{d\widetilde{\nu}}{d\mu}\right] \\ &= E_\nu\left[-\delta^W v - \frac{|v|^2_H}{2}\right] - E_\nu\left[-\delta^W \widetilde{v} - \frac{|\widetilde{v}|^2_H}{2}\right] \end{aligned}$$

$$= E_\nu\left[-\delta^V v + \frac{|v|_H^2}{2}\right] - E_\nu\left[-\delta^V \widetilde{v} + < v, \widetilde{v} >_H - \frac{|\widetilde{v}|_H^2}{2}\right]$$

$$= E_\nu\left[-\delta^V (v - \widetilde{v}) + \frac{|v - \widetilde{v}|_H^2}{2}\right].$$

If $v - \widetilde{v} \in L_a^2(\nu, H)$ we have $E_\nu[-\delta^V (v - \widetilde{v})] = 0$ so that equality (3.5) holds. Conversely we assume that $H(\nu|\widetilde{\nu}) < \infty$ and for each $n \in \mathbb{N}$ we set $\tau_n = \inf(\{t \in [0,1] : |\pi_t(v - \widetilde{v})|_H > n\})$ with the convention $\inf(\emptyset) = 1$. Since $v - \widetilde{v} \in L_a^0(\mu, H)$ we have $\mu - a.s.\ \tau_n \uparrow 1$. Therefore the monotone convergence theorem implies

$$\begin{aligned} E_\nu\left[|v - \widetilde{v}|_H^2\right] &= E_\nu\left[\lim_{n\to\infty} |\pi_{\tau_n}(v - \widetilde{v})|_H^2\right] \\ &= \lim_{n\to\infty} E_\nu\left[|\pi_{\tau_n}(v - \widetilde{v})|_H^2\right] \\ &= \lim_{n\to\infty} H(\nu|_{\mathcal{F}_{\tau_n}} | \widetilde{\nu}|_{\mathcal{F}_{\tau_n}}) \\ &\leq H(\nu|\widetilde{\nu}), \end{aligned}$$

where $\widetilde{\nu}|_{\mathcal{F}_{\tau_n}}$ (resp. $\nu|_{\mathcal{F}_{\tau_n}}$) is the measure induced by $\widetilde{\nu}$ (resp. by ν) on $\mathcal{F}_{\tau_n}$. Hence we proved that $v - \widetilde{v} \in L_a^2(\nu, H)$ if and only if $H(\nu|\widetilde{\nu}) < \infty$, and that we always have Eq. (3.5). □

Remark 3.1. Consider the function $g^\lambda(\eta) := H(\eta|\lambda)$ where η is any Borelian probability on $\mathbb{R}^d$ which is equivalent to Lebesgue measure λ on $\mathbb{R}^d$. For any $h \in \mathbb{R}^d$ the translation $T_h : x \in \mathbb{R}^d \to x + h \in \mathbb{R}^d$ is invertible and λ is invariant under the action of T_h (i.e., $T_h\lambda = \lambda$). Thus for any such η and any $h \in \mathbb{R}^d$, $g^\lambda(T_h\eta) = H(T_h\eta|\lambda) = H(T_h\eta|T_h\lambda) = H(\eta|\lambda) = g^\lambda(\eta)$. On the path space the Lebesgue measure is no more defined and one often consider $g^\mu(\nu) := H(\nu|\mu)$ where ν is a Borelian measure. For any $h \in H$ let $\tau_h : \omega \in W \to \omega + h \in W$. Since $\tau_h\mu \neq \mu$ for any $h \neq 0$, we generally don't have $g^\mu(\nu) := g^\mu(\tau_h\nu)$. However the function $f(\nu) := H(\nu|\mu_\nu)$ has the nice property to be invariant under translations along H (just as g^λ was on $\mathbb{R}^d$). Indeed an easy application of the Cameron–Martin theorem shows that for any $h \in H$

$$\frac{d\mu_{\tau_h\nu}}{d\mu} = \rho(-\delta^W(\tau_{-h}(E_\nu[v])), \tag{3.7}$$

where v is the Girsanov drift associated with ν and that

$$\tau_h\mu_\nu = \mu_{\tau_h\nu}.$$

Hence for any $h \in H$ we have

$$\begin{aligned} f(\tau_h \nu) &= H(\tau_h \nu | \mu_{\tau_h \nu}) \\ &= H(\tau_h \nu | \tau_h \mu_\nu) \\ &= H(\nu | \mu_\nu) \\ &= f(\nu), \end{aligned}$$

where the last equalities hold since τ_h is invertible.

Let ν be a probability equivalent to μ and let $V := I_W + v$ be its Girsanov shift. Since V is a $(\mathcal{F}_t)$-Wiener process on $(W, \mathcal{F}, \nu)$ we know that (I_W, V) is a weak solution of Eq. (3.8) with law ν. However, the uniqueness in law may not hold for Eq. (3.8) and some weak solutions of Eq. (3.8) may exist with a law on W which is not equal to ν. Proposition 3.2 shows that when the uniqueness in law does not hold for Eq. (3.8) there exists a weak solution whose law on W is not equivalent to μ.

Proposition 3.2. *Let $v \in L^0_a(\mu, H)$ be such that $E_\mu[\rho(-\delta^W v)] = 1$ and consider the following stochastic differential equation:*

$$\mathrm{d}X_t = \mathrm{d}B_t - \dot{v}_t \circ X \mathrm{d}t; \quad X_0 = 0. \tag{3.8}$$

The uniqueness in the sense of probability law does not hold for Eq. (3.8) [i.e., there exist two weak solutions of Eq. (3.8) whose laws on W are not equal] if and only if there exist two weak solutions of Eq. (3.8) whose laws on W are not equivalent.

Proof. The sufficiency is obvious. Hence we just have to prove that if the uniqueness in the sense of probability law does not hold for Eq. (3.8) there exist two weak solutions of Eq. (3.8) whose laws on W are not equivalent. We set $\nu := \rho(-\delta^W v).\mu$. Since $(I_W, V := I_W + v)$ is a weak solution on $(W, \mathcal{F}, \nu)$, ν is the law of a solution. Thus it suffices to prove that if there is a weak solution of Eq. (3.8) whose law is not equal to ν, it is not equivalent to ν neither. Let $(\Omega, \mathcal{G}, \mathbb{P})$ be a complete probability space and $(\mathcal{G}_t)_{t \in [0,1]}$ a complete and continuous filtration on it. Further assume that (U, B) is a weak solution of Eq. (3.8) on that space with that filtration and note $u := U - B$. If $\mathbb{P}(|u|_H = \infty) > 0$ then $U\mathbb{P}$ is not absolutely continuous with respect to μ (see Theorem 2.4.1 of [34]) and in particular $U\mathbb{P}$ is not equivalent to ν. We still have to prove that if $\mathbb{P}(|u|_H = \infty) = 0$, then $U\mathbb{P} \neq \nu$ implies that μ is not equivalent to $U\mathbb{P}$. By contraposition we suppose that $\mathbb{P}(|u|_H = \infty) = 0$ and $\mu \sim U\mathbb{P}$, and we have to show that it implies $U\mathbb{P} = \nu$. The hypothesis $\mu \sim U\mathbb{P}$ yields the existence of a $v^i \in L^0_a(\mu, H)$ such that $V^i := I_W + v^i$ is a $(\mathcal{F}_t)$-Wiener process on $(W, \mathcal{F}, U\mathbb{P})$ and $\mu - a.s.$

$$\frac{\mathrm{d}U\mathbb{P}}{\mathrm{d}\mu} = \rho(-\delta^W v^i).$$

On the other hand (U, B) is a weak solution. Hence for any $l \in W^\star$ and for any $\theta \in C^b(W)$ which is $\mathcal{F}_s$ measurable we have:

$$E_{U\mathbb{P}}[(\delta^V \pi_t l - \delta^V \pi_s l)\theta] = E_{\mathbb{P}}[(\delta^B \pi_t l - \delta^B \pi_s l)\theta \circ U]$$
$$= 0,$$

where the last equality holds since $\mathcal{U}_s \subset \mathcal{G}_s$ where $(\mathcal{U}_s)$ is the filtration generated by $t \to U_t$. From Paul Levy's criterion it implies that $t \to V_t$ is an abstract $(\mathcal{F}_t)$-Wiener process on $(W, \mathcal{F}, U\mathbb{P})$. Hence for any $l \in W^\star$ $t \to \delta^{V^i - V}\pi_t l =< \pi_t l, v^i - v >_H$ is a continuous martingale with finite variations. Therefore $\mu - a.s.$ $v = v^i$, and $U\mathbb{P} = \rho(-\delta^W v) = \nu$. This achieves the proof. □

Two straightforward consequences of Proposition 3.1 which we will use in the sequel are the following path space version of two inequalities: the Talagrand inequality (see [27,31], and references therein) and the Sobolev inequality [11,31]. The associated proofs are well known and can be found in [30,31]. However, before we do this we have to recall the definition of the Wasserstein distance.

Definition 3.1. Let ρ and ν be two probabilities on a Wiener space $\widetilde{W}$ (in the sequel we shall consider $\widetilde{W} = S$ or $\widetilde{W} = W$). We then note $\Sigma(\rho, \nu)$ be the set of the measures on $(\widetilde{W} \times \widetilde{W}, \mathcal{B}(\widetilde{W} \times \widetilde{W}))$ whose first (resp. second) marginal is ρ (resp. ν). A measure $\gamma \in \Sigma(\rho, \nu)$ is said to be the solution of the Monge–Kantorovitch problem if

$$J(\gamma) = \int_{W\times W} |x-y|_H^2 d\gamma(x,y) = \inf\left(\left\{\int_{W\times W} |x-y|_H^2 d\beta(x,y) : \beta \in \Sigma(\rho,\nu)\right\}\right).$$

Let $d(e, v) := J(\gamma)$, the wasserstein distance between e and v is $\sqrt{d(e, v)}$.

Proposition 3.3. *For any probability ν equivalent to μ we have*

$$d(\nu, \mu) \leq 2H(\nu|\mu). \tag{3.9}$$

Proposition 3.4. *For any probability ν equivalent to μ which is such that $\frac{d\nu}{d\mu} \in \mathbb{D}_{2,1}$, we have*

$$H(\nu|\mu) \leq J(\nu|\mu),$$

where

$$J(\nu|\mu) := E_\nu\left[\left|\nabla \ln \frac{d\nu}{d\mu}\right|_H^2\right].$$

4 Invertibility of Adapted Shifts

In this section we first recall our definition of invertibility for adapted shifts on W. Then we recall two propositions which we already proved elsewhere [17] and which we shall use in the sequel.

Definition 4.1. Let $v \in L_a^0(\mu, H)$ be such that $E_\mu\left[\rho(-\delta^W v)\right] = 1$. Then $V := I_W + v$ is said to be (globally) invertible with (a global) inverse $U := I_W + u$ where $u \in L_a^0(\mu, H)$, if and only if $\mu - a.s.$

$$V \circ U = I_W$$

and

$$U \circ V = I_W.$$

The next proposition, which was proved in [17], enlighten the hypothesis of [30] and is also very useful to get the invertibility from the right invertibility.

Proposition 4.1. *Let ν be a probability equivalent to μ, and let $V = I_W + v$ be the Girsanov shift associated with ν. Further assume that there is a $u \in L_a^0(\mu, H)$ such that $U := I_W + u$ is the right inverse of V, i.e., $\mu - a.s.$ $V \circ U = I_W$. Then the following assertions are equivalent:*

1. $E_\mu[\rho(-\delta^W u)] = 1$.
2. $U\mu \sim \mu$.
3. *V is invertible with inverse U (see Definition 4.1).*
4. $U\mu = \nu$.

Proposition 4.2 was proved in [17]. It is an improvement of Theorem 3.1 of [32]. Contrary to the latter, Proposition 4.2 provides an explicit formula for the inverse of the stopped shift.

Proposition 4.2. *Let $v \in L_a^0(\mu, H)$ be such that $E_\mu\left[\rho(-\delta^W v)\right] = 1$ and let σ and τ be two $(\mathcal{F}_t)$-optional times such that $\mu - a.s.$ $\sigma \leq \tau$. Further assume that $V^\tau := I_W + v^\tau$ is invertible with inverse $U = I_W + u$ (see Definition 4.1) where $u \in L_a^0(\mu, H)$. Then $V^\sigma := I_W + v^\sigma$ is invertible with inverse*

$$\widetilde{U} := I_W + \pi_{\sigma \circ U} u.$$

Moreover, we have $\mu - a.s.$

$$v^\sigma \circ U = v^\sigma \circ \widetilde{U}.$$

5 Local Invertibility: General Case

We introduced explicitly the notion of local invertibility of adapted shifts in [17] for a $v \in L^0_a(\mu, H)$ such that $E_\mu[\rho(-\delta^W v)] = 1$. We showed that under a finite energy condition (i.e., $v \in L^2_a(\nu, H)$ where $\nu = \rho(-\delta^W v).\mu$), it was equivalent to the invertibility. As a matter of fact it seems that this notion already existed implicitly in the literature [35] and that the equivalence between invertibility and local invertibility was known under the condition $v \in L^2_a(\mu, H)$. These two results suggest that the equivalence between invertibility and local invertibility may be more general. Theorem 5.1 completely solves this problem and shows that this equivalence holds in full generality.

Definition 5.1. Let $v \in L^0_a(\mu, H)$ be such that $E_\mu[\rho(-\delta^W v)] = 1$ and let $V := I_W + v$. V is said to be locally invertible if there is a sequence $(u^n)_{n\in\mathbb{N}} \subset L^0_a(\mu, H)$ and a sequence $(\tau_n)_{n\in\mathbb{N}}$ of $(\mathcal{F}_t)$-optional times such that $\mu - a.s.$ $\tau_n \uparrow 1$ and for each $n \in \mathbb{N}$, $V^n := I_W + \pi_{\tau_n} v$ is invertible (see Definition 4.1) with inverse $U^n := I_W + u^n$.

Lemma 5.1. *Let $v \in L^0_a(\mu, H)$ be such that $E_\mu\left[\rho(-\delta^W v)\right] = 1$, and let $(\tau_n)_{n\in\mathbb{N}}$ be a sequence of optional times such that $\mu - a.s.$ $\tau_n \uparrow 1$. Further assume that there is a sequence $(u^n) \subset L^0_a(\mu, H)$ such that for each $n \in \mathbb{N}$ $V^n := I_W + \pi_{\tau_n} v$ is invertible with inverse $U^n := I_W + u^n$. Then (u^n) converges in $L^0_a(\mu, H)$.*

Proof. For convenience of notations we note $v^n := \pi_{\tau_n} v$. By definition and by Proposition 4.2, for each $n \geq m$ we have $\mu - a.s.$

$$v^m \circ U^m = v^m \circ U^n.$$

This yields

$$\begin{aligned}
\mu\left(|u^n - u^m|_H > \epsilon\right) &= \mathrm{E}_\mu\left[1_{|u^n - u^m|_H > \epsilon}\right] \\
&= \mathrm{E}_\mu\left[1_{|v^n \circ U^n - v^m \circ U^m|_H > \epsilon}\right] \\
&= \mathrm{E}_\mu\left[1_{|v^n \circ U^n - v^m \circ U^n|_H > \epsilon}\right] \\
&= \mathrm{E}_\mu\left[1_{|v^n \circ U^n \circ V^n - v^m \circ U^n \circ V^n|_H > \epsilon}\rho(-\delta^W v^n)\right] \\
&= \mathrm{E}_\mu\left[1_{|v^n - v^m|_H > \epsilon}\rho(-\delta^W v^n)\right].
\end{aligned}$$

On the other hand $(\rho(-\delta^W v^n))_{n\in\mathbb{N}}$ (resp. $(v^n)_{n\in\mathbb{N}}$) converges almost surely to $\rho(-\delta^W v)$ (resp. to v). Since $(\rho(-\delta^W v^n))_{n\in\mathbb{N}}$ is uniformly integrable, the dominated convergence theorem implies that (u^n) converges in $L^0(\mu, H)$. □

Remark 5.1. Of course since the norm $|.|_W$ is weaker than $|.|_H$ the convergence of (u^n) in $L^0_a(\mu, H)$ to a $u \in L^0_a(\mu, H)$ also implies the convergence in probability of (U^n) to $U := I_W + u$.

The next theorem completely solves the problem of the equivalence between invertibility and local invertibility.

Theorem 5.1. *Let* $v \in L_a^0(\mu, H)$ *be such that* $E_\mu\left[\rho(-\delta^W v)\right] = 1$. *Then* $V := I_W + v$ *is locally invertible if and only if it is invertible.*

Proof. The sufficiency is obvious by taking $(\tau_n := 1 - 1/n)$ and applying Proposition 4.2. Conversely we henceforth assume that V is locally invertible. By hypothesis there is a sequence (τ_n) of optional times and a sequence $(u^n) \subset L_a^0(\mu, H)$ such that for each $n \in \mathbb{N}$ $V^n := I_W + \pi_{\tau_n} v$ is invertible with inverse $U^n := I_W + u^n$ and $\mu - a.s.$ $\tau_n \uparrow 1$. By Lemma 5.1 (u^n) converges in $L_a^0(\mu, H)$. We note $u \in L_a^0(\mu, H)$ this limit and $U := I_W + u$. We will show that V is invertible with inverse U. For convenience of notations we set $v^n := \pi_{\tau_n} v$, $L_n := \rho(-\delta^W v^n)$, $L := \rho(-\delta^W v)$, and $\nu := L.\mu$. From Doob's optional stopping theorem $L_n := E_\mu[L|\mathcal{F}_{\tau_n}]$ so that (L_n) is uniformly integrable and converges to L in $L^1(\mu)$. On the other hand (see Remark 5.1), U^n converges to U in probability. Therefore the dominated convergence theorem yields

$$
\begin{aligned}
E_\nu\left[e^{il}\right] &= E_\mu\left[Le^{il}\right] \\
&= \lim_{n\to\infty} E_\mu\left[L_n e^{il}\right] \\
&= \lim_{n\to\infty} E_\mu\left[e^{il\circ U^n}\right] \\
&= E_\mu\left[e^{il\circ U}\right]
\end{aligned}
$$

for any $l \in W^\star$, i.e., $U\mu = \nu$. Thus from Proposition 4.1 we know that V is invertible with inverse U if and only if U is the almost sure right inverse of V (i.e., $\mu - a.s.$ $V \circ U = I_W$ or equivalently $\mu - a.s.$ $u + v \circ U = 0$). As we shall see we can show this last result thanks to Lusin's theorem. Let $c > 0$; we have

$$
\begin{aligned}
\mu\left(|v \circ U + u^n|_W > c\right) &= \mu\left(|v \circ U - v^n \circ U^n|_W > c\right) \\
&\leq \mu\left(|v \circ U - v \circ U^n|_H > \frac{c}{2}\right) + E_\mu\left[L_n 1_{|v - v^n|_H > \frac{c}{2}}\right].
\end{aligned}
$$

Let $\alpha > 0$; the dominated convergence theorem implies the existence of a N_1 such that for any $n > N_1$, $E_\mu[L_n 1_{|v - v^n|_H > \frac{c}{2}}] < \alpha/2$. To control $\nu\left(|v \circ U - v \circ U^n|_H > \frac{c}{2}\right)$ we use Lusin's theorem from which we know the existence of a compact set $K_\alpha \subset W$ such that $\nu(K_\alpha) \geq 1 - \alpha/8$, and v is uniformly continuous on K_α. By setting

$$
\Omega_n = \left\{\omega : |v \circ U - v \circ U^n|_H > \frac{c}{2}, (U, U^n) \in K_\alpha \times K_\alpha\right\},
$$

we then have

$$\mu\left(|v\circ U - v\circ U^n|_H > \frac{c}{2}\right) \le \mu(\Omega_n) + \mu(U \notin K_\alpha) + \mu(U^n \notin K_\alpha)$$
$$= \mu(\Omega_n) + \nu(\omega \notin K_\alpha) + E_\mu\left[\rho(-\delta v^n)1_{\omega\notin K_\alpha}\right].$$

Moreover the very definition of K_α yields

$$\nu(\omega \notin K_\alpha) < \alpha/8$$

and the dominated convergence theorem implies

$$\lim_{n\to\infty} E_\mu[\rho(-\delta v^n)1_{\omega\notin K_\alpha}] = \nu(\omega \notin K_\alpha) < \frac{\alpha}{8}.$$

Thus there is an N_2 such that for any $n > N_2$, $E_\mu[\rho(-\delta v^n)1_{\omega\notin K_\alpha}] < \alpha/4$. We then have for any $n > \sup(\{N_1, N_2\})$

$$\mu(|v\circ U + u^n|_W > c) \le \frac{7\alpha}{8} + \mu(\Omega_n). \tag{3.10}$$

On the other hand, the uniform continuity of v on K_α yields the existence of a $\beta_{\alpha,c}$ such that $|u^n - u|_H < \beta_{\alpha,c}$ and $(U^n, U) \in K_\alpha \times K_\alpha$ imply $|v\circ U - v\circ U^n|_H < \frac{c}{2}$. To control the last term of Eq. (3.10) we then set

$$\widetilde{\Omega}_n = \{\omega \in W : |u^n - u| < \beta_{\alpha,c}\}.$$

In particular $\Omega_n \cap \widetilde{\Omega}_n = \emptyset$ so that we get

$$\begin{aligned}\mu(\Omega_n) &\le \mu(\Omega_n \cap \widetilde{\Omega}_n) + \mu(\Omega_n \cap (\widetilde{\Omega}_n^c))\\ &= \mu(\Omega_n \cap (\widetilde{\Omega}_n^c))\\ &\le \mu(\widetilde{\Omega}_n^c)\\ &= \mu(|u_n - u| > \beta_{\alpha,c}).\end{aligned}$$

Since $u_n \to u$ in $L^0_a(\mu, H)$ there is an N_3 such that for any $n > N_3$, $\mu(\Omega_n) < \alpha/8$. Therefore for any $n > \sup(\{N_1, N_2, N_3\})$ we have

$$\mu(|v\circ U + u^n|_W > c) < \alpha$$

This proves that $u^n \to -v\circ U$ in $L^0(\mu, H)$. By uniqueness of the limit $u + v\circ U = 0$, i.e., U is the right inverse of V. But we already showed that this implies the invertibility of V which is the result. □

We now state the result which will be used in the applications:

Corollary 5.1. *Let* $v \in L^0_a(\mu, H)$ *be such that* $E_\mu[\rho(-\delta^W v)] = 1$. *Assume that there is a sequence* $(v^n) \subset L^0_a(\mu, H)$ *with the property that for each* $n \in \mathbb{N}$

$$E_\mu[\rho(-\delta^W v^n)] = 1$$

and such that $V^n := I_W + v^n$ *is invertible. Further assume that there is a sequence of* $(\mathcal{F}_t)$*-stopping times* (τ_n) *such that* $\mu - a.s.$ $\tau_n \uparrow 1$ *and such that for each* $n \in \mathbb{N}$ *we have* $\mu - a.s.$

$$\pi_{\tau_n} v = \pi_{\tau_n} v^n.$$

Then $V := I_W + v$ *is invertible.*

Proof. By Proposition 4.2 for each $n \in \mathbb{N}$ the shift defined by $\widetilde{V^n} := I_W + \pi_{\tau_n} v^n$ is invertible. Therefore V is locally invertible, and its invertibility follows from Theorem 5.1. □

6 A Sufficient Condition for the Invertibility of Markovian Shifts with States in $\mathbb{R}^d$

In this section we will only consider the case where $S = \mathbb{R}^d$ for a d $\in \mathbb{N}$. The main result of this section is Theorem 6.1 which is a sufficient condition of invertibility for Markovian shifts. From the main result of [36] any shift which is both Markovian and bounded is invertible. Here we give a local version of this fact. Note that this extension is different from those of [14].

Definition 6.1. A $v := \int_0^\cdot \dot{v}_s ds \in L^0_a(\mu, H)$ is said to be locally bounded if there is a sequence of $(\mathcal{F}_t)$-stopping times (τ_n) such that $\mu - a.s.$, $\tau_n \uparrow 1$ and such that for each $n \in \mathbb{N}$ we have $\mu - a.s.$

$$\sup_{s \leq \tau_n} |\dot{v}_s|_{H_S} < \infty.$$

Proposition 6.1 will enable us to use the notion of Definition 6.1.

Proposition 6.1. *Let* $v \in L^0_a(\mu, H)$ *and* $\sigma_n := \inf(\{t : \sup_{s\in[0,t]} |\dot{v}_s| > n\}) \wedge 1$. *Then* v *is locally bounded (see Definition 6.1) if and only if* $\mu - a.s.$ $\sigma_n \uparrow 1$.

Proof. If $\sigma_n \uparrow 1$ $\mu{-}a.s.$ the sequence (σ_n) satisfies the hypothesis of Definition 6.1. Conversely we assume that v is locally bounded, and we define Ω to be the set of the $\omega \in W$ such that $\tau_n(\omega) \uparrow 1$ and such that for all $n \in \mathbb{N}$, $\sup_{s \leq \tau_n(\omega)} |\dot{v}_s(\omega)| < \infty$. From the hypothesis $\mu(\Omega) = 1$. Given $\omega \in \Omega$ and $\epsilon \in [0, 1)$ there is an $n_0 \in \mathbb{N}$ and a $K > 0$ such that $\tau_{n_0}(\omega) > \epsilon$ and $\sup_{s\in[0,\tau_{n_0}(\omega)]} |\dot{v}_s| < K$. Let $m_0 \in \mathbb{N}$ be such that $m_0 > K$. Then $\sup_{s\in[0,\tau_{n_0}(\omega))} |\dot{v}_s| < K < m_0$ so that $\sigma_{m_0}(\omega) \geq \tau_{n_0}(\omega) > \epsilon$. Since $\sigma_n(\omega)$ increases, this implies that $\sigma_n(\omega) \uparrow 1$. □

Under a mild condition, the next proposition shows that a Markovian shift which is locally bounded in the sense of Definition 6.1 is invertible.

Theorem 6.1. *Let $v \in L_a^0(\mu, H)$ be such that $E_\mu[\rho(-\delta^W v)] = 1$. Further assume that there is a measurable $b : [0,1)\times\mathbb{R}^d \to \mathbb{R}^d$ such that $\mu-a.s.$ for each $t \in [0,1)$*

$$v_t = \int_0^t b(s, W_s)\mathrm{d}s$$

and that v is locally bounded in the sense of Definition 6.1. Then $V := I_W + v$ is invertible.

Proof. Let (σ_n) be as in Proposition 6.1. Since v is locally bounded $\mu - a.s.$ $\sigma_n \uparrow 1$. For each $T \in [0,1)$ and $n \in \mathbb{N}$ we set

$$v^T := \int_0^{\cdot} b(t, W_t)1_{t\leq T}\mathrm{d}t$$

and

$$v^{n,T} := \int_0^{\cdot} b^{n,T}(t, W_t)\mathrm{d}t,$$

where

$$b^{n,T}(t,x) := b(t,x)1_{|b(t,x)|<n}1_{t\leq T}.$$

Since $b^{n,T}(t,x)$ is both measurable and bounded the main result of [36] yields the existence of a strong solution for the equation

$$\mathrm{d}X_t = \mathrm{d}B_t - b^{n,T}(t, W_t)\mathrm{d}t. \tag{3.11}$$

Thus $V^{n,T} := I_W + v^{n,T}$ is right invertible with an inverse $U^{n,T} := I_W + u^{n,T}$ (note that since $b^{n,T}$ is bounded the condition $u^{n,T} \in L_a^0(\mu, H)$ is filled). Moreover the fact that $b^{n,T}$ is bounded, together with the Novikov criterion, yields $E_\mu[\rho(-\delta^W u^{n,T})] = 1$. From Proposition 4.1 it yields the invertibility of each $V^{n,T}$. On the other hand from the hypothesis we obviously have $\pi_{\sigma_n} v^T = \pi_{\sigma_n} v^{n,T}$. Hence Corollary 5.1 implies the invertibility of $V^T := I_W + v^T$ for each $T < 1$. In particular V is locally invertible (take $\tau_n = 1 - 1/n$) and therefore invertible (Theorem 5.1). □

7 Invertibility of Free Schrödinger Shifts

Let $\widehat{\nu}$ be a probability equivalent to $\widehat{\mu}$, it is well known [9] that

$$H(\widehat{\nu}|\widehat{\mu}) = \inf(\{H(\nu|\mu) : W_1\nu = \widehat{\nu}\}) \tag{3.12}$$

and that the optimum is attained by the probability ν which is defined by

$$\frac{d\nu}{d\mu} = \frac{d\widehat{\nu}}{d\widehat{\mu}} \circ W_1.$$

Moreover ν is an h-path process in the sense of Doob (see [9] and the references therein) so that the Girsanov drift associated with the optimal ν is Markovian. Such a process may be seen as a particular Schrödinger bridge (see [9] and alternatively [39] and references therein). Thus it is connected to stochastic mechanics [23–25] and to stochastic control problems (see [20–22] and references therein). The connection of Schrödinger bridges with stochastic mechanics was clearly shown in [39, 40], and is trivial to see in the special case of h-path processes. Moreover it is known for a long time that such mechanics are related to stochastic control both through Yasue's approach [38] and through the Guerra–Morato(–Nelson) approach [12, 24]. Consider now the equation

$$dU_t = dW_t - \dot{v}_t \circ U dt; \quad U_0 = 0, \tag{3.13}$$

where v is the Girsanov drift associated with the optimal probability ν. As it appears clearly from [10, 39] a solution of Eq. (3.13) may been interpreted physically as a free euclidean quantum (time imaginary) particle starting from the origin whose final marginal is empirically estimated by $\widehat{\nu}$. In this context the equivalences investigated in [10] show that the relative entropy $H(\nu|\mu)$ is the analogous of the Guerra–Morato action associated with the free euclidean particle (or field). Furthermore as it is stressed in [9, 10] the formula (3.12) is also related to the large deviation theory through Sanov's theorem which yields a very concrete intuition of the experiment. Since the reader may be not familiar with these notions, it seems necessary to recall here briefly and formally the main lines of this stochastic picture of euclidean quantum mechanics. Let $\mathcal{V} : S \to \mathbb{R}$ be a smooth potential which is such that

$$\frac{d\mu_{\mathcal{V}}}{d\mu} := e^{-\int_0^1 \mathcal{V}(W_s)ds}$$

defines a probability equivalent to μ. Jensen's inequality easily implies that the infimum of

$$\{H(\nu|\mu_{\mathcal{V}})|W_1\nu = \widehat{\nu}\}$$

is attained by the probability ν defined by

$$\frac{d\nu}{d\mu_{\mathcal{V}}} = \frac{1}{E_\mu[\frac{d\mu_{\mathcal{V}}}{d\mu}|\sigma(W_1)]} \frac{d\widehat{\nu}}{d\widehat{\mu}} \circ W_1.$$

Let

$$H = -\frac{\Delta}{2} + \mathcal{V}.$$

Then at least formally in the case $\dim S < \infty$, by noting λ the Lebesgue measure on S, it is straightforward to check that we have

$$\frac{\mathrm{d}W_t\nu}{\mathrm{d}\lambda} = \theta(t,x)\theta^\star(t,x), \tag{3.14}$$

where θ solves

$$\partial_t\theta = H\theta$$

and where $\theta^\star$ is the fundamental solution of the time-reversed equation

$$-\partial_t\theta^\star = H\theta^\star.$$

By substituting formally $t \to it$ (this procedure is usually called the rotation of Wick) we would have density $\rho(t,x)$ with the shape

$$\rho(t,x) = \Psi(t,x)\Psi^\star(t,x),$$

where Ψ (resp. $\Psi^\star$) solves the Schrödinger equation (its conjugate). For that reason ν is said to model imaginary–time quantum mechanics. By considering the associated space-time metric, it is also called euclidean quantum mechanics. Note that within this framework, ν is the law of a solution to

$$\mathrm{d}X_t = \mathrm{d}W_t - \dot{v}_t \circ X\mathrm{d}t; \quad X_0 = 0,$$

where v is the Girsanov drift associated with ν. The possibility to deal with euclidean quantum mechanics through stochastic mechanics in such a way was first showed in [39]. For that reason it seems relevant to call the Girsanov shift $V := I_W + v$ associated with the above measure ν, the Schrödinger shifts associated with $\widehat{\nu}$ under the potential $\mathcal{V}$. Although our results and methods may extend to the case of potentials, we preferred to focus on the free case (i.e., $\mathcal{V} = 0$) and to treat both the finite- and the infinite-dimensional cases. Henceforth we allow S to be of infinite dimensions, unless otherwise stated. In the next section we will also see that these processes are also involved in information theory. Hence these processes are involved in several fields in which it would be relevant to prove the pathwise uniqueness. For instance, in the point of view of stochastic mechanics, the pathwise uniqueness for Eq. (3.13) means that the stochastic description of free euclidean quantum mechanics fits with the classical picture of determinism. This is the main motivation of this section in which we give some (very large) sufficient conditions for the Girsanov shift associated with such probabilities to be invertible. This motivates the following definition:

Definition 7.1. Let $\widehat{\nu}$ be a probability such that $\widehat{\nu} << \widehat{\mu}$. Further note ν the probability on W which is defined by

$$\frac{\mathrm{d}\nu}{\mathrm{d}\mu} = \frac{\mathrm{d}\widehat{\nu}}{\mathrm{d}\widehat{\mu}} \circ W_1$$

and note v the Girsanov drift associated with ν. We say that ν is the (optimal) measure associated with $\widehat{\nu}$ (on the space on the path) and that $V := I_W + v$ (resp. v) is the (free) Schrödinger shift (resp. drift) associated with the final marginal $\widehat{\nu}$.

We will show below that when $\dim S < \infty$ (resp. $\dim S = \infty$) free Schrödinger shifts with a final marginal of finite entropy (resp. with a bounded density with respect to the Wiener measure on S) are always invertible. In particular we don't assume any regularity conditions on the density.

Theorem 7.1. *Let $S = \mathbb{R}^d$ and let $\widehat{\nu}$ be a probability equivalent to the Gaussian measure $\widehat{\mu}$ on $\mathbb{R}^d$ with finite entropy ($H(\widehat{\nu}|\widehat{\mu}) < \infty$). Then the free Schrödinger shift with final marginal $\widehat{\nu}$ is invertible.*

Proof. Let ν be the optimal measure associated with $\widehat{\nu}$ by Definition 7.1 and let v be the Girsanov drift associated with ν so that $V := I_W + v$ is the Schrödinger shift associated with $\widehat{\nu}$. It is well known [9] and straightforward to see that the Itô formula yields the following expression for the Schrödinger drift v. For each $t \in [0, 1)$

$$\dot{v}_t(\omega) := b(t, W_t), \tag{3.15}$$

where $b : (t, x) \in [0, 1)\times\mathbb{R}^d \to -D \ln P_{1-t}\frac{d\widehat{\nu}}{d\widehat{\mu}}(x)$ and where $(t, x) \in [0, 1)\times\mathbb{R}^d \to P_{1-t}\frac{d\widehat{\nu}}{d\widehat{\mu}}(x)$ is the heat kernel defined by

$$P_{1-t}\frac{d\widehat{\nu}}{d\widehat{\mu}}(x) = \int_W d\mu(\omega)\frac{d\widehat{\nu}}{d\widehat{\mu}}(W_{1-t} + x).$$

We will show that the finite entropy condition implies that v is locally bounded (see Definition 6.1) so that the shift V will be invertible. First note that for each integer $i \in [1, d]$ $(\dot{v}^i_s, s \in [0, 1))$ is a $(\mathcal{F}_t)_{t\in[0,1)}$ martingale on $(W, \mathcal{F}, \nu)$. Indeed for each $t < 1$ and $s \leq t$, let $\theta_s \in C_b(W)$ be $\mathcal{F}_s$ measurable. We then have

$$\begin{aligned}
E_\nu\left[-\dot{v}^i_t\theta_s\right] &= E_\nu\left[D^i \ln P_{1-t}\frac{d\widehat{\nu}}{d\widehat{\mu}}(W_t)\theta_s\right] \\
&= E_\mu\left[\frac{d\nu}{d\mu}D^i \ln P_{1-t}\frac{d\widehat{\nu}}{d\widehat{\mu}}(W_t)\theta_s\right] \\
&= E_\mu\left[E_\mu\left[\frac{d\nu}{d\mu}|\mathcal{F}_t\right] D^i \ln P_{1-t}\frac{d\widehat{\nu}}{d\widehat{\mu}}(W_t)\theta_s\right] \\
&= E_\mu\left[P_{1-t}\frac{d\widehat{\nu}}{d\widehat{\mu}}(W_t)D^i \ln P_{1-t}\frac{d\widehat{\nu}}{d\widehat{\mu}}(W_t)\theta_s\right] \\
&= E_\mu\left[D^i P_{1-t}\frac{d\widehat{\nu}}{d\widehat{\mu}}(W_t)\theta_s\right]
\end{aligned}$$

$$
\begin{aligned}
&= E_\mu\left[P_{t-s} D^i P_{1-s}\frac{d\widehat{\nu}}{d\widehat{\mu}}(W_s)\theta_s\right]\\
&= E_\mu\left[D^i P_{1-s}\frac{d\widehat{\nu}}{d\widehat{\mu}}(W_s)\theta_s\right]\\
&= E_\nu\left[D^i \ln P_{1-s}\frac{d\widehat{\nu}}{d\widehat{\mu}}(W_s)\theta_s\right]\\
&= E_\nu\left[-\dot{v}_s^i\theta_s\right].
\end{aligned}
$$

Hence $t \in [0,1) \to E_\nu[(\dot{v}_t^i)^2]$ is increasing. Together with the finite energy condition $v \in L_a^2(\nu, H)$ it yields $\dot{v}_t^i \in L^2(\nu)$ for each $t \in [0,1)$. Therefore by Doob's inequality we get that for each $t \in [0,1)$

$$\sup_{s\in[0,t]} |\dot{v}_s|^2_{\mathbb{R}^d} \in L^1(\nu). \tag{3.16}$$

Since $\nu \sim \mu$ Eq. (3.16) implies that the hypothesis of Theorem 6.1 is satisfied with $\tau_n := 1 - 1/n$. Therefore V is invertible. □

In the case where $\dim S = \infty$ it is harder to get a clean expression of the Schrödinger shift since we need an Itô formula. To avoid the use of such a formula we give an elementary proof in Proposition 7.1 which is in the spirit of the proof of the Clark–Ocone formula.

Proposition 7.1. *Let $\widehat{\nu}$ be a probability such that $\widehat{\nu} \sim \widehat{\mu}$ and $\frac{d\widehat{\nu}}{d\widehat{\mu}} \in L^2(\widehat{\mu})$. We further note v the free Schrödinger drift associated with $\widehat{\nu}$ (see Definition 7.1). We then have for each $t \in [0,1)$*

$$\dot{v}_t = b(t, W_t) \tag{3.17}$$

$d\mu \times dt$ *a.s. where*

$$b(t,x) := -D \ln Q_{1-t}\frac{d\widehat{\nu}}{d\widehat{\mu}}(x)$$

and where Q_t is the heat kernel on S, i.e., $Q_t\frac{d\widehat{\nu}}{d\widehat{\mu}}(x) = E_\mu[\frac{d\widehat{\nu}}{d\widehat{\mu}}(x + W_t)]$ for each $x \in S$.

Proof. Let ν be the measure associated with $\widehat{\nu}$ by Definition 7.1 and let v be the Girsanov drift associated with ν which is also the Schrödinger drift of $\widehat{\nu}$. When $\frac{d\widehat{\nu}}{d\widehat{\mu}} \in L^2(\widehat{\mu})$ the same proof as Lemma 3.3.2 of [34] applies and we know that for each $t < 1$ there is a modification of $Q_{1-t}\frac{d\widehat{\nu}}{d\widehat{\mu}}$ such that for each $\omega \in S$ the map $h \in H_S \to Q_{1-t}\frac{d\widehat{\nu}}{d\widehat{\mu}}(\omega + h)$ is real analytic on H_S. We then chose this modification. Hence, it is straightforward to check that for any $k \in H_S$ $Q_t D_k Q_s = D_k Q_t Q_s = D_k Q_{t+s}$ and that $E_\mu[\frac{d\widehat{\nu}}{d\widehat{\mu}} \circ W_1 | \mathcal{F}_t] = Q_{1-t}\frac{d\widehat{\nu}}{d\widehat{\mu}}(W_t)$. In order to avoid the use of the theory of Watanabe distributions, or of an Itô formula on abstract Wiener space, we give an elementary proof. For any $t < 1$ let $L_a^2(\mu, H, t)$ be the subset of the $u \in L_a^2(\mu, H)$ for which the support of $s \in [0,1] \to \dot{u}_s \in H_S$ is in $[0,t] \subset [0,1)$. We recall that from the martingale representation theorem which holds on W (see [34] Chap. 2), $\{\delta^W \alpha, \alpha \in L_a^2(\mu, H, t)\}$ is dense in $\{X - E_\mu[X] : X \in L^2(\mu)\ ,\ X \text{ is } \mathcal{F}_t \text{ measurable}\}$. Then we have for each $h \in L_a^2(\mu, H, t)$

$$\begin{aligned}
E_\mu\left[\frac{\mathrm{d}\nu}{\mathrm{d}\mu} \circ \tau_h \rho(-\delta h)\right] &= E_\mu\left[E_\mu\left[\frac{\mathrm{d}\nu}{\mathrm{d}\mu} \circ \tau_h | \mathcal{F}_t\right] \rho(-\delta h)\right] \\
&= E_\mu\left[E_\mu\left[\frac{\mathrm{d}\widehat{\nu}}{\mathrm{d}\widehat{\mu}} \circ (W_1 + h_t) | \mathcal{F}_t\right] \rho(-\delta h)\right] \\
&= E_\mu\left[(Q_{1-t}\frac{\mathrm{d}\widehat{\nu}}{\mathrm{d}\widehat{\mu}})(W_t + h_t)\rho(-\delta h)\right].
\end{aligned}$$

Hence

$$\begin{aligned}
&\frac{\mathrm{d}}{\mathrm{d}\lambda} E_\mu\left[\left[\frac{\mathrm{d}\nu}{\mathrm{d}\mu} \circ \tau_{\lambda h} - 1\right] \rho(-\delta\lambda h)\right] |_{\lambda=0} \\
&= \frac{\mathrm{d}}{\mathrm{d}\lambda} E_\mu\left[\left[\left(Q_{1-t}\frac{\mathrm{d}\widehat{\nu}}{\mathrm{d}\widehat{\mu}}\right)(W_t + \lambda h_t) - 1\right] \rho(-\delta\lambda h)\right] |_{\lambda=0} \\
&= E_\mu\left[D_{h_t} Q_{1-t}\frac{\mathrm{d}\widehat{\nu}}{\mathrm{d}\widehat{\mu}} \circ (W_t)\right] - E_\mu\left[\left[Q_{1-t}\frac{\mathrm{d}\widehat{\nu}}{\mathrm{d}\widehat{\mu}}(W_t) - 1\right] \delta h\right].
\end{aligned}$$

On the other hand the Cameron–Martin theorem yields $E_\mu\left[\left[\frac{\mathrm{d}\nu}{\mathrm{d}\mu} \circ \tau_{\lambda h} - 1\right] \rho(-\delta\lambda h)\right] = 0$. By setting $\alpha^t_\cdot = \int_0^\cdot 1_{s \le t} D Q_{1-s}\left[\frac{\mathrm{d}\widehat{\nu}}{\mathrm{d}\widehat{\mu}}\right](W_s)\mathrm{d}s$ we get

$$\begin{aligned}
E_\mu\left[\left(E_\mu\left[\frac{\mathrm{d}\nu}{\mathrm{d}\mu} | \mathcal{F}_t\right] - 1\right) \delta h\right] &= E_\mu\left[\left[Q_{1-t}\frac{\mathrm{d}\widehat{\nu}}{\mathrm{d}\widehat{\mu}}(W_t) - 1\right] \delta h\right] \\
&= E_\mu\left[< D Q_{1-t}\frac{\mathrm{d}\widehat{\nu}}{\mathrm{d}\widehat{\mu}}(W_t), h_t >_{H_S}\right] \\
&= \int_0^t E_\mu\left[< D Q_{1-t}\frac{\mathrm{d}\widehat{\nu}}{\mathrm{d}\widehat{\mu}}(W_t), \dot{h}_s >_{H_S}\right] \mathrm{d}s \\
&= \int_0^t E_\mu\left[D_{\dot{h}_s} Q_{1-t}\frac{\mathrm{d}\widehat{\nu}}{\mathrm{d}\widehat{\mu}}(W_t)\right] \mathrm{d}s \\
&= \int_0^t E_\mu\left[E_\mu\left[D_{\dot{h}_s} Q_{1-t}\frac{\mathrm{d}\widehat{\nu}}{\mathrm{d}\widehat{\mu}}(W_t) | \mathcal{F}_s\right]\right] \mathrm{d}s \\
&= \int_0^t E_\mu\left[Q_{t-s} D_{\dot{h}_s} Q_{1-t}\frac{\mathrm{d}\widehat{\nu}}{\mathrm{d}\widehat{\mu}}(W_s)\right] \mathrm{d}s
\end{aligned}$$

$$= \int_0^t E_\mu \left[D_{\dot{h}_s} Q_{1-s} \frac{d\widehat{\nu}}{d\widehat{\mu}}(W_s) \right] ds$$

$$= \int_0^t E_\mu \left[< DQ_{1-s} \frac{d\widehat{\nu}}{d\widehat{\mu}} \circ (W_s), \dot{h}_s >_{H_S} \right] ds$$

$$= E_\mu \left[< \alpha^t, h >_H \right]$$

$$= E_\mu \left[\delta^W \alpha^t \delta^W h \right]$$

which means that $E_\mu \left[\frac{d\nu}{d\mu} | \mathcal{F}_t \right] - 1 = \delta^W \alpha^t$, i.e.,

$$E_\mu \left[\frac{d\nu}{d\mu} | \mathcal{F}_t \right] = 1 + \int_0^t DQ_{1-s} \frac{d\widehat{\nu}}{d\widehat{\mu}} \circ (W_s) dW_s$$

By construction of ν (see Sect. 3) it yields Eq. (3.17). □

Lemma 7.1. *Let $L \in L^\infty(\widehat{\mu})$, $\epsilon \in (0,1)$ and $T \in [0,1)$. Let $b : [0,1] \times S \to H_S$ be the mapping defined by*

$$b(t,x) := -D \ln P_{1-t} L(x) 1_{t<T} 1_{P_{1-t}L(x)>\epsilon}$$

and let v be defined by

$$v := \int_0^\cdot b(s, W_s) ds.$$

Then $V := I_W + v$ is invertible.

Proof. For any $h \in H_S$, $x \in S$, and $t < 1$, we have [15]

$$|D_h \ln Q_{1-t} L(x)| = \left| \frac{1}{\sqrt{1-t}} \frac{\int_S (< h, y >) L(x + \sqrt{1-t} y) \widehat{\mu}(dy)}{\int_S L(x + \sqrt{1-t} y) \widehat{\mu}(dy)} \right|$$
$$\leq \frac{|h|_H K}{\sqrt{1-t} P_{1-t} L(x)},$$

where K is the essential supremum of L. Therefore $|D \ln Q_{1-t} L(x)|_H \leq \frac{K}{\sqrt{1-t} P_{1-t} L(x)}$ and

$$|b(t,x)| = |D \ln P_{1-t} L(x) 1_{t<T} 1_{P_{1-t}L(x)>\epsilon}| \leq \frac{K}{\epsilon \sqrt{1-T}}.$$

In particular $v \in L_a^0(\mu, H)$ and together with Novikov's criterion it yields $E_\mu \left[\rho(-\delta^W v) \right] = 1$. Moreover if we assume that V is right invertible with an inverse $U := I_W + u$, the boundedness of b will imply $E_\mu \left[\rho(-\delta^W u) \right] = 1$ so that the invertibility will automatically follow from Proposition 4.1. Hence, to get the invertibility of V, it suffices to prove that V is right invertible. Specifically, since b is bounded it suffices to prove that

$$dX_t = dB_t - \dot{v}_t \circ X dt \tag{3.18}$$

has a unique strong solution. Since S is a Polish space the Yamada–Watanabe criterion [13] also applies for Eq. (3.18). Therefore to prove the existence of a strong solution it suffices to check that a weak solution of Eq. (3.18) exists, and that the pathwise uniqueness holds. Since b is bounded, we already have the existence of a weak solution and the uniqueness in law for Eq. (3.18) by transformation of the drift [13]. Hence we only have to prove the pathwise uniqueness. Before we do this, we have made some preparations and to prove that $\mu - a.s.$ for any $t < 1$ and $(h,k) \in \mathcal{H}_{t,\omega} \times \mathcal{H}_{t,\omega}$ we have

$$|D \ln P_{1-t}L(W_t + h) - D \ln P_{1-t}L(W_t + k)|_{H_S} \leq C|h - k|_{H_S} \tag{3.19}$$

for a $C > 0$ which depends on T. For any $(t,\omega) \in [0,1) \times W$ we set

$$\mathcal{H}_{t,\omega} := \{h \in H_S : P_{1-t}L(W_t + h) > \epsilon\}.$$

Since $x \to P_{1-t}L(x)$ is H_S-continuous (and even $H_S - C^\infty$; see [15]) $\mathcal{H}_{t,\omega}$ is an open set. On the other hand we have for any $h, k \in H_S$, and $x \in S$

$$D^2_{h,k} \ln Q_{1-t}L(x) = \frac{D_h D_k Q_{1-t}L}{Q_{1-t}L} - \frac{(D_h Q_{1-t}L)(D_k Q_{1-t}L)}{(Q_{1-t}L)^2}$$

and

$$\begin{aligned} |D^2_{h,k} Q_{1-t}L(x)| &= \frac{1}{1-t}\left|\int (<h,y><k,y>-<h,k>_{H_S})L(x+\sqrt{1-t}\,y)\widehat{\mu}(dy)\right| \\ &\leq \frac{K}{1-t}|h|_{H_S}|k|_{H_S}, \end{aligned}$$

where the last inequality follows from

$$\int (<h,y><k,y> - <h,k>_{H_S})^2 \widehat{\mu}(dy) = |h|^2_{H_S}|k|^2_{H_S}.$$

Hence, by taking $C := \frac{K}{(1-T)\epsilon} + (\frac{K}{\sqrt{1-T}\epsilon})^2$, we get Eq. (3.19). We can now begin the proof of pathwise uniqueness. Let U and $\widetilde{U}$ be two solutions with initial distribution α, which are defined on the same space $(\Omega, \mathcal{G}, \mathbb{P})$, with the same filtration $(\mathcal{G}_t)$ defined on it, and with the same $(\mathcal{G}_t)$-Brownian motion(B_t). We define two $(\mathcal{G}_t)$-optional times:

$$\tau := \inf(\{t : |u_t - \widetilde{u}_t|_{H_S} > 0\}) \wedge 1 \tag{3.20}$$

and

$$\sigma := \inf(\{t : (P_{1-t}L \circ U_t - \epsilon)(P_{1-t}L \circ \widetilde{U}_t - \epsilon) < 0\}) \wedge 1. \tag{3.21}$$

Let $\widetilde{\Omega}$ be the set of the $\omega \in \Omega$ for which both $t \to U_t$ and $t \to \widetilde{U}_t$ are continuous. By construction it is such that $\mathbb{P}(\widetilde{\Omega}) = 1$. Moreover for any $\omega \in \widetilde{\Omega}$ the continuity implies that we have $\sigma(\omega) \leq \tau(\omega)$ if and only if $\tau(\omega) = 1$. Therefore, to prove pathwise uniqueness, it suffices to prove that $\mu - a.s.$ $\sup_{t\in[0,\sigma(\omega)]} |u_t - \widetilde{u}_t|_{H_S} = 0$. For any $t \in [0, \sigma(\omega)]$ the definition of σ implies:

$$\begin{aligned}|u_t - \widetilde{u}_t|_{H_S} &\leq \int_0^t |b(s, U_s) - b(s, \widetilde{U}_s)|\mathrm{d}s \\ &\leq A + B,\end{aligned}$$

where

$$A := \int_0^t |b(s, U_s) - b(s, \widetilde{U}_s)| 1_{\mathcal{H}_{s,\omega}\times\mathcal{H}_{s,\omega}}(u_s, \widetilde{u}_s)$$

and

$$B := \int_0^t |b(s, U_s) - b(s, \widetilde{U}_s)| 1_{(\mathcal{H}_{s,\omega})^c\times(\mathcal{H}_{s,\omega})^c}(u_s, \widetilde{u}_s)$$

and where $(\mathcal{H}_{s,\omega})^c$ is the complement of $\mathcal{H}_{s,\omega}$ in H_S. From the definitions of $\mathcal{H}_{t,\omega}$ and of $b(t,x)$ $\mu - a.s.$ $B = 0$. On the other hand Eq. (3.19) implies

$$\begin{aligned}A &\leq C \int_0^t |u_s - \widetilde{u}_s|_{H_S} 1_{\mathcal{H}_{s,\omega}\times\mathcal{H}_{s,\omega}}(u_s, \widetilde{u}_s) \\ &\leq C \int_0^t |u_s - \widetilde{u}_s|_{H_S}.\end{aligned}$$

Hence, up to a negligible set, Gronwall's lemma implies that $u_t(\omega) = \widetilde{u}_t(\omega)$ for every $t < \sigma(\omega)$. This proves that $\mu - a.s.$ $\sigma \leq \tau$, from which we know that $\mu - a.s.$ $\tau = 1$. □

Theorem 7.2. *Let $\widehat{\nu}$ be a probability equivalent to $\widehat{\mu}$ such that $\frac{\widehat{d\nu}}{d\mu} \in L^\infty(\mu)$. Then the free Schrödinger shift with final marginal $\widehat{\nu}$ is invertible.*

Proof. Let v be the Girsanov drift associated with the probability $\nu := \frac{\widehat{d\nu}}{d\mu} \circ W_1$. From Definition 7.1 $V := I_W + v$ is the free Schrödinger shift associated with $\widehat{\nu}$. For any $(n, T) \in \mathbb{N} \times [0, 1)$ we set

$$v^{n,T} := \int_0^{\cdot} b^{n,T}(s, W_s)\mathrm{d}s,$$

where

$$b^{n,T}(t,x) = -D \ln P_{1-t}\frac{d\widehat{\nu}}{d\widehat{\mu}}(x)1_{t\leq T}1_{P_{1-t}\frac{d\widehat{\nu}}{d\mu}(x)>\frac{1}{n}}$$

and

$$v^T := -\int_0^{\cdot} D \ln P_{1-t}\frac{d\widehat{\nu}}{d\widehat{\mu}}(x)1_{t\leq T}dt.$$

By Lemma 7.1 all the $V^{n,T} := I_W + v^{n,T}$ are invertible. Let $\tau_n := \inf(\{t : P_{1-t}L(W_t) < \frac{1}{n}\} \wedge 1$. Since $\widehat{\nu} \sim \widehat{\mu}$ we have $\mu - a.s.$ $(\tau_n) \uparrow 1$. On the other hand from the definitions $\pi_{\tau_n}v^T = \pi_{\tau_n}v^{n,T}$. From Corollary 5.1 it implies that V^T is invertible for any $T < 1$. Taking $\tau_n := 1 - \frac{1}{n}$ yields the local invertibility of V which is therefore invertible (Theorem 5.1). □

8 Information Loss on the Path Space and Shannon's Inequality

The idea to use h-path processes in information theory is not new and we found it implicitly in an original but somehow misleading unpublished paper [18] in the case of the classical Wiener space. We generalized and clarified some of these results. By completing that work we were acquainted that the author of [18] also took the same way. Nevertheless our results are still more general and we find it interesting enough to be presented here. Although the essential ideas of the proofs are not so new there are several original contributions in this section. First we give a version of a Brascamp–Lieb inequality which holds on any abstract Wiener space. Then we give an abstract Wiener space version of Shannon's inequality which holds on any abstract Wiener space. Since the Lebesgue measure is no more well defined when $\dim S = \infty$ we had to write it in terms of Gaussian measure. We succeeded in this task by making a change of measure from which we get a formulation of Shannon's inequality which seems to be new. By making an analogous change of measure on the path space W we show that Üstünel's criterion (the main result of [30]) of which we present a generalization here (Theorem 8.2) may be written in terms of variance. This generalization was first given in Theorem 6 of [16] in the case where the underlying probability space was the Wiener space, and it stresses the connection between information loss and invertibility. Its interpretation in terms of variance is new. Within this framework we reduce the proof of [18] as a consequence of the additive properties of variance. Moreover our precise results shows that Shannon's inequality as well as the Brascamp–Lieb inequality may be seen as the result of information loss in a Gaussian channel.

8.1 Talagrand's Inequality and Sobolev's Inequality on Any Abstract Wiener Space

The Monge–Kantorovich problem on abstract Wiener space has been investigated for instance in [5] or [6] (see also [37] for a general overview on this topic). The results of this section are almost trivial and may be well known. However it seems relevant to give it here for pedagogical reasons. As a matter of fact here are the two simplest cases in which h-path processes can be used to yield inequalities on any abstract Wiener space from inequalities on the path space. We first recall the following result which is a particular case of Theorem 3.2 of [6] in order to achieve the proof of Proposition 8.1.

Theorem 8.1. *Let $\widehat{\nu}$ be a probability such that $\widehat{\nu} << \widehat{\mu}$ (i.e., absolutely continuous). Assume that $d(\widehat{\nu},\widehat{\mu}) < \infty$. Then there is a measurable mapping $T^S : S \to S$ which is solution to the original Monge problem. Moreover its graph supports the unique solution of the Monge–Kantorovitch problem γ, i.e.,*

$$(I_S \times T^S)\widehat{\mu} = \gamma$$

In particular $T^S\widehat{\mu} = \widehat{\nu}$, $T^S - I_S \in L^2(\widehat{\mu}, H)$, and there is a mapping $\widetilde{S}^S := (T^S)^{-1}$ such that

$$\mu\left(\{\omega|\widetilde{S}^S \circ T^S = I_S\}\right) = \nu\left(\{\omega|T^S \circ \widetilde{S}^S = I_S\}\right) = 1.$$

The next proposition sums up basic properties of the optimal measure associated with a marginal.

Proposition 8.1. *Let $\widehat{\nu}$ be a probability such that $\widehat{\nu} \sim \widehat{\mu}$ with $H(\widehat{\nu}|\widehat{\mu}) < \infty$. Let ν be the measure associated with $\widehat{\nu}$ by Definition 7.1. We then have*

- $H(\widehat{\nu}|\widehat{\mu}) = H(\nu|\mu)$.
- $d(\widehat{\nu},\widehat{\mu}) \leq d(\nu,\mu)$.

where $d(\nu,\mu)$ is given by Definition 3.1. If we further assume that $\frac{d\widehat{\nu}}{d\widehat{\mu}} \in \mathbb{D}_{2,1}(\widehat{\mu})$ we have

$$J(\widehat{\nu},\widehat{\mu}) = J(\nu,\mu),$$

where

$$J(\widehat{\nu},\widehat{\mu}) := E_{\widehat{\nu}}\left[\left|D \ln \frac{d\widehat{\nu}}{d\widehat{\mu}}\right|^2_{H_S}\right]$$

and where $J(\nu,\mu)$ is defined in Proposition 3.4.

Proof. Let v be the Girsanov drift associated with ν. The fact that $H(\widehat{\nu}|\widehat{\mu}) = H(\nu|\mu)$ is obvious. Let $s \to T_s := W_s + \int_0^s \dot{t}_s ds$ be the optimum given by Theorem 8.1 which attains $d(\nu,\mu)$. By definition we have $d(\nu,\mu) = E_\mu[\int_0^1 |\dot{t}_s|^2_{H_S} ds]$.

On the other hand $T_1\mu = W_1 T\mu = W_1\nu = \widehat{\nu}$ and $W_1\mu = \widehat{\mu}$ so that:

$$d(\widehat{\nu},\widehat{\mu}) \le E_\mu[|T_1 - W_1|^2_{H_S}]$$

Hence Jensen's inequality implies

$$\begin{aligned} d(\widehat{\nu},\widehat{\mu}) &\le E_\mu[|T_1 - W_1|^2_{H_S}] \\ &\le E_\mu\left[\left|\int_0^1 \dot{t}_s \mathrm{d}s\right|^2_{H_S}\right] \\ &\le E_\mu\left[\int_0^1 |\dot{t}_s|^2_{H_S} \mathrm{d}s\right] \\ &= d(\nu,\mu). \end{aligned}$$

We now state the last part of the claim. From Proposition 3.1 we have

$$\begin{aligned} \nabla_h \frac{\mathrm{d}\nu}{\mathrm{d}\mu} &= \nabla_h \frac{\mathrm{d}\widehat{\nu}}{\mathrm{d}\widehat{\mu}} \circ W_1 \\ &= < h_1, D\frac{\mathrm{d}\widehat{\nu}}{\mathrm{d}\widehat{\mu}} \circ W_1 >_{H_S} \\ &= \int_0^1 < \dot{h}_s, D\frac{\mathrm{d}\widehat{\nu}}{\mathrm{d}\widehat{\mu}} \circ W_1 >_{H_S} \mathrm{d}s \end{aligned}$$

so that $D_s \ln \frac{\mathrm{d}\nu}{\mathrm{d}\mu} = D \ln \widehat{\frac{\mathrm{d}\nu}{\mathrm{d}\mu}} \circ W_1$. In Sect. 3 we recalled that the Clark–Ocone formula yields $\dot{v}_1 = D_1 \ln \frac{\mathrm{d}\nu}{\mathrm{d}\mu}$. However in that case it can also be seen directly with the same proof as Proposition 7.1, but in that case with $t = 1$. Since $W_1\nu = \widehat{\nu}$ we get

$$\begin{aligned} J(\nu,\mu) &= E_\nu\left[\left|\nabla \ln \frac{\mathrm{d}\nu}{\mathrm{d}\mu}\right|^2_H\right] \\ &= \int_0^1 E_\nu\left[\left|D_s \ln \frac{\mathrm{d}\nu}{\mathrm{d}\mu}\right|^2_{H_S}\right] \mathrm{d}s \\ &= \int_0^1 E_\nu\left[\left|D \ln \frac{\mathrm{d}\widehat{\nu}}{\mathrm{d}\widehat{\mu}} \circ W_1\right|^2_{H_S}\right] \mathrm{d}s \\ &= E_\nu\left[\left|D \ln \frac{\mathrm{d}\widehat{\nu}}{\mathrm{d}\widehat{\mu}} \circ W_1\right|^2_{H_S}\right] \end{aligned}$$

$$= E_{\widehat{\nu}}\left[\left|D \ln \frac{d\widehat{\nu}}{d\widehat{\mu}}\right|_{H_S}^2\right]$$
$$= J(\widehat{\nu}, \widehat{\mu}).$$

□

Proposition 8.2 shows how we can use h-path processes to get inequalities on S from inequalities on the path space W.

Proposition 8.2. *Let $\widehat{\nu}$ be a probability equivalent to $\widehat{\mu}$ with $H(\widehat{\nu}|\widehat{\mu}) < \infty$. Let $\sqrt{d(\widehat{\mu}, \widehat{\nu})}$ be the Wasserstein distance, then we have*

$$d(\widehat{\mu}, \widehat{\nu}) \leq 2H(\widehat{\nu}|\widehat{\mu}).$$

Assume henceforth that $\frac{d\widehat{\nu}}{d\widehat{\mu}} \in \mathbb{D}_{2,1}(\widehat{\mu})$ and let $J(\widehat{\nu}|\widehat{\mu})$ be the Fisher information, which is defined by

$$J(\widehat{\nu}|\widehat{\mu}) = E_{\widehat{\nu}}\left[\left|D \ln \frac{d\widehat{\nu}}{d\widehat{\mu}}\right|_{H_S}^2\right].$$

Then we have

$$J(\widehat{\nu}|\widehat{\mu}) = E_{\widehat{\nu}}\left[\frac{|\dot{v}_1|_{H_S}^2}{2}\right],$$

where v is the Schrödinger drift associated with ν. In particular we get the logarithmic Sobolev inequality

$$H(\widehat{\nu}|\widehat{\mu}) \leq J(\widehat{\nu}|\widehat{\mu}).$$

Proof. Let ν be the probability associated with $\widehat{\nu}$ by Definition 7.1 and v be the Girsanov drift of ν which is the free Schrödinger drift associated with $\widehat{\nu}$. Then Propositions 8.1 and 3.4 yield

$$H(\widehat{\nu}|\widehat{\mu}) = H(\nu|\mu) \leq J(\nu|\mu) \leq J(\widehat{\nu}|\widehat{\mu}).$$

Furthermore Propositions 8.1 and 3.3 directly imply

$$d(\widehat{\mu}, \widehat{\nu}) \leq d(\nu, \mu) \leq 2H(\nu|\mu) = 2H(\widehat{\nu}|\widehat{\mu}).$$

□

8.2 Information Loss on the Path Space

Let $(\Omega, \mathcal{F}^{\mathbb{P}}, \mathbb{P})$ be a complete probability space, and let $(\mathcal{F}_t^{\mathbb{P}})_{t\in[0,1]}$ be a continuous filtration which satisfies the usual conditions. On this space let $(B_t)_{t\in[0,1]}$ be a $(\mathcal{F}_t^{\mathbb{P}})_{t\in[0,1]}$ and S-valued Brownian motion starting from the origin. In the point of view of information theory one may think of B as a Gaussian noise in a transmission channel. Assume now that one sends a signal u through this channel and that the receptor observes $U = B + u$ but not u. Let $(\mathcal{F}_t^u)$ be the augmentation with respect to $\mathbb{P}$ of the filtration generated by the observed signal $t \to U_t$. If the receptor tries to estimate dynamically u, his best estimation will be

$$\widehat{u}_t = \int_0^t E_{\mathbb{P}}[\dot{u}_s | \mathcal{F}_s^u] \mathrm{d}s$$

which is usually called the causal estimate of u. Hence the estimated signal will be $\widehat{u}$, while the emitted signal is u. Since $\widehat{u}$ is a projection of u, the energy of u is always bigger that the energy of $\widehat{u}$. In other words, some energy will dissipate in the channel, and the value of this dissipated energy is

$$E_{\mathbb{P}}\left[\frac{|u|_H^2}{2}\right] - E_{\mathbb{P}}\left[\frac{|\widehat{u}|_H^2}{2}\right].$$

As a matter of fact the value of this dissipated energy may be seen equivalently as an error or a loss of information. Indeed

$$E_{\mathbb{P}}\left[\frac{|u|_H^2}{2}\right] - E_{\mathbb{P}}\left[\frac{|\widehat{u}|_H^2}{2}\right] = \frac{1}{2} E_{\mathbb{P}}\left[|u - \widehat{u}|_H^2\right]$$

is equal to (half) the error of the causal estimate. Theorem 8.2 (which was first proved in [16] in the case of the Wiener space) states that the dissipated energy (or information) only depends on two parameters: the energy of the signal and the law of the observed signal through its relative entropy. This loss of information relies on the fact that the observer only gets the information of the filtration $(\mathcal{F}_t^u)$ generated by $t \to U_t \in S$ which is smaller than $(\mathcal{F}_t^{\mathbb{P}})_{t\in[0,1]}$. Moreover one expects the equality to occur if and only if one can reconstruct the Brownian path until t from $\mathcal{F}_t^u$. This is exactly what shows the equality case in Theorem 8.2. Before going further we have to set some notations. We note $L^2((\Omega, \mathcal{F}^{\mathbb{P}}, \mathbb{P}), H))$ or when there are no ambiguity on the underlying filtered space $L^2(\mathbb{P}, H)$ the set of the measurable mapping $u : \Omega \to H$, such that $E_{\mathbb{P}}[|u|_H^2] < \infty$. We also define $L_a^2(\mathbb{P}, H)$ the subset of the $u \in L^2(\mathbb{P}, H)$ such that the mapping $t \to \dot{u}_t$ is adapted to $(\mathcal{F}_t^{\mathbb{P}})_{t\in[0,1]}$. We recall that we defined $\mathcal{F}$ (resp. $(\mathcal{F}_t)$) as the completion of the sigma field $\mathcal{B}(W)$ with respect to μ (resp. the augmentation with respect to μ of the sigma field generated by the coordinate process $t \to W_t$).

Definition 8.1. Let $(B_t)_{t\in[0,1]}$ be an S-valued $(\mathcal{F}_t^{\mathbb{P}})_{t\in[0,1]}$ Brownian motion starting from the origin on a complete probability space $(\Omega, \mathcal{F}^{\mathbb{P}}, \mathbb{P})$ with a continuous filtration $(\mathcal{F}_t^{\mathbb{P}})_{t\in[0,1]}$ which satisfies the usual conditions. Given a $u \in L_a^2(\mathbb{P}, H)$ such that $U\mathbb{P} \sim \mu$ where $U := I_W + u$, we define $(\mathcal{F}_t^u)$ as being the filtration $(\sigma(U_s, s \leq t))_{t\in[0,1]}$ augmented with respect to $\mathbb{P}$. We also note $L_u^2(\mathbb{P}, H)$ the subset of the $\widetilde{u} \in L_a^2(\mathbb{P}, H)$ such that $t \to \dot{\widetilde{u}}_t$ is adapted to $(\mathcal{F}_t^u)$. Moreover $\widehat{u}$ will denote the projection of u on $L_u^2(\mathbb{P}, H)$ which is a closed subspace of $L_a^2(\mathbb{P}, H)$.

As a matter of fact $\widehat{u}$ is the dual predicable projection of u [4] on the augmentation with respect to $\mathbb{P}$ of the filtration generated by $t \to U_t = B_t + u_t$. We recall that physically $\widehat{u}$ may be seen as the causal estimator of a signal u and is written:

$$\widehat{u}_t = \int_0^t E_{\mathbb{P}}[\dot{u}_s | \mathcal{F}_s^u] \mathrm{d}s.$$

Lemma 8.1. *Let $(\Omega, \mathcal{F}^{\mathbb{P}}, \mathbb{P})$ be a complete probability space with a continuous filtration $(\mathcal{F}_t^{\mathbb{P}})_{t\in[0,1]}$ on it which satisfies the usual conditions. Let $(B_t)_{t\in[0,1]}$ be an S-valued $\mathcal{F}_t^P$-Brownian motion on $(\Omega, \mathcal{F}^{\mathbb{P}}, \mathbb{P})$ and $U : t \in [0,1] \to U_t \in S$ be any $(\mathcal{F}_t)$-adapted, continuous process. Further assume that $U\mathbb{P} \sim \mu$. Then we have $\mathbb{P} - a.s.$*

$$\widehat{u} + v \circ U = 0, \tag{3.22}$$

where v is the Girsanov drift associated with $U\mathbb{P}$.

Proof. Since $t \to V_t$ is an abstract Wiener process on $(W, \mathcal{F}, U\mathbb{P})$, for any $\theta \in L_a^2(U\mathbb{P}, H)$ we have.

$$\begin{aligned}
E_{\mathbb{P}}[< \theta \circ U, u >_H] &= E_{\mathbb{P}}\left[\int_0^1 < \dot{\theta}_s \circ U, \dot{u}_s >_{H_S} \mathrm{d}s\right] \\
&= E_{\mathbb{P}}\left[\int_0^1 \dot{\theta}_s \circ U \mathrm{d}B_s + \int_0^1 < \dot{\theta}_s \circ U, \dot{u}_s >_{H_S} \mathrm{d}s - \int_0^1 \dot{\theta}_s \circ U \mathrm{d}B_s\right] \\
&= E_{\mathbb{P}}\left[\left(\int_0^1 \dot{\theta}_s \mathrm{d}W_s\right) \circ U - \int_0^1 \dot{\theta}_s \circ U \mathrm{d}B_s\right] \\
&= E_{\mathbb{P}}\left[\left(\int_0^1 \dot{\theta}_s \mathrm{d}W_s\right) \circ U\right] \\
&= E_{U\mathbb{P}}\left[\int_0^1 \dot{\theta}_s \mathrm{d}W_s\right] \\
&= E_{U\mathbb{P}}\left[\int_0^1 \dot{\theta}_s \mathrm{d}W_s + \int_0^1 < \dot{\theta}, \dot{v}_s >_{H_S} \mathrm{d}s - \int_0^1 < \dot{\theta}_s, \dot{v}_s >_{H_S} \mathrm{d}s\right] \\
&= E_{U\mathbb{P}}\left[\int_0^1 \dot{\theta}_s \mathrm{d}V_s - \int_0^1 < \dot{\theta}, \dot{v}_s >_{H_S} \mathrm{d}s\right]
\end{aligned}$$

$$= -E_{U\mathbb{P}}\left[\int_0^1 <\dot{\theta}, \dot{v}_s>_{H_S} \mathrm{d}s\right]$$
$$= -E_{\mathbb{P}}[< v \circ U, \theta \circ U >_H].$$

This shows Eq. (3.22). □

Theorem 8.2. *Let $(\Omega, \mathcal{F}^{\mathbb{P}}, \mathbb{P})$ be a complete probability space with a continuous filtration $(\mathcal{F}_t^{\mathbb{P}})_{t\in[0,1]}$ which satisfies the usual conditions. Let $(B_t)_{t\in[0,1]}$ be an S-valued $(\mathcal{F}_t^P)$-Brownian motion on $(\Omega, \mathcal{F}^{\mathbb{P}}, \mathbb{P})$ and $U : [0,1] \to S$ be any $(\mathcal{F}_t^P)$-adapted, continuous process such that $U\mathbb{P} \sim \mu$. Then*

$$2H(U\mathbb{P}|B\mathbb{P}) \leq E_{\mathbb{P}}[|U - B|_H^2]. \tag{3.23}$$

We further note $u := U - B$ and henceforth assume that $u \in L^2(\mathbb{P}, H)$. If we note $\widehat{u}$ the dual predicable projection of u on $(\mathcal{F}_t^u)$ which is the filtration $\sigma(U_s, s \leq t)$ augmented with respect to $\mathbb{P}$, we then have

$$2H(U\mathbb{P}|\mu) = E_{\mathbb{P}}[|U - B|_H^2] - \epsilon_U^2, \tag{3.24}$$

where

$$\epsilon_U = \sqrt{E_{\mathbb{P}}[|u - \widehat{u}|_H^2]}$$

Moreover the following assertions are equivalent:

- $\epsilon_U = 0$.
- *$V \circ U = B$ on $(\Omega, \mathcal{F}^{\mathbb{P}}, \mathbb{P})$ where $V = I_W + v$ is the Girsanov shift associated with the measure $U\mathbb{P}$ on the Wiener space.*

Finally for any probability ν which is equivalent to the Wiener measure μ

$$2H(\nu|\mu) = \inf\left(\{E_{\mathbb{P}}[|U - B|_H^2]\}\right), \tag{3.25}$$

where the infimum is taken on all the $(\Omega, \mathcal{F}^{\mathbb{P}}, (\mathcal{F}_t^{\mathbb{P}})_{t\in[0,1]}, \mathbb{P})$ and all the (U, B) defined on it as above and such that $U\mathbb{P} = \nu$. Moreover, we can always find (at least) one space $(\Omega, \mathcal{F}^{\mathbb{P}}, (\mathcal{F}_t^{\mathbb{P}})_{t\in[0,1]}, \mathbb{P})$ and a (U, B) defined on it with the same properties as in the first part of the claim, which attains the infimum.

Proof. If $U - B \notin L^2(\mathbb{P}, H)$ then Eq. (3.23) holds. Otherwise let $u := U - I_W$ and let v be the Girsanov shift associated to $U\mathbb{P}$. By Proposition 3.1 and Lemma 8.1 we have:

$$\begin{aligned} 2H(U\mathbb{P}|\mu) &= E_{U\mathbb{P}}[|v|_H^2] \\ &= E_{\mathbb{P}}[|v \circ U|_H^2] \\ &= E_{\mathbb{P}}[|\widehat{u}|_H^2] \\ &= E_{\mathbb{P}}[|u|_H^2] - E_{\mathbb{P}}[|u - \widehat{u}|_H^2], \end{aligned}$$

where the last line follows from the fact that $\widehat{u}$ is an orthogonal projection of u. This clearly yields Eq. (3.24) and the equivalences in case of equality. The inequality (3.23) clearly yields the inequality of Eq. (3.25). By taking $(\Omega, \mathcal{F}^{\mathbb{P}}, (\mathcal{F}_t^{\mathbb{P}})_{t\in[0,1]}, \mathbb{P}) = (W, \mathcal{F}, (\mathcal{F}_t)_{t\in[0,1]}, \nu)$ and $(U, B) = (I_W, V)$, where $V = I_W + v$ is the Girsanov shift associated with ν, we know (see Proposition 3.1) that the optimum is attained in Eq. (3.25). □

Remark 8.1. In Theorem 8.2 we wanted to stress that the physical origin of the inequality in Eq. (3.23) is the information loss. This is the reason why we need Lemma 8.1. We already showed this result in [16] (the underlying space was the Wiener space but the proof is the same). However the inequality (3.23) as well as the fact that the equality in Eq. (3.23) occurs if and only if $V \circ U = B$ may be showed directly. Indeed these results follow easily from Proposition 3.1 and the Cauchy–Schwarz inequality as we showed it in [17]. Henceforth, let us assume that the underlying probability space is the Wiener space W. As we seen, under the condition $E_\mu[\rho(-\delta^W u)] = 1, U := I_W + u$ is the inverse of V if and only if it is the right inverse of V. Hence the equality in Eq. (3.23) is equivalent to the invertibility and we recover the main result of [30]. For that reason we refer to this result as Üstünel's criterion.

Proposition 8.3 is a path space version of the Brascamp–Lieb inequality. In Proposition 8.4 we apply it to an h-path process, so that we get a Brascamp–Lieb inequality which holds on any abstract Wiener space. The proof of it is a generalization of the one given in [18]; however the ideas are essentially the same. As Theorem 8.2 enlightens it the next inequalities are involved by information loss in the Gaussian channel.

Proposition 8.3. *Let $(\widetilde{\pi}^i)$ be a family of projections on H_S. Further assume that for any i, there is an $(e^i_j)_{j=1}^\infty$ which is an Hilbert basis of H_S and an $(I_i) \subset \mathbb{N}$ such that $\widetilde{\pi}_i = \sum_{j\in I_i} < x, e^i_j >_{H_S} e^i_j$ and that*

$$\sum_i \alpha_i \widetilde{\pi}_i = I_{H_S}$$

for a sequence of positive numbers $(\alpha_i) \subset \mathbb{R}^+$. We then set

$$\widetilde{T}_i x = \sum_{j\in I_i} < e^i_j, x > e^i_j$$

which is well defined as a measurable mapping from $L^0(\widehat{\mu}, S) \to L^0(\widehat{\mu}, S)$. We also note $\pi_i : H \to H$ the mapping such that $(\pi_i h)(t) = \widetilde{\pi}_i(h_t)$ for any $(h, t) \in H \times [0, 1]$ and $T_i : \omega \in W \to T_i(\omega) \in W$ the mapping defined pathwise by $T_i(\omega) : t \in [0, 1] \to \widetilde{T}_i(W_t)$. Then for any measure ν equivalent to μ with $H(\nu|\mu) < \infty$ we have

$$H(\nu|\mu) \geq \sum_i \alpha_i H(T_i \nu | T_i \mu). \tag{3.26}$$

In particular for every $t \in [0,1]$ *we have*

$$H(\nu|\mu) \geq \sum_i \alpha_i H(\widetilde{T}_i \nu_t | \widetilde{T}_i \mu_t), \tag{3.27}$$

where $\nu_t := W_t \nu$ *(resp.* $\mu_t := W_t \mu$*) is the marginal at* t*.*

Proof. From the definitions it is straightforward to check that (π^i) is a family of projections on H such that $\sum_i \alpha_i \pi^i = I_H$. Let (U, B) be the pair defined on a space $(\Omega, \mathcal{F}^{\mathbb{P}}, \mathbb{P})$ which attains the optimum in the variational problem given in Theorem 8.2, and set $u := U - B$. Theorem 8.2 yields

$$\begin{aligned}
2H(\nu|\mu) &= E_{\mathbb{P}}[|u|_H^2] \\
&= E_{\mathbb{P}}\left[< \sum_i \alpha_i \pi_i u, u >_H \right] \\
&= \sum_i \alpha_i E_{\mathbb{P}}[|\pi_i u|_H^2] \\
&= \sum_i \alpha_i E_{\mathbb{P}}[|B + \pi_i u - B|_H^2] \\
&\geq \sum_i \alpha_i H((B + \pi_i u)\mathbb{P}|\mu) \\
&\geq \sum_i \alpha_i H(T_i(B + \pi_i u)\mathbb{P}|T_i \mu),
\end{aligned}$$

where the last equality comes from the fact that $H(\nu|\mu) \geq H(X\nu|X\mu)$ for any measurable $X : W \to E$ where E is a Polish space. By definition, we also have $T^i \circ \pi^i h = \pi^i h$ for any $h \in H$ so that $H(T_i(B + \pi_i u)\mathbb{P}|T_i \mu) = H(T_i U \mathbb{P}|T_i \mu) = H(T_i \nu|T_i \mu)$. Therefore we have Eq. (3.26). Since by definition for any $t \in [0,1]$ $\mu - a.s.$ $W_t \circ T^i = \widetilde{T}_i(W_t)$, we also have

$$H(T_i \nu|T_i \mu) \geq H(W_t T_i \nu|W_t T_i \mu) = H(\widetilde{T}^i \nu_t | \widetilde{T}_i \mu_t).$$

□

Proposition 8.4. *Let* $(\widehat{\pi}^i)$ *be a family of projections on* H_S*. Further assume that for any* i *there is a* $(e^i_j)_{j=1}^{\infty}$ *which is an Hilbert basis of* H_S *and a* $(I_i) \subset \mathbb{N}$ *such that* $\widehat{\pi}_i = \sum_{j \in I_i} < x, e^i_j > e^i_j$ *and that*

$$\sum_i \alpha_i \widehat{\pi}_i = I_{H_S}.$$

For a sequence of positive numbers $(\alpha_i) \subset \mathbb{R}^+$. *We define*

$$\widehat{T}_i x = \sum_{j \in I_i} < e^i_j, x > e^i_j$$

which is well defined as a mapping from $L^0(\widehat{\mu}, S) \to L^0(\widehat{\mu}, S)$. *Then for any measure* $\widehat{\nu}$ *equivalent to* $\widehat{\mu}$ *such that* $H(\widehat{\nu}|\widehat{\mu}) < \infty$ *we have*

$$H(\widehat{\nu}|\widehat{\mu}) \geq \sum_i \alpha_i H(\widehat{T}_i \widehat{\nu} | \widehat{T}_i \widehat{\mu}).$$

Proof. Let ν be the measure associated with $\widehat{\nu}$ by Definition 7.1 so that $W_1 \nu = \widehat{\nu}$. By Proposition 8.1 $H(\widehat{\nu}|\widehat{\mu}) = H(\nu|\mu)$. Hence Proposition 8.3 with $t = 1$ implies

$$H(\widehat{\nu}|\widehat{\mu}) = H(\nu|\mu) \geq \sum_i \alpha_i H(\widehat{T}_i \widehat{\nu} | \widehat{T}_i \widehat{\mu}).$$

□

Remark 8.2. Note that in the case $S = \mathbb{R}^d$ we have $\widehat{\pi}_i = \widehat{T}_i$ and if we note λ the Lebesgue measure and if we assume that $\widehat{\nu} = X\lambda$ for an X, we have $H(X\lambda|\widehat{\mu}) \geq \sum_i \alpha_i H((\widehat{\pi}^i X)\lambda | \widehat{\pi}^i \widehat{\mu})$ where $\widehat{\pi}^i \widehat{\mu}$ is the law of a standard Gaussian vector with range in $\widehat{\pi}(H_S)$. The relationship of this equality with the Brascamp–Lieb inequality was shown in [2] as it is also recalled in [18].

8.3 Üstünel's Criterion in Terms of Variance and Shannon's Inequality

In finite dimension, Shannon's inequality involves some entropies with respect of the Lebesgue measure. When we seek to write it in terms of Gaussian measures some correlation terms appear since we then lose the invariance under translations. Here we use a trick to recover the property of invariance under translation by performing a change of measure. Under this change of measure Theorem 8.2 takes the form of Corollary 8.1. The variational formulation of the entropy is then written in terms of variance instead of in terms of energy. In Theorem 8.3 we get the abstract Wiener space version of the Shannon inequality as a consequence of two facts: the information loss on the path space and the addition property of the variances of independent random variables.

Corollary 8.1. *Let* $(\Omega, \mathcal{F}^{\mathbb{P}}, \mathbb{P})$ *be a complete probability space with a continuous filtration* $(\mathcal{F}^{\mathbb{P}}_t)_{t \in [0,1]}$ *which satisfies the usual conditions. Let* $(B_t)_{t \in [0,1]}$ *be a* $\mathcal{F}^P_t$ S*-valued Brownian motion on that space, and* $U : [0,1] \to S$ *be any* $(\mathcal{F}_t)$*-adapted continuous process such that* $U\mathbb{P} \sim \mu$. *Further assume that* U *is of the form* $U := B + u$ *where* $u \in L^2_a(\mathbb{P}, H)$, *and let* μ_U *be the probability defined by*

$$\frac{d\mu_U}{d\mu} := \rho(\delta^W m_U),$$

where $m_U(t) = \int_0^t E_{\mathbb{P}}[\dot{u}_s]ds$. *Then*

$$2H(U\mathbb{P}|\mu_U) \leq \mathcal{V}ar_{\mathbb{P}}(u),$$

where

$$\mathcal{V}ar_{\mathbb{P}}(u) = E_{\mathbb{P}}[|u - m_U|_H^2].$$

Let $\widehat{u}$ *be the dual predicable projection of* $u := U - B$ *on* $(\mathcal{F}_t^u)$ *which is the filtration* $\sigma(U_s, s \leq t)$ *augmented with respect to* $\mathbb{P}$. *Then we have*

$$2H(U\mathbb{P}|\mu_U) = \mathcal{V}ar_{\mathbb{P}}(u) - \epsilon_U^2,$$

where

$$\epsilon_U = (E_{\mathbb{P}}[|u - \widehat{u}|_H^2])^{\frac{1}{2}}.$$

Moreover, the following assertions are equivalent:

- $u = \widehat{u}$.
- $V \circ U = B$ *where* $V = I_W + v$ *is the Girsanov shift associated with* $U\mathbb{P}$.
- $2H(U\mathbb{P}|\mu_U) = \mathcal{V}ar_{\mathbb{P}}(u)$.
- $2H(U\mathbb{P}|\mu) = E_{\mathbb{P}}[|U - B|_H^2]$.

Finally, for any probability ν *which is equivalent to the Wiener measure* μ,

$$\frac{d\mu_\nu}{d\mu} := \rho(\delta^W m_\nu),$$

where $m_\nu(t) = -\int_0^t E_\nu[\dot{v}_s]ds = -E_\nu[v_t]$. *Then*

$$2H(\nu|\mu_\nu) = \inf\left(\{\mathcal{V}ar_{\mathbb{P}}(U - B)\}\right),$$

where the infimum is taken on all the $(\Omega, \mathcal{F}^{\mathbb{P}}, (\mathcal{F}_t^{\mathbb{P}})_{t\in[0,1]}, \mathbb{P})$ *and all the* (U, B) *defined on it as above and such that* $U\mathbb{P} = \nu$. *Moreover, we can always find (at least) one space* $(\Omega, \mathcal{F}^{\mathbb{P}}, (\mathcal{F}_t^{\mathbb{P}})_{t\in[0,1]}, \mathbb{P})$ *and a* (U, B) *defined on it with the same properties as in the first part of the claim, which attains the infimum. Finally the optimum is attained by the same shifts as the variational problem of Theorem 8.2.*

Proof. From the definitions and Theorem 8.2, we have

$$\begin{aligned} H(U\mathbb{P}|\mu_U) &= H(U\mathbb{P}|\mu) - E_{\mathbb{P}}\left[\ln \frac{d\mu_U}{d\mathbb{P}} \circ U\right] \\ &= H(U\mathbb{P}|\mu) - \frac{|m_U|_H^2}{2} \end{aligned}$$

$$= E_{\mathbb{P}}\left[\frac{|\widehat{u}|_H^2}{2}\right] - \frac{|m_U|_H^2}{2}$$
$$= E_{\mathbb{P}}\left[\frac{|\widehat{u}|_H^2}{2}\right] - E_{\mathbb{P}}\left[\frac{|u|_H^2}{2}\right] - \left(E_{\mathbb{P}}\left[\frac{|u|_H^2}{2}\right] - \frac{|m_U|_H^2}{2}\right)$$

which is the main part of the result. Note that if $U\mathbb{P} = \nu$ we have

$$\begin{aligned} m_U &= \int_0^{\cdot} E_{\mathbb{P}}[\dot{u}_s]\mathrm{d}s \\ &= \int_0^{\cdot} E_{\mathbb{P}}[\dot{\widehat{u}}_s]\mathrm{d}s \\ &= -\int_0^{\cdot} E_{\mathbb{P}}[\dot{v}_s \circ U]\mathrm{d}s \\ &= -\int_0^{\cdot} E_{\nu}[\dot{v}_s \circ U]\mathrm{d}s \\ &= m_\nu \end{aligned}$$

so that $\mu_\nu = \mu_U$. By taking $(\Omega, \mathcal{F}^{\mathbb{P}}, (\mathcal{F}_t^{\mathbb{P}})_{t\in[0,1]}, \mathbb{P}) = (W, \mathcal{F}, (\mathcal{F}_t)_{t\in[0,1]}, \nu)$, and $(U, B) = (I_W, V)$, where $V = I_W + v$ is the Girsanov shift associated with ν, we know (see Proposition 3.1) that the optimum is attained. □

The next lemma is the price to pay for working with Gaussian measures instead of the Lebesgue measure (which is not defined when $\dim S = \infty$).

Lemma 8.2. *Let ν be a probability equivalent to μ such that $H(\nu|\mu) < \infty$, and let $\widehat{\nu}$ a probability equivalent to $\widehat{\mu}$. Further assume that*

$$W_1\nu = \widehat{\nu}$$

and let

$$\frac{\mathrm{d}\mu_\nu}{\mathrm{d}\mu} := \rho(\delta^W m_\nu),$$

where $m_\nu(t) = -\int_0^t E_\nu[\dot{v}_s]\mathrm{d}s = -E_\nu[v_t]$ where v is the Girsanov drift associated with ν, $\nu_t = W_t\nu$ and

$$\frac{\mathrm{d}\widehat{\mu}_{\widehat{\nu}}}{\mathrm{d}\widehat{\mu}}(x) := \exp\left(< x, m_{\widehat{\nu}} > - \frac{|m_{\widehat{\nu}}|_{H_S}^2}{2}\right),$$

where $m_{\widehat{\nu}}$ is the mean of $\widehat{\nu}_i$. Then

$$H(\widehat{\nu}|\widehat{\mu}_{\widehat{\nu}}) \le H(\nu|\mu_\nu) \tag{3.28}$$

Moreover, if ν is the measure associated with $\widehat{\nu}$ by Definition 7.1, the inequality (3.28) is an equality.

Proof. Since $\widehat{\mu}$ is centered $V_1\nu = \widehat{\mu}$ and $W_1\nu = \widehat{\nu}$ we have $m_\nu(1) = -E_\nu[v_1] = \int d\widehat{\nu}(x)x = m_{\widehat{\nu}}$. Hence by applying two times the Cameron–Martin theorem [15] once on W, once on S, $W_1\mu_\nu = (W_1 + m_\nu(1))\mu = (W_1 + m_{\widehat{\nu}})\mu = (I_S + m_{\widehat{\nu}})\widehat{\mu} = \widehat{\mu}_{\widehat{\nu}}$. Since we also have $W_1\nu = \widehat{\nu}$, we then get $H(\widehat{\nu}|\widehat{\mu}_{\widehat{\nu}}) = H(W_1\nu|W_1\mu_\nu) \leq H(\nu|\mu_\nu)$. Moreover if ν is the measure associated with $\widehat{\nu}$ by Definition 7.1 it is straightforward to check that we have an equality. □

We now give an abstract Wiener space version of Shannon's inequality.

Theorem 8.3. *Let $(\widehat{\nu}^i)$ be a sequence of probabilities equivalent to $\widehat{\mu}$ such that $H(\widehat{\nu}^i|\widehat{\mu}) < \infty$. For a sequence (p_i) of positive reals such that*

$$\sum_i p_i = 1,$$

we set $\widehat{\nu}^\Sigma := (\sum_i \sqrt{p_i}\pi_i) \otimes_i \widehat{\nu}^i$ where π_i is the projection on the ith coordinate of the product space $S^{\mathbb{N}}$. We further define a family of measure $(\widehat{\mu}^i)$ by

$$\frac{\mathrm{d}\widehat{\mu}^i}{\mathrm{d}\widehat{\mu}} = \exp\left(< x, \widehat{m}_i > -\frac{|\widehat{m}_i|^2_{H_S}}{2} \right),$$

where $\widehat{m}_i$ is the mean of $\widehat{\nu}^i$ and we set

$$\frac{\mathrm{d}\widehat{\mu}^\Sigma}{d\widehat{\mu}} = \exp\left(< x, \sum_i \sqrt{p_i}\widehat{m}_i > -\frac{|\sum_i \sqrt{p_i}\widehat{m}_i|^2_{H_S}}{2} \right).$$

Then we have

$$H(\widehat{\nu}^\Sigma|\widehat{\mu}^\Sigma) \leq \sum_i p_i H(\widehat{\nu}^i|\widehat{\mu}^i).$$

Proof. For any $i \in \mathbb{N}$ let ν^i be the optimal measure associated with $\widehat{\nu}^i$ by Definition 7.1, and let V^i be the Girsanov shift associated with ν^i. For any i we also note μ^i the measure associated with ν^i on W by Lemma 8.2. From Corollary 8.1 we have

$$H(\nu^i|\mu^i) = \mathcal{V}ar_{\nu^i}[(V^i - I_W)].$$

Thus Lemma 8.2 yields

$$H(\widehat{\nu}^i|\widehat{\mu}^i) = \mathcal{V}ar_{\nu^i}[(V^i - I_W)]. \tag{3.29}$$

We set $\Omega^\Sigma = W^{\mathbb{N}}$, $\mathbb{P}^\Sigma := \otimes_i \nu^i$, and we define the filtration $(\mathcal{G}_t^\Sigma)$ by $\mathcal{G}_t^\Sigma = \sigma(\{\{V_s^i \circ Pr_i\}_{i\in\mathbb{N}}, s \leq t\})$ for any $t \in [0,1]$, where Pr_i is the projection on the

ith coordinate of $W^{\mathbb{N}}$. We also set $\mathcal{G}^{\Sigma} := \mathcal{G}_1^{\Sigma}$. Then from Paul Levy's theorem, $B^{\Sigma} := \sum_i \sqrt{p_i} V^i \circ Pr_i$ is a $(\mathcal{G}_t^{\Sigma})$-Brownian motion on $(\Omega^{\Sigma}, \mathcal{G}^{\Sigma}, \mathbb{P}^{\Sigma})$. On the other hand $U^{\Sigma} := \sum_i \sqrt{p_i} Pr_i$ is adapted to $(\mathcal{G}_t^{\Sigma})$. Hence Corollary 8.1 applies and we get

$$H(U^{\Sigma}\mathbb{P}^{\Sigma}|\mu_{\Sigma}) \leq \mathcal{V}ar_{\mathbb{P}^{\Sigma}}(U^{\Sigma} - B^{\Sigma}), \tag{3.30}$$

where μ^{Σ} is the measure defined in Corollary 8.1. Since $W_1(U^{\Sigma}\mathbb{P}^{\Sigma}) = W_1 \sum_i \sqrt{p_i} Pr_i \mathbb{P}^{\Sigma} = \widehat{\nu}^{\Sigma}$ Lemma 8.2 yields

$$H(\widehat{\nu}^{\Sigma}|\widehat{\mu}_{\Sigma}) \leq \mathcal{V}ar_{\mathbb{P}^{\Sigma}}(U^{\Sigma} - B^{\Sigma}). \tag{3.31}$$

The property of the variance of a sum of independent variables writes

$$\mathcal{V}ar_{\mathbb{P}^{\Sigma}}(U^{\Sigma} - B^{\Sigma}) = \sum_i p_i \mathcal{V}ar_{\nu^i}(V^i - I_W). \tag{3.32}$$

By gathering Eqs. (3.29), (3.31), and (3.32) we get the result. □

Remark 8.3. In the finite-dimensional case $S = \mathbb{R}^n$ with the Lebesgue measure λ on it, let (X_i) be a sequence of independent random elements with values in $\mathbb{R}^n$ defined on a space $(\Omega, \mathcal{F}, \mathbb{P})$ such that $H(X_i\mathbb{P}|\lambda) < \infty$. Shannon's inequality can be written:

$$\sum_i p_i H(X^i\mathbb{P}|\lambda) \geq H\left(\sum_i \sqrt{p_i} X_i \mathbb{P}|\lambda\right), \tag{3.33}$$

where (p_i) is a sequence of positive numbers such that $\sum_i p_i = 1$. However, the Lebesgue measure is not defined in infinite dimensions and we had to write it in terms of Gaussian measure (which still makes sense in infinite dimension as a Wiener measure). Let γ be the standard Gaussian measure on $\mathbb{R}^n$. The trick to keep a formula as simple as possible is then to introduce the following measures: for any i we set

$$\frac{d\gamma^i}{d\lambda}(x) := \frac{1}{(2\pi)^{\frac{n}{2}}} \exp\left(-\frac{(x - E[X_i])^2}{2}\right)$$

and

$$\frac{d\gamma^{\Sigma}}{d\lambda}(x) := \frac{1}{(2\pi)^{\frac{n}{2}}} \exp\left(-\frac{(x - \sum_i \sqrt{p_i} E[X_i])^2}{2}\right).$$

We then have

$$\begin{aligned} H(X^i\mathbb{P}|\gamma^i) &= E_{X^i\mathbb{P}}\left[\ln \frac{dX^i\mathbb{P}}{d\lambda}\right] - E_{X^i\mathbb{P}}\left[\ln \frac{d\gamma^i}{d\lambda}\right] \\ &= H(X^i\mathbb{P}|\lambda) - E_{X^i\mathbb{P}}\left[\ln \frac{d\gamma^i}{d\lambda}\right] \end{aligned}$$

$$= H(X^i\mathbb{P}|\lambda) - \int d\lambda(x)\frac{(X^i(x) - E[X^i])^2}{2} - \frac{n}{2}\ln(2\pi)$$
$$= H(X^i\mathbb{P}|\lambda) - \mathcal{V}ar_{\mathbb{P}}(X^i) - \frac{n}{2}\ln(2\pi).$$

Since the X^i are independent we have $\mathcal{V}ar_{\mathbb{P}}(\sum_i \sqrt{p_i}X^i) = \sum_i p_i \mathcal{V}ar_{\mathbb{P}}(X_i)$. Hence if we set $\nu^\Sigma = \sum_i \sqrt{p_i}X_i\mathbb{P}$ we get

$$\sum_i p_i H(X^i\mathbb{P}|\gamma^i) - H(\nu^\Sigma|\gamma^\Sigma) = \sum_i p_i H(X^i\mathbb{P}|\lambda) - H(\nu^\Sigma|\lambda).$$

Hence Shannon's inequality may be written $H(\nu^\Sigma|\gamma^\Sigma) \leq \sum_i p_i H(X^i\mathbb{P}|\gamma^i)$. This is exactly what we proved in this proposition.

9 Invertibility of the Free Schrödinger Shifts in the Perspective of Optimal Transport

In this section we will handle the following sets.

Definition 9.1. Let $\widehat{\nu}$ be a probability equivalent to $\widehat{\mu}$ then we set

$$\mathcal{R}(\mu, \widehat{\nu}) := \left\{u \in L^0(\mu, H) : U_1\mu = \widehat{\nu} \; where \; U := I_W + u\right\}.$$

and

$$\mathcal{R}_a(\mu, \widehat{\nu}) := L^0_a(\mu, H) \cap \mathcal{R}(\mu, \widehat{\nu}).$$

In Sect. 7 we have given mild sufficient conditions for the invertibility of free Schrödinger shifts. It also involves a generalization for the representation formula of [18] which is given in Proposition 9.1. We recall that in [18] Proposition 9.1 is proved on the classical Wiener space under the condition that $\frac{d\widehat{\nu}}{d\mu} = e^f$ where f is C^2 with all its derivatives bounded. Moreover the latter does not relate the equality case to invertibility but rather to right invertibility.

Proposition 9.1. *Let $\widehat{\nu}$ be a probability equivalent to $\widehat{\mu}$ with finite entropy, i.e., $H(\widehat{\nu}|\widehat{\mu}) < \infty$. If $\dim S = \infty$ further assume that $\frac{d\widehat{\nu}}{d\mu} \in L^\infty(\widehat{\mu})$. Then we have*

$$2H(\widehat{\nu}|\widehat{\mu}) = \min\left(\left\{E_\mu[|u|_H^2] : u \in \mathcal{R}_a(\mu, \widehat{\nu})\right\}\right). \tag{3.34}$$

Moreover the infimum is attained by a $U := I_W + u$ which is the inverse of the Schrödinger shift (see Definition 7.1) associated with $\widehat{\nu}$.

Proof. Let $\widehat{\nu}$ be a probability equivalent to $\widehat{\mu}$. From Eq. (3.12) for any $u \in \mathcal{R}(\mu, \widehat{\nu})$ we have

$$H(\widehat{\nu}|\widehat{\mu}) \leq H(U\mu|\mu). \tag{3.35}$$

Then Theorem 8.2 implies that

$$2H(U\mu|\mu) \leq E_\mu[|u|_H^2]. \tag{3.36}$$

Together with Eq. (3.35) it yields the inequality in Eq. (3.34). Moreover from formula (3.12) the equality in Eq. (3.35) is attained by the optimal measure associated with ν. By applying Theorem 7.1 (or Theorem 7.2 if $\dim S = \infty$) we get the existence of a u which attains the equality in Eq. (3.36). Hence the inverse of the free Schrödinger shift associated with $\widehat{\nu}$ attains the optimum in Eq. (3.34). □

Naturally related questions are whether a similar representation formula holds for $d(\widehat{\nu}, \widehat{\mu})$ (see definition 3.1) and whether the infimum is also attained by an invertible shift T with a given marginal $\widehat{\nu}$. The answer is given by Proposition 9.2. Also note that the law of the nonadapted shift T which appears in the statement of Proposition 9.2 is the analogous in the nonadapted case to the optimal law $\nu := \frac{\widehat{d\nu}}{\widehat{d\mu}} \circ W_1$.

Proposition 9.2. *Let $\widehat{\nu}$ be a probability equivalent to $\widehat{\mu}$ of finite Wasserstein distance with respect to $\widehat{\mu}$. Then $d(\widehat{\nu}, \widehat{\mu})$ is given by*

$$d(\widehat{\nu}, \widehat{\mu}) = \inf\left(\left\{E_\mu[|u|_H^2] : u \in \mathcal{R}(\mu, \widehat{\nu})\right\}\right) \tag{3.37}$$

and the infimum is attained by an invertible shift T such that

$$W_1 \circ T = T^S \circ W_1. \tag{3.38}$$

Moreover, let $\widetilde{S}$ be the inverse of T, i.e., $\mu - a.s.$ $T \circ S = I_W$ and $S \circ T = I_W$. We also have

$$W_1 \circ \widetilde{S} = \widetilde{S}^S \circ W_1, \tag{3.39}$$

where T^S and $\widetilde{S}^S$ are the solutions of the Monge problem (resp. its inverse) on S defined in Theorem 8.1.

Proof. We recall that I_S denotes the identity map on S. We note $\mathcal{R}(\widehat{\mu}, \widehat{\nu})$ the set of the mappings $u^S \in L^2(\widehat{\mu}, H_S)$ such that $U^S\widehat{\mu} = \widehat{\nu}$ where $U^S = I_S + u^S$. It may be seen as a subset of $\mathcal{R}(\mu, \widehat{\nu})$. Indeed for any such u^S we can set $i(u^S)_. = \int_0^{\cdot} u^S \circ W_1 ds$. Obviously $i(u^s) \in \mathcal{R}(\mu, \widehat{\nu})$ so that $i(\mathcal{R}(\widehat{\mu}, \widehat{\nu})) \subset \mathcal{R}(\mu, \widehat{\nu})$ and

$$\begin{aligned}\inf\left(\left\{E_\mu[|i(u^S)|_H^2 : u^S \in \mathcal{R}(\widehat{\mu}, \widehat{\nu})\right\}\right) &= \inf\left(\left\{E_\mu[|u|_H^2 : u \in i(\mathcal{R}(\widehat{\mu}, \widehat{\nu}))\right\}\right) \\ &\geq \inf\left(\left\{E_\mu[|u|_H^2 : u \in \mathcal{R}(\mu, \widehat{\nu})\right\}\right).\end{aligned}$$

Of course we have

$$E_\mu\left[\left|i(u^S)\right|_H^2\right] = E_{\widehat{\mu}}\left[\left|u^S\right|_{H_S}^2\right] \tag{3.40}$$

and

$$\begin{aligned} d(\widehat{\nu},\widehat{\mu}) &= \inf\left(\left\{E_{\widehat{\mu}}[|u^S|_{H_S}^2 : u^S \in \mathcal{R}(\widehat{\mu},\widehat{\nu})\right\}\right) \\ &= \inf\left(\left\{E_\mu[|i(u^S)|_H^2 : u^S \in \mathcal{R}(\widehat{\mu},\widehat{\nu})\right\}\right). \end{aligned}$$

Hence we get

$$d(\widehat{\nu},\widehat{\mu}) \geq \inf\left(\left\{E_\mu[|u|_H^2] : u \in \mathcal{R}(\mu,\widehat{\nu})\right\}\right). \tag{3.41}$$

On the other hand for any $u \in \mathcal{R}(\mu,\widehat{\nu})$ Jensen's inequality yields

$$E_\mu\left[|u_1|_{H_S}^2\right] \leq E_\mu\left[|u|_H^2\right] \tag{3.42}$$

and since $(U_1 \times W_1)\mu \in \Sigma(\widehat{\nu},\widehat{\mu})$ we have

$$d(\widehat{\nu},\widehat{\mu}) \leq E_\mu\left[\frac{|u_1|_{H_S}^2}{2}\right]. \tag{3.43}$$

The inequalities (3.42) and (3.43) clearly yield

$$d(\widehat{\nu},\widehat{\mu}) \leq \inf\left(\left\{E_\mu\left[\frac{|u|_H^2}{2}\right] : u \in \mathcal{R}(\mu,\widehat{\nu})\right\}\right). \tag{3.44}$$

Together with Eq. (3.41) the inequality (3.44) yields Eq. (3.37). Now we set $T := I_W + i(t^S)$ where $t^S := T^S - I_S$, i.e.,

$$T : (\sigma,\omega) \in [0,1] \times W \to T_\sigma := W_\sigma + \sigma t^S \circ W_1$$

and we want to show that it attains the infimum of Eq. (3.37). Indeed by hypothesis we have Eq. (3.38) so that $T_1\mu = T^S\widehat{\mu} = \widehat{\nu}$. Hence

$$t := T - I_W \in \mathcal{R}(\mu,\widehat{\nu})$$

On the other hand Eq. (3.40) yields

$$\begin{aligned} E_\mu[|T - I_W|_H^2] &= E_{\widehat{\mu}}[|T^S - I_S|_{H_S}^2] \\ &= d(\widehat{\nu},\widehat{\mu}). \end{aligned}$$

Hence t attains the infimum of Eq. (3.37). We now prove the last part of the claim. It is easy to see that if we set $\widetilde{S}_\sigma := W_\sigma + \sigma\widetilde{s}^S \circ W_1$ where $\widetilde{s}^S := \widetilde{S}^S - I_S$ we then have

$$
\begin{aligned}
(T \circ \widetilde{S})_\sigma &= W_\sigma + \sigma \widetilde{s}^S \circ W_1 + \sigma T^S \circ (W_1 + \widetilde{s}^S \circ W_1) \\
&= W_\sigma + \sigma (T^S \circ \widetilde{S}^S - I_S) \circ W_1 \\
&= W_\sigma
\end{aligned}
$$

which shows that $T := I_W + t$ is invertible with inverse $\widetilde{S}$. □

Under the hypothesis of Proposition 9.1, we then have

$$d(\widehat{\mu}, \widehat{\nu}) = \inf\left(\{E_\mu[|u|_H^2] : u \in \mathcal{R}(\mu, \widehat{\nu})\}\right)$$

and

$$2H(\widehat{\nu}, \widehat{\mu}) = \inf\left(\{E_\mu[|u|_H^2] : u \in \mathcal{R}_a(\mu, \widehat{\nu})\}\right).$$

Since

$$\mathcal{R}_a(\mu, \widehat{\nu}) \subset \mathcal{R}(\mu, \widehat{\nu})$$

we get again the Talagrand inequality. Moreover in the proof of Proposition 9.2 the existence of T clearly follows from the existence of an invertible shift on S which solves the Monge problem. This suggests to investigate the connection between the invertibility of the Schrödinger shifts and the problem of invertibility on S. For that reason henceforth and until the end of this section we assume that $S = C_0([0,1], \mathbb{R}^d)$ (however our results extend to the case where S is an abstract Wiener space with a time structure as in [33]). Let $\widehat{\nu}$ be a probability equivalent to $\widehat{\mu}$, we then have the existence of a Girsanov shift $V^S := I_S + v^S$ such that $\widehat{\mu} - a.s.$

$$\frac{d\widehat{\nu}}{d\widehat{\mu}} = \exp\left(-\delta^{W^S} v^S - \frac{|v^S|_{H_S}}{2}\right)$$

and such that $t \to V_t^S$ is a Wiener process under $\widehat{\nu}$ on S. We recall that in that case $\delta^{W^S} v^S$ denotes the stochastic integral of v^S with respect to the coordinate process $t \to W_t^S$ on S. $L_a^0(\widehat{\mu}, H_S)$ is the subset of the elements of $L^0(\widehat{\mu}, H_S)$ which are adapted to the filtration generated by the coordinate process on S. We call v^S (resp. $V^S := I_S + v^S$ where I_S is the identity map on S) the S-Girsanov drift (resp. shift) associated with $\widehat{\nu}$. In this case we also define

$$\mathcal{R}(\widehat{\mu}, \widehat{\nu}) := \left\{u \in L^0(\mu, H_S) : U\widehat{\mu} = \widehat{\nu} \; where \; U := I_S + u\right\}$$

and $\mathcal{R}_a(\widehat{\mu}, \widehat{\nu}) = \mathcal{R}(\widehat{\mu}, \widehat{\nu}) \cap L_a^0(\widehat{\mu}, H)$. Propositions 9.3 and 9.4 complete the analogy between Eqs. (3.34) and (3.37): in particular Eq. (3.48) has to be compared with Eq. (3.38).

Proposition 9.3. *Let $\widehat{\nu}$ be a probability equivalent to $\widehat{\mu}$ on $S = C_0([0,1], \mathbb{R}^d)$ which is such that $\frac{d\widehat{\nu}}{d\widehat{\mu}} \in L^\infty(\widehat{\mu})$, and let $V := I_W + v$ be the Schrödinger shift associated with $\widehat{\nu}$ (see Definition 7.1). Then we have*

$$W_1 \circ V = V^S \circ W_1, \tag{3.45}$$

where $V^S := I_S + v^S$ *is the* S*-Girsanov shift associated with* $\widehat{\nu}$*. If we denote by* $U := I_W + u$ *the inverse of the Schrödinger shift we also have equivalently*

$$V^S \circ U_1 = W_1 \tag{3.46}$$

and

$$E_\mu[|u_1|^2_{H_S}] = E_\mu[|u|^2_H].$$

In particular the following variational formula holds:

$$\begin{aligned} H(\widehat{\nu}|\widehat{\mu}) &= E_\nu\left[\frac{|v_1|^2_{H_S}}{2}\right] \\ &= \inf\left(\left\{E_\mu\left[\frac{|a_1|^2_{H_S}}{2}\right] : a \in \mathcal{R}_a(\mu,\widehat{\nu})\right\}\right). \end{aligned}$$

Proof. Let $a \in \mathcal{R}_a(\mu,\widehat{\nu})$ and let $\widehat{a_1}^{A_1}$ be the projection of a_1 on the closed subspace

$$\{\theta \circ A_1 : \theta \in L^2_a(\widehat{\nu}, H_S)\}.$$

Since

$$(\delta^{W^S}\theta)\circ A_1 - \delta^{W_1}(\theta\circ A_1) =< \theta\circ A_1, a_1 >_{H_S},$$

we get

$$\begin{aligned} E_\mu\left[< a_1, \theta\circ A_1 >_{H_S}\right] &= E_\mu\left[(\delta^{W^S}\theta)\circ A_1\right] - E_\mu\left[\delta^{W_1}(\theta\circ A_1)\right] \\ &= E_{\widehat{\nu}}\left[\delta^{W^S}\theta\right] \\ &= E_{\widehat{\nu}}\left[\delta^{V^S}\theta\right] - E_{\widehat{\nu}}\left[< v^S, \theta >_{H_S}\right] \\ &= -E_\mu\left[< v^S\circ A_1, \theta\circ A_1 >_{H_S}\right]. \end{aligned}$$

Hence

$$\widehat{a_1}^{A_1} + v^S \circ A_1 = 0$$

$\widehat{\mu} - a.s.$ In particular together with Proposition 3.1 it yields

$$\begin{aligned} 2H(\widehat{\nu}|\widehat{\mu}) &= E_{\widehat{\nu}}\left[|v^S|^2_{H_S}\right] \\ &= E_\mu\left[|v^S\circ A_1|^2_{H_S}\right] \\ &= E_\mu\left[|\widehat{a_1}^{A_1}|^2_{H_S}\right] \\ &\le E_\mu\left[|a_1|^2_{H_S}\right]. \end{aligned}$$

Then Jensen's inequality implies

$$2H(\widehat{\nu}|\widehat{\mu}) \leq E_{\mu}[|a|_H^2] \tag{3.47}$$

for any $a \in \mathcal{R}_a(\mu, \widehat{\nu})$. If we apply this result to the optimal shift $U := I_W + u$ which is the inverse of V and Theorem 8.2 we get

$$2H(\widehat{\nu}|\widehat{\mu}) \leq E_{\mu}[|u|_H^2] = 2H(\nu|\mu) = 2H(\widehat{\nu}|\widehat{\mu}),$$

where the last equality is a consequence of Proposition 8.1 and where ν is the optimal probability associated with $\widehat{\nu}$ by Definition 7.1. Hence in that case the inequalities are equalities and $\widehat{u_1}^{U_1} = u_1$ so that $u_1 = -v^S \circ U_1$ and Eq. (3.46) is proved. By applying V to both terms of Eq. (3.46) we get Eq. (3.45). From this the result comes easily. □

Proposition 9.4. *With the same hypothesis and notations as in Proposition 9.3, let $\widehat{\nu}$ be a probability equivalent to $\widehat{\mu}$ such that $\frac{\mathrm{d}\widehat{\nu}}{\mathrm{d}\mu} \in L^{\infty}(\widehat{\mu})$. Moreover, let U be the optimal shift given by Proposition 9.1 which is the inverse of the Schrödinger shift associated with $\widehat{\nu}$. Then the following assertions are equivalent:*

- *There is a measurable mapping $U^S : S \to S$ such that $\mu - a.s.$*

$$W_1 \circ U = U^S \circ W_1. \tag{3.48}$$

- *There is a $u^S \in \mathcal{R}_a(\widehat{\mu}, \widehat{\nu})$ such that $U^S = I_S + u^S$ is the both sided $\widehat{\mu}$ almost sure inverse of $V^S := I_S + v^S$ where v^S is the S-Girsanov drift associated with $\widehat{\nu}$.*

Moreover, in that case, both the U^S are the same and

$$2H(\widehat{\nu}|\widehat{\mu}) = E_{\widehat{\mu}}[|u^S|_{H^S}^2].$$

Proof. Assume that there is a mapping U^S such that Eq. (3.48) holds. We have

$$U^S\widehat{\mu} = U^S \circ W_1\mu = U_1\mu = \widehat{\nu}.$$

Moreover, by Eq. (3.46) of Proposition 9.3, we get

$$\begin{aligned}\widehat{\mu}\left(\{\omega \in S : V^S \circ U^S = I_S\}\right) &= \mu\left(\{\omega \in W : V^S \circ U^S \circ W_1 = W_1\}\right)\\ &= \mu\left(\{\omega \in W : V^S \circ U_1 = W_1\}\right)\\ &= 1.\end{aligned}$$

Hence $V^S \circ U^S = I_S \; \widehat{\mu} - a.s.$ Since $U^S\widehat{\mu} = \widehat{\nu} \sim \widehat{\mu}$ Proposition 4.1 also implies that $\widehat{\mu} - a.s.$ we have $U^S \circ V^S = I_S$. Hence U^S is the both sided inverse of V^S.

This implies that $t \to U_t^S$ generates the same filtration as the coordinate process on S. Let $u^S = U^S - I_S$. Since $u^S = -v^S \circ U^S$, $E_{\widehat{\mu}}[|u^S|^2_{H_S}] = E_\nu[|v^S|^2_H] = H(\widehat{\nu}|\widehat{\mu})$. Moreover, since $v^S \in L^0_a(\widehat{\mu}, H_S)$ is adapted to the filtration generated by the coordinate process on S, u^S is also adapted to the filtration generated by U^S and hence to the filtration generated by the coordinate process on S, i.e., $u^S \in L^0_a(\widehat{\mu}, H_S)$. Conversely assume that $U^S = I_W + u^S$ where $u^S \in \mathcal{R}_a(\widehat{\mu}, \widehat{\nu})$ is the inverse of V^S. By applying U^S to both terms of formula Eq. (3.46) in Proposition 9.3 we finally get Eq. (3.48). □

10 An Extension of Üstünel's Criterion for Some Stochastic Differential Equation with Dispersion

We adopt special notations throughout this last section. We note $\mathcal{A}^{d,r}$ the set of the progressively measurable processes (with respect to the canonical filtrations) $(\alpha_s, s \in [0,1])$ with values in $\mathbb{R}^d \otimes \mathbb{R}^r$. Let $(\sigma_s, s \in [0,1]) \in \mathcal{A}^{d,d}$. We first consider the following stochastic differential equation:

$$\mathrm{d}X_t = \sigma_t(X)\mathrm{d}B_t; \quad X_0 = 0. \tag{3.49}$$

We further assume that $(\sigma_s, s \in [0,1]) \in \mathcal{A}^{d,d}$ satisfies the following condition:

(H1) σ is such that Eq. (3.49) has a unique solution η in the sense of the probability law.

Let $\mathbb{P} \in \mathcal{P}(W)$ be any Borelian probability on W. We note $\mathcal{F}^{\mathbb{P}}$ be the completion of $\mathcal{B}(W)$ with respect to $\mathbb{P}$. We recall that $\mathbb{P}$ has a unique extension on $\mathcal{F}^{\mathbb{P}}$ which we still note $\mathbb{P}$. We also note $(\mathcal{F}_t^{\mathbb{P}})$ the usual augmentation [3] with respect to $\mathbb{P}$ of the filtration generated by the coordinate process $t \to W_t$. We now focus on equations of the form

$$\mathrm{d}X_t = \sigma_t(X)(\mathrm{d}B_t - \dot{\beta}_t \circ X\mathrm{d}t); \quad X_0 = 0. \tag{3.50}$$

Without loss of generality we may always assume that there is a predicable $(\dot{\xi}_s, s \in [0,1]) \in \mathcal{A}^{d,1}$ such that $\dot{\beta}_t = \sigma_t^\star \dot{\xi}_t$, where $\sigma_t^\star$ is the transpose of σ_t. We have to introduce some other notations in order to set our second hypothesis. For a $\mathbb{P} \in \mathcal{P}(W)$ $\mathcal{M}_2^c(\mathbb{P})$ (resp. $\mathcal{M}_2^{c,\mathrm{loc}}(\mathbb{P})$) will denote the set of the continuous square integrable $(\mathcal{F}_t^{\mathbb{P}})$-martingale (resp. continuous locally square integrable $(\mathcal{F}_t^{\mathbb{P}})$-martingale) on $(W, \mathcal{F}^{\mathbb{P}}, \mathbb{P})$, vanishing at zero. For any $N \in \mathcal{M}_2^c(\mathbb{P})$ (resp.in $\mathcal{M}_2^{c,\mathrm{loc}}(\mathbb{P})$) we set $\mathcal{L}^2(\mathbb{P}, <N>) := \{(\dot{a}_s, s \in [0,1]) \in \mathcal{A}^{1,1} : E_\nu[\int_0^1 \dot{a}_s^2 d <N,N>_s] < \infty\}$ (resp. $\mathcal{L}^2_{\mathrm{loc}}(\mathbb{P}, <N>) := \{(\dot{a}_s, s \in [0,1]) \in \mathcal{A}^{1,1} : \mathbb{P}-a.s. \int_0^1 \dot{a}_s^2 d <N,N>_s < \infty\}$). Let $t \to M_t \in \mathbb{R}^d$ and $(\dot{u}_s, s \in [0,1] | \dot{u}_s \in \mathbb{R}^d)$ be such that for any $i \in \mathbb{R}^d$ $M^i \in \mathcal{M}_2^{c,\mathrm{loc}}(\mathbb{P})$ and $(\dot{u}_s^i) \in \mathcal{L}^2_{\mathrm{loc}}(\mathbb{P}, <M^i>)$ and let $t \to A_t \in \mathbb{R}^d$ be a continuous and adapted process such that for any i, $t \to A_t^i$ is of finite variation. For $U := M + A$, we set the following notation of the stochastic integral:

$$\delta^U u := \sum_i \int_0^1 \dot{u}_s^i \mathrm{d}M_s^i + \sum_i \int_0^1 \dot{u}_s^i \mathrm{d}A_s^i.$$

In particular on $(W, \mathcal{F}^\eta, \eta)$ the coordinate process is a semimartingale which enables us to set

$$\rho^\sigma(-\delta^W \xi) := \exp\left(-\delta^W \xi - \frac{< \delta^W \xi, \delta^W \xi >_1}{2}\right).$$

Note that it is straightforward to check that $< \delta^W \xi, \delta^W \xi >_1 = |\beta|_H^2$. With this notation our second hypothesis will be:

(H2) The uniqueness in law holds for Eq. (3.50); $\beta \in L_a^0(\eta, H)$ and

$$E_\eta\left[\rho^\sigma(-\delta^W \xi)\right] = 1.$$

We recall that by the Girsanov theorem (H1) and (H2) imply the existence of a weak solution for Eq. (3.50) [13]. We note ν the law of this solution which is unique by $(H2)$. Proposition 10.1 is nothing but an extension of the old result of Föllmer we recalled in Sect. 3. It is probably well known, but we prefer to recall the proof for the sake of completeness.

Proposition 10.1. *Assume that $(H1)$ and $(H2)$ hold and that ν (resp. η) still denotes the law of the unique solution to Eq. (3.50) (resp. to Eq. (3.49)). Then we have $\eta - a.s.$*

$$\frac{\mathrm{d}\nu}{\mathrm{d}\eta} = \rho^\sigma(-\delta^W \xi) \tag{3.51}$$

(in particular $\nu \sim \eta$) and

$$2H(\nu|\eta) = E_\nu[|\beta|_H^2]. \tag{3.52}$$

Proof. We set

$$\frac{\mathrm{d}\widetilde{\nu}}{\mathrm{d}\eta} := \rho^\sigma(-\delta^W \xi).$$

Since for any $i \in [1, d]$ $t \to W_t^i$ is a continuous local martingale $N := -\delta^W \pi.\xi \in \mathcal{M}_2^{c,\text{loc}}$ and $< N, N >_1 = |\beta|_H^2$. By applying the Girsanov theorem for any $f \in C_b^2(\mathbb{R}^d)$ we have

$$f(W_t) - f(W_0) - \int_0^t (Af)(s, \omega)\mathrm{d}s \in \mathcal{M}_2^{c,\text{loc}}(\widetilde{\nu}),$$

where (repeated indices are implicitly summed over)

$$(Af)(t, \omega) := \frac{(\sigma_s(\omega)\sigma_s^\star(\omega))^{i,j}}{2}\partial_{i,j} f(W_t) - (\sigma_s(\omega)\beta_s(\omega))^i \partial_i f(W_t).$$

Hence $\widetilde{\nu}$ is the law of a solution to Eq. (3.50) and from the uniqueness $\nu = \widetilde{\nu}$ which implies Eq. (3.51). We now turn to the proof of Eq. (3.52). First assume that

$H(\nu|\eta) < \infty$. Then we set for any $n \in \mathbb{N}$ $\tau_n = \inf(\{t :< N, N >_t > n\} \wedge 1$. By the Girsanov theorem, $t \to N_t^{\tau_n} + < N^{\tau_n}, N^{\tau_n} >_t$ is a martingale under ν so that $E_\nu[N_{\tau_n} + < N^{\tau_n}, N^{\tau_n} >] = 0$. Therefore

$$\begin{aligned}
E_\nu\left[< N, N >_1\right] &= E_\nu\left[\lim_{n\to\infty} < N, N >_{\tau_n}\right] \\
&\leq \liminf_{n\to\infty} E_\nu\left[< N, N >_{\tau_n}\right] \\
&= \liminf_{n\to\infty} E_\nu\left[-2N_{\tau_n} - 2 < N, N >_{\tau_n} + < N, N >_{\tau_n}\right] \\
&= 2\liminf_{n\to\infty} E_\nu\left[-N_{\tau_n} - \frac{< N, N >_{\tau_n}}{2}\right] \\
&= 2\liminf_{n\to\infty} E_\nu\left[E_\eta[\frac{d\nu}{d\eta}|\mathcal{F}_{\tau_n}]\right] \\
&= 2\liminf_{n\to\infty} E_\eta\left[E_\eta[\frac{d\nu}{d\eta}|\mathcal{F}_{\tau_n}]\ln E_\eta[\frac{d\nu}{d\eta}|\mathcal{F}_{\tau_n}]\right] \\
&\leq 2H(\nu|\eta),
\end{aligned}$$

where the last line follows from Jensen's inequality. Hence $H(\nu|\eta) < \infty$ implies $E_\nu[< N, N >_1] < \infty$. Conversely, if $E_\nu[< N, N >_1] < \infty$ $t \to N_t$ is a martingale under η for the filtration $(\mathcal{F}_t^\eta)$. Hence the Girsanov theorem implies $E_\nu[\widetilde{N}_1] = 0$, where $\widetilde{N}_. := N_. + < N, N >_.$. We then get

$$\begin{aligned}
H(\nu|\eta) &= E_\nu\left[-\widetilde{N}_1 + \frac{< N, N >_1}{2}\right] \\
&= E_\nu\left[\frac{< N, N >_1}{2}\right].
\end{aligned}$$

Hence $H(\nu|\eta) < \infty$ if and only if $E_\nu[< N, N >_1] < \infty$, and we always have

$$2H(\nu|\eta) = E_\nu[< N, N >_1]$$

which is Eq. (3.52). □

We still assume that $(H1)$ and $(H2)$ hold and recall that η (resp. ν) denotes the law of the unique solution to Eq. (3.49) [resp. Eq. (3.50)]. We further assume that the drift term β satisfies the following finite energy condition.

(H3) $E_\nu[|\beta|_H^2] < \infty$.

The next theorem provides an entropy-based criterion for strong solutions. The notion of perturbation of the identity (cf. [30]) has to be generalized by some U of the form

$$U_. := \int_0^{\cdot} \sigma_t(U)(\mathrm{d}W_t + \dot{\beta}_t^u \mathrm{d}t).$$

Theorem 10.1 is a generalization of Theorem 7 of [30] (take $\sigma_t = \delta^{i,j}$ for any $t \in [0,1]$).

Theorem 10.1. *Assume that $(H1)$,$(H2)$, and $(H3)$ hold. Let $U : W \to W$ be such that*

$$U\mu = \nu \tag{3.53}$$

and $t \to U_t$ is adapted to $(\mathcal{F}_t^\mu)$ which is the usual augmentation of the coordinate process $t \to W_t$ with respect to the Wiener measure μ. Further assume that U solves

$$U_. = \int_0^{\cdot} \sigma_s(U)(\mathrm{d}W_s + \dot{\beta}_s^u \mathrm{d}s). \tag{3.54}$$

Then we have

$$2H(\nu|\eta) \leq E_\mu[|\beta^u|_H^2] \tag{3.55}$$

with equality if and only if $\mathrm{d}t \times \mathrm{d}\mu - a.s.$ $\beta^u + \dot{\beta}_t \circ U = 0$, *i.e., if and only if U is a strong solution of Eq. (3.50) on $(W, \mathcal{F}^\mu, \mu)$ with the filtration $(\mathcal{F}_t^\mu)$, and the Brownian motion $t \to W_t$, i.e.,*

$$U_. = \int_0^{\cdot} \sigma_t \circ U(\mathrm{d}W_t - \dot{\beta}_t \circ U\mathrm{d}t).$$

Proof. We set $N := \int_0^{\cdot} \sigma_s(U)\mathrm{d}W_s$ and $u = \int_0^{\cdot} \sigma_s(U)\dot{\beta}_s^u \mathrm{d}s$ so that from the hypothesis $U_. := N + u$. Note that for any $i \in [1,d]$ $M_.^i := W_.^i - \int_0^{\cdot}[\sigma_t \dot{\beta}_t]^i \mathrm{d}t \in \mathcal{M}_2^{c,\mathrm{loc}}(\nu)$. We then have by definition

$$< \delta^M \xi, \delta^M \xi >_1 = < \delta^W \xi, \delta^W \xi >_1 = |\beta|_H^2. \tag{3.56}$$

Together with $(H3)$, Eq. (3.56) yields

$$E_\nu[\delta^M \xi] = 0. \tag{3.57}$$

Moreover, from Eq. (3.53) we also have

$$< \delta^N(\xi \circ U), \delta^N(\xi \circ U) >_1 = < \delta^W \xi, \delta^W \xi >_1 \circ U = |\beta \circ U|_H^2. \tag{3.58}$$

Since the finite energy condition also reads $E_\mu[|\xi \circ U|_H^2] < \infty$, together with Eq. (3.58) it yields

$$E_\mu[\delta^N(\xi \circ U)] = 0. \tag{3.59}$$

We then have

$$
\begin{aligned}
E_\nu\left[|\beta|_H^2\right] &= E_\nu\left[<\beta,\beta>_H\right]\\
&= E_\nu\left[\int_0^1 <\sigma_s^\star \dot{\xi}_s, \dot{\beta}_s>_{\mathbb{R}^d}\, \mathrm{d}s\right]\\
&= E_\nu\left[\int_0^1 <\dot{\xi}_s, \sigma_s \dot{\beta}_s>_{\mathbb{R}^d}\, \mathrm{d}s\right]\\
&= -E_\nu\left[\delta^W \xi\right] - E_\nu\left[\delta^M \xi\right]\\
&= -E_\nu\left[\delta^W \xi\right],
\end{aligned}
$$

where the last equality follows from Eq. (3.57). On the other hand Eqs. (3.53) and (3.54) yield

$$
\begin{aligned}
E_\nu\left[|\beta|_H^2\right] &= -E_\mu\left[\delta^U(\xi\circ U)\right]\\
&= -E_\mu\left[\delta^N(\xi\circ U)\right] - E_\mu\left[<u,\xi\circ U>_H\right]\\
&= -E_\mu\left[\delta^N(\xi\circ U)\right] - E_\mu\left[\int_0^1 <\sigma_s\circ U \dot{\beta}_s^u, \dot{\xi}_s\circ U>_{\mathbb{R}^d}\, \mathrm{d}s\right].
\end{aligned}
$$

Finally Eq. (3.59) yields

$$
\begin{aligned}
E_\nu[|\beta|_H^2] &= -E_\mu\left[\int_0^1 <\sigma_s\circ U \dot{\beta}_s^u, \dot{\xi}_s\circ U>_{\mathbb{R}^d}\, \mathrm{d}s\right]\\
&= -E_\mu\left[\int_0^1 <\dot{\beta}_s^u, (\sigma_s^\star \dot{\xi}_s)\circ U>_{\mathbb{R}^d}\, \mathrm{d}s\right]\\
&= -E_\mu\left[\int_0^1 <\dot{\beta}_s^u, \dot{\beta}_s\circ U>_{\mathbb{R}^d}\, \mathrm{d}s\right]\\
&= -E_\mu\left[<\beta^u, \beta\circ U>_H\right].
\end{aligned}
$$

Proposition 10.1 then implies

$$
2H(\nu|\eta) = E_\nu[|\beta|_H^2] = -E_\mu[<\beta^u,\beta\circ U>_H]. \tag{3.60}
$$

The result follows directly from Eq. (3.60). Indeed the Cauchy–Schwarz inequality yields

$$
\begin{aligned}
2H(\nu|\eta) &= -E_\mu[<\beta^u,\beta\circ U>_H]\\
&\le \sqrt{E_\mu[|\beta^u|_H^2]}\sqrt{E_\mu[|\beta\circ U|_H^2]}\\
&\le \sqrt{E_\mu[|\beta^u|_H^2]}\sqrt{E_\nu[|\beta|_H^2]}\\
&= \sqrt{E_\mu[|\beta^u|_H^2]}\sqrt{2H(\nu|\eta)}
\end{aligned}
$$

which implies inequality Eq. (3.55). Moreover, from the case of equality in the Cauchy–Schwarz inequality, we have an equality in Eq. (3.55) if and only if $\mu - a.s.$ $\beta^u + \beta \circ U = 0$. □

References

1. Artstein, S., Ball, K.M., Barthe, F., Naor, A.: Solution of Shannon's problem on the monotonicity of entropy. J. Amer. Math. Soc. **17**(4), 975–982 (2004)
2. Carlen, E., Cordero-Erausquin, D.: Subadditivity of the entropy and its relation to Brascamp–Lieb type inequalities. Geom. Funct. Anal. **19**, 373–405 (2009)
3. Dellacherie, C., Meyer, P.A.: Probabilités et Potentiel Ch. I à IV. Hermann, Paris (1975)
4. Dellacherie, C., Meyer, P.A.: Probabilités et Potentiel Ch. V à VIII. Hermann, Paris (1980)
5. Feyel, D., Üstünel, A.S.: Monge-Kantorovitch measure transportation and Monge-Ampère equation on Wiener space. Probab. Theor. Relat. Fields **128**(3), 347–385 (2004)
6. Feyel, D., Üstünel, A.S.: Solution of the Monge-Ampère equation on Wiener space for general log-concave measures. J. Funct. Anal. **232**(1), 29–55 (2006)
7. Föllmer, H.: Time reversal on Wiener space. In: Lecture Notes in Mathematics. Springer, Berlin (1986)
8. Föllmer, H.: Time reversal of infinite-dimensional diffusions. Stoch. Proc. Appl. **22**(1), 59–77 (1986)
9. Föllmer, H.: Random fields and diffusion processes. Ecole d'Été de Saint Flour XV-XVII. In: Lecture Notes in Mathematics, vol. 1362, pp. 101–203. Springer, New York (1988)
10. Föllmer, H., Gandert, N.: Entropy minimization and Schrödinger process in infinite demensions. Ann. Probab. **25**(2), 901–926 (1997)
11. Gross, L.: Logarithmic Sobolev inequalities. Amer. J. Math. **97**, 1061–1083 (1975)
12. Guerra, F., Morato, L.: Quantization of dynamical systems and stochastic control theory. Phys. Rev. D **27**(8) 1774–1786 (1983)
13. Ikeda, N., Watanabe, S.: Stochastic Differential Equations and Diffusion Processes. North Holland, Amsterdam (Kodansha Ltd., Tokyo) (1981)
14. Krylov, N.V., Röckner, M.R.: Strong solutions of stochastic equations with singular time dependent drift. Probab. Theory Relat. Fields **131**(2), 154–196 (2005)
15. Kuo, H.: Gaussian measures in Banach spaces. In: Lecture Notes in Mathematics, vol. 463. Springer, Berlin (1975)
16. Lassalle, R.: Invertibility of adapted perturbations of the identity on abstract Wiener space. J. Func. Anal. **262**(6), 2734–2776 (2012)
17. Lassalle, R.: Local invertibility of adapted shifts on Wiener space, under finite energy condition, Stochastics, An International Journal of Probability and Stochastics Processes, (2012) DOI: 10.1080/17442508.2012.720257
18. Lehec, J.: A stochastic formula for the entropy and applications. (2010) (Preprint)
19. Malliavin, P.: Stochastic Analysis. Springer, New York (1997)
20. Mikami, T.: Dynamical systems in the variational formulation of the Fokker-Planck equation by the Wasserstein metric, Series 464 (1999). Available at http://eprints3.math.sci.hokudai.ac.jp/655/1/464.PDF
21. Mikami, T.: Optimal control for absolutely continuous stochastic processes and the mass transportation problem. Elect. Comm. Probab. **7**, 199–213 (2002)
22. Mikami, T., Thieullen, M.: Duality theorem for stochastic optimal control problem. Stoch. Proc. Appl. **116**, 1815–1835 (2006). MR230760
23. Nelson, E.: Derivation of the Schrödinger equation from Newtonian mechanics. Phys. Rev. **150**(4) (1966)

24. Nelson, E.: Quantum Fluctuations. Princeton Series in Physics. Princeton University Press, NJ (1985)
25. Nelson, E.: Dynamical Theories of Brownian Motion. Princeton university Press, NJ (1967)
26. Nualart, D.: The Malliavin calculus and related topics. Probability and its Applications, vol. 21. Springer, Berlin (1995)
27. Talagrand, M.: Transportation cost for Gaussian and other product measures. Geom. Funct. Anal. **6**, 587–600 (1996)
28. Üstünel, A.S.: Introduction to analysis on Wiener space. In: Lecture Notes in Mathematics, vol. 1610. Springer, Berlin (1995)
29. Üstünel, A.S.: A necessary and sufficient condition for the invertibility of adapted perturbations of identity on the Wiener space. C.R. Acad. Sci. Paris, Ser. I **346**, 97–900 (2008)
30. Üstünel, A.S.: Entropy, invertibility and variational calculus of adapted shifts on Wiener space. J. Funct. Anal. **257**(11), 3655–3689 (2009)
31. Üstünel, A.S.: Analysis on Wiener Space and Applications (2010) arxiv
32. Üstünel, A.S.: Persistence of invertibility on the Wiener space. Commun. Stoch. Anal. **4**(2), 201–213 (2010)
33. Üstünel, A.S., Zakai, M.: The construction of filtrations on abstract Wiener space. J. Funct. Anal. **143**, 10–32 (1997)
34. Üstünel, A.S., Zakai, M.: Transformation of Measure on Wiener Space. Springer, Berlin (1999)
35. Üstünel, A.S., Zakai, M.: Sufficient conditions for the invertibility of adapted perturbations of identity on the Wiener space. Probab. Theory Relat. Fields **139**, 207–234 (2007)
36. Veretennikov, A.Yu.: On strong solutions and explicit formulas for solutions of stochastic integral equations. Mat. Sb. (N.S.) **111**(153)(3), 434–452 (1980)
37. Villani, C.: Topics in optimal transportation. Graduate Series in Mathematics, vol. 58. American Mathematical Society, Providence (2003)
38. Yasue, K.: Stochastic calculus of variations. J. Funct. Anal. (1981)
39. Zambrini, J.C.: Stochastic mechanics according to E. Schrödinger. Phys. Rev. A **33**(3) (1986)
40. Zambrini, J.C.: Variational processes and stochastic versions of mechanics. J. math. Phys. **27** (9), 2307–2330 (1986)

Chapter 4
Dilation Vector Field on Wiener Space*

Hélène Airault

Abstract We consider the heat operator Δ_H, heat equation, and heat kernel measures $(\nu_t)_{t\geq 0}$ on Wiener space Ω as explained in Driver (Contemp. Math. 338:101–141, 2003). We define the notion of heat dilation vector field associated to a family of probability measures $(\mu_t)_{t\geq 0}$ on Ω. Let $\omega \in \Omega$. The vector field V on Ω is expressed for $F(\omega) = f(\omega(t_1), \omega(t_2), \ldots, \omega(t_n))$ as $VF(\omega) = (vf)(\omega(t_1), \omega(t_2), \ldots, \omega(t_n))$ where $vf = \sum_{k=1}^{n} x_k \frac{\partial}{\partial x_k}$. The vector field V is shown to be a heat vector field for the heat kernel measures $(\nu_t)_{t\geq 0}$. We project down "through a nondegenerate map Z", Ornstein–Uhlenbeck operators defined on Ω by $\mathcal{L}_t F = t\Delta_H F - VF$. We obtain a first-order partial differential equation for the density of the random vector Z. We compare this differential equation to the heat equation and to Stein's equation for the density.

Keywords Wiener space • Heat kernel measures • Dilation vector fields • Ornstein–Uhlenbeck operators • Malliavin calculus

MSC Classification 2010 60H07 (60H30, 35R60)

Received 7/7/2011; Accepted 2/14/2012; Final 6/7/2012

---

*International Conference on Malliavin Calculus and Stochastic Analysis in honor of David Nualart: University of Kansas, 19–21 March 2011.

H. Airault (✉)
INSSET, UMR 7132 CNRS, Université de Picardie Jules Verne,
48 rue Raspail, 02100 Saint Quentin (Aisne), France
e-mail: helene.airault@u-picardie.fr

F. Viens et al. (eds.), *Malliavin Calculus and Stochastic Analysis: A Festschrift in Honor of David Nualart*, Springer Proceedings in Mathematics & Statistics 34,
DOI 10.1007/978-1-4614-5906-4_4, © Springer Science+Business Media New York 2013

1 Introduction

On the path space of a Riemannian manifold, see [2], different Wiener type measures (μ_t) exist corresponding to the variance t. A heat operator $(\mathcal{A}_t)$ associated to (μ_t) is a family $(\mathcal{A}_t)_{|t\in]0;+\infty[}$ of elliptic operators on functions ϕ defined on the path space and such that

$$\int \mathcal{A}_t \phi \mathrm{d}\mu_t = \frac{\mathrm{d}}{\mathrm{d}t}\int \phi \mathrm{d}\mu_t \tag{4.1}$$

whenever the integrals in Eq. (4.1) exist. A dilation vector field associated to (μ_t) is a vector field V defined on the path space such that

$$t\,\frac{\mathrm{d}}{\mathrm{d}t}\int \phi \mathrm{d}\mu_t = \int (V\phi)\,\mathrm{d}\mu_t. \tag{4.2}$$

In [2], a dilatation vector field on the path space is obtained by constructing a Laplacian on the path space and integrating by parts. This shows a rescaling of Wiener measure under dilations. In [9], heat dilation vector fields have been defined on loop groups. For the space $\mathbb{R}^n$, let $x = (x_1, x_2, \ldots, x_n) \in \mathbb{R}^n$, we define

$$Yf(x) = \frac{1}{2}x\cdot(\text{grad } f)(x) = \frac{1}{2}\sum_{j=1}^{n} x_j \frac{\partial}{\partial x_j} f. \tag{4.3}$$

Then Y is a dilation vector field for the gaussian measures

$$\mu_t = (2\pi t)^{-n/2} \exp(-(x_1^2 + x_2^2 + \cdots + x_n^2)/2t)\mathrm{d}x_1\mathrm{d}x_2\cdots \mathrm{d}x_n. \tag{4.4}$$

Of course, it is not the only one. Moreover, fixing $t_1, t_2, \ldots$ positive real numbers, it can be proved by integrating by parts that Y is a dilation vector field for the measures $(\mu_t)_{t\geq 0}$ on $\mathbb{R}^n$,

$$\mu_t = R_t(x_1, x_2, \ldots, x_n)\mathrm{d}x_1\mathrm{d}x_2\cdots \mathrm{d}x_n, \tag{4.5}$$

where

$$\begin{aligned} R_t(x_1, x_2, \ldots, x_n) = & \frac{1}{\sqrt{2\pi t_1 t}} \exp\left(-\frac{x_1^2}{2t_1 t}\right) \\ & \times \frac{1}{\sqrt{2\pi(t_2 - t_1)t}} \exp\left(-\frac{(x_2 - x_1)^2}{2(t_2 - t_1)t}\right) \times \cdots \\ & \times \frac{1}{\sqrt{2\pi(t_n - t_{n-1})t}} \exp\left(-\frac{(x_n - x_{n-1})^2}{2(t_n - t_{n-1})t}\right). \end{aligned} \tag{4.6}$$

The objective of this work is to relate Laplace operator on Wiener space, via a dilation map, to a system of PDEs given in [1] for the density of a smooth random vector on Wiener space. This system of PDEs has been inspired by the analysis of Nourdin and Peccati, [10]. While the system of PDEs in [1] may not determine the density of a smooth vector on Wiener space, by using the device of dilation, interpreting the dilation parameter t as a time parameter, we establish in the present work that a time-dependent version of this system of PDEs, coupled with a single dilation PDE whose coefficients are determined by the *dilation* vector field on Wiener space, takes to the heat equation on Wiener space which was defined by Driver and whose solution is known to be unique. This situation is in stark contrast with the uniqueness obtained by Nourdin and Viens in [11], for a density equation for Malliavin differentiable scalar random variables on Wiener space. Our point of view can be observed in finite-dimensional space by the following considerations. The gaussian density on the real line,

$$p_{t\,t_1}(x) = \frac{1}{\sqrt{2\pi t t_1}} \exp\left(-\frac{x^2}{2\,t t_1}\right) \tag{4.7}$$

is solution of any of the three differential equations

$$\frac{\mathrm{d}}{\mathrm{d}t} p_{tt_1}(x) = \frac{t_1}{2}\frac{\mathrm{d}^2}{\mathrm{d}x^2}\, p_{tt_1}(x), \tag{4.8}$$

$$\frac{\mathrm{d}}{\mathrm{d}t} p_{tt_1}(x) = -\frac{1}{2t}\frac{\mathrm{d}}{\mathrm{d}x}(x\, p_{tt_1}(x)), \tag{4.9}$$

$$\frac{\mathrm{d}}{\mathrm{d}x}(t t_1\, p_{t\,t_1}(x)) = -x\, p_{tt_1}(x). \tag{4.10}$$

These three equations are related: Consider a differentiable function $f(t,x)$ defined for $t > 0$, $x \in \mathbb{R}$ and such that $\frac{\partial}{\partial x}_{|x=0} f(t,x) = 0$. If $f(t,x)$ satisfies two of the three Eqs. (4.8)–(4.10) then it satisfies the third one. For example, if $f(t,x)$ is solution of Eqs. (4.9) and (4.10), we replace the right hand side of Eq. (4.10) into the right hand side of Eq. (4.9), then we find Eq. (4.8).

Similarly, in dimension n, let $t_1, t_2, \ldots, t_n$, and t be positive real numbers. We put $x = (x_1, x_2, \ldots, x_n)$. The function defined on $\mathbb{R}^n$ by Eq. (4.6) is solution of the heat equation

$$\frac{\mathrm{d}}{\mathrm{d}t} R_t(x) = \frac{1}{2}\sum_{j,k=1}^{n} t_j \wedge t_k \left(\frac{\partial^2}{\partial x_j \partial x_k} R_t\right)(x). \tag{4.11}$$

On the other hand, $R_t(x)$ is also solution of

$$\frac{\mathrm{d}}{\mathrm{d}t} R_t(x) = -\frac{1}{2t}\left[\frac{\partial}{\partial x_1}(x_1\, R_t) + \frac{\partial}{\partial x_2}(x_2\, R_t) + \cdots + \frac{\partial}{\partial x_n}(x_n\, R_t)\right] \tag{4.12}$$

just as well as a solution of the first-order system:

$$t_1 \frac{\partial R}{\partial x_1} + t_1 \wedge t_2 \frac{\partial R}{\partial x_2} + t_1 \wedge t_3 \frac{\partial R}{\partial x_3} + \cdots = -x_1\, R \times \frac{1}{t},$$

$$t_1 \wedge t_2 \frac{\partial R}{\partial x_1} + t_2 \frac{\partial R}{\partial x_2} + t_2 \wedge t_3 \frac{\partial R}{\partial x_3} + \cdots = -x_2\, R \times \frac{1}{t},$$

$$t_1 \wedge t_3 \frac{\partial R}{\partial x_1} + t_2 \wedge t_3 \frac{\partial R}{\partial x_2} + t_3 \frac{\partial R}{\partial x_3} + \cdots = -x_3\, R \times \frac{1}{t},$$

$$\cdots. \tag{4.13}$$

Moreover, if we carry in Eq. (4.12), the $x_j R \times \frac{1}{t}$ given by the right hand side of Eq. (4.13), we find Eq. (4.11). Consider a differentiable function $f(t, x)$ defined on $]0, +\infty[\times R^n$. If $f(t, x)$ satisfies Eqs. (4.12) and (4.13), then it satisfies Eq. (4.11). In the same way, if $f(t, x)$ satisfies Eqs. (4.11)–(4.13), then it satisfies Eq. (4.12). Our purpose is to use Malliavin's projecting down and lifting up through a nondegenerate map, see [6], p. 75, to analyze the differential equations giving the density of a random variable. We shall lift Eqs. (4.8)–(4.10) and (4.11)–(4.13) to Wiener space Ω. Let $t \in [0, 1]$. Let $F = \{F(\omega) : \omega \in \Omega\}$ be a smooth random variable on Ω. Let ν be the Wiener measure on Ω. We define the dilated Wiener measure ν_t on Ω by

$$\int F(\omega) \mathrm{d}\nu_t(\omega) = \int F(\sqrt{t}\, \omega) \mathrm{d}\nu(\omega). \tag{4.14}$$

The dilation vector field V on Ω extends the formula

$$VF(\omega) = \sum_{j=1}^{n} \omega(t_j) \frac{\partial f}{\partial x_j}(\omega(t_1), \omega(t_2), \ldots, \omega(t_n)) \tag{4.15}$$

for simple random variables $F(\omega) = f(\omega(t_1), \omega(t_2), \ldots, \omega(t_n))$, where f is a bounded differentiable function on $\mathbb{R}^n$. On Wiener space, the dilation vector field is solution of

$$t \frac{\mathrm{d}}{\mathrm{d}t} \nu_t(F) = \frac{1}{2} \nu_t(VF). \tag{4.16}$$

The Laplacian (or heat) operator on Wiener space is defined by

$$\Delta_H F(\omega) = \sum_{h \in \mathcal{B}} D_h^2 F(\omega), \tag{4.17}$$

where $\mathcal{B}$ is an orthonormal system in $L^2([0, 1])$ and D_h is Malliavin derivative. By [4], we have

$$\frac{\mathrm{d}}{\mathrm{d}t} \nu_t(F) = \frac{1}{2} \nu_t(\Delta_H F) \quad \text{with} \quad \lim_{t \to 0} \nu_t(F) = F(0). \tag{4.18}$$

Let $\mathcal{L}$ be the classical Ornstein–Uhlenbeck operator on Wiener space. We define the dilated $\mathcal{L}_t$ by

$$(\mathcal{L}_t F)(\omega) = (\mathcal{L} F_t)\left(\frac{\omega}{\sqrt{t}}\right) \quad \text{and} \quad F_t(\omega) = F(\sqrt{t}\omega). \tag{4.19}$$

We prove (Proposition 2.3) that

$$\mathcal{L}_t = t\Delta_H - V. \tag{4.20}$$

Let $Z = (Z_1, \ldots, Z_n)$ be a smooth nondegenerate random variable on Wiener space. The conditional expectation (relative to ν_t) of a random vector U with respect to the random variable Z taking the value x is denoted $E_{\nu_t}[U \mid Z = x]$. In Sects. 3 and 4, following [1], taking conditional expectations, we project on $\mathbb{R}^n$, the differential operators on Wiener space given in Eqs. (4.16), (4.20), (4.18). For that purpose, we define the conditional expectations

$$\begin{aligned} &\alpha_j(t,x) = E_{\nu_t}[VZ_j \mid Z = x], \quad \lambda_j(t,x) = E_{\nu_t}[\mathcal{L}_t Z_j \mid Z = x] \\ &\beta_{kj}(t,x) = E_{\nu_t}[(DZ_j|DZ_k)_{L^2([0,1])} \mid Z = x], \quad \gamma_j(t,x) = E_{\nu_t}[\Delta_H Z_j \mid Z = x], \end{aligned} \tag{4.21}$$

where $x \in \mathbb{R}^n$. From Eq. (4.20), we deduce for $j = 1, \ldots, n$,

$$\lambda_j(t,x) = t\gamma_j(t,x) - \alpha_j(t,x). \tag{4.22}$$

For fixed t, the density of the vector Z under ν_t is denoted by ρ_t. As a function of $(t,x) \in [0,1] \times \mathbb{R}^n$, the density $\rho = \{\rho_t(x)\}$ satisfies the *dilation* PDE (projection of the differential equation (4.16) solved by the dilation vector field on Wiener space):

$$\frac{\mathrm{d}}{\mathrm{d}t}\rho_t = -\frac{1}{2t}\sum_{j=1}^{n} \frac{\partial}{\partial x_j}(\alpha_j(t,x) \times \rho_t(x)). \tag{4.23}$$

See Proposition 3.3. As a function of $x \in \mathbb{R}^n$, for every fixed t, ρ_t satisfies the dilated version of the system of PDEs in [1]: for all $j = 1, \ldots, n$, and every fixed t,

$$\sum_{k=1}^{n} \frac{\partial}{\partial x_k}(\beta_k^j(t,x)\rho_t(x)) = \frac{1}{t}\,\rho_t(x)\,E_{\nu_t}[\mathcal{L}_t Z_j \mid Z = x]. \tag{4.24}$$

From Eq. (4.20), we prove that this system of $n+1$ PDEs given by Eqs. (4.23) and (4.24) implies that ρ as a function of $(t,x) \in [0,1] \times \mathbb{R}^n$, also solves the equation which is obtained as projection of the heat equation (4.18) on Wiener space through the random vector Z, namely,

$$\frac{\mathrm{d}}{\mathrm{d}t}\rho_t(x) = -\frac{1}{2}\sum_{j=1}^{n} \frac{\partial}{\partial x_j}(\gamma_j(t,x)\rho_t(x)) + \frac{1}{2}\sum_{j,k} \frac{\partial^2}{\partial x_j \partial x_k}(\beta_{kj}(t,x)\rho_t(x)). \tag{4.25}$$

See Proposition 3.2. If $Z(\omega) = (\omega(t_1), \omega(t_2), \ldots, \omega(t_n))$, then $\alpha_j(t,x) = x_j$, $\beta_{jk}(t,x) = t_j \wedge t_k$, $\gamma_j(t,x) = 0$, we find that Eq. (4.25) becomes Eqs. (4.11) and (4.23) becomes Eq. (4.12).

2 Heat Measures and Laplacian, Dilation Vector Field, Ornstein–Uhlenbeck Operators on the Wiener Space

We consider vector fields and differential operators on Wiener space. We recall some definitions relative to Malliavin calculus, see [12]. To be self-contained and harmonize the notations, we write identities like Eq. (4.45) that can be found in [12], but we give a different proof. The reason is that in this work, we do not develop the Wiener chaos approach and we restrict mainly to integrations by parts for Wiener integrals. In Lemma 3.1, for example, we again keep the point of view of integration by parts. Likewise, we restrict to the infinitesimal version of Cameron–Martin formula, see [3,4], p. 124. Our only account to chaos will be as a complement in Sect. 5 to relate them to the *dilated* Ornstein–Uhlenbeck operator $\mathcal{L}_t$.

Let Ω be the Wiener space of continuous real valued maps ω defined on $[0, 1]$ and such that $\omega(0) = 0$. It is a Banach space with the norm $||\omega|| = \sup_{t\in[0,1]} |\omega(t)|$. Let H be the Cameron–Martin space of continuous, real valued differentiable maps defined on $[0, 1]$. The space H is a Hilbert space with the scalar product

$$(h_1|h_2) = \int_0^1 h_1'(s)h_2'(s)\mathrm{d}s. \tag{4.26}$$

Let $\mathcal{B}$ be an orthonormal basis of H, then $\sum_{h\in\mathcal{B}} h(t_1)h(t_2) = t_1 \wedge t_2$. The Malliavin differentiation operator, see [5], is given for $F : \Omega \to R$ by

$$D_h F(\omega) = \lim_{\epsilon\to 0} \frac{1}{\epsilon}[F(\omega + \epsilon h) - F(\omega)]. \tag{4.27}$$

The differentiation operators D_s on Wiener space, see [12], p. 24, satisfy

$$D_h F(\omega) = \int_0^1 D_s F(\omega)\, h'(s)\, \mathrm{d}s \quad \forall h \in H. \tag{4.28}$$

We define $DF(\omega)$ as element of the Cameron–Martin space such that for any $h \in H$,

$$(DF(\omega)|h) = D_h F(\omega). \tag{4.29}$$

See [12]. Then the inner product of two Malliavin derivatives DF and DG is expressed as

$$(DF|DG) = \int_0^1 D_s F(\omega)\, D_s G(\omega)\, \mathrm{d}s. \tag{4.30}$$

Consider the evaluation functions $\epsilon_{t_1,t_2,\ldots,t_n} : \Omega \to \mathbb{R}^n$, defined by

$$\epsilon_{t_1,t_2,\ldots,t_n}(\omega) = (\omega(t_1), \omega(t_2), \ldots, \omega(t_n)). \tag{4.31}$$

If $F = f \, o \, \epsilon_{t_1,t_2,\ldots,t_n}$ where $f : \mathbb{R}^n \to R$ is differentiable, then by composition of differentiation,

$$D_h F(\omega) = \sum_{j=1}^{n} \left(\frac{\partial}{\partial x_j} f \right) (\omega(t_1), \omega(t_2), \ldots, \omega(t_n)) \times h(t_j), \tag{4.32}$$

$$D_h^2 F(\omega) = \sum_{j,k=1}^{n} \left(\frac{\partial^2}{\partial x_j \partial x_k} f \right) (\omega(t_1), \omega(t_2), \ldots, \omega(t_n)) \times h(t_j)\, h(t_k). \tag{4.33}$$

2.1 *Laplacian and Heat Kernel Measures*

The Laplacian or heat operator Δ_H on Wiener space is defined by Eq. (4.17). We have

$$\sum_{h \in \mathcal{B}} h(t_j) h(t_k) = t_j \wedge t_k. \tag{4.34}$$

Thus, if $F = f \, o \, \epsilon_{t_1,t_2,\ldots,t_n}$, we deduce

$$\Delta_H F(\omega) = \sum_{j,k=1}^{n} t_j \wedge t_k \left(\frac{\partial^2}{\partial x_j \partial x_k} f \right) (\omega(t_1), \omega(t_2), \ldots, \omega(t_n)). \tag{4.35}$$

Let ν be Wiener measure on Ω and let F be a measurable function on Ω. For $t > 0$, we define the measures ν_t with Eq. (4.14). We have

$$\int f(\omega_{t_1}) \mathrm{d}\nu_t(\omega) = \int f(x)\, p_{tt_1}(x)\, \mathrm{d}x, \tag{4.36}$$

where p_{tt_1} is given by Eq. (4.7), see [13] p. 135 for the physical interpretation of the parameter t. If $F_1(\omega) = F(\sqrt{t}\,\omega)$, then

$$D_s F_1(\omega) = \sqrt{t}\; D_s F(\sqrt{t}\,\omega). \tag{4.37}$$

Proposition 2.1. *The measures ν_t defined by Eq. (4.14) satisfy the heat equation (4.18). Such measures ν_t on the Wiener space Ω are uniquely determined by Eq. (4.18) and ν_1 is the Wiener measure. See, for example, [4] p. 125.*

2.2 *Dilation Vector Field on Wiener Space* Ω

Definition 2.1. We define the **dilation vector field** on Ω by

$$(VF)(\omega) = \sum_{h \in \mathcal{B}} \int_0^1 h'(s)\, \mathrm{d}\omega(s)\, .\, D_h F(\omega). \tag{4.38}$$

If $F(\omega) = f(\omega(t_1), \omega(t_2), \ldots, \omega(t_n))$, we have

$$VF(\omega) = (V_n f)(\omega(t_1), \omega(t_2), \ldots, \omega(t_n)) \tag{4.39}$$

with

$$(V_n f)(x_1, x_2, \ldots, x_n) = \sum_{j=1}^{n} x_j \frac{\partial}{\partial x_j} f. \tag{4.40}$$

Proposition 2.2. *The vector field V is a dilation vector field associated to (ν_t). For any function $F : \Omega \to \mathbb{R}$, we have Eq. (4.16).*

Proof. We prove Eq. (4.16) when $F = f o \epsilon_{t_1, t_2, \ldots, t_n}$. □

2.3 Ornstein–Uhlenbeck Operators $\mathcal{L}_t$

Let δ the adjoint of D with respect to the scalar product on the Cameron–Martin space, see [12] p. 35. Let ν be the Wiener measure. We denote $E_\nu[\Phi] = \int \Phi(\omega) \mathrm{d}\nu(\omega)$. For a real valued functional u defined on $[0, 1] \times \Omega$, such that $\int \mathrm{d}\nu(\omega) \int_0^1 u^2(s, \omega) \mathrm{d}s < +\infty$ and u is adapted, we have

$$E_\nu \left[\int_0^1 u(s, \omega) D_s G(\omega) \mathrm{d}s \right] = E_\nu[\delta(u)(\omega)\, G(\omega)] \tag{4.41}$$

for any measurable functional G such that the integrals in Eq. (4.41) exist. We define the Ornstein–Uhlenbeck $\mathcal{L}$ on Ω by

$$\mathcal{L} = -\delta D. \tag{4.42}$$

See Proposition 1.4.3 p. 54 in [12]. From Eq. (4.42), we deduce that for real valued functionals F, G defined on Ω,

$$-E_\nu[\mathcal{L}F(\omega)\, .\, G(\omega)] = E_\nu \left[\int_0^1 D_s F(\omega)\, .D_s G(\omega)\, \mathrm{d}s \right]. \tag{4.43}$$

If $F(\omega) = \omega(\tau)$, then $\mathcal{L}F(\omega) = -\omega(\tau)$ and a particular case of this last equation is the "energy identity" $E_\nu[\omega_s\, .\omega_\tau] = s \wedge \tau$. From Eq. (4.43), we deduce that

$$\int (\mathcal{L}F)(\omega)\, \mathrm{d}\nu(\omega) = 0. \tag{4.44}$$

Lemma 2.1 (See Proposition 1.4.5 in [12]). *Let $Z = (Z_1, Z_2, \ldots, Z_n) : \Omega \to \mathbb{R}^n$ be a nondegenerate map and $\phi : \mathbb{R}^n \to \mathbb{R}$. Assume that $F = \phi o Z$, then*

$$(\mathcal{L}F)(\omega) = \sum_{j=1}^{n} (\mathcal{L}Z_j)(\omega)\, \frac{\partial \phi}{\partial x_j}(Z(\omega)) + \sum_{j,k} (DZ_j | DZ_k)\, \frac{\partial^2 \phi}{\partial x_j \partial x_k}(Z(\omega)). \quad (4.45)$$

Proof. For $G : \Omega \to \mathbb{R}$,

$$\begin{aligned} J &= \int G(\omega)(\mathcal{L}(\phi\, o Z))(\omega) \mathrm{d}\nu(\omega) \\ &= -\int (DG | D(\phi\, o Z)) \mathrm{d}\nu(\omega) \\ &= -\sum_k \int \frac{\partial \phi}{\partial \xi_k}(Z(\omega))(DG | DZ_k) \mathrm{d}\nu(\omega). \end{aligned}$$

Thus $J = J_1 + J_2$ with

$$\begin{aligned} J_1 &= -\sum_k \int (D(G \frac{\partial \phi}{\partial \xi_k}(Z(\omega))) | DZ_k) \mathrm{d}\nu(\omega) \\ &= \sum_k \int \mathcal{L}Z_k \times G(\omega) \frac{\partial \phi}{\partial \xi_k}(Z(\omega)) \mathrm{d}\nu(\omega), \\ J_2 &= \sum_k \int G(\omega)\, (\, D(\frac{\partial \phi}{\partial \xi_k}(Z(\omega)) | DZ_k\,)\, \mathrm{d}\nu(\omega) \\ &= \sum_{j,k} \int G(\omega)\, \frac{\partial^2 \phi}{\partial \xi_k \partial \xi_j}(Z(\omega))(DZ_j | DZ_k)\,)\, \mathrm{d}\nu(\omega). \end{aligned}$$

□

Proposition 2.3. *Let $F : \Omega \to \mathbb{R}$. We define $\mathcal{L}_t F$ by Eq. (4.19). For $t > 0$, we have Eq. (4.20).*

Proof. Assume that $F(\omega) = f(\omega(t_1), \omega(t_2), \ldots, \omega(t_n))$, we shall prove that

$$(\mathcal{L}_t F)(\omega) = t\, (\Delta_H F)(\omega) - (VF).$$

Let $F_1(\omega) = f(\sqrt{t}\omega(t_1), \sqrt{t}\omega(t_2), \ldots, \sqrt{t}\omega(t_n))$. By Lemma 2.1,

$$\begin{aligned} (\mathcal{L}F_1)(\omega) = &\sum_{j=1}^{n} (-\omega(t_j))\, \sqrt{t}\, \frac{\partial f}{\partial x_j}(\sqrt{t}\omega(t_1), \sqrt{t}\omega(t_2), \ldots, \sqrt{t}\omega(t_n)) \\ &+ t\, (t_j \wedge t_k) \sum_{j,k} \frac{\partial^2 f}{\partial x_j \partial x_k}(\sqrt{t}\omega(t_1), \sqrt{t}\omega(t_2), \ldots, \sqrt{t}\omega(t_n)) \end{aligned}$$

and

$$(\mathcal{L}F_1)\left(\frac{\omega}{\sqrt{t}}\right) = \sum_{j=1}^{n}(-\omega(t_j))\,\frac{\partial f}{\partial x_j}(\omega(t_1),\omega(t_2),\ldots,\omega(t_n))$$
$$+t\,(t_j \wedge t_k)\sum_{j,k}\frac{\partial^2 f}{\partial x_j \partial x_k}(\omega(t_1),\omega(t_2),\ldots,\omega(t_n)).$$

□

Remark 2.1. If $t = 1$, $\mathcal{L}_1$ is the operator $\mathcal{L}$; see [5–8, 12]. In Sect. 5, we calculate $\mathcal{L}_t F$ for various functionals F. From Eq. (4.20), we see that our choice for the operators $\mathcal{L}_t$ differs up to multiplication by t from the Ornstein–Uhlenbeck operators $\Delta_H-(1/t)V$, see [8] p. 168. Let $L = \frac{\mathrm{d}^2}{\mathrm{d}x^2} - x\frac{\mathrm{d}}{\mathrm{d}x}$ be the classical Ornstein–Uhlenbeck operator on $\mathbb{R}$. For a differentiable function $f : \mathbb{R} \to \mathbb{R}$, similarly to Eq. (4.19), we put $f_t(x) = f(\sqrt{t}\,x)$ and we define $(L_t f)(x) = (Lf_t)(x/\sqrt{t})$. We have $(L_t f)(x) = tf''(x) - xf'(x)$ which is in accordance with Eq. (4.20). On the other hand, $\int (L_t f)(x)\,\exp(-(x^2/2t))\,(1/\sqrt{2\pi\,t})\,\mathrm{d}x = 0$ is the one-dimensional analogue of Eq. (4.44).

The next proposition extends Eq. (4.43) to the measure ν_t and to the operator $\mathcal{L}_t$. This is also a consequence of Proposition 1.4.5 p. 55 in [12].

Proposition 2.4. *We have*

$$-\frac{1}{t}\,E_{\nu_t}[(\mathcal{L}_t F)(\omega)\,.\,G(\omega)] = E_{\nu_t}\left[\int_0^1 D_s F(\omega)\,.\,D_s G(\omega)\,\mathrm{d}s\right]. \tag{4.46}$$

Proof. By Eq. (4.14),

$$J = E_{\nu_t}\left[\int_0^1 D_s F(\omega)\,.\,D_s G(\omega)\,\mathrm{d}s\right] = E_{\nu_1}\left[\int_0^1 D_s F(\sqrt{t}\,\omega)\,.\,D_s G(\sqrt{t}\,\omega)\,\mathrm{d}s\right].$$

We put $F_1(\omega) = F(\sqrt{t}\,\omega)$ and $G_1(\omega) = G(\sqrt{t}\,\omega)$. Then by Eq. (4.37),

$$(D_s F)(\sqrt{t}\,\omega)\,.\,(D_s G)(\sqrt{t}\,\omega) = \frac{1}{t}(D_s F_1)(\omega)\,.\,(D_s G_1)(\omega).$$

This gives using Eq. (4.43),

$$J = \frac{1}{t}E_{\nu_1}\left[\int_0^1 (D_s F_1)(\omega)\,.\,(D_s G_1)(\omega)\,\mathrm{d}s\right] = -\frac{1}{t}E_{\nu_1}[(\mathcal{L}F_1)(\omega)\,.\,G_1(\omega)]$$
$$= -\frac{1}{t}E_{\nu_t}\left[(\mathcal{L}F_1)\left(\frac{\omega}{\sqrt{t}}\right)\,.\,G(\omega)\right].$$

□

3 Taking Conditional Expectation and Projecting Down

Given a measure ν on the Wiener space and a differentiable map $Z : \Omega \to R^n$, we denote $Z * \nu$ the image measure on $\mathbb{R}^n$ through the map Z, or equivalently the law of Z. For $\phi : \mathbb{R}^n \to \mathbb{R}$, we have

$$\int \phi(Z(\omega))\mathrm{d}\nu(\omega) = \int \phi(x)\,\mathrm{d}(Z * \nu)(x) \quad \text{for} \quad x = (x_1, x_2, \ldots, x_n). \tag{4.47}$$

Assume that ν is the Wiener measure on Ω. If the map Z is nondegenerate, see [6], p. 77, then $Z * \nu$, the law of Z, is absolutely continuous with respect to the n-dimensional Lebesgue measure. On the other hand, let $G : \Omega \to \mathbb{R}$ be a nondegenerate map; we denote $G\nu$ the measure on Ω which has density G with respect to ν:

$$\int F(\omega)\mathrm{d}(G\nu)(\omega) = \int F(\omega)G(\omega)\mathrm{d}\nu(\omega). \tag{4.48}$$

The conditional expectation of G conditioned with $Z(\omega) = x$ is defined as the ratio of the two densities of the measures $Z * (G\,\nu)$ and $Z * \nu$:

$$E_\nu[G \mid Z = x] = \frac{\mathrm{d}(Z * (G\,\nu))}{\mathrm{d}(Z * \nu)}(x). \tag{4.49}$$

Let $m = Z * \nu$ the law of Z; then for $\psi : \mathbb{R}^n \to \mathbb{R}$ and $F : \Omega \to \mathbb{R}$, there holds

$$\int E_\nu[F \mid Z = x]\ \psi(x)\,\mathrm{d}m(x) = \int F(\omega)\psi(Z(\omega))\mathrm{d}\nu(\omega). \tag{4.50}$$

3.1 Stein Equation (4.59)

Following [6], Sect 2.4 p. 70, we project vector fields and operators with conditional expectations. The Nourdin–Peccati analysis [10] and [1] is obtained by projection of vector fields from Wiener space to $\mathbb{R}^n$ by the conditional expectation $E_\nu[U \mid Z = x]$ where ν is Wiener measure and U is a random vector. This extends to the projection by conditional expectation $E_{\nu_t}[U \mid Z = x]$ where we take the measure ν_t instead of Wiener measure ν.

Proposition 3.1. *Let $F : \Omega \to \mathbb{R}$ and let $Z = (Z_1, Z_2, \ldots, Z_n) : \Omega \to \mathbb{R}^n$. Assume that Z is nondegenerate so that the law of Z is absolutely continuous with respect to the n-dimensional Lebesgue measure* $\mathrm{d}x$. *Let $\rho_t(x)$ be the density of $Z * \nu_t$ with respect to* $\mathrm{d}x$. *We consider the conditional expectation*

$$f_t(x) = -E_{\nu_t}[(\mathcal{L}_t F)(\omega) \mid Z = x]. \tag{4.51}$$

On $\mathbb{R}^n$, we define the vector field v_t by

$$v_t(\phi)(x) = E_{\nu_t}[(DF \mid D(\phi \, o \, Z)) \mid Z = x]$$
$$= \sum_{k=1}^{n} E_{\nu_t}[(DF|DZ_k) \mid Z = x] \frac{\partial \phi}{\partial x_k}(x), \quad \text{where} \quad \phi : \mathbb{R}^n \to \mathbb{R}. \tag{4.52}$$

We have

$$E_{\nu_t}[(DF|D(\phi \, oZ))] = E_{\nu_t}[(v_t\phi)(Z(\omega))]. \tag{4.53}$$

Moreover, there holds "Stein equation"

$$\mathrm{div}_{\rho_t(x)\mathrm{d}x}(v_t) = (1/t) \, f_t(x). \tag{4.54}$$

or equivalently,

$$\frac{1}{t} \int f_t(x)\phi(x)\rho_t(x) \, \mathrm{d}x = \int (v_t\phi)(x)\rho_t(x) \, \mathrm{d}x. \tag{4.55}$$

Proof. Equation (4.53) is a consequence of Eq. (4.50). To prove Eq. (4.55), we verify that

$$-\frac{1}{t} E_{\nu_t}[(\mathcal{L}_t F)(\omega)\phi(Z(\omega))] = E_{\nu_t}[(DF|D(\phi \, oZ))]. \tag{4.56}$$

This last identity comes from Eq. (4.46). □

In the case of nonrandom elements h in Cameron–Martin space, we can formulate Proposition 3.1 as follows.

Lemma 3.1. *Let h be in Cameron–Martin space and put*

$$\delta h(\omega) = \int_0^1 h'(s) \mathrm{d}\omega(s), \tag{4.57}$$

where Eq. (4.57) is a classical Ito stochastic integral (see, for example, [12], p. 14); in particular since h is not random, $\delta h = I_1(h)$ is in the first Wiener chaos. We have

$$\frac{1}{t} E_{\nu_t}[\delta h \mid Z = x] = \mathrm{div}_{\rho_t(x)\mathrm{d}x} \, v(t,x), \tag{4.58}$$

where $v(t,x)$ is the vector field

$$v(t,x) = \sum_{k=1}^{n} E_{\nu_t}[D_h Z_k \mid Z = x] \frac{\partial}{\partial x_k}. \tag{4.59}$$

Proof. We verify Eq. (4.55) with $f_t(x) = E_{\nu_t}[\delta h \mid Z = x]$. □

3.2 Projection of the Heat Equation Through a Nondegenerate Map

Let $Z = (Z_1, Z_2, \ldots, Z_n) : \Omega \to \mathbb{R}^n$. If $F(\omega) = (\phi \, o \, Z)(\omega)$, then by composition of differentiations

$$
\begin{aligned}
(\Delta_H F)(\omega) &= \sum_{h \in \mathcal{B}} D_h^2 F(\omega) \\
&= \sum_{j=1}^{n} (\Delta_H Z_j)(\omega) \, \frac{\partial \phi}{\partial x_j}(Z(\omega)) \\
&\quad + \sum_{j,k} (DZ_j | DZ_k) \, \frac{\partial^2 \phi}{\partial x_j \, \partial x_k}(Z(\omega)).
\end{aligned}
\tag{4.60}
$$

When Z is an evaluation function, Eq. (4.35) is deduced from Eq. (4.60). For $x = (x_1, x_2, \ldots, x_n)$ in $\mathbb{R}^n$, consider the conditional expectation

$$
(\Delta_t^Z \phi)(x) = E_{\nu_t}[(\Delta_H F)(\omega) \mid Z = x] \tag{4.61}
$$

We have

$$
\begin{aligned}
(\Delta_t^Z \phi)(x) &= \sum_{j=1}^{n} E_{\nu_t}[\Delta_H Z_j(\omega) \mid Z = x] \, \frac{\partial \phi}{\partial x_j}(x) \\
&\quad + \sum_{j,k} E_{\nu_t}[(DZ_j | DZ_k) \mid Z = x] \, \frac{\partial^2 \phi}{\partial x_j \, \partial x_k}(x).
\end{aligned}
\tag{4.62}
$$

Proposition 3.2. *Let $\rho_t(x)$ be the density of the law $Z * \nu_t$ with respect to the Lebesgue measure. Then $\rho_t(x)$ is the unique solution of*

$$
\begin{aligned}
\frac{\mathrm{d}}{\mathrm{d}t} \rho_t(x) &= -\frac{1}{2} \sum_{j=1}^{n} \frac{\partial}{\partial x_j} (E_{\nu_t}[\Delta_H Z_j(\omega) \mid Z = x] \, \rho_t(x)) \\
&\quad + \frac{1}{2} \sum_{j,k} \frac{\partial^2}{\partial x_j \, \partial x_k} (E_{\nu_t}[(DZ_j | DZ_k) \mid Z = x] \, \rho_t(x))
\end{aligned}
\tag{4.63}
$$

satisfying the condition $\lim_{t \to 0} \int \rho_t(u) \phi(u) \, \mathrm{d}u = \phi(0)$ *for any integrable function* $\phi : \mathbb{R}^n \to \mathbb{R}$.

Proof. From the heat equation (4.18), we have

$$\frac{\mathrm{d}}{\mathrm{d}t}\int \phi(x)\rho_t(x)\mathrm{d}x = \frac{1}{2}\int (\Delta_t^Z \phi)(x)\,\rho_t(x)\mathrm{d}x.$$

Integrating by parts, the density $\rho_t(x)$ satisfies Eq. (4.63). □

3.3 Image of the Dilation Vector Field V Through a Nondegenerate Map

Proposition 3.3. *Let V be the dilation vector field Eq. (4.38).*
The density $\rho_t(x)$ for the law of Z satisfies

$$\frac{\mathrm{d}}{\mathrm{d}t}\rho_t = -\frac{1}{2t}\sum_{j=1}^{n}\frac{\partial}{\partial x_j}(E_{\nu_t}[VZ_j \mid Z = x] \times \rho_t(x)). \tag{4.64}$$

Proof. For any differentiable function $\phi : \mathbb{R}^n \to \mathbb{R}$, we have

$$V(\phi\, o\, Z)(\omega) = \sum_{j=1}^{n}\frac{\partial \phi}{\partial x_j}(Z(\omega))\,(VZ_j)(\omega). \tag{4.65}$$

The relation (4.65) comes from the differentiation for composition of functions. From Eq. (4.16), there holds

$$\frac{\mathrm{d}}{\mathrm{d}t}\nu_t(\phi\, oZ) = \frac{1}{2t}\nu_t(\,V(\phi\, oZ)\,). \tag{4.66}$$

We replace Eq. (4.65) into Eq. (4.66). This gives

$$\begin{aligned}\frac{\mathrm{d}}{\mathrm{d}t}\nu_t(\phi\, oZ) &= \frac{1}{2t}\sum_{j=1}^{n} E_{\nu_t}\left[\frac{\partial \phi}{\partial x_j}(Z(\omega))\,(VZ_j)(\omega)\right]\\ &= \frac{1}{2t}\sum_{j=1}^{n}\int \frac{\partial \phi}{\partial x_j}(x)\,E_{\nu_t}[VZ_j \mid Z = x]\,\rho_t(x)\mathrm{d}x.\end{aligned}$$

We obtain Eq. (4.64) after integration by parts. □

4 From Stein's Equation to Heat Equation: Density of a Random Variable

We assume that $Z = (Z_1, Z_2, \dots, Z_n) : \Omega \to \mathbb{R}^n$ is nondegenerate. Let $\rho_t(x)$ be the density of $Z * \nu_t$ with respect to the n-dimensional Lebesgue measure:

$$Z * \nu_t = \rho_t(x)\,\mathrm{d}x. \tag{4.67}$$

Let $\alpha_j(t,x)$, $\lambda_j(t,x)$, $\beta_{kj}(t,x)$, and $\gamma_j(t,x)$ as in Eq. (4.21). By Proposition 2.3, we have Eq. (4.22). By Propositions 3.2 and 3.3, the density $\rho_t(x)$ satisfies each one of Eqs. (4.23)–(4.25).

Lemma 4.1. *Let*

$$w_j(t,x) = \lambda_j(t,x)\rho_t(x) - \sum_{k=1}^{n} \frac{\partial}{\partial x_k}(\beta_{jk}(t,x)\rho_t(x)) \tag{4.68}$$

then

$$\sum_{j=1}^{n} \frac{\partial}{\partial x_j} w_j(t,x) = 0. \tag{4.69}$$

Proof. We eliminate $\frac{\partial}{\partial t}\rho_t$ between Eqs. (4.25) and (4.23) □

Proposition 4.1 (**n-dimensional Stein's equation**). *As in Eq. (4.21), for $j = 1, \dots, n$ and $k = 1, \dots, n$, let*

$$\beta_k^j(t,x) = E_{\nu_t}[(DZ_j|DZ_k) \mid Z = x].$$

Then the density $\rho_t(x)$ of the random vector Z satisfies the system of n Eq. (4.24).

Proof. We apply Proposition 3.1 taking $F = Z_j$. As in Eq. (4.51), we put

$$f_t^j(x) = -E_{\nu_t}[\mathcal{L}_t Z_j \mid Z = x].$$

Integrating by parts the right hand side of Eq. (4.55), we obtain Eq. (4.24). □

Remark 4.1. With Proposition 4.1, we show that taking $F = Z_j$, $j = 1, \dots, n$ in Proposition 3.1, then all functions $w_j(t,x)$, $j = 1, \dots, n$ in Eq. (4.68) are equal to zero. This gives a n-dimensional system. If $Z(\omega) = (\omega(t_1), \omega(t_2), \dots, \omega(t_n))$, we find the system (4.13).

Main Theorem 4.1. If $\rho_t(x)$ is solution of the system (4.24) as well as solution of the dilation equation (4.23), then $\rho_t(x)$ is solution of the heat equation (4.25).

Proof. In Eq. (4.24), we have

$$\frac{1}{t}\rho_t(x)E_{\nu_t}[\mathcal{L}_t Z_j \mid Z = x] \tag{i}$$

and in Eq. (4.23), we have

$$-\frac{1}{t}\rho_t(x)E_{\nu_t}[V Z_j \mid Z = x]. \tag{ii}$$

Since [see Eqs. (4.20)–(4.22)]

$$E_{\nu_t}[V Z_j \mid Z = x] = -E_{\nu_t}[\mathcal{L}_t Z_j \mid Z = x] + t\, E_{\nu_t}[\Delta_H Z_j \mid Z = x]. \tag{iii}$$

We replace $E_{\nu_t}[V Z_j \mid Z = x]$ by this expression in Eq. (4.23). We obtain that $\rho_t(x)$ satisfies

$$\begin{aligned}\frac{\mathrm{d}}{\mathrm{d}t}\rho_t &= \frac{1}{2t}\sum_{j=1}^{n}\frac{\partial}{\partial x_j}(E_{\nu_t}[\mathcal{L}_t Z_j \mid Z = x] \times \rho_t(x))\\ &\quad -\frac{1}{2}\sum_{j=1}^{n}\frac{\partial}{\partial x_j}(E_{\nu_t}[\Delta_H Z_j \mid Z = x] \times \rho_t(x)).\end{aligned} \tag{iv}$$

In this expression, we replace $\frac{1}{t}\rho_t(x)E_{\nu_t}[\mathcal{L}_t Z_j \mid Z = x]$ by its expression given in the system (4.24). We obtain that $\rho_t(x)$ satisfies the heat equation (4.25). □

5 Hermite Polynomials and $\mathcal{L}_t$

We have defined $\mathcal{L}_t$ with Eq. (4.19). In this section, $t > 0$ is fixed and we calculate $\mathcal{L}_t$ on Hermite polynomials. Our Hermite polynomials $H_n^t(x,\tau)$ depend on t. Given $x \in \mathbb{R}$ and $\tau > 0$, we consider the Hermite polynomial of two variables

$$H_n^t(x,\tau) = (-1)^n\,(t\tau)^n\,\exp\left(\frac{x^2}{2t\tau}\right)\frac{\mathrm{d}^n}{\mathrm{d}x^n}\exp\left(-\frac{x^2}{2t\tau}\right). \tag{4.70}$$

See, for example, [12] p. 22. It is classical that $H_1^t(x,\tau) = x$, $H_2^t(x,\tau) = x^2 - t\tau$, $H_3^t(x,\tau) = x^3 - 3t\tau\, x$, $H_4^t(x,\tau) = x^4 - 6t\tau\, x^2 + 3t^2\tau^2, \ldots$.

$$\sum_{n=0}^{\infty} H_n^t(x,\tau)\,\frac{z^n}{n!} = \exp\left(-\frac{t\tau}{2}z^2 + xz\right), \tag{4.71}$$

$$t\,\tau\,\frac{\mathrm{d}^2}{\mathrm{d}x^2}H_n^t(x,\tau) - x\,\frac{\mathrm{d}}{\mathrm{d}x}H_n^t(x,\tau) = -n\,H_n^t(x,\tau), \tag{4.72}$$

$$\frac{\mathrm{d}}{\mathrm{d}x}H_n^t(x,\tau) = nH_{n-1}^t(x,\tau). \tag{4.73}$$

Proposition 5.1.

$$\mathcal{L}_t[H_n^t(\omega(\tau),\tau)] = -n\, H_n^t(\omega(\tau),\tau) \tag{4.74}$$

Proof. We put $F_n(x,\tau) = H_n^t(\sqrt{t}x,\tau)$. According to Eq. (4.45),

$$\mathcal{L}[H_n^t(\sqrt{t}\omega(\tau),\tau)] = -\sqrt{t}\omega(\tau)\left(\frac{\mathrm{d}}{\mathrm{d}x}H_n^t\right)(\sqrt{t}\omega(\tau),\tau)+t\tau\left(\frac{\mathrm{d}^2}{\mathrm{d}x^2}H_n^t\right)(\sqrt{t}\omega(\tau),\tau).$$

From Eq. (4.19),

$$\mathcal{L}_t[H_n^t(\omega(\tau),\tau)] = -\omega(\tau)\left(\frac{\mathrm{d}}{\mathrm{d}x}H_n^t\right)(\omega(\tau),\tau) + t\tau\left(\frac{\mathrm{d}^2}{\mathrm{d}x^2}H_n^t\right)(\omega(\tau),\tau).$$

Then we use Eq. (4.72). □

Corollary 5.1. *For the following conditional expectations, there holds*

$$E_{\nu_t}[\omega(t_1) \mid \omega(t_2) = x] = x\,\frac{t_1 \wedge t_2}{t_2},$$

$$E_{\nu_t}[\omega(t_1)^2 - t\,t_1 \mid \omega(t_2) = x] = \left(\frac{t_1 \wedge t_2}{t_2}\right)^2 (x^2 - t\,t_2),$$

$$E_{\nu_t}[H_n^t(w(t_1),\, t_1) \mid \omega(t_2) = x] = \left(\frac{t_1 \wedge t_2}{t_2}\right)^n H_n^t(x,\, t_2). \tag{4.75}$$

Proof. From Eq. (4.74),

$$J = E_{\nu_t}[H_n^t(w(t_1),t_1)\psi(w(t_2))] = -\frac{1}{n}E_{\nu_t}[\,\mathcal{L}_t[H_n^t(w(t_1),t_1)]\psi(w(t_2))].$$

Then from Eq. (4.56),

$$J = \frac{t(t_1 \wedge t_2)}{n}E_{\nu_t}\left[\left(\frac{\mathrm{d}}{\mathrm{d}x}H_n^t\right)(w(t_1),t_1)\psi'(w(t_2))\right].$$

With Eq. (4.73), this gives

$$E_{\nu_t}[H_n^t(w(t_1),t_1)\psi(w(t_2))] = t(t_1 \wedge t_2)E_{\nu_t}[H_{n-1}^t(w(t_1),t_1)\psi'(w(t_2))].$$

Writing this last identity with H_{n-1}^t, then H_{n-2}^t, ..., we obtain

$$E_{\nu_t}[H_n^t(w(t_1),t_1)\psi(w(t_2))] = t^n(t_1 \wedge t_2)^n E_{\nu_t}[\psi^{(n)}(w(t_2))].$$

By Eq. (4.36)

$$E_{\nu_t}[\psi^{(n)}(w(t_2))] = \int \left(\frac{\mathrm{d}^n}{\mathrm{d}x^n}\psi\right)(x)p_{t\,t_2}(x)\mathrm{d}x. \tag{4.76}$$

We integrate by parts the right hand side of Eq. (4.76). With Eq. (4.63), we obtain

$$E_{\nu_t}[\psi^{(n)}(w(t_2))] = \frac{1}{t^n\,t_2^n}E_{\nu_t}[H_n^t(w(t_2),t_2)\psi(w(t_2))].$$

This gives Eq. (4.75). □

Acknowledgements A preliminary version of this note was presented in June 2011 at the "Journée mathématique d'Amiens". I thank Frederi Viens for friendly and useful discussions in July 2011. I also thank the referee for significant suggestions for the writing.

References

1. Airault, H., Malliavin, P., Viens, F.: Stokes formula on the Wiener space and n-dimensional Nourdin-Peccati analysis. J. Funct. Anal. **258**(5), 1763–1783 (2010)
2. Airault, H., Malliavin, P.: Integration by parts formulas and dilatation vector fields on elliptic probability spaces. Probab. Theor. Rel. Fields. **106**(4), 447–494 (1996)
3. Cameron, R.H., Martin, W.T.: Transformations of Wiener integrals under translations. Ann. Math. **45**(2), 386–396 (1944)
4. Driver, B.K.: Heat kernels measures and infinite dimensional analysis. Heat kernels and analysis on manifolds, graphs, and metric spaces (Paris, 2002). Contemp. Math. **338**, 101–141 (2003) (American Mathematical Society, Providence)
5. Malliavin, P.: Analyse différentielle sur l'espace de Wiener. In: Proceedings of the International Congress of Mathematicians, vol. 1, 2 (Warsaw, 1983), pp. 1089–1096. Warsaw (1984)
6. Malliavin, P.: Stochastic analysis. Grundlehren der Mathematischen Wissenschaften, vol. 313. Springer, Berlin (1997)
7. Malliavin, P.: Calcul des variations, intégrales stochastiques et complexes de De Rham sur l'espace de Wiener. C.R. Acad. Sci. Paris Série I, **299**(8), 347–350 (1984)
8. Malliavin, P.: Calcul des variations stochastiques subordonné au processus de la chaleur. C. R. Acad. Sci. Paris Sér. I Math. **295**(2), 167–172 (1982)
9. Mancino, M.E.: Dilatation vector fields on the loop group. J. Funct. Anal. **166**(1), 130–147 (1999)
10. Nourdin, I., Peccati, G.: Stein's method on Wiener chaos. Probab. Theor. Rel. Fields **145**(1–2), 75–118 (2009)
11. Nourdin, I., Viens, F.G.: Density formula and concentration inequalities with Malliavin calculus. Electron. J. Probab. **14**(78), 2287–2309 (2009)
12. Nualart, D.: The Malliavin calculus and related topics. In: Probability and its Applications. Springer, Berlin (1995)
13. Wiener, N.: Differential space. J. Math. Phys. **2**, 131–174 (1923)

Chapter 5
The Calculus of Differentials for the Weak Stratonovich Integral

Jason Swanson

Abstract The weak Stratonovich integral is defined as the limit, in law, of Stratonovich-type symmetric Riemann sums. We derive an explicit expression for the weak Stratonovich integral of $f(B)$ with respect to $g(B)$, where B is a fractional Brownian motion with Hurst parameter 1/6, and f and g are smooth functions. We use this expression to derive an Itô-type formula for this integral. As in the case where g is the identity, the Itô-type formula has a correction term which is a classical Itô integral and which is related to the so-called signed cubic variation of $g(B)$. Finally, we derive a surprising formula for calculating with differentials. We show that if $\mathbf{d}M = X\,\mathbf{d}N$, then $Z\,\mathbf{d}M$ can be written as $ZX\,\mathbf{d}N$ minus a stochastic correction term which is again related to the signed cubic variation.

Keywords Stochastic integration • Stratonovich integral • Fractional Brownian motion • Weak convergence

AMS: Primary 60H05; secondary 60F17, 60G22

Received 4/5/2011; Accepted 9/3/2011; Final 1/4/2012

1 Introduction

If X and Y are stochastic processes, then the Stratonovich integral of X with respect to Y can be defined as the ucp (uniformly on compacts in probability) limit, if it exists, of the process

J. Swanson
University of Central Florida, Orlando, FL, USA
e-mail: jason@swansonsite.com

F. Viens et al. (eds.), *Malliavin Calculus and Stochastic Analysis: A Festschrift in Honor of David Nualart*, Springer Proceedings in Mathematics & Statistics 34,
DOI 10.1007/978-1-4614-5906-4_5,

$$t \mapsto \sum_{t_j \le t} \frac{X(t_{j-1}) + X(t_j)}{2}(Y(t_j) - Y(t_{j-1})),$$

as the mesh of the partition $\{t_j\}$ goes to zero. If we specialize to the uniformly spaced partition, $t_j = j/n$, then we are interested in the Stratonovich-type symmetric Riemann sums,

$$\sum_{j=1}^{\lfloor nt \rfloor} \frac{X(t_{j-1}) + X(t_j)}{2}(Y(t_j) - Y(t_{j-1})), \tag{5.1}$$

where $\lfloor x \rfloor$ denotes the greatest integer less than or equal to x.

It is well-known (see [2,4]) that if $Y = B^H$, a fractional Brownian motion with Hurst parameter H, and $X = f(B^H)$ for a sufficiently differentiable function f, then the Stratonovich integral of X with respect to Y exists for all $H > 1/6$ but does not exist for $H = 1/6$. Moreover, if $H > 1/6$, then the Stratonovich integral satisfies the classical Stratonovich change-of-variable formula, which corresponds to the usual fundamental theorem of calculus.

In [6], we studied the case $H = 1/6$. There we showed that if $Y = B = B^{1/6}$ and $X = f(B)$, where $f \in C^\infty(\mathbb{R})$, then the sequence of processes Eq. (5.1) converges in law. We let $\int_0^t f(B(s))\,\mathbf{d}B(s)$ denote a process with this limiting law, and we referred to this as the weak Stratonovich integral. We also showed that the weak Stratonovich integral with respect to B does not satisfy the classical Stratonovich change-of-variable formula. Rather, it satisfies an Itô-type formula with a correction term that is a classical Itô integral. Namely,

$$f(B(t)) = f(B(0)) + \int_0^t f'(B(s))\,\mathbf{d}B(s) - \frac{1}{12}\int_0^t f'''(B(s))\,\mathbf{d}[\![B]\!]_s, \tag{5.2}$$

where $[\![B]\!]$ is what we called the signed cubic variation of B. That is, $[\![B]\!]$ is the limit in law of the sequence of processes $\sum_{j=1}^{\lfloor nt \rfloor}(B(t_j) - B(t_{j-1}))^3$. It is shown in [7] that $[\![B]\!] = \kappa W$, where W is a standard Brownian motion, independent of B, and κ is an explicitly defined constant whose approximate numerical value is $\kappa \simeq 2.322$ [see Eq. (5.7) for the precise definition of κ.]. The correction term above is a standard Itô integral with respect to Brownian motion. Similar Itô-type formulas with an Itô integral correction term were developed in [1, 5]. There, the focus was on quartic variation processes and midpoint-style Riemann sums. A formula similar to Eq. (5.2), but with an ordinary integral correction term, was established in [3] for the Russo–Vallois symmetric integral with respect to finite cubic variation processes.

The precise results in [1, 6], as well as in this paper, involve demonstrating the joint convergence of all of the processes involved, with the type of convergence being weak convergence as processes in the Skorohod space of càdlàg functions. In Sect. 2, we establish the formal definition of the weak Stratonovich integral as an equivalence class of sequences of càdlàg step functions, and we demonstrate in

Theorem 2.1 the joint convergence in law of such sequences. For simplicity, we omit discussion of these details in this introduction and only summarize the results of Sect. 3, in which we derive our various change-of-variable formulas.

In Sect. 3, we extend the Itô-type formula (5.2) to the case $Y = g(B)$. We show that the sequence of processes (5.1) converges in law to an integral satisfying the Itô-type formula

$$\varphi(Y(t)) = \varphi(Y(0)) + \int_0^t \varphi'(Y(s))\,\mathbf{d}Y(s) - \frac{1}{12}\int_0^t \varphi'''(Y(s))\,\mathbf{d}[\![Y]\!]_s, \tag{5.3}$$

where

$$[\![Y]\!]_t = \int_0^t (g'(B(s)))^3\,\mathbf{d}[\![B]\!]_s$$

is the limit, in law, of $\sum_{j=1}^{\lfloor nt \rfloor}(Y(t_j) - Y(t_{j-1}))^3$. That is, $[\![Y]\!]$ is the signed cubic variation of Y.

This result is actually just one of the two main corollaries of our central result (see Corollary 3.1). To motivate the other results, consider the following. Formulas such as Eqs. (5.2) and (5.3) are typically referred to as change-of-variable formulas. They have the same structure as Itô's rule, which is also generally referred to as a change-of-variable formula. In elementary calculus, we perform a change-of-variable when we convert an integral with respect to one variable into an integral with respect to another. In Itô's stochastic calculus, we may wish to convert an integral with respect to one semimartingale into an integral with respect to another. Strictly speaking, Itô's rule is not sufficient for this purpose. Itô's rule simply tells us how to expand a function of a semimartingale into a sum of integrals. In order to convert one integral into another, we must combine Itô's rule with a theorem that says

$$\text{if } M = \int X\,dY, \text{ then } \int Z\,dM = \int ZX\,dY.$$

Or, in differential form,

$$\text{if } dM = X\,dY, \text{ then } Z\,dM = ZX\,dY. \tag{5.4}$$

For Itô integrals, this theorem is usually proved very early on in the construction of the integral. It is also true for the classical Stratonovich integral for semimartingales as well as for ordinary Lebesgue–Stieltjes integrals. In fact, in the theory of Lebesgue–Stieltjes integration, it is often this result which is called the change-of-variable formula.

In terms of the calculus of differentials, Itô's rule tells us that if $M = f(Y)$, then $dM = f'(Y)\,dY + \frac{1}{2}f''(Y)\,d\langle Y\rangle$, where $\langle Y\rangle$ is the quadratic variation of Y, and Eq. (5.4) tells us that it is permissible to substitute this expression into $Z\,dM$, so that $Z\,dM = Zf'(Y)\,dY + \frac{1}{2}Zf''(Y)\,d\langle Y\rangle$.

In this paper, we will show that Eq. (5.4) is not true for the weak Stratonovich integral. A very simple example which illustrates this is the following. First, let us note that when the integral is defined as a limit of Stratonovich-type symmetric Riemann sums, it is always the case that $\int \theta\, d\theta = \frac{1}{2}\theta^2$, for any process θ. Let us therefore define $M = \frac{1}{2}B^2$, so that $\mathbf{d}M = B\,\mathbf{d}B$. On the other hand,

$$\int M\,\mathbf{d}M = \frac{1}{2}M^2 = \frac{1}{8}B^4.$$

Using Eq. (5.2), we have

$$\frac{1}{8}B^4 = \int \frac{1}{2}B^3\,\mathbf{d}B - \frac{1}{12}\int 3B\,\mathbf{d}[\![B]\!] = \int MB\,\mathbf{d}B - \frac{1}{4}\int B\,\mathbf{d}[\![B]\!].$$

It follows that, in this example, Eq. (5.4) does not hold for the weak Stratonovich integral. Instead, we have that $\mathbf{d}M = B\,\mathbf{d}B$, whereas $M\,\mathbf{d}M = MB\,\mathbf{d}B - \frac{1}{4}B\,\mathbf{d}[\![B]\!]$.

The second main corollary of our central result is that the weak Stratonovich integral satisfies a rule analogous to Eq. (5.4) but with a correction term (see Corollary 3.2). Namely, suppose $X = f(B)$, $Y = g(B)$, and $Z = h(B)$, where $f, g, h \in C^\infty(\mathbb{R})$. Then the weak Stratonovich integral satisfies the following rule for calculating with differentials:

$$\text{If } \mathbf{d}M = X\,\mathbf{d}Y,\text{ then } Z\,\mathbf{d}M = ZX\,\mathbf{d}Y - \frac{1}{4}(f'g'h')(B)\,\mathbf{d}[\![B]\!]. \tag{5.5}$$

We actually prove a slightly more general rule; see Eq. (5.18).

Both Eqs. (5.3) and (5.5) will be demonstrated as corollaries of the following general result. With X and Y as above,

$$\int_0^t X(s)\,\mathbf{d}Y(s) = \Phi(B(t)) - \Phi(B(0)) + \frac{1}{12}\int_0^t (f''g' - f'g'')(B(s))\,\mathbf{d}[\![B]\!]_s, \tag{5.6}$$

where $\Phi \in C^\infty(\mathbb{R})$ is chosen to satisfy $\Phi' = fg'$. See Theorem 3.1 for the precise statement. Theorem 3.1 is actually formulated more generally for integrators of the form $Y + V$, where $V = \int \theta(B)\,\mathbf{d}[\![B]\!]$. This generalization is necessary to make sense of $\int Z\,\mathbf{d}M$ in Eq. (5.5), since if $M = \int X\,\mathbf{d}Y$, then according to Eq. (5.6), M is not a function of B but is rather the sum of a function of B and a process V which is in an integral against $[\![B]\!]$.

2 Notation and Definitions

2.1 Basic Notation

Let $B = B^{1/6}$ be a fractional Brownian motion with Hurst parameter $H = 1/6$. That is, B is a centered Gaussian process, indexed by $t \geq 0$, such that

$$E[B(s)B(t)] = \frac{1}{2}(t^{1/3} + s^{1/3} - |t-s|^{1/3}).$$

For compactness of notation, we will sometimes write B_t instead of $B(t)$ and similarly for other processes. Given a positive integer n, let $t_j = t_{j,n} =\cdot j/n$. We shall frequently have occasion to deal with the quantity

$$\beta_j = \beta_{j,n} = \frac{B(t_{j-1}) + B(t_j)}{2}.$$

Let $\Delta B_{j,n} = B(t_j) - B(t_{j-1})$ and $B^*(T) = \sup_{0\le t\le T} |B(t)|$.

Let $\kappa > 0$ be defined by

$$\kappa^2 = \frac{3}{4}\sum_{r\in\mathbb{Z}}(|r+1|^{1/3} + |r-1|^{1/3} - 2|r|^{1/3})^3. \tag{5.7}$$

Let $D_{\mathbb{R}^d}[0,\infty)$ denote the Skorohod space of càdlàg functions from $[0,\infty)$ to $\mathbb{R}^d$. Throughout the paper, "$\Rightarrow$" will denote convergence in law. The phrase "uniformly on compacts in probability" will be abbreviated "ucp." If X_n and Y_n are càdlàg processes, we shall write $X_n \approx Y_n$ or $X_n(t) \approx Y_n(t)$ to mean that $X_n - Y_n \to 0$ ucp.

2.2 *The Space* $[\mathcal{S}]$

Recall that for fixed n, we defined $t_k = k/n$. Let $\mathcal{S}_n$ denote the vector space of stochastic processes $\{L(t) : t \ge 0\}$ of the form $L = \sum_{k=0}^{\infty} \lambda_k 1_{[t_k,t_{k+1})}$, where each $\lambda_k \in \mathcal{F}_\infty^B$. Note that $\lambda_k = L(t_k)$. Given $L \in \mathcal{S}_n$, let $\delta_j(L) = L(t_j) - L(t_{j-1})$, for $j \ge 1$. Since $t \in [t_k, t_{k+1})$ if and only if $\lfloor nt \rfloor = k$, we may write

$$L(t) = L(0) + \sum_{j=1}^{\lfloor nt \rfloor} \delta_j(L).$$

Definition 2.1. Let $\mathcal{S}$ denote the vector space of sequences $\Lambda = \{\Lambda_n\}_{n=1}^{\infty}$ such that:

(i) $\Lambda_n \in \mathcal{S}_n$.
(ii) $\Lambda_n(0)$ converges in probability.
(iii) There exist $\varphi_1, \varphi_3, \varphi_5 \in C^\infty(\mathbb{R})$ such that

$$\delta_j(\Lambda_n) = \varphi_1(\beta_j)\Delta B_{j,n} + \varphi_3(\beta_j)\Delta B_{j,n}^3 + \varphi_5(\beta_j)\Delta B_{j,n}^5 + R_{j,n}, \tag{5.8}$$

where for each $T, K > 0$, there exists a finite constant $C_{T,K}$ such that

$$|R_{j,n}| 1_{\{B^*(T)\le K\}} \le C_{T,K}|\Delta B_{j,n}|^7,$$

whenever $j/n \le T$.

If $X = f(B)$, where $f \in C^\infty(\mathbb{R})$, then we define

$$\Lambda_n^X = \sum_{k=0}^{\infty} X(t_k) 1_{[t_k, t_{k+1})},$$

and $\Lambda^X = \{\Lambda_n^X\}_{n=1}^\infty$. Note that the map $X \mapsto \Lambda^X$ is linear.

Lemma 2.1. *If $X = f(B)$, where $f \in C^\infty(\mathbb{R})$, then $\Lambda^X \in \mathcal{S}$ and $\Lambda_n^X \to X$ uniformly on compacts a.s.*

Proof. Since X is continuous a.s., we have that $\Lambda_n^X \to X$ uniformly on compacts a.s. Clearly, $\Lambda_n^X \in \mathcal{S}_n$ and $\Lambda_n^X(0) = X(0)$ for all n, so that Definition 2.1(i) and (ii) hold. For $a, b \in \mathbb{R}$, we use the Taylor expansion

$$f(b) - f(a) = f'(x)(b-a) + \frac{1}{24} f'''(x)(b-a)^3 + \frac{1}{5!2^4} f^{(5)}(x)(b-a)^5 + h(a,b)(b-a)^7,$$

where $x = (a+b)/2$ and $|h(a,b)| \le M(a,b) = \sup_{x \in [a \wedge b, a \vee b]} |g^{(7)}(x)|$. For a derivation of this Taylor expansion, see the proof of Lemma 5.2 in [6].

Taking $a = B(t_{j-1})$ and $b = B(t_j)$ gives

$$\begin{aligned} \delta_j(\Lambda_n^X) &= f(B(t_j)) - f(B(t_{j-1})) \\ &= f'(\beta_j)\Delta B_{j,n} + \frac{1}{24} f'''(\beta_j)\Delta B_{j,n}^3 + \frac{1}{5!2^4} f^{(5)}(\beta_j)\Delta B_{j,n}^5 + R_{j,n}, \end{aligned} \tag{5.9}$$

where $|R_{j,n}| \le M(B(t_{j-1}), B(t_j))|\Delta B_{j,n}|^7$. If $j/n \le T$ and $B^*(T) \le K$, then $B(t_{j-1}), B(t_j) \in [-K, K]$, which implies $M(B(t_{j-1}), B(t_j)) \le \sup_{x \in [-K,K]} |g^{(7)}(x)| < \infty$, and this verifies Definition 2.1 (iii) showing that $\Lambda^X \in \mathcal{S}$. □

We may now identify $X = f(B)$ with $\Lambda^X \in \mathcal{S}$ and will sometimes abuse notation by writing $X \in \mathcal{S}$. In this way, we identify the space of smooth functions of B with a space of sequences in such a way that each sequence converges a.s. to its corresponding process. What we see next is that every sequence in $\mathcal{S}$ converges to a stochastic process, at least in law.

Theorem 2.1. *Let $\Lambda^{(1)}, \dots, \Lambda^{(m)} \in \mathcal{S}$. For $1 \le k \le m$, choose $\varphi_{1,k}, \varphi_{3,k}, \varphi_{5,k} \in C^\infty(\mathbb{R})$ satisfying Eq. (5.8) for $\Lambda^{(k)}$ and let $\mathcal{I}^{(k)}(0)$ be the limit in probability of $\Lambda_n^{(k)}(0)$ as $n \to \infty$. Let $\Phi_k \in C^\infty(\mathbb{R})$ satisfy $\Phi_k' = \varphi_{1,k}$ and $\Phi_k(0) = 0$. Let W be a Brownian motion independent of B, and let $\kappa > 0$ be given by Eq. (5.7). Define*

$$\mathcal{I}^{(k)}(t) = \mathcal{I}^{(k)}(0) + \Phi_k(B(t)) + \kappa \int_0^t \left(\varphi_{3,k} - \frac{1}{24}\varphi_{1,k}'' \right)(B(s))\, dW(s),$$

where this last integral is an Itô integral. Then $(B, \Lambda_n^{(1)}, \dots, \Lambda_n^{(m)}) \Rightarrow (B, \mathcal{I}^{(1)}, \dots, \mathcal{I}^{(m)})$ in $D_{\mathbb{R}^{m+1}}[0,\infty)$ as $n \to \infty$.

Proof. By Definition 2.1, we may write

$$\Lambda_n^{(k)}(t) = \Lambda_n^{(k)}(0)+\sum_{j=1}^{\lfloor nt\rfloor} \varphi_{1,k}(\beta_j)\Delta B_{j,n}+\sum_{j=1}^{\lfloor nt\rfloor} \varphi_{3,k}(\beta_j)\Delta B_{j,n}^3+\sum_{j=1}^{\lfloor nt\rfloor} \varphi_{5,k}(\beta_j)\Delta B_{j,n}^5+R_n(t),$$

where $R_n(t) = \sum_{j=1}^{\lfloor nt\rfloor} R_{j,n}$. Let $R_n^*(T) = \sup_{0\le t\le T} |R_n(t)| \le \sum_{j=1}^{\lfloor nT\rfloor} |R_{j,n}|$. Let $\varepsilon > 0$ and choose K such that $P(B^*(T) > K) < \varepsilon$. Then

$$P(R_n^*(T) > \varepsilon) \le P(B^*(T) > K) + P\left(C_{T,K} \sum_{j=1}^{\lfloor nT\rfloor} |\Delta B_{j,n}|^7 > \varepsilon\right).$$

Since B has a nontrivial 6-variation (see Theorem 2.11 in [6]), we have $\sum_{j=1}^{\lfloor nT\rfloor} |\Delta B_{j,n}|^7 \to 0$ a.s. Hence, for n sufficiently large, we have $P(R_n^*(T) > \varepsilon) < 2\varepsilon$, which gives $R_n \to 0$ ucp.

As in the proof of Theorem 2.13 in [6], we may assume without loss of generality that each $\varphi_{i,k}$ has compact support. By Lemma 5.1 in [6], if $\varphi \in C^1(\mathbb{R})$ has compact support, then $\sum_{j=1}^{\lfloor nt\rfloor} \varphi(\beta_j)\Delta B_{j,n}^5 \to 0$ ucp. Thus,

$$\Lambda_n^{(k)}(t) \approx \mathcal{I}^{(k)}(0) + \sum_{j=1}^{\lfloor nt\rfloor} \varphi_{1,k}(\beta_j)\Delta B_{j,n} + \sum_{j=1}^{\lfloor nt\rfloor} \varphi_{3,k}(\beta_j)\Delta B_{j,n}^3.$$

Similarly, by Eq. (5.9),

$$\begin{aligned}\Phi_k(B(t)) &\approx \sum_{j=1}^{\lfloor nt\rfloor}(\Phi_k(B(t_j)) - \Phi_k(B(t_{j-1})))\\ &\approx \sum_{j=1}^{\lfloor nt\rfloor} \varphi_{1,k}(\beta_j)\Delta B_{j,n} + \frac{1}{24}\sum_{j=1}^{\lfloor nt\rfloor} \varphi_{1,k}''(\beta_j)\Delta B_{j,n}^3.\end{aligned}$$

Therefore,

$$\Lambda_n^{(k)}(t) \approx \mathcal{I}^{(k)}(0) + \Phi_k(B(t)) + \sum_{j=1}^{\lfloor nt\rfloor} \psi_k(\beta_j)\Delta B_{j,n}^3,$$

where $\psi_k = \varphi_{3,k} - \frac{1}{24}\varphi_{1,k}''$. Let $V_n(\psi, t) = \sum_{j=1}^{\lfloor nt\rfloor} \psi(\beta_j)\Delta B_{j,n}^3$ and $J_k(t) = \kappa \int_0^t \psi_k(B(s))\, dW(s)$. By Lemma 5.2 and Theorem 2.13 in [6], we have $(B, V_n(\psi_1), \dots, V_n(\psi_m)) \Rightarrow (B, J_1, \dots, J_m)$, in $D_{\mathbb{R}^{m+1}}[0,\infty)$ as $n \to \infty$, which implies $(B, \Lambda_n^{(1)}, \dots, \Lambda_n^{(m)}) \Rightarrow (B, \mathcal{I}^{(1)}, \dots, \mathcal{I}^{(m)})$. □

We now define an equivalence relation on $\mathcal{S}$ by $\Lambda \equiv \Theta$ if and only if $\Lambda_n - \Theta_n \to 0$ ucp.

Lemma 2.2. *If $\Lambda \in \mathcal{S}$, then there exist unique functions φ_1, φ_3 which satisfy Eq. (5.8). If we denote these unique functions by $\varphi_{1,\Lambda}$ and $\varphi_{3,\Lambda}$, then $\Lambda \equiv \Theta$ and only if both of the following conditions hold:*

(i) $\Lambda_n(0) - \Theta_n(0) \to 0$ *in probability.*
(ii) $\varphi_{1,\Lambda} = \varphi_{1,\Theta}$ *and* $\varphi_{3,\Lambda} = \varphi_{3,\Theta}$.

Proof. Let $\Lambda \in \mathcal{S}$. Let $\{\varphi_1, \varphi_3, \varphi_5\}$ and $\{\widetilde{\varphi}_1, \widetilde{\varphi}_3, \widetilde{\varphi}_5\}$ be two sets of functions, each of which satisfies Eq. (5.8). Let $\mathcal{I}(0)$ be the limit in probability of $\Lambda_n(0)$ as $n \to \infty$. Let $\Phi, \widetilde{\Phi} \in C^\infty(\mathbb{R})$ satisfy $\Phi' = \varphi_1$, $\widetilde{\Phi}' = \widetilde{\varphi}_1$, and $\Phi(0) = \widetilde{\Phi}(0) = 0$. Then, by Theorem 2.1, Λ_n converges in law in $D_\mathbb{R}[0,\infty)$ to

$$\begin{aligned}\mathcal{I}(t) &= \mathcal{I}(0) + \Phi(B(t)) + \kappa \int_0^t \left(\varphi_3 - \frac{1}{24}\varphi_1''\right)(B(s))\,dW(s)\\ &= \mathcal{I}(0) + \widetilde{\Phi}(B(t)) + \kappa \int_0^t \left(\widetilde{\varphi}_3 - \frac{1}{24}\widetilde{\varphi}_1''\right)(B(s))\,dW(s).\end{aligned}$$

Hence, $E[\mathcal{I}(t) - \mathcal{I}(0) \mid \mathcal{F}_\infty^B] = \Phi(B(t)) = \widetilde{\Phi}(B(t))$ a.s. for all $t \geq 0$, which implies $\Phi = \widetilde{\Phi}$, and hence, $\varphi_1 = \widetilde{\varphi}_1$. It follows that

$$\mathcal{M}(t) = \int_0^t (\varphi_3 - \widetilde{\varphi}_3)(B(s))\,dW(s) = 0.$$

Hence, $E[\mathcal{M}(t)^2 \mid \mathcal{F}_\infty^B] = \int_0^t |(\varphi_3 - \widetilde{\varphi}_3)(B(s))|^2\,ds = 0$ a.s. for all $t \geq 0$, which implies $\varphi_3 = \widetilde{\varphi}_3$. This shows that there exist unique functions $\varphi_{1,\Lambda}, \varphi_{3,\Lambda}$ which satisfy Eq. (5.8).

Let $\Lambda, \Theta \in \mathcal{S}$ and define $\Gamma = \Lambda - \Theta$. Note that $\Lambda_n - \Theta_n \to 0$ ucp if and only if $\Gamma_n \Rightarrow 0$ in $D_\mathbb{R}[0,\infty)$.

First assume (i) and (ii) hold. Then $\Gamma_n(0) \to 0$ in probability, so by Theorem 2.1, Γ_n converges in law in $D_\mathbb{R}[0,\infty)$ to

$$\Phi_\Gamma(B(t)) + \kappa \int_0^t \left(\varphi_{3,\Gamma} - \frac{1}{24}\varphi_{1,\Gamma}''\right)(B(s))\,dW(s),$$

where $\Phi_\Gamma' = \varphi_{1,\Gamma}$ and $\Phi_\Gamma(0) = 0$. But from Eq. (5.8), we see that $\varphi_{1,\Gamma} = \varphi_{1,\Lambda} - \varphi_{1,\Theta} = 0$ and $\varphi_{3,\Gamma} = \varphi_{3,\Lambda} - \varphi_{3,\Theta} = 0$. Hence, $\Gamma_n \Rightarrow 0$ and $\Lambda \equiv \Theta$.

Now assume $\Lambda \equiv \Theta$. Then $\Gamma_n \to 0$ ucp, so by Theorem 2.1, for all $t \geq 0$,

$$\mathcal{I}(t) = \mathcal{I}(0) + \Phi_\Gamma(B(t)) + \kappa \int_0^t \left(\varphi_{3,\Gamma} - \frac{1}{24}\varphi_{1,\Gamma}''\right)(B(s))\,dW(s) = 0,$$

where $\mathcal{I}(0)$ is the limit in probability of $\Lambda_n(0) - \Theta_n(0)$ as $n \to \infty$ and $\Phi_\Gamma' = \varphi_{1,\Gamma}$ with $\Phi_\Gamma(0) = 0$. Thus, $\mathcal{I}(0) = 0$, which shows that (i) holds. And as above, we obtain $\varphi_{1,\Gamma} = \varphi_{3,\Gamma} = 0$, which shows that (ii) holds. $\square$

Let $[\Lambda]$ denote the equivalence class of Λ under this relation, and let $[\mathcal{S}]$ denote the set of equivalence classes. If $N = [\Lambda] \in [\mathcal{S}]$, then we define $\varphi_{1,N} = \varphi_{1,\Lambda}$, $\varphi_{3,N} = \varphi_{3,\Lambda}$, $\mathcal{I}_N(0) = \lim \Lambda_n(0)$, and

$$\mathcal{I}_N(t) = \mathcal{I}_N(0) + \Phi_N(B(t)) + \kappa \int_0^t \left(\varphi_{3,N} - \frac{1}{24}\varphi''_{1,N}\right)(B(s))\,dW(s), \quad (5.10)$$

where $\Phi'_N = \varphi_{1,N}$ and $\Phi_N(0) = 0$. Notice that by Theorem 2.1, if $N_1, \dots, N_m \in [\mathcal{S}]$ and $\Lambda^{(k)} \in N_k$ are arbitrary, then $(B, \Lambda_n^{(1)}, \dots, \Lambda_n^{(m)}) \Rightarrow (B, \mathcal{I}_{N_1}, \dots, \mathcal{I}_{N_m})$ in $D_{\mathbb{R}^{m+1}}[0,\infty)$.

It is easily verified that $[\mathcal{S}]$ is a vector space under the operations $c[N] = [cN]$ and $[M] + [N] = [M + N]$ and that $N \mapsto \mathcal{I}_N$ is linear and injective. This gives us a one-to-one correspondence between $[\mathcal{S}]$ and processes of the form Eq. (5.10).

If $X = f(B)$, where $f \in C^\infty(\mathbb{R})$, then we define $N^X = [\Lambda^X] \in [\mathcal{S}]$. We may now identify X with N^X and will sometimes abuse notation by writing $X \in [\mathcal{S}]$. It may therefore be necessary to deduce from context whether X refers to the process $f(B)$, the sequence $\Lambda^X = \{\Lambda_n^X\}$, or the equivalence class $N^X = [\Lambda^X]$. Typically, there will be only one sensible interpretation, but when ambiguity is possible, we will be specific.

Note that, using Eq. (5.9), we obtain $\varphi_{1,X} = f'$, $\varphi_{3,X} = \frac{1}{24} f'''$, $\mathcal{I}_X(0) = X(0) = f(0)$, and $\Phi_X = f - f(0)$. Hence, by Eq. (5.10), we have $\mathcal{I}_X(t) = X(t)$. Because of this and because of the one-to-one correspondence between $N \in [\mathcal{S}]$ and the process $\mathcal{I}_N(t)$ in Eq. (5.10), we will sometimes abuse notation and write $N(t) = N_t = \mathcal{I}_N(t)$. Again, when there is a possible ambiguity as to whether N refers to an element of $[\mathcal{S}]$ or to the process $\mathcal{I}_N$, we will be specific.

2.3 *The Signed Cubic Variation*

If $\Lambda \in \mathcal{S}$, we define $V_n^\Lambda(t) = \sum_{j=1}^{\lfloor nt \rfloor} (\delta_j(\Lambda_n))^3$ and $V^\Lambda = \{V_n^\Lambda\}$. Since $\delta_j(V_n^\Lambda) = (\delta_j(\Lambda_n))^3$, it is easy to see from Eq. (5.8) that $V^\Lambda \in \mathcal{S}$, $\varphi_{1,V^\Lambda} = 0$ and $\varphi_{3,V^\Lambda} = \varphi_{1,\Lambda}^3$. Hence, if $\Lambda \equiv \Theta$, then $V^\Lambda \equiv V^\Theta$. We may therefore define the signed cubic variation of $N = [\Lambda] \in [\mathcal{S}]$ to be $[V^\Lambda] \in [\mathcal{S}]$. We denote the signed cubic variation of N by $[\![N]\!]$. We then have $\varphi_{1,[\![N]\!]} = 0$, $\varphi_{3,[\![N]\!]} = \varphi_{1,N}^3$, and $\mathcal{I}_{[\![N]\!]}(0) = 0$, so that by Eq. (5.10),

$$[\![N]\!]_t = \mathcal{I}_{[\![N]\!]}(t) = \kappa \int_0^t (\varphi_{1,N}(B(s)))^3\,dW(s).$$

For example, suppose $X = f(B)$, where $f \in C^\infty(\mathbb{R})$. Then $[\![X]\!] = [\![N^X]\!]$. Since $N^X = [\Lambda^X]$, we have $[\![N^X]\!] = [V^{\Lambda^X}]$. Note that $V^{\Lambda^X} = \{V_n^{\Lambda^X}\}$ and

$$V_n^{\Lambda^X}(t) = \sum_{j=1}^{\lfloor nt \rfloor} (\delta_j(\Lambda_n^X))^3 = \sum_{j=1}^{\lfloor nt \rfloor} (X(t_j) - X(t_{j-1}))^3.$$

In other words, $[\![X]\!]$ is the equivalence class in $\mathcal{S}$ of the above sequence of sums of cubes of increments of X. By Theorem 2.1, $[\![X]\!]_t = \mathcal{I}_{[\![X]\!]}(t)$ is the stochastic process which is the limit in law of this sequence. Since $\varphi_{1,X} = f'$, we have $\varphi_{1,[\![X]\!]} = 0$ and $\varphi_{3,[\![X]\!]} = (f')^3$, so that

$$[\![X]\!]_t = \mathcal{I}_{[\![X]\!]}(t) = \kappa \int_0^t (f'(B(s)))^3 \, dW(s).$$

In particular, taking $f(x) = x$ gives $[\![B]\!]_t = \kappa W$.

2.4 The Weak Stratonovich Integral

If $\Lambda_n, \Theta_n \in \mathcal{S}_n$, then we define

$$(\Lambda_n \circ \Theta_n)(t) = \sum_{j=1}^{\lfloor nt \rfloor} \frac{\Lambda_n(t_{j-1}) + \Lambda_n(t_j)}{2} \delta_j(\Theta_n).$$

If $\Lambda, \Theta \in \mathcal{S}$, then we define $\Lambda \circ \Theta = \{\Lambda_n \circ \Theta_n\}_{n=1}^{\infty}$.

Lemma 2.3. *If $X = f(B)$, where $f \in C^\infty(\mathbb{R})$ and $\Lambda \in \mathcal{S}$, then $\Lambda^X \circ \Lambda \in \mathcal{S}$. Moreover, if $\Lambda \equiv \Theta$, then $\Lambda^X \circ \Lambda \equiv \Lambda^X \circ \Theta$.*

Proof. Clearly, $\Lambda_n^X \circ \Lambda_n \in \mathcal{S}_n$ and $\Lambda_n^X \circ \Lambda_n(0) = 0$ for all n, so that Definition 2.1(i) and (ii) hold. For $a, b \in \mathbb{R}$, we use the Taylor expansion

$$\frac{f(b) + f(a)}{2} = f(x) + \frac{1}{8} f''(x)(b-a)^2 + \frac{1}{4!2^4} f^{(4)}(x)(b-a)^4 + h(a,b)(b-a)^6,$$

where $x = (a+b)/2$ and $|h(a,b)| \le M(a,b) = \sup_{x \in [a \wedge b, a \vee b]} |g^{(6)}(x)|$. For a derivation of this Taylor expansion, see the proof of Lemma 5.2 in [6].

Taking $a = B(t_{j-1})$ and $b = B(t_j)$ gives

$$\begin{aligned}\frac{\Lambda_n^X(t_{j-1}) + \Lambda_n^X(t_j)}{2} &= \frac{f(B(t_{j-1})) + f(B(t_j))}{2} \\ &= f(\beta_j) + \frac{1}{8} f''(\beta_j) \Delta B_{j,n}^2 + \frac{1}{4!2^4} f^{(4)}(\beta_j) \Delta B_{j,n}^4 + R_{j,n},\end{aligned}$$

where for each $T, K > 0$, there exists a finite constant $C_{T,K}$ such that

$$|R_{j,n}| 1_{\{B^*(T) \le K\}} \le C_{T,K} |\Delta B_{j,n}|^6,$$

whenever $j/n \le T$. Choose $\varphi_5 \in C^\infty(\mathbb{R})$ such that

$$\delta_j(\Lambda_n) = \varphi_{1,\Lambda}(\beta_j) \Delta B_{j,n} + \varphi_{3,\Lambda}(\beta_j) \Delta B_{j,n}^3 + \varphi_5(\beta_j) \Delta B_{j,n}^5 + \widetilde{R}_{j,n},$$

where for each $T, K > 0$, there exists a finite constant $\widetilde{C}_{T,K}$ such that

$$|\widetilde{R}_{j,n}|1_{\{B^*(T)\le K\}} \le \widetilde{C}_{T,K}|\Delta B_{j,n}|^7,$$

whenever $j/n \le T$. Then

$$\begin{aligned}\delta_j(\Lambda_n^X \circ \Lambda_n) &= \frac{\Lambda_n^X(t_{j-1}) + \Lambda_n^X(t_j)}{2}\delta_j(\Lambda_n)\\ &= (f\varphi_{1,\Lambda})(\beta_j)\Delta B_{j,n} + \left(\frac{1}{8}f''\varphi_{1,\Lambda} + f\varphi_{3,\Lambda}\right)(\beta_j)\Delta B_{j,n}^3\\ &\quad + h(\beta_j)\Delta B_{j,n}^5 + \widehat{R}_{j,n},\end{aligned}$$

for an appropriately chosen smooth function h, and with $\widehat{R}_{j,n}$ satisfying Definition 2.1(iii).

It follows that $\Lambda^X \circ \Lambda \in \mathcal{S}$ and that $\varphi_{1,\Lambda^X\circ\Lambda} = f\varphi_{1,\Lambda}$ and $\varphi_{3,\Lambda^X\circ\Lambda} = \frac{1}{8}f''\varphi_{1,\Lambda} + f\varphi_{3,\Lambda}$. This implies that if $\Lambda \equiv \Theta$, then $\Lambda^X \circ \Lambda \equiv \Lambda^X \circ \Theta$. □

If $X = f(B)$, where $f \in C^\infty(\mathbb{R})$, and $N = [\Lambda] \in [\mathcal{S}]$, we may now define $X \circ N = [\Lambda^X \circ \Lambda]$. Note that if $Y = g(B)$, where $g \in C^\infty$, and $M \in [\mathcal{S}]$, then $(X + Y) \circ N = X \circ N + Y \circ N$ and $X \circ (N + M) = X \circ N + X \circ M$. From the proof of Lemma 2.3, we have

$$\mathcal{I}_{X\circ N}(0) = 0, \tag{5.11}$$

$$\varphi_{1,X\circ N} = f\varphi_{1,N}, \tag{5.12}$$

$$\varphi_{3,X\circ N} = \frac{1}{8}f''\varphi_{1,N} + f\varphi_{3,N}. \tag{5.13}$$

We may use these formulas, together with Eq. (5.10), to calculate $\mathcal{I}_{X\circ N}$, given f, $\varphi_{1,N}$, and $\varphi_{3,N}$.

We now adopt some more traditional notation. If $X = f(B)$, where $f \in C^\infty$, and $N \in [\mathcal{S}]$, then

$$\int X\,\mathbf{d}N = X \circ N \in [\mathcal{S}],$$

and

$$\int_0^t X(s)\,\mathbf{d}N(s) = (X \circ N)_t = \mathcal{I}_{X\circ N}(t).$$

As we noted earlier, there is a one-to-one correspondence between $[\mathcal{S}]$ and processes of the form Eq. (5.10). We may therefore go back and forth between the above two objects according to what is more convenient at the time. We will use the shorthand notation $\mathbf{d}M = X\,\mathbf{d}N$ to denote the equality $M = \int X\,\mathbf{d}N$.

Before investigating our change-of-variable formulas, let us first consider some examples.

Example 2.1. Let $X = f(B)$ and $Y = g(B)$, where $f, g \in C^\infty(\mathbb{R})$. Then

$$\int X\,\mathbf{d}Y = X \circ Y = X \circ N^Y = [\Lambda^X \circ \Lambda^Y],$$

and $\Lambda^X \circ \Lambda^Y = \{\Lambda_n^X \circ \Lambda_n^Y\}$, where

$$\begin{aligned}(\Lambda_n^X \circ \Lambda_n^Y)(t) &= \sum_{j=1}^{\lfloor nt \rfloor} \frac{\Lambda_n^X(t_{j-1}) + \Lambda_n^X(t_j)}{2} \delta_j(\Lambda_n^Y) \\ &= \sum_{j=1}^{\lfloor nt \rfloor} \frac{X(t_{j-1}) + X(t_j)}{2} (Y(t_j) - Y(t_{j-1})).\end{aligned}$$

In other words, $\int X\,\mathbf{d}Y$ is the equivalence class in $\mathcal{S}$ of the above sequence of Stratonovich-type symmetric Riemann sums. Also, $\int_0^t X(s)\,\mathbf{d}Y(s) = \mathcal{I}_{X \circ Y}(t)$, so that by Theorem 2.1, $\int_0^t X(s)\,\mathbf{d}Y(s)$ is the stochastic process which is the limit in law of this sequence.

Example 2.2. Again let $X = f(B)$ and $Y = g(B)$, where $f, g \in C^\infty(\mathbb{R})$. Then

$$\int X\,\mathbf{d}[\![Y]\!] = X \circ [\![Y]\!] = [\Lambda^X \circ V^{\Lambda^Y}]$$

and $\Lambda^X \circ V^{\Lambda^Y} = \{\Lambda_n^X \circ V_n^{\Lambda^Y}\}$, where

$$\begin{aligned}(\Lambda_n^X \circ V_n^{\Lambda^Y})(t) &= \sum_{j=1}^{\lfloor nt \rfloor} \frac{\Lambda_n^X(t_{j-1}) + \Lambda_n^X(t_j)}{2} \delta_j(V_n^{\Lambda^Y}) \\ &= \sum_{j=1}^{\lfloor nt \rfloor} \frac{X(t_{j-1}) + X(t_j)}{2} (Y(t_j) - Y(t_{j-1}))^3.\end{aligned}$$

In other words, $\int X\,\mathbf{d}[\![Y]\!]$ is the equivalence class in $\mathcal{S}$ of the above sequence of sums, and $\int_0^t X(s)\,\mathbf{d}[\![Y]\!]_s = \mathcal{I}_{X \circ [\![Y]\!]}(t)$ is the limit in law of this sequence. Recall that $\varphi_{1,[\![Y]\!]} = 0$ and $\varphi_{3,[\![Y]\!]} = (g')^3$. Hence, by Eqs. (5.12) and (5.13), we have $\varphi_{1,X \circ [\![Y]\!]} = f\varphi_{1,[\![Y]\!]} = 0$ and $\varphi_{3,X \circ [\![Y]\!]} = \frac{1}{8} f'' \varphi_{1,[\![Y]\!]} + f\varphi_{3,[\![Y]\!]} = f(g')^3$, so that by Eq. (5.10), we have

$$\int_0^t X(s)\,\mathbf{d}[\![Y]\!]_s = \kappa \int_0^t f(B(s))(g'(B(s)))^3\,dW(s). \tag{5.14}$$

Example 2.3. For one last example, let $X = f(B)$, $Y = g(B)$, and $Z = h(B)$, where $f, g, h \in C^\infty(\mathbb{R})$, and let $N = \int Y\,\mathbf{d}Z$. Then

$$\int X\,\mathbf{d}N = X \circ N = X \circ [\Lambda^Y \circ \Lambda^Z] = [\Lambda^X \circ (\Lambda^Y \circ \Lambda^Z)],$$

and

$$\begin{aligned}(\Lambda_n^X \circ (\Lambda^Y \circ \Lambda^Z)_n)(t) &= (\Lambda_n^X \circ (\Lambda_n^Y \circ \Lambda_n^Z))(t) \\ &= \sum_{j=1}^{\lfloor nt \rfloor} \frac{X(t_{j-1}) + X(t_j)}{2} \frac{Y(t_{j-1})+Y(t_j)}{2}(Z(t_j)-Z(t_{j-1})).\end{aligned}$$

Hence, $\int X \, \mathbf{d}N$ is the equivalence class in $\mathcal{S}$ of the above sequence of sums, and $\int_0^t X(s) \, \mathbf{d}N(s)$ is the limit in law of this sequence.

3 Change-of-Variable Formulas

We have already identified smooth functions of B with their corresponding sequences in $\mathcal{S}$, as well as with their equivalence classes in $[\mathcal{S}]$. In this section, it will be helpful to do the same for $\mathcal{F}_\infty^B$-measurable random variables, which can serve as initial values for the stochastic processes we are considering.

Let η be an $\mathcal{F}_\infty^B$-measurable random variable, let $\Lambda_n^\eta(t) = \eta$ for all $t \geq 0$, and let $\Lambda^\eta = \{\Lambda_n^\eta\}$. Since $\delta_j(\Lambda_n^\eta) = 0$ for all j and n, we have that $\Lambda^\eta \in \mathcal{S}$. We may therefore identify η with $\Lambda^\eta \in \mathcal{S}$ and also with $N^\eta = [\Lambda^\eta] \in [\mathcal{S}]$. Note, then, that $\varphi_{1,\eta} = \varphi_{3,\eta} = 0$ and $\eta(t) = N^\eta(t) = \mathcal{I}_{N^\eta}(t) = \eta$ for all $t \geq 0$. Note also that $\int X \, \mathbf{d}\eta = 0$.

We begin with the following result, which tells us that every element of $[\mathcal{S}]$ has a unique decomposition into the sum of a smooth function of B and an integral against $[\![B]\!]$.

Lemma 3.1. *Each $N \in [\mathcal{S}]$ can be written as $N = \eta + Y + V$, where η is an $\mathcal{F}_\infty^B$-measurable random variable, $Y = g(B)$ for some $g \in C^\infty(\mathbb{R})$, and $V = \int \theta(B) \, \mathbf{d}[\![B]\!]$ for some $\theta \in C^\infty(\mathbb{R})$.*

Suppose $N = \widetilde{\eta} + \widetilde{Y} + \widetilde{V}$ is another such representation, with $\widetilde{Y} = \widetilde{g}(B)$ and $\widetilde{V} = \int \widetilde{\theta}(B) \, \mathbf{d}[\![B]\!]$. Let $c = g(0) - \widetilde{g}(0)$. Then $\widetilde{\eta} = \eta + c$, $\widetilde{g} = g - c$, and $\widetilde{\theta} = \theta$. In particular, there is a unique such representation with $g(0) = 0$.

An explicit representation is given by $\eta = N(0) = \mathcal{I}_N(0)$, $\theta = \varphi_{3,N} - \frac{1}{24}\varphi''_{1,N}$ and g chosen so that $g' = \varphi_{1,N}$ and $g(0) = 0$.

Proof. Let $N \in [\mathcal{S}]$. Let $\eta = N(0)$ and $\theta = \varphi_{3,N} - \frac{1}{24}\varphi''_{1,N}$ and choose g so that $g' = \varphi_{1,N}$ and $g(0) = 0$. Let $Y = g(B)$ and $V = \int \theta(B) \, \mathbf{d}[\![B]\!]$. To prove that $N = \eta + Y + V$, it will suffice to show that

$$\begin{aligned}N(t) &= \eta(t) + Y(t) + V(t) \\ &= N(0) + g(B(t)) + \int_0^t \theta(B(s)) \, \mathbf{d}[\![B]\!]_s.\end{aligned}$$

But this follows immediately from Eqs. (5.10) and (5.14).

Now suppose $N(t) = \widetilde{\eta} + \widetilde{g}(B(t)) + \int_0^t \widetilde{\theta}(B(s))\,\mathbf{d}[\![B]\!]_s$. Then $E[N(t) \mid \mathcal{F}_\infty^B] = \eta + g(B(t)) = \widetilde{\eta} + \widetilde{g}(B(t))$ a.s., which gives $\eta - \widetilde{\eta} + (g - \widetilde{g})(B(t)) = 0$ a.s. for all $t \geq 0$. Hence, there exists a constant $c \in \mathbb{R}$ such that $g - \widetilde{g} = c$, and it follows that $\widetilde{\eta} = \eta + c$. We then have $\mathcal{M}(t) = \int_0^t (\theta - \widetilde{\theta})(B(s))\,dW(s) = 0$ a.s., so that $E[\mathcal{M}(t)^2 \mid \mathcal{F}_\infty^B] = \int_0^t |(\theta - \widetilde{\theta})(B(s))|^2\,ds = 0$ a.s. for all $t \geq 0$, which implies $\theta = \widetilde{\theta}$. □

We next verify that processes of the form $V = \int \theta(B)\,\mathbf{d}[\![B]\!]$ behave as we would expect them to in regards to integration.

Lemma 3.2. *Let $X = f(B)$, where $f \in C^\infty(\mathbb{R})$, and let $\theta \in C^\infty(\mathbb{R})$. If $\mathbf{d}V = \theta(B)\,\mathbf{d}[\![B]\!]$, then $X\,\mathbf{d}V = X\theta(B)\,\mathbf{d}[\![B]\!]$.*

Proof. Let $V = \int \theta(B)\,\mathbf{d}[\![B]\!]$, $U = \int X\theta(B)\,\mathbf{d}[\![B]\!]$, and $N = \int X\,\mathbf{d}V$. Since $N(0) = U(0) = 0$, it will suffice to show that $\varphi_{1,U} = \varphi_{1,N}$ and $\varphi_{3,U} = \varphi_{3,N}$. By Example 2.2, $\varphi_{1,V} = \varphi_{1,U} = 0$, $\varphi_{3,V} = \theta$, and $\varphi_{3,U} = f\theta$. On the other hand, by Eqs. (5.12) and (5.13), we have $\varphi_{1,N} = f\varphi_{1,V} = 0$ and $\varphi_{3,N} = \frac{1}{8}f''\varphi_{1,V} + f\varphi_{3,V} = f\theta$. □

We finally present our main result for doing calculations with the weak Stratonovich integral.

Theorem 3.1. *Let $N \in [S]$ and write $N = \eta + Y + V$, where η is an $\mathcal{F}_\infty^B$-measurable random variable, $Y = g(B)$, and $V = \int \theta(B)\,\mathbf{d}[\![B]\!]$ for some $g, \theta \in C^\infty(\mathbb{R})$. Let $X = f(B)$, where $f \in C^\infty(\mathbb{R})$. Then*

$$\int X\,\mathbf{d}N = \Phi(B) + \frac{1}{12}\int (f''g' - f'g'')(B)\,\mathbf{d}[\![B]\!] + \int X\,\mathbf{d}V, \tag{5.15}$$

where $\Phi \in C^\infty(\mathbb{R})$ is chosen so that $\Phi' = fg'$ and $\Phi(0) = 0$.

Remark 3.1. Since $M = \int X\,\mathbf{d}N \in [S]$, Lemma 3.1 tells us that M has a unique decomposition into the sum of a smooth function of B and an integral against $[\![B]\!]$. Theorem 3.1 gives us a convenient formula for this decomposition.

Remark 3.2. Theorem 3.1 and the corollaries that are to follow express equalities in the space $[S]$. Each side of Eq. (5.15) is an equivalence class of sequences of Riemann sums that converge in law. The equivalence relation is such that if we choose any sequence from the class on the left and any sequence from the class on the right, then their difference will converge to zero ucp. Note that this is a stronger statement than simply asserting that the two sequences have the same limiting law.

Proof of Theorem 3.1. Since $\int X\,\mathbf{d}N = \int X\,\mathbf{d}\eta + \int X\,\mathbf{d}Y + \int X\,\mathbf{d}V$ and $\int X\,\mathbf{d}\eta = 0$, it follows from Eq. (5.14) that we need only show

$$\int_0^t X(s)\,\mathbf{d}Y(s) = \Phi(B(t)) + \frac{\kappa}{12}\int_0^t (f''g' - f'g'')(B(s))\,dW(s). \tag{5.16}$$

By Eq. (5.10), we have

$$\int_0^t X(s)\,\mathbf{d}Y(s) = \Phi_M(B(t)) + \kappa \int_0^t \left(\varphi_{3,M} - \frac{1}{24}\varphi_{1,M}''\right)(B(s))\,dW(s),$$

where $M = X \circ Y$. Recall that $\varphi_{1,Y} = g'$ and $\varphi_{3,Y} = \frac{1}{24}g'''$. By Eqs. (5.12) and (5.13), we have $\varphi_{1,M} = fg'$ and $\varphi_{3,M} = \frac{1}{8}f''g' + \frac{1}{24}fg'''$. Since $\Phi_M(0) = 0$ and $\Phi_M' = \varphi_{1,M} = fg'$, we have $\Phi_M = \Phi$, and we also have

$$\begin{aligned}\varphi_{3,M} - \frac{1}{24}\varphi_{1,M}'' &= \frac{1}{8}f''g' + \frac{1}{24}fg''' - \frac{1}{24}(fg')'' \\ &= \frac{1}{8}f''g' + \frac{1}{24}fg''' - \frac{1}{24}f''g' - \frac{1}{12}f'g'' - \frac{1}{24}fg''' \\ &= \frac{1}{12}(f''g' - f'g''),\end{aligned}$$

and this verifies Eq. (5.16). □

Corollary 3.1. *Let $Y = g(B)$, where $g \in C^\infty(\mathbb{R})$, and let $\varphi \in C^\infty$. Then*

$$\varphi(Y(t)) = \varphi(Y(0)) + \int_0^t \varphi'(Y(s))\,\mathbf{d}Y(s) - \frac{1}{12}\int_0^t \varphi'''(Y(s))\,\mathbf{d}[\![Y]\!]_s. \qquad (5.17)$$

Proof. Let $X = \varphi'(Y) = f(B)$, where $f = \varphi' \circ g$. By Theorem 3.1,

$$\int X\,\mathbf{d}Y = \Phi(B) + \frac{1}{12}\int (f''g' - f'g'')(B)\,\mathbf{d}[\![B]\!],$$

where $\Phi \in C^\infty(\mathbb{R})$ is chosen so that $\Phi' = fg'$ and $\Phi(0) = 0$. Since $(\varphi \circ g)' = fg'$, we have $\Phi = (\varphi \circ g) - (\varphi \circ g)(0)$. Also,

$$f''g' - f'g'' = ((\varphi''' \circ g)(g')^2 + (\varphi'' \circ g)g'')g' - (\varphi'' \circ g)g'g'' = (\varphi''' \circ g)(g')^3.$$

Thus,

$$\begin{aligned}\int_0^t \varphi'(Y(s))\,\mathbf{d}Y(s) &= \int_0^t X(s)\,\mathbf{d}Y(s) \\ &= (\varphi \circ g)(B(t)) - (\varphi \circ g)(0) \\ &\quad + \frac{1}{12}\int_0^t (\varphi''' \circ g)(B(s))(g'(B(s)))^3\,\mathbf{d}[\![B]\!]_s \\ &= \varphi(Y(t)) - \varphi(Y(0)) + \frac{\kappa}{12}\int_0^t \varphi'''(Y(s))(g'(B(s)))^3\,dW(s).\end{aligned}$$

By Eq. (5.14), this gives

$$\int_0^t \varphi'(Y(s))\,\mathbf{d}Y(s) = \varphi(Y(t)) - \varphi(Y(0)) + \frac{1}{12}\int_0^t \varphi'''(Y(s))\,\mathbf{d}[\![Y]\!]_s,$$

which is Eq. (5.17). □

Corollary 3.2. *Let $N \in [\mathcal{S}]$ and write $N = \eta + Y + V$, where η is an $\mathcal{F}^B_\infty$-measurable random variable, $Y = g(B)$, and $V = \int \theta(B)\,\mathbf{d}[\![B]\!]$ for some $g, \theta \in C^\infty(\mathbb{R})$. Let $X = f(B)$ and $Z = h(B)$, where $f, h \in C^\infty(\mathbb{R})$. Then*

$$\textit{if } \mathbf{d}M = X\,\mathbf{d}N, \textit{ then } Z\,\mathbf{d}M = ZX\,\mathbf{d}N - \frac{1}{4}(f'g'h')(B)\,\mathbf{d}[\![B]\!]. \tag{5.18}$$

Moreover, the above correction term is a "weak triple covariation" in the following sense: If $\mathcal{V} = \{\mathcal{V}_n\}$, where

$$\mathcal{V}_n(t) = \sum_{j=1}^{\lfloor nt \rfloor}(X(t_j) - X(t_{j-1}))(Y(t_j) - Y(t_{j-1}))(Z(t_j) - Z(t_{j-1})),$$

then $\mathcal{V} \in \mathcal{S}$ and $[\mathcal{V}] = \int (f'g'h')(B)\,\mathbf{d}[\![B]\!]$.

Proof. Let N, X, and Z be as in the hypotheses, and let $M = \int X\,\mathbf{d}N$. By Theorem 3.1,

$$M = \Phi(B) + \frac{1}{12}\int (f''g' - f'g'')(B)\,\mathbf{d}[\![B]\!] + \int X\,\mathbf{d}V,$$

where $\Phi \in C^\infty(\mathbb{R})$ is chosen so that $\Phi' = fg'$ and $\Phi(0) = 0$. Hence, by Lemma 3.2,

$$\int Z\,\mathbf{d}M = \int Z\,\mathbf{d}\Phi(B) + \frac{1}{12}\int (f''g'h - f'g''h)(B)\,\mathbf{d}[\![B]\!] + \int ZX\,\mathbf{d}V. \tag{5.19}$$

By Theorem 3.1,

$$\int Z\,\mathbf{d}\Phi(B) = \Psi(B) + \frac{1}{12}\int (h''\Phi' - h'\Phi'')(B)\,\mathbf{d}[\![B]\!],$$

where $\Psi \in C^\infty(\mathbb{R})$ is chosen so that $\Psi' = h\Phi'$ and $\Psi(0) = 0$. Theorem 3.1 also gives

$$\int ZX\,\mathbf{d}Y = \widetilde{\Psi}(B) + \frac{1}{12}\int ((fh)''g' - (fh)'g'')(B)\,\mathbf{d}[\![B]\!],$$

where $\widetilde{\Psi} \in C^\infty(\mathbb{R})$ is chosen so that $\widetilde{\Psi}' = fhg'$ and $\widetilde{\Psi}(0) = 0$. Note, however, that this implies $\Psi = \widetilde{\Psi}$, which gives

$$\int Z\,\mathbf{d}\Phi(B) = \int ZX\,\mathbf{d}Y + \frac{1}{12}\int (h''\Phi' - h'\Phi'' - (fh)''g' + (fh)'g'')(B)\,\mathbf{d}[\![B]\!].$$

Substituting $\Phi' = fg'$ into the above and simplifying gives

$$\int Z\,\mathbf{d}\Phi(B) = \int ZX\,\mathbf{d}Y + \frac{1}{12}\int (f'g''h - f''g'h - 3f'g'h')(B)\,\mathbf{d}[\![B]\!].$$

Substituting this into Eq. (5.19) gives

$$\begin{aligned}\int Z\,\mathbf{d}M &= \int ZX\,\mathbf{d}Y - \frac{1}{4}\int (f'g'h')(B)\,\mathbf{d}[\![B]\!] + \int ZX\,\mathbf{d}V\\ &= \int ZX\,\mathbf{d}N - \frac{1}{4}\int (f'g'h')(B)\,\mathbf{d}[\![B]\!],\end{aligned}$$

and this verifies Eq. (5.18).

Finally, if $\mathcal{V} = \{\mathcal{V}_n\}$, then $\delta_j(\mathcal{V}_n) = \delta_j(\Lambda_n^X)\delta_j(\Lambda_n^Y)\delta_j(\Lambda_n^Z)$. From Eq. (5.8), we see that $\mathcal{V} \in \mathcal{S}$, $\varphi_{1,\mathcal{V}} = 0$, and $\varphi_{3,\mathcal{V}} = \varphi_{1,X}\varphi_{1,Y}\varphi_{1,Z} = f'g'h'$. Since $[\mathcal{V}]_0 = 0$, it follows from Example 2.2 that $[\mathcal{V}] = \int (f'g'h')(B)\,\mathbf{d}[\![B]\!]$. □

Acknowledgements Thanks go to Tom Kurtz and Frederi Viens for stimulating and helpful comments, feedback, and discussions. Jason Swanson was supported in part by NSA grant H98230-09-1-0079.

References

1. Burdzy, K., Swanson, J.: A change of variable formula with Itô correction term. Ann. Probab. **38**(5), 1817–1869 (2010)
2. Cheridito, P., Nualart, D.: Stochastic integral of divergence type with respect to fractional Brownian motion with Hurst parameter $H \in (0, \frac{1}{2})$. Ann. Inst. H. Poincaré Probab. Statist. **41**(6), 1049–1081 (2005)
3. Errami, M., Russo, F.: n-covariation, generalized Dirichlet processes and calculus with respect to finite cubic variation processes. Stoch. Process Appl. **104**(2), 259–299 (2003)
4. Gradinaru, M., Nourdin, I., Russo, F., Vallois, P.: m-order integrals and generalized Itô's formula: the case of a fractional Brownian motion with any Hurst index. Ann. Inst. H. Poincaré Probab. Statist. **41**(4), 781–806 (2005)
5. Nourdin, I., Réveillac, A.: Asymptotic behavior of weighted quadratic variations of fractional Brownian motion: the critical case $H = 1/4$. Ann. Probab. **37**(6), 2200–2230 (2009)
6. Nourdin, I., Réveillac, A., Swanson, J.: The weak Stratonovich integral with respect to fractional Brownian motion with Hurst parameter 1/6. Electron. J. Probab. **15**, 2087–2116 (2010)
7. Nualart, D., Ortiz-Latorre, S.: Central limit theorems for multiple stochastic integrals and Malliavin calculus. Stoch. Process Appl. **118**(4), 614–628 (2008)

Part II
Stochastic Differential Equations

Chapter 6
Large Deviations for Hilbert-Space-Valued Wiener Processes: A Sequence Space Approach

Andreas Andresen, Peter Imkeller, and Nicolas Perkowski

Dedicated to David Nualart on the occasion of his 60th birthday

Abstract Ciesielski's isomorphism between the space of α-Hölder continuous functions and the space of bounded sequences is used to give an alternative proof of the large deviation principle (LDP) for Wiener processes with values in Hilbert space.

Keywords Large deviations • Schilder's theorem • Hilbert space valued Wiener process • Ciesielski's isomorphism

MSC subject classifications 2010: 60F10; 60G15.

Received 9/23/2011; Accepted 3/5/2012; Final 3/20/2012

1 Introduction

The large deviation principle (LDP) for Brownian motion β on $[0, 1]$—contained in Schilder's theorem [11]—describes the exponential decay of the probabilities with which $\sqrt{\varepsilon}\beta$ takes values in closed or open subsets of the path space of continuous functions in which the trajectories of β live. The path space is equipped with the topology generated by the uniform norm. The decay is dominated by a rate function capturing the "energy" $\frac{1}{2}\int_0^1(\dot{f}(t))^2 \mathrm{d}t$ of functions f on the Cameron–Martin space

A. Andresen • P. Imkellera (✉) • N. Perkowski
Institut für Mathematik, Humboldt-Universität zu Berlin,
Rudower Chaussee 25, 12489 Berlin, Germany
e-mail: imkeller@mathematik.hu-berlin.de

F. Viens et al. (eds.), *Malliavin Calculus and Stochastic Analysis: A Festschrift in Honor of David Nualart*, Springer Proceedings in Mathematics & Statistics 34,
DOI 10.1007/978-1-4614-5906-4_6, © Springer Science+Business Media New York 2013

for which a square-integrable derivative exists. Schilder's theorem is of central importance to the theory of large deviations for randomly perturbed dynamical systems or diffusions taking their values in spaces of continuous functions (see [6,9], and references therein, [10]). A version of Schilder's theorem for a Q-Wiener processes W taking values in a separable Hilbert space H is well known (see [5]; Theorem 12.7 gives an LDP for Gaussian laws on Banach spaces). Here Q is a self-adjoint positive trace-class operator on H. If $(\lambda_i)_{i\geq 0}$ are its summable eigenvalues with respect to an eigenbasis $(e_k)_{k\geq 0}$ in H, W may be represented with respect to a sequence of one-dimensional Wiener processes $(\beta_k)_{k\geq 0}$ by $W = \sum_{k=0}^{\infty} \lambda_k \beta_k \, e_k$. The LDP in this framework can be derived by means of techniques of reproducing kernel Hilbert spaces (see [5], Chap. 12.1). The rate function is then given by an analogous energy functional for which $\dot{f}^2$ is replaced by $\|Q^{-\frac{1}{2}} \dot{F}\|^2$ for continuous functions F possessing square-integrable derivatives $\dot{F}$ on $[0, 1]$.

Schilder's theorem for β may for instance be derived via approximation of β by random walks from LDP principles for discrete processes (see [6]). Baldi and Roynette [1] give a very elegant alternative proof of Schilder's theorem, the starting point of which is a Fourier decomposition of β by a complete orthonormal system (CONS) in $L^2([0,1])$. The rate function for β is then simply calculated by the rate functions of one-dimensional Gaussian unit variables. In this approach, the LDP is first proved for balls of the topology, and then generalized by means of exponential tightness to open and closed sets of the topology. As a special feature of the approach, Schilder's theorem is obtained in a stricter sense on all spaces of Hölder continuous functions of order $\alpha < \frac{1}{2}$. This enhancement results quite naturally from a characterization of the Hölder topologies on function spaces by appropriate infinite sequence spaces (see [4]). Representing the one-dimensional Brownian motions β_k for instance by the CONS of Haar functions on $[0, 1]$, we obtain a description of the Hilbert space valued Wiener process W in which a double sequence of independent standard normal variables describes randomness. Starting with this observation, in this paper we extend the direct proof of Schilder's theorem by [1] to Q-Wiener spaces W with values on H. On the way, we also retrieve the enhancement of the LDP to spaces of Hölder continuous functions on $[0, 1]$ of order $\alpha < \frac{1}{2}$. The idea of approaching problems related to stochastic processes with values in function spaces by sequence space methods via Ciesielski's isomorphism is not new: it has been employed in [2] to give an alternative treatment of the support theorem for Brownian motion, in [3] to enhance the Freidlin–Wentzell theory from the uniform to Hölder norms, and in [7, 8] further to Besov–Orlicz spaces.

In Sect. 2 we first give a generalization of Ciesielski's isomorphism of spaces of Hölder continuous functions and sequence spaces to functions with values on Hilbert spaces. We briefly recall the basic notions of Gaussian measures and Wiener processes on Hilbert spaces. Using Ciesielski's isomorphism we give a Schauder representation of Wiener processes with values in H. Additionally we give a short overview of concepts and results from the theory of LDP needed in the derivation of Schilder's theorem for W. In main Sect. 3 the alternative proof of the LDP for W is given. We first introduce a new norm on the space of Hölder continuous functions $C_\alpha([0, 1], H)$ with values in H which is motivated by the sequence space representation in Ciesielski's isomorphism and generates a coarser topology. We adapt the description of the rate function to the Schauder series setting and then prove the LDP for a basis of the coarser topology using Ciesielski's isomorphism. We finally establish the last ingredient, the crucial property of exponential tightness, by construction of appropriate compact sets in sequence space.

2 Preliminaries

In this section we collect some ingredients needed for the proof of a LDP for Hilbert space valued Wiener processes. We first prove Ciesielski's theorem for Hilbert space valued functions which translates properties of functions into properties of the sequences of their Fourier coefficients with respect to complete orthonormal systems in $L^2([0,1])$. We summarize some basic properties of Wiener processes W with values in a separable Hilbert space H. We then discuss Fourier decompositions of W, prove that its trajectories lie almost surely in $C^0_\alpha([0,1],H)$, and describe its image under the Ciesielski isomorphism. We will always denote by H a separable Hilbert space equipped with a symmetric inner product $\langle\cdot,\cdot\rangle$ that induces the norm $\|\cdot\|_H$ and a countable CONS (e_k) $k \in \mathbb{N}$.

2.1 Ciesielski's Isomorphism

The **Haar functions** $(\chi_n, n \geq 0)$ are defined as $\chi_0 \equiv 1$:

$$\chi_{2^k+l}(t) := \begin{cases} \sqrt{2^k}, & \frac{2l}{2^{k+1}} \leq t < \frac{2l+1}{2^{k+1}}, \\ -\sqrt{2^k}, & \frac{2l+1}{2^{k+1}} \leq t \leq \frac{2l+2}{2^{k+1}}, \\ 0, & \text{otherwise.} \end{cases} \tag{6.1}$$

The Haar functions form a CONS of $L^2([0,1],\mathrm{d}x)$. Note that because of their wavelet structure, the integral $\int_{[0,1]} \chi_n \mathrm{d}f$ is well defined for all functions f. For $n = 2^k + l$ where $k \in \mathbb{N}$ and $0 \leq l \leq 2^k - 1$ we have $\int_{[0,1]} \chi_n \mathrm{d}F = \sqrt{2^k}[2F(\frac{2l+1}{2^{k+1}}) - F(\frac{2l+2}{2^{k+1}}) - F(\frac{2l}{2^{k+1}})]$, and it does not matter whether F is a real or Hilbert space valued function.

The primitives of the Haar functions are called **Schauder functions**, and they are given by

$$\phi_n(t) = \int_0^t \chi_n(s)\mathrm{d}s,\ t \in [0,1],\ n \geq 0.$$

Slightly abusing notation, we denote the α-Hölder seminorms on $C_\alpha([0,1];H)$ and on $C_\alpha([0,1];\mathbb{R})$ by the same symbols

$$\|F\|_\alpha := \sup_{0\leq s<t\leq 1} \frac{\|F(t)-F(s)\|_H}{|t-s|^\alpha}, \quad F \in C_\alpha([0,1];H),$$

$$\|f\|_\alpha := \sup_{0\leq s<t\leq 1} \frac{|f(t)-f(s)|}{|t-s|^\alpha}, \quad f \in C_\alpha([0,1];\mathbb{R}).$$

$C_\alpha([0,1];H)$ is of course the space of all functions $F : [0,1] \to H$ such that $\|F\|_\alpha < \infty$ and similarly for $C_\alpha([0,1];\mathbb{R})$. We also denote the supremum norm on $C([0,1];H)$ and $C([0,1];\mathbb{R})$ by the same symbol $\|\cdot\|_\infty$.

Denote in the sequel for an H-valued function F its orthogonal component with respect to e_k by $F_k = \langle F, e_k\rangle, k \geq 0$. Further denote by P_k (resp. R_k) the orthogonal projectors on $\text{span}(e_1,\ldots,e_k)$ (resp. its orthogonal complement), $k \geq 0$. For every $F \in C_\alpha([0,1];H)$ and every $k \geq 0, s,t \in [0,1]$ we have

$$|\langle F(t), e_k\rangle - \langle F(s), e_k\rangle| \leq \|F(t)-F(s)\|_H.$$

More generally, for any $k \geq 0, s,t \in [0,1]$, we have

$$\|P_k F(t) - P_k F(s)\|_H \leq \|F(t)-F(s)\|_H, \quad \|R_k F(t) - R_k F(s)\|_H \leq \|F(t)-F(s)\|_H.$$

Our approach starts with the observation that we may decompose functions $F \in C_\alpha([0,1];H)$ by double series with respect to the system $(\phi_n\, e_k : n,k \geq 0)$.

Lemma 2.1. *Let $\alpha \in (0,1)$ and $F \in C_\alpha([0,1];H)$. Then we have*

$$F = \sum_n \int_{[0,1]} \chi_n \mathrm{d}F \phi_n = \sum_{n=0}^\infty \sum_{k=0}^\infty \int_{[0,1]} \chi_n \mathrm{d}F_k e_k \phi_n$$

with convergence in the uniform norm on $C([0,1];H)$.

Proof. For the real-valued functions $F_k, k \geq 0$, the representation

$$F_k = \sum_{n=0}^\infty \int_{[0,1]} \chi_n \mathrm{d}F_k\, \phi_n$$

is well known from [4]. Therefore we may write for $F \in C_\alpha([0,1];H)$

$$\begin{aligned} F &= \sum_{k=0}^\infty F_k e_k \\ &= \sum_{k=0}^\infty e_k \sum_{n=0}^\infty \int_{[0,1]} \chi_n \mathrm{d}F_k \phi_n \end{aligned}$$

$$= \sum_{n=0}^{\infty}\sum_{k=0}^{\infty}\int_{[0,1]} \chi_n \mathrm{d}F_k e_k \phi_n$$

$$= \sum_{n=0}^{\infty}\int_{[0,1]} \chi_n \mathrm{d}F \phi_n.$$

To justify the exchange in the order of summation and the convergence in the uniform norm, we have to show

$$\lim_{N,m\to\infty} \left\| \sum_{n\geq N} \int_{[0,1]} \chi_n \mathrm{d}R_m F \phi_n \right\|_\infty = 0.$$

For this purpose, note first that by definition of the Haar system for any $n, m \geq 0, n = 2^k + l$, where $0 \leq l \leq 2^k - 1$

$$\left\| \int_{[0,1]} \chi_n dR_m F \right\|_H = \sqrt{2^k} \left\| 2R_m F\left(\frac{2l+1}{2^{k+1}}\right) - R_m F\left(\frac{2l+2}{2^{k+1}}\right) - R_m F\left(\frac{2l}{2^{k+1}}\right) \right\|_H$$

$$\leq 2\| R_m F \|_\alpha 2^{-\alpha(k+1)} 2^{\frac{1}{2}k}$$

$$= \| R_m F \|_\alpha 2^{-\alpha(k+1)+\frac{1}{2}k+1}.$$

Therefore, for $K \geq 0$ such that $2^K \leq N \leq 2^{K+1}$, using the fact that $\phi_{2^k+l}, 0 \leq l \leq 2^k - 1$ have disjoint support and that $\|\phi_{2^k+l}\|_\infty \leq 2^{-\frac{k}{2}-1}$, we obtain

$$\left\| \sum_{n\geq N} \int_{[0,1]} \chi_n \mathrm{d}R_m F \phi_n \right\|_\infty \leq \sum_{k\geq K} \left\| \sum_{0\leq l\leq 2^k-1} \int_{[0,1]} \chi_{2^k+l} \mathrm{d}F \phi_{2^k+l} \right\|_\infty$$

$$\leq \sum_{k\geq K} \sup_{0\leq l\leq 2^k-1} \left\| \int_{[0,1]} \chi_{2^k+l} \mathrm{d}R_m F \right\|_\infty 2^{-\frac{k}{2}-1}$$

$$\leq \sum_{k\geq K} \| R_m F \|_\alpha 2^{-\alpha(k+1)}$$

$$\leq \| R_m F \|_\alpha \sum_{k\geq K} (2^\alpha)^{-k} \xrightarrow[K,m\to\infty]{} 0.$$

Here we use $\| R_m F \|_\alpha \leq \| F \|_\alpha < \infty$ for all $m \geq 0$, the fact that $\lim_{m\to\infty} R_m F(t) = 0$ for any $t \in [0, 1]$, and dominated convergence to obtain $\lim_{m\to\infty} \| R_m F \|_\alpha = 0$.

□

A closer inspection of the coefficients in the decomposition of Lemma 2.1 leads us to the following isomorphism, described by [4] in the one-dimensional case. To formulate it, denote by C_0^H the space of H-valued sequences $(\eta_n)_{n\in\mathbb{N}}$ such that $\lim_{n\to\infty}\|\eta_n\|_H = 0$. If we equip C_0^H with the supremum norm (using again the symbol $\|\cdot\|_\infty$), it becomes a Banach space.

Theorem 2.1 (Ciesielski's isomorphism for Hilbert spaces). *Let* $0 < \alpha < 1$. *Let* (χ_n) *denote the Haar functions and* (ϕ_n) *denote the for Schauder functions. Let for* $0 \le n = 2^k + l \ge 0$, *where* $0 \le l \le 2^k - 1$

$$c_0(\alpha) := 1, \quad c_n(\alpha) := 2^{k(\alpha-1/2)+\alpha-1}.$$

Define

$$T_\alpha^H : C_\alpha^0([0,1];H) \to C_0^H \qquad F \mapsto \left(c_n(\alpha)\int_{[0,1]} \chi_n \mathrm{d}F\right)_{n\in\mathbb{N}}.$$

Then T_α^H *is continuous and bijective, its operator norm is 1, and its inverse is given by*

$$(T_\alpha^H)^{-1} : C_0^H \to C_\alpha^0([0,1];H), \quad (\eta_n) \mapsto \sum_{n=0}^{\infty} \frac{\eta_n}{c_n(\alpha)}\phi_n.$$

The norm of $(T_\alpha^H)^{-1}$ *is bounded by*

$$\|(T_\alpha^H)^{-1}\| \le \frac{2}{(2^\alpha - 1)(2^{1-\alpha} - 1)}.$$

Proof. Observe that for $n \in \mathbb{N}$ with $n = 2^k + l$, $0 \le l \le 2^k - 1$

$$\begin{aligned}
&\left\|\int_{[0,1]} \chi_n \mathrm{d}F\right\|_H \\
&= \sqrt{2^k}\left\|2F\left(\frac{2l+1}{2^{k+1}}\right) - F\left(\frac{2l+2}{2^{k+1}}\right) - F\left(\frac{2l}{2^{k+1}}\right)\right\|_H \\
&\le \frac{1}{2c_\alpha(n)}\left(\frac{\left\|F(\frac{2l+2}{2^{k+1}}) - F(\frac{2l+1}{2^{k+1}})\right\|_H}{2^{-\alpha(k+1)}} + \frac{\left\|F(\frac{2l+1}{2^{k+1}}) - F(\frac{2l}{2^{k+1}})\right\|_H}{2^{-\alpha(k+1)}}\right) \\
&\le \frac{1}{c_\alpha(n)} \sup_{t,s\in[0,1],\ |t-s|\le 2^{-k-1}} \frac{\|F(t) - F(s)\|_H}{|t-s|^\alpha} \\
&\le \frac{1}{c_\alpha(n)}\|F\|_\alpha.
\end{aligned}$$

This gives the desired bound on the norm. Moreover, since $F \in C_\alpha^0([0,1], H)$ we have

$$\lim_{n\to\infty} c_\alpha(n) \left\| \int_{[0,1]} \chi_n \mathrm{d}F \right\|_H \leq \lim_{n\to\infty} \sup_{t,s\in[0,1],\ |t-s|\leq 2^{-k-1}} \frac{\|F(t)-F(s)\|_H}{|t-s|^\alpha} = 0.$$

Thus the range of T_α^H is indeed contained in C_0^H. Taking $F : [0,1] \to H$ with $F(s) = se_1$ for $s \in [0,1]$ we find that $T_\alpha^H(F) = (e_1, 0, 0, \ldots)$; thus $\|F\|_\alpha = \|T_\alpha^H(F)\|_\infty$. Therefore $\|T_\alpha^H\| = 1$. Clearly T_α^H is injective.

To see that T_α^H is bijective and that the inverse is bounded as claimed, define

$$A : C_0^H \to C_\alpha^0([0,1]; H), \quad (\eta_n) \mapsto \sum_{n=0}^{\infty} \frac{\eta_n}{c_n(\alpha)} \phi_n.$$

Now a straightforward calculation using the orthogonality of the $(\chi_n)_{n\geq 0}$ gives for any $(\eta_n)_{n\geq 0} \subset C_0^H$

$$\begin{aligned}
T_\alpha^H \circ A((\eta_n)_{n\geq 0}) &= T_\alpha^H \left(\sum_{n=0}^{\infty} \frac{\eta_n}{c_n(\alpha)} \phi_n \right) \\
&= \left(\sum_{n,m=0}^{\infty} \eta_n \int_{[0,1]} \chi_m \mathrm{d}\phi_n \right)_{m\in\mathbb{N}} \\
&= \left(\sum_{n,m=0}^{\infty} \eta_n \int \chi_n(t)\chi_m(t)\mathrm{d}t \right)_{m\in\mathbb{N}} \\
&= (\eta_m)_{m\geq 0}.
\end{aligned}$$

Consequently we can infer that $A = (T_\alpha^H)^{-1}$.

We still have to show that $(T_\alpha^H)^{-1}$ satisfies the claimed norm inequality and maps every sequence $(\eta_n)_{n\geq 0} \in C_0^H$ to an element of $C_\alpha^0([0,1], H)$. For this purpose let $(\eta_n)_{n\geq 0} \in C_0^H$, set $F = (T_\alpha^H)^{-1}((\eta_n))$ and let $s, t \in [0,1]$ be given. Then we have

$$\|F(t) - F(s)\|_H \leq \|(\eta_n)_{n\geq 0}\|_\infty \left(|t-s| + \sum_{k=0}^{\infty} \sum_{l=0}^{2^k-1} \frac{|\phi_{2^k+l}(t) - \phi_{2^k+l}(s)|}{c_{2^k}(\alpha)} \right).$$

The term in brackets on the right-hand side is exactly the one appearing in the real-valued case [4]. Consequently we have the same bound, given by

$$\|(T_\alpha^H)^{-1}\| \leq \frac{1}{(2^\alpha - 1)(2^{\alpha-1} - 1)}.$$

A more careful estimation yields

$$\|F(t) - F(s)\|_H \leq \|\eta_0\| |t - s| + \sum_{k=0}^{\infty} \sum_{l=0}^{2^k-1} \frac{1}{c_{2^k}(\alpha)} \|\eta_{2^k+l}\| |\phi_{2^k+l}(t) - \phi_{2^k+l}(s)|.$$

This is the same expression as in the real-valued case. Its well-known treatment implies

$$\lim_{|t-s|\to 0} \frac{\|F(t) - F(s)\|_H}{|t - s|^{\alpha}} = 0.$$

This finishes the proof. □

2.2 Wiener Processes on Hilbert Spaces

We recall some basic concepts of Gaussian random variables and Wiener processes with values in a separable Hilbert space H. Especially we will derive a Fourier sequence decomposition of Wiener processes. Our presentation follows [5].

Definition 2.1. Let $(\Omega, \mathcal{F}, \mathbb{P})$ be a probability space, $m \in H$ and $Q : H \to H$ a positive self-adjoint operator. An H-valued random variable X such that for every $h \in H$

$$E[\exp(i \langle h, X \rangle)] = \exp\left(i \langle h, m \rangle - \frac{1}{2} \langle Qh, h \rangle \right)$$

is called Gaussian with covariance operator Q and mean $m \in H$. We denote the law of X by $\mathcal{N}(m, Q)$.

By Proposition 2.15 of [5], Q has to be a positive, self-adjoint trace-class operator, i.e., a bounded operator from H to H that satisfies:

1. $\langle Qx, x \rangle \geq 0$ for every $x \in H$
2. $\langle Qx, x \rangle = \langle x, Qx \rangle$ for every $x \in H$
3. $\sum_{k=0}^{\infty} \langle Qe_k, e_k \rangle < \infty$ for every CONS $(e_k)_{k\geq 0}$

If Q is a positive, self-adjoint trace-class operator on H, then there exists a CONS $(e_k)_{k\geq 0}$ such that $Qe_k = \lambda_k e_k$, where $\lambda_k \geq 0$ for all k and $\sum_{k=0}^{\infty} \lambda_k < \infty$. Note that for such a Q, an operator $Q^{1/2}$ can be defined by setting $Q^{1/2} e_k := \sqrt{\lambda_k} e_k, k \in \mathbb{N}_0$. Then $Q^{1/2} Q^{1/2} = Q$.

Definition 2.2. Let Q be a positive, self-adjoint trace-class operator on H. A Q-Wiener process $(W(t) : t \in [0, 1])$ is a stochastic process with values in H such that:

1. $W(0) = 0$.
2. W has continuous trajectories.

3. W has independent increments.
4. $\mathcal{L}(W(t) - W(s)) = \mathcal{N}(0, (t-s)Q)$.

In this case $(W(t_1), \ldots, W(t_n))$ is H^n-valued Gaussian for all $t_1, \ldots, t_n \in [0,1]$. By Proposition 4.2 of [5] we know that such a process exists for every positive, self-adjoint trace-class operator Q on H. To get the Fourier decomposition of a Q-Wiener process along the Schauder basis we use a different standard characterization.

Lemma 2.2. *A stochastic process Z on $(H, \mathcal{B}(H))$ is a Q-Wiener process if and only if:*

- $Z_0 = 0$ $\mathbb{P}$*-a.s..*
- Z *has continuous trajectories.*
- $cov(\langle v, Z_t\rangle\langle w, Z_s\rangle) = (t \wedge s)\langle v, Qw\rangle\ \forall v, w \in H,\ \forall 0 \le s \le t < \infty.$
- $\forall (v_1, \ldots, v_n) \in H^n$ $(\langle v_1, Z\rangle, \ldots, \langle v_n, Z\rangle)$ *is an* $\mathbb{R}^n$*-valued Gaussian process.*

Independent Gaussian random variables with values in a Hilbert space asymptotically allow the following bounds.

Lemma 2.3. *Let $Z_n \sim \mathcal{N}(0, Q)$, $n \in \mathbb{N}$, be independent. Then there exists an a.s. finite real-valued random variable C such that*

$$\|Z_n\|_H \le C\sqrt{\log n}\ \mathbb{P}\ a.s..$$

Proof. By using the exponential integrability of $\lambda\|Z_n\|_H^2$ for small enough λ and Markov's inequality, we obtain that there exist $\lambda, c \in \mathbb{R}_+$ such that for any $a > 0$

$$\mathbb{P}(\|Z\|_H > a) \le ce^{-\lambda a^2}.$$

Thus for $\alpha > 1$ and n big enough

$$\mathbb{P}\left(\|Z_n\|_H \ge \sqrt{\lambda^{-1}\alpha \log n}\right) \le cn^{-\alpha}.$$

We set $A_n = \left\{\|Z_n\|_H \ge \sqrt{\lambda^{-1}\alpha \log n}\right\}$ and have

$$\sum_{n=0}^{\infty} \mathbb{P}(A_n) < \infty.$$

Hence the lemma of Borel–Cantelli gives that $\mathbb{P}(\limsup_n A_n) = 0$, i.e., $\mathbb{P} - a.s.$ for almost all $n \in \mathbb{N}$ we have $\|Z_n\|_H \le \sqrt{\lambda^{-1}\alpha \log n}$. In other words

$$C := \sup_{n \ge 0} \frac{\|Z_n\|_H}{\sqrt{\log n}} < \infty\ \mathbb{P} - a.s.$$

□

Using Lemma 2.3 and the characterization of Q-Wiener processes of Lemma 2.2, we now obtain its Schauder decomposition which can be seen as a Gaussian version of Lemma 2.1.

Proposition 2.1. *Let $\alpha \in (0, 1/2)$; let $(\phi_n)_{n\geq 0}$ be the Schauder functions and $(Z_n)_{n\geq 0}$ a sequence of independent, $\mathcal{N}(0, Q)$-distributed Gaussian variables, where Q is a positive self-adjoint trace-class operator on H. The series-defined process*

$$W_t = \sum_{n=0}^{\infty} \phi_n(t) Z_n, \quad t \in [0, 1],$$

converges $\mathbb{P}$-a.s. with respect to the $\|\cdot\|_\alpha$-norm on $[0, 1]$ and is an H-valued Q-Wiener process.

Proof. We have to show that the process defined by the series satisfies the conditions given in Lemma 2.2. The first and the two last conditions concerning the covariance structure and Gaussianity of scalar products have standard verifications. Let us just argue for absolute and $\|\cdot\|_\alpha$-convergence of the series, thus proving Hölder continuity of the trajectories.

Since T_α^H is an isomorphism and since any single term of the series is even Lipschitz continuous, it suffices to show that

$$\left(T_\alpha^H \left(\sum_{n=0}^{m} \phi_n Z_n \right) : m \in \mathbb{N} \right)$$

is a Cauchy sequence in C_0^H. Let us first calculate the image of term N under T_α^H. We have

$$(T_\alpha^H \phi_n Z_n)_N = 1_{\{n=N\}} c_N(\alpha) Z_N.$$

Therefore for $m_1, m_2 \geq 0, m_1 \leq m_2$

$$\sum_{n=m_1}^{m_2} (T_\alpha^H \phi_n Z_n)_N = 1_{\{m_1 \leq N \leq m_2\}} c_N(\alpha) Z_N = \left(T_\alpha^H \left(\sum_{n=m_1}^{m_2} \phi_n Z_n \right) \right)_N.$$

So if we can prove that $c_N(\alpha) Z_N$ a.s. converges to 0 in H as $N \to \infty$, the proof is complete. But this follows immediately from Lemma 2.3: $c_N(\alpha)$ decays exponentially fast, and $\|Z_N\|_H \leq C\sqrt{\log N}$. □

In particular we showed that for $\alpha < 1/2$ W a.s. takes its trajectories in

$$C_\alpha^0([0, 1]; H) := \left\{ F : [0, 1] \to H, F(0) = 0, \lim_{\delta \to 0} \sup_{\substack{t \neq s, \\ |t-s| < \delta}} \frac{\|F(t) - F(s)\|_H}{|t - s|^\alpha} = 0 \right\}.$$

By Lipschitz continuity of the scalar product, we also have $\langle F, e_k \rangle \in C_\alpha^0([0,1];\mathbb{R})$.

Since P_k and R_k are orthogonal projectors and therefore Lipschitz continuous, we obtain that for $F \in C_\alpha^0([0,1];H)$,

$$\sup_{k\geq 0} \|\langle F, e_k \rangle\|_\alpha \leq \|F\|_\alpha.$$

We also saw that $T_\alpha^H(W)$ is well defined almost surely. As a special case this is also true for the real-valued Brownian motion. We have by Proposition 2.1

$$T_\alpha^H(W) = (c_n(\alpha) Z_n)$$

where $(Z_n)_{n\geq 0}$ is a sequence of i.i.d. $\mathcal{N}(0, Q)$-variables.

Plainly, the representation of the preceding lemma can be used to prove the representation formula for Q-Wiener processes by scalar Brownian motions according to [5], Theorem 4.3.

Proposition 2.2. *Let W be a Q-Wiener process. Then*

$$W(t) = \sum_{k=0}^{\infty} \sqrt{\lambda_k} \beta_k(t) e_k, \; t \in [0,1],$$

where the series on the right-hand side $\mathbb{P}$-a.s. converges uniformly on $[0,1]$ and $(\beta_k)_{k\geq 0}$ is a sequence of independent real-valued Brownian motions.

Proof. Using arguments as in the proof of Theorem 2.1 and Lemma 2.3 to justify changes in the order of summation we get

$$W = \sum_{n=0}^{\infty} \phi_n Z_n = \sum_{k\geq 0}\sum_{n\geq 0} \phi_n \langle Z_n, e_k \rangle e_k = \sum_{k\geq 0} \sqrt{\lambda_k} \sum_{n\geq 0} \phi_n N_{n,k} e_k = \sum_{k\geq 0} \sqrt{\lambda_k} \beta_k e_k,$$

where the equivalences are $\mathbb{P}$-a.s. and $(N_{n,k})_{n,k\geq 0}, (\beta_k)_{k\geq 0}$ are real-valued iid $\mathcal{N}(0,1)$ random variables, resp., Brownian motions. For the last step we applied Proposition 2.1 for the one-dimensional case. □

2.3 Large Deviations

Let us recall some basic notions of the theory of large deviations that will suffice to prove the LDP for Hilbert space valued Wiener processes. We follow [6]. Let X be a topological Hausdorff space. Denote its Borel σ-algebra by $\mathcal{B}$.

Definition 2.3 (Rate function). A function $I : X \to [0, \infty]$ is called a rate function if it is lower semi-continuous, i.e., if for every $C \geq 0$ the set

$$\Psi_I(C) := \{x \in X : I(x) \leq C\}$$

is closed. It is called a good rate function, if $\Psi_I(C)$ is compact. For $A \in \mathcal{B}$ we define $I(A) := \inf_{x \in A} I(x)$.

Definition 2.4 (LDP). Let I be a rate function. A family of probability measures $(\mu_\varepsilon)_{\varepsilon>0}$ on $(X, \mathcal{B})$ is said to satisfy the LDP with rate function I if for any closed set $F \subset X$ and any open set $G \subset X$ we have

$$\limsup_{\varepsilon \to 0} \varepsilon \log \mu_\varepsilon(F) \leq -I(F) \text{ and}$$

$$\liminf_{\varepsilon \to 0} \varepsilon \log \mu_\varepsilon(G) \geq -I(G).$$

Definition 2.5 (Exponential tightness). A family of probability measures $(\mu_\varepsilon)_{\varepsilon>0}$ is said to be exponentially tight if for every $a > 0$ there exists a compact set $K_a \subset X$ such that

$$\limsup_{\varepsilon \to 0} \varepsilon \log \mu_\varepsilon(K_a^c) < -a.$$

In our approach to Schilder's theorem for Hilbert space valued Wiener processes we shall mainly use the following proposition which basically states that the rate function has to be known for elements of a subbasis of the topology.

Proposition 2.3. *Let $\mathcal{G}_0$ be a collection of open sets in the topology of X such that for every open set $G \subset X$ and for every $x \in G$ there exists $G_0 \in \mathcal{G}_0$ such that $x \in G_0 \subset G$. Let I be a rate function and let $(\mu_\varepsilon)_{\varepsilon>0}$ be an exponentially tight family of probability measures. Assume that for every $G \in \mathcal{G}_0$ we have*

$$-\inf_{x \in G} I(x) = \lim_{\varepsilon \to 0} \varepsilon \log \mu_\varepsilon(G).$$

Then I is a good rate function, and $(\mu_\varepsilon)_\varepsilon$ satisfies an LDP with rate function I.

Proof. Let us first establish the lower bound. In fact, let G be an open set. Choose $x \in G$ and a basis set G_0 such that $x \in G_0 \subset G$. Then evidently

$$\liminf_{\varepsilon \to 0} \varepsilon \ln \mu_\varepsilon(G) \geq \liminf_{\varepsilon \to 0} \varepsilon \ln \mu_\varepsilon(G_0) = -\inf_{y \in G_0} I(y) \geq -I(x).$$

Now the lower bound follows readily by taking the sup of $-I(x), x \in G$, on the right-hand side, the left-hand side not depending on x.

For the upper bound, fix a compact subset K of X. For $\delta > 0$ denote

$$I^\delta(x) = (I(x) - \delta) \wedge \frac{1}{\delta}, \quad x \in X.$$

For any $x \in K$, use the lower semicontinuity of I, more precisely that $\{y \in X : I(y) > I^\delta(x)\}$ is open to choose a set $G_x \in \mathcal{G}_0$ such that

$$-I^\delta(x) \geq \limsup_{\varepsilon \to 0} \varepsilon \ln \mu_\varepsilon(G_x).$$

Use compactness of K to extract from the open cover $K \subset \cup_{x \in K} G_x$ a finite subcover $K \subset \cup_{i=1}^n G_{x_i}$. Then with a standard argument we obtain

$$\limsup_{\varepsilon \to 0} \varepsilon \ln \mu_\varepsilon(K) \leq \max_{1 \leq i \leq n} \limsup_{\varepsilon \to 0} \varepsilon \ln \mu_\varepsilon(G_{x_i}) \leq - \min_{1 \leq i \leq n} I^\delta(x_i) \leq - \inf_{x \in K} I^\delta(x).$$

Now let $\delta \to 0$. Finally use exponential tightness to show that I is a good rate function (see [6], Sect. 4.1). □

The following propositions show how LDPs are transferred between different topologies on a space, or via continuous maps to other topological spaces.

Proposition 2.4 (Contraction principle). *Let X and Y be topological Hausdorff spaces, and let $I : X \to [0, \infty]$ be a good rate function. Let $f : X \to Y$ be a continuous mapping. Then*

$$I' : Y \to [0, \infty], I'(y) = \inf\{I(x) : f(x) = y\}$$

is a good rate function, and if $(\mu_\varepsilon)_{\varepsilon>0}$ satisfies an LDP with rate function I on X, then $(\mu_\varepsilon \circ f^{-1})_{\varepsilon>0}$ satisfies an LDP with rate function I' on Y.

Proposition 2.5. *Let $(\mu_\varepsilon)_{\varepsilon>0}$ be an exponentially tight family of probability measures on $(X, \mathcal{B}_{\tau_2})$ where $\mathcal{B}_{\tau_2}$ are the Borel sets of τ_2. Assume (μ_ε) satisfies an LDP with rate function I with respect to some Hausdorff topology τ_1 on X which is coarser than τ_2, i.e., $\tau_2 \subset \tau_1$. Then $(\mu_\varepsilon)_{\varepsilon>0}$ satisfies the LDP with respect to τ_2, with good rate function I.*

The main idea of our sequence space approach to Schilder's theorem for Hilbert space valued Wiener processes will just extend the following LDP for a standard normal variable with values in $\mathbb{R}$ to sequences of i.i.d. variables of this kind.

Proposition 2.6. *Let Z be a standard normal variable with values in $\mathbb{R}$,*

$$I : \mathbb{R} \to [0, \infty), \; x \mapsto \frac{x^2}{2},$$

and for Borel sets B in $\mathbb{R}$ let $\mu_\varepsilon(B) := \mathbb{P}(\sqrt{\varepsilon} Z \in B)$. Then $(\mu_\varepsilon)_{\varepsilon>0}$ satisfies an LDP with good rate function I.

3 Large Deviations for Hilbert Space Valued Wiener Processes

Ciesielski's isomorphism and the Schauder representation of Brownian motion yield a very elegant and simple method of proving LDPs for the Brownian motion. This was first noticed by [1] who gave an alternative proof of Schilder's theorem based on this isomorphism. We follow their approach and extend it to Wiener processes with values on Hilbert spaces. In this entire section we always assume $0 < \alpha < 1/2$. By further decomposing the orthogonal one-dimensional Brownian motions in the representation of an H-valued Wiener process by its Fourier coefficients with respect to the Schauder functions, we describe it by double sequences of real-valued normal variables.

3.1 Appropriate Norms

We work with new norms on the spaces of α-Hölder continuous functions given by

$$\|F\|'_\alpha := \|T^H_\alpha F\|_\infty = \sup_{k,n} \left| c_n(\alpha) \int_{[0,1]} \chi_n(s) \mathrm{d}\langle F, e_k\rangle(s) \right|, F \in C^0_\alpha([0,1]; H),$$

$$\|f\|'_\alpha := \|T_\alpha f\|_\infty = \sup_n \left| c_n(\alpha) \int_{[0,1]} \chi_n(s) \mathrm{d} f(s) \right|, f \in C^0_\alpha([0,1]; \mathbb{R}).$$

Since T^H_α is one-to-one, $\|.\|'_\alpha$ is indeed a norm. Also, we have $\|.\|'_\alpha \leq \|.\|_\alpha$. Hence the topology generated by $\|.\|'_\alpha$ is coarser than the usual topology on $C^0_\alpha([0,1], H)$.

Balls with respect to the new norms $U^\delta_\alpha(F) := \{G \in C^0_\alpha([0,1]; H) : \|G - F\|'_\alpha < \delta\}$ for $F \in C^0_\alpha([0,1]; H), \delta > 0$, have a simpler form for our reasoning, since the condition that for $\delta > 0$ a function $G \in C^0_\alpha([0,1], H)$ lies in $U^\delta_\alpha(F)$ translates into the countable set of one-dimensional conditions $|\langle T^H_\alpha(F)_n - T^H_\alpha(G)_n, e_k\rangle| < \delta$ for all $n, k \geq 0$. This will facilitate the proof of the LDP for the basis of open balls of the topology generated by $\|.\|'_\alpha$. We will first prove the LDP in the topologies generated by these norms and then transfer the result to the finer sequence space topologies using Proposition 2.5 and finally to the original function space using Ciesielski's isomorphism and Proposition 2.4.

3.2 The Rate Function

Recall that Q is supposed to be a positive self-adjoint trace-class operator on H. Let $H_0 := (Q^{1/2}H, \|\cdot\|_0)$, equipped with the inner product

$$\langle x, y\rangle_{H_0} := \langle Q^{-1/2}x, Q^{-1/2}y\rangle_H,$$

that induces the norm $\|\cdot\|_0$ on H_0. We define the Cameron–Martin space of the Q-Wiener process W by

$$\mathcal{H} := \left\{ F \in C([0,1]; H) : F(\cdot) = \int_0^\cdot U(s)\mathrm{d}s \text{ for some } U \in L^2([0,1]; H_0) \right\}.$$

Here $L^2([0,1]; H_0)$ is the space of measurable functions U from $[0,1]$ to H_0 such that $\int_0^1 \|U\|_{H_0}^2 dx < \infty$. Define the function I via

$$I : C([0,1]; H) \to [0, \infty],$$
$$F \mapsto \inf\left\{ \frac{1}{2} \int_0^1 \|U(s)\|_{H_0}^2 \mathrm{d}s : U \in L^2([0,1]; H_0),\ F(\cdot) = \int_0^{\cdot} U(s)\mathrm{d}s \right\},$$

where by convention $\inf \emptyset = \infty$. In the following we will denote any restriction of I to a subspace of $C([0,1]; H)$ (e.g., to $(C_\alpha([0,1]; H))$ by I as well. We will use the structure of H to simplify our problem. It allows us to compute the rate function I from the rate function of the one-dimensional Brownian by the following lemma.

Lemma 3.1. *Let $\tilde{I} : C([0,1]; \mathbb{R})$ be the rate function of the Brownian motion, i.e.,*

$$\tilde{I}(f) := \begin{cases} \int_0^1 |\dot{f}(s)|^2 \mathrm{d}s, & f(\cdot) = \int_0^{\cdot} \dot{f}(s)\mathrm{d}s \text{ for a square-integrable function } \dot{f}, \\ \infty, & \text{otherwise.} \end{cases}$$

Let $(\lambda_k)_{k \geq 0}$ be the sequence of eigenvalues of Q. Then for all $F \in C([0,1]; H)$ we have

$$I(F) = \sum_{k=0}^{\infty} \frac{1}{\lambda_k} \tilde{I}(\langle F, e_k \rangle)$$

where we convene that $c/0 = \infty$ for $c > 0$ and $0/0 = 0$.

Proof. Let $F \in C([0,1]; H)$.

1. First assume $I(F) < \infty$. Then there exists $U \in L^2([0,1]; H_0)$ such that $F = \int_0^{\cdot} U(s)\mathrm{d}s$ and thus $\langle F, e_k \rangle = \int_0^{\cdot} \langle U(s), e_k \rangle \mathrm{d}s$ for $k \geq 0$. Consequently we have by monotone convergence

$$\begin{aligned} \frac{1}{2} \int_0^1 \|U(s)\|_{H_0}^2 \mathrm{d}s &= \frac{1}{2} \int_0^1 \left\| \sum_{k=0}^{\infty} \langle U(s), e_k \rangle e_k \right\|_{H_0} \mathrm{d}s \\ &= \frac{1}{2} \int_0^1 \sum_{k=0}^{\infty} \langle U(s), e_k \rangle^2 \langle Q^{-\frac{1}{2}} e_k, Q^{-\frac{1}{2}} e_k \rangle \mathrm{d}s \\ &= \frac{1}{2} \int_0^1 \sum_{k=0}^{\infty} \frac{1}{\lambda_k} \langle U(s), e_k \rangle^2 \mathrm{d}s \\ &= \sum_{k=0}^{\infty} \frac{1}{\lambda_k} \tilde{I}(\langle F, e_k \rangle). \end{aligned}$$

   The last expression does not depend on the choice of U. Hence we get that $I(F) < \infty$ implies $I(F) = \sum_{k=0}^{\infty} \frac{1}{\lambda_k} \tilde{I}(\langle F, e_k \rangle)$.

2. Conversely assume $\sum_{k=0}^{\infty} \frac{1}{\lambda_k} \tilde{I}(\langle F, e_k \rangle) < \infty$. Since $\tilde{I}(\langle F, e_k \rangle) < \infty$ for all $k \geq 0$, we know that there exists a sequence $(U_k)_{k \geq 0}$ of square-integrable real-valued functions such that $\langle F, e_k \rangle = \int_0^{\cdot} U_k(s) \mathrm{d}s$. Further, those functions U_k satisfy by monotone convergence

$$\int_0^1 \sum_{k=0}^{\infty} \frac{1}{\lambda_k} |U_k(s)|^2 \mathrm{d}s = \sum_{k=0}^{\infty} \frac{1}{\lambda_k} \int_0^1 |U_k(s)|^2 \mathrm{d}s = \sum_{k=0}^{\infty} \frac{2}{\lambda_k} \tilde{I}(\langle F, e_k \rangle) < \infty.$$

So if we define $U(s) := \sum_{k=0}^{\infty} U_k(s) e_k, s \in [0,1]$, then $U \in L^2([0,1]; H_0)$. This follows from

$$U \in L^2([0,1]; H_0) \text{ iff } \int_0^1 \|U(s)\|_{H_0}^2 \mathrm{d}s = \int_0^1 \sum_{k=0}^{\infty} \frac{1}{\lambda_k} |U_k(s)|^2 \mathrm{d}s < \infty.$$

Finally we obtain by dominated convergence ($\|F(t)\|_H < \infty$)

$$F(t) = \sum_{k=0}^{\infty} \langle F(t), e_k \rangle e_k = \sum_{k=0}^{\infty} e_k \int_0^t U_k(s) \mathrm{d}s = \int_0^t U(s) \mathrm{d}s,$$

such that

$$I(F) \leq \frac{1}{2} \int_0^1 \|U(s)\|_{H_0}^2 \mathrm{d}s = \frac{1}{2} \int_0^1 \sum_{k=0}^{\infty} \frac{1}{\lambda_k} |U_k(s)|^2 \mathrm{d}s < \infty.$$

Combining the two steps we obtain $I(F) < \infty$ iff $\sum_{k=0}^{\infty} \frac{1}{\lambda_k} \tilde{I}(\langle F, e_k \rangle) < \infty$ and in this case

$$I(F) = \sum_{k=0}^{\infty} \frac{1}{\lambda_k} \tilde{I}(\langle F, e_k \rangle).$$

This completes the proof. □

Lemma 3.1 allows us to show that I is a rate function.

Lemma 3.2. *I is a rate function on $(C_\alpha^0([0,1]; H), \|.\|'_\alpha)$.*

Proof. For a constant $C \geq 0$ we have to prove that if $(F_n)_{n \geq 0} \subset \Psi_I(C) \cap C_\alpha^0([0,1]; H)$ converges in $C_\alpha^0([0,1]; H)$ to F, then F is also in $\Psi_I(C)$.

It was observed in [1] that $\tilde{I}$ is a rate function for the $\|.\|'_\alpha$-topology on $C_0^\alpha([0,1|; \mathbb{R})$. By our assumption we know that for every $k \in \mathbb{N}$, $(\langle F_n, e_k \rangle)_{n \geq 0}$ converges in $(C_0^\alpha([0,1|; \mathbb{R}), \|.\|'_\alpha)$ to $\langle F, e_k \rangle$. Therefore

$$\tilde{I}(\langle F, e_k \rangle) \leq \liminf_{n \to \infty} \tilde{I}(\langle F_n, e_k \rangle),$$

so by Lemma 3.1 and by Fatou's lemma,

$$C \geq \liminf_{n\to\infty} I(F_n) = \liminf_{n\to\infty} \sum_{k=0}^{\infty} \frac{1}{\lambda_k} \tilde{I}(\langle F_n, e_k\rangle) \geq \sum_{k=0}^{\infty} \frac{1}{\lambda_k} \liminf_{n\to\infty} \tilde{I}(\langle F_n, e_k\rangle)$$

$$\geq \sum_{k=0}^{\infty} \frac{1}{\lambda_k} \tilde{I}(\langle F, e_k\rangle) = I(F).$$

Hence $F \in \Psi_I(C)$. □

3.3 LDP for a Subbasis of the Coarse Topology

To show that the Q-Wiener process $(W(t) : t \in [0,1])$ satisfies an LDP on $(C_\alpha([0,1]; H), \|.\|_\alpha)$ with good rate function I as defined in the last section we now show that the LDP holds for open balls in our coarse topology induced by $\|.\|'_\alpha$. The proof is an extension of the version of [1] for the real-valued Wiener process.

For $\varepsilon > 0$ denote by μ_ε the law of $\sqrt{\varepsilon} W$, i.e., $\mu_\varepsilon(A) = \mathbb{P}(\sqrt{\varepsilon} W \in A)$, $A \in \mathcal{B}(H)$.

Lemma 3.3. *For every $\delta > 0$ and every $F \in C^0_\alpha([0,1]; H)$ we have*

$$\lim_{\varepsilon\to 0} \varepsilon \log \mu_\varepsilon(U^\delta_\alpha(F)) = - \inf_{G\in U^\delta_\alpha(F)} I(G).$$

Proof. 1. Write $T^H_\alpha F = (\sum_{k=0}^{\infty} F_{n,k} e_k)_{n\in\mathbb{N}}$. Then $\sqrt{\varepsilon} W$ is in $U^\delta_\alpha(F)$ if and only if

$$\sup_{k,n\geq 0} \left| \sqrt{\varepsilon} c_n(\alpha) \int_0^1 \chi_n \mathrm{d}\langle W, e_k\rangle - F_{k,n} \right| < \delta.$$

Now for $k \geq 0$ we recall $\langle W, e_k\rangle = \sqrt{\lambda_k} \beta_k$, where $(\beta_k)_{k\geq 0}$ is a sequence of independent standard Brownian motions. Therefore for $n, k \geq 0$

$$\left| \int_0^1 \chi_n \mathrm{d}\langle W, e_k\rangle \right| = \left| \sqrt{\lambda_k} Z_{k,n} \right|,$$

where $(Z_{k,n})_{k,n\geq 0}$ is a double sequence of independent standard normal variables. Therefore by independence

$$\mu_\varepsilon(U^\delta_\alpha(F)) = \mathbb{P}\left(\bigcap_{k,n\in\mathbb{N}_0} \left| c_n(\alpha) \sqrt{\varepsilon\lambda_k} Z_{k,n} - F_{k,n} \right| < \delta \right)$$

$$= \prod_{k=0}^{\infty} \prod_{n=0}^{\infty} \mathbb{P}\left(c_n(\alpha) \sqrt{\varepsilon\lambda_k} Z_{k,n} \in (F_{k,n} - \delta, F_{k,n} + \delta) \right).$$

To abbreviate, we introduce the notation

$$\mathbb{P}_{k,n}(\varepsilon) = \mathbb{P}\left(c_n(\alpha)\sqrt{\varepsilon\lambda_k}Z_{k,n} \in (F_{k,n} - \delta, F_{k,n} + \delta)\right), \varepsilon > 0, n, k \in \mathbb{N}_0.$$

For every $k \geq 0$ we split $\mathbb{N}_0$ into subsets Λ_i^k, $i = 1, 2, 3, 4$, for each of which we will calculate $\prod_{k=0}^{\infty} \prod_{n\in\Lambda_i^k} \mathbb{P}_{n,k}(\varepsilon)$ separately. Let

$$\begin{aligned}
\Lambda_1^k &= \{n \geq 0 : 0 \notin [F_{k,n} - \delta, F_{k,n} + \delta]\} \\
\Lambda_2^k &= \{n \geq 0 : F_{k,n} = \pm\delta\} \\
\Lambda_3^k &= \{n \geq 0 : [-\delta/2, \delta/2] \subset [F_{k,n} - \delta, F_{k,n} + \delta]\} \\
\Lambda_4^k &= (\Lambda_1^k \cup \Lambda_2^k \cup \Lambda_3^k)^c.
\end{aligned}$$

By applying Ciesielski's isomorphism to the real-valued functions $\langle F, e_k\rangle$, we see that for every fixed k, Λ_3^k contains nearly all n. Since $(T_\alpha^H F)_n$ converges to zero in H, in particular $\sup_{k\geq 0} |F_{k,n}|$ converges to zero as $n \to \infty$. But for every fixed n, $(F_{k,n})_k$ is in l^2 and therefore converges to zero. This shows that for large enough k we must have $\Lambda_3^k = \mathbb{N}_0$, and therefore $\cup_k (\Lambda_3^k)^c$ is finite.

2. First we examine $\prod_{k=0}^{\infty} \prod_{n\in\Lambda_3^k} \mathbb{P}_{k,n}(\varepsilon)$. Note that for $n \in \Lambda_3^k$ we have

$$[-\delta/2, \delta/2] \subset [F_{k,n} - \delta, F_{k,n} + \delta],$$

and therefore

$$\begin{aligned}
\prod_{k=0}^{\infty} \prod_{n\in\Lambda_3^k} \mathbb{P}_{k,n}(\varepsilon) &\geq \prod_{k=0}^{\infty} \prod_{n\in\Lambda_3^k} \mathbb{P}\left(Z_{k,n} \in \left(-\frac{\delta}{2c_n(\alpha)\sqrt{\varepsilon\lambda_k}}, \frac{\delta}{2c_n(\alpha)\sqrt{\varepsilon\lambda_k}}\right)\right) \\
&= \prod_{k=0}^{\infty} \prod_{n\in\Lambda_3^k} \left(1 - \sqrt{\frac{2}{\pi}} \int_{\delta/(2c_n(\alpha)\sqrt{\varepsilon\lambda_k})}^{\infty} e^{-u^2/2} du\right).
\end{aligned}$$

For $a > 1$ we have $\int_a^\infty e^{-x^2/2} dx \leq e^{-a^2/2}$. Thus for small enough ε:

$$\prod_{k=0}^{\infty} \prod_{n\in\Lambda_3^k} \mathbb{P}_{k,n}(\varepsilon) \geq \prod_{k=0}^{\infty} \prod_{n\in\Lambda_3^k} \left(1 - \sqrt{\frac{2}{\pi}} \exp\left(-\frac{\delta^2}{8c_n^2(\alpha)\varepsilon\lambda_k}\right)\right).$$

This amount will tend to 1 if and only if its logarithm tends to 0 as $\varepsilon \to 0$. Since $\log(1 - x) \leq -x$ for $x \in (0, 1)$, it suffices to prove that

$$\lim_{\varepsilon\to 0} \sum_{k=0}^{\infty} \sum_{n\geq 0} \exp\left(-\frac{\delta^2}{8c_n^2(\alpha)\varepsilon\lambda_k}\right) = 0. \tag{6.2}$$

This is true by dominated convergence, because $c_n(\alpha) = 2^{n(\alpha-1/2)+\alpha-1}$, and since $(\lambda_k) \in l_1$.

We will make this more precise. First observe that for $a > 0$

$$\mathrm{e}^{-a} \leq \frac{1}{a}\mathrm{e}^{-1}$$

$$\text{if } \log(a) - a \leq -1.$$

For $k, n \geq 0$ we write $\eta_{n,k} = \frac{\delta^2}{8c_n^2(\alpha)\varepsilon\lambda_k}$. Clearly there exists a finite set $T \subset \mathbb{N}_0^2$ such that $\log(\eta_{n,k}) - \eta_{n,k} \leq -1$ for all $(n,k) \in T^c$. We set $C = \sum_{(n,k)\in T} \mathrm{e}^{-\eta_{n,k}}$ and get

$$\begin{aligned}\sum_{k=0}^{\infty}\sum_{n=0}^{\infty}\exp\left(-\frac{\delta^2}{8c_n^2(\alpha)\varepsilon\lambda_k}\right) &= C + \sum_{(n,k)\in T^c}^{\infty} \mathrm{e}^{-\eta_{n,k}} \\ &\leq C + \sum_{(n,k)\in T^c}^{\infty} \frac{1}{\eta_{n,k}}\mathrm{e}^{-1} \\ &\leq C + \frac{8\varepsilon\mathrm{e}^{-1}}{\delta^2}\sum_{k\geq 0}\lambda_k \sum_{n\geq 0} c_n(\alpha)^2 < \infty.\end{aligned}$$

3. Since $\cup_{k\geq 0}\Lambda_4^k$ is finite and since for every n in Λ_4^k the interval $(\mathcal{F}_{k,n}-\delta, \mathcal{F}_{k,n}+\delta)$ contains a small neighborhood of 0, we have

$$\lim_{\varepsilon\to 0}\prod_{k=0}^{\infty}\prod_{n\in\Lambda_4^k} \mathbb{P}_{k,n}(\varepsilon) = 1. \tag{6.3}$$

4. Again because $\cup_{k\geq 0}\Lambda_2^k$ is finite, we obtain from its definition that

$$\lim_{\varepsilon\to 0}\prod_{k=0}^{\infty}\prod_{n\in\Lambda_2^k} \mathbb{P}_{k,n}(\varepsilon) = 2^{-|\cup_k \Lambda_2^k|}. \tag{6.4}$$

5. Finally we calculate $\lim_{\varepsilon\to 0}\prod_{k=0}^{\infty}\prod_{n\in\Lambda_1^k}\mathbb{P}_{k,n}(\varepsilon)$. For given k, n define

$$\bar{F}_{k,n} = \begin{cases} F_{k,n} - \delta, \ F_{k,n} > \delta, \\ F_{k,n} + \delta, \ F_{k,n} < -\delta. \end{cases}$$

We know that $Z_{k,n}$ is standard normal, so that by Proposition 2.6 for $n \in \Lambda_1^k$

$$\lim_{\varepsilon\to 0}\varepsilon\log\mathbb{P}_{k,n}^0(\varepsilon) = -\frac{\bar{F}_{k,n}^2}{2c_n^2(\alpha)\lambda_k},$$

and therefore again by the finiteness of $\cup_k \Lambda_1^k$

$$\lim_{\varepsilon\to 0} \varepsilon \log \prod_{k=0}^{\infty} \prod_{n\in\Lambda_1^k} \mathbb{P}_{k,n}^0(\varepsilon) = -\sum_{k=0}^{\infty} \sum_{n\in\Lambda_1^k} \frac{\bar{F}_{k,n}^2}{2c_n^2(\alpha)\lambda_k}. \tag{6.5}$$

6. Combining Eqs. (6.2)–(6.5) we obtain

$$\lim_{\varepsilon\to 0} \varepsilon \log \mu_\varepsilon(U_\alpha^\delta(F)) = -\sum_{k=0}^{\infty} \frac{1}{\lambda_k} \sum_{n\in\Lambda_1^k} \frac{\bar{F}_{k,n}^2}{2c_n^2(\alpha)}.$$

So if we manage to show

$$-\sum_{k=0}^{\infty} \frac{1}{\lambda_k} \sum_{n\in\Lambda_1^k} \frac{\bar{F}_{k,n}^2}{2c_n^2(\alpha)} = -\inf_{G\in U_\alpha^\delta(F)} I(G),$$

the proof is complete. By Ciesielski's isomorphism, every $G \in C_\alpha^0([0,1]; H)$ has the representation

$$G = \sum_{k=0}^{\infty} e_k \sum_{n=0}^{\infty} \frac{G_{k,n}}{c_n(\alpha)} \phi_n.$$

Its derivative fulfills (if it exists) for any $k \geq 0$

$$\langle \dot{G}, e_k \rangle = \sum_{n=0}^{\infty} \frac{G_{k,n}}{c_n(\alpha)} \chi_n.$$

Since the Haar functions $(\chi_n)_{n\geq 0}$ are a CONS for $L^2([0,1])$, we see that $\tilde{I}(\langle G, e_k \rangle) < \infty$ if and only if $(G_{k,n}/c_n(\alpha)) \in l_2$, and in this case

$$\tilde{I}(\langle G, e_k \rangle) = \frac{1}{2} \int_0^1 \langle \dot{G}(s), e_k \rangle^2 \mathrm{d}s = \sum_{n=0}^{\infty} \frac{G_{k,n}^2}{2c_n^2(\alpha)}.$$

So we finally obtain with Lemma 3.1 the desired equality

$$\begin{aligned}\inf_{G\in U_\alpha^\delta(F)} I(G) &= \inf_{G\in U_\alpha^\delta(F)} \sum_{k=0}^{\infty} \frac{1}{\lambda_k} \tilde{I}(\langle G, e_k \rangle) = \inf_{G\in U_\alpha^\delta(F)} \sum_{k=0}^{\infty} \frac{1}{\lambda_k} \sum_{n=0}^{\infty} \frac{G_{k,n}^2}{2c_n^2(\alpha)} \\ &= \sum_{k=0}^{\infty} \frac{1}{\lambda_k} \sum_{n\in\Lambda_1^k} \frac{\bar{F}_{k,n}^2}{2c_n^2(\alpha)}.\end{aligned}$$

□

3.4 Exponential Tightness

The final ingredient needed in the proof of the LDP for Hilbert space valued Wiener processes is exponential tightness. It will be established in two steps. The first step claims exponential tightness for the family of laws of $\sqrt{\varepsilon}Z, \varepsilon > 0$, where Z is an H-valued $\mathcal{N}(0, Q)$-variable.

Lemma 3.4. *Let $\varepsilon > 0$ and $\nu_\varepsilon = \mathbb{P} \circ (\sqrt{\varepsilon}Z)^{-1}$ for a centered Gaussian random variable Z with values in the separable Hilbert space H and covariance operator Q. Then $(\nu_\varepsilon)_{\varepsilon\in(0,1]}$ is exponentially tight. More precisely for every $a > 0$ there exists a compact subset K_a of H, such that for every $\varepsilon \in (0, 1]$*

$$\nu_\varepsilon(K_a^c) \leq \mathrm{e}^{-a/\varepsilon}$$

Proof. We know that for a sequence $(b_k)_{k\geq 0}$ converging to 0, the operator $T_{(b_k)} := \sum_{k=0}^{\infty} b_k \langle \cdot, e_k\rangle e_k$ is compact. That is, for bounded sets $A \subset H$, the set $T_{(b_k)}(A)$ is precompact in H. Since H is complete, this means that $cl(T_{(b_k)}(A))$ is compact. Let $a' > 0$ to be specified later. Denote by $B(0, \sqrt{a'}) \subset H$ the ball of radius $\sqrt{a'}$ in H. We will show that there exists a zero sequence $(b_k)_{k\geq 0}$, such that the compact set $K_{a'} = cl(T_{(b_k)}(B(0, \sqrt{a'})))$ satisfies for all $\varepsilon \in (0, 1]$:

$$\mathbb{P}(\sqrt{\varepsilon}Z \in (K_{a'})^c) \leq c\mathrm{e}^{-a'/\varepsilon} \tag{6.6}$$

with a constant $c > 0$ that does not depend on a'. Thus, for given a, we can choose $a' > a$ such that for every $\varepsilon \in (0, 1]$

$$c \leq \mathrm{e}^{(a'-a)/\varepsilon}$$

and therefore the proof is complete once we proved Eq. (6.6).

Since Z is Gaussian, $\mathrm{e}^{\lambda\|Z\|_H}$ is integrable for small λ, and we can apply Markov's inequality to obtain constants $\lambda(Q), c(Q) > 0$ such that $\mathbb{P}(\|Z\|_H \geq \sqrt{a'}) \leq c(Q)\mathrm{e}^{-\lambda(Q)a'}$.

Note that if $(\lambda_k)_{k\geq 0} \in l^1$, we can always find a sequence $(c_k)_{k\geq 0}$ such that $\lim_{k\to\infty} c_k = \infty$ and $\sum_{k\geq 0} c_k\lambda_k < \infty$. For $\beta > 0$ that will be specified later, we set $b_k = \sqrt{\frac{\beta}{c_k}}$ for all $k \geq 0$. We can define $(T_{(b_k)})^{-1} = \sum_{k=0}^{\infty} \frac{1}{b_k}\langle \cdot, e_k\rangle e_k$. This gives

$$\begin{aligned}
\mathbb{P}(\sqrt{\varepsilon}Z \in (K_{a'})^c) &\leq \mathbb{P}(\sqrt{\varepsilon}(T_{(b_k)})^{-1}(Z) \notin B(0, \sqrt{a'})) \\
&= \mathbb{P}(\|(T_{(b_k)})^{-1}(Z)\|_H^2 \geq \frac{a'}{\varepsilon}) \\
&= \mathbb{P}\left(\sum_{k=0}^{\infty} c_k|\langle Z, e_k\rangle|^2 \geq \frac{\beta a'}{\varepsilon}\right) \\
&= \mathbb{P}\left(\|\tilde{Z}\|_H \geq \sqrt{\frac{\beta a'}{\varepsilon}}\right),
\end{aligned}$$

where $\tilde{Z}$ is a centered Gaussian random variable with trace-class covariance operator

$$\tilde{Q} = \sum_{k=0}^{\infty} c_k \lambda_k \langle \cdot, e_k \rangle e_k.$$

Consequently we obtain

$$\mathbb{P}(\sqrt{\varepsilon} Z \in (K_{a'})^c) \leq c(\tilde{Q}) \mathrm{e}^{-\frac{\lambda(\tilde{Q})\beta a'}{\varepsilon}}.$$

Choosing $\beta = \frac{1}{\lambda(\tilde{Q})}$ proves the claim (6.6). □

With the help of Lemma 3.4 we are now in a position to prove exponential tightness for the family $(\mu_\varepsilon)_{\varepsilon\in(0,1]}$.

Lemma 3.5. $(\mu_\varepsilon)_{\varepsilon\in(0,1]}$ *is an exponentially tight family of probability measures on* $(C^0_\alpha([0,1];H), \|.\|_\alpha)$.

Proof. Let $a > 0$. We will construct a suitable set of the form

$$\tilde{K^a} = \prod_{n=0}^{\infty} K^a_n$$

such that

$$\limsup_{\varepsilon\to 0} \varepsilon \log \mu_\varepsilon \left[\left(\left(T^H_\alpha\right)^{-1} \tilde{K^a} \right)^c \right] \leq -a.$$

Here each K^a_n is a compact subset of H, such that the diameter of K^a_n tends to 0 as n tends to ∞. Then $\tilde{K^a}$ will be sequentially compact in C^H_0 by a diagonal sequence argument. Since C^H_0 is a metric space, $\tilde{K^a}$ will be compact. As we saw in Theorem 2.1, $(T^H_\alpha)^{-1}$ is continuous, so that then $K^a := (T^H_\alpha)^{-1}(\tilde{K^a})$ is compact in $(C^0_\alpha([0,1],H), \|\cdot\|_\alpha)$.

Let $\nu_\varepsilon = \mathbb{P} \circ (\sqrt{\varepsilon} Z)^{-1}$ for a random variable Z on H with $Z \sim \mathcal{N}(0,Q)$. By Lemma 3.4, we can find a sequence of compact sets $(K^a_n)_{n\in\mathbb{N}} \subset H$ such that for all $\varepsilon \in (0,1]$,

$$\nu_\varepsilon((K^a_n)^c) \leq \exp\left(\frac{-(n+1)a}{\varepsilon} \right).$$

To guarantee that the diameter of K^a_n converges to zero, denoting by $\overline{B}(0,d)$ the closed ball of radius d around 0, we set

$$\tilde{K^a} := \prod_{n=0}^{\infty} c_n(\alpha) \left(\overline{B}\left(0, \sqrt{\frac{a(n+1)}{\lambda}}\right) \cap K^a_n \right).$$

Since $c_n(\alpha)\sqrt{a(n+1)/\lambda} \to 0$ as $n \to \infty$, this is a compact set in C^H_0. Thus $K^a := (T^H_\alpha)^{-1}(\tilde{K^a})$ is compact in $(C^0_\alpha([0,1],H), \|\cdot\|_\alpha)$.

Remember that by Lemma 2.1 we have $W = \sum_{n=0}^{\infty} \phi_n Z_n$, where $(Z_n)_{n\geq 0}$ is an i.i.d. sequence of $\mathcal{N}(0,Q)-$ variables. This implies $T_\alpha^H(W) = (c_n(\alpha)Z_n)_{n\geq 0}$ and thus for any $\varepsilon \in (0,1]$

$$\mu_\varepsilon((K^a)^c) = \mathbb{P}\left[\cup_{n\in\mathbb{N}_0}\left\{c_n(\alpha)\sqrt{\varepsilon}Z_n \in \left(c_n(\alpha)\left(B\left(0,\sqrt{\frac{a(n+1)}{\lambda}}\right)\cap K_n^a\right)\right)^c\right\}\right]$$

$$\leq \sum_{n=0}^{\infty}\left(\nu_\varepsilon((K_n^a)^c) + \mathbb{P}\left(\|Z_n\| \geq \sqrt{\frac{a(n+1)}{\varepsilon\lambda}}\right)\right)$$

$$\leq \sum_{n=0}^{\infty}\left(\mathrm{e}^{\frac{-(n+1)a}{\varepsilon}} + c\mathrm{e}^{\frac{-a(n+1)}{\varepsilon}}\right)$$

$$= (1+c)\frac{\mathrm{e}^{\frac{-a}{\varepsilon}}}{1-\mathrm{e}^{\frac{-a}{\varepsilon}}}.$$

So we have

$$\limsup_{\varepsilon\to 0} \varepsilon \log \mu_\varepsilon((K^a)^c) \leq -a.$$

□

We now combine the arguments given so far to obtain an LDP in the Hölder spaces.

Lemma 3.6. $(\mu_\varepsilon)_{\varepsilon\in(0,1]}$ *satisfies an LDP on* $(C_\alpha^0([0,1];H), \|.\|_\alpha)$ *with good rate function* I.

Proof. We know $\|.\|_\alpha' \leq \|.\|_\alpha$. Therefore the $\|.\|_\alpha'$-topology is coarser, which in turn implies that every compact set in the $\|.\|_\alpha$-topology is also a compact set in the $\|.\|_\alpha'$-topology. From Lemma 3.5 we thus obtain that $(\mu_\varepsilon)_{\varepsilon\in(0,1]}$ is also exponentially tight on $(C_\alpha^0([0,1];H), \|.\|_\alpha^0)$.

Proposition 2.3 implies that $(\mu_\varepsilon)_{\varepsilon\in(0,1]}$ satisfies an LDP with good rate function I on $(C_\alpha^0([0,1];H), \|.\|_\alpha^0)$.

Finally we obtain from Proposition 2.5 and from Lemma 3.5 that $(\mu_\varepsilon)_{\varepsilon\in(0,1]}$ satisfies an LDP with good rate function I on $(C_\alpha^0([0,1];H), \|.\|_\alpha)$. □

We may now extend the LDP from $(C_\alpha^0([0,1];H), \|.\|_\alpha)$ to $(C_\alpha([0,1];H), \|.\|_\alpha)$. This is an immediate consequence of the contraction principle (Proposition 2.4), since the inclusion map from $C_\alpha^0([0,1];H)$ to $C_\alpha([0,1];H)$ is continuous. Similarly we can transfer the LDP from $C_\alpha^0([0,1];H)$ to $C([0,1];H)$, the space of continuous functions on $[0,1]$ with values in H, equipped with the uniform norm.

Theorem 3.1. *Let* $(W(t) : t \in [0,1])$ *be a* Q*-Wiener process and for* $\varepsilon \in (0,1]$, *let* μ_ε *be the law of* $\sqrt{\varepsilon}W$. *Then* $(\mu_\varepsilon)_{\varepsilon\in(0,1]}$ *satisfies an LDP on* $(C([0,1];H), \|.\|_\infty)$ *with rate function* I.

Proof. First we can transfer the LDP from $(C_\alpha^0([0,1];H), \|.\|_\alpha)$ to $(C_\alpha^0([0,1];H), \|.\|_\infty)$. This is because on $C_\alpha^0([0,1];H)$, $\|.\|_\infty \leq \|.\|_\alpha$, whence the $\|.\|_\infty$-topology is coarser. Therefore I is a good rate function for the $\|.\|_\infty$-topology as well, and $(\mu_\varepsilon)_{\varepsilon\in(0,1]}$ satisfies an LDP on $(C_\alpha^0([0,1];H), \|.\|_\infty)$ with good rate function I.

The inclusion map from $(C_\alpha^0([0,1];H), \|.\|_\infty)$ to $(C([0,1];H), \|.\|_\infty)$ is continuous, so that an application of the contraction principle (Proposition 2.4) finishes the proof. □

Acknowledgment Nicolas Perkowski is supported by a Ph.D. scholarship of the Berlin Mathematical School.

References

1. Baldi, P., Roynette, B.: Some exact equivalents for the Brownian motion in Hölder norm. Probab. Theory Relat. Fields **93**, 457–484 (1992)
2. Ben Arous, G., Gradinaru, M.: Hölder norms and the support theorem for diffusions. Ann. Inst. H. Poincaré **30**, 415–436 (1994)
3. Ben Arous, G., Ledoux, M.: Grandes déviations de Freidlin-Wentzell en norme Hölderienne. Séminaire de Probabilités **28**, 293–299 (1994)
4. Ciesielski, Z.: On the isomorphisms of the spaces H_α and m. Bull. Acad. Pol. Sci. **8**, 217–222 (1960)
5. Da Prato, G., Zabczyk, J.: Stochastic Equations in Infinite Dimensions. Cambridge University Press, Cambridge (1992)
6. Dembo, A., Zeitouni, O.: Large Deviation Techniques and Applications. Springer, New York (1998)
7. Eddahbi, M., Ouknine, Y.: Large deviations of diffusions on Besov-Orlicz spaces. Bull. Sci. Math. **121**, 573–584 (1997)
8. Eddahbi, M., N'zi, M., Ouknine, Y.: Grandes déviations des diffusions sue les espaces de Besov-Orlicz et application. Stoch. Stoch. Rep. **65**, 299–315 (1999)
9. Freidlin, M., Wentzell, A.: Random Perturbations of Dynamical Systems, 2nd edn. Springer, New York (1998)
10. Galves, A., Olivieri, E., Vares, M.: Metastability for a class of dynamical systems subject to small random perturbations. Ann. Probab. **15**, 1288–1305 (1987)
11. Schilder, M.: Asymptotic formulas for Wiener integrals. Trans. Amer. Math. Soc. **125**, 63–85 (1966)

Chapter 7
Stationary Distributions for Jump Processes with Inert Drift

K. Burdzy, T. Kulczycki, and R.L. Schilling

Dedicated to David Nualart

Abstract We analyze jump processes Z with "inert drift" determined by a "memory" process S. The state space of (Z, S) is the Cartesian product of the unit circle and the real line. We prove that the stationary distribution of (Z, S) is the product of the uniform probability measure and a Gaussian distribution.

Received 4/8/2011; Accepted 9/6/2011; Final 1/30/2012

1 Introduction

We are going to find stationary distributions for jump processes with inert drift. We will first review various sources of inspiration for this project, related models, and

---

K. Burdzy (✉)
Department of Mathematics, University of Washington, Box 354350, Seattle, WA 98195, USA
e-mail: burdzy@math.washington.edu

T. Kulczycki
Institute of Mathematics, Polish Academy of Sciences,
ul. Kopernika 18, 51–617 Wrocław, Poland

Institute of Mathematics and Computer Science, Wrocław University of Technology,
Wybrzeze Wyspianskiego 27, 50–370 Wrocław, Poland
e-mail: t.kulczycki@impan.pl

R.L. Schilling
Institut für Stochastik, TU Dresden, 01062 Dresden, Germany
e-mail: rene.schilling@tu-dresden.de

F. Viens et al. (eds.), *Malliavin Calculus and Stochastic Analysis: A Festschrift in Honor of David Nualart*, Springer Proceedings in Mathematics & Statistics 34,
DOI 10.1007/978-1-4614-5906-4_7,

results. Then we will discuss some technical aspects of the paper that may have independent interest.

This paper is concerned with the following system of stochastic differential equations (the precise statement is in the next section):

$$dY_t = dX_t + W(Y_t)S_t\, dt, \tag{7.1}$$

$$dS_t = W'(Y_t)\, dt, \tag{7.2}$$

where X is a stable Lévy process and W is a C^4 function. This equation is similar to equation [1, (4.1)], driven by Brownian motion, but in Eq. (7.1) the term $\frac{1}{2}(A\nabla V)(X_t)\, dt$ from the first line of [1, (4.1)] is missing. An explanation for this can be found in heuristic calculations in [7, Example 3.7]. The paper [7] deals with Markov processes with finite state spaces and (continuous-space) inert drifts. This class of processes is relatively easy to analyze from the technical point of view. It can be used to generate conjectures, for example, [7, Example 3.7] contains a conjecture about the process defined by Eqs. (7.1) and (7.2). We want to point out that the function W used in the present paper corresponds to W' in [7, Example 3.7]. This means that the assumptions made in the present article are weaker than those in [7] and hence our result is stronger than that conjectured in [7].

The main result of this paper, that is, Theorem 2.2, is concerned with the stationary distribution of a transformation of (Y, S). In order to obtain non-trivial results, we "wrap" Y on the unit circle, so that the state space for the transformed process Y is compact. In other words, we consider $(Z_t, S_t) = (e^{iY_t}, S_t)$. The stationary distribution for (Z_t, S_t) is the product of the uniform distribution on the circle and the normal distribution.

The product form of the stationary distribution for a two-component Markov process is obvious if the two components are independent Markov processes. The product form is far from obvious if the components are not independent, but it does appear in a number of contexts, from queuing theory to mathematical physics. The paper [7] was an attempt to understand this phenomenon for a class of models.

One expects to encounter a Gaussian distribution as (a part of) the stationary distribution in some well-understood situations. First, Gaussian distributions arise in the context of the central limit theorem (CLT) and continuous limits of CLT-based models. Another class of examples of processes with Gaussian stationary measures comes from mathematical physics. The Gibbs measure is given by $c_1 \exp(-c_2 \sum_{i,j}(x_i - x_j)^2)$ in some models, such as the Gaussian free field; see [17]. In such models, the Gaussian nature of the stationary measure arises because the strength of the potential between two elements of the system is proportional to their "distance" (as in Hooke's law for springs) and, therefore, the potential energy is proportional to the square of the distance between two elements. Our model is different in that the square in the exponential function represents the "kinetic energy" (square of the drift magnitude) and not potential energy of a force. The unexpected appearance of the Gaussian distribution in some stationary measures was noticed in [6] before it was explored more deeply in [1,7].

This paper has a companion [8] in which we analyze a related jump process with "memory." In that model, the memory process affects the rate of jumps but it does not add a drift to the jump process. The stationary distribution for that model is also the product of uniform probability measure and a Gaussian distribution.

An ongoing research project of one of the authors is concerned with Markov processes with inert drift when the noise [represented by X in Eq. (7.1)] goes to 0. In other words, one can regard the process (Y, S) as a trajectory of a dynamical system perturbed by a small noise. No matter how small the noise is, the second component of the stationary measure will always be Gaussian. Although we do not study small noise asymptotics in this paper, it is clear from our results that the Gaussian character of the stationary distribution for the perturbed dynamical system does not depend on the Gaussian character of the noise—it holds for the stable noise.

Models of Markov processes with inert drift can represent the motion of an inert particle in a potential, with small noise perturbing the motion. Although such models are related to the Langevin equation [13], they are different. There are several recent papers devoted to similar models; see, e.g., [2–5].

We turn to the technical aspects of the paper. The biggest effort is directed at determining a core of the generator of the process. This is done by showing that the semigroup T_t of the process (Y_t, S_t) preserves C_b^2; see Theorem 3.1. The main idea is based on an estimate of the smoothness of the stochastic flow of solutions to Eqs. (7.1) and (7.2). This result, proved in greater generality than that needed for our main results, is presented in Sect. 3; see Proposition 3.1. This proposition actually makes an assertion on the *pathwise* smoothness of the flow. It seems that Theorem 3.1 and Proposition 3.1 are of independent interest.

We are grateful to the referee for very helpful suggestions.

1.1 Notation

Since the paper uses a large amount of notation, we collect most frequently used symbols in Table 7.1, for easy reference.

2 A Jump Process with a Smooth Drift

Let $\mathbb{S} = \{z \in \mathbb{C} : |z| = 1\}$ be the unit circle in $\mathbb{C}$. Consider a C^4 function $V : \mathbb{S} \to \mathbb{R}$ which is not identically constant and put $W(x) = V(\mathrm{e}^{ix})$, $x \in \mathbb{R}$. Let X_t be a symmetric α-stable Lévy process on $\mathbb{R}$ which has the jump density $\mathcal{A}_\alpha |x - y|^{-1-\alpha}$, $\alpha \in (0, 2)$. Let (Y, S) be a Markov process with the state space $\mathbb{R}^2$ satisfying the following SDE:

$$\begin{cases} \mathrm{d}Y_t = \mathrm{d}X_t + W(Y_t)S_t\,\mathrm{d}t, \\ \mathrm{d}S_t = W'(Y_t)\,\mathrm{d}t. \end{cases} \tag{7.3}$$

Table 7.1 Frequently used notation

$a \vee b, a \wedge b$	$\max(a,b)$, $\min(a,b)$
a_+, a_-	$\max(a,0)$, $-\min(a,0)$
$\lvert x\rvert_{\ell^1}$	$\sum_{j=1}^m \lvert x_j\rvert$ where $x=(x_1,\dots,x_m)\in\mathbb{R}^m$
c	generic constant (without sub- or superscript) which may change its value from line to line
e_k	The kth unit base vector in the usual orthonormal basis for $\mathbb{R}^n$
$\mathcal{A}_\alpha$	$\alpha\Gamma\left(\frac{1+\alpha}{2}\right)\frac{2^{\alpha-1}}{\sqrt{\pi}\,\Gamma(1-\frac{\alpha}{2})}$, $\alpha\in(0,2)$
D^α	$\frac{\partial^{\lvert\alpha\rvert}}{\partial x_1^{\alpha_1}\cdots\partial x_d^{\alpha_d}}$, $\alpha=(\alpha_1,\dots,\alpha_d)\in\mathbb{N}_0^d$
C^k	k-times continuously differentiable functions
C_b^k, C_c^k, C_0^k	Functions in C^k which, together with all their derivatives up to order k, are "bounded," are "compactly supported," and "vanish at infinity," respectively
$\lVert f\rVert_{\infty,B}$	$\sup_{x\in B}\lvert f(x)\rvert$ for $f:\mathbb{R}^n\to\mathbb{R}$
$\lVert D^{(j)}f\rVert_{\infty,B}$	$\sum_{\lvert\alpha\rvert=j}\lVert D^\alpha f\rVert_{\infty,B}$
$\lVert f\rVert_{(j),B}, \lVert f\rVert_{(j)}$	$\sum_{\lvert\alpha\rvert\le j}\sup_{x\in B}\lvert D^\alpha f(x)\rvert$, resp., $\sum_{\lvert\alpha\rvert\le j}\lVert D^\alpha f\rVert_\infty$
$\lVert D^{(j)}V\rVert_{\infty,B}, \lVert D^{(j)}V\rVert_\infty$	$\sum_{\lvert\alpha\rvert=j}\sum_{k=1}^n\sup_{x\in B}\lvert D^\alpha V_k(x)\rvert$, resp., $\sum_{\lvert\alpha\rvert=j}\sum_{k=1}^n\lVert D^\alpha V_k\rVert_\infty$ for any function $V:\mathbb{R}^n\to\mathbb{R}^n$
$\lVert V\rVert_{(j),B}, \lVert V\rVert_{(j)}$	$\sum_{i=0}^j\lVert D^{(i)}V\rVert_{\infty,B}$, resp., $\sum_{i=0}^j\lVert D^{(i)}V\rVert_\infty$
$\mathbb{S}$	$\{z\in\mathbb{C}:\lvert z\rvert=1\}$ unit circle in $\mathbb{C}$

Lemma 2.1. *The SDE* (7.3) *has a unique strong solution which is a strong Markov process with càdlàg paths.*

Proof. For every $n\in\mathbb{N}$ define the function $f_n:\mathbb{R}\to\mathbb{R}$ by $f_n(s):=(-n)\vee s\wedge n$. We consider for fixed $n\in\mathbb{N}$ the following SDE:

$$\begin{cases} \mathrm{d}Y_t^{(n)} = \mathrm{d}X_t + W(Y_t^{(n)})f_n(S_t^{(n)})\,\mathrm{d}t, \\ \mathrm{d}S_t^{(n)} = W'(Y_t^{(n)})\,\mathrm{d}t. \end{cases} \tag{7.4}$$

Note that $\mathbb{R}^2\ni(y,s)\mapsto W(y)f_n(s)$ is a Lipschitz function. By [14, Theorem V.7] and [14, Theorems V.31, V.32] the SDE (7.4) has a unique strong solution which has the strong Markov property and càdlàg paths for every fixed $n\in\mathbb{N}$.

Now fix $t_0<\infty$ and a starting point $\mathbb{R}^2\ni(y,s)=(Y_0^{(n)},S_0^{(n)})$. Note that for any $t\le t_0$ we have

$$\left|S_t^{(n)}\right| = \left|S_0^{(n)} + \int_0^t W'(Y_s^{(n)})\,\mathrm{d}s\right| \le |s| + t_0\lVert W'\rVert_\infty.$$

Pick $n > |s| + t_0 \|W'\|_\infty$, $n \in \mathbb{N}$. For such n and any $t \le t_0$, the process defined by $(Y_t, S_t) := (Y_t^{(n)}, S_t^{(n)})$ is a solution to Eq. (7.3) with starting point (y, s). This shows that for any fixed starting point $(y, s) = (Y_0, S_0)$ and fixed $t_0 < \infty$ the SDE (7.3) has a unique strong solution up to time t_0. The solution is strong Markov and has càdlàg paths. Since $t_0 < \infty$ and the starting point (y, s) are arbitrary, the lemma follows. □

We will now introduce some notation. Let $\mathbb{N}$ be the positive integers and denote by $\mathbb{N}_0 = \mathbb{N} \cup \{0\}$. For any $f : \mathbb{S} \to \mathbb{R}$ we set

$$\tilde{f}(x) := f(e^{ix}), \quad x \in \mathbb{R}.$$

We say that $f : \mathbb{S} \to \mathbb{R}$ is *differentiable* at $z = e^{ix}$, $x \in \mathbb{R}$, if and only if $\tilde{f}$ is differentiable at x and we put

$$f'(z) := (\tilde{f})'(x), \quad \text{where} \quad z = e^{ix}, \quad x \in \mathbb{R}.$$

Analogously, we say that $f : \mathbb{S} \to \mathbb{R}$ is *n times differentiable* at $z = e^{ix}$, $x \in \mathbb{R}$, if and only if $\tilde{f}$ is n times differentiable at x and we write

$$f^{(n)}(z) = (\tilde{f})^{(n)}(x), \quad \text{where} \quad z = e^{ix}, \quad x \in \mathbb{R}.$$

In a similar way we define for $f : \mathbb{S} \times \mathbb{R} \to \mathbb{R}$

$$\tilde{f}(y, s) = f(e^{iy}, s), \quad y, s \in \mathbb{R}. \tag{7.5}$$

We say that $D^\alpha f(z, s)$, $z = e^{iy}$, $y, s \in \mathbb{R}$, $\alpha \in \mathbb{N}_0^2$, exists if and only if $D^\alpha \tilde{f}(y, s)$ exists and we set

$$D^\alpha f(z, s) = D^\alpha \tilde{f}(y, s), \quad \text{where} \quad z = e^{iy}, \quad y, s \in \mathbb{R}.$$

When writing $C^2(\mathbb{S})$, $C_c^2(\mathbb{S} \times \mathbb{R})$, etc., we are referring to the derivatives defined above.

Let

$$Z_t = e^{iY_t}. \tag{7.6}$$

Then (Z, S) is "a symmetric α-stable process with inert drift wrapped on the unit circle." In general, a function of a (strong) Markov process is not any longer a Markov process. We will show that the "wrapped" process $(Z_t, S_t) = (e^{iY_t}, S_t)$ is a strong Markov process because the function $W(x) = V(e^{ix})$ is periodic.

Lemma 2.2. *Let* (Y_t, S_t) *be the solution of the SDE* (7.3). *Then*

$$\mathbb{P}^{(y+2\pi, s)}(Y_t \in A + 2\pi,\ S_t \in B) = \mathbb{P}^{(y,s)}(Y_t \in A,\ S_t \in B)$$

holds for all $(y, s) \in \mathbb{R}^2$ *and all Borel sets* $A, B \subset \mathbb{R}$.

Proof. Denote by (Y_t^y, S_t^s) the unique solution of the SDE (7.3) with initial value $(Y_0^y, S_0^s) = (y, s)$. We assume without loss of generality that $X_0 \equiv 0$. By definition, the process $(Y_t^{y+2\pi}, S_t^s)$ solves

$$\begin{cases} \hat{Y}_t = y + 2\pi + X_t + \int_0^t W(\hat{Y}_r)\hat{S}_r \, dr, \\ \hat{S}_t = s + \int_0^t W'(\hat{Y}_r) \, dr. \end{cases}$$

Since the function W is periodic with period 2π, we have $W(\hat{Y}_r) = W(\hat{Y}_r - 2\pi)$ and $W'(\hat{Y}_r) = W'(\hat{Y}_r - 2\pi)$. Therefore, $(Y_t^{y+2\pi}, S_t^s)$ solves the system

$$\begin{cases} \hat{Y}_t = y + 2\pi + X_t + \int_0^t W(\hat{Y}_r - 2\pi)\hat{S}_r \, dr, \\ \hat{S}_t = s + \int_0^t W'(\hat{Y}_r - 2\pi) \, dr. \end{cases}$$

By subtracting 2π from both sides of the first equation we get

$$\begin{cases} \hat{Y}_t - 2\pi = y + X_t + \int_0^t W(\hat{Y}_r - 2\pi)\hat{S}_r \, dr, \\ \hat{S}_t = s + \int_0^t W'(\hat{Y}_r - 2\pi) \, dr. \end{cases}$$

Since the solutions are unique, this shows that $(Y_t^{y+2\pi}, S_t) = (Y_t^y + 2\pi, S_t)$ from which the claim follows. □

We can now use a rather general result on transformations of the state space due to Dynkin [9, 10.25, Theorem 10.13]; see also Glover [11] and Sharpe [16, Sect. 13].

Corollary 2.1. *Let* $\gamma : \mathbb{R}^2 \to \mathbb{S} \times \mathbb{R}$, $\gamma(y, s) := (e^{iy}, s)$ *and* (Y_t, S_t) *be the unique, càdlàg strong Markov solution of the SDE* (7.3). *Then* $(Z_t, S_t) = (e^{iY_t}, S_t)$ *is also a strong Markov process. Let* $P_t((y, s), A \times B)$ *denote the transition function of* (Y, S) *and* $P_t^{\mathbb{S}}((y, s), A \times B)$ *the transition function of* (Z, S). *Then for* $y, s \in \mathbb{R}$ *and Borel sets* $A, B \subset \mathbb{R}$,

$$P_t^{\mathbb{S}}(\gamma(y, s), A \times B) = P_t((y, s), \gamma^{-1}(A \times B)).$$

Proof. All we have to do is to verify Dynkin's condition [9, 10.25.A] saying that

$$P_t((y, s), \gamma^{-1}(A \times B)) = P_t((y', s'), \gamma^{-1}(A \times B))$$

holds for all Borel sets $A \subset \mathbb{S}$, $B \subset \mathbb{R}$ and all points $(y, s), (y', s') \in \mathbb{R}^2$ such that $\gamma(y, s) = \gamma(y', s')$. Clearly, $s = s'$ and $y - y' = 2j\pi$ for some $j \in \mathbb{Z}$. Denote $f(y) = e^{iy}$. Applying Lemma 2.2 repeatedly we find

$$\begin{aligned}\mathbb{P}^{(y,s)}\left((Y_t,S_t)\in\gamma^{-1}(A\times B)\right) &= \mathbb{P}^{(y,s)}\left(Y_t\in f^{-1}(A),\ S_t\in B\right)\\ &= \mathbb{P}^{(y+2\pi j,s)}\left(Y_t\in f^{-1}(A)+2\pi j,\ S_t\in B\right)\\ &= \mathbb{P}^{(y+2\pi j,s)}\left(Y_t\in f^{-1}(A),\ S_t\in B\right)\\ &= \mathbb{P}^{(y+2\pi j,s)}\left((Y_t,S_t)\in\gamma^{-1}(A\times B)\right). \qquad \square\end{aligned}$$

We are going to calculate the generators of the processes X_t, (Y_t, S_t) and (Z_t, S_t).

By $\mathcal{G}^X$ let us denote the generator of the semigroup, defined on the Banach space $(C_b(\mathbb{R}), \|\cdot\|_\infty)$, of the process X_t. By $\mathcal{D}(\mathcal{G}^X)$ we denote the domain of $\mathcal{G}^X$. It is well known that $C_b^2(\mathbb{R}) \subset \mathcal{D}(\mathcal{G}^X)$ and for $f \in C_b^2(\mathbb{R})$ we have $\mathcal{G}^X f = -(-\Delta)^{\alpha/2} f$, where

$$-(-\Delta)^{\alpha/2} f(x) = \mathcal{A}_\alpha \lim_{\varepsilon\to 0+} \int_{|y-x|>\varepsilon} \frac{f(y)-f(x)}{|x-y|^{1+\alpha}}\, dy, \quad x \in \mathbb{R}.$$

If $f \in C_b^2(\mathbb{R})$ is periodic with period 2π then we have

$$\begin{aligned}-(-\Delta)^{\alpha/2} f(x) &= \mathcal{A}_\alpha \lim_{\varepsilon\to 0+} \int_{\pi>|y-x|>\varepsilon} \frac{f(y)-f(x)}{|x-y|^{1+\alpha}}\, dy\\ &\quad + \mathcal{A}_\alpha \sum_{n\in\mathbb{Z}\setminus\{0\}} \int_{\pi>|y-x|} \frac{f(y)-f(x)}{|x-y+2n\pi|^{1+\alpha}}\, dy.\end{aligned} \tag{7.7}$$

In the sequel we will need the following auxiliary notation:

Definition 2.1.

$$\begin{aligned}C_*(\mathbb{R}^2) &:= \big\{f:\mathbb{R}^2\to\mathbb{R} \;:\; \exists N>0\ \operatorname{supp}(f)\subset\mathbb{R}\times[-N,N],\\ &\qquad f \text{ is bounded and uniformly continuous on } \mathbb{R}^2\big\},\\ C_*^2(\mathbb{R}^2) &:= C_*(\mathbb{R}^2)\cap C_b^2(\mathbb{R}^2).\end{aligned}$$

Let us define the transition semigroup $\{T_t\}_{t\ge 0}$ of the process (Y_t, S_t) by

$$T_t f(y,s) = \mathbb{E}^{(y,s)} f(Y_t, S_t), \quad y, s \in \mathbb{R}, \tag{7.8}$$

for functions $f \in C_b(\mathbb{R}^2)$. Let $\mathcal{G}^{(Y,S)}$ be the generator of $\{T_t\}_{t\ge 0}$ and let $\mathcal{D}(\mathcal{G}^{(Y,S)})$ be the domain of $\mathcal{G}^{(Y,S)}$.

Lemma 2.3. *We have $C_*^2(\mathbb{R}^2) \subset \mathcal{D}(\mathcal{G}^{(Y,S)})$ and for $f \in C_*^2(\mathbb{R}^2)$, and $y, s \in \mathbb{R}$*

$$\mathcal{G}^{(Y,S)} f(y,s) = -(-\Delta_y)^{\alpha/2} f(y,s) + W(y) s f_y(y,s) + W'(y) f_s(y,s), \quad (7.9)$$

Proof. Let $f \in C_*(\mathbb{R}^2)$. Throughout we assume that $\operatorname{supp}(f) \subset \mathbb{R} \times (-M_0, M_0)$ for some $M_0 > 0$. Note that for any starting point $(Y_0, S_0) = (y, s) \in \mathbb{R} \times [-M_0, M_0]$ and all $0 \le t \le 1$,

$$|S_t| = \left| S_0 + \int_0^t W'(Y_r)\,dr \right| \le M_0 + \|W'\|_\infty.$$

Put

$$M_1 = M_0 + \|W'\|_\infty.$$

Note that if $(y,s) \notin \mathbb{R} \times [-M_1, M_1]$ and $(Y_0, S_0) = (y, s)$ then for any $0 \le t \le 1$ we have

$$|S_t| = \left| S_0 + \int_0^t W'(Y_r)\,dr \right| > M_1 - \|W'\|_\infty = M_0,$$

and, therefore, $f(Y_t, S_t) = 0$. It follows that for any $(y,s) \notin \mathbb{R} \times [-M_1, M_1]$ and $0 < h \le 1$ we have

$$\frac{\mathbb{E}^{(y,s)} f(Y_h, S_h) - f(y,s)}{h} = 0.$$

We may, therefore, assume that $(y,s) \in \mathbb{R} \times [-M_1, M_1]$. We will also assume that $0 < h \le 1$.

As above we see that for any starting point $(Y_0, S_0) = (y, s) \in \mathbb{R} \times [-M_1, M_1]$ and all $0 \le t \le 1$ we have $|S_t| \le M_1 + \|W'\|_\infty$. Set $M_2 := M_1 + \|W'\|_\infty$. We assume without loss of generality that $X_0 \equiv 0$. Then

$$Y_t = y + X_t + \int_0^t W(Y_r) S_r\,dr,$$

$$S_t = s + \int_0^t W'(Y_r)\,dr.$$

It follows that

$$\begin{aligned}\frac{T_h f(y,s) - f(y,s)}{h} &= \frac{\mathbb{E}^{(y,s)} f(Y_h, S_h) - f(y,s)}{h} \\ &= \frac{1}{h}\,\mathbb{E}^{(y,s)}[f(Y_h, S_h) - f(Y_h, s)] + \frac{1}{h}\,\mathbb{E}^{(y,s)}[f(Y_h, s) - f(y,s)] \\ &= \mathrm{I} + \mathrm{II}.\end{aligned}$$

Using Taylor's theorem we find

$$\begin{aligned}
\mathrm{I} &= \mathbb{E}^{(y,s)}\left[\frac{1}{h}\frac{\partial f}{\partial s}(Y_h,s)\int_0^h W'(Y_r)\,\mathrm{d}r + \frac{1}{2h}\frac{\partial^2 f}{\partial s^2}(Y_h,\xi)\left(\int_0^h W'(Y_r)\,\mathrm{d}r\right)^2\right] \\
&= \mathbb{E}^{(y,s)}\left[\frac{1}{h}\frac{\partial f}{\partial s}(Y_h,s)\int_0^h W'(y)\,\mathrm{d}r + \frac{1}{h}\frac{\partial f}{\partial s}(Y_h,s)\int_0^h (W'(Y_r)-W'(y))\,\mathrm{d}r\right. \\
&\qquad \left. + \frac{1}{2h}\frac{\partial^2 f}{\partial s^2}(Y_h,\xi)\left(\int_0^h W'(Y_r)\,\mathrm{d}r\right)^2\right],
\end{aligned}$$

where ξ is a point between s and S_h. Note that

$$\begin{aligned}
&\mathbb{E}^{(y,s)}\left[\left|\frac{1}{h}\frac{\partial f}{\partial s}(Y_h,s)\int_0^h (W'(Y_r)-W'(y))\,\mathrm{d}r\right|\right] \\
&\le \mathbb{E}^{(y,s)}\left[\frac{1}{h}\left\|\frac{\partial f}{\partial s}\right\|_\infty \int_0^h\left\{\left(\|W''\|_\infty\left|X_r + \int_0^r W(Y_t)S_t\,\mathrm{d}t\right|\right)\wedge 2\|W'\|_\infty\right\}\mathrm{d}r\right] \\
&\le \left\|\frac{\partial f}{\partial s}\right\|_\infty \mathbb{E}^{(y,s)}\left[\left\{\|W''\|_\infty\left(\sup_{0\le r\le h}|X_r| + h\|W\|_\infty M_2\right)\right\}\wedge 2\|W'\|_\infty\right] \\
&\xrightarrow[h\to 0+]{} 0,
\end{aligned}$$

uniformly for all $(y,s)\in\mathbb{R}\times[-M_1,M_1]$. The convergence follows from the right continuity of X_t and our assumption that $X_0=0$. We also have

$$\mathbb{E}^{(y,s)}\left[\left|\frac{1}{2h}\frac{\partial^2 f}{\partial s^2}(Y_h,\xi)\left(\int_0^h W'(Y_r)\,\mathrm{d}r\right)^2\right|\right] \le \left\|\frac{\partial^2 f}{\partial s^2}\right\|_\infty \frac{h}{2}\|W'\|_\infty^2 \xrightarrow[h\to 0+]{} 0,$$

uniformly for all $(y,s)\in\mathbb{R}\times[-M_1,M_1]$. Because Y_h is right continuous it is easy to see that

$$\mathbb{E}^{(y,s)}\left[\frac{1}{h}\frac{\partial f}{\partial s}(Y_h,s)\int_0^h W'(y)\,\mathrm{d}r\right]\xrightarrow[h\to 0+]{}\frac{\partial f}{\partial s}(y,s)W'(y),$$

uniformly for all $(y,s)\in\mathbb{R}\times[-M_1,M_1]$. It follows that

$$\mathrm{I}\xrightarrow[h\to 0+]{}\frac{\partial f}{\partial s}(y,s)W'(y),$$

uniformly for all $(y,s)\in\mathbb{R}\times[-M_1,M_1]$.

Now let us consider II. We have

$$\begin{aligned}\mathrm{II} &= \frac{1}{h}\,\mathbb{E}^{(y,s)}[f(y+X_h,s)-f(y,s)] + \frac{1}{h}\,\mathbb{E}^{(y,s)}[f(Y_h,s)-f(y+X_h,s)]\\ &= \mathrm{II}_1 + \mathrm{II}_2.\end{aligned}$$

It is well known that

$$\mathrm{II}_1 \xrightarrow[h\to 0+]{} -(-\Delta_y)^{\alpha/2} f(y,s),$$

uniformly for all (y,s). We also have

$$\begin{aligned}\mathrm{II}_2 &= \mathbb{E}^{(y,s)}\left[\frac{1}{h}\frac{\partial f}{\partial y}(y+X_h,s)\int_0^h W(Y_r)S_r\,\mathrm{d}r + \frac{1}{2h}\frac{\partial^2 f}{\partial y^2}(\xi,s)\left(\int_0^h W(Y_r)S_r\,\mathrm{d}r\right)^2\right]\\ &= \mathbb{E}^{(y,s)}\left[\frac{1}{h}\frac{\partial f}{\partial y}(y+X_h,s)\left(\int_0^h W(y)s\,\mathrm{d}r + \int_0^h W(Y_r)(S_r-s)\,\mathrm{d}r\right.\right.\\ &\qquad \left.\left. + \int_0^h (W(Y_r)-W(y))s\,\mathrm{d}r\right) + \frac{1}{2h}\frac{\partial^2 f}{\partial y^2}(\xi,s)\left(\int_0^h W(Y_r)S_r\,\mathrm{d}r\right)^2\right],\end{aligned}$$

where ξ is a point between $y+X_h$ and Y_h. Using similar arguments as above we obtain

$$\mathrm{II}_2 \xrightarrow[h\to 0+]{} \frac{\partial f}{\partial y}(y,s)W(y)s,$$

uniformly for all $(y,s) \in \mathbb{R}\times[-M_1,M_1]$.

It follows that

$$\frac{T_h f(y,s)-f(y,s)}{h} \xrightarrow[h\to 0+]{} -(-\Delta_y)^{\alpha/2} f(y,s) + W(y)s\frac{\partial f}{\partial y}(y,s) + W'(y)\frac{\partial f}{\partial s}(y,s),$$

uniformly for all $(y,s) \in \mathbb{R}\times[-M_1,M_1]$. This means that $f \in \mathcal{D}(\mathcal{G}^{(Y,S)})$ and Eq. (7.9) holds. □

Remark 2.1. A weaker version of Lemma 2.3 can be proved as follows. If we rewrite the SDE (7.3) in the form

$$d\begin{pmatrix}Y_t\\ S_t\end{pmatrix} = \begin{pmatrix}1 & W(Y_t)S_t\\ 0 & W'(Y_t)\end{pmatrix} d\begin{pmatrix}X_t\\ t\end{pmatrix} = \Phi(Y_t,S_t)\,d\begin{pmatrix}X_t\\ t\end{pmatrix}$$

and notice that $(X_t,t)^\top$ is a two-dimensional Lévy process with characteristic exponent $\psi(\xi,\tau) = |\xi|^\alpha + i\tau$, we can use [15, Theorem 3.5, Remark 3.6] to deduce that $C_c^\infty(\mathbb{R}^2) \subset \mathcal{D}(\mathcal{G}^{(Y,S)})$. This argument uses the fact that the SDE has only jumps in the direction of the α-stable process, while it is local in the other direction. Theorem 3.1 of [15] now applies and shows that $\mathcal{G}^{(Y,S)}$ is a pseudo-differential operator $\mathcal{G}^{Y,S}u(x,s) = (2\pi)^{-2}\int_{\mathbb{R}^2} p(x,s;\xi,\tau)\,\mathcal{F}u(\xi,\tau)\,e^{ix\xi+is\tau}\,d\xi\,d\tau$, where $\mathcal{F}$ denotes the Fourier transform, with symbol

$$p(x,s;\xi,\tau) = \psi(\Phi(y,s)^{\top}(\xi,\tau)^{\top}) = |\xi|^{\alpha} + i\xi W(x)s.$$

A Fourier inversion argument now shows that Eq. (7.9) holds for $f \in C_c^{\infty}(\mathbb{R}^2)$ and by a standard closure argument we deduce from this that Eq. (7.9) also holds for $f \in C_0^2(\mathbb{R}^2)$.

We say that $f \in C_0(\mathbb{S}\times\mathbb{R})$ if and only if for every $\varepsilon > 0$ there exists a compact set $K \subset \mathbb{S}\times\mathbb{R}$ such that $|f(u)| < \varepsilon$ for $u \in K^c$. Let us define the semigroup $\{T_t^{\mathbb{S}}\}_{t\geq 0}$ of the process (Z_t, S_t) by

$$T_t^{\mathbb{S}} f(z,s) = \mathbb{E}^{(z,s)} f(Z_t,S_t), \quad z\in\mathbb{S}, \quad s\in\mathbb{R}, \tag{7.10}$$

for f belonging to $C_0(\mathbb{S}\times\mathbb{R})$. Let $z = e^{iy}$, $y \in \mathbb{R}$. For future reference, we note the following consequences of Corollary 2.1:

$$\begin{aligned} T_t^{\mathbb{S}} f(z,s) = \mathbb{E}^{(z,s)} f(Z_t,S_t) = \mathbb{E}^{(y,s)} f(e^{iY_t},S_t) &= \mathbb{E}^{(y,s)} \tilde{f}(Y_t,S_t) \\ &= T_t\tilde{f}(y,s), \end{aligned} \tag{7.11}$$

and

$$\widetilde{T_t^{\mathbb{S}} f}(y,s) = T_t\tilde{f}(y,s). \tag{7.12}$$

By $\operatorname{Arg}(z)$ we denote the argument of $z \in \mathbb{C}$ contained in $(-\pi,\pi]$. For $g \in C^2(\mathbb{S})$ let us put

$$\begin{aligned} Lg(z) = {} & \mathcal{A}_\alpha \lim_{\varepsilon\to 0+}\int_{\mathbb{S}\cap\{|\operatorname{Arg}(w/z)|>\varepsilon\}} \frac{g(w)-g(z)}{|\operatorname{Arg}(w/z)|^{1+\alpha}}\,dw \\ & + \mathcal{A}_\alpha \sum_{n\in\mathbb{Z}\setminus\{0\}} \int_{\mathbb{S}} \frac{g(w)-g(z)}{|\operatorname{Arg}(w/z)+2n\pi|^{1+\alpha}}\,dw, \end{aligned} \tag{7.13}$$

where $\mathcal{A}_\alpha$ is the constant appearing in Eq. (7.7) and dw denotes the arc length measure on $\mathbb{S}$; note that $\int_{\mathbb{S}} dw = 2\pi$.

Let $\mathcal{G}$ be the generator of the semigroup $\{T_t^{\mathbb{S}}\}_{t\geq 0}$ and let $\mathcal{D}(\mathcal{G})$ be its domain.

Lemma 2.4. *We have $C_c^2(\mathbb{S}\times\mathbb{R}) \subset \mathcal{D}(\mathcal{G})$, and for $f \in C_c^2(\mathbb{S}\times\mathbb{R})$,*

$$\mathcal{G}f(z,s) = L_z f(z,s) + V(z)sf_z(z,s) + V'(z)f_s(z,s), \quad z\in\mathbb{S}, \quad s\in\mathbb{R}.$$

Proof. Let $f \in C_c^2(\mathbb{S}\times\mathbb{R})$. Note that $\tilde{f} \in C_*^2(\mathbb{R}^2)$. We obtain from Eq. (7.9), for $z = e^{iy}$, $y, s \in \mathbb{R}$,

$$\lim_{t\to 0+} \frac{T_t^{\$} f(z,s) - f(z,s)}{t} = \lim_{t\to 0+} \frac{T_t \tilde{f}(y,s) - \tilde{f}(y,s)}{t} \\ = -(-\Delta)^{\alpha/2} \tilde{f}(y,s) + W(y) s \tilde{f}_y(y,s) + W'(y) \tilde{f}_s(y,s). \tag{7.14}$$

By Lemma 2.3 this limit exists uniformly in z and s, that is, $f \in \mathcal{D}(\mathcal{G})$.

We get from Eq. (7.7)

$$-(-\Delta_y)^{\alpha/2} \tilde{f}(y,s) = L_z f(z,s). \tag{7.15}$$

Recall that we have $W(y) = V(e^{iy})$, $y \in \mathbb{R}$. Using our definitions we get $V'(z) = W'(y)$ for $z = e^{iy}$, $y \in \mathbb{R}$. Hence Eq. (7.14) equals

$$L_z f(z,s) + V(z) s f_z(z,s) + V'(z) f_s(z,s),$$

which gives the assertion of the lemma. □

We will need the following auxiliary lemma.

Lemma 2.5. *For any $f \in C^2(\mathbb{S})$ we have*

$$\int_{\mathbb{S}} Lf(z)\, dz = 0.$$

Proof. Recall that $\mathrm{Arg}(z)$ denotes the argument of $z \in \mathbb{C}$ belonging to $(-\pi, \pi]$. First we will show that

$$\iint_{\mathbb{S}\times\mathbb{S}} \mathbb{1}_{\{w : |\mathrm{Arg}(w/z)| > \varepsilon\}}(w) \frac{f(w) - f(z)}{|\mathrm{Arg}(w/z)|^{1+\alpha}}\, dw\, dz = 0. \tag{7.16}$$

We interchange the integration variables z and w, use Fubini's theorem, and observe that $|\mathrm{Arg}(z/w)| = |\mathrm{Arg}(w/z)|$,

$$\begin{aligned}
&\iint_{\mathbb{S}\times\mathbb{S}} \mathbb{1}_{\{w : |\mathrm{Arg}(w/z)| > \varepsilon\}}(w) \frac{f(w) - f(z)}{|\mathrm{Arg}(w/z)|^{1+\alpha}}\, dw\, dz \\
&= \iint_{\mathbb{S}\times\mathbb{S}} \mathbb{1}_{\{z : |\mathrm{Arg}(z/w)| > \varepsilon\}}(z) \frac{f(z) - f(w)}{|\mathrm{Arg}(z/w)|^{1+\alpha}}\, dz\, dw \\
&= \iint_{\mathbb{S}\times\mathbb{S}} \mathbb{1}_{\{z : |\mathrm{Arg}(z/w)| > \varepsilon\}}(z) \frac{f(z) - f(w)}{|\mathrm{Arg}(z/w)|^{1+\alpha}}\, dw\, dz \\
&= -\iint_{\mathbb{S}\times\mathbb{S}} \mathbb{1}_{\{w : |\mathrm{Arg}(w/z)| > \varepsilon\}}(w) \frac{f(w) - f(z)}{|\mathrm{Arg}(w/z)|^{1+\alpha}}\, dw\, dz,
\end{aligned}$$

which proves Eq. (7.16).

By interchanging z and w we also get that

$$\begin{aligned}\sum_{n\in\mathbb{Z}\setminus\{0\}} \int_{\mathbb{S}}\int_{\mathbb{S}} \frac{f(w)-f(z)}{|\operatorname{Arg}(w/z)+2n\pi|^{1+\alpha}}\,\mathrm{d}w\,\mathrm{d}z \\ = \sum_{n\in\mathbb{Z}\setminus\{0\}} \int_{\mathbb{S}}\int_{\mathbb{S}} \frac{f(z)-f(w)}{|\operatorname{Arg}(z/w)+2n\pi|^{1+\alpha}}\,\mathrm{d}z\,\mathrm{d}w.\end{aligned} \tag{7.17}$$

Note that for $\operatorname{Arg}(w/z) \neq \pi$ we have $|\operatorname{Arg}(z/w)+2n\pi| = |\operatorname{Arg}(w/z)-2n\pi|$. Hence the expression in Eq. (7.17) equals 0.

Set

$$L_\varepsilon f(z) := \int_{\mathbb{S}\cap\{|\operatorname{Arg}(w/z)|>\varepsilon\}} \frac{f(w)-f(z)}{|\operatorname{Arg}(w/z)|^{1+\alpha}}\,\mathrm{d}w.$$

What is left is to show that

$$\int_{\mathbb{S}} \lim_{\varepsilon\to 0+} L_\varepsilon f(z)\,\mathrm{d}z = \lim_{\varepsilon\to 0+} \int_{\mathbb{S}} L_\varepsilon f(z)\,\mathrm{d}z. \tag{7.18}$$

By the Taylor expansion we have for $f \in C^2(\mathbb{S})$

$$f(w)-f(z) = \operatorname{Arg}(w/z) f'(z) + \operatorname{Arg}^2(w/z) r(w,z), \quad w, z \in \mathbb{S},$$

where $|r(w,z)| \le c(f)$. Hence,

$$\begin{aligned}|L_\varepsilon f(z)| &= \left| \int_{\mathbb{S}\cap\{|\operatorname{Arg}(w/z)|>\varepsilon\}} r(w,z)\operatorname{Arg}^{1-\alpha}(w/z)\,\mathrm{d}w \right| \\ &\le c(f) \int_{\mathbb{S}} |\operatorname{Arg}^{1-\alpha}(w/z)|\,\mathrm{d}w = c(f,\alpha).\end{aligned}$$

Therefore, we get Eq. (7.18) by the bounded convergence theorem. □

We will identify the stationary measure for (Z_t, S_t).

Proposition 2.1. *For $z \in \mathbb{S}$ and $s \in \mathbb{R}$ let*

$$\rho_1(z) \equiv \frac{1}{2\pi}, \quad \rho_2(s) = \frac{1}{\sqrt{2\pi}} \mathrm{e}^{-s^2/2}, \quad \pi(dz, ds) = \rho_1(z)\rho_2(s)\,dz\,ds.$$

Then for any $f \in C_c^2(\mathbb{S}\times\mathbb{R})$ we have

$$\int_{\mathbb{S}}\int_{\mathbb{R}} \mathcal{G} f(z,s)\,\pi(\mathrm{d}z,\mathrm{d}s) = 0.$$

Proof. We have

$$\int_{\mathbb{S}}\int_{\mathbb{R}} \mathcal{G}f(z,s)\,\pi(\mathrm{d}z,\mathrm{d}s)$$
$$= \frac{1}{2\pi}\int_{\mathbb{S}}\int_{\mathbb{R}} \big(L_z f(z,s) + V(z)sf_z(z,s) + V'(z)f_s(z,s)\big)\rho_2(s)\,\mathrm{d}s\,\mathrm{d}z.$$

Integrating by parts, we see that this is equal to

$$\frac{1}{2\pi}\int_{\mathbb{S}}\int_{R} L_z f(z,s)\rho_2(s)\,\mathrm{d}s\,\mathrm{d}z - \frac{1}{2\pi}\int_{\mathbb{S}}\int_{R} V'(z)sf(z,s)\rho_2(s)\,\mathrm{d}s\,\mathrm{d}z$$
$$-\frac{1}{2\pi}\int_{\mathbb{S}}\int_{R} V'(z)f(z,s)\rho_2{}'(s)\,\mathrm{d}s\,\mathrm{d}z = \mathrm{I} + \mathrm{II} + \mathrm{III}.$$

Since $\rho_2{}'(s) = -s\rho_2(s)$ we find that $\mathrm{II} + \mathrm{III} = 0$, while $\mathrm{I} = 0$ by Lemma 2.5. The claim follows. □

Proposition 2.2. *For any $t \geq 0$ we have*

$$T_t^{\mathbb{S}} : C_c^2(\mathbb{S}\times\mathbb{R}) \to C_c^2(\mathbb{S}\times\mathbb{R}).$$

The proof of this proposition is quite difficult. It is deferred to the next section in which we prove this result in much greater generality for solutions of SDEs driven by Lévy processes.

Theorem 2.1. *Let*

$$\pi(dz,ds) = \frac{1}{(2\pi)^{3/2}}\,\mathrm{e}^{-s^2/2}\,dz\,ds, \quad z\in\mathbb{S}, \quad s\in\mathbb{R}. \tag{7.19}$$

Then π is a stationary distribution of the process (Z_t, S_t).

Proof. Let (Y_t, S_t) be a Markov process satisfying the SDE (7.3) and denote by $(Z_t, S_t) = (\mathrm{e}^{iY_t}, S_t)$. Recall that $\{T_t^{\mathbb{S}}\}_{t\geq 0}$ is the semigroup on $C_0(\mathbb{S}\times\mathbb{R})$ defined by Eq. (7.10) and $\mathcal{G}$ is its generator. Let $\mathcal{P}(\mathbb{R}\times\mathbb{R})$ and $\mathcal{P}(\mathbb{S}\times\mathbb{R})$ denote the sets of all probability measures on $\mathbb{R}\times\mathbb{R}$ and $\mathbb{S}\times\mathbb{R}$, respectively. In this proof, for any $\tilde{\mu}\in\mathcal{P}(\mathbb{S}\times\mathbb{R})$, we define $\mu\in\mathcal{P}(\mathbb{R}\times\mathbb{R})$ by $\mu([0,2\pi)\times\mathbb{R}) = 1$ and $\mu(A\times B) = \tilde{\mu}(\mathrm{e}^{iA}\times B)$ for Borel sets $A\subset[0,2\pi)$, $B\subset\mathbb{R}$.

Consider any $\tilde{\mu}\in\mathcal{P}(\mathbb{S}\times\mathbb{R})$ and the corresponding $\mu\in\mathcal{P}(\mathbb{R}\times\mathbb{R})$.

For this μ there exists a Markov process (Y_t, S_t) given by Eq. (7.3) such that (Y_0, S_0) has the distribution μ. It follows that for any $\widetilde{\mu}\in\mathcal{P}(\mathbb{S}\times\mathbb{R})$ there exists a Markov process (Z_t, S_t) given by Eq. (7.3) and $Z_t = \mathrm{e}^{iY_t}$ such that (Z_0, S_0) has the distribution $\widetilde{\mu}$. By Proposition 4.1.7 [10], (Z_t, S_t) is a solution of the martingale problem for $(\mathcal{G}, \widetilde{\mu})$. The Hille–Yosida theorem shows that the assumptions of Theorem 4.4.1 [10] are satisfied if we take $A = A' = \mathcal{G}$. Thus

Theorem 4.4.1 [10] implies that for any $\widetilde{\mu} \in \mathcal{P}(\$ \times \mathbb{R})$, uniqueness holds for the martingale problem for $(\mathcal{G}, \widetilde{\mu})$. Hence the martingale problem for $\mathcal{G}$ is well posed.

Note that $C_c^2(\$ \times \mathbb{R})$ is dense in $C_0(\$ \times \mathbb{R})$, that is, in the set on which the semigroup $\{T_t^{\$}\}_{t\geq 0}$ is defined. It follows from Proposition 2.2 and Proposition 1.3.3 from [10] that $C_c^2(\$ \times \mathbb{R})$ is a core for $\mathcal{G}$. Now using Proposition 2.1 and Proposition 4.9.2 from [10] we get that π is a stationary measure for $\mathcal{G}$. This means that (Z_t, S_t) has a stationary distribution π. □

Theorem 2.2. *The measure π defined in Eq.* (7.19) *is the unique stationary distribution of the process (Z_t, S_t).*

Proof. Step 1. Suppose that for some càdlàg processes X^1 and X^2, processes (Y_t^1, S_t^1) and (Y_t^2, S_t^2) satisfy

$$Y_t^1 = y + X_t^1 + \int_0^t W(Y_r^1) S_r^1 \, dr, \tag{7.20}$$

$$S_t^1 = s + \int_0^t W'(Y_r^1) \, dr, \tag{7.21}$$

$$Y_t^2 = y + X_t^2 + \int_0^t W(Y_r^2) S_r^2 \, dr, \tag{7.22}$$

$$S_t^2 = s + \int_0^t W'(Y_r^2) \, dr. \tag{7.23}$$

Then

$$|S_t^1 - S_t^2| \leq \int_0^t |W'(Y_r^1) - W'(Y_r^2)| \, dr \leq \|W''\|_\infty \int_0^t |Y_r^1 - Y_r^2| \, dr, \tag{7.24}$$

and, therefore, for $t \leq 1$,

$$\begin{aligned}
|Y_t^1 - Y_t^2| &\leq |X_t^1 - X_t^2| + \int_0^t |W(Y_r^1)S_r^1 - W(Y_r^2)S_r^2| \, dr \\
&\leq |X_t^1 - X_t^2| + \int_0^t |W(Y_r^1)(S_r^1 - S_r^2)| \, dr + \int_0^t |(W(Y_r^1) - W(Y_r^2))S_r^2| \, dr \\
&\leq |X_t^1 - X_t^2| + \|W\|_\infty \sup_{0\leq r\leq t} |S_r^1 - S_r^2| t + \|W'\|_\infty \sup_{0\leq r\leq t} |S_r^2| \int_0^t |Y_r^1 - Y_r^2| \, dr \\
&\leq |X_t^1 - X_t^2| + \|W\|_\infty t \, \|W''\|_\infty \int_0^t |Y_r^1 - Y_r^2| \, dr \\
&\quad + \|W'\|_\infty \left(|s| + \|W'\|_\infty t\right) \int_0^t |Y_r^1 - Y_r^2| \, dr
\end{aligned}$$

$$\le |X_t^1 - X_t^2| + (c_1 t + c_2|s|) \int_0^t |Y_r^1 - Y_r^2| \, dr$$

$$\le |X_t^1 - X_t^2| + (c_1 + c_2|s|) \int_0^t |Y_r^1 - Y_r^2| \, dr.$$

By Gronwall's inequality,

$$\sup_{0 \le r \le t} |Y_r^1 - Y_r^2| \le \sup_{0 \le r \le t} |X_r^1 - X_r^2| + \int_0^t |X_r^1 - X_r^2|(c_1 + c_2|s|) \exp\{(c_1 + c_2|s|)\, t\} \, dr$$
$$\le \sup_{0 \le r \le t} |X_r^1 - X_r^2| \big(1 + t(c_1 + c_2|s|) \exp\{(c_1 + c_2|s|)t\}\big).$$

For $t = 1$, the inequality becomes

$$\sup_{0 \le r \le 1} |Y_r^1 - Y_r^2| \le \sup_{0 \le r \le 1} |X_r^1 - X_r^2| \big(1 + (c_1 + c_2|s|) \exp\{(c_1 + c_2|s|)\}\big). \tag{7.25}$$

We substitute Eq. (7.21) into Eq. (7.20) and rearrange terms to obtain

$$X_t^1 = -y + Y_t^1 - \int_0^t W(Y_r^1) \left(s + \int_0^r W'(Y_u^1) \, du \right) dr.$$

We substitute the (non-random) number y for Y_t^1 in the above formula to obtain

$$X_t^1 = -y + y - \int_0^t W(y) \left(s + \int_0^r W'(y) \, du \right) dr$$
$$= -W(y)(ts + t^2 W'(y)/2). \tag{7.26}$$

From now on, X^1 will denote the process defined in Eq. (7.26). It is easy to see that X_t^1 is well defined for all $t \ge 0$. If we substitute this X^1 into Eqs. (7.20) and (7.21) then $Y_t \equiv y$.

It follows from [18, Theorem II, p. 9] that every continuous function is in the support of the distribution of the symmetric α-stable Lévy process on $\mathbb{R}$. We will briefly outline how to derive the last claim from the much more general result in [18, Theorem II, p. 9]. One should take $a(\cdot) \equiv 0$ and $b(\cdot, z) \equiv z$. Note that the "skeleton" functions in [18, (5), p. 9] can have jumps at any times and of any sizes so the closure of the collection of all such functions in the Skorokhod topology contains the set of all continuous functions. Standard arguments then show that every continuous function is in the support of the distribution of the stable process also in the topology of uniform convergence on compact time intervals. We see that if X^1 is the continuous function defined in Eq. (7.26) and X_t^2 is a stable process as in Eq. (7.3) then for every $\varepsilon > 0$ there exists $\delta > 0$ such that

$$\mathbb{P}\left(\sup_{0 \le r \le 1} |X_r^1 - X_r^2| \le \varepsilon \right) \ge \delta.$$

This and Eq. (7.25) show that for any $y, s \in \mathbb{R}$ and $\varepsilon > 0$ there exists $\delta > 0$ such that

$$\mathbb{P}^{y,s}\left(\sup_{0 \le r \le 1} |X_r^1 - X_r^2| \le \varepsilon, \ \sup_{0 \le r \le 1} |Y_r^2 - y| \le \varepsilon \right) \ge \delta.$$

Note that S can change by at most $\|W'\|_\infty$ on any interval of length 1. This, the Markov property, and induction show that for any $\varepsilon > 0$ there exist $\delta_k > 0$, $k \ge 1$, such that

$$\mathbb{P}^{y,s}\left(\sup_{k \le r \le k+1} |X_r^1 - X_r^2| \le 2^{-k}\varepsilon, \ \sup_{k \le r \le k+1} |Y_r^2 - Y_k^2| \le 2^{-k}\varepsilon \right) \ge \delta_k,$$

where X^1 is defined in Eq. (7.26). This implies that for any $\tau < \infty$, $y, s \in \mathbb{R}$ and $\varepsilon > 0$ there exists $\delta' > 0$ such that

$$\mathbb{P}^{y,s}\left(\sup_{0 \le r \le \tau} |X_r^1 - X_r^2| \le 2\varepsilon, \ \sup_{0 \le r \le \tau} |Y_r^2 - y| \le 2\varepsilon \right) \ge \delta'. \tag{7.27}$$

Step 2. Recall that V is not identically constant. This and the fact that $V \in C^4$ easily imply that W' is strictly positive on some interval and it is strictly negative on some other interval. We fix some $a_1, a_2 \in (-\pi, \pi)$, $b_1 > 0$, $b_2 < 0$, and $\varepsilon_0 \in (0, \pi/100)$, such that $V'(z) > b_1$ for $z \in \mathbb{S}$, $\operatorname{Arg}(z) \in [a_1 - 4\varepsilon_0, a_1 + 4\varepsilon_0]$, and $V'(z) < b_2$ for $z \in \mathbb{S}$, $\operatorname{Arg}(z) \in [a_2 - 4\varepsilon_0, a_2 + 4\varepsilon_0]$.

Suppose that there exist two stationary probability distributions π and $\widehat{\pi}$ for (Z, S). Let $((Z_t, S_t))_{t \ge 0}$ and $((\widehat{Z}_t, \widehat{S}_t))_{t \ge 0}$ be processes with (Z_0, S_0) and $(\widehat{Z}_0, \widehat{S}_0)$ distributed according to π and $\widehat{\pi}$, respectively. The transition probabilities for these processes are the same as for the processes defined by Eqs. (7.3) and (7.6). Let X denote the driving stable Lévy process for Z.

Let A be an open set such that $W'(y) > c > 0$ for all $y \in A$. In view of the relationship between V and W, we can assume that A is periodic, that is, $y \in A$ if and only if $y + 2\pi \in A$. It follows easily from Eq. (7.3) that there exist $q_1 > 0$ and $s_1 < \infty$ such that for any (Y_0, S_0), the process Y enters A at some random time $T_1 \le s_1$ with probability greater than q_1. Since Y is right continuous, if $Y_{T_1} \in A$, then Y_t stays in A for all t in some interval (T_1, T_2), with $T_2 \le 2s_1$. Then Eq. (7.3) implies that $S_t \ne 0$ for some $t \in (T_1, T_2)$. A repeated application of the Markov property at the times $2s_1, 4s_1, 6s_1, \ldots$ shows that the probability that $S_t = 0$ for all $t \le 2ks_1$ is less than $(1-q_1)^k$. Letting $k \to \infty$, we see that $S_t \neq 0$ for some $t > 0$, a.s.

Suppose without loss of generality that there exist $\varepsilon_1 > 0$, $t_2 > 0$, and $p_1 > 0$ such that $\mathbb{P}^\pi(S_{t_2} > \varepsilon_1) > p_1$. Let $F_1 = \{S_{t_2} > \varepsilon_1\}$ and $t_3 = \varepsilon_1/(2\|W'\|_\infty)$. It is easy to see that for some $p_2 > 0$,

$$\mathbb{P}^\pi\left(\exists t \in [t_2, t_2+t_3] \ : \ \mathrm{Arg}(Z_t) \in [a_2 - \varepsilon_0, a_2 + \varepsilon_0] \,\big|\, F_1\right) > p_2.$$

This implies that there exist $\varepsilon_1 > 0$, $t_2 > 0$, $t_4 \in [t_2, t_2+t_3]$, and $p_3 > 0$ such that

$$\mathbb{P}^\pi(S_{t_2} > \varepsilon_1, \mathrm{Arg}(Z_{t_4}) \in [a_2 - 2\varepsilon_0, a_2 + 2\varepsilon_0]) > p_3.$$

Note that $|S_{t_4} - S_{t_2}| \le \|W'\|_\infty t_3 < \varepsilon_1/2$. Hence,

$$\mathbb{P}^\pi(S_{t_4} > \varepsilon_1/2, \mathrm{Arg}(Z_{t_4}) \in [a_2 - 2\varepsilon_0, a_2 + 2\varepsilon_0]) > p_3.$$

Let $\varepsilon_2 \in (\varepsilon_1/2, \infty)$ be such that

$$\mathbb{P}^\pi(S_{t_4} \in [\varepsilon_1/2, \varepsilon_2], \mathrm{Arg}(Z_{t_4}) \in [a_2 - 2\varepsilon_0, a_2 + 2\varepsilon_0]) > p_3/2.$$

Let $t_5 = 2\varepsilon_2/|b_2|$ and $t_6 = t_4 + t_5$. By Eq. (7.27), for any $\varepsilon_3 > 0$ and some $p_4 > 0$,

$$\mathbb{P}^\pi\Big(\sup_{t_4 \le r \le t_6} |X_r^1 - X_r| \le \varepsilon_3, S_{t_4} \in [\varepsilon_1/2, \varepsilon_2],$$
$$\mathrm{Arg}(Z_t) \in [a_2 - 3\varepsilon_0, a_2 + 3\varepsilon_0] \ \text{ for all } \ t \in [t_4, t_6]\Big) > p_4,$$

where X^1 is the function defined in Eq. (7.26). Observe that $V'(z) < b_2 < 0$ for $\mathrm{Arg}\, z \in [a_2 - 3\varepsilon_0, a_2 + 3\varepsilon_0]$, if the event in the last formula holds, then

$$S_{t_6} = S_{t_4} + \int_{t_4}^{t_6} V'(Z_s)\,\mathrm{d}s \le \varepsilon_2 + b_2 t_5 \le -\varepsilon_2.$$

This implies that

$$\mathbb{P}^\pi\Big(\sup_{t_4 \le r \le t_6} |X_r^1 - X_r| \le \varepsilon_3, S_{t_4} \ge \varepsilon_1/2, S_{t_6} \le -\varepsilon_2\Big) > p_4. \tag{7.28}$$

Step 3. By the Lévy–Itô representation we can write the stable Lévy process X in the form $X_t = J_t + \widetilde{X}_t$, where J is a compound Poisson process comprising all jumps of X which are greater than ε_0 and $\widetilde{X} = X - J$ is an independent Lévy process (accounting for all small jumps of X). Denote $\lambda = \lambda(\alpha, \varepsilon_0)$ the rate of the compound Poisson process J.

Let $(\widetilde{Y}, \widetilde{S})$ be the solution to Eq. (7.3), with X_t replaced by $\widetilde{X}_t$ for $t \geq t_4$. Take $\varepsilon_3 < \varepsilon_0/2$. Then $\sup_{t_4 \leq r \leq t_6} |X_r^1 - \widetilde{X}_r| \leq \varepsilon_3$ entails that $\sup_{t_4 \leq r \leq t_6} |J_{t_4} - J_r| = 0$. Thus, Eq. (7.28) becomes

$$\begin{aligned}
&\mathbb{P}^{\pi}\Big(\sup_{t_4 \leq r \leq t_6} |X_r^1 - \widetilde{X}_r| \leq \varepsilon_3,\ \widetilde{S}_{t_4} \geq \tfrac{\varepsilon_1}{2},\ \widetilde{S}_{t_6} \leq -\varepsilon_2\Big) \\
&\quad \geq \mathbb{P}^{\pi}\Big(\sup_{t_4 \leq r \leq t_6} |X_r^1 - \widetilde{X}_r| \leq \varepsilon_3,\ \sup_{t_4 \leq r \leq t_6} |J_{t_4} - J_r| = 0,\ \widetilde{S}_{t_4} \geq \tfrac{\varepsilon_1}{2},\ \widetilde{S}_{t_6} \leq -\varepsilon_2\Big) \\
&\quad > p_4 > 0.
\end{aligned}$$

Let τ be the time of the first jump of J in the interval $[t_4, t_6]$; we set $\tau = t_6$ if there is no such jump. We can represent $\{(Y_t, S_t), 0 \leq t \leq \tau\}$ in the following way: $(Y_t, S_t) = (\widetilde{Y}_t, \widetilde{S}_t)$ for $0 \leq t < \tau$, $S_\tau = \widetilde{S}_\tau$, and $Y_\tau = \widetilde{Y}_\tau + J_\tau - J_{\tau-}$.

We say that a non-negative measure μ_1 is a component of a non-negative measure μ_2 if $\mu_2 = \mu_1 + \mu_3$ for some non-negative measure μ_3. Let $\mu(dz, ds) = \mathbb{P}^{\pi}(Z_\tau \in dz, S_\tau \in ds)$. We will argue that $\mu(dz, ds)$ has a component with a density bounded below by $c_2 > 0$ on $\mathbb{S} \times (-\varepsilon_2, \varepsilon_1/2)$. We find for every Borel set $A \subset \mathbb{S}$ of arc length $|A|$ and every interval $(s_1, s_2) \subset (-\varepsilon_2, \varepsilon_1/2)$

$$\begin{aligned}
&\mu(A \times (s_1, s_2)) \\
&\quad = \mathbb{P}^{\pi}\left(Z_\tau \in A,\ S_\tau \in (s_1, s_2)\right) \\
&\quad \geq \mathbb{P}^{\pi}\Big(Z_\tau \in A, S_\tau \in (s_1, s_2),\ \sup_{t_4 \leq r \leq t_6} |X_r^1 - \widetilde{X}_r| \leq \varepsilon_3,\ \widetilde{S}_{t_4} \geq \tfrac{\varepsilon_1}{2},\ \widetilde{S}_{t_6} \leq -\varepsilon_2\Big) \\
&\quad \geq \mathbb{P}^{\pi}\Big(\mathrm{e}^{i(J_\tau - J_{\tau-})} \in \mathrm{e}^{-i\widetilde{X}_{\tau-}}A,\ \widetilde{S}_\tau \in (s_1, s_2), \\
&\qquad \sup_{t_4 \leq r \leq t_6} |X_r^1 - \widetilde{X}_r| \leq \varepsilon_3, \widetilde{S}_{t_4} \geq \varepsilon_1/2,\ \widetilde{S}_{t_6} \leq -\varepsilon_2,\ N^J = 1\Big).
\end{aligned}$$

Here N^J counts the number of jumps of the process J occurring during the interval $[t_4, t_6]$. Without loss of generality we can assume that $\varepsilon_0 < 2\pi$. In this case the density of the jump measure of J is bounded below by $c_3 > 0$ on $(2\pi, 4\pi)$. Observe that the processes $(\widetilde{X}, \widetilde{S})$ and J are independent. Conditional on $\{N^J = 1\}$, τ is uniformly distributed on $[t_4, t_6]$, and the probability of the event $\{N^J = 1\}$ is $\lambda(t_6 - t_4)\mathrm{e}^{-\lambda(t_6 - t_4)}$. Thus,

$$\begin{aligned}
&\mu(A \times (s_1, s_2)) \\
&\geq c_3|A|\,\mathbb{P}^{\pi}\Big(\widetilde{S}_\tau \in (s_1, s_2)\Big|\ \sup_{t_4 \leq r \leq t_6} |X_r^1 - \widetilde{X}_r| \leq \varepsilon_3, \widetilde{S}_{t_4} \geq \varepsilon_1/2, \widetilde{S}_{t_6} \leq -\varepsilon_2, N^J = 1\Big) \\
&\quad \times p_4 \cdot \lambda(t_6 - t_4)\mathrm{e}^{-\lambda(t_6 - t_4)}.
\end{aligned}$$

Since the process $\widetilde{S}$ spends at least $(s_2 - s_1)/\|W'\|_\infty$ units of time in (s_1, s_2) we finally arrive at

$$\mu(A, (s_1, s_2)) \geq p_4 \lambda e^{-\lambda(t_6 - t_4)} c_3 |A| (s_2 - s_1)/\|W'\|_\infty.$$

This proves that $\mu(dz, ds)$ has a component with a density bounded below by the constant $c_2 = p_4 \lambda e^{-\lambda(t_6-t_4)} c_3/\|W'\|_\infty$ on $\mathbb{S} \times (-\varepsilon_2, \varepsilon_1/2)$.

Step 4. Let $\varepsilon_4 = \varepsilon_1/2 \wedge \varepsilon_2 > 0$. We have shown that for some stopping time τ, $\mathbb{P}^\pi(Z_\tau \in dz, S_\tau \in ds)$ has a component with a density bounded below by $c_2 > 0$ on $\mathbb{S} \times (-\varepsilon_4, \varepsilon_4)$. We can prove in an analogous way that for some stopping time $\widehat{\tau}$ and $\widehat{\varepsilon}_4 > 0$, $\mathbb{P}^{\widehat{\pi}}(\widehat{Z}_{\widehat{\tau}} \in dz, \widehat{S}_{\widehat{\tau}} \in ds)$ has a component with a density bounded below by $\widehat{c}_2 > 0$ on $\mathbb{S} \times (-\widehat{\varepsilon}_4, \widehat{\varepsilon}_4)$.

Since $\pi \neq \widehat{\pi}$, there is a Borel set $A \subset \mathbb{S} \times \mathbb{R}$ such that $\pi(A) \neq \widehat{\pi}(A)$. Moreover, since any two stationary probability measures are either mutually singular or identical (cf. [19, Chap. 2, Theorem 4]), we have $\pi(A) > 0$ and $\widehat{\pi}(A) = 0$ for some A. By the strong Markov property applied at τ and the ergodic theorem (see [19, Chap. 1, p. 12]), we have $\mathbb{P}^\pi$-a.s.

$$\lim_{t\to\infty} (1/t) \int_\tau^t \mathbb{1}_{\{(Z_s, S_s) \in A\}} \, ds = \pi(A) > 0.$$

Similarly, we see that $\mathbb{P}^{\widehat{\pi}}$-a.s.

$$\lim_{t\to\infty} (1/t) \int_{\widehat{\tau}}^t \mathbb{1}_{\{(\widehat{Z}_s, \widehat{S}_s) \in A\}} \, ds = \widehat{\pi}(A) = 0.$$

Since the distributions of (Z_τ, S_τ) and $(\widehat{Z}_{\widehat{\tau}}, \widehat{S}_{\widehat{\tau}})$ have mutually absolutely continuous components, the last two statements contradict each other. This shows that we must have $\pi = \widehat{\pi}$. □

Remark 2.2. It is not hard to show that Theorem 2.1 holds even if we take $\alpha = 2$ in Eq. (7.3), that is, if X_t is Brownian motion. It seems that for $\alpha = 2$ uniqueness of the stationary distribution can be proved using techniques employed in Proposition 4.8 in [1]. A close inspection of the proofs in this section reveals that our results remain also valid if X_t is a symmetric Lévy process with jump measure having full support.

3 Smoothness of $T_t f$

In this section, we will show that if $f \in C_b^2$ then $T_t f \in C_b^2$ where $\{T_t\}_{t \geq 0}$ is the semigroup of a process defined by a stochastic differential equation driven by a Lévy process. We use this result to show Proposition 2.2, but it may well be of independent interest. We found some related results in the literature, but none of them was sufficiently strong for our purposes. The key element of the proof is explicit bounds for derivatives of the flow of solutions to the SDE. This is done in Proposition 3.1. We provide a direct and elementary proof of this proposition. Note that our bounds are non-random and do not depend on the sample path. This is a new feature in this type of analysis since usually (see, e.g., Kunita [12]), the constants are random since they are derived with the Kolmogorov–Chentsov–Totoki lemma or a Borel–Cantelli argument. Let us, however, point out that there is an alternative way of proving Proposition 3.1. It is possible to use [14, Theorems V.39, V.40] and [14, formula (D), p. 305] to obtain bounds for derivatives of the flow. Since this alternative approach demands similar arguments and is not shorter than our proof of Proposition 3.1, we decided to prove Proposition 3.1 directly.

Consider the following system of stochastic differential equations in $\mathbb{R}^n$:

$$\begin{cases} \mathrm{d}Y_1(t) = \mathrm{d}X_1(t) + \mathcal{V}_1(Y(t))\,\mathrm{d}t, \\ \quad\vdots \\ \mathrm{d}Y_n(t) = \mathrm{d}X_n(t) + \mathcal{V}_n(Y(t))\,\mathrm{d}t, \end{cases} \tag{7.29}$$

where $Y(t) = (Y_1(t), \dots, Y_n(t)) \in \mathbb{R}^n$, $X(t) = (X_1(t), \dots, X_n(t)) \in \mathbb{R}^n$. We assume that $X(0) = 0$, $X_1, \dots, X_n$ are Lévy processes on $\mathbb{R}$ and $\mathcal{V}_i : \mathbb{R}^n \to \mathbb{R}$ are locally Lipschitz. We allow $X_1, \dots, X_n$ to be degenerate, that is, some or all X_i may be identically equal to 0.

By [14, Theorem V.38] it follows that if $Y(0) = x$ then there exists a stopping time $\zeta(x,\omega) : \mathbb{R}^n \times \Omega \to [0,\infty]$ and there exists a unique solution of Eq. (7.29) with $Y(0) = x$ with $\limsup_{t\to\zeta(x,\cdot)} |Y(t)| = \infty$ a.s. on $\zeta < \infty$; ζ is called the *explosion time*. In order to apply [14, Theorem V.38] we take in the equations marked ($\otimes$) in [14, p. 302] $m = n+1$, $X_t^i = Y_i(t)$, $x^i = Y_i(0)$, $Z_t^\alpha = X_\alpha(t)$ for $\alpha \in \{1,\dots,n\}$, $Z_t^{n+1} = t$, and $f_\alpha^i = \delta_{\alpha i}$ for $\alpha, i \in \{1,\dots,n\}$ and $f_{n+1}^i(x) = \mathcal{V}_i(x)$ for $i \in \{1,\dots,n\}$.

By $Y^x(t)$ we denote the process with starting point $Y^x(0) = x$. In the rest of this section, we will assume that Eq. (7.29) holds not only a.s. but for all $\omega \in \Omega$. More precisely, we can and will assume that the solution to Eq. (7.29) is constructed on a probability space Ω such that $X(0) = 0$ and

$$Y^x(t) = x + X(t) + \int_0^t \mathcal{V}(Y(s))\,\mathrm{d}s,$$

for all $t \ge 0$ and all $\omega \in \Omega$.

Set

$$\|x\| = \max\{|x_1|, \dots, |x_n|\}, \quad x = (x_1, \dots, x_n),$$

and

$$B^*(x,r) = \{y \in \mathbb{R}^n : \|y - x\| < r\}, \quad x \in \mathbb{R}^n,\ r > 0.$$

For $f : \mathbb{R}^n \to \mathbb{R}$ and $A \subset \mathbb{R}^n$ we write $D^{(1)} f = \nabla f$,

$$\|f\|_{\infty,A} = \sup_{x\in A}|f(x)|,\ \|D^{(j)}f\|_{\infty,A} = \sum_{|\alpha|=j}\sup_{x\in A}|D^{\alpha}f(x)|,$$

$$\|f\|_{(j),A} = \|f\|_{\infty,A} + \|D^{(1)}f\|_{\infty,A} + \ldots + \|D^{(j)}f\|_{\infty,A}.$$

When $A = \mathbb{R}^n$ we drop A from this notation. For $\mathcal{V} = (\mathcal{V}_1, \ldots, \mathcal{V}_n)$ from Eq. (7.29) and $A \subset \mathbb{R}^n$ we put

$$\|\mathcal{V}\|_{\infty,A} = \sum_{i=1}^{n}\|\mathcal{V}_i\|_{\infty,A}, \quad \|D^{(j)}\mathcal{V}\|_{\infty,A} = \sum_{i=1}^{n}\|D^{(j)}\mathcal{V}_i\|_{\infty,A},$$

$$\|\mathcal{V}\|_{(j),A} = \|\mathcal{V}\|_{\infty,A} + \|D^{(1)}\mathcal{V}\|_{\infty,A} + \ldots + \|D^{(j)}\mathcal{V}\|_{\infty,A}.$$

For $f : \mathbb{R}^n \to \mathbb{R}$, $x \in \mathbb{R}^n$ and $0 \le t < \infty$ we define the operator T_t by

$$T_t f(x) = \mathbb{E}\left[f(Y^x(t)); t < \zeta(x)\right]. \tag{7.30}$$

Before formulating the results for the process $Y(t)$ let us go back for a moment to the original problem (7.3), that is,

$$\begin{cases} \mathrm{d}Y_t = \mathrm{d}X_t + W(Y_t)S_t\,\mathrm{d}t, \\ \mathrm{d}S_t = W'(Y_t)\,\mathrm{d}t. \end{cases}$$

This SDE is of type Eq. (7.29) because we can rewrite it as

$$\begin{cases} \mathrm{d}Y_1(t) = \mathrm{d}X_1(t) + \mathcal{V}_1(Y(t))\,\mathrm{d}t, \\ \mathrm{d}Y_2(t) = \mathrm{d}X_2(t) + \mathcal{V}_2(Y(t))\,\mathrm{d}t, \end{cases} \tag{7.31}$$

where $X_1(t) = X_t$ is a symmetric α-stable Lévy process on $\mathbb{R}$, with index $\alpha \in (0, 2)$, $X_2(t) \equiv 0$, $\mathcal{V}_1(y_1, y_2) = W(y_1)y_2$, $\mathcal{V}_2(y_1, y_2) = W'(y_1)$. By Lemma 2.1 there exists a unique solution to this SDE and the explosion time for this process is infinite a.s. We want to show that $T_t f \in C_b^2$ whenever $f \in C_b^2$. Our proof of Theorem 3.1 requires that $\mathcal{V}_i$ and its derivatives up to order 3 are bounded. However, $\mathcal{V}_1(y_1, y_2) = W(y_1)y_2$ is not bounded on $\mathbb{R}^2$. We will circumvent this difficulty by proving in Proposition 3.2 that $T_t f \in C_*^2(\mathbb{R}^2)$ whenever $f \in C_*^2(\mathbb{R}^2)$, where $C_*^2(\mathbb{R}^2)$ is given by Definition 2.1.

Let us briefly discuss the reasons that made us choose this particular set of functions, $C_*^2(\mathbb{R}^2)$. This discussion gives also an explanation for the specific assumptions in the main result of this section, Theorem 3.1.

Assume that $f \in C^2(\mathbb{R}^2)$ and $\operatorname{supp} f \subset K_0 = \mathbb{R} \times [-r, r]$, $r > 0$. Fix $t_0 < \infty$. If $|s| = |S_0| > r + t_0\|W'\|_\infty$ then for $t \le t_0$,

$$\left|S_t^{(y,s)}\right| = \left|s + \int_0^t W'(Y_u^{(y,s)})\,\mathrm{d}u\right| > r$$

and, therefore,

$$T_t f(y,s) = \mathbb{E} f\big(Y_t^{(y,s)}, S_t^{(y,s)}\big) = 0.$$

It follows that if $t \le t_0$ then

$$\operatorname{supp}(T_t f) \subset K = \mathbb{R} \times \big[-r - t_0\|W'\|_\infty,\, r + t_0\|W'\|_\infty\big]. \tag{7.32}$$

For technical reasons, we enlarge K as follows:

$$K_3 = \mathbb{R} \times \big(-r - t_0\|W'\|_\infty - 3,\, r + t_0\|W'\|_\infty + 3\big).$$

In view of Eq. (7.32), we have to consider only starting points $(y,s) \in K$ in order to prove that $T_t f \in C_*^2(\mathbb{R}^2)$. Note that for the starting point $(y,s) \in K_3$ and $t \le t_0$ we have

$$\left|S_t^{(y,s)}\right| = \left|s + \int_0^t W'(Y_u^{(y,s)})\,\mathrm{d}u\right| \le r + 2t_0\|W'\|_\infty + 3.$$

Thus, for all starting points $(y,s) \in K_3$ and $t \le t_0$,

$$\big(Y_t^{(y,s)}, S_t^{(y,s)}\big) \in M := \mathbb{R} \times \big[-r - 2t_0\|W'\|_\infty - 3,\, r + 2t_0\|W'\|_\infty + 3\big]. \tag{7.33}$$

But the function $\mathcal{V}_1(y_1,y_2) = W(y_1)y_2$ is bounded on M. Using our assumptions on W, namely, periodicity of W and $W \in C^4$, we obtain also that the derivatives of $\mathcal{V}_1(y_1,y_2) = W(y_1)y_2$ up to order 3 are bounded on M.

Now we return to the general process $Y(t)$. Let us formulate the main result for this process.

Theorem 3.1. *Let $f : \mathbb{R}^n \to \mathbb{R}$ be a function in C_b^2. Fix $0 < t_0 < \infty$. Let $Y^x(t)$ be a solution of Eq. (7.29). Assume that the explosion time $\zeta(x,\omega) \equiv \infty$ for all $x \in \mathbb{R}^n$ and all $\omega \in \Omega$. Let $T_t f$ be defined by Eq. (7.30). Assume that $K \subset \mathbb{R}^n$, for every $t \le t_0$ $\operatorname{supp}(T_t f) \subset K$, and that there exists a convex set $M \subset \mathbb{R}^n$ such that $Y^x(t,\omega) \in M$ for all $x \in K_3 := \bigcup_{x \in K} B^*(x,3)$, $t \le t_0$, and $\omega \in \Omega$. Assume that $\|\mathcal{V}\|_{\infty,M} < \infty$ and $\|D^{(j)}\mathcal{V}\|_{\infty,M} < \infty$ for $j = 1,2,3$. Then we have*

$$T_t f \in C_b^2 \quad \textit{for all} \quad t \le t_0.$$

Remark 3.1. When $\|\mathcal{V}\|_{(3)} < \infty$ (i.e., when the assumptions of Theorem 3.1 hold with $K = M = \mathbb{R}^n$) then the above theorem implies that we have for any $f \in C_b^2$

$$T_t f \in C_b^2 \quad \text{for all} \quad t > 0.$$

The first step in proving Theorem 3.1 will be the following proposition.

Proposition 3.1. *Fix $0 < t_0 < \infty$. Let $Y^x(t)$ be a solution of Eq. (7.29). Assume that the explosion time $\zeta(x,\omega) \equiv \infty$ for all $x \in \mathbb{R}^n$ and all $\omega \in \Omega$. Let $K \subset \mathbb{R}^n$.*

Assume that there exists a convex set $M \subset \mathbb{R}^n$ such that $Y^x(t,\omega) \in M$ for all $x \in K_3 := \bigcup_{x\in K} B^*(x,3)$, $t \le t_0$, and $\omega \in \Omega$. Assume that $\|\mathcal{V}\|_{(3),M} < \infty$. Put

$$\tau := \frac{1}{2\,\|D^{(1)}\mathcal{V}\|_{\infty,M}} \wedge t_0, \qquad \left(\frac{1}{0} := \infty\right). \tag{7.34}$$

For every $\omega \in \Omega$ we have the following:

(i) *For all $0 < t \le \tau$, $x \in K_2 = \bigcup_{x\in K} B^*(x,2)$, $h \in \mathbb{R}^n$, $\|h\| < 1$,*

$$\|Y^{x+h}(t,\omega) - Y^x(t,\omega)\| \le 2\|h\|. \tag{7.35}$$

(ii) *Recall that e_i is the ith unit vector in the usual orthonormal basis for $\mathbb{R}^n$. For all $0 < t \le \tau$, $x \in K_2$, $i \in \{1,\dots,n\}$,*

$$D_i Y^x(t,\omega) := \lim_{u\to 0} \frac{Y^{x+ue_i}(t,\omega) - Y^x(t,\omega)}{u}$$

exists and

$$\|D_i Y^x(t,\omega)\| \le 2. \tag{7.36}$$

We will write $D_i Y^x(t,\omega) = (D_i Y_1^x(t,\omega),\dots,D_i Y_n^x(t,\omega))$.

(iii) *For all $0 < t \le \tau$, $x \in K_1 = \bigcup_{x\in K} B^*(x,1)$, $h \in \mathbb{R}^n$, $\|h\| < 1$, and any $i \in \{1,\dots,n\}$,*

$$\|D_i Y^{x+h}(t,\omega) - D_i Y^x(t,\omega)\| \le 8\,\|D^{(2)}\mathcal{V}\|_{\infty,M}\,\tau\,\|h\|. \tag{7.37}$$

(iv) *For all $0 < t \le \tau$, $x \in K_1$, $i,k \in \{1,\dots,n\}$,*

$$D_{ik} Y^x(t,\omega) := \lim_{u\to 0} \frac{D_i Y^{x+ue_k}(t,\omega) - D_i Y^x(t,\omega)}{u}$$

exists and

$$\|D_{ik} Y^x(t,\omega)\| \le 8\,\|D^{(2)}\mathcal{V}\|_{\infty,M}\,\tau. \tag{7.38}$$

We will write $D_{ik} Y^x(t,\omega) = (D_{ik} Y_1^x(t,\omega),\dots,D_{ik} Y_n^x(t,\omega))$.

(v) *For all $0 < t \le \tau$, $x \in K$, $h \in \mathbb{R}^n$, $\|h\| < 1$, $i,k \in \{1,\dots,n\}$,*

$$\begin{aligned}&\|D_{ik} Y^{x+h}(t,\omega) - D_{ik} Y^x(t,\omega)\|\\ &\quad \le 96\,\|D^{(2)}\mathcal{V}\|_{\infty,M}^2\,\tau^2\,\|h\| + 16\,\|D^{(3)}\mathcal{V}\|_{\infty,M}\,\tau\,\|h\|.\end{aligned}$$

Remark 3.2. The existence of $D_i Y^x(t)$ and $D_{ik} Y^x(t)$ follows from [14, Theorem V.40]. What is new here are the explicit bounds for $D_i Y^x(t)$ and $D_{ik} Y^x(t)$ which are needed in the proof of Theorem 3.1; see Lemma 3.1. The proof of Proposition 3.1 is self-contained. We do not use [14, Theorem V.40].

Proof of Proposition 3.1. The proof has a structure that might be amenable to presentation as a case of mathematical induction. After careful consideration we came to the conclusion that setting up an inductive argument would not shorten the proof.

Recall that we assume that Eq. (7.29) holds for all $\omega \in \Omega$, not only a.s. Throughout this proof we fix one path $\omega \in \Omega$.

(i) Let $x \in K_2$, $h \in \mathbb{R}^n$, $\|h\| < 1$, and $0 < t \leq \tau$. Recall that $X(0) = 0$. For any $1 \leq j \leq n$ we have

$$Y_j^{x+h}(t) - Y_j^x(t) = h_j + \int_0^t \left[\mathcal{V}_j(Y^{x+h}(s)) - \mathcal{V}_j(Y^x(s))\right] \mathrm{d}s. \tag{7.39}$$

Let

$$c_1 := c_1(x,h) := \sup_{0<t\leq\tau} \|Y^{x+h}(t) - Y^x(t)\|.$$

Note that for $0 < t \leq \tau$ we have $Y^x(t) \in M$ and $Y^{x+h}(t) \in M$. By Eq. (7.39) and $\|\mathcal{V}\|_{\infty,M} < \infty$ we get that c_1 is finite. Moreover,

$$\begin{aligned}\|Y_j^{x+h}(t) - Y_j^x(t)\| &\leq \|h\| + \int_0^t \|D^{(1)}\mathcal{V}_j\|_{\infty,M} \|Y^{x+h}(s) - Y^x(s)\| \,\mathrm{d}s \\ &\leq \|h\| + \tau \, \|D^{(1)}\mathcal{V}_j\|_{\infty,M} \, c_1.\end{aligned}$$

Hence,

$$c_1 \leq \|h\| + \tau \, \|D^{(1)}\mathcal{V}\|_{\infty,M} \, c_1,$$

which, when combined with Eq. (7.34), gives

$$\sup_{0<t\leq\tau} \|Y^{x+h}(t) - Y^x(t)\| = c_1 \leq \frac{\|h\|}{1 - \tau \, \|D^{(1)}\mathcal{V}\|_{\infty,M}} \leq 2\|h\|.$$

(ii) Denote

$$R_j^{x,h}(t) = Y_j^{x+h}(t) - Y_j^x(t)$$

and $R^{x,h}(t) = (R_1^{x,h}(t), \ldots, R_n^{x,h}(t))$. Using the Taylor expansion we get from Eq. (7.39),

$$R_j^{x,h}(t) = h_j + \int_0^t D^{(1)}\mathcal{V}_j(Y^x(s)) \cdot R^{x,h}(s) \,\mathrm{d}s + O(\|h\|^2). \tag{7.40}$$

For $i \in \{1, \ldots, n\}$ and $h = ue_i$, let

$$c_2 = c_2(x,i) = \max_{1\le j\le n} \sup_{0<t\le\tau} \left(\limsup_{u\to 0} \frac{R_j^{x,h}(t)}{u} - \liminf_{u\to 0} \frac{R_j^{x,h}(t)}{u} \right).$$

Note that c_2 is finite because for $u \in (-1,1)$ we have $|R_j^{x,h}(t)| \le 2u$, by Eq. (7.35). Consider $0 < t \le \tau$, $x \in K_2$, $i, j \in \{1,\dots,n\}$. From Eq. (7.40) we obtain for $u, u' \in (-1,1) \setminus \{0\}$, $h = ue_i$, and $h' = u'e_i$,

$$\frac{R_j^{x,h}(t)}{u} - \frac{R_j^{x,h'}(t)}{u'} = \int_0^t \sum_{k=1}^n D_k \mathcal{V}_j(Y^x(s)) \left(\frac{R_k^{x,h}(s)}{u} - \frac{R_k^{x,h'}(s)}{u'} \right) \mathrm{d}s$$
$$+ O(u) + O(u').$$

Letting $u, u' \to 0$ leads to

$$\limsup_{u\to 0} \frac{R_j^{x,h}(t)}{u} - \liminf_{u'\to 0} \frac{R_j^{x,h'}(t)}{u'} \le \tau \, \|D^{(1)}\mathcal{V}\|_{\infty,M} \cdot c_2,$$

and since $0 < t \le \tau$ and $j \in \{1,\dots,n\}$ are arbitrary, we get

$$c_2 \le \tau \, \|D^{(1)}\mathcal{V}\|_{\infty,M} \cdot c_2.$$

So $c_2 = 0$ which means that $D_i Y^x(t)$ exists. Estimate (7.36) is now an easy consequence of Eq. (7.35).

(iii) From Eq. (7.40) and the bounded convergence theorem, we obtain

$$D_i Y_j^x(t) = \delta_{ij} + \int_0^t D^{(1)}\mathcal{V}_j(Y^x(s)) \cdot D_i Y^x(s)\,\mathrm{d}s. \tag{7.41}$$

Let $x \in K_1$, $h \in \mathbb{R}^n$, $\|h\| < 1$, and $i \in \{1,\dots,n\}$. Set

$$c_3 := c_3(x,h,i) := \sup_{0<t\le\tau} \|D_i Y^{x+h}(t) - D_i Y^x(t)\|.$$

Because of Eq. (7.36), c_3 is finite. For any $0 < t \le \tau$ we have

$$D_i Y_j^{x+h}(t) - D_i Y_j^x(t)$$
$$= \int_0^t \left[D^{(1)}\mathcal{V}_j(Y^{x+h}(s)) \cdot D_i Y^{x+h}(s) - D^{(1)}\mathcal{V}_j(Y^x(s)) \cdot D_i Y^x(s) \right] \mathrm{d}s$$
$$= \int_0^t \Big(\left[D^{(1)}\mathcal{V}_j(Y^{x+h}(s)) - D^{(1)}\mathcal{V}_j(Y^x(s)) \right] \cdot D_i Y^{x+h}(s)$$
$$+ D^{(1)}\mathcal{V}_j(Y^x(s)) \cdot \left[D_i Y^{x+h}(s) - D_i Y^x(s) \right] \Big) \mathrm{d}s, \tag{7.42}$$

so

$$\left|D_i Y_j^{x+h}(t) - D_i Y_j^x(t)\right| \leq \int_0^t \left[\sum_{k=1}^n |D_k \mathcal{V}_j(Y^{x+h}(s)) - D_k \mathcal{V}_j(Y^x(s))|\,|D_i Y_k^{x+h}(s)| + \sum_{k=1}^n |D_k \mathcal{V}_j(Y^x(s))|\,|D_i Y_k^{x+h}(s) - D_i Y_k^x(s)|\right] \mathrm{d}s.$$

In view of Eqs. (7.35) and (7.36), we have for $0 < s \leq \tau$,

$$\sum_{k=1}^n |D_k \mathcal{V}_j(Y^{x+h}(s)) - D_k \mathcal{V}_j(Y^x(s))| \leq \|D^{(2)}\mathcal{V}\|_{\infty,M} \|Y^{x+h}(s) - Y^x(s)\| \leq 2\,\|D^{(2)}\mathcal{V}\|_{\infty,M}\,\|h\|,$$

$$\|D_i Y^{x+h}(s)\| \leq 2, \qquad \sum_{k=1}^n |D_k \mathcal{V}_j(Y^x(s))| \leq \|D^{(1)}\mathcal{V}\|_{\infty,M}.$$

It follows that

$$|D_i Y_j^{x+h}(t) - D_i Y_j^x(t)| \leq 4\,\|D^{(2)}\mathcal{V}\|_{\infty,M}\,\tau\,\|h\| + \tau\,\|D^{(1)}\mathcal{V}\|_{\infty,M} \cdot c_3,$$

so,

$$c_3 \leq 4\,\|D^{(2)}\mathcal{V}\|_{\infty,M}\,\tau\,\|h\| + \tau\,\|D^{(1)}\mathcal{V}\|_{\infty,M} \cdot c_3.$$

By definition, $\tau \leq 1/(2\|D^{(1)}\mathcal{V}\|_{\infty,M})$, so

$$c_3 \leq 4\,\|D^{(2)}\mathcal{V}\|_{\infty,M}\,\tau\,\|h\| + c_3/2.$$

This gives

$$\sup_{0<t\leq\tau} \|D_i Y^{x+h}(t) - D_i Y^x(t)\| = c_3 \leq 8\,\|D^{(2)}\mathcal{V}\|_{\infty,M}\,\tau\,\|h\|.$$

(iv) Set

$$Q_{i,j}^{x,h}(t) := D_i Y_j^{x+h}(t) - D_i Y_j^x(t)$$

and $Q_i^{x,h}(t) = (Q_{i,1}^{x,h}(t), \ldots, Q_{i,n}^{x,h}(t))$. Using the Taylor expansion we get from Eq. (7.42),

$$Q_{i,j}^{x,h}(t) = \int_0^t \sum_{l=1}^n D_i Y_l^{x+h}(s) \sum_{m=1}^n D_{lm}\mathcal{V}_j(Y^x(s)) R_m^{x,h}(s)\,\mathrm{d}s + O(\|h\|^2) + \int_0^t D^{(1)}\mathcal{V}_j(Y^x(s)) \cdot Q_i^{x,h}(s)\,\mathrm{d}s$$

$$= \int_0^t \sum_{l=1}^n D_i Y_l^{x+h}(s) D^{(1)} D_l \mathcal{V}_j(Y^x(s)) \cdot R^{x,h}(s)\,\mathrm{d}s + O(\|h\|^2)$$

$$+ \int_0^t D^{(1)} \mathcal{V}_j(Y^x(s)) \cdot Q_i^{x,h}(s)\,\mathrm{d}s. \tag{7.43}$$

Consider $k \in \{1, \ldots, n\}$ and let $h = ue_k$. Define

$$c_4 := c_4(x,i,k) := \max_{1\le j\le n} \sup_{0<t\le\tau} \left(\limsup_{u\to 0} \frac{Q_{i,j}^{x,h}(t)}{u} - \liminf_{u\to 0} \frac{Q_{i,j}^{x,h}(t)}{u} \right).$$

Note that c_4 is finite because we have $|Q_{i,j}^{x,h}(t)| \le 8\,\|D^{(2)}\mathcal{V}\|_{\infty,M}\,\tau\,u$ for $u \in (-1,1)$, by Eq. (7.37). For $u, u' \in (-1,1)\setminus\{0\}$, $h = ue_k$, and $h' = u'e_k$, Eq. (7.43) implies that

$$\frac{Q_{i,j}^{x,h}(t)}{u} - \frac{Q_{i,j}^{x,h'}(t)}{u'}$$

$$= \int_0^t \sum_{l=1}^n D_i Y_l^{x+h}(s) D^{(1)} D_l \mathcal{V}_j(Y^x(s)) \cdot \frac{R^{x,h}(s)}{u}\,\mathrm{d}s + O(u)$$

$$- \int_0^t \sum_{l=1}^n D_i Y_l^{x+h'}(s) D^{(1)} D_l \mathcal{V}_j(Y^x(s)) \cdot \frac{R^{x,h'}(s)}{u'}\,\mathrm{d}s + O(u')$$

$$+ \int_0^t D^{(1)} \mathcal{V}_j(Y^x(s)) \left(\frac{Q_i^{x,h}(s)}{u} - \frac{Q_i^{x,h'}(s)}{u'} \right) \mathrm{d}s.$$

The first two integrals cancel in the limit as $u, u' \to 0$. To see that we can pass to the limit, we use the bounded convergence theorem. This theorem is applicable because Eq. (7.35) provides a bound for $u^{-1} R^{x,h}(s)$, Eq. (7.36) provides a bound for $D_i Y_l^{x+h}(s)$, and we also have $\|D^{(2)}\mathcal{V}\|_{\infty,M} < \infty$, by assumption. Letting $u, u' \to 0$ we get

$$\limsup_{u\to 0} \frac{Q_{i,j}^{x,h}(t)}{u} - \liminf_{u'\to 0} \frac{Q_{i,j}^{x,h'}(t)}{u'} \le \tau\,\|D^{(1)}\mathcal{V}\|_{\infty,M} \cdot c_4.$$

Since $0 < t \le \tau$ and $j \in \{1, \ldots, n\}$ are arbitrary we see that

$$c_4 \le \tau\,\|D^{(1)}\mathcal{V}\|_{\infty,M} \cdot c_4,$$

so $c_4 = 0$; this proves that $D_{ik} Y^x(t)$ exists. The estimate (7.38) follows now from Eq. (7.37).

(v) By Eq. (7.43) we get for $h = ue_k$

$$D_{ik}Y_j^x(t) = \lim_{u\to 0}\frac{Q_{i,j}^{x,h}(t)}{\|h\|} = \int_0^t \sum_{l=1}^n D_i Y_l^x(s)\, D^{(1)} D_l \mathcal{V}_j(Y^x(s)) \cdot D_k Y^x(s)\,\mathrm{d}s$$
$$+ \int_0^t D^{(1)}\mathcal{V}_j(Y^x(s)) D_{ik} Y^x(s)\,\mathrm{d}s.$$

Let $x \in K$, $h \in \mathbb{R}^n$, $\|h\| < 1$, and $i, k \in \{1, \ldots, n\}$. Put

$$c_5 := c_5(x,h,i,k) := \sup_{0<t\le\tau} \|D_{ik}Y^{x+h}(t) - D_{ik}Y^x(t)\|.$$

Because of Eq. (7.38), c_5 is finite. For any $0 < t \le \tau$ and $j \in \{1,\ldots,n\}$ we have

$$\begin{aligned}
&D_{ik}Y_j^{x+h}(t) - D_{ik}Y_j^x(t)\\
&= \int_0^t \sum_{l=1}^n \sum_{m=1}^n \Big[D_i Y_l^{x+h}(s) D_{lm}\mathcal{V}_j(Y^{x+h}(s)) D_k Y_m^{x+h}(s)\\
&\quad - D_i Y_l^x(s) D_{lm}\mathcal{V}_j(Y^x(s)) D_k Y_m^x(s)\Big]\mathrm{d}s\\
&\quad + \int_0^t \sum_{l=1}^n \Big[D_l\mathcal{V}_j(Y^{x+h}(s)) D_{ik} Y_l^{x+h}(s) - D_l\mathcal{V}_j(Y^x(s)) D_{ik} Y_l^x(s)\Big]\mathrm{d}s\\
&= \mathrm{I} + \mathrm{II}.
\end{aligned}$$

We obtain from Eqs. (7.36)–(7.38)

$$\begin{aligned}
|\mathrm{I}| \le &\int_0^t \sum_{l=1}^n \sum_{m=1}^n \Big[\Big| D_{lm}\mathcal{V}_j(Y^{x+h}(s)) D_i Y_l^{x+h}(s) \big[D_k Y_m^{x+h}(s) - D_k Y_m^x(s)\big]\Big|\\
&+ \Big| D_{lm}\mathcal{V}_j(Y^{x+h}(s)) D_k Y_m^x(s) \Big[D_i Y_l^{x+h}(s) - D_i Y_l^x(s)\Big]\Big|\\
&+ \big| D_i Y_l^x(s) D_k Y_m^x(s) \big[D_{lm}\mathcal{V}_j(Y^{x+h}(s)) - D_{lm}\mathcal{V}_j(Y^x(s))\big]\big| \Big]\mathrm{d}s\\
\le &\;\tau \Big[\|D^{(2)}\mathcal{V}\|_{\infty,M}^2\, 32\,\tau\,\|h\| + 8\,\|D^{(3)}\mathcal{V}\|_{\infty,M}\,\|h\|\Big]
\end{aligned}$$

as well as

$$\begin{aligned}
|\mathrm{II}| \le &\int_0^t \sum_{l=1}^n \Big[\Big| D_l\mathcal{V}_j(Y^{x+h}(s)) \Big[D_{ik} Y_l^{x+h}(s) - D_{ik} Y_l^x(s)\Big]\Big|\\
&+ \big| D_{ik} Y_l^x(s) \big[D_l\mathcal{V}_j(Y^{x+h}(s)) - D_l\mathcal{V}_j(Y^x(s))\big]\big| \Big]\mathrm{d}s\\
\le &\;\tau \Big[\|D^{(1)}\mathcal{V}\|_{\infty,M} \cdot c_5 + 16\,\|D^{(2)}\mathcal{V}\|_{\infty,M}^2 \tau\,\|h\|\Big].
\end{aligned}$$

Combining these two estimates we find for all $0 < t \le \tau$ and $1 \le j \le n$,

$$|D_{ik}Y_j^{x+h}(t) - D_{ik}Y_j^x(t)|$$
$$\le 48\,\|D^{(2)}\mathcal{V}\|^2_{\infty,M}\,\tau^2\|h\| + 8\,\|D^{(3)}\mathcal{V}\|_{\infty,M}\tau\,\|h\| + \tau\,\|D^{(1)}\mathcal{V}\|_{\infty,M}\cdot c_5.$$

Hence,

$$c_5 \le 48\,\|D^{(2)}\mathcal{V}\|^2_{\infty,M}\,\tau^2\,\|h\| + 8\,\|D^{(3)}\mathcal{V}\|_{\infty,M}\,\tau\,\|h\| + \tau\,\|D^{(1)}\mathcal{V}\|_{\infty,M}\cdot c_5,$$

so, recalling Eq. (7.34),

$$c_5 \le 96\,\|D^{(2)}\mathcal{V}\|^2_{\infty,M}\,\tau^2\,\|h\| + 16\,\|D^{(3)}\mathcal{V}\|_{\infty,M}\,\tau\,\|h\|,$$

which finishes the proof. □

The next step in proving Theorem 3.1 is the following lemma.

Lemma 3.1. *Let $g : \mathbb{R}^n \to \mathbb{R}$ be a function in C_b^2. Fix $0 < t_1 < \infty$ and let $Y^x(t)$ be the solution of Eq. (7.29). Assume that the explosion time $\zeta(x,\omega) \equiv \infty$ for all $x \in \mathbb{R}^n$ and all $\omega \in \Omega$. Let $T_t g$ be defined by Eq. (7.30). Assume that $K \subset \mathbb{R}^n$, for every $t \le t_1$ $\operatorname{supp} T_t g \subset K$, and there exists a convex set $M \subset \mathbb{R}^n$ such that $Y^x(t,\omega) \in M$ for all $x \in K_3 := \bigcup_{x\in K} B^*(x,3)$, $t \le t_1$, and $\omega \in \Omega$. Assume that $\|\mathcal{V}\|_{(3),M} < \infty$ and let*

$$\tilde{\tau} = \frac{1}{2\,\|D^{(1)}\mathcal{V}\|_{\infty,M}} \wedge t_1 \qquad \left(\frac{1}{0} := \infty\right).$$

Then we have

(i) *For all $0 < t \le \tilde{\tau}$, $x \in K$, and $i \in \{1,\dots,n\}$, the derivative $D_i T_t g(x)$ exists and*

$$D_i T_t g(x) = \mathbb{E}\left(D^{(1)}g(Y^x(t))D_iY^x(t)\right). \tag{7.44}$$

(ii) *For all $0 < t \le \tilde{\tau}$, $x \in K$ and $i,k \in \{1,\dots,n\}$, the derivative $D_{ik}T_t g(x)$ exists and*

$$D_{ik}T_t g(x) = \mathbb{E}\Bigg(D^{(1)}g(Y^x(t))\cdot D_{ik}Y^x(t) + \sum_{j=1}^{n} D_iY_j^x(t)D^{(1)}(D_jg)(Y^x(t))\cdot D_kY^x(t)\Bigg). \tag{7.45}$$

(iii) *For all $0 < t \le \tilde{\tau}$ and $i,k \in \{1,\dots,n\}$, the derivative $D_{ik}T_t g(x)$ is continuous for $x \in K$.*

Proof. (i) Let $0 < t \le \tilde{\tau}$, $x \in K$, fix $i \in \{1,\dots,n\}$ and let $h = ue_i$. By Taylor's theorem and Eq. (7.35), we get

$$
\begin{aligned}
D_i T_t g(x) &= \lim_{u\to 0} \frac{T_t g(x+h) - T_t g(x)}{u} \\
&= \lim_{u\to 0} \mathbb{E}\left(\frac{g(Y^{x+h}(t)) - g(Y^x(t))}{u}\right) \\
&= \lim_{u\to 0} \mathbb{E}\left(\frac{D^{(1)}g(Y^x(t))\cdot(Y^{x+h}(t) - Y^x(t))}{u}\right) \\
&\quad + \lim_{u\to 0} \mathbb{E}\left(\frac{\sum_{1\le l,m\le n} D_{lm}g(\xi)(Y_l^{x+h}(t) - Y_l^x(t))(Y_m^{x+h}(t) - Y_m^x(t))}{2u}\right) \\
&= \mathbb{E}\left(D^{(1)}g(Y^x(t))\cdot D_i Y^x(t)\right) + \lim_{u\to 0} \mathbb{E}\left(o\left(\frac{\|Y^{x+h}(t) - Y^x(t)\|^2}{u}\right)\right) \\
&= \mathbb{E}\left(D^{(1)}g(Y^x(t))\cdot D_i Y^x(t)\right),
\end{aligned}
$$

where $\xi = \xi_{x,h,t,l,m}$ is an intermediate point between $Y^x(t)$ and $Y^{x+h}(t)$. This yields Eq. (7.44).

(ii) Fix $i, k \in \{1, \ldots, n\}$ and let $h = ue_k$. We have, using (i),

$$
\begin{aligned}
D_{ik} T_t g(x) &= \lim_{u\to 0} \frac{D_i T_t g(x+h) - D_i T_t g(x)}{u} \\
&= \lim_{u\to 0} \mathbb{E}\left(\frac{D^{(1)}g(Y^{x+h}(t))\cdot D_i Y^{x+h}(t) - D^{(1)}g(Y^x(t))\cdot D_i Y^x(t)}{u}\right) \\
&= \lim_{u\to 0} \mathbb{E}\left(\frac{D^{(1)}g(Y^{x+h}(t))\cdot(D_i Y^{x+h}(t) - D_i Y^x(t))}{u}\right) \\
&\quad + \lim_{u\to 0} \mathbb{E}\left(\frac{D_i Y^x(t)\cdot(D^{(1)}g(Y^{x+h}(t)) - D^{(1)}g(Y^x(t)))}{u}\right) \\
&= \mathrm{I} + \mathrm{II}.
\end{aligned}
$$

By Eq. (7.37) and bounded convergence theorem,

$$
\mathrm{I} = \mathbb{E}\left(D^{(1)}g(Y^x(t))\cdot D_{ik} Y^x(t)\right).
$$

We apply the Taylor theorem Eq. (7.35) and the bounded convergence theorem to see that

$$
\begin{aligned}
\mathrm{II} &= \lim_{u\to 0} \mathbb{E}\left(\frac{\sum_{j=1}^n D_i Y_j^x(t)(D_j g(Y^{x+h}(t)) - D_j g(Y^x(t)))}{u}\right) \\
&= \lim_{u\to 0} \mathbb{E}\left(\frac{\sum_{j=1}^n D_i Y_j^x(t) D^{(1)}(D_j g)(Y^x(t))\cdot(Y^{x+h}(t) - Y^x(t))}{u}\right)
\end{aligned}
$$

$$+\lim_{u\to 0}\mathbb{E}\left(o\left(\frac{\|Y^{x+h}(t)-Y^{x}(t)\|^{2}}{u}\right)\right)$$

$$=\mathbb{E}\left(\sum_{j=1}^{n} D_i Y_j^x(t) D^{(1)}(D_j g)(Y^x(t))\cdot D_k Y^x(t)\right).$$

This proves Eq. (7.45).

(iii) By Proposition 3.1, all derivatives on the right-hand side of Eq. (7.45) are continuous. Thus the functions $D_{ik}T_t g(x)$, $i,k\in\{1,\ldots,n\}$ are continuous for $x\in K$, and each $0<t\le\tilde{\tau}$. This proves (iii). □

Proof of Theorem 3.1. We set

$$\tau := \frac{1}{2\,\|D^{(1)}\mathcal{V}\|_{\infty,M}}\wedge t_0.$$

We will use induction. The induction step is the following. Assume that $T_s f\in C_b^2$ for some $s\in[0,t_0]$. We will show that for all $r\le\tau$ such that $s+r\le t_0$ we have $T_{s+r}f\in C^2$ and $\|T_{s+r}f\|_{(2)}<\infty$. To show this we use Lemma 3.1. Put $g=T_s f$ and $t_1=t_0-s$. Note that $r\le\tau\wedge t_1=\tilde{\tau}$ and $g=T_s f$ satisfy the assumptions of Lemma 3.1. Hence we obtain that $T_{r+s}f=T_r g\in C^2$. A combination of the estimates (7.44) and (7.45), the fact that $\operatorname{supp} T_r g\subset K$, and the estimates from Proposition 3.1 yield $\|T_r g\|_{(2)}<\infty$.

An assumption of Theorem 3.1 states that $f\in C_b^2$. Hence, $T_0 f=f\in C_b^2$. The induction step shows that $T_s f\in C_b^2$ for all $s\le\tau\wedge t_0$. Subsequent induction steps extend this claim to $T_s f\in C_b^2$ for all $s\le j\tau\wedge t_0$, $j=2,3,\ldots$. Therefore, $T_s f\in C_b^2$ for all $s\le t_0$. □

Proposition 3.2. *Let $\{T_t\}_{t\ge 0}$ be the semigroup given by Eq. (7.8) of the process (Y_t,S_t) defined by Eq. (7.3). Let $C_*^2(\mathbb{R}^2)$ be the class of functions given by Definition 2.1. We have*

$$T_t : C_*^2(\mathbb{R}^2)\to C_*^2(\mathbb{R}^2).$$

Proof. We will repeat some of the arguments given before the statement of Theorem 3.1. Note that the SDE (7.3) is of the form Eq. (7.29). By Lemma 2.1 there exists a unique solution of Eq. (7.3) with explosion time $\zeta((y,s),\omega)\equiv\infty$ for all $(y,s)\in\mathbb{R}^2$ and $\omega\in\Omega$. Suppose that $f\in C_*^2(\mathbb{R}^2)$. Then $\operatorname{supp} f\subset\mathbb{R}\times[-r,r]$, for some $r>0$. Fix $t_0>0$. By Eq. (7.32), for any $t\le t_0$, we have

$$\operatorname{supp} T_t f\subset K := \mathbb{R}\times\big[-r-t_0\|W'\|_\infty,\, r+t_0\|W'\|_\infty\big]. \tag{7.46}$$

We have

$$K_3=\bigcup_{(y,s)\in K} B^*((y,s),3)=\mathbb{R}\times\big(-r-t_0\|W'\|_\infty-3,\, r+t_0\|W'\|_\infty+3\big).$$

Let

$$M = \mathbb{R} \times \big[-r - 2t_0 \|W'\|_\infty - 3,\, r + 2t_0 \|W'\|_\infty + 3 \big].$$

By Eq. (7.33) we have $(Y_t^{(y,s)}, S_t^{(y,s)}) \in M$ for all $(y,s) \in K_3$. Rewriting Eq. (7.3) as Eq. (7.31) we have $\mathcal{V}_1(y_1, y_2) = W(y_1)y_2, \mathcal{V}_2(y_1, y_2) = W'(y_1)$. Since $W \in C^4$ and since it is periodic, we get $\|\mathcal{V}\|_{(3),M} < \infty$. Therefore, the solution of Eq. (7.31) satisfies the assumptions of Theorem 3.1. It follows that for any $t \le t_0$ we have

$$T_t f \in C^2 \quad \text{and} \quad \|T_t f\|_{(2)} < \infty.$$

This and Eq. (7.46) yield $T_t f \in C_*^2(\mathbb{R}^2)$. □

Proof of Proposition 2.2 Suppose that $f \in C_c^2(\mathbb{S} \times \mathbb{R})$. Then $\tilde{f} \in C_*^2(\mathbb{R}^2)$ where $\tilde{f}$ is given by Eq. (7.5). By Proposition 3.2, $T_t \tilde{f} \in C_*^2(\mathbb{R}^2)$. Using this and Eq. (7.12) we get $T_t^{\mathbb{S}} f \in C_c^2(\mathbb{S} \times \mathbb{R})$. □

Acknowledgements K. Burdzy was supported in part by NSF grant DMS-0906743 and by grant N N201 397137, MNiSW, Poland. T. Kulczycki was supported in part by grant N N201 373136, MNiSW, Poland. R.L. Schilling was supported in part by DFG grant Schi 419/5–1.

References

1. Bass, R., Burdzy, K., Chen, Z., Hairer, M.: Stationary distributions for diffusions with inert drift. Probab. Theory Rel. Fields **146**, 1–47 (2010)
2. Benaïm, M., Ledoux, M., Raimond, O.: Self-interacting diffusions. Probab. Theory Rel. Fields **122**, 1–41 (2002)
3. Benaïm, M., Raimond, O.: Self-interacting diffusions. II. Convergence in law. Ann. Inst. H. Poincaré Probab. Statist. **39**, 1043–1055 (2003)
4. Benaïm, M., Raimond, O.: Self-interacting diffusions. III. Symmetric interactions. Ann. Probab. **33**, 1717–1759 (2005)
5. Bou-Rabee, N., Owhadi, H.: Ergodicity of Langevin processes with degenerate diffusion in momentums. Int. J. Pure Appl. Math. **45**, 475–490 (2008)
6. Burdzy, K., White, D.: A Gaussian oscillator. El. Comm. Probab. **9**(10), 92–95 (2004)
7. Burdzy, K., White, D.: Markov processes with product-form stationary distribution. El. Comm. Probab. **13**, 614–627 (2008)
8. Burdzy, K., Kulczycki, T., Schilling, R.: Stationary distributions for jump processes with memory. Ann. Inst. Henri Poincaré Probab. & Statist. **48**, 609–630 (2012)
9. Dynkin, E.B.: Markov Processes, vol. 1. Springer, Berlin (1965)
10. Ethier, S.N., Kurtz, T.G.: Markov Processes: Characterization and Convergence. Wiley, New York (1986)
11. Glover, J.: Markov functions. Ann. Inst. Poincaré B **27**, 221–238 (1991)
12. Kunita, H.: SDEs based on Lévy processes and stochastic flows of diffeomorphisms. In: Rao, M.M. (ed.) Real and Stochastic Analysis, pp. 305–374. Birkhäuser, New York (2005)
13. Lachal, A.: Applications de la théorie des excursions à l'intégrale du mouvement brownien. Sém. Probab. XXXVIII. Lecture Notes in Math, vol. 1801, pp. 109–195. Springer, Berlin (2003)
14. Protter, P.: Stochastic Integration and Differential Equations, 2nd edn. Springer, Berlin (2004)

15. Schilling, R.L., Schnurr, A.: The symbol associated with the solution of a stochastic differential equation. Electron. J. Probab. **15**(43), 1369–1393 (2010)
16. Sharpe, M.: General Theory of Markov Processes. Academic, Boston (1988)
17. Sheffield, S.: Gaussian free fields for mathematicians. Probab. Theory Rel. Fields **139**, 521–541 (2007)
18. Simon, T.: Support theorem for jump processes. Stoch. Proc. Appl. **89**, 1–30 (2000)
19. Sinai, Ya.G.: Topics in Ergodic Theory. Princeton University Press, Princeton (1994)

Chapter 8
An Ornstein-Uhlenbeck-Type Process Which Satisfies Sufficient Conditions for a Simulation-Based Filtering Procedure

Arturo Kohatsu-Higa and Kazuhiro Yasuda

Abstract In this article, we verify all the conditions stated in [8] in order for a filtering/estimation procedure based on Monte Carlo simulations of unknown densities of diffusion processes to converge to its theoretical values. In order to verify these hypotheses one needs to use extensively various properties of the diffusion processes and its Euler–Maruyama approximation. In particular, we need to study flow properties, upper and lower bounds for densities and existence of invariant measures and α-mixing properties.

As a consequence one obtains that there is a tuning procedure which chooses the number of steps in the Euler–Maruyama scheme, the window size of the kernel estimation method and the Monte Carlo simulation size in function of the number of available data.

Received 10/30/2011; Accepted 3/13/2012; Final 4/6/2012

1 Introduction

In many statistical estimation/filtering problems one needs to estimate quantities of the following type:

A. Kohatsu-Higa (✉)
Department of Mathematical Sciences, Ritsumeikan University, Kyoto,
Japan Japan Science and Technology Agency, Tokyo, Japan
e-mail: arturokohatsu@gmail.com

K. Yasuda
Faculty of Science and Engineering, Hosei University, Tokyo, Japan
e-mail: k_yasuda@hosei.ac.jp

F. Viens et al. (eds.), *Malliavin Calculus and Stochastic Analysis: A Festschrift in Honor of David Nualart*, Springer Proceedings in Mathematics & Statistics 34,
DOI 10.1007/978-1-4614-5906-4_8,

$$E_N[f] := E_\theta[f|Y_0,\ldots,Y_N] := \frac{\int f(\theta)\phi_\theta(Y_0^N)\pi(\theta)\mathrm{d}\theta}{\int \phi_\theta(Y_0^N)\pi(\theta)\mathrm{d}\theta}, \tag{8.1}$$

where $\{Y_i;\ i = 0,\ldots,N\}$ are observed data at times $\{i\Delta;\ i = 0,\ldots,N\}$ from a process which has the same law as the solution to the following stochastic differential equation:

$$Y_t = Y_0 + \int_0^t b(\theta_0, Y_s)\mathrm{d}s + \int_0^t \sigma(Y_s)\mathrm{d}W_s,\ t \geq 0. \tag{8.2}$$

Here in order to simplify the situation, we consider the case where the prior density, π, is concentrated in a one-dimensional compact interval $[\theta^l, \theta^u] \subseteq \mathbb{R}$. Furthermore, $b : [\theta^l, \theta^u] \times \mathbb{R} \to \mathbb{R}$ and $\sigma : \mathbb{R} \to \mathbb{R}$ are smooth functions with bounded derivatives. σ is also bounded and uniformly elliptic. Suppose that the diffusion possesses an invariant measure μ_θ such that $\int e^{c_1 y^2}\mu_\theta(dy) < \infty$ for some positive constant c_1 (see, e.g., [4] for a much more general situation) and that it is α-mixing. We furthermore assume that Y forms a stationary α-mixing Markov chain.

Furthermore, $\phi_\theta(Y_0^N) = \phi_\theta(Y_0,\ldots,Y_N) = \mu_\theta(Y_0)\prod_{j=1}^N p_\theta(Y_{j-1}, Y_j)$ denotes the joint density of $(Y_0, Y_1,\ldots,Y_N)$ where p_θ denotes the transition density for Y_Δ.

Clearly as the transition density and the invariant measure of Y are unknown the above estimation procedure is at best a theoretical formula. In general, one has to resort to simulation procedures in order to approximate the filtering formula (8.1). In this article, we assume that the transition density is approximated using an Euler–Maruyama scheme with n partition points and h is the window size in the kernel density estimation method.

In this setting we have that there are three parameters (N, n, h) which need to be tuned in order for the overall estimation procedure to work well.

In a recent article (see [8]), we have discussed a theoretical framework where one can study such estimation procedure and provided a proof which clearly states that a correct tuning is needed. Due to the lack of space, we have not considered explicit examples and the objective of this paper together with another explicit example (see [9]) is to provide the reader with explicit cases where the theory is applicable.

The list of conditions in order to achieve the correct tuning is long, albeit every condition being natural. For briefness, we refer the reader to [8] for exact statements. In this article, we explain the meaning of each condition and prove that in the case of one-dimensional diffusions with Ornstein–Uhlenbeck type behavior all the required conditions are satisfied.

The way we verify the requirements follows a different order than in [8]. This is done for pedagogical reasons. While in [8] the conditions are introduced as one deepens into the tuning problem. Here the order of exposition is in the order of easy verification.

Our final goal in this article, is to show the reader an application area where the full strength of Malliavin calculus is needed in order to verify various conditions so as to obtain a practical result. The order of exposition will take us first into the verification of the ergodic property and the α-strongly mixing condition. Then we will proceed verifying various flow properties. After this, we will quote and deduce results about the upper and lower bounds for densities. Finally, we will verify the identifiability conditions and at the end we state our main theorem.

2 Framework

We will verify that all hypotheses related to the approximation process in [8] are satisfied by the Euler–Maruyama scheme under enough regularity conditions. So in our present example, Y_i is the random variable associated with $X_{i\Delta}$, where X is a one-dimensional diffusion process with regular drift $b(\theta, \cdot)$. θ takes values in a compact set $\Theta = [\theta^l, \theta^u]$ and regular diffusion coefficient $\sigma(\cdot)$ (independent of θ). We assume that the diffusion coefficient is uniformly elliptic which guarantees the existence of a smooth strictly positive density. A copy of X starting at y is denoted by $X^y(\theta)$. Furthermore $X^y_{(m)}(\theta)$ denotes the Euler–Maruyama scheme of step size $\frac{\Delta}{m}$ with $m \equiv m(N) \in \mathbb{N}$. Then $\tilde{p}^N_\theta(y, z)$ denotes the density of $X^y_{(m)}(\theta)$ which is regular and strictly positive given the uniformly elliptic condition.

Let $K : \mathbb{R} \to \mathbb{R}_+$ be a twice continuously differentiable kernel which satisfies $\int K(x)dx = 1$. Denote by $\hat{p}^N_\theta(y, z; \hat{\omega})$, $\hat{\omega} \in \hat{\Omega}$, where $\hat{\Omega}$ denotes the sample space where simulations are carried out. The kernel density estimate of $\tilde{p}^N_\theta(y, z)$ based on n simulated i.i.d. copies of $X^y_{(m)}(\theta)$ which are defined on $(\hat{\Omega}, \hat{\mathcal{F}}, \hat{P}_\theta)$ and denoted by $X^{y,(k)}_{(m)}(\theta, \cdot)$, $k = 1, \ldots, n$; for $h \in (0, 1)$, is given by

$$\hat{p}^N_\theta(y, z; \hat{\omega}) := \hat{p}^N_\theta(y, z; \hat{\omega}; m(N), h(N), n(N))$$
$$:= \frac{1}{n(N)h(N)} \sum_{k=1}^{n(N)} K\left(\frac{X^{y,(k)}_{(m(N))}(\theta, \hat{\omega}) - z}{h(N)}\right).$$

For given m, we introduce the "average" approximative transition density over all trajectories with respect to the kernel K as

$$\bar{p}^N_\theta(y, z) := \bar{p}^N_\theta(y, z; m(N), h(N)) := \hat{E}\left[\hat{p}^N_\theta(y, z; \cdot)\right]$$
$$= \hat{E}\left[\frac{1}{h(N)} K\left(\frac{X^{y,(1)}_{(m(N))}(\theta, \cdot) - z}{h(N)}\right)\right],$$

where $\hat{E}$ means the expectation with respect to $\hat{P}$.

Then we consider the following approximation of Eq. (8.1);

$$\hat{E}^n_{N,m}[f] = \frac{\int f(\theta)\hat{\phi}^N_\theta(Y^N_0)\pi(\theta)\mathrm{d}\theta}{\int \hat{\phi}^N_\theta(Y^N_0)\pi(\theta)\mathrm{d}\theta},$$

where set $\hat{\phi}_\theta^N(Y_0^N) = \mu_\theta(Y_0)\prod_{j=1}^N \hat{p}_\theta^N(Y_{j-1}, Y_j)$. Strong convergence of $\hat{E}_{N,m}^n[f]$ to $E_N[f]$ and the rate of convergence with respect to the number of data N under a general diffusion case has been studied in Kohatsu-Higa et al. [8].

3 Invariant Measure and α-Mixing Condition w.r.t. Diffusion Processes

In this section we give sufficient conditions for a one-dimensional diffusion of the type (8.2) to have an invariant measure which satisfies the α-mixing condition.

In fact, from pp.213 in Bibby, Jacobsen, and Sorensen [3], we give sufficient conditions for the existence of an invariant measure of a one-dimensional stochastic differential equation. We define the density function of the scale measure for $X^y(\theta)$ as

$$s(x;\theta) = \exp\left(-2\int_{x^\sharp}^x \frac{b(\theta,y)}{\sigma(y)^2}\mathrm{d}y\right),$$

where $x^\sharp$ is an arbitrary point in $(-\infty,\infty)$.

Assumption 3.1. *The following holds for all $\theta \in \Theta$:*

$$\int_{x^\sharp}^\infty s(x;\theta)\mathrm{d}x = \int_{-\infty}^{x^\sharp} s(x;\theta)\mathrm{d}x = \infty$$

and

$$A(\theta) := \int_{-\infty}^\infty \frac{1}{s(x;\theta)\sigma(x)^2}\mathrm{d}x < \infty.$$

Under Assumption 3.1, the process $X_t^y(\theta)$ is ergodic with an invariant probability measure that has density

$$\mu_\theta(x) = \frac{1}{A(\theta)s(x;\theta)\sigma(x)^2}, \quad x \in (-\infty,\infty). \tag{8.3}$$

Example 3.1. Set $b(\theta,x) = -x+\theta$ and $\sigma(x) = \sigma > 0$ (constant). Then we have

$$s(x;\theta) = \exp\left(2\int_{x^\sharp}^x \frac{y-\theta}{\sigma^2}\mathrm{d}y\right) = \exp\left(\frac{2}{\sigma^2}\left(\frac{1}{2}x^2 - \theta x + C\right)\right),$$

where C is a constant. Obviously we have, for all $\theta \in \Theta$,

$$\int_{x^\sharp}^\infty s(x;\theta)\mathrm{d}x = \int_{-\infty}^{x^\sharp} s(x;\theta)\mathrm{d}x = \infty$$

and

$$A(\theta) = \int_{-\infty}^{\infty} \frac{1}{s(x;\theta)\sigma^2} \mathrm{d}x = \frac{1}{\sigma} \exp\left(-\frac{2}{\sigma^2}C + \frac{\theta^2}{\sigma^2}\right) < \infty.$$

And the invariant measure is given as

$$\mu_\theta(x) = \frac{1}{A(\theta)s(x;\theta)\sigma^2} = \frac{1}{\sigma} \exp\left(-\frac{(x-\theta)^2}{\sigma^2}\right).$$

Next, from Corollary 2.1 in Genon-Catalot, Jeantheau, and Larédo [5], for fixed θ, we give some sufficient conditions so that $X^y(\theta)$ is α-mixing.

Assumption 3.2. *For fixed* $\theta \in \Theta$,

(i) *The function* b *is continuously differentiable with respect to* x *and* σ *is twice continuously differentiable with respect to* x, $\sigma(x) > 0$ *for all* $x \in (-\infty, \infty)$. *We also assume that there exists constants* C_1^θ, $C_2 > 0$ *such that*

$$|b(\theta, x)| \leq C_1^\theta(1 + |x|) \ \ and \ \ \sigma(x)^2 \leq C_2(1 + x^2) \ \ for\ all\ x \in (-\infty, \infty).$$

(ii) $\sigma(x)\mu_\theta(x) \to 0$ *as* $x \downarrow -\infty$ *and* $x \uparrow \infty$.

(iii) $\frac{1}{\gamma(\theta,x)}$ *has a finite limit as* $x \downarrow -\infty$ *and* $x \uparrow \infty$, *where*

$$\gamma(\theta;x) := \frac{\partial}{\partial x}\sigma(x) - \frac{2b(\theta,x)}{\sigma(x)}.$$

Under Assumptions 3.1 and 3.2, for fixed $\theta \in \Theta$, the process $X^y(\theta)$ is geometrically α-mixing, i.e., α-mixing with mixing coefficients that tend to zero geometrically fast. Therefore our data $\{Y_i\}$, which is in the case $\theta = \theta_0$, is geometrically α-mixing.

Example 3.2 (cont. of Example 3.1). For $b(\theta, x) = -x + \theta$ and $\sigma(x) = \sigma$, (i) and (ii) clearly hold since θ takes values in a compact set $\Theta = [\theta^l, \theta^u]$ and the diffusion coefficient σ is constant. For (iii), now

$$\gamma(\theta;x) = \frac{2}{\sigma}(x - \theta)$$

and for all $\theta \in \Theta$, $\frac{1}{\gamma(\theta,x)}$ converges to 0 as $x \to \pm\infty$.

Assumption 3.3. *We assume that the invariant measure* μ_θ *satisfies the following integrability condition for some strictly positive constant* c_1

$$E[\mathrm{e}^{c_1 Y_1^2}] = \int \mathrm{e}^{c_1 y^2} \mu_\theta(\mathrm{d}y) < \infty.$$

Note that if we assume that $b(\theta, x)$ is a function with linear growth with respect to x uniformly in θ and continuity in θ and $\sigma(x)$ is a bounded and continuous function with uniformly ellipticity, then the invariant measure $\mu_\theta(x)$, which is defined in Eq. (8.3), is continuous and bounded and for all $x \in \mathbb{R}$ and $\theta \in \Theta$, $\mu_\theta(x) > 0$ holds.

4 Flow-Related Properties

In this section we verify some of the conditions given in Sect. 6.5 in [8] that are related with the explicit tuning procedure. These conditions appear when one has to determine the uniform (wrt parameter and space values in compact sets, see the definition of B^N below) rates of convergence of the Monte Carlo simulation of the approximative density. In particular, the conditions required follow from arguments that rely on the Borel–Cantelli lemma.

For this, consider $c_2 > \frac{2}{c_1}$, $n = C_1 N^{\alpha_1}$ for $\alpha_1, C_1 > 0$, and $h = C_2 N^{-\alpha_2}$ for $\alpha_2, C_2 > 0$. Furthermore, let $B^N := \left\{(\mathbf{x}, \theta) = (x, y, \theta) \in \mathbb{R}^2 \times \Theta;\ \|\mathbf{x}\| < a_N := \sqrt{c_2 \ln N}\right\}$, where we assume $a_N \geq 1$. Then, we define

$$Z_{3,N}^{(k)}(\omega) := a_N^{-2} \left(\sup_{(\mathbf{x},\theta)\in B^N} \left| X_{(m)}^{y,(k)}(\theta, \omega) \right| + 1 \right) \sup_{(\mathbf{x},\theta)\in B^N} \left| \partial_\theta X_{(m)}^{y,(k)}(\theta, \omega) \right|, \quad (8.4)$$

$$Z_{4,N}^{(k)}(\omega) := a_N^{-1} \left(\sup_{(\mathbf{x},\theta)\in B^N} \left| \partial_y X_{(m)}^{y,(k)}(\theta; \omega) \right| + \sup_{(\mathbf{x},`)\in B^N} \left| \partial_\theta X_{(m)}^{y,(k)}(\theta; \omega) \right| \right),$$

$$\dot{Z}_{4,N}^{(k)}(\omega) := a_N^{-1} \Bigg(h \sup_{(\mathbf{x},\theta)\in B^N} \left| \partial_y \partial_\theta X_{(m)}^{y,(k)}(\theta; \omega) \right| + h \sup_{(\mathbf{x},\theta)\in B^N} \left| \partial_\theta \partial_\theta X_{(m)}^{y,(k)}(\theta; \omega) \right|$$

$$+ \left(Z_{4,N}^{(k)} + 1 \right) \sup_{(\mathbf{x},\theta)\in B^N} \left| \partial_\theta X_{(m)}^{y,(k)}(\theta; \omega) \right| \Bigg),$$

$$\dot{Z}_{6,N}^{(j)}(\omega) := a_N^{-1} \sup_{(\mathbf{x},\theta)\in B^N} \left\{ \left| \partial_\theta X_{(m)}^{y,(j)}(\theta; \omega) \right| + E\left[\left| \partial_\theta X_{(m)}^{y,(1)}(\theta; \cdot) \right| \right] \right\}. \quad (8.5)$$

Then the goal in this section is to prove that the conditions below are satisfied.

(iii) (Borel–Cantelli for $Z_{3,N}^{(k)}(\omega)$, **(H3)**) For some $r_3 > 0$ and $b_{3,N} := \frac{C_3 (N^{\gamma_3} n)^{\frac{1}{r_3}} c_2 \ln N}{h^2}. \geq 1$,

$$\sum_{N=1}^{\infty} \frac{n a_N^{2r_3}}{(h^2 b_{3,N})^{r_3}} < +\infty \quad \text{and} \quad \sup_{N\in\mathbb{N}} E\left[|Z_{3,N}(\cdot)|^{r_3} \right] < +\infty.$$

(iv) (Borel–Cantelli for $Z_{4,N}^{(k)}(\omega)$, **(H4)**) For some $r_4 > 0$ and $b_{4,N} \geq 1$,

$$\sum_{N=1}^{\infty} \frac{n}{(b_{4,N})^{r_4}} < +\infty \quad \text{and} \quad \sup_{N \in \mathbb{N}} E\left[\left|Z_{4,N}\left(\cdot\right)\right|^{r_4}\right] < +\infty.$$

(vi) (Borel–Cantelli for $\dot{Z}^{(k)}_{4,N}(\omega)$, **(H4')**) For some $\dot{r}_4 > 0$ and $\dot{b}_{4,N} \geq 1$,

$$\sum_{N=1}^{\infty} \frac{n}{(\dot{b}_{4,N})^{\dot{r}_4}} < +\infty \quad \text{and} \quad \sup_{N \in \mathbb{N}} E\left[\left|\dot{Z}_{4,N}\left(\cdot\right)\right|^{\dot{r}_4}\right] < +\infty.$$

(viii) (Borel–Cantelli for $\dot{Z}^{(k)}_{6,N}(\omega)$, **(H6a')**) For some $\dot{r}_6 > 0$ and $\dot{b}_{6,N} = \left(\dot{C}_6 n N^{\dot{\gamma}_6}\right)^{\frac{1}{\dot{r}_6}} \geq 1$,

$$\sum_{N=1}^{\infty} \frac{n}{(\dot{b}_{6,N})^{\dot{r}_6}} < +\infty \quad \text{and} \quad \sup_{N} E\left[\left|\dot{Z}_{6,N}(\cdot)\right|^{\dot{r}_6}\right] < +\infty.$$

Remark 4.1. The summability conditions and the other conditions not quoted here (i.e., (i), (ii), (v), and (vii) in [8]) are finally simplified by requesting that the inequalities (8.7) and (8.8) which appear below, be satisfied.

The conditions that remain are the above finite moment conditions in the conditions above. These conditions are satisfied due to the regularity conditions on the coefficients b and σ and flow properties. The proofs are tedious but the essential technique already exists so we will only sketch the result in the following lemma.

Lemma 4.1. *Assume that b and σ are smooth and at most linear growth in x with all derivatives bounded by constants which are independent of θ. Then the moment conditions stated in* (iii), (iv), (vi), *and* (viii) *are satisfied.*

Proof. For the proof, we will only indicate how to prove that the uniform moments are finite for one of the terms of Eq. (8.4). This should point to reader how to proceed in the other cases in a similar fashion.

We can estimate $\sup_{(\mathbf{x},\theta)\in B_N} |X^y_{(m)}(\theta)|$ as follows. Let (y_0, θ_0) be a point in $(-a_N, a_N) \times \Theta$, where set $\Theta = [\theta^l, \theta^u]$, then we have

$$\begin{aligned}
&\sup_{(\mathbf{x},\theta)\in B_N} \left|X^y_{(m)}(\theta)\right| \\
&\leq \sup_{(\mathbf{x},\theta)\in B_N} \left|X^y_{(m)}(\theta) - X^y_{(m)}(\theta_0)\right| + \sup_{(\mathbf{x},\theta)\in B_N} \left|X^y_{(m)}(\theta_0) - X^{y_0}_{(m)}(\theta_0)\right| + \left|X^{y_0}_{(m)}(\theta_0)\right| \\
&= \sup_{(\mathbf{x},\theta)\in B_N} \left|\int_{\theta_0}^{\theta} \partial_\theta X^y_{(m)}(\kappa) d\kappa\right| + \sup_{(\mathbf{x},\theta)\in B_N} \left|\int_{y_0}^{y} \partial_y X^{\tilde{y}}_{(m)}(\theta_0) d\tilde{y}\right| + \left|X^{y_0}_{(m)}(\theta_0)\right| \\
&\leq \int_{\theta_l}^{\theta_u} \sup_{y\in(-a_N, a_N)} \left|\partial_\theta X^y_{(m)}(\kappa) - \partial_\theta X^{y_0}_{(m)}(\kappa)\right| d\kappa + \int_{\theta_l}^{\theta_u} \left|\partial_\theta X^{y_0}_{(m)}(\kappa)\right| d\kappa
\end{aligned}$$

$$
\begin{aligned}
&+ \int_{-a_N}^{a_N} \left|\partial_y X_{(m)}^{\tilde{y}}(\theta_0)\right| \mathrm{d}\tilde{y} + \left|X_{(m)}^{y_0}(\theta_0)\right| \\
&\leq \int_{\theta_l}^{\theta_u} \int_{-a_N}^{a_N} \left|\partial_{\theta y}^2 X_{(m)}^{\tilde{y}}(\kappa)\right| \mathrm{d}\tilde{y}\mathrm{d}\kappa + \int_{\theta_l}^{\theta_u} \left|\partial_\theta X_{(m)}^{y_0}(\kappa)\right| \mathrm{d}\kappa \\
&+ \int_{-a_N}^{a_N} \left|\partial_y X_{(m)}^{\tilde{y}}(\theta_0)\right| \mathrm{d}\tilde{y} + \left|X_{(m)}^{y_0}(\theta_0)\right|. \qquad (8.6)
\end{aligned}
$$

From this calculation, we see that we need to find uniform $L^p(\Omega)$ estimates in θ and y of the above derivatives. These estimates are essentially quoted in Lemma 4.3 in [1] which uses the same method as in the proof of Lemma 2.2 in the same paper for the derivatives wrt y. The proof is based in general derivative formula for the composition of functions (Faà di Bruno formula). □

(ix). ((**H6b'**)) For some $\dot{q}_6 > 1$,

$$
\left(\frac{\eta_N h^2}{(\|K'\|_\infty (\dot{b}_{6,N})^2 a_N} \exp\left(- \frac{(\eta_N)^2}{2(\frac{\|K'\|_\infty}{h^2} \dot{b}_{6,N} a_N)^2} \right) \right)^{\dot{q}_6} \leq \frac{\dot{C}_6}{n^{1+\dot{\alpha}_6}}
$$

$$
\text{and} \quad \sup_{N \in \mathbb{N}} E\left[\left| \dot{Z}_{6,N}(\cdot) \right|^{\dot{q}_6} \right] < +\infty,
$$

where $\dot{C}_6$ and $\dot{\alpha}_6$ are the same as (vii) of Sect. 6.5 in [8].

In order for the above condition to be satisfied, we choose $\eta_N := \frac{C_{K,\Delta,\Theta} h^2}{N^{\frac{\varphi_2 c_2}{\Delta} + \frac{1}{2}} (N^{\gamma_3} n)^{\frac{1}{r_3}} c_2 \ln N}$ with

$$
\left(4\alpha_2 + 2\frac{\alpha_1 + \dot{\gamma}_6}{\dot{r}_6} + \frac{\varphi_2 c_2}{\Delta} + \frac{1}{2} + \frac{\gamma_3}{r_3} + \frac{\alpha_1}{r_3} \right) \dot{q}_6 > \alpha_1, \qquad (8.7)
$$

which has to be satisfied together with

$$
\alpha_1 \left(1 - \frac{2}{r_3} - \frac{2}{\dot{r}_6} \right) > 8\alpha_2 + 1 + \frac{2\varphi_2 c_2}{\Delta} + \frac{2\gamma_3}{r_3} + 2\frac{\dot{\gamma}_6}{\dot{r}_6}. \qquad (8.8)
$$

All the above constants have been already defined with the exception of φ_2 (which appears in (**H1**)).

5 Regularity of the Densities

In this section, we discuss conditions (**H5**) and (**H5'**) in [8] which are related with conditions on the density of the approximation processes. This is the section where the core of Malliavin calculus has to be used in order to obtain uniform regularity of the density of the Euler–Maruyama scheme with upper and lower bounds.

(H5) Assume that there exists some positive constant $C_5 > 0$ such that for all $y, z \in \mathbb{R}$, $m \in \mathbb{N}$, and $\theta \in \Theta$,

$$\left|\partial_y \bar{p}_\theta^N(y,z)\right|, \ \left|\partial_z \bar{p}_\theta^N(y,z)\right|, \ \left|\partial_\theta \bar{p}_\theta^N(x,y)\right| \leq C_5 < +\infty.$$

(H5') Assume that there exists some positive constant $\dot{C}_5 > 0$ such that for all $y, z \in \mathbb{R}$, $m \in \mathbb{N}$, and $\theta \in \Theta$,

$$\left|\partial_y \partial_\theta \bar{p}_\theta^N(y,z)\right|, \ \left|\partial_z \partial_\theta \bar{p}_\theta^N(y,z)\right|, \ \left|\partial_\theta^2 \bar{p}_\theta^N(y,z)\right| \leq \dot{C}_5 < +\infty.$$

All the above conditions are regularity properties of the Euler scheme which are well known and are proved using Malliavin calculus techniques.

For example, in [7], Lemma 3.3, it is proved that the process $X_{(m)}^y + hZ$ where Z is a standard Gaussian random variable has a Malliavin covariance matrix whose inverse determinant belong to all the spaces $L^p(\Omega)$ if $h = m^{-1/2}$. It is also mentioned (see equation 3.1) and the comments thereafter) that the process $X_{(m)}^y(\theta) + hZ$ belongs uniformly (in m) to all the spaces $\mathbb{D}^{k,p}$. Therefore all the necessary integration by parts formulas can be carried out in order to obtain that the above hypotheses **(H5)** and **(H5')** are satisfied. Therefore, from now on, we will take $h = m^{-1/2}$.

We remark here that the derivatives with respect to the starting point and the parameter θ of the flow defined by the Euler–Maruyama scheme are differentiable in the stochastic sense as they can be rewritten as solutions of linear equations with random bounded coefficients.

One can also explicitly write the above upper bounds as Gaussian type upper bounds by using the technique in section A of [6]. We will do this without further mentioning it.

In the next results, we need to prove upper and lower bounds of Gaussian type. In fact, we verify the following hypotheses **(H1)** and **(H2)**.

(H1) Assume that there exist some positive constants φ_1, φ_2, where φ_1 is independent of N and φ_2 is independent of N and Δ, such that the following holds:

$$\inf_{(\mathbf{x},\theta)\in B^N} \bar{p}_\theta^N(y,z) \geq \varphi_1 \exp\left(-\frac{\varphi_2 a_N^2}{\Delta}\right).$$

(H2) Assume that the kernel K is the Gaussian kernel:

$$K(z) := \frac{1}{\sqrt{2\pi}} \exp\left(-\frac{1}{2}z^2\right).$$

Recall that $\tilde{p}_\theta^N(y,z)$ denotes the transition density function of $X_{(m)}^y(\theta)$.

Lemma 5.1. *Assume hypothesis* **(H2)**. *Let X denote a standard normal random variable, then we have*

$$\bar{p}_\theta^N(y,z) = E\left[\tilde{p}_\theta^N(y, hX + z)\right].$$

Proof. Note that K is the density function of the random variable X. By using a change of variables ($u = \frac{x-z}{h}$),

$$\begin{aligned}\bar{p}_\theta^N(y,z) &= E\left[\frac{1}{h}K\left(\frac{X_{(m)}^y(\theta)-z}{h}\right)\right]\\ &= \int_{-\infty}^{\infty}\frac{1}{h}K\left(\frac{x-z}{h}\right)\tilde{p}_\theta^N(y,x)\mathrm{d}x\\ &= \int_{-\infty}^{\infty}K(u)\,\tilde{p}_\theta^N(y,hu+z)\mathrm{d}u = E\left[\tilde{p}_\theta^N(y,hX+z)\right].\end{aligned}$$

□

Now we consider the lower bound **(H1)**.

Proposition 5.1. *Assume that σ is a uniformly elliptic, bounded, and smooth function with bounded derivatives. Similarly assume that b is a smooth function with at most linear growth in x uniformly in θ and bounded derivatives. Then under the hypothesis* **(H2)**, **(H1)** *is satisfied.*

Proof. First, we obtain a lower bound for $\tilde{p}_\theta^N(y,z)$. We somewhat abuse the notation using the delta distribution function $\delta_z(x)$. The formal argument can be obtained by proper approximation arguments which are left to the reader. If we denote by $t_j = \frac{\Delta j}{m}$, $j = 1,\cdots,m$ the partition points of the Euler–Maruyama scheme, $s_j = \frac{\Delta j}{Rm}$, $j = 1,\cdots,Rm$, $R \in \mathbb{N}$ the partition points that will be used in the argument for the proof of the lower bound. We denote by $X_m^y(t,\theta)$, its corresponding Euler–Maruyama scheme with Rm time steps in $[0,\Delta]$ at time t which starts from y at time 0, and by $\varphi : \mathbb{R} \to [0,1]$ a smooth function such that it takes the value 1 in the interval $[-2a_N-1, 2a_N+1]$ and 0 in the complement of $[-2a_N-2, 2a_N+2]$. Then, we have by the Chapman–Kolmogorov equation and the fact that the range of φ is $[0,1]$ that

$$\begin{aligned}\tilde{p}_\theta^N(y,z) &= \int_{\mathbb{R}^{Rm-1}} E\left[\left(\prod_{j=1}^{Rm-1}\delta_{x_j}\left(X_m^y(s_j,\theta)\right)\right)\delta_z\left(X_m^y(s_{Rm},\theta)\right)\right]\\ &\quad\times \mathrm{d}x_1\cdots\mathrm{d}x_{Rm-1}\\ &\geq \int_{\mathbb{R}^{Rm-1}} E\left[\left(\prod_{j=1}^{Rm-1}\delta_{x_j}\left(X_m^y(s_j,\theta)\right)\varphi\left(X_m^y(s_j,\theta)\right)\right)\delta_z\left(X_m^y(s_{Rm},\theta)\right)\right]\\ &\quad\times \mathrm{d}x_1\cdots\mathrm{d}x_{Rm-1}.\end{aligned}$$

Next we use the Markov property so that (from now on, we let $x_0 = y$ and $x_{Rm} = z$, in order to simplify the notation)

$$\tilde{p}_\theta^N(y,z) \geq \int_{\mathbb{R}^{Rm-1}} E\left[\left(\prod_{j=1}^{Rm-1} \delta_{x_j}\left(X_m^{s_{j-1},x_{j-1}}(s_j,\theta)\right)\varphi\left(X_m^{s_{j-1},x_{j-1}}(s_j,\theta)\right)\right)\right.$$
$$\left. \times\, \delta_z\left(X_m^{s_{Rm-1},x_{Rm-1}}(s_{Rm},\theta)\right)\right] dx_1\cdots dx_{Rm-1}.$$

Then we can evaluate the expectations by conditioning obtaining

$$\tilde{p}_\theta^N(y,z)$$
$$\geq \int_{\mathbb{R}^{Rm-1}} \left(\prod_{i=1}^{R}\prod_{j=1}^{m} \frac{\exp\left(-\frac{(x_{R(j-1)+i}-x_{R(j-1)+i-1}-b(\theta,x_{R(j-1)+i-1})\Delta(Rm)^{-1})^2}{2\sigma^2(x_{R(j-1)+i-1})\Delta(Rm)^{-1}}\right)}{\sqrt{2\pi\sigma^2(x_{R(j-1)}+i-1)\Delta(Rm)^{-1}}}\right.$$
$$\left. \times\, \varphi(x_{R(j-1)+i})\right) dx_1\cdots dx_{Rm-1}. \tag{8.9}$$

Next we restrict the integration regions using for each integral above a "tube" that will go from y to z as follows. Define $z_i = y + (z-y)\frac{i}{Rm}$, $i = 0,\cdots,Rm$. Then around each of these points, we consider the regions of integration $A = \{\mathbf{x} = (x_1,\cdots,x_{Rm-1}) \in \mathbb{R}^{Rm-1} |\ x_i \in A_i$ for all $i = 1,2,\cdots,Rm-1\}$, where we let $A_i := (z_{i-1} - M\sqrt{\Delta(Rm)^{-1}}, z_{i-1} + M\sqrt{\Delta(Rm)^{-1}})$, $i = 1,\cdots,Rm-1$. Here M is a positive constant, chosen so that $M\sqrt{\Delta(Rm)^{-1}} \leq 1$. If we restrict the above $Rm-1$ integrals to these regions, we will obtain a lower bound. On these regions, we have that the following inequalities are satisfied for $j = 1,\cdots,R$, $i = 1,\cdots,m$, and $\varphi(x_{R(j-1)}) > 0$:

$$\left|x_{R(j-1)+i} - x_{R(j-1)+i-1} - b(\theta,x_{R(j-1)+i-1})\Delta(Rm)^{-1}\right|$$
$$\leq \left|x_{R(j-1)+i} - z_{R(j-1)+i-1}\right| + \left|z_{R(j-1)+i-1} - z_{R(j-1)+i-2}\right|$$
$$+ \left|z_{R(j-1)+i-2} - x_{R(j-1)+i-1}\right| + \left|b(\theta,x_{R(j-1)+i-1})\right|\Delta(Rm)^{-1}$$
$$\leq 2M\sqrt{\Delta(Rm)^{-1}} + |y-z|(Rm)^{-1} + C_0(1+2a_N)\Delta(Rm)^{-1}.$$

We now choose $R = \frac{|2a_N|^2}{m\Delta}$. Using that $|y-z||2a_N|^{-1} \leq 1$, we have that

$$\left|x_{R(j-1)+i} - x_{R(j-1)+i-1} - b(\theta,x_{R(j-1)+i-1})\Delta(Rm)^{-1}\right|$$
$$\leq (2M+1)\sqrt{\Delta(Rm)^{-1}} + C_0(1+(2a_N)^{-1})\Delta^{\frac{3}{2}}(Rm)^{-\frac{1}{2}}.$$

Assuming without loss of generality that $\Delta \leq 1$, we have that

$$\exp\left(-\frac{(x_{R(j-1)+i} - x_{R(j-1)+i-1} - b(\theta, x_{R(j-1)+i-1})\Delta(Rm)^{-1})^2}{2\sigma^2(x_{R(j-1)+i-1})\Delta(Rm)^{-1}}\right)$$
$$\geq \exp\left(-\frac{2(M+C_0)+1}{2c_0^2}\right).$$

Here c_0 is the constant of uniform ellipticity of σ. Replacing this in Eq. (8.9), we obtain for a new constant $K > 1$ that

$$\bar{p}_\theta^N(y,z) \geq \int_A \left(\prod_{i=1}^R \prod_{j=1}^m \frac{K^{-1}}{\sqrt{2\pi C_0^2 \Delta(Rm)^{-1}}} \varphi(x_{R(j-1)+i})\right) \mathrm{d}x_1 \cdots \mathrm{d}x_{Rm-1}.$$

Next we remark that for any $\mathbf{x} = (x_1, \cdots, x_{Rm-1}) \in A$, the $(R(j-1)+i)$-th element of $\mathbf{x}$ satisfies $\varphi(x_{R(j-1)+i}) = 1$ for all $j = 1, \ldots, R$, $i = 1, \ldots, m$ and therefore as the integrands will be constants, we obtain for some constant $K_1 > 0$ that

$$\tilde{p}_\theta^N(y,z) \geq (2M\sqrt{\Delta(Rm)^{-1}})^{Rm-1} \left(\frac{K^{-1}}{\sqrt{2\pi C_0^2 \Delta(Rm)^{-1}}}\right)^{Rm}$$
$$= \frac{c_1}{\sqrt{\Delta}} \exp\left(-Rm \log K_1 - \frac{1}{2}\log(Rm)\right).$$

where c_1 is a suitable positive constant. Therefore using that $Rm = \frac{|2a_N|^2}{\Delta}$ and as

$$-Rm \log K_1 - \frac{1}{2}\log(Rm) \geq -K_2 \frac{|a_N|^2}{\Delta}.$$

for an appropriately chosen positive constant K_2 which is independent of a_N and Δ, finally we obtain

$$\inf_{|y|,|z| \leq 2a_N} \tilde{p}_\theta^N(y,z) \geq \frac{c_1}{\sqrt{\Delta}} \exp\left(-\frac{K_2 a_N^2}{\Delta}\right). \tag{8.10}$$

Next we consider the lower bound of $\tilde{p}_\theta^N(y,z)$. From $0 < h < 1$, for all z such that $|z| \leq a_N$, we have

$$\left\{ u \,\middle|\, \frac{-2a_N - z}{h} \leq u \leq \frac{2a_N - z}{h} \right\} \supset \{ u \mid -a_N \leq u \leq a_N \}.$$

And from $a_N \geq 1$ and the Chernoff bound , we have

$$\int_{-a_N}^{a_N} \frac{1}{\sqrt{2\pi}} \mathrm{e}^{-\frac{u^2}{2}} \mathrm{d}u = 2\left(\frac{1}{2} - \int_{a_N}^{\infty} \frac{1}{\sqrt{2\pi}} \mathrm{e}^{-\frac{u^2}{2}} \mathrm{d}u\right) \geq 2\left(\frac{1}{2} - \frac{1}{2}\mathrm{e}^{-\frac{a_N^2}{2}}\right) \geq 1 - \mathrm{e}^{-\frac{1}{2}}.$$

From Lemma 5.1, Eq. (8.10), and the above results, we have

$$\begin{aligned}
\inf_{(\mathbf{x},\theta)\in B^N} \bar{p}_\theta^N(y,z) &= E\left[\tilde{p}_\theta^N(y, hX + z)\mathbf{1}(|hX + z| \leq 2a_N)\right. \\
&\quad \left. + \tilde{p}_\theta^N(y, hX + z)\mathbf{1}(|hX + z| > 2a_N)\right] \\
&\geq \int_{-\infty}^{\infty} \frac{1}{\sqrt{2\pi}} \mathrm{e}^{-\frac{u^2}{2}} \tilde{p}_\theta^N(y, hu + z)\mathbf{1}_{[-2a_N, 2a_N]}(hu + z)\mathrm{d}u \\
&\geq \int_{-a_N}^{a_N} \frac{1}{\sqrt{2\pi}} \mathrm{e}^{-\frac{u^2}{2}} \mathrm{d}u \frac{c_1}{\sqrt{\Delta}} \mathrm{e}^{-K_2 \frac{a_N^2}{\Delta}} \\
&\geq \left(1 - \mathrm{e}^{-\frac{1}{2}}\right) \frac{c_1}{\sqrt{\Delta}} \mathrm{e}^{-K_2 \frac{a_N^2}{\Delta}}.
\end{aligned}$$

□

Remark 5.1. A similar lower estimation of the density is given in [2]. Here we give a simpler proof in our settings and as pointed out in [2], the uniformly elliptic condition can be weakened with an elliptic condition around a "tube."

We now discuss condition 5 of Assumption 2.2 in Kohatsu-Higa et al. [8] on the regularity of the log density. The technical problem in these estimates is that derivatives of the logarithm will make appear the density in the denominator of various expressions. One may control these by using the Gaussian lower bounds obtained in Proposition 5.1.

Lemma 5.2 (Regularity of the log density). *Let $b(\theta, x)$ be a smooth function with at most linear growth in x uniformly in θ and bounded derivatives, and also $\sigma(x)$ be a uniformly elliptic, bounded, and smooth function with bounded derivatives. For $q_\theta = p_\theta$, $\bar{p}_\theta^N$ and Δ small enough, we have*

$$\sup_N \sup_{\theta\in\Theta} \iint \left(\frac{\partial^i}{\partial\theta^i} \ln q_\theta(y,z)\right)^{12} p_{\theta_0}(y,z)\mu_{\theta_0}(y)\mathrm{d}y\mathrm{d}z < +\infty, \quad for\ i = 0, 1, 2,$$

$$\sup_N \sup_{\theta\in\Theta} \left|\frac{\partial^2}{\partial\theta^2} \iint (\ln q_\theta(y,z))\, \bar{p}_{\theta_0}^N(y,z)\mu_{\theta_0}(y)\mathrm{d}y\mathrm{d}z\right| < +\infty,$$

$$\sup_N \sup_{\theta\in\Theta} \iint \left|\frac{\partial^i}{\partial\theta^i} \ln q_\theta(y,z)\right| \bar{p}_{\theta_0}^N(y,z)\mu_{\theta_0}(y)\mathrm{d}y\mathrm{d}z < +\infty, \quad for\ i = 0, 1,$$

where $\frac{\partial^0}{\partial\theta^0} q_\theta = q_\theta$.

Proof. We will give the ideas in order to prove one of these inequalities and leave the others for the reader as they are all similar. Consider the last one: first note that in the case $i = 0$ one only uses the upper bounds for densities result either for the diffusion and its approximation which can be obtained in a variety of ways, e.g., using Malliavin calculus (see, e.g, section A in [6]). Also note that the lower bound is obtained here for the approximation in Proposition 5.1. In fact, applying (with a slight modification to the formula of H) Theorem 2.1.4 in [10] we have the following expression for the density of $X_\Delta^x(\theta)$:

$$p_\theta(x,y) = E\left[\mathbf{1}\left(X_\Delta^x(\theta) \le y\right) H\left(X_\Delta^x(\theta);1\right)\right],$$

where for any $p > 1$ there exists positive constants (which can be made explicit) C, k_1, and p_1 so that

$$H\left(X_\Delta^x(\theta);1\right) = \int_0^\Delta \frac{A_\Delta}{A_s\sigma(X_s^x)}\mathrm{d}W_s,$$

$$D_s X_\Delta^x(\theta) = \sigma\left(X_s^x(\theta)\right)\exp\left(\int_s^\Delta \left\{\partial_x b\left(\theta, X_t^x(\theta)\right) - \frac{\partial_x\sigma(X_t^x(\theta))^2}{2}\right\}\mathrm{d}t + \int_s^\Delta \partial_x\sigma\left(X_t^x(\theta)\right)\mathrm{d}W_t\right) =: \sigma\left(X_s^x(\theta)\right)A_\Delta A_s^{-1},$$

$$\|H\left(X_\Delta^x(\theta);1\right)\|_p \le C\|X_\Delta^x(\theta)\|_{k_1,p_1}.$$

For $i = 1$, we proceed in a similar fashion (using the integration by parts formula of Malliavin calculus) with the addition of the following extra ingredient:

$$\left|\frac{\partial}{\partial\theta}\ln p_\theta(y,z)\right| = \left|E\left[H(X^y(\theta), \partial_\theta X^y(\theta))/\, X^y(\theta) = z\right]\right| \le E\left[\left|H(X^y(\theta), \partial_\theta X^y(\theta))\right|/\, X^y(\theta) = z\right].$$

Then for $\alpha \in (0,1)$ and $p^{-1} + q^{-1} = 1$ we have

$$\begin{aligned}
&\int\left|\frac{\partial}{\partial\theta}\ln p_\theta(y,z)\right|\bar{p}_{\theta_0}^N(y,z)\mathrm{d}z \\
&\le \int E\left[\left|H(X_\Delta^y(\theta), \partial_\theta X_\Delta^y(\theta))\right|/\, X_\Delta^y(\theta) = z\right]\bar{p}_{\theta_0}^N(y,z)\mathrm{d}z \\
&\le \left(\int E\left[\left|H(X_\Delta^y(\theta), \partial_\theta X_\Delta^y(\theta))\right|^p/\, X_\Delta^y(\theta) = z\right]\bar{p}_{\theta_0}^N(y,z)^{\alpha p}\mathrm{d}z\right)^{1/p} \\
&\times\left(\int \bar{p}_{\theta_0}^N(y,z)^{(1-\alpha)q}\mathrm{d}z\right)^{1/q}.
\end{aligned}$$

Note that as we have upper Gaussian estimates for $\bar{p}_{\theta_0}^N(y,z)$ then the second term above is finite. Furthermore for the first term we have that

$$\int E\left[\left|H(X_\Delta^y(\theta),\partial_\theta X_\Delta^y(\theta))\right|^p / X_\Delta^y(\theta)=z\right]\bar{p}_{\theta_0}^N(y,z)^{\alpha p}\mathrm{d}z$$

$$=\int E\left[\left|H(X_\Delta^y(\theta),\partial_\theta X_\Delta^y(\theta))\right|^p \delta_z\left(X_\Delta^y(\theta)\right)\right]p_\theta(y,z)^{-1}\bar{p}_{\theta_0}^N(y,z)^{\alpha p}\mathrm{d}z.$$

Therefore by choosing p big enough, we will have that $p_\theta(y,z)^{-1}\bar{p}_{\theta_0}^N(y,z)^{\alpha p}$ is bounded by a Gaussian term and then the finiteness will follow from estimates for H. □

Next we will verify Assumption 2.2 6-(b) of Kohatsu-Higa et al. [8] leaving 6-(a) for the end as this requires various conditions as explained in the paper.

Lemma 5.3. *Let $b(\theta,x)$ be a smooth function with at most linear growth in x uniformly in θ and bounded derivatives and also $\sigma(x)$ be a uniformly elliptic, bounded, and smooth function with bounded derivatives. Then for each $y,z\in\mathbb{R}$ and Δ small enough, there exist factors $C_1^N(y,z)$ and $c_1(y,z)$ such that*

$$\left|p_{\theta_0}(y,z)-\bar{p}_{\theta_0}^N(y,z)\right|\le C_1^N(y,z)a_1(N),$$

where $\sup_N C_1^N(y,z)<+\infty$ and $a_1(N):=m(N)^{-1}+h(N)^2\to 0$ as $N\to\infty$, and

$$C_1^N(y,z)a_1(N)\sqrt{N}<c_1(y,z), \tag{8.11}$$

where c_1 satisfies

$$\sup_N\sup_{\theta\in\Theta}\iint\left|\frac{\partial}{\partial\theta}\ln\bar{p}_\theta^N(y,z)\right|c_1(y,z)\mu_{\theta_0}(y)\mathrm{d}y\mathrm{d}z<+\infty. \tag{8.12}$$

Proof. It is well known that the rate of convergence of the Euler scheme with the kernel modification is of the order $m^{-1}+h^2=:a_1(N)$ (see, e.g., [7]). Therefore the idea to prove the bound (8.12) is similar as in the proof of Lemma 5.2. In fact,

$$C_1^N(y,z):=a_1(N)^{-1}\left|p_{\theta_0}(y,z)-\bar{p}_{\theta_0}^N(y,z)\right|.$$

As before, we consider

$$\int\left|\frac{\partial}{\partial\theta}\ln\bar{p}_\theta^N(y,z)\right|\sup_N C_1^N(y,z)\mathrm{d}z$$

$$\le\left(\int E\left[\left|H(X_{(m)}^y(\theta),\partial_\theta X_{(m)}^y(\theta))\right|^p\delta_z\left(X_{(m)}^y(\theta)\right)\right]\bar{p}_\theta^N(y,z)^{-1}\sup_N C_1^N(y,z)^{p\alpha}\mathrm{d}z\right)^{1/p}$$

$$\times\left(\int\sup_N C_1^N(y,z)^{q(1-\alpha)}\mathrm{d}z\right)^{1/q}.$$

The above quantity will be finite by using upper Gaussian estimates for $\sup_N C_1^N(y,z)$ with appropriate values for p and α as shown in, e.g., [7]. □

Now we verify Assumption 2.2 6-(c) of Kohatsu-Higa et al. [8].

Lemma 5.4. *Let $b(\theta, x)$ be a smooth function with at most linear growth in x uniformly in θ and bounded derivatives and also $\sigma(x)$ be a uniformly elliptic, bounded, and smooth function with bounded derivatives. And $K(x)$ is in $C^2(\mathbb{R})$. Then for Δ small enough, there exist some function $g^N : \mathbb{R}^2 \to \mathbb{R}$ and constant $a_2(N)$, which depends on N, such that for all $y, z \in \mathbb{R}$*

$$\sup_{\theta\in\Theta}\left|\frac{\partial}{\partial\theta}\ln\bar{p}_\theta^N(y,z) - \frac{\partial}{\partial\theta}\ln p_\theta(y,z)\right| \le |g^N(y,z)|a_2(N),$$

where $\sup_N E_{\theta_0}[|g^N(Y_0,Y_1)|^4] < +\infty$ *and* $a_2(N) \to 0$ *as* $N \to \infty$.

Proof. This condition is similar to Lemma 5.3 with the exception that the approximation is of the logarithmic derivative. Again, one can also prove alternatively

$$a_2(N)^{-4}\int\int\sup_{\theta\in\Theta}\left|\frac{\partial}{\partial\theta}\ln\bar{p}_\theta^N(y,z) - \frac{\partial}{\partial\theta}\ln p_\theta(y,z)\right|^4 p_{\theta_0}(y,z)\mu_{\theta_0}(y)\mathrm{d}y\mathrm{d}z < \infty.$$

Using a Sobolev embedding inequality (or a simple argument like in the beginning of the proof of Lemma 4.1) to deal with the supremum by requiring higher derivatives with respect to θ. The arguments closes as in the proof of the previous lemmas. □

5.1 Identifiability Conditions

We verify now that the following identifiability condition is satisfied.

Lemma 5.5. (i) *Assume that there exists some $x \in \mathbb{R}$ such that for all $\theta \neq \theta_0$, there exists some $y \in \mathbb{R}$ such that $p_\theta(x,y) \neq p_{\theta_0}(x,y)$ and $\partial_\theta p_{\theta_0}(x,y) \neq 0$. Then there exist $c_1 : \mathbb{R} \to (0,\infty)$ such that for all $\theta \in \Theta$,*

$$\int |p_\theta(y,z) - p_{\theta_0}(y,z)|\,\mathrm{d}z \ge c_1(y)|\theta - \theta_0|,$$

and $C_1(\theta_0) := \int c_1(y)^2\mu_{\theta_0}(y)\mathrm{d}y \in (0,+\infty)$.

(ii) *Let $b(\theta, x)$ be a smooth function with at most linear growth in x uniformly in θ and bounded derivatives, and also $\sigma(x)$ be a uniformly elliptic, bounded, and smooth function with bounded derivatives. Assume the same hypotheses as in (i). Furthermore we assume that there exists some N_0 such that for all $N \ge N_0$,* *the same conditions as in (i) are satisfied with* p *replaced by* $\bar{p}^N$. *Then there exist* $c_2 : \mathbb{R} \to (0, \infty)$ *and* $N_0 \in \mathbb{N}$ *such that for all* $\theta \in \Theta$,

$$\inf_{N \geq N_0} \int \left| \bar{p}_{\theta}^N(x, y) - \bar{p}_{\theta_0}^N(x, y) \right| \mathrm{d}y \geq c_2(x)|\theta - \theta_0|,$$

and $C_2(\theta_0) := \int c_2(x)^2 \mu_{\theta_0}(x)\mathrm{d}x \in (0, +\infty)$ *with* N *big enough.*

Proof of (i). First note that the identifiability condition for p is equivalent to

$$\infty > \int \left(\inf_{\theta \in \Theta} \int \frac{|p_\theta(x, y) - p_{\theta_0}(x, y)|}{|\theta - \theta_0|} \mathrm{d}y \right)^2 \mu_{\theta_0}(x)\mathrm{d}x \geq \int c(x)^2 \mu_{\theta_0}(x)\mathrm{d}x > 0.$$

By using the fundamental theorem of calculus and changing variables, set

$$\beta = \alpha\theta + (1 - \alpha)\theta_0,$$

we have

$$\infty > \int \left(\inf_{\theta \in \Theta} \int \left| \int_0^1 \partial_\theta p_{\alpha\theta + (1-\alpha)\theta_0}(x, y)\mathrm{d}\alpha \right| \mathrm{d}y \right)^2 \mu_{\theta_0}(x)\mathrm{d}x$$
$$\geq \int c(x)^2 \mu_{\theta_0}(x)\mathrm{d}x > 0.$$

The integrability (upper estimation) is easily obtained from Gaussian upper estimates of derivatives of densities of uniformly elliptic diffusions. One may alternatively use the method of proof in the proof of Lemma 5.2.

Now $\mu_{\theta_0}(x) > 0$ for all $x \in \mathbb{R}$. Therefore it is enough to prove that

$$\inf_{\theta \in \Theta} \int \left| \int_0^1 \partial_\theta p_{\alpha\theta + (1-\alpha)\theta_0}(x, y)\mathrm{d}\alpha \right| \mathrm{d}y > 0,$$

for x in a set of positive measure. We will prove this statement by contradiction. Therefore, we assume that for almost all x

$$\inf_{\theta \in \Theta} \int \left| \int_0^1 \partial_\theta p_{\alpha\theta + (1-\alpha)\theta_0}(x, y)\mathrm{d}\alpha \right| \mathrm{d}y = 0.$$

Due to the continuity of $\partial_\theta p$ we have that for almost all $x \in \mathbb{R}$, there exists some $\theta^* = \theta^*(x)$ such that

$$\int \left| \int_0^1 \partial_\theta p_{\alpha\theta^* + (1-\alpha)\theta_0}(x, y)\mathrm{d}\alpha \right| \mathrm{d}y = 0.$$

As $\theta^* = \theta_0$ contradicts the assumption we have that for all $x \in \mathbb{R}$, there exists some $\theta^* = \theta^*(x) \neq \theta_0$ such that for all $y \in \mathbb{R}$,

$$(\theta^* - \theta) \int_0^1 \partial_\theta p_{\alpha\theta^*+(1-\alpha)\theta_0}(x, y)\mathrm{d}\alpha = p_{\theta^*}(x, y) - p_{\theta_0}(x, y) = 0, \tag{8.13}$$

which also contradicts the assumption.

(Proof of (ii)*)* By using a similar argument, we obtain the identifiability condition for $\bar{p}^N$. Set $B := \int \{\inf_\theta \inf_{N \geq N_0} \int \frac{|\bar{p}_\theta^N(x,y) - \bar{p}_{\theta_0}^N(x,y)|}{|\theta - \theta_0|} \mathrm{d}y\}^2 \mu_{\theta_0}(x)\mathrm{d}x \in [0, +\infty)$. As before, it is easy to prove $B < +\infty$.

Here we also use proof by contradiction. If $B = 0$, then from the assumption of the coefficients, we have $\mu_\theta(x) > 0$ for all $x \in \mathbb{R}$ and $\theta \in \Theta$. So that we have, for all $N_0 \in \mathbb{N}$ and almost all $x \in \mathbb{R}$,

$$\inf_\theta \inf_{N \geq N_0} \int \frac{|\bar{p}_\theta^N(x, y) - \bar{p}_{\theta_0}^N(x, y)|}{|\theta - \theta_0|} \mathrm{d}y = 0.$$

Then for all $N_0 \in \mathbb{N}$ and all $x \in \mathbb{R}$, there exists some sequence $\theta_n = \theta_n(x, N_0)$ such that

$$\lim_{n \to \infty} \inf_{N \geq N_0} \int \frac{|\bar{p}_{\theta_n}^N(x, y) - \bar{p}_{\theta_0}^N(x, y)|}{|\theta_n - \theta_0|} \mathrm{d}y = 0.$$

And also, for all $x \in \mathbb{R}$, there exists some sequence $\theta_n = \theta_n(x, N_0)$ such that there exists some sequence $N_n = N_n(x, \theta_n) \geq N_0$ such that

$$\lim_{n \to \infty} \int \frac{|\bar{p}_{\theta_n}^{N_n}(x, y) - \bar{p}_{\theta_0}^{N_n}(x, y)|}{|\theta_n - \theta_0|} \mathrm{d}y = 0.$$

By using the mean value theorem, we consider the following: for all $x \in \mathbb{R}$, there exists some sequence $\theta_n = \theta_n(x, N_0)$ such that there exists some sequence $N_n = N_n(x, \theta_n)$ such that

$$\lim_{n \to \infty} \int \left| \int_0^1 \partial_\theta \bar{p}_{\alpha\theta_n+(1-\alpha)\theta_0}^{N_n}(x, y)\mathrm{d}\alpha \right| \mathrm{d}y = 0. \tag{8.14}$$

If N_n has a subsequence converging to infinity then as θ_n is a sequence in a compact set there is also an accumulation point. In that case, the proof finishes as in the proof of (i) by taking limits in n.

In the contrary if N_n is a bounded sequence then one obtains that there exists N' such that for all $x, y \in \mathbb{R}$,

$$\bar{p}_{\theta^*}^{N'}(x, y) = \bar{p}_{\theta_0}^{N'}(x, y),$$

if $\theta_n \to \theta^* \neq \theta_0$ or $\partial_\theta \bar{p}_{\theta_0}^{N'}(x, y) = 0$. □

Example 5.1. (i). Here we give an example which satisfies the conditions in Lemma 5.5 (i). If we assume that $|\partial_\theta b(\theta, x)| > c > 0$ for all $x \in \mathbb{R}$ and $\partial_x b$ and $\partial_x \sigma$ are bounded, then the assumption of Lemma 5.5 holds. In fact, suppose without loss of generality that $\partial_\theta b_\theta(x) > c$ for all $x \in \mathbb{R}$, we have

$$\partial_\theta X_t^x(\theta) = \varepsilon_t \int_0^t \varepsilon_s^{-1} \partial_\theta b\left(\theta, X_s^x(\theta)\right) \mathrm{d}s > c\varepsilon_t \int_0^t \varepsilon_s^{-1} \mathrm{d}s > 0 \quad a.s.,$$

where set

$$\varepsilon_t = \exp\left\{ \int_0^t \left(\partial_x b\left(\theta, X_u^x(\theta)\right) - \frac{1}{2} \partial_x \sigma \left(X_s^x(\theta)\right)^2 \right) \mathrm{d}u \right. \\ \left. + \int_0^t \partial_x \sigma\left(X_u^x(\theta)\right) \mathrm{d}W_u \right\}.$$

Therefore $X_t^x(\theta)$ is almost surely strictly increasing with respect to θ.

Let $F_{x,\theta}(y)$ be the distribution function of $X_\Delta^x(\theta)$:

$$F_{x,\theta}(y) = \int_{-\infty}^y p_\theta(x, z) \mathrm{d}z.$$

From the monotonicity of $X_\Delta^x(\theta)$ and supp $p_\theta(x, \cdot) = \mathbb{R}$, we have $F_{x,\theta}(y) \neq F_{x,\theta_0}(y)$ for all $y \in \mathbb{R}$ and $\theta (\neq \theta_0)$. Therefore there exists some $y \in \mathbb{R}$ such that $p_\theta(x, y) \neq p_{\theta_0}(x, y)$.

(ii) Next we give an example which satisfies the conditions in Lemma 5.5(ii). Here we consider the case $\sigma(x) = \sigma > 0$ and assume that $|\partial_x b(\theta, x)|$ is bounded by a positive constant M uniformly in θ, and there exists some positive constant c such that $|\partial_\theta b(\theta, x)| > c$ for all $x \in \mathbb{R}$. From Eq. (8.14), we have $\lim_{n\to\infty} \int_0^1 \partial_\theta \bar{p}^{N_n}_{\alpha\theta_n + (1-\alpha)\theta_0}(x, y) \mathrm{d}\alpha = 0$ for all $x, y \in \mathbb{R}$. And we integrate both sides from $-\infty$ to y with respect to the second variable to obtain

$$\begin{aligned} 0 &= \int_{-\infty}^y \lim_{n\to\infty} \int_0^1 \partial_\theta \bar{p}^{N_n}_{\alpha\theta_n + (1-\alpha)\theta_0}(x, z) \mathrm{d}\alpha \mathrm{d}z \\ &= \lim_{n\to\infty} \int_0^1 \partial_\theta \bar{F}^{N_n}_{x,\alpha\theta_n + (1-\alpha)\theta_0}(y) \mathrm{d}\alpha, \end{aligned} \tag{8.15}$$

where set

$$\bar{F}^N_{x,\theta}(y) = \int_{-\infty}^y \bar{p}^N_\theta(x, z) \mathrm{d}z.$$

Note that we have the following expression:

$$\begin{aligned} \bar{F}^N_{x,\theta}(y) &= \int_{-\infty}^y E\left[\tilde{p}^N_\theta(x, hX + z)\right] \mathrm{d}z = E\left[\tilde{F}^N_{x,\theta}(hX + y)\right] \\ &= E\left[\mathbf{1}\left(X^x_{(m)}(\theta) \leq hX + y\right)\right], \end{aligned} \tag{8.16}$$

where X is a random variable with the standard normal distribution and $\tilde{F}^N_{x,\theta}(y)$ is the distribution function for $\tilde{p}^N_\theta(x, y)$.

Set $X^x_l(\theta) = X^x_{\frac{l}{m}}(\theta)$ with the Euler–Maruyama approximation. From the above assumptions (case $\partial_\theta b(\theta, x) > c$), we have

$$\partial_\theta X^x_1(\theta) = \partial_\theta b(\theta, x)\, \Delta t > c\Delta t,$$

and if $\partial_\theta X^x_l(\theta) > 0$ and Δt is small enough (e.g., $1 - \partial_x b(\theta, X^x_l(\theta))\Delta t > 0$, i.e., $\frac{1}{M} > \Delta t$), then we have

$$\partial_\theta X^x_{l+1}(\theta) = \partial_\theta X^x_l(\theta) + \left\{\partial_\theta b\left(\theta, X^x_l(\theta)\right) + \partial_x b\left(\theta, X^x_l(\theta)\right) \partial_\theta X^x_l(\theta)\right\} \Delta t > 0.$$

Finally by induction, we obtain $\partial_\theta X^x_{(m)}(\theta) > 0$.

From the expression of the distribution function and the above calculation, we have, for all $x, y \in \mathbb{R}$ and $\theta \in \Theta$,

$$\partial_\theta \bar{F}^N_{x,\theta}(y) = E\left[\delta_{hX+y}\left(X^x_{(m)}(\theta)\right) \partial_\theta X^x_{(m)}(\theta)\right] > 0$$

Hence the conclusion follows as in (i).

6 Main Result

Finally by summing up all the hypotheses of the previous sections, and all the verification of various hypotheses, we obtain the following consequence from Theorem 3.1 in Kohatsu-Higa et al. [8].

Theorem 6.1. *Assume hypothesis* **(H2)**, *Assumptions 3.1, 3.2, and 3.3 with* $c_1c_2 > 2$, *and the assumptions in Lemma 5.5.*

Let $\alpha_1 > 0$, $\alpha_2 > \frac{1}{4}$ *and* $m \in \mathbb{N}$ *be such that* $\alpha_1 > 8\alpha_2 + 1 + \frac{4\varphi_2 c_2}{\Delta}$ *and* $m \geq \sqrt{N}$, *where number of the Monte–Carlo simulation* $n = C_1 N^{\alpha_1}$ *and the bandwidth size* $h = C_2 N^{-\alpha_2}$, *where* C_1 *and* C_2 *are positive constants. Assume that* $b(\theta, x)$ *is smooth and at most linear growth in* x *uniformly in* θ *with bounded derivatives. And assume that* $\sigma(x)$ *is a smooth bounded function with bounded derivatives and uniformly elliptic. Then for* Δ *small enough, there exist some positive finite random variable* Ξ_1 *and* Ξ_2 *such that for* $f \in C^1(\Theta)$, *we have*

$$|E_N[f] - f(\theta_0)| \leq \frac{\Xi_1}{\sqrt{N}} \quad a.s. \quad and \quad |\hat{E}^n_{N,m}[f] - f(\theta_0)| \leq \frac{\Xi_2}{\sqrt{N}} \quad a.s.,$$

and therefore $|E_N[f] - \hat{E}^n_{N,m}[f]| \leq \frac{\Xi_1+\Xi_2}{\sqrt{N}}$ *a.s.*

In fact, we remark that we are able to simplify the inequalities (8.7) and (8.8) to the above $\alpha_1 > 8\alpha_2 + 1 + \frac{4\varphi_2 c_2}{\Delta}$ since one can freely choose the constants r_3, r_4, $\dot{r}_4$, $\dot{r}_6$, and $\dot{q}_6$ due to the existence of all moments associated with the processes in the hypotheses (iii), (iv), (vi), (viii), and (ix). Remember that φ_2 is the constant which was introduced in the lower bound of $\bar{p}_\theta^N$ in assumption **(H1)** and c_1 is the constant related to the integrability condition in Assumption 3.3. Hence from the assumptions $c_1 c_2 > 2$ and $\alpha_1 > 8\alpha_2 + 1 + \frac{4\varphi_2 c_2}{\Delta}$, we can find that c_1 and φ_2 are connected through the parameter c_2. Finally, the assumption $\alpha_2 > \frac{1}{4}$ is needed as the bandwidth h has to satisfy $h^2 < \frac{1}{\sqrt{N}}$ in Eq. (8.11) of Lemma 5.3.

Acknowledgements This research was supported by grants from the Japan Ministry of Education and Science and the Japan Science and Technology Agency. The authors would like to thank all the people that gave us information about related results.

References

1. Ahn, H., Kohatsu-Higa, A.: The Euler scheme for Anticipating Stochastic Differential Equations. Stoch. Stoch. Rep. **54**, 247–269 (1995)
2. Bally, V.: Lower Bounds for the Density of Locally Elliptic Ito Processes. Ann. Prob. **36**(6), 2406–2440 (2006)
3. Bibby, B.M., Jacobsen, M., Sorensen, M.: Estimating Functions for Discretely Sample Diffusion-Type Models. Hand. Financ. Economet. Vol.1-Tools and Techniques, 203–268 (2010)
4. Da Prato, G., Zabczyk, J.: Ergodicity for Infinite Dimensional Systems. London Mathematical Society. Lecture Notes Series, vol. 229. Cambridge University Press, Cambridge (1996)
5. Genon-Catalot, V., Jeantheau, T. : Stochastic Volatility Models as Hidden Markov Models and Statistical Applications. Bernoulli. **6**(6), 1051–1079 (2000)
6. Gobet, E.: LAN property for ergodic diffusions with discrete observations. Annales de l'IHP (B) Probabilite et statistiques, **38**(5), 711–737 (2002)
7. Kohatsu-Higa, A.: High order Itô-Taylor approximations to heat kernels. J. Math. Kyoto Univ. **37**(1), 129–151 (1997)
8. Kohatsu-Higa, A., Vayatis, N., Yasuda, K.: Tuning of a Bayesian Estimator under Discrete Time Observations and Unknown Transition Density, Preprint.
9. Kohatsu-Higa, A., Vayatis, N., Yasuda, K.: Strong consistency of Bayesian estimator for the Ornstein-Uhlenbeck process. To appear in the Proceedings of the Metabief Conference.
10. Nualart, D.: The Malliavin Calculus and Related Topics (2nd). Springer, Berlin (2006)

Chapter 9
Escape Probability for Stochastic Dynamical Systems with Jumps

Huijie Qiao, Xingye Kan, and Jinqiao Duan

Dedicated to Professor David Nualart on the occasion of his 60th birthday

Abstract The escape probability is a deterministic concept that quantifies some aspects of stochastic dynamics. This issue has been investigated previously for dynamical systems driven by Gaussian Brownian motions. The present work considers escape probabilities for dynamical systems driven by non-Gaussian Lévy motions, especially symmetric α-stable Lévy motions. The escape probabilities are characterized as solutions of the Balayage-Dirichlet problems of certain partial differential-integral equations. Differences between escape probabilities for dynamical systems driven by Gaussian and non-Gaussian noises are highlighted. In certain special cases, analytic results for escape probabilities are given.

Keywords Escape probability • Balayage–Dirichlet problem • Lévy processes • Discontinuous stochastic dynamical systems • Nonlocal differential equation • Non-Gaussian noise

AMS Subject Classification (2010): 60H10, 60J75, 35S15, 31C05.

Received 10/31/2011; Accepted 4/3/2012; Final 5/1/2012

H. Qiao
Department of Mathematics, Southeast University, Nanjing, Jiangsu 211189, China
e-mail: hjqiaogean@yahoo.com.cn

X. Kan • J. Duan (✉)
Institute for Pure and Applied Mathematics, University of California,
Los Angeles, CA 90095, USA

Department of Applied Mathematics, Illinois Institute of Technology, Chicago, IL 60616, USA
e-mail: xkan@hawk.iit.edu; duan@iit.edu

F. Viens et al. (eds.), *Malliavin Calculus and Stochastic Analysis: A Festschrift in Honor of David Nualart*, Springer Proceedings in Mathematics & Statistics 34,
DOI 10.1007/978-1-4614-5906-4_9,

1 Introduction

Stochastic dynamical systems arise as mathematical models for complex phenomena in biological, geophysical, physical, and chemical sciences, under random fluctuations. A specific orbit (or trajectory) for such a system could vary wildly from one realization to another, unlike the situation for deterministic dynamical systems. It is desirable to have different concepts for quantifying stochastic dynamical behaviors. The escape probability is such a concept.

Brownian motions are Gaussian stochastic processes and thus are appropriate for modeling Gaussian random fluctuations. Almost all sample paths of Brownian motions are continuous in time. For a dynamical system driven by Brownian motions, almost all orbits (or paths or trajectories) are thus continuous in time. The *escape probability* is the likelihood that an orbit, starting inside an open domain D, exits this domain first through a specific part Γ of the boundary ∂D. This concept helps understand various phenomena in sciences. One example is in molecular genetics [23]. The frequency of collisions of two single strands of long helical DNA molecules that leads to a double-stranded molecule is of interest and can be computed by virtue of solving an escape probability problem. It turns out that the escape probability satisfies an elliptic partial differential equation with properly chosen boundary conditions [4, 16, 22, 23].

Non-Gaussian random fluctuations are widely observed in various areas such as physics, biology, seismology, electrical engineering, and finance [14, 18, 26]. Lévy motions are a large class of non-Gaussian stochastic processes whose sample paths are discontinuous in time. For a dynamical system driven by Lévy motions, almost all the orbits X_t are discontinuous in time. In fact, these orbits are càdlàg (right continuous with left limit at each time instant), that is, each of these orbits has countable jumps in time. Due to these jumps, an orbit could escape an open domain without passing through its boundary. In this case, the *escape probability* is the likelihood that an orbit, starting inside an open domain D, exits this domain first by landing in a target domain U in D^c (the complement of domain D).

As we see, the escape probability is defined slightly differently for dynamical systems driven by Gaussian or non-Gaussian processes. Although the escape probability for the former has been investigated extensively, the characterization for the escape probability for the latter has not been well documented as a dynamical systems analysis tool for applied mathematics and science communities. See our recent works [5, 10] for numerical analysis of escape probability and mean exit time for dynamical systems driven by symmetric α-stable Lévy motions.

In this paper, we carefully derive a partial differential–integral equation to be satisfied by the escape probability for a class of dynamical systems driven by Lévy motions, especially symmetric α-stable Lévy motions. Namely the escape probability is a solution of a nonlocal differential equation. We highlight the differences between escape probabilities for dynamical systems driven by Gaussian and non-Gaussian processes. These are illustrated in a few examples.

More precisely, let $\{X_t, t \geqslant 0\}$ be a $\mathbb{R}^d$-valued Markov process defined on a complete filtered probability space $(\Omega, \mathcal{F}, \{\mathcal{F}_t\}_{t\geqslant 0}, \mathbb{P})$. Let D be an open domain in $\mathbb{R}^d$. Define the *exit time*

$$\tau_{D^c} := \inf\{t > 0 : X_t \in D^c\},$$

where D^c is the complement of D in $\mathbb{R}^d$. Namely, τ_{D^c} is the first time when X_t hits D^c.

When X_t has almost surely continuous paths, that is, X_t is either a Brownian motion or a solution process for a dynamical system driven by Brownian motions, a path starting at $x \in D$ will hit D^c by hitting ∂D first (assume for the moment that ∂D is smooth). Thus $\tau_{D^c} = \tau_{\partial D}$. Let Γ be a subset of the boundary ∂D. The likelihood that X_t, starting at x, exits from D first through Γ is called the escape probability from D to Γ, denoted as $p(x)$. That is,

$$p(x) = \mathbb{P}\{X_{\tau_{\partial D}} \in \Gamma\}.$$

We will verify that (Sect. 3.2) the escape probability $p(x)$ solves the following Dirichlet boundary value problem:

$$\begin{cases} \mathcal{L}p = 0, & x \in D, \\ p|_{\partial D} = \psi, \end{cases} \tag{9.1}$$

where $\mathcal{L}$ is the infinitesimal generator of the process X_t and the boundary data ψ is defined as follows:

$$\psi(x) = \begin{cases} 1, & x \in \Gamma, \\ 0, & x \in \partial D \setminus \Gamma. \end{cases}$$

When X_t has càdlàg paths which have countable jumps in time, that is, X_t could be either a Lévy motion or a solution process of a dynamical system driven by Lévy motions, the first hitting of D^c may occur somewhere in D^c. For this reason, we take a subset U of D^c and define the likelihood that X_t exits firstly from D by landing in the target set U as the escape probability from D to U, also denoted by $p(x)$. That is,

$$p(x) = \mathbb{P}\{X_{\tau_{D^c}} \in U\}.$$

We will demonstrate that (Sect. 3.4) the escape probability $p(x)$ solves the following Balayage–Dirichlet boundary value problem:

$$\begin{cases} Ap = 0, & x \in D, \\ p|_{D^c} = \varphi, \end{cases} \tag{9.2}$$

where A is the characteristic operator of X_t and φ is defined as follows:

$$\varphi(x) = \begin{cases} 1, & x \in U, \\ 0, & x \in D^c \setminus U. \end{cases}$$

Therefore by solving a deterministic boundary value problem (9.1) or (9.2), we obtain the escape probability $p(x)$.

This paper is arranged as follows. In Sect. 2, we introduce Balayage–Dirichlet problem for discontinuous Markov processes and also define Lévy motions. The main result is stated and proved in Sect. 3. In Sect. 4, we present analytic solutions for escape probabilities in a few special cases.

2 Preliminaries

In this section, we recall basic concepts and results that will be needed throughout the paper.

2.1 Balayage–Dirichlet Problem for Discontinuous Markov Processes

The following materials are from [3, 6, 11, 15, 17, 24]. Let $\mathcal{G}$ be a locally compact space with a countable base and $\mathscr{G}$ be the Borel σ-field of $\mathcal{G}$. Also, ς is adjoined to $\mathcal{G}$ as the point at infinity if $\mathcal{G}$ is noncompact and as an isolated point if $\mathcal{G}$ is compact. Furthermore, let $\mathscr{G}_\varsigma$ be the σ-field of Borel sets of $\mathcal{G}_\varsigma = \mathcal{G} \cup \{\varsigma\}$.

Definition 2.1. A Markov process Y with state space $(\mathcal{G}, \mathscr{G})$ is called a Hunt process provided:

(i) The path functions $t \to Y_t$ are right continuous on $[0, \infty)$ and have left-hand limits on $[0, \zeta)$ almost surely, where $\zeta := \inf\{t : Y_t = \varsigma\}$.
(ii) Y is strong Markov.
(iii) Y is quasi-left-continuous: whenever $\{\tau_n\}$ is an increasing sequence of $\mathcal{F}_t$-stopping times with limit τ, then almost surely $Y_{\tau_n} \to Y_\tau$ on $\{\tau < \infty\}$.

Definition 2.2. Let G be an open subset of $\mathcal{G}$ and $Y_t(x)$ be a Hunt process starting at $x \in G$. A nonnegative function h defined on $\mathcal{G}$ is said to be harmonic with respect to Y_t in G if for every compact set $K \subset G$,

$$\mathbb{E}[h(Y_{\tau_{K^c}}(x))] = h(x), \quad x \in G.$$

Definition 2.3. Let f be nonnegative on G^c. We say h defined on $\mathcal{G}$ solves the Balayage–Dirichlet problem for G with "boundary value" f, denoted by (G, f), if $h = f$ on G^c, h is harmonic with respect to Y_t in G and further satisfies the following boundary condition:

$$\forall z \in \partial G, \quad h(y) \to f(z), \quad \text{as } y \to z \text{ from inside } G.$$

A point $z \in \partial G$ is called regular for G^c with respect to $Y_t(z)$ if

$$\mathbb{P}\{\tau_{G^c} = 0\} = 1.$$

Here G is said to be regular if any $z \in \partial G$ is regular for G^c.

Let ρ be a metric on $\mathcal{G}$ compatible with the given topology. Let $\mathcal{I}_G$ be the family of functions $g \geqslant 0$ bounded on $\mathcal{G}$ and lower semicontinuous in G such that $\forall x \in G$, there is a number $Ag(x)$ satisfying

$$\frac{\mathbb{E}[g(Y_{\tau_\varepsilon}(x))] - g(x)}{\mathbb{E}[\tau_\varepsilon]} \to Ag(x), \text{ as } \varepsilon \downarrow 0,$$

where $\tau_\varepsilon := \inf\{t > 0 : \rho(Y_t(x), x) > \varepsilon\}$. We call A with domain $\mathcal{I}_G$ the characteristic operator of Y_t relative to G. If $\mathcal{L}$ with domain $\mathcal{D}_G$ is the infinitesimal generator of Y_t relative to G, $\mathcal{D}_G \subseteq \mathcal{I}_G$, and

$$Af = \mathcal{L}f, \quad f \in \mathcal{D}_G.$$

(cf. [9])

We quote the following result about the existence and regularity of the solution for the Balayage–Dirichlet problem.

Theorem 2.1 ([15]). *Suppose that G is relatively compact and regular and f is nonnegative and bounded on G^c. If f is continuous at any $z \in \partial G$, then $h(x) = \mathbb{E}[f(Y_{\tau_{G^c}}(x))]$ is the unique solution to the Balayage–Dirichlet problem (G, f), and $Ah(x) = 0$ for $h \in I_G$.*

2.2 Lévy Motions

Definition 2.4. A process L_t, with $L_0 = 0$ a.s. is a d-dimensional Lévy process or Lévy motion if:

(i) L_t has independent increments; that is, $L_t - L_s$ is independent of $L_v - L_u$ if $(u, v) \cap (s, t) = \emptyset$.
(ii) L_t has stationary increments; that is, $L_t - L_s$ has the same distribution as $L_v - L_u$ if $t - s = v - u > 0$.
(iii) L_t is stochastically continuous.
(iv) L_t is right continuous with left limit.

The characteristic function for L_t is given by

$$\mathbb{E}\left(\exp\{i\langle z, L_t\rangle\}\right) = \exp\{t\Psi(z)\}, \quad z \in \mathbb{R}^d,$$

where $\langle \cdot, \cdot \rangle$ is the scalar product in $\mathbb{R}^d$. The function $\Psi : \mathbb{R}^d \to C$ is called the characteristic exponent of the Lévy process L_t. By the Lévy–Khintchine formula, there exist a nonnegative definite $d \times d$ matrix Q, a measure ν on $\mathbb{R}^d$ satisfying

$$\nu(\{0\}) = 0 \text{ and } \int_{\mathbb{R}^d \setminus \{0\}} (|u|^2 \wedge 1)\nu(du) < \infty,$$

and $\gamma \in \mathbb{R}^d$ such that

$$\Psi(z) = i\langle z, \gamma\rangle - \frac{1}{2}\langle z, Qz\rangle + \int_{\mathbb{R}^d \setminus \{0\}} \left(e^{i\langle z,u\rangle} - 1 - i\langle z, u\rangle 1_{|u|\le 1}\right)\nu(du). \tag{9.3}$$

The measure ν is called the Lévy measure of L_t, Q is the diffusion matrix, and γ is the drift vector.

We now introduce a special class of Lévy motions, that is, the symmetric α-stable Lévy motions L_t^α.

Definition 2.5. For $\alpha \in (0, 2)$, a d-dimensional symmetric α-stable Lévy motion L_t^α is a Lévy process with characteristic exponent

$$\Psi(z) = -C|z|^\alpha, \quad z \in \mathbb{R}^d, \tag{9.4}$$

where

$$C = \pi^{-1/2}\frac{\Gamma((1+\alpha)/2)\Gamma(d/2)}{\Gamma((d+\alpha)/2)}.$$

(cf. [21, p. 115] for the above formula of C.)

Thus, for a d-dimensional symmetric α-stable Lévy motion L_t^α, the diffusion matrix $Q = 0$, the drift vector $\gamma = 0$, and the Lévy measure ν is given by

$$\nu(du) = \frac{C_{d,\alpha}}{|u|^{d+\alpha}}du,$$

where

$$C_{d,\alpha} = \frac{\alpha\Gamma((d+\alpha)/2)}{2^{1-\alpha}\pi^{d/2}\Gamma(1-\alpha/2)}.$$

(cf. [7, p. 1312] for the above formula of $C_{d,\alpha}$.) Moreover, comparing Eq. (9.4) with Eq. (9.3), we obtain

$$-C|z|^\alpha = \int_{\mathbb{R}^d \setminus \{0\}} \left(e^{i\langle z,u\rangle} - 1 - i\langle z, u\rangle 1_{|u|\le 1}\right)\frac{C_{d,\alpha}}{|u|^{d+\alpha}}du.$$

Let $C_0(\mathbb{R}^d)$ be the space of continuous functions f on $\mathbb{R}^d$ satisfying $\lim_{|x|\to\infty} f(x) = 0$ with norm $\|f\|_{C_0(\mathbb{R}^d)} = \sup_{x\in\mathbb{R}^d} |f(x)|$. Let $C_0^2(\mathbb{R}^d)$ be the set of

$f \in \mathcal{C}_0(\mathbb{R}^d)$ such that f is two times differentiable and the first and second order partial derivatives of f belong to $\mathcal{C}_0(\mathbb{R}^d)$. Let $\mathcal{C}_c^\infty(\mathbb{R}^d)$ stand for the space of all infinitely differentiable functions on $\mathbb{R}^d$ with compact supports. Define

$$(\mathcal{L}_\alpha f)(x) := \int_{\mathbb{R}^d\setminus\{0\}} \big(f(x+u) - f(x) - \langle \partial_x f(x), u\rangle 1_{|u|\le 1}\big)\frac{C_{d,\alpha}}{|u|^{d+\alpha}}\mathrm{d}u$$

on $\mathcal{C}_0^2(\mathbb{R}^d)$. And then for $\xi \in \mathbb{R}^d$

$$(\mathcal{L}_\alpha \mathrm{e}^{i\langle\cdot,\xi\rangle})(x) = \mathrm{e}^{i\langle x,\xi\rangle}\int_{\mathbb{R}^d\setminus\{0\}} \big(\mathrm{e}^{i\langle u,\xi\rangle} - 1 - i\langle \xi, u\rangle 1_{|u|\le 1}\big)\frac{C_{d,\alpha}}{|u|^{d+\alpha}}\mathrm{d}u.$$

By Courrège's second theorem [1, Theorem 3.5.5, p. 183], for every $f \in \mathcal{C}_c^\infty(\mathbb{R}^d)$

$$\begin{aligned}
&(\mathcal{L}_\alpha f)(x)\\
&= \frac{1}{(2\pi)^{d/2}}\int_{\mathbb{R}^d} \mathrm{e}^{i\langle z,x\rangle}\left[\mathrm{e}^{-i\langle x,z\rangle}(\mathcal{L}_\alpha \mathrm{e}^{i\langle\cdot,z\rangle})(x)\right]\hat{f}(z)\mathrm{d}z\\
&= \frac{1}{(2\pi)^{d/2}}\int_{\mathbb{R}^d} \mathrm{e}^{i\langle z,x\rangle}\left[\int_{\mathbb{R}^d\setminus\{0\}} \big(\mathrm{e}^{i\langle u,z\rangle} - 1 - i\langle z, u\rangle 1_{|u|\le 1}\big)\frac{C_{d,\alpha}}{|u|^{d+\alpha}}\mathrm{d}u\right]\hat{f}(z)\mathrm{d}z\\
&= -\frac{C}{(2\pi)^{d/2}}\int_{\mathbb{R}^d} \mathrm{e}^{i\langle z,x\rangle}|z|^\alpha \hat{f}(z)\mathrm{d}z\\
&= C\cdot[-(-\Delta)^{\alpha/2} f](x).
\end{aligned}$$

Set $p_t := L_t - L_{t-}$. Then p_t defines a stationary $(\mathcal{F}_t)$-adapted Poisson point process with values in $\mathbb{R}^d\setminus\{0\}$ [12]. And the characteristic measure of p is the Lévy measure ν. Let $N_p((0,t],\mathrm{d}u)$ be the counting measure of p_t, that is, for $B \in \mathcal{B}(\mathbb{R}^d\setminus\{0\})$

$$N_p((0,t],B) := \#\{0 < s \le t : p_s \in B\},$$

where # denotes the cardinality of a set. The compensator measure of N_p is given by

$$\tilde{N}_p((0,t],\mathrm{d}u) := N_p((0,t],\mathrm{d}u) - t\nu(\mathrm{d}u).$$

The Lévy–Itô theorem states that for a symmetric α-stable process L_t:

1. For $1 \le \alpha < 2$,

$$L_t = \int_0^t\int_{|u|\le 1} u\tilde{N}_p(\mathrm{d}s,\mathrm{d}u) + \int_0^t\int_{|u|>1} uN_p(\mathrm{d}s,\mathrm{d}u).$$

2. For $0 < \alpha < 1$,

$$L_t = \int_0^t\int_{\mathbb{R}^d\setminus\{0\}} uN_p(\mathrm{d}s,\mathrm{d}u).$$

3 Boundary Value Problems for Escape Probability

In this section, we formulate boundary value problems for the escape probability associated with Brownian motions, SDEs driven by Brownian motions, Lévy motions, and SDEs driven by Lévy motions. For Lévy motions, in particular, we consider symmetric α-stable Lévy motions. We will see that the escape probability can be found by solving deterministic partial differential equations or partial differential–integral equations, with properly chosen boundary conditions.

3.1 Boundary Value Problem for Escape Probability of Brownian Motions

Suppose that a particle executes an unbiased random walk on a straight line. Let $D = (a, b)$. Figure 9.1 shows the random walk scenario. That is, a particle moves according to the following rules [16]:

1. During the passage of a certain fixed time interval, a particle takes 1 step of a certain fixed length δ along the x axis.
2. It is equally probable that the step is to the right or to the left.

If the particle starting from $x \in D$ eventually escapes D by crossing the boundary b, then it must have moved to one of the two points adjacent to x first and then crossed the boundary. Thus

$$p(x) = \frac{1}{2}[p(x-\delta) + p(x+\delta)],$$

for $x \in D$. By Taylor expansion on the right-hand side to the second order, we have

$$\frac{1}{2}p''(x) = 0.$$

The boundary conditions are

$$\lim_{x \to b} p(x) = 1, \quad \lim_{x \to a} p(x) = 0,$$

since the nearer the particle starts to b, the more likely it will first cross the boundary through b.

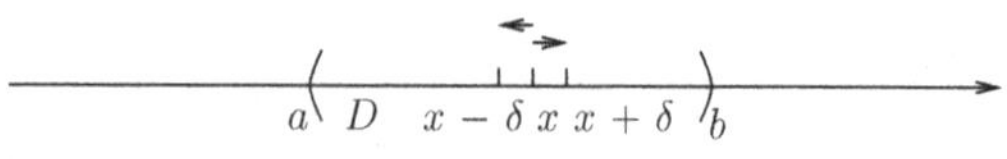

Fig. 9.1 A particle executing unbiased random walk in a bounded interval

Note that the limit of the random walk is a standard Brownian motion W_t, that is:

1. W has independent increments.
2. For $0 < s < t$, $W_t - W_s$ is a Gaussian random variable with mean zero and variance $(t - s)$.

Thus, the escape probability $p(x)$ of a standard Brownian motion from D through the boundary b satisfies

$$\begin{cases} \frac{1}{2}\Delta p(x) = 0, \\ p(b) = 1, \\ p(a) = 0, \end{cases}$$

where $\frac{1}{2}\Delta = \frac{1}{2}\partial_{xx}$ is the infinitesimal generator for a scalar standard Brownian motion W_t.

3.2 *Boundary Value Problem for Escape Probability of* SDEs *Driven by Brownian Motions*

Some results in this section can be found in [19, Chap. 9].

Let $\{W(t)\}_{t \geqslant 0}$ be an m-dimensional standard $\mathcal{F}_t$-adapted Brownian motion. Consider the following stochastic differential equation (SDE) in $\mathbb{R}^d$:

$$X_t(x) = x + \int_0^t b(X_s(x))\,\mathrm{d}s + \int_0^t \sigma(X_s(x))\,\mathrm{d}W_s. \tag{9.5}$$

We make the following assumptions about the drift $b : \mathbb{R}^d \mapsto \mathbb{R}^d$ and the diffusion coefficient $\sigma : \mathbb{R}^d \mapsto \mathbb{R}^d \times \mathbb{R}^m$:

$(\mathbf{H}^1_{b,\sigma})$

$$|b(x) - b(y)| \leq \lambda(|x - y|),$$
$$|\sigma(x) - \sigma(y)| \leq \gamma(|x - y|).$$

Here λ and γ are increasing concave functions with the properties $\lambda(0) = \gamma(0) = 0$, and $\int_{0+} \frac{1}{\lambda(u)}\mathrm{d}u = \int_{0+} \frac{1}{\gamma^2(u)}\mathrm{d}u = \infty$.

Under $(\mathbf{H}^1_{b,\sigma})$, it is well known that there exists a unique strong solution to Eq. (9.5) ([27]). This solution is denoted by $X_t(x)$.

We also make the following assumption.

$(\mathbf{H}^2_{\sigma})$ There exists a $\xi > 0$ such that for any $x, y \in D$

$$\langle y, \sigma\sigma^*(x)y\rangle \geqslant \xi|y|^2.$$

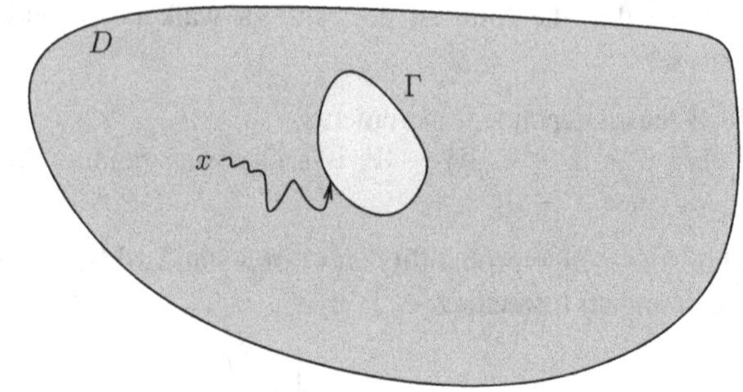

Fig. 9.2 Escape probability for SDEs driven by Brownian motions: an annular open domain D with a subset Γ of its boundary ∂D

This condition guarantees that the infinitesimal generator

$$\mathcal{L} := \sum_{i=1}^{d} b_i(x)\frac{\partial}{\partial x_i} + \sum_{i,j=1}^{d} a_{ij}(x)\frac{\partial^2}{\partial x_i \partial x_j}$$

for Eq. (9.5) is uniformly elliptic in D, since then the eigenvalues of $\sigma\sigma^*$ are away from 0 in D. Here the matrix $[a_{ij}] := \frac{1}{2}\sigma(x)\sigma^*(x)$.

Let D be an open annular domain as in Fig. 9.2. In one-dimensional case, it is just an open interval. Let Γ be its inner (or outer) boundary. Taking

$$\psi(x) = \begin{cases} 1, & x \in \Gamma, \\ 0, & x \in \partial D \setminus \Gamma, \end{cases} \tag{9.6}$$

we have

$$\begin{aligned} \mathbb{E}[\psi(X_{\tau_{\partial D}}(x))] &= \int_{\{\omega: X_{\tau_{\partial D}}(x)\in\Gamma\}} \psi(X_{\tau_{\partial D}}(x))\mathrm{d}\mathbb{P}(\omega) \\ &\quad + \int_{\{\omega: X_{\tau_{\partial D}}(x)\in\partial D\setminus\Gamma\}} \psi(X_{\tau_{\partial D}}(x))\mathrm{d}\mathbb{P}(\omega) \\ &= \mathbb{P}\{\omega : X_{\tau_{\partial D}}(x) \in \Gamma\} \\ &= p(x). \end{aligned}$$

This means that, for this specific ψ, $\mathbb{E}[\psi(X_{\tau_{\partial D}}(x))]$ is the escape probability $p(x)$, which we are looking for.

We need to use [19, Theorem 9.2.14] or [8] in order to see that the escape probability $p(x)$ is closely related to a harmonic function with respect to X_t. This requires that the boundary data ψ to be bounded and continuous on ∂D. For the domain D taken as in Fig. 9.2, with Γ the inner boundary (or outer) boundary, the above chosen ψ in Eq. (9.6) is indeed bounded and continuous on ∂D. Thus, we have the following result by [19, Theorem 9.2.14].

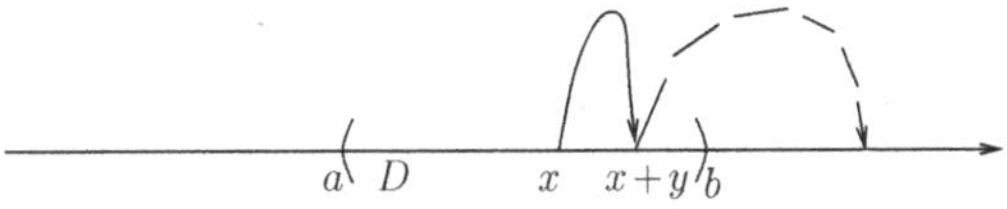

Fig. 9.3 A particle executing Lévy motion in a bounded interval

Theorem 3.1. *The escape probability $p(x)$ from an open annular domain D to its inner (or outer) boundary Γ, for the dynamical system driven by Brownian motions (9.5), is the solution to the following Dirichlet boundary value problem:*

$$\begin{cases} \mathcal{L}p = 0, \\ p|_{\Gamma} = 1, \\ p|_{\partial D\setminus\Gamma} = 0. \end{cases}$$

3.3 Boundary Value Problem for Escape Probability of Symmetric α-Stable Lévy Motions

Assume a particle is taking a one-dimensional Lévy flight, where the distribution of step sizes is a symmetric α-stable distribution (Fig. 9.3). Let $p(x)$ denote the escape probability of the particle starting at x in $D = (a,b)$ and then first escapes D over the right boundary b. It could first move to somewhere inside D, say $x + y \in D$, and then achieve its goal by jumping over the right boundary b from the new starting point $x + y$. More precisely,

$$p(x) = \int_{\mathbb{R}\setminus\{0\}} P\{\text{the first step length is } y\} p(x+y)\mathrm{d}y. \tag{9.7}$$

According to [2], the symmetric α-stable probability density function is the following:

$$f_{\alpha,0}(y) = \begin{cases} -\frac{1}{\pi}\sum_{k=1}^{\infty}\frac{(-1)^k}{k!}\frac{\Gamma(\alpha k+1)}{y|y|^{\alpha k}}\sin[k(\frac{\alpha\pi}{2}-\alpha\arg y)], & 0<\alpha<1, \\ \frac{1}{\pi}\sum_{k=0}^{\infty}(-1)^k\frac{\Gamma(\frac{k+1}{\alpha})}{k!\alpha}y^k\cos[k(\frac{\pi}{2})], & 1<\alpha<2, \end{cases}$$

where $\arg y = \pi$ when $y < 0$.

For $0 < \alpha < 2$, the asymptotic expansion has also been given by [2] as follows:

$$\begin{aligned} f_{\alpha,0}(y) &= \begin{cases} -\frac{1}{\pi}\sum_{k=1}^{n}\frac{(-1)^k}{k!}\frac{\Gamma(\alpha k+1)}{y|y|^{\alpha k}}\sin\left[k(\frac{\alpha\pi}{2}-\alpha\arg y)\right]+o(|y|^{-\alpha(n+1)-1}), & |y|\to\infty, \\ \frac{1}{\pi}\sum_{k=0}^{n}(-1)^k\frac{\Gamma(\frac{k+1}{\alpha})}{k!\alpha}y^k\cos\left[k(\frac{\pi}{2})\right]+o(|y|^{n+1}), & |y|\to 0, \end{cases} \\ &= \begin{cases} C_1(\alpha)/|y|^{1+\alpha}+o(|y|^{-2\alpha-1}), & |y|\to\infty, \\ C_2(\alpha)+o(|y|^2), & |y|\to 0, \end{cases} \end{aligned}$$

where $C_1(\alpha) = \frac{1}{\pi}\sin(\frac{\pi\alpha}{2})\Gamma(1+\alpha)$ and $C_2(\alpha) = \frac{1}{\pi}\frac{\Gamma(1/\alpha)}{\alpha}$. Take $N > 0$ large enough and fix it. Thus,

$$\begin{aligned} 0 &= \int_{\mathbb{R}\setminus\{0\}} f_{\alpha,0}(y)[p(x+y)-p(x)]\mathrm{d}y \\ &= \int_{(-N,N)\setminus\{0\}} f_{\alpha,0}(y)[p(x+y)-p(x)]\mathrm{d}y \\ &\quad + \int_{\mathbb{R}\setminus(-N,N)} f_{\alpha,0}(y)[p(x+y)-p(x)]\mathrm{d}y \\ &=: I_1 + I_2. \end{aligned}$$

For I_1, by self-affine property in [25], we obtain

$$\begin{aligned} I_1 &= \int_{(-N,N)\setminus\{0\}} f_{\alpha,0}\left(N\cdot\frac{y}{N}\right)[p(x+y)-p(x)]\mathrm{d}y \\ &= \int_{(-N,N)\setminus\{0\}} \frac{f_{\alpha,0}(N)}{(|y|/N)^{1+\alpha}}[p(x+y)-p(x)]\mathrm{d}y \\ &= \int_{(-N,N)\setminus\{0\}} \frac{C_1(\alpha)}{N^{1+\alpha}(|y|/N)^{1+\alpha}}[p(x+y)-p(x)]\mathrm{d}y \\ &= \int_{(-N,N)\setminus\{0\}} \frac{C_1(\alpha)}{|y|^{1+\alpha}}[p(x+y)-p(x)]\mathrm{d}y. \end{aligned} \tag{9.8}$$

For I_2, we calculate

$$\begin{aligned} I_2 &= \int_N^{\infty} f_{\alpha,0}(y)[1-p(x)]dy + \int_{-\infty}^{-N} f_{\alpha,0}(y)[0-p(x)]\mathrm{d}y \\ &= \int_N^{\infty}\left[\frac{C_1(\alpha)}{y^{1+\alpha}} + o\left(\frac{1}{y^{1+2\alpha}}\right)\right][1-p(x)]\mathrm{d}y \\ &\quad - \int_{-\infty}^{-N}\left[\frac{C_1(\alpha)}{(-y)^{1+\alpha}} + o\left(\frac{1}{(-y)^{1+2\alpha}}\right)\right]p(x)\mathrm{d}y \\ &= \int_N^{\infty}\frac{C_1(\alpha)}{y^{1+\alpha}}\mathrm{d}y - \int_{\mathbb{R}\setminus[-N,N]}\frac{C_1(\alpha)}{|y|^{1+\alpha}}p(x)\mathrm{d}y \\ &= \int_{\mathbb{R}\setminus[-N,N]}\frac{C_1(\alpha)p(x+y)}{|y|^{1+\alpha}}\mathrm{d}y - \int_{\mathbb{R}\setminus[-N,N]}\frac{C_1(\alpha)}{|y|^{1+\alpha}}p(x)\mathrm{d}y. \end{aligned} \tag{9.9}$$

Note that for $0 < \alpha < 1$, by the fact that the integral of an odd function on a symmetric interval is zero, it holds that

$$\int_{\mathbb{R}\setminus\{0\}} p'(x)yI_{\{|y|\le 1\}}\frac{C_1(\alpha)}{|y|^{1+\alpha}}\mathrm{d}y = p'(x)C_1(\alpha)\int_{\{|y|\le 1\}\setminus\{0\}}\frac{y}{|y|^{1+\alpha}}\mathrm{d}y = 0. \tag{9.10}$$

Thus, putting Eqs. (9.8), (9.9), and (9.10) together, we have for $0 < \alpha < 1$

$$\int_{\mathbb{R}\backslash\{0\}} \left[p(x+y) - p(x) - p'(x)yI_{\{|y|\leq 1\}}\right] \frac{C_1(\alpha)}{|y|^{1+\alpha}} \mathrm{d}y = 0.$$

Moreover, $C_1(\alpha) = C_{1,\alpha}$.

For $\alpha \in [1,2)$, we only divide I_1 into two parts I_{11} and I_{12}, where

$$I_{11} := \int_{\{|y|\leq \varrho\}\backslash\{0\}} f_{\alpha,0}(y)[p(x+y) - p(x)]\mathrm{d}y,$$

$$I_{12} := \int_{(-N,N)\backslash(-\varrho,\varrho)} f_{\alpha,0}(y)[p(x+y) - p(x)]\mathrm{d}y,$$

and $\varrho > 0$ is a small enough constant.

For I_{11}, by Taylor expansion and self-affine property in [23], we get

$$\begin{aligned}
&\int_{\{|y|\leq \varrho\}\backslash\{0\}} f_{\alpha,0}(y)[p(x+y) - p(x)]\mathrm{d}y \\
&= \int_{\{|y|\leq \varrho\}\backslash\{0\}} f_{\alpha,0}\left(\frac{1}{\varrho}\cdot \varrho y\right) p'(x) y \mathrm{d}y \\
&= \int_{\{|y|\leq \varrho\}\backslash\{0\}} \frac{f_{\alpha,0}(\frac{1}{\varrho})}{(\varrho|y|)^{1+\alpha}} p'(x) y \mathrm{d}y \\
&= \int_{\{|y|\leq \varrho\}\backslash\{0\}} \frac{C_1(\alpha)}{(\frac{1}{\varrho})^{1+\alpha}\cdot(\varrho|y|)^{1+\alpha}} p'(x) y \mathrm{d}y \\
&= \int_{\mathbb{R}\backslash\{0\}} p'(x) y I_{\{|y|\leq \varrho\}} \frac{C_1(\alpha)}{|y|^{1+\alpha}} \mathrm{d}y.
\end{aligned}$$

For I_{12}, we apply the same technique as that in dealing with I_1 for $\alpha \in (0,1)$. Next, by the similar calculation to that for $\alpha \in (0,1)$, we obtain for $\alpha \in [1,2)$

$$\int_{\mathbb{R}\backslash\{0\}} \left[p(x+y) - p(x) - p'(x)yI_{\{|y|\leq 1\}}\right] \frac{C_1(\alpha)\mathrm{d}y}{|y|^{1+\alpha}} = 0.$$

Since the limit of the Lévy flight is a symmetric α-stable Lévy motion L_t^α, the escape probability $p(x)$ of a symmetric α-stable Lévy motion, from D to $[b,\infty)$ satisfies

$$\begin{cases}
-(-\Delta)^{\frac{\alpha}{2}} p(x) = 0, \\
p(x)|_{[b,\infty)} = 1, \\
p(x)|_{(-\infty,a]} = 0.
\end{cases}$$

Note that $-(-\Delta)^{\frac{\alpha}{2}}$ is the infinitesimal generator for a scalar symmetric α-stable Lévy motion L_t^α.

3.4 Boundary Value Problem for Escape Probability of SDEs Driven by General Lévy Motions

Let L_t be a Lévy process independent of W_t. Consider the following SDE in $\mathbb{R}^d$:

$$X_t(x) = x + \int_0^t b(X_s(x))\,\mathrm{d}s + \int_0^t \sigma(X_s(x))\,\mathrm{d}W_s + L_t. \tag{9.11}$$

Assume that the drift b and the diffusion σ satisfy the following conditions:

($\mathbf{H}_b$) There exists a constant $C_b > 0$ such that for $x, y \in \mathbb{R}^d$

$$|b(x) - b(y)| \leq C_b|x - y| \cdot \log(|x - y|^{-1} + \mathrm{e}).$$

($\mathbf{H}_\sigma$) There exists a constant $C_\sigma > 0$ such that for $x, y \in \mathbb{R}^d$

$$|\sigma(x) - \sigma(y)|^2 \leq C_\sigma|x - y|^2 \cdot \log(|x - y|^{-1} + \mathrm{e}).$$

Under ($\mathbf{H}_b$) and ($\mathbf{H}_\sigma$), it is well known that there exists a unique strong solution to Eq. (9.11) (see [20]). This solution will be denoted by $X_t(x)$. Moreover, $X_t(x)$ is continuous in x.

Lemma 3.1. *The solution process $X_t(x)$ of the SDE* (9.11) *is a strong Markov process.*

Proof. Let η be a $(\mathcal{F}_t)_{t\geqslant 0}$-stopping time. Set

$$\mathcal{G}_t := \sigma\{W_{\eta+t} - W_\eta, L_{\eta+t} - L_\eta\} \cup \mathcal{N}, \quad t \geqslant 0,$$

where $\mathcal{N}$ is of all P-zero sets. That is, $\mathcal{G}_t$ is a completed σ-algebra generated by $W_{\eta+t} - W_\eta$ and $L_{\eta+t} - L_\eta$. Besides, $\mathcal{G}_t$ is independent of $\mathcal{F}_t$. Let $X(x, \eta, \eta + t)$ denote the unique solution of the following SDE:

$$X(x,\eta,\eta+t) = x + \int_\eta^{\eta+t} b(X(x,\eta,s))\,\mathrm{d}s + \int_\eta^{\eta+t} \sigma(X(x,\eta,s))\,\mathrm{d}W_s + L_{\eta+t} - L_\eta. \tag{9.12}$$

Moreover, $X(x, \eta, \eta + t)$ is $\mathcal{G}_t$-measurable and $X(x, 0, t) = X_t(x)$. By the uniqueness of the solution to Eq. (9.12), we have

$$X(x, 0, \eta + t) = X(X(x, 0, \eta), \eta, \eta + t), \quad a.s..$$

For any bounded measurable function g,

$$\begin{aligned}\mathbb{E}[g(X_{\eta+t}(x))|\mathcal{F}_\eta] &= \mathbb{E}[g(X(x,0,\eta+t))|\mathcal{F}_\eta] \\ &= \mathbb{E}[g(X(X(x,0,\eta),\eta,\eta+t))|\mathcal{F}_\eta] \\ &= \mathbb{E}[g(X(y,\eta,\eta+t))]|_{y=X(x,0,\eta)} \\ &= \mathbb{E}[g(X(y,0,t))]|_{y=X(x,0,\eta)}. \end{aligned} \tag{9.13}$$

Here the last equality holds because the distribution of $X(y,\eta,\eta+t)$ is the same to that of $X(y,0,t)$. The proof is completed since Eq. (9.13) implies that

$$\mathbb{E}[g(X_{\eta+t}(x))|\mathcal{F}_\eta] = \mathbb{E}[g(X_{\eta+t}(x))|X_\eta(x)].$$

□

Because L_t has càdlàg and quasi-left-continuous paths ([21]), $X_t(x)$ also has càdlàg and quasi-left-continuous paths. Thus by Lemma 3.1 and Definition 2.1, we see that $X_t(x)$ is a Hunt process. Let D be a relatively compact and regular open domain (Fig. 9.4 or Fig. 9.5). Theorem 2.1 implies that $\mathbb{E}[\varphi(X_{\tau_{D^c}}(x))]$ is the unique solution to the Balayage–Dirichlet problem (D,φ), under the condition that φ is nonnegative and bounded on D^c. Set

$$\varphi(x) = \begin{cases} 1, & x \in U, \\ 0, & x \in D^c \setminus U. \end{cases}$$

Then φ is nonnegative and bounded on D^c. We observe that

$$\begin{aligned}\mathbb{E}[\varphi(X_{\tau_{D^c}}(x))] &= \int_{\{\omega: X_{\tau_{D^c}}(x)\in U\}} \varphi(X_{\tau_{D^c}}(x))\mathrm{d}\mathbb{P}(\omega) \\ &\quad + \int_{\{\omega: X_{\tau_{D^c}}(x)\in D^c\setminus U\}} \varphi(X_{\tau_{D^c}}(x))\mathrm{d}\mathbb{P}(\omega) \\ &= \mathbb{P}\{\omega : X_{\tau_{D^c}}(x) \in U\} \\ &= p(x). \end{aligned}$$

This means that, for this specific φ, $\mathbb{E}[\varphi(X_{\tau_{D^c}}(x))]$ is the escape probability $p(x)$ that we are looking for. By the definition of the characteristic operator, $p \in I_D$, and by Theorem 2.1, $Ap(x) = 0$. Thus we obtain the following theorem.

Theorem 3.2. *Let D be a relatively compact and regular open domain, and let U be a set in D^c. Then the escape probability $p(x)$, for the dynamical system driven by Lévy motions* (9.11), *from D to U, is the solution of the following Balayage–Dirichlet problem:*

$$\begin{cases} Ap = 0, \\ p|_U = 1, \\ p|_{D^c\setminus U} = 0, \end{cases}$$

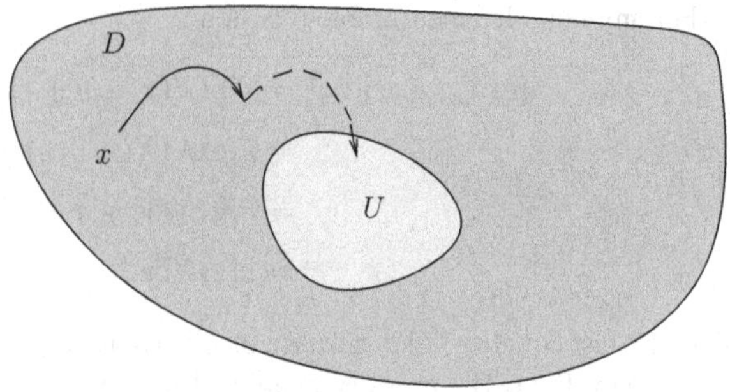

Fig. 9.4 Escape probability for SDEs driven by Lévy motions: an open annular domain D, with its inner part U (which is in D^c) as a target domain

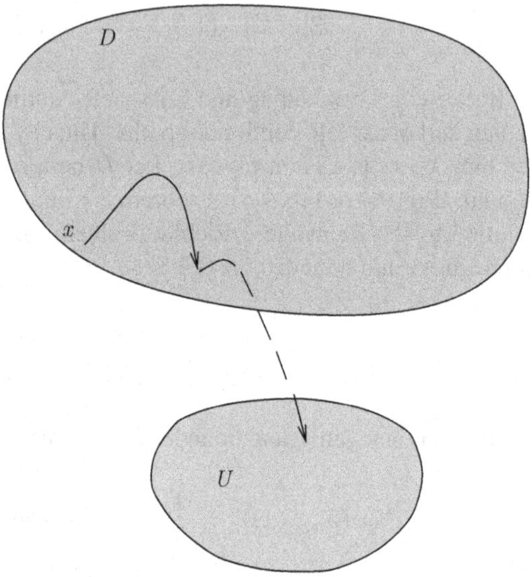

Fig. 9.5 Escape probability for SDEs driven by Lévy motions: a general open domain D, with a target domain U in D^c

where A is the characteristic operator for this system.

Remark 3.1. Unlike the SDEs driven by Brownian motions, a typical open domain D here could be a quite general open domain (Fig. 9.5), as well as an annular domain (Fig. 9.4). This is due to the jumping properties of the solution paths. It is also due to the fact that, in Theorem 2.1, the function f is only required to be continuous on the boundary ∂D (not on the domain D^c).

Finally we consider the representation of the characteristic operator A, for an SDE driven by a symmetric α-stable Lévy process L_t^α, with $\alpha \in (0, 2)$:

$$X_t(x) = x + \int_0^t b(X_s(x))\,\mathrm{d}s + \int_0^t \sigma(X_s(x))\,\mathrm{d}W_s + L_t^\alpha. \tag{9.14}$$

Let us first consider the case of $1 \leq \alpha < 2$. For $f \in C_0^2(\mathbb{R}^d)$, applying the Itô formula to $f(X_{\tau_\varepsilon}(x))$, we obtain

$$
\begin{aligned}
f(X_{\tau_\varepsilon}(x)) - f(x) &= \int_0^{\tau_\varepsilon} \langle \partial_y f(X_s), b(X_s)\rangle \mathrm{d}s + \int_0^{\tau_\varepsilon} \langle \partial_y f(X_s), \sigma(X_s)\mathrm{d}W_s\rangle \\
&+ \int_0^{\tau_\varepsilon} \int_{|u|\le 1} (f(X_s+u) - f(X_s))\, \tilde{N}_p(\mathrm{d}s, \mathrm{d}u) \\
&+ \int_0^{\tau_\varepsilon} \int_{|u|>1} (f(X_s+u) - f(X_s))\, N_p(\mathrm{d}s, \mathrm{d}u) \\
&+ \frac{1}{2}\int_0^{\tau_\varepsilon} \left(\frac{\partial^2}{\partial y_i \partial y_j} f(X_s)\right) \sigma_{ik}(X_s)\sigma_{kj}(X_s)\mathrm{d}s \\
&+ \int_0^{\tau_\varepsilon} \int_{|u|\le 1} (f(X_s+u) - f(X_s) \\
&- \langle \partial_y f(X_s), u\rangle) \frac{C_{d,\alpha}}{|u|^{d+\alpha}} \mathrm{d}u\mathrm{d}s.
\end{aligned}
$$

Here and hereafter, we use the convention that repeated indices imply summation from 1 to d. Taking expectation on both sides, we get

$$
\begin{aligned}
&\mathbb{E}[f(X_{\tau_\varepsilon}(x))] - f(x) \\
&= \mathbb{E}\int_0^{\tau_\varepsilon} \langle \partial_y f(X_s), b(X_s)\rangle \mathrm{d}s + \frac{1}{2}\mathbb{E}\int_0^{\tau_\varepsilon} \left(\frac{\partial^2}{\partial y_i \partial y_j} f(X_s)\right) \sigma_{ik}(X_s)\sigma_{kj}(X_s)\mathrm{d}s \\
&+ \mathbb{E}\int_0^{\tau_\varepsilon} \int_{\mathbb{R}^d\setminus\{0\}} (f(X_s+u) - f(X_s) - \langle \partial_y f(X_s), u\rangle) \frac{C_{d,\alpha}}{|u|^{d+\alpha}} \mathrm{d}u\mathrm{d}s.
\end{aligned}
$$

The infinitesimal generator $\mathcal{L}$ of Eq. (9.11) is as follows [1]:

$$
\begin{aligned}
(\mathcal{L}f)(x) &= \langle \partial_x f(x), b(x)\rangle + \frac{1}{2}\left(\frac{\partial^2}{\partial x_i \partial x_j} f(x)\right) \sigma_{ik}(x)\sigma_{kj}(x) \\
&+ \int_{\mathbb{R}^d\setminus\{0\}} (f(x+u) - f(x) - \langle \partial_x f(x), u\rangle) \frac{C_{d,\alpha}}{|u|^{d+\alpha}} \mathrm{d}u.
\end{aligned}
$$

So,

$$
\begin{aligned}
\left|\frac{\mathbb{E}[f(X_{\tau_\varepsilon}(x))] - f(x)}{\mathbb{E}[\tau_\varepsilon]} - (\mathcal{L}f)(x)\right| &= \left|\frac{\mathbb{E}\int_0^{\tau_\varepsilon} (\mathcal{L}f)(X_s)\mathrm{d}s}{\mathbb{E}[\tau_\varepsilon]} - \frac{\mathbb{E}\int_0^{\tau_\varepsilon} (\mathcal{L}f)(x)\mathrm{d}s}{\mathbb{E}[\tau_\varepsilon]}\right| \\
&\le \frac{\mathbb{E}\int_0^{\tau_\varepsilon} |(\mathcal{L}f)(X_s) - (\mathcal{L}f)(x)|\mathrm{d}s}{\mathbb{E}[\tau_\varepsilon]} \\
&\le \sup_{|y-x|<\varepsilon} |(\mathcal{L}f)(y) - (\mathcal{L}f)(x)|.
\end{aligned}
$$

Because $(\mathcal{L}f)(x)$ is continuous in x,

$$Af(x) = \lim_{\varepsilon \downarrow 0} \frac{\mathbb{E}[f(X_{\tau_\varepsilon}(x))] - f(x)}{\mathbb{E}[\tau_\varepsilon]} = (\mathcal{L}f)(x).$$

Similarly, we also have $A = \mathcal{L}$ for $0 < \alpha < 1$.

Remark 3.2. The above deduction tells us $Af = \mathcal{L}f$ for $f \in C_0^2(\mathbb{R}^d)$. If the considered driving process is not a symmetric α-stable Lévy motion, the domain of $\mathcal{L}$ is unclear and thus $A = \mathcal{L}$ may not be true. The corresponding escape probability $p(x)$ is the solution of the following Balayage–Dirichlet problem (in terms of operator $\mathcal{L}$, instead of A):

$$\begin{cases} \mathcal{L}p = 0, \\ p|_U = 1, \\ p|_{D^c \setminus U} = 0. \end{cases}$$

4 Examples

In this section we consider a few examples.

Example 4.1. In one-dimensional case, take $D = (-r, r)$ and $\Gamma = \{r\}$. For each $x \in D$, the escape probability $p(x)$ of $X_t = x + W_t$ from D to Γ satisfies the following differential equation:

$$\begin{cases} \frac{1}{2}p''(x) = 0, \quad x \in (-r, r), \\ p(r) = 1, \\ p(-r) = 0. \end{cases}$$

We obtain that $p(x) = \frac{x+r}{2r}$ for $x \in [-r, r]$. It is a straight line (see Fig. 9.6).

In two-dimensional case, take $D = \{x \in \mathbb{R}^2; r < |x| < R\}$ and $\Gamma = \{x \in \mathbb{R}^2; |x| = r\}$. For every $x \in D$, the escape probability $p(x)$ of $X_t = x + W_t$ from D to Γ satisfies the following elliptic partial differential equation:

$$\begin{cases} \frac{1}{2}\Delta p(x) = 0, \quad x \in D, \\ p(x)|_{|x|=r} = 1, \\ p(x)|_{|x|=R} = 0. \end{cases}$$

By solving this equation, we obtain that $p(x) = \frac{\log R - \log|x|}{\log R - \log r}$. It is plotted in Fig. 9.7.

Example 4.2. Consider the following SDE driven by Brownian motions:

$$\mathrm{d}X_t = b(X_t)\mathrm{d}t + \sigma(X_t)\mathrm{d}W_t,$$

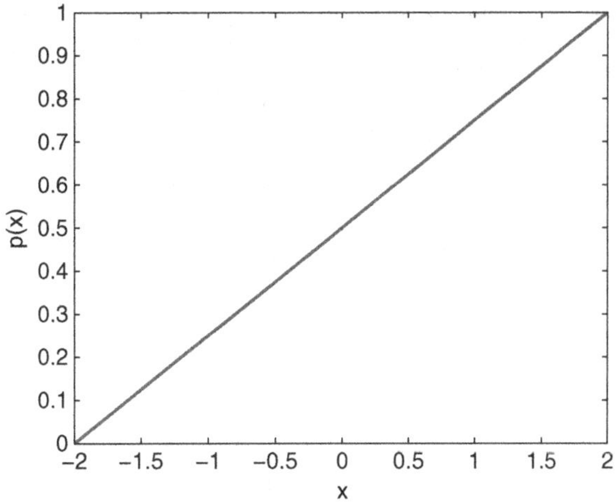

Fig. 9.6 Escape probability for one-dimensional Brownian motion in Example 4.1, $r = 2$

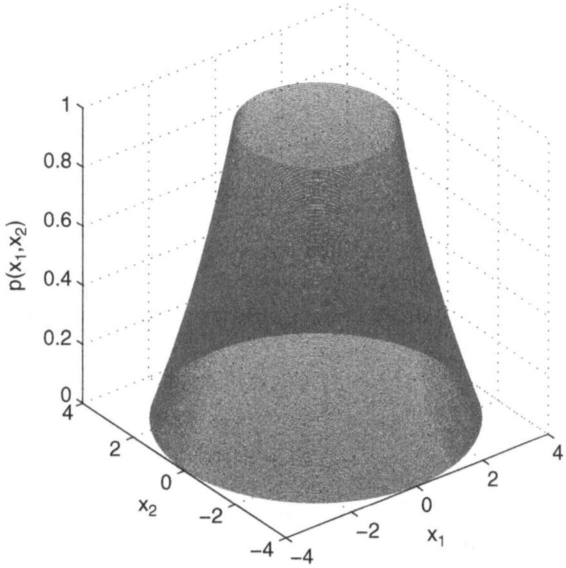

Fig. 9.7 Escape probability for two-dimensional Brownian motion in Example 4.1: $r = 2$, $R = 4$

where b and (nonzero) σ are real functions. When b and σ satisfy $(\mathbf{H}^1_{b,\sigma})$, the equation has a unique solution which is denoted as X_t. We take $D = (-r, r)$ and $\Gamma = \{r\}$. For each $x \in D$, under the condition $(\mathbf{H}^2_{\sigma})$, the escape probability $p(x)$ satisfies

$$\begin{cases} \frac{1}{2}\sigma^2(x)p''(x) + b(x)p'(x) = 0, & x \in (-r, r), \\ p(r) = 1, \\ p(-r) = 0. \end{cases}$$

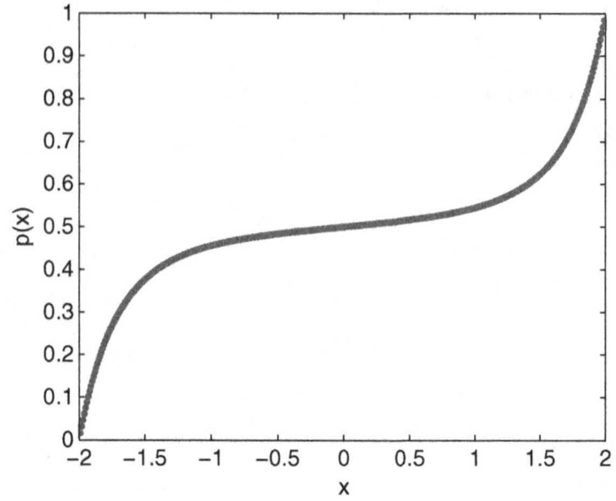

Fig. 9.8 Escape probability in Example 4.2: $b(x) = -x, \sigma(x) = 1, r = 2$

The solution is

$$p(x) = \frac{\int_{-r}^{x} e^{-2\int_{-r}^{y} \frac{b(z)}{\sigma^2(z)} \mathrm{d}z} \mathrm{d}y}{\int_{-r}^{r} e^{-2\int_{-r}^{y} \frac{b(z)}{\sigma^2(z)} \mathrm{d}z} \mathrm{d}y}$$

for $x \in [-r, r]$. See Fig. 9.8.

Example 4.3. In one-dimensional case, take $D = (-r, r)$ and $U = [r, \infty)$. For each $x \in D$ and a symmetric α-stable Lévy process L_t^α, the escape probability $p(x)$ of $X_t = x + L_t^\alpha$ from D to U satisfies the following differential–integral equation:

$$\begin{cases} -(-\Delta)^{\frac{\alpha}{2}} p(x) = 0, & x \in (-r, r), \\ p(x)|_{[r,\infty)} = 1, \\ p(x)|_{(-\infty,-r]} = 0. \end{cases}$$

It is difficult to deal with this equation because of the fractional Laplacian operator. But we can solve it via Poisson kernel. From [13], for $x \in (-r, r)$,

$$p(x) = \frac{\sin\frac{\pi\alpha}{2}}{\pi} \int_r^\infty \frac{(r^2 - x^2)^{\alpha/2}}{(y^2 - r^2)^{\alpha/2}} \frac{1}{(y - x)} \mathrm{d}y.$$

Obviously, $p(-r) = 0$. To justify $p(r) = 1$, we apply the substitution $y = (r^2 - xv)(x - v)^{-1}$ to obtain

$$\begin{aligned} p(r) &= \frac{\sin\frac{\pi\alpha}{2}}{\pi} \int_{-r}^{r} (r - v)^{\alpha-1} (r^2 - v^2)^{-\frac{\alpha}{2}} \mathrm{d}v \\ &= \frac{\sin\frac{\pi\alpha}{2}}{\pi} \int_0^1 (1 - v)^{\frac{\alpha}{2}-1} v^{1-\frac{\alpha}{2}-1} \mathrm{d}v \end{aligned}$$

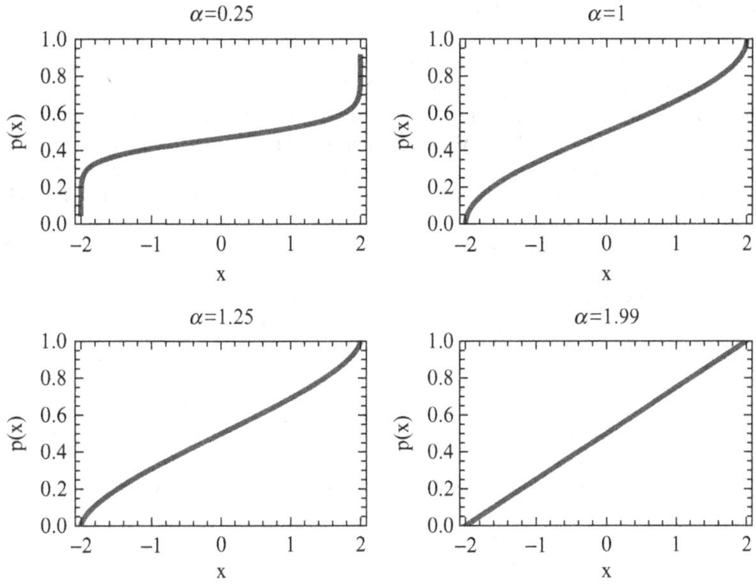

Fig. 9.9 Escape probability in Example 4.3: $r = 2$

$$
\begin{aligned}
&= \frac{\sin\frac{\pi\alpha}{2}}{\pi} B\left(\frac{\alpha}{2}, 1-\frac{\alpha}{2}\right) \\
&= \frac{\sin\frac{\pi\alpha}{2}}{\pi} \Gamma\left(\frac{\alpha}{2}\right)\Gamma\left(1-\frac{\alpha}{2}\right) \\
&= 1,
\end{aligned}
$$

where the beta and gamma functions and their properties are used in the last two steps. The escape probability $p(x)$ is plotted in Fig. 9.9 for various α values.

Acknowledgment We have benefited from our previous collaboration with Ting Gao, Xiaofan Li, and Renming Song. We thank Ming Liao, Renming Song, and Zhen–Qing Chen for helpful discussions. This work was done while Huijie Qiao was visiting the Institute for Pure and Applied Mathematics (IPAM), Los Angeles. This work is partially supported by the NSF of China (No. 11001051 and No. 11028102) and the NSF grant DMS-1025422.

References

1. Applebaum, D.: Lévy Processes and Stochastic Calculus, 2nd edn. Cambridge Univ. Press, Cambridge (2009)
2. Bergström, H.: On some expansions of stable distribution functions. Ark. Math. **2**, 375–378 (1952)

3. Blumenthal, R.M., Getoor, R.K.: Markov Processes and Potential Theory. Academic, New York (1968)
4. Brannan, J., Duan, J., Ervin, V.: Escape probability, mean residence time and geophysical fluid particle dynamics. Physica D **133**, 23–33 (1999)
5. Chen, H., Duan, J., Li, X., Zhang, C.: A computational analysis for mean exit time under non-Gaussian Levy noises. Appl. Math. Comput. **218**(5), 1845–1856 (2011)
6. Chen, Z.Q.: On notions of harmonicity. P. Am. Math. Soc. **137**, 3497–3510 (2009)
7. Chen, Z., Kim, P., Song, R.: Heat kernel estimates for the Dirichlet fractional Laplacian. J. Eur. Math. Soc. **12**, 1307–1329 (2010)
8. Chen, Z.Q., Zhao, Z.: Diffusion processes and second order elliptic operators with singular coefficients for lower order terms. Math. Ann. **302**, 323–357 (1995)
9. Dynkin, E.B.: Markov Processes, vol. I. Springer, Berlin (1965)
10. Gao, T., Duan, J., Li, X., Song, R.: Mean exit time and escape probability for dynamical systems driven by Lévy noise. Preprint (2011) arXiv:1201.6015v1 [math.NA]
11. Guan, Q., Ma, Z.: Boundary problems for fractional Laplacians. Stochast. Dynam. **5**, 385–424 (2005)
12. Ikeda, N., Watanabe, S.: Stochastic Differential Equations and Diffusion Processes, 2nd edn. North-Holland, Amsterdam (1989)
13. Jacob, N.: Pseudo differential operators and Markov processes. Markov Processes and Applications, vol. III. Imperial College Press, London (2005)
14. Koren, T., Chechkin, A.V., Klafter, J.: On the first passage time and leapover properties of Lévy motions. Physica A **379**, 10–22 (2007)
15. Liao, M.: The Dirichlet problem of a discontinuous Markov process. Acta. Math. Sin. **5**, 9–15 (1989)
16. Lin, C.C., Segel, L.A.: Mathematics Applied to Deterministic Problems in the Natural Sciences. SIAM, Philadelphia (1988)
17. Ma, Z., Zhu, R., Zhu, X.: On notions of harmonicity for non-symmetric Dirichlet form. Sci. China (Mathematics) **53**, 1407–1420 (2010)
18. Matkowsky, B., Schuss, Z., Tier, C.: Asymptotic methods for Markov jump processes. AMS Lect. Appl. Math. **27**, 215–240 (1991)
19. Oksendal, B.: Stochastic Differential Equations: An Introduction with Applications, 6th edn. Springer, Berlin (2003)
20. Qiao, H., Zhang, X.: Homeomorphism flows for non-Lipschitz stochastic differential equations with jumps. Stoch. Proc. Appl. **118**, 2254–2268 (2008)
21. Sato, K.: Lévy Processes and Infinitely Divisible Distributions. Cambridge University Press, Cambridge (1999)
22. Schuss, Z.: Theory and Applications of Stochastic Differential Equations. Wiley, New York (1980)
23. Smith, W., Watson, G.S.: Diffusion out of a triangle. J. Appl. Prob. **4**, 479–488 (1967)
24. Song, R.: Probabilistic approach to the Dirichlet problem of perturbed stable processes. Probab. Theory Rel. **95**(3), 371–389 (1993)
25. Viswanathan, G.M., Afanasyev, V., Buldyrev, S.V., Havlin, S., Daluz, M., Raposo, E., Stanley, H.: Lévy flights in random searches. Physica A **282**, 1–12 (2000)
26. Woyczynski, W.A.: Lévy processes in the physical sciences. In: Barndorff-Nielsen, O.E., Mikosch, T., Resnick, S.I. (eds.) Lévy Processes: Theory and Applications, pp. 241–266. Birkhäuser, Boston (2001)
27. Yamada, T., Watanabe, S.: On the uniqueness of solutions of stochastic differential equations. J. Math. Kyoto Univ. **11**, 155–167 (1971)

Part III
Stochastic Partial Differential Equations

Chapter 10
On the Stochastic Navier–Stokes Equation Driven by Stationary White Noise

Chia Ying Lee and Boris Rozovskii

Abstract We consider an unbiased approximation of stochastic Navier–Stokes equation driven by spatial white noise. This perturbation is unbiased in that the expectation of a solution of the perturbed equation solves the deterministic Navier–Stokes equation. The nonlinear term can be characterized as the highest stochastic order approximation of the original nonlinear term $u\nabla u$. We investigate the analytical properties and long-time behavior of the solution. The perturbed equation is solved in the space of generalized stochastic processes using the Cameron–Martin version of the Wiener chaos expansion and generalized Malliavin calculus. We also study the accuracy of the Galerkin approximation of the solutions of the unbiased stochastic Navier–Stokes equations.

Received 12/12/2011; Accepted 5/10/2012; Final 5/26/2012

1 Introduction

Stochastic perturbations of the Navier–Stokes equation have received much attention over the past few decades. Among the early studies of the stochastic Navier–Stokes equations are those by Bensoussan and Temam [1], Foias et al. [4–6], and Flandoli [7, 8]. Traditionally, the types of perturbations that were proposed include stochastic forcing by a noise term such as a Gaussian random

C.Y. Lee
Statistical and Applied Mathematical Sciences Institute,
19 T.W. Alexander Drive, P.O. Box 14006, Research Triangle Park, NC 27709, USA
e-mail: cylee@samsi.info

B. Rozovskii
Division of Applied Mathematics, Brown University,
182 George St., Providence, RI 02912, USA
e-mail: Boris_Rozovsky@brown.edu

F. Viens et al. (eds.), *Malliavin Calculus and Stochastic Analysis: A Festschrift in Honor of David Nualart*, Springer Proceedings in Mathematics & Statistics 34,
DOI 10.1007/978-1-4614-5906-4_10,

field or a cylindrical Wiener process and are broadly accepted as a natural way to incorporate stochastic effects into the system. The stochastic Navier–Stokes equation

$$\begin{aligned}
&u_t + u^i u_{x_i} + \nabla P = \nu \Delta u + f(t,x) + \left(\sigma^i(t,x) u_{x_i} + g(t,x)\right) \dot{W}(t,x), \\
&\operatorname{div} u \equiv 0, \\
&u(0,x) = w(x), \quad u|_{\partial D} = 0
\end{aligned} \tag{10.1}$$

is underpinned by a familiar physical basis, because it can be derived from Newton's second law via the fluid flow map, using a particular assumption on the stochasticity of the governing SODE of the flow map, known as the Kraichnan turbulence. (See [12, 13] and the references therein.) However, due to the nonlinearity, stochastic Navier–Stokes equation (10.1) is a *biased* perturbation of the underlying deterministic Navier–Stokes equation. That is, the mean of the solution of the stochastic equation does not coincide with the solution of the underlying deterministic Navier–Stokes equation. Of course, this observation is also true for other nonlinear equations such as the stochastic Burgers equation and Ginzburg–Landau equation. In fact, the mean of Eq. (10.1) solves the famous Reynolds equation.

An unbiased version of stochastic Navier–Stokes equation (10.1)

$$\begin{aligned}
u_t + u^i \diamond u_{x_i} + \nabla P = \nu \Delta u + f(t,x) + \left(\sigma^i(t,x) u_{x_i} + g(t,x)\right) \dot{W}(t,x), \\
\operatorname{div} u \equiv 0, \\
u(0,x) = w(x), \quad u|_{\partial D} = 0.
\end{aligned} \tag{10.2}$$

has been introduced and studied in Eq.[14]. The unbiased version, Eq. (10.2) differs from Eq. (10.1) by the nonlinear term: the product $u^i u_{x_i}$ is replaced by the Wick product $u^i \diamond u_{x_i}$. In fact, Wick product $u^i \diamond u_{x_i}$ can be interpreted as Malliavin integral of u_{x_i} with respect to u (see [11]). An important property of Wick product is that

$$\mathbb{E}[u^i \diamond u_{x_i}] = \mathbb{E}u^i \, \mathbb{E}u_{x_i}. \tag{10.3}$$

Due to this property, stochastic Navier–Stokes equation (10.2) with Wick nonlinearity is an unbiased perturbation of stochastic Navier–Stokes equation (10.1). In the future, we will refer to unbiased perturbations of stochastic Navier–Stokes equation as *unbiased stochastic Navier–Stokes equation.*

In this paper we will study an unbiased stochastic Navier–Stokes equations on an open bounded smooth domain $D \in \mathbb{R}^d$, $d = 2, 3$, driven by purely spatial noise. In particular, we will study equation

$$\begin{aligned}
&u_t + u^i \diamond u_{x_i} + \nabla P = \nu \Delta u + f(t,x) + \left(\sigma^i(x) u_{x_i} + g(t,x)\right) \diamond \dot{W}(x), \\
&\operatorname{div} u \equiv 0, \\
&u(0,x) = w(x), \quad u|_{\partial D} = 0,
\end{aligned} \tag{10.4}$$

where the diffusivity constant is $\nu > 0$ and the functions f, g, σ are given deterministic $\mathbb{R}^d$-valued functions. Here, the driving noise $\dot{W}(x) = \sum_k \mathfrak{u}_l(x)\xi_l$ is a stationary Gaussian white noise on $L_2(D)$, and we assume that $\sup_l \|\mathfrak{u}_l\|_{L^\infty} < \infty$.

We will also study the stationary (elliptic) version of Eq. (10.4)

$$\begin{aligned} &\bar{u}^i \diamond \bar{u}_{x_i} + \nabla \bar{P} = \nu \Delta \bar{u} + \bar{f}(x) + \left(\bar{\sigma}^i(x)\bar{u}_{x_i} + \bar{g}(x)\right) \diamond \dot{W}(x), \\ &\operatorname{div} \bar{u} \equiv 0, \\ &\bar{u}|_{\partial D} = 0, \end{aligned} \tag{10.5}$$

where $\bar{f}(x), \bar{g}(x), \bar{\sigma}(x)$ are given deterministic $\mathbb{R}^d$-valued functions. It will be shown that $u(t,x) \to \bar{u}(x)$ as $t \to \infty$.

Solutions of Eqs. (10.4) and (10.5) will be defined by their respective Wiener chaos expansions:

$$u(t,x) = \sum_{\alpha} u_\alpha(t,x)\, \xi_\alpha \text{ and } \bar{u}(x) = \sum_{\alpha} \bar{u}_\alpha(x)\, \xi_\alpha, \tag{10.6}$$

where $\{\xi_\alpha, \alpha \in J\}$ is the Cameron–Martin basis generated by $\dot{W}(x)$, $v_\alpha := E(v\xi_\alpha)$, and J is the set of multi-indices $\alpha = \{\alpha_k, k \geq 1\}$ such that for every k, $\alpha_k \in \mathbf{N}_0(\mathbf{N}_0 = \{0, 1, 2, \ldots\})$ and $|\alpha| = \sum_k \alpha_k < \infty$. It will be shown that Wiener chaos coefficients $u_\alpha(t,x)$ and $\bar{u}_\alpha(x)$ solve lower triangular systems of deterministic equations. We will refer to these systems as *propagators* of $u_\alpha(t,x)$ and $\bar{u}_\alpha(x)$, respectively.

In fact, Eqs. (10.4) and (10.5) could be viewed as the *highest stochastic order approximations* of similar equations with standard nonlinearities $u^i u_{x_i}$ and $\bar{u}^i \bar{u}_{x_i}$, respectively. Indeed, it was shown in [14] that under certain natural assumptions, the following equality holds:

$$v\nabla v = \sum_{n=0}^{\infty} \frac{\mathcal{D}^n v \diamondsuit \mathcal{D}^n \nabla v}{n!} \tag{10.7}$$

where $\mathcal{D}^n$ is the nth power of Malliavin derivative $\mathcal{D} = \mathcal{D}_{\dot{W}}$. Taking into account expansion (10.7),

$$v\nabla v \approx v \diamondsuit \nabla v. \tag{10.8}$$

This approximation is the *highest stochastic order* approximation of $v\nabla v$ in that $v\diamondsuit\nabla v$ contains the highest-order Hermite polynomials of the driving noise, while the remaining terms of the right hand side of Eq. (10.7) include only lower-order elements of the Cameron–Martin basis. This fact could be illustrated by the following simple fact:

$$\xi_\alpha \xi_\beta = \xi_{\alpha+\beta} + \sum_{\gamma < \alpha+\beta} k_\gamma \xi_\gamma,$$

where k_γ are constants.

As a side note, we remark that in comparison, the usual stochastic Navier–Stokes equation has a propagator system that is a full system of equations which, comparatively, is a much tougher beast to tackle. Additionally, apart from the zero-th chaos mode which, being the mean, solves the deterministic Navier–Stokes equation, all higher modes in the propagator system solve a linearized

Stokes equation. Thus, where a result is known for the deterministic Navier–Stokes equation, it is sometimes the case that an analogous result may be shown for the unbiased approximation of the stochastic Navier-Stokes equation. For instance, the existence of a unique stationary solution of Eq. (10.5) requires the same condition on the largeness of the viscosity ν as does the existence of a unique steady solution of the deterministic equation (10.13a).

There is substantial theory on the steady solutions of the deterministic Stokes and Navier–Stokes equations, the long-time convergence of a time-dependent solution to the steady solution, as well as other dynamical behavior of the solution. In the subsequent sections, we begin to study some of these same questions for the unbiased Navier–Stokes equation, focusing on the large viscosity case where the uniqueness of steady solutions and long-time convergence has been established in the deterministic setting. We will study the existence of a unique stationary solution of Eq. (10.5) as well as the existence of a unique time-dependent solution of Eq. (10.4) on a finite time interval. The Wiener chaos expansion and the propagator system will be the central tool in obtaining a generalized solution, but to place the solution in a Kondratiev space involves a useful result invoking the Catalan numbers. The Catalan numbers arise naturally from the convolution of the Wiener chaos modes in the nonlinear term. It was used to study the Wick versions of the stochastic Burgers [10] and Navier–Stokes [14] equations.

2 Generalized Random Variables and Functional Analytic Framework

To study Eqs. (10.4) and (10.5), we will give the basic definitions for the generalized stochastic spaces that will be used. The definitions of the generalized solution will be defined in the variational/weak sense such as described in [15, 16], and before stating those definitions, we first state some standard notation and facts about the vector spaces.

Let $d = 2, 3$ be the dimension. Denote the vector spaces $\mathbb{L}^2(D) = (L^2(D))^d$ with the norm $|\cdot|$ and $\mathbb{H}^m(D) = (H^m(D))^d$ with the norm $\|\cdot\|_{H^m}$. Denote the following spaces:

$$\mathcal{V} := \{v \in (C_0^\infty(D))^d : \operatorname{div} v = 0\}.$$

$$V := \text{closure of } \mathcal{V} \text{ in the } \mathbb{H}_0^1(D) \text{ norm} \equiv \{u \in \mathbb{H}_0^1(D) : \operatorname{div} u = 0\}.$$

$$H := \text{closure of } \mathcal{V} \text{ in the } \mathbb{L}^2(D) \text{ norm.}$$

$$V' := \text{dual space of } V \text{ w.r.t. inner product in } H.$$

Also denote the norms in V and V' by $\|\cdot\|_V$ and $\|\cdot\|_{V'}$, respectively. In particular, we have $\|\cdot\|_V := |\nabla \cdot|$.

The operator[1] $-\Delta$ on H, defined on the domain dom$(-\Delta)$, is symmetric positive definite and thus defines a norm $|\cdot|_2$ via $|\cdot|_2 = |\Delta \cdot|$, which is equivalent to the norm $\|w\|_{H^2}$. For $m > 0$, the spaces $V_m := \text{dom}((-\Delta)^{m/2})$ are closed subspaces of $\mathbb{H}^m(D)$ with the norms $|\cdot|_m = |(-\Delta)^{m/2} \cdot|$. In this paper, we will commonly use $m = 1/2, 3/2$, and 2. Note that $|\cdot|_1 = \|\cdot\|_V$ and the norms $|\cdot|_m$ and $\|\cdot\|_{H^m}$ are equivalent. We thus have a constant c_1 so that

$$c_1\|w\|_{H^1}^2 \leq |w|_1^2 \leq \frac{1}{c_1}\|w\|_{H^1}^2, \quad \text{for all } w \in V.$$

Denote $\lambda_1 > 0$ to be the smallest eigenvalue of $-\Delta$; then we have a Poincaré inequality

$$\lambda_1|v|^2 \leq \|v\|_V^2, \quad \text{for } v \in V. \tag{10.9}$$

Define the trilinear continuous form b on $V \times V \times V$ by

$$b(u, v, w) = \int_D u^k \partial_{x_k} v^j w^j \, \mathrm{d}x$$

and the mapping $B : V \times V \to V'$ by

$$\langle B(u, v), w\rangle = b(u, v, w).$$

It is easy to check that

$$b(u, v, w) = -b(u, w, v) \quad \text{and} \quad b(u, v, v) = 0$$

for all $u, v, w \in V$. B and b have many useful properties that follow from the following lemma.

Lemma 2.1 (Lemma 2.1 in [15]). *The form b is defined and is trilinear continuous on $H^{m_1} \times H^{m_2+1} \times H^{m_3}$, where $m_i \geq 0$ and*

$$\begin{aligned} m_1 + m_2 + m_3 &\geq \frac{d}{2} \text{ if } m_i \neq \frac{d}{2}, \; i = 1, 2, 3, \\ m_1 + m_2 + m_3 &> \frac{d}{2} \text{ if } m_i = \frac{d}{2}, \; \text{some } i. \end{aligned} \tag{10.10}$$

In view of Lemma 2.1, let c_b be the constant in

$$|b(u, v, w)| \leq c_b |u|_{m_1} |v|_{m_2+1} |w|_{m_3},$$

where m_i satisfies (10.10). Also let c_d, $d = 2, 3$, be the constants in

[1] Technically, the correct operator is $Au := -P\Delta u$, where P is the orthogonal projection onto H. We abuse notation here and continue writing $-\Delta$.

$$|b(u,v,w)| \le c_2|u|^{1/2}\|u\|_V^{1/2}\|v\|_V^{1/2}|\Delta v|^{1/2}|w| \qquad \text{if } d = 2$$
$$|b(u,v,w)| \le c_3\|u\|_V\|v\|_V^{1/2}|\Delta v|^{1/2}|w| \qquad \text{if } d' = 3$$

for all $u \in V$, $v \in \mathrm{dom}(-\Delta)$, and $w \in H$ (Eqs. (2.31–32) in [15]). Other useful consequences of Lemma 2.1 are that $B(\cdot,\cdot)$ is a bilinear continuous operator from $V \times H^2 \to L^2$ and also from $H^2 \times V \to L^2$.

Next, we introduce the basic notation that will be used to define the generalized stochastic spaces and the generalized solution. Let $(\Omega, \mathcal{F}, P)$ be a probability space where the σ-algebra $\mathcal{F}$ is generated by $\{\xi_k, k = 1, 2, \ldots\}$, where ξ_k are independent and identically distributed $N(0,1)$ random variables. Let $\mathcal{U} = L^2(D)$ and let $\{\mathrm{u}_k(x), k = 1, 2, \ldots\}$ be a complete orthonormal basis for $\mathcal{U}$. Then the Gaussian white noise on $\mathcal{U}$ is

$$\dot{W}(x) = \sum_{k\ge 1} \mathrm{u}_k(x)\xi_k.$$

Let $\mathcal{J} = \{\alpha = (\alpha_1, \alpha_2, \ldots), \alpha_k \in \mathbb{N}_0\}$ be the set of multi-indices of finite length. Denote $|\alpha| = \sum_{k\ge 1}\alpha_k < \infty$ and ϵ_k is the unit multi-index with $|\epsilon_k| = 1$ and kth entry $(\epsilon_k)_k = 1$. For $\alpha, \beta \in \mathcal{J}$,

$$\alpha + \beta = (\alpha_1 + \beta_1, \alpha_2 + \beta_2, \cdots), \quad \text{and} \quad \alpha! = \prod_{k\ge 1}\alpha_k!.$$

For a sequence $\rho = (\rho_1, \rho_2, \ldots)$, set $\rho^\alpha = \prod \rho_k^{\alpha_k}$.

For each $\alpha \in \mathcal{J}$, let

$$\xi_\alpha = \prod_{k\ge 1} \frac{H_{\alpha_k}(\xi_k)}{\sqrt{\alpha_k}}$$

where H_n is the nth Hermite polynomial given by $H_n(x) = (-1)^n(\frac{d^n e^{-x^2/2}}{dx^n} e^{x^2/2})$. It is a well-known fact that the set $\Xi = \{\xi_\alpha, \alpha \in \mathcal{J}\}$ forms an orthonormal basis in $L^2(\Omega)$ [2]. Thus, for a Hilbert space X, if $f \in L^2(\Omega; X)$ and $f_\alpha = E[f\xi_\alpha]$, then the *Wiener chaos expansion* of f is $f = \sum_{k\ge 1} f_\alpha \xi_\alpha$, and moreover $E|f|_X^2 = \sum_{\alpha\in\mathcal{J}} |f_\alpha|_X^2$. The set Ξ is the Cameron–Martin basis of $L^2(\Omega)$.

For a Hilbert space X, define the (stochastic) test function and distribution spaces:

$$\mathcal{D}(X) = \left\{ v = \sum_\alpha v_\alpha \xi_\alpha : v_\alpha \in X \text{ and only finitely many } v_\alpha \text{ are non-zero} \right\},$$

$$\mathcal{D}'(X') = \left\{ \text{All formal series } u = \sum_\alpha u_\alpha \xi_\alpha \text{ with } u_\alpha \in X' \right\}.$$

Random variables in $\mathcal{D}(X)$ serve as test functions for the distributions in $\mathcal{D}'(X')$. If $\langle\cdot,\cdot\rangle$ is the duality pairing between X' and X, then the duality pairing between $u \in \mathcal{D}'(X')$ and $v \in \mathcal{D}(X)$ is

$$\langle\langle u, v\rangle\rangle = \sum_\alpha \langle u_\alpha, v_\alpha\rangle.$$

The space $\mathcal{D}'$ is a very large space. To quantify the asymptotic growth of the Wiener chaos coefficients, we introduce the Kondratiev spaces. For $q > 0$, denote the sequence $(2\mathbb{N})^{-q} = ((2k)^{-q})_{k=1,2,\ldots}$, and let the weights $r_\alpha^2 = (2\mathbb{N})^{-q\alpha}/\alpha!$. The Kondratiev space $\mathcal{S}_{-1,-q}(X)$ is

$$\mathcal{S}_{-1,-q}(X) = \left\{ u = \sum_\alpha u_\alpha \xi_\alpha : u_\alpha \in X \text{ and } \sum_\alpha |u_\alpha|_X^2 r_\alpha^2 < \infty \right\}.$$

$\mathcal{S}_{-1,-q}(X)$ is a Hilbert space with the norm $\|u\|^2_{\mathcal{S}_{-1,-q}(X)} = \sum_\alpha |u_\alpha|_X^2 r_\alpha^2$.

Definition 2.1. For $\alpha, \beta \in \mathcal{J}$, the *Wick product* is defined as

$$\xi_\alpha \diamond \xi_\beta = \sqrt{\binom{\alpha+\beta}{\alpha}} \xi_{\alpha+\beta}.$$

Extending by linearity, for $u, v \in \mathcal{D}'(\mathbb{R})$, the Wick product $u \diamond v$ is a $\mathcal{D}'(\mathbb{R})$ element with

$$u \diamond v = \sum_\alpha \left(\sum_{0 \le \gamma \le \alpha} \sqrt{\binom{\alpha}{\gamma}} u_\gamma v_{\alpha-\gamma} \right) \xi_\alpha.$$

In particular, for $G \in \mathcal{S}_{-1,-q}(L^2(D))$,

$$(G(x) \diamond \dot{W}(x))_\alpha = \sum_{k \ge 1} \sqrt{\alpha_k} G_{\alpha-\epsilon_k}(x) \mathfrak{u}_k(x).$$

We now proceed to define the weak solution of Eq. (10.4). Recall that for a smooth function p, $(\nabla p, v) = 0$ for all $v \in V$. This leads us to define the weak solution by taking the test function space V, so that the pressure term drops out.

Definition 2.2. Let $T < \infty$. A generalized weak solution of Eq. (10.4) is a generalized random element $u \in \mathcal{D}'(L^2(0, T; V))$ such that

$$\langle\langle u_t + u^i \diamond u_{x_i}, \phi \rangle\rangle = \langle\langle \nu \Delta u + f + \left(\sigma^i u_{x_i} + g\right) \diamond \dot{W}(x), \phi \rangle\rangle \qquad (10.11)$$

for all test functions $\phi \in \mathcal{D}(V)$.

The pressure term can be recovered from the generalized weak solution in the standard way.

Using the Wiener chaos expansion, we will study Eqs. (10.4) and (10.5) through the analysis of the propagator system—an equivalent infinite system of deterministic PDE that gives the coefficients u_α of the solution, thereby equivalently characterizing the solution u. Recalling the definition of the Wick product, the propagator system of Eq. (10.4) is, for $\alpha = (0)$,

$$\begin{aligned} &\partial_t u_0 + B(u_0, u_0) = \nu \Delta u_0 + f, \\ &\operatorname{div} u_{(0)} = 0, \\ &u_0(0, x) = w(x), \quad u_0|_{\partial D} = 0 \end{aligned} \tag{10.12a}$$

and for $|\alpha| \geq 1$,

$$\begin{aligned} &\partial_t u_\alpha + B(u_\alpha, u_0) + B(u_0, u_\alpha) + \textstyle\sum_{0<\gamma<\alpha} \sqrt{\binom{\alpha}{\gamma}}\, B(u_\gamma, u_{\alpha-\gamma}) \\ &\quad = \nu \Delta u_\alpha + \textstyle\sum_l \sqrt{\alpha_l} \mathfrak{u}_l(x) \big(\sigma^i \partial_{x_i} u_{\alpha-\epsilon_l} + \mathbf{1}_{\alpha=\epsilon_l} g\big), \\ &\quad \operatorname{div} u_\alpha = 0, \\ &\quad u_\alpha(0, x) = 0, \quad u_\alpha|_{\partial D} = 0 \end{aligned} \tag{10.12b}$$

with equality holding in V'. Note that each equation in the propagator system involves only the divergence-free part; the pressure term P_α can be recovered from each equation by a standard technique (see, e.g., [16]). Hereon, we will focus only on studying the velocity field u.

Similarly, the propagator system of Eq. (10.5) is

$$\begin{aligned} &B(\bar{u}_0, \bar{u}_0) = \nu \Delta \bar{u}_0 + \bar{f}, \\ &\operatorname{div} \bar{u}_0 = 0, \quad \bar{u}_0|_{\partial D} = 0, \end{aligned} \tag{10.13a}$$

$$\begin{aligned} &B(\bar{u}_\alpha, \bar{u}_0) + B(\bar{u}_0, \bar{u}_\alpha) + \textstyle\sum_{0<\gamma<\alpha} \sqrt{\binom{\alpha}{\gamma}}\, B(\bar{u}_\gamma, \bar{u}_{\alpha-\gamma}) \\ &\quad = \nu \Delta u_\alpha + \textstyle\sum_l \sqrt{\alpha_l} \mathfrak{u}_l(x) \big(\bar{\sigma}^i \partial_{x_i} \bar{u}_{\alpha-\epsilon_l} + \bar{g}_{\alpha-\epsilon_l}\big), \\ &\operatorname{div} \bar{u}_\alpha = 0, \quad \bar{u}_\alpha|_{\partial D} = 0 \end{aligned} \tag{10.13b}$$

with equality holding in V'.

The zeroth mode $u_0 = \mathbb{E}u$ is the mean of Eq. (10.4) and solves the unperturbed Navier–Stokes equations (10.12a).

3 The Stationary Unbiased Stochastic Navier–Stokes Equation

Given deterministic functions $\bar{f}, \bar{g}, \bar{\sigma} \in L_2(D)$, we seek a weak/variational solution $\bar{u} \in \mathcal{D}'(V)$ satisfying

$$-\nu \langle\!\langle \Delta \bar{u}, \varphi \rangle\!\rangle + \langle\!\langle \bar{u}^i \diamond \partial_{x_i} \bar{u}, \varphi \rangle\!\rangle = \langle\!\langle \bar{f}, \varphi \rangle\!\rangle + \langle\!\langle \big(\bar{\sigma}^i \partial_{x_i} \bar{u} + \bar{g}\big) \diamond \dot{W}(x), \varphi \rangle\!\rangle$$

for all test random elements $\varphi \in \mathcal{D}(V)$.

We will first show the existence and uniqueness of a generalized strong solution.

Proposition 3.1. *Assume the dimension $d = 2, 3$. Assume $\bar{f}, \bar{g}, \bar{\sigma}$ are deterministic functions satisfying*

$$\bar{f}, \bar{g}, \bar{\sigma} \in H, \tag{A.0}$$

$$\nu^2 > c_b \|\bar{f}\|_{V'}, \tag{A.1}$$

$$\bar{g} \in \mathbb{H}^1(D), \quad \bar{\sigma} \in (W^{1,\infty}(D))^d. \tag{A.2}$$

Then there exists a unique generalized strong solution $u \in \mathcal{D}'(\mathbb{H}^2(D)) \cap V$ *of Eq.* (10.5).

Remark. It is interesting to note that condition (A.1) in Proposition 3.1, which ensures the existence of a generalized strong solution, is the same condition that ensures the uniqueness of the strong solution of the deterministic Navier–Stokes equation. Thus, Proposition 3.1 generalizes the analogous result in the deterministic Navier–Stokes theory, which is the subcase when $\bar{g} = \bar{\sigma} = 0$.

Proof. **Solution for** $\alpha = (0)$.

The equation for $\bar{u}_0$ is the deterministic stationary Navier–Stokes equation, for which the existence and uniqueness of weak solutions is well known [15, 16]. From (A.1), there exists a unique weak solution $\bar{u}_0 \in V$ of Eq. (10.13a) satisfying

$$\|\bar{u}_0\|_V \leq \frac{1}{\nu} \|\bar{f}\|_{V'} < \frac{\nu}{c_b}. \tag{10.14}$$

Moreover, since $\bar{f} \in L_2(D)$, then $\bar{u}_0 \in \mathrm{dom}(-\Delta)$, with

$$|\Delta \bar{u}_0| \leq \frac{2}{\nu} |\bar{f}| + \frac{c_d^2}{\nu^5 \lambda_1^{3/2}} |\bar{f}|^3.$$

THE BILINEAR FORM $\bar{a}_0(\cdot,\cdot)$. Define the bilinear continuous form $\bar{a}_0$ on $V \times V$ by

$$\bar{a}_0(u, v) = \nu(\nabla u, \nabla v) + b(u, \bar{u}_0, v) + b(\bar{u}_0, u, v), \tag{10.15}$$

where $\bar{u}_0(x)$ is the solution of the stationary (deterministic) Navier–Stokes equation (10.13a) just found. Also define the mapping $\bar{A}_0 : V \to V'$, by

$$\langle \bar{A}_0(u), v \rangle = \bar{a}_0(u, v), \quad \text{for all } v \in V.$$

Then Eq. (10.13b) can be written as

$$\bar{A}_0(\bar{u}_\alpha) = -\sum_{0<\gamma<\alpha} \sqrt{\binom{\alpha}{\gamma}}\, B(\bar{u}_\gamma, \bar{u}_{\alpha-\gamma}) + \sum_l \sqrt{\alpha_l}\, \mathrm{u}_l(x) \big(\bar{\sigma}^i \partial_{x_i} \bar{u}_{\alpha-\epsilon_l} + \mathbf{1}_{\alpha=\epsilon_l} \bar{g}\big)$$

for $|\alpha| \geq 1$.

To obtain the existence and uniqueness of u_α, we intend to apply the Lax–Milgram lemma to the bilinear form $\bar{a}_0(\cdot,\cdot)$. To do this, we first check the coercivity of $\bar{a}_0(\cdot,\cdot)$ on V.

Lemma 3.1. *Assume (A.1), and assume u_0 solves (10.13a) with $f \in V'$. Then $\bar{a}_0(\cdot,\cdot)$ defined in Eq.* (10.15) *is coercive and bounded on V.*

Proof. Indeed, for any $v \in V$,

$$\begin{aligned}\bar{a}_0(v,v) &= \nu|\nabla v|^2 + b(v,\bar{u}_0,v) + b(\bar{u}_0,v,v)\\ &\geq \nu|\nabla v|^2 - c_b\|\bar{u}_0\|_V\|v\|_V^2\\ &= \big(\nu - c_b\|\bar{u}_0\|_V\big)\|v\|_V^2 = \bar{\beta}\|v\|_V^2,\end{aligned}$$

where $\bar{\beta} := \nu - c_b\|\bar{u}_0\|_V > 0$ by Eq. (10.14). Next, $\bar{a}_0(\cdot,\cdot)$ is bounded, because

$$\begin{aligned}|\bar{a}_0(v,w)| &\leq \nu\|v\|_V\|w\|_V + |b(v,\bar{u}_0,w)| + |b(\bar{u}_0,v,w)|\\ &\leq \big(\nu + c_b\|\bar{u}_0\|_V\big)\|v\|_V\|w\|_V\end{aligned}$$

for any $v, w \in V$. □

We continue with the proof of Proposition 3.1.

Solutions for $\alpha = \epsilon_l$. Equation (10.13b) in variational form reduces to finding $\bar{u}_{\epsilon_l} \in V$ such that

$$\bar{a}_0(\bar{u}_{\epsilon_l}, v) = \langle \mathrm{u}_l(\bar{\sigma}^i\partial_{x_i}\bar{u}_0 + \bar{g}), v\rangle =: \langle G_{\epsilon_l}, v\rangle$$

for all $v \in V$. To apply the Lax–Milgram lemma to Eq. (10.13b), we check that the term

$$G_{\epsilon_l} := \mathrm{u}_l(\bar{\sigma}^i\partial_{x_i}\bar{u}_0 + \bar{g})$$

belongs to V'. In fact, we have that G_{ϵ_l} belongs to $\mathbb{L}^2(D)$. Indeed, due to assumption (A.2), $|\bar{\sigma}^i\partial_{x_i}\bar{u}_0| \leq \|\bar{\sigma}\|_{L^\infty}\|\bar{u}_0\|_V$, and from Eq. (10.14),

$$\begin{aligned}|G_{\epsilon_l}| &\leq C\|\mathrm{u}_l\|_{L^\infty}\Big(\|\bar{\sigma}\|_{W^{1,\infty}}\|\bar{u}_0\|_V + \|\bar{g}\|_{H^1}\Big)\\ &\leq C\|\mathrm{u}_l\|_{L^\infty}\Big(\frac{\nu}{c_b}\|\bar{\sigma}\|_{W^{1,\infty}} + \|\bar{g}\|_{H^1}\Big).\end{aligned}$$

By the Lax–Milgram lemma, there exists a unique variational solution $\bar{u}_{\epsilon_l} \in V$ with the estimate

$$\|\bar{u}_{\epsilon_l}\|_V \leq \frac{1}{\bar{\beta}}C\|\mathrm{u}_l\|_{L^\infty}\Big(\frac{\nu}{c_b}\|\bar{\sigma}\|_{W^{1,\infty}} + \|\bar{g}\|_{H^1}\Big).$$

Additionally, by a standard technique in [16], there exists $\bar{P}_{\epsilon_l} \in L^2(D)$ such that Eq. (10.13b) holds in V'.

Next, observe that by the continuity property of the bilinear form $B : V \times \mathbb{H}^2 \to \mathbb{L}^2$,

$$-\nu \Delta \bar{u}_{\epsilon_l} = G_{\epsilon_l} - B(\bar{u}_{\epsilon_l}, \bar{u}_0) - B(\bar{u}_0, \bar{u}_{\epsilon_l}) \in L^2(D).$$

Hence, $\bar{u}_{\epsilon_l} \in \text{dom}(-\Delta)$, and we have the estimate

$$\begin{aligned}
|\Delta \bar{u}_{\epsilon_l}| &\leq \frac{1}{\nu}\Big(|G_{\epsilon_l}| + |B(\bar{u}_{\epsilon_l}, \bar{u}_0)| + |B(\bar{u}_0, \bar{u}_{\epsilon_l})|\Big) \\
&\leq \frac{1}{\nu}\left(|G_{\epsilon_l}| + 2c_b |\Delta \bar{u}_0| \, \|\bar{u}_{\epsilon_l}\|_V\right) \\
&\leq \frac{C \sup_l \|\mathrm{u}_l\|_{L^\infty}}{\nu \bar{\beta}} \left(\frac{\nu}{c_b}\|\bar{\sigma}\|_{W^{1,\infty}} + \|\bar{g}\|_{H^1}\right)\left(1 + \frac{2c_b}{\bar{\beta}}|\Delta \bar{u}_0|\right) \\
&= \bar{K},
\end{aligned}$$

and $\bar{K} = \bar{K}(\nu, \bar{f}, \bar{g}, \bar{\sigma})$ does not depend on l.

Solutions for $|\alpha| \geq 2$. Denote

$$G_\alpha := \sum_l \sqrt{\alpha_l}\, \mathrm{u}_l \left(\bar{\sigma}^i \partial_{x_i} \bar{u}_{\alpha - \epsilon_l}\right),$$

$$F_\alpha := - \sum_{0<\gamma<\alpha} \sqrt{\binom{\alpha}{\gamma}}\; B(\bar{u}_\gamma, \bar{u}_{\alpha-\gamma}).$$

We first find $\bar{u}_\alpha \in V$ such that

$$\bar{a}_0(\bar{u}_\alpha, v) = \langle F_\alpha + G_\alpha, v \rangle$$

for all $v \in V$.

We prove by induction. In the above, we have shown the existence of a unique solution $\bar{u}_{\epsilon_l} \in \text{dom}(-\Delta)$. Assume that for some α with $|\alpha| = n$, we have shown the existence of a unique solution $\bar{u}_\gamma \in \text{dom}(-\Delta)$ for all $|\gamma| \leq n - 1$. We now show that $\bar{u}_\alpha \in \text{dom}(-\Delta)$. By a similar argument as above, we have $G_\alpha \in L^2(D)$ with

$$|G_\alpha| \leq C \sum_l \sqrt{\alpha_l} \|\mathrm{u}_l\|_{L^\infty} \|\bar{\sigma}\|_{W^{1,\infty}} \|\bar{u}_{\alpha-\epsilon_l}\|_V < \infty.$$

Also, since $B(\cdot, \cdot)$ is a bilinear continuous form $\mathbb{H}^2 \times \mathbb{H}^2 \to \mathbb{L}^2$, we deduce that $F_\alpha \in \mathbb{L}^2(D)$ with

$$|F_\alpha| \leq c_b \sum_{0<\gamma<\alpha} \sqrt{\binom{\alpha}{\gamma}}\; |\Delta \bar{u}_\gamma| \, |\Delta \bar{u}_{\alpha-\gamma}| < \infty.$$

Applying the Lax–Milgram lemma, there exists a unique solution $\bar{u}_\alpha \in V$ with the estimates

$$\|\bar{u}_\alpha\|_V \le \frac{1}{\bar{\beta}}(|G_\alpha| + |F_\alpha|).$$

Finally, since

$$-\nu\Delta\bar{u}_\alpha = F_\alpha + G_\alpha - B(\bar{u}_\alpha, \bar{u}_0) - B(\bar{u}_0, \bar{u}_\alpha) \in \mathbb{L}^2(D),$$

we deduce that $u_\alpha \in \operatorname{dom}(-\Delta)$, with

$$\begin{aligned}
|\Delta\bar{u}_\alpha| &\le \frac{1}{\nu}\left(|F_\alpha| + |G_\alpha| + |B(\bar{u}_\alpha, \bar{u}_0)| + |B(\bar{u}_0, \bar{u}_\alpha)|\right) \\
&\le \frac{1}{\nu}\left(|F_\alpha| + |G_\alpha| + 2c_b\|\bar{u}_\alpha\|_V|\Delta\bar{u}_0|\right) \\
&\le \frac{1}{\nu}(|F_\alpha| + |G_\alpha|)\left(1 + \frac{2c_b}{\bar{\beta}}|\Delta\bar{u}_0|\right) < \infty.
\end{aligned}$$

Hence, we have found a solution $\bar{u} \in \mathcal{D}'(\mathbb{H}^2(D) \cap V)$. □

Next, we find the appropriate Kondratiev space to which the solution u belongs. As described previously, the estimation of the Kondratiev norm makes use of the recursion properties of the Catalan numbers. The Catalan number rescaling technique used in our estimates has been described in [10], and is detailed in Appendix A.

Proposition 3.2. *Assume (A.0–10.2) hold. Then there exists $q_0 > 2$, depending on ν, $\bar{f}$, $\bar{g}$, $\bar{\sigma}$ such that $\bar{u}$ belongs to the Kondratiev space $S_{-1,-q}(\mathbb{H}^2(D) \cap V)$, for $q > q_0$.*

Proof. For $|\alpha| \ge 1$, we have found in the proof of Proposition 3.1 estimates for $|\Delta\bar{u}_\alpha|$:

$$|\Delta\bar{u}_{\epsilon_l}| \le \bar{K}$$

$$\frac{1}{\sqrt{\alpha!}}|\Delta\bar{u}_\alpha| \le \bar{B}_0\left(\sum_{0<\gamma<\alpha}\frac{|\Delta\bar{u}_\gamma|}{\sqrt{\gamma!}}\frac{|\Delta\bar{u}_{\alpha-\gamma}|}{\sqrt{(\alpha-\gamma)!}} + \mathbf{1}_{\sigma\neq 0}\sum_l \mathbf{1}_{\alpha_l\neq 0}\frac{\|\bar{u}_{\alpha-\epsilon_l}\|_V}{\sqrt{(\alpha-\epsilon_l)!}}\right),$$

where $\bar{B}_0$ depends on $\nu, \bar{f}, \bar{\sigma}$. Also from the proof of Proposition 3.1, the same bounds hold for $\|\bar{u}_\alpha\|_V$ in the LHS of the above inequalities. Thus the above inequalities can be rewritten as

$$\hat{L}_{\epsilon_l} \le \bar{K},$$

$$\hat{L}_\alpha \le \bar{B}_0\left(\sum_{0<\gamma<\alpha}\hat{L}_\gamma\,\hat{L}_{\alpha-\gamma} + \mathbf{1}_{\sigma\neq 0}\sum_{\substack{\gamma<\alpha\\|\gamma|=|\alpha|-1}}\hat{L}_\gamma\right),$$

where $\hat{L}_\alpha = \frac{1}{\sqrt{\alpha!}}(|\Delta\bar{u}_\alpha| + \|\bar{u}_\alpha\|_V)$ for $|\alpha| \geq 1$. Now let $L_{\epsilon_l} = 1 + \hat{L}_{\epsilon_l}$ and $L_\alpha = \hat{L}_\alpha$ for $|\alpha| \geq 2$. Then for $|\alpha| \geq 2$,

$$
\begin{aligned}
L_\alpha &\leq \bar{B}_0\Bigg(\sum_{\substack{0<\gamma<\alpha \\ 2\leq|\gamma|\leq|\alpha|-1}} L_\gamma L_{\alpha-\gamma} + \sum_{\substack{0<\gamma<\alpha \\ |\gamma|=1}} (L_\gamma - 1)\hat{L}_{\alpha-\gamma} + \sum_{\substack{0<\gamma<\alpha \\ |\gamma|=|\alpha|-1}} \hat{L}_\gamma (L_{\alpha-\gamma} - 1) \\
&\qquad + \mathbf{1}_{\sigma\neq 0} \sum_{\substack{\gamma<\alpha \\ |\gamma|=|\alpha|-1}} \hat{L}_\gamma \Bigg) \\
&= \bar{B}_0\Bigg(\sum_{0<\gamma<\alpha} L_\gamma L_{\alpha-\gamma} + \sum_{\substack{0<\gamma<\alpha \\ |\gamma|=|\alpha|-1}} (-2 + \mathbf{1}_{\sigma\neq 0})\hat{L}_\gamma \Bigg) \\
&\leq \bar{B}_0\Bigg(\sum_{0<\gamma<\alpha} L_\gamma L_{\alpha-\gamma} \Bigg).
\end{aligned}
$$

By the Catalan numbers method in Appendix A

$$
|\Delta\bar{u}_\alpha|^2 \leq \alpha! C_{|\alpha|-1}^2 \binom{|\alpha|}{\alpha} (2\mathbb{N})^\alpha \bar{B}_0^{2(|\alpha|-1)} \bar{K}^{2|\alpha|} \tag{10.16}
$$

for $|\alpha| \geq 1$. The result holds with q_0 satisfying

$$
\bar{B}_0^2 \bar{K}^2 2^{5-q_0} \sum_{i=1}^{\infty} i^{1-q_0} = 1. \tag{10.17}
$$

□

4 The Time-Dependent Case

In this section, we consider for simplicity equation (10.4) with $\sigma(t,x) = 0$. We will consider the time-dependent solution $u(t)$ of Eq. (10.4) on a finite time interval $[0,T]$ if $d = 2,3$, and also study its uniform boundedness on $[0,\infty)$ for d=2. The former result allows an arbitrarily large time interval, thereby ensuring a global-in-time solution. On the other hand, the latter result will become useful for showing the long-time convergence of the solution to a steady-state solution.

For any $T < \infty$, it is known that a strong solution $u_0(t)$ of the deterministic Navier–Stokes equation (10.12a) exists on the finite interval $[0,T]$ if $d = 2$, and exists on $[0,(T \wedge T_1)]$ for a specific $T_1 = T_1(u_0(0))$ depending on $u_0(0)$ if $d = 3$.

Without further conditions, we have the following result for a generalized strong solution of the unbiased Navier–Stokes equation.

Lemma 4.1. *For $d = 2, 3$, let $T < \infty$ if $d = 2$ or $T \leq T_1$ if $d = 3$. Assume the forcing terms f, g and initial condition $u(0)$ are deterministic functions satisfying*

$$f, g \in L^2(0, T; H), \qquad u(0) \in V. \tag{A0$'$}$$

Then there exists a unique generalized strong solution $u(t) \in \mathcal{D}'(\mathbb{H}^2(D) \cap V)$ for a.e. $t \in [0, T]$. Moreover, $u_\alpha \in C([0, T], V)$ for all α.

Proof. For $\alpha = (0)$, it is well known (see, e.g., [15]) that Eq. (10.12a) has a unique solution u_0 and

$$u_0 \in L^2([0, T]; \mathrm{dom}(-\Delta)), \quad u_0 \in C([0, T]; V).$$

THE BILINEAR FORM $a_0(t)$. For $t \in [0, T]$, define the bilinear continuous form $a_0(t)$ on $V \times V$ by

$$a_0(u, v; t) = \nu(\nabla u, \nabla v) + b(u, u_0(t), v) + b(u_0(t), u, v),$$

where $u_0(t, x)$ is the solution of the time-dependent (deterministic) Navier–Stokes equations given in Eq. (10.12a) just found. Also define the mapping $A_0(t) : V \to V'$, for $t \in [0, T]$, by

$$\langle A_0(t)u, v\rangle = a_0(u, v; t), \quad \text{for all } v \in V.$$

Then Eq. (10.12b) can be written as

$$\partial_t u_\alpha + A_0(t)u_\alpha + \sum_{0<\gamma<\alpha} \sqrt{\binom{\alpha}{\gamma}}\, B(u_\gamma, u_{\alpha-\gamma}) = \sum_l \sqrt{\alpha_l}\, \mathrm{u}_l(x)\big(\sigma^i \partial_{x_i} u_{\alpha-\epsilon_l} + \mathbf{1}_{\alpha=\epsilon_l} g\big).$$

This is a linear Stokes equation of the form

$$\begin{aligned} &\partial_t U + A_0(t)U = F, \\ &U|_{\partial D} = 0, \quad U(0) = w. \end{aligned}$$

Since $u_0 \in L^2(0, T; \mathrm{dom}(-\Delta))$, it can be shown by standard compactness techniques (see, e.g., [3]) that if $F \in L^2([0, T]; H)$ and $w \in V$, then there exists a unique strong solution $U \in L^2([0, T]; \mathrm{dom}(-\Delta)) \cap C(0, T; V)$ and $U_t \in L^2([0, T]; H)$ with the estimates

$$\sup_{t\leq T} \|U(t)\|_V + \|U\|_{L^2([0,T];\mathrm{dom}(-\Delta))} + \|U_t\|_{L^2([0,T];H)} \leq C\left(\|U(0)\|_V + \|F\|_{L^2([0,T];H)}\right), \tag{10.18}$$

where the constant C depends only on T, ν, D and $\|u_0\|_{L^2([0,T];\mathrm{dom}(-\Delta))}$.

We prove the lemma by induction. We have earlier shown that $u_{(0)} \in L^2([0,T]; \text{dom}(-\Delta)) \cap C(0,T;V)$. Assume for some α that $u_\gamma \in L^2([0,T]; \text{dom}(-\Delta)) \cap C(0,T;V)$ for all $\gamma < \alpha$. We now show that $u_\alpha \in L^2([0,T]; \text{dom}(-\Delta)) \cap C(0,T;V)$ also. We check for the RHS of Eq. (10.12b):

$$-\sum_{0<\gamma<\alpha} \sqrt{\binom{\alpha}{\gamma}}\, B(u_\gamma, u_{\alpha-\gamma}) + \mathbf{1}_{\alpha=\epsilon_l} \mathfrak{u}_l g \in L^2([0,T]; H).$$

This follows from (A0′) and the fact that $|B(u_\gamma, u_{\alpha-\gamma})| \le c_b |\Delta u_\gamma| \, \|u_{\alpha-\gamma}\|$. It follows from Eq. (10.18) that there exists a unique solution u_α of Eq. (10.12b) with

$$u_\alpha \in L^2([0,T]; \text{dom}(-\Delta)), \quad \partial_t u_\alpha \in L^2([0,T]; H), \quad \text{and } u_\alpha \in C([0,T]; V).$$

□

Remark. If $\sigma \neq 0$, then in addition to (A0′), we must require that $g \in L^2([0,T]; \mathbb{H}^1(D))$ and $\sigma \in L^2([0,T]; (W^{1,\infty}(D))^d)$. (Compare with (A.2).)

Next, we study $\|u(t)\|_{-1,-q;\mathbb{H}^2}$ on a finite interval $[0,T]$ as well as the uniform boundedness of $\|u(t)\|_{-1,-q;V}$ for all time $t \in [0,\infty)$. We recall the following established result on the uniform bounds of u_0 in the V and $\mathbb{H}^2(D)$ norms.

Lemma 4.2 (Lemma 11.1 in [15]; see also [9]). *Assume for the initial condition that $u_0(0,\cdot) \in V$, and assume*

$$f \text{ is continuous and bounded from } [0,\infty) \text{ into } H,$$

$$\frac{\partial f}{\partial t} \text{ is continuous and bounded from } [0,\infty) \text{ into } V'.$$

Let $u_0(t)$ be the strong solution of the deterministic Navier–Stokes equations (10.12a), defined on $[0,\infty)$ if $d = 2$ or on $[0,T_1]$ if $d = 3$. Then

$$\sup_{t\ge 0} \|u_0(t)\|_V \le c'(\|u_{(0)}(0,\cdot)\|_V, \nu, f, D), \tag{10.19a}$$

$$\sup_{t\ge \tau} |\Delta u_0(t)| \le c''(\tau, \|u_{(0)}(0,\cdot)\|_V, \nu, f, D), \tag{10.19b}$$

for any $\tau > 0$. In the case of $d = 3$, the suprema are taken over $0 \le t \le T_1$ and $0 \le \tau \le t \le T_1$, respectively.

Proposition 4.1. *(i) For $d = 2, 3$, assume the same conditions as in Lemma 4.1. Then there exists some $q_1 > 2$ depending on ν, c', c_b, and T, such that for $q > q_1$,*

$$u \in \mathcal{S}_{-1,-q}(L^2(0,T; dom(-\Delta))) \cap \mathcal{S}_{-1,-q}(L^\infty(0,T;V)).$$

(ii) For $d = 2$, assume the hypothesis of Lemma 4.2, and assume g is bounded from $[0, \infty)$ into H. Also assume

$$\nu^4 > \frac{2^7 c_b^4 c'^4}{\lambda_1} \tag{A1'}$$

where $c' = c'(\|u(0,\cdot)\|_V, \nu, \bar{f}, D)$ in Eq. (10.19a).
Then there exists $q_2 > 2$, depending on ν, c', and c_b, such that for $q > q_2$,

$$\sup_{t\geq 0} \|u(t)\|_{-1,-q;V} < \infty \quad \text{and} \quad \sup_{t\geq\tau} \|u(t)\|_{-1,-q;dom(-\Delta)} < \infty$$

for any $\tau > 0$, as in Eq. (10.19b). *In fact,*

$$u \in \mathcal{S}_{-1,-q}(L^\infty([0,\infty); V)) \quad \text{and} \quad u \in \mathcal{S}_{-1,-q}(L^\infty([\tau,\infty); V)).$$

Remark. Part (ii) of the equation asserts a uniform-in-time bound of the $\mathcal{S}_{-1,-q}(V)$ norm of the solution on the infinite time interval. Unfortunately, this result does not follow from part (i) because, under the present proof, the estimates for the $\mathcal{S}_{-1,-q}(\mathbb{H}^2(D) \cap V)$ norm of the solution on the finite time interval increase to infinity as the terminal time $T \to \infty$.

Proof. (i) The proof of this result is identical to the proof of Proposition 3.2, by using the estimates (10.18). For $\alpha = (0)$, Eq. (10.19a) and the usual deterministic theory imply that $u_0 \in L^2(0,T; \text{dom}(-\Delta)) \cap L^\infty(0,T;V)$. Let $\tilde{L}_\alpha = \|u_\alpha\|_{L^2(0,T;H^2)} + \|u_\alpha\|_{L^2(0,T;V)}$, and let $L_\alpha = \frac{1}{\sqrt{\alpha!}}\tilde{L}_\alpha$ for $|\alpha| \geq 1$. For $\alpha = \epsilon_l$, the estimates (10.18) yield

$$L_{\epsilon_l} \leq C \sup_l \|\mathrm{u}_l\|_{L^\infty(D)} |g| =: K_1,$$

where K_1 does not depend on l. For $|\alpha| \geq 2$,

$$\tilde{L}_\alpha \leq C \left(\sum_{0<\gamma<\alpha} \sqrt{\tbinom{\alpha}{\gamma}} \|u_\gamma\|_{L^2(0,T;H^2)} \|u_{\alpha-\gamma}\|_{L^2(0,T;V)} \right)$$
$$\leq C \sum_{0<\gamma<\alpha} \sqrt{\tbinom{\alpha}{\gamma}} \tilde{L}_\gamma \tilde{L}_{\alpha-\gamma}.$$

Then for $L_\alpha := \frac{1}{\sqrt{\alpha!}}\tilde{L}_\alpha$

$$L_\alpha \leq B_1 \sum_{0<\gamma<\alpha} L_\gamma L_{\alpha-\gamma}.$$

By the Catalan numbers method as per Appendix A,

$$\|u_\alpha\|_{L^\infty(0,T;V)} + \|\Delta u_\alpha\|_{L^2(0,T;H)} \leq \sqrt{\alpha!} C_{|\alpha|-1} \binom{|\alpha|}{\alpha} B_1^{|\alpha|-1} K_1^{|\alpha|}$$

and the statement of the proposition holds with q_1 satisfying

$$B_1^2 K_1^2 2^{5-q_1} \sum_{i=1}^{\infty} i^{1-q_1} = 1.$$

(ii) We now show the uniform boundedness of each mode u_α on the infinite time interval. For $\alpha = (0)$, this is shown in the estimates of Eqs. (10.19a) and (10.19b). For $|\alpha| = 1$, $\alpha = \epsilon_l$, choose in Eq. (10.12b) the test function $v = (-\Delta)u_\alpha$:

$$\begin{aligned}
\frac{1}{2}\frac{\mathrm{d}}{\mathrm{d}t}\|u_{\epsilon_l}\|_V^2 + \nu|\Delta u_{\epsilon_l}|^2 &\leq |b(u_{\epsilon_l}, u_0, \Delta u_{\epsilon_l})| + |b(u_0, u_{\epsilon_l}, \Delta u_{\epsilon_l})| + |\langle \mathfrak{u}_l g, \Delta u_{\epsilon_l}\rangle| \\
&\leq 2c_b\|u_0\|_V \|u_{\epsilon_l}\|_V^{1/2} |\Delta u_{\epsilon_l}|^{3/2} + |\mathfrak{u}_l g|\,|\Delta u_{\epsilon_l}| \\
&\leq \frac{\varepsilon}{2}|\Delta u_{\epsilon_l}|^2 + \frac{1}{2\varepsilon}\left(2c_b\|u_0\|_V\|u_{\epsilon_l}\|_V^{1/2}|\Delta u_{\epsilon_l}|^{1/2} + |\mathfrak{u}_l g|\right)^2 \\
&\leq \frac{\varepsilon}{2}|\Delta u_{\epsilon_l}|^2 + \frac{2c_b^2\|u_0\|_V^2}{2\varepsilon}\|u_{\epsilon_l}\|_V|\Delta u_{\epsilon_l}| + \frac{1}{\varepsilon}|\mathfrak{u}_l g|^2 \\
&\leq \varepsilon|\Delta u_{\epsilon_l}|^2 + \frac{2^3 c_b^4\|u_0\|_V^4}{\varepsilon^3}\|u_{\epsilon_l}\|_V^2 + \frac{1}{\varepsilon}|\mathfrak{u}_l g|^2.
\end{aligned}$$

Taking $\varepsilon = \frac{\nu}{2}$,

$$\frac{\mathrm{d}}{\mathrm{d}t}\|u_{\epsilon_l}\|_V^2 + \nu|\Delta u_{\epsilon_l}|^2 \leq \frac{2^7 c_b^4}{\nu^3}\|u_0\|_V^2\|u_{\epsilon_l}\|_V^2 + \frac{4}{\nu}|\mathfrak{u}_l g|^2,$$

and from Eqs. (10.9) and (10.19a),

$$\begin{aligned}
\frac{\mathrm{d}}{\mathrm{d}t}\|u_{\epsilon_l}\|_V^2 &\leq \left(\frac{2^7 c_b^4 c'^4}{\nu^3} - \nu\lambda_1\right)\|u_{\epsilon_l}\|_V^2 + \frac{4}{\nu}|\mathfrak{u}_l g|^2 \\
&\leq -\beta\|u_{\epsilon_l}\|_V^2 + \frac{4}{\nu}|\mathfrak{u}_l g|^2,
\end{aligned}$$

where $\beta := -(\frac{2^7 c_b^4 c'^4}{\nu^3} - \nu\lambda_1) > 0$ by (A1′). By Gronwall's inequality,

$$\|u_{\epsilon_l}(T)\|_V^2 \leq \int_0^T \frac{4}{\nu}|\mathfrak{u}_l g|^2 \mathrm{e}^{-\beta(T-s)}\mathrm{d}s \leq \frac{4}{\nu\beta}\|\mathfrak{u}_l\|_{L^\infty}^2\|g\|_{L^\infty(0,\infty;H)}^2\left(1 - \mathrm{e}^{-\beta T}\right)$$

for any $T > 0$. Also,

$$|\Delta u_{\epsilon_l}(t)|^2 \le \frac{2^7 c_b^4 c'^2}{\nu^4} \|u_{\epsilon_l}(t)\|_V^2 + \frac{4}{\nu^2} |u_l g(t)|^2.$$

It follows that

$$L_{\epsilon_l} := \sup_{t \ge 0} \left(\|u_{\epsilon_l}(t)\|_V + |\Delta u_{\epsilon_l}(t)| \right) \le K_2,$$

for all l, where the constant K_2 is independent of l and t. For $|\alpha| \ge 2$, let $L_\alpha := \frac{1}{\sqrt{\alpha!}} \sup_{t \ge 0}(\|u_\alpha(t)\|_V + |\Delta u_\alpha(t)|)$. Then

$$\begin{aligned}
&\frac{1}{2}\frac{\mathrm{d}}{\mathrm{d}t} \|u_\alpha\|_V^2 + \nu |\Delta u_\alpha|^2 \\
&\quad \le |b(u_\alpha, u_0, \Delta u_\alpha)| + |b(u_0, u_\alpha, \Delta u_\alpha)| + \sum_{0<\gamma<\alpha} \sqrt{\binom{\alpha}{\gamma}}\, |b(u_\gamma, u_{\alpha-\gamma}, \Delta u_\alpha)| \\
&\quad \le 2c_b \|u_0\|_V \|u_\alpha\|_V^{1/2} |\Delta u_\alpha|^{3/2} + \sum_{0<\gamma<\alpha} \sqrt{\binom{\alpha}{\gamma}}\, c_b \|u_\gamma\|_V\, |\Delta u_{\alpha-\gamma}|\, |\Delta u_\alpha|.
\end{aligned}$$

By similar computations,

$$\begin{aligned}
&\frac{1}{2}\frac{\mathrm{d}}{\mathrm{d}t} \|u_\alpha\|_V^2 + \nu |\Delta u_\alpha|^2 \\
&\quad \le \frac{2^7 c_b^4}{\nu^3} \|u_0\|_V^4 \|u_\alpha\|_V^2 + \frac{4c_b^2}{\nu} \left(\sum_{0<\gamma<\alpha} \sqrt{\binom{\alpha}{\gamma}}\, \|u_\gamma\|_V\, |\Delta u_{\alpha-\gamma}| \right)^2 \\
&\quad \le \frac{2^7 c_b^4}{\nu^3} \|u_0\|_V^4 \|u_\alpha\|_V^2 \\
&\qquad + \frac{4c_b^2}{\nu} \left(\sum_{0<\gamma<\alpha} \sqrt{\binom{\alpha}{\gamma}} \left(\sup_{s \ge 0} \|u_\gamma(s)\|_V \right) \left(\sup_{s \ge 0} |\Delta u_{\alpha-\gamma}(s)| \right) \right)^2
\end{aligned}$$

and so

$$\frac{\mathrm{d}}{\mathrm{d}t} \|u_\alpha\|_V^2 \le -\beta \|u_\alpha\|_V^2 + \frac{4c_b^2}{\nu} \left(\sum_{0<\gamma<\alpha} \sqrt{\alpha!} L_\gamma L_{\alpha-\gamma} \right)^2.$$

By Gronwall's inequality and triangle inequality,

$$\|u_\alpha(T)\|_V^2 \le \frac{4c_b^2}{\nu}\int_0^T \left(\sum_{0<\gamma<\alpha} \sqrt{\alpha!}L_\gamma L_{\alpha-\gamma}\, \mathrm{e}^{-\beta(T-s)/2}\right)^2 \mathrm{d}s$$

$$\le \frac{4c_b^2}{\nu}\left(\sum_{0<\gamma<\alpha} \sqrt{\alpha!}L_\gamma L_{\alpha-\gamma}\Big(\int_0^T \mathrm{e}^{-\beta(T-s)}\mathrm{d}s\Big)^{1/2}\right)^2$$

so

$$\frac{1}{\sqrt{\alpha!}}\sup_{T\ge 0}\|u_\alpha(T)\|_V \le \frac{2c_b^2}{\sqrt{\nu\beta}}\sum_{0<\gamma<\alpha} L_\gamma L_{\alpha-\gamma}.$$

We have also

$$|\Delta u_\alpha(t)|^2 \le \frac{2^7 c_b^4 c'^4}{\nu^4}\|u_\alpha(t)\|_V^2 + \frac{4c_b^2}{\nu^2}\left(\sum_{0<\gamma<\alpha}\sqrt{\alpha!}L_\gamma L_{\alpha-\gamma}\right)^2$$

for any $t \ge 0$. Hence, it follows that

$$L_\alpha \le B_2 \sum_{0<\gamma<\alpha} L_\gamma L_{\alpha-\gamma}$$

where B_2 depends on ν, c', and c_b but is independent of t. By the Catalan method in Appendix A,

$$\sup_{t\ge 0}\left(\|u_\alpha(t)\|_V + |\Delta u_\alpha(t)|\right) \le \sqrt{\alpha!}C_{|\alpha|-1}\binom{|\alpha|}{\alpha}B_2^{|\alpha|-1}K_2^{|\alpha|}$$

for $|\alpha| \ge 1$, and the statement of the proposition holds with q_2 satisfying

$$B_2^2 K_2^2 2^{5-q_2}\sum_{i=1}^{\infty} i^{1-q_2} = 1.$$

□

5 Long Time Convergence to the Stationary Solution

In this section, we study the solutions $u(t,x)$ of Eq. (10.4) and $\bar{u}(x)$ of Eq. (10.5) with $\sigma(t,x) = \bar{\sigma}(x) = 0$ and for simplicity consider the case with $f(t,x) = \bar{f}(x)$ and $g(t,x) = \bar{g}(x)$. We study the convergence of $u(t,x)$ to the stationary solution

$\bar{u}(x)$ as $t \to \infty$, first in a weak sense (in the generalized space $\mathcal{D}'(H)$ with some exponential rate of convergence in each mode), then in a strong sense (in some Kondratiev space $\mathcal{S}_{-1,-q}(H)$) using a compact embedding argument. The latter proof, unfortunately, does not provide a rate of convergence. For time-dependent f, g, similar results can be obtained under suitable assumptions, but the exponential convergence of each mode is not guaranteed.

Let $z(t) := u(t) - \bar{u}$. The propagator system for z is

$$z_{0,t} + B(u_0, u_0) - B(\bar{u}_0, \bar{u}_0) = \nu \Delta z_0, \tag{10.20a}$$

$$z_{\alpha,t} + A_0(t; u_\alpha) - \bar{A}_0(\bar{u}_\alpha) = -\sum_{0<\gamma<\alpha} \sqrt{\tbinom{\alpha}{\gamma}} \left(B(u_\gamma, u_{\alpha-\gamma}) - B(\bar{u}_\gamma, \bar{u}_{\alpha-\gamma}) \right) \tag{10.20b}$$

with $z_\alpha(0,x) = u_\alpha(0,x) - \bar{u}_\alpha(x)$, $z|_{\partial D} = 0$, and $\operatorname{div} z_\alpha \equiv 0$, for all α.

Proposition 5.1. *Let $d = 2$. Assume (A.0), (A0′) and (A.2), and assume*

$$\nu \left(\frac{\lambda_1}{c_2'} \right)^{3/4} > \frac{2}{\nu} |\bar{f}| + \frac{c_2^2}{\nu^5 \lambda_1^{3/2}} |\bar{f}|^3, \tag{A3}$$

where c_2, c_2' are specific constants depending only on D.

Then the solution $u(t)$ of Eq. (10.4) converges in $\mathcal{D}'(H)$ to the sólution $\bar{u}$ of Eq. (10.5):

$$u(t) \overset{\mathcal{D}'(H)}{\longrightarrow} \bar{u}, \qquad \text{as } t \to \infty.$$

Remark. In the following proof, all computations follow through even when $d = 3$. So, a similar statement to Proposition 5.1 can be made for $d = 3$, provided a strong solution $u(t)$ exists in $\mathcal{D}'(\mathbb{H}^2 \cap V)$ for all $t > 0$, and the zero-th mode $u_0(t)$ satisfies the energy inequality (cf. [15])

$$\frac{1}{2} \frac{\mathrm{d}}{\mathrm{d}t} |u_0(t)|^2 + \nu \|u_0(t)\|_V^2 \le \langle \bar{f}, u_0(t) \rangle.$$

Remark. If $f(t,x)$ and $g(t,x)$ depend on time, then an additional condition for the proposition to hold is that $f(t), g(t)$ converge to $\bar{f}, \bar{g}$ in H.

Proof. For $\alpha = (0)$, the convergence for the deterministic Navier–Stokes equation is well known due to [15], Theorem 10.2: if $u_0(t)$ is any weak solution of Eq. (10.12a) with initial condition $u_0(0) \in H$, then $u_0(t) \longrightarrow \bar{u}_{(0)}$ in H as $t \to \infty$, provided (A3) holds. Moreover, $|z_0(t)|$ decays exponentially:

$$|z_0(t)| \le |z_0(0)| \, \mathrm{e}^{-\bar{\nu} t}, \tag{10.21}$$

where $\bar{\nu} := \nu\lambda_1 - \frac{c_2'}{\nu^{1/3}}|\Delta\bar{u}_0|^{4/3} > 0$. (The positivity of $\bar{\nu}$ follows from the fact that $|\Delta\bar{u}_0|$ can be majorized by the RHS of (A3).)

For $\alpha = \epsilon_l$, choosing the test function $v = z_{\epsilon_l}$ in the weak formulation of Eq. (10.20b),

$$
\begin{aligned}
&\frac{1}{2}\frac{\mathrm{d}}{\mathrm{d}t}|z_{\epsilon_l}|^2 + \nu\|z_{\epsilon_l}\|_V^2 \\
&\quad \le |b(z_{\epsilon_l}, \bar{u}_0, z_{\epsilon_l})| + |b(z_{\epsilon_l}, z_0, z_{\epsilon_l})| + |b(z_0, \bar{u}_{\epsilon_l}, z_{\epsilon_l})| + |b(\bar{u}_{\epsilon_l}, z_0, z_{\epsilon_l})| \\
&\quad \le c_b\|\bar{u}_0\|_V\|z_{\epsilon_l}\|_V^2 + c_b\|z_0\|_{L^\infty}|z_{\epsilon_l}|\,\|z_{\epsilon_l}\|_V + 2c_b|\Delta\bar{u}_{\epsilon_l}|\,|z_0|\,\|z_{\epsilon_l}\|_V \\
&\quad \le c_b\|\bar{u}_0\|_V\|z_{\epsilon_l}\|_V^2 + \frac{c_b^2}{2\varepsilon}\|z_0\|_{L^\infty}^2|z_{\epsilon_l}|^2 + \varepsilon\|z_{\epsilon_l}\|_V^2 + \frac{2c_b^2}{\varepsilon}|\Delta\bar{u}_{\epsilon_l}|^2|z_0|^2,
\end{aligned}
$$

where we have used the Young's inequality in the last line with any $0 < \varepsilon < \bar{\beta}$. So,

$$
\frac{1}{2}\frac{\mathrm{d}}{\mathrm{d}t}|z_{\epsilon_l}|^2 + (\bar{\beta} - \varepsilon)\|z_{\epsilon_l}\|_V^2 \le \frac{c_b^2}{2\varepsilon}\|z_0\|_{L^\infty}^2|z_{\epsilon_l}|^2 + \frac{2c_b^2}{\varepsilon}|\Delta\bar{u}_{\epsilon_l}|^2|z_0|^2. \tag{10.22}
$$

Using the Poincaré inequality (10.9) and taking $\varepsilon = \frac{\bar{\beta}}{2}$,

$$
\frac{\mathrm{d}}{\mathrm{d}t}|z_{\epsilon_l}|^2 + \bar{\beta}\lambda_1|z_{\epsilon_l}|^2 \le \frac{2c_b^2}{\bar{\beta}}\|z_0\|_{L^\infty}^2|z_{\epsilon_l}|^2 + \frac{8c_b^2}{\bar{\beta}}|\Delta\bar{u}_{\epsilon_l}|^2|z_0|^2.
$$

For some appropriately chosen $t_0 \in (0, \infty)$ to be discussed next, we apply Gronwall's inequality:

$$
|z_{\epsilon_l}(T)|^2 \le \mathrm{e}^{\int_{t_0}^T \varphi(t)\mathrm{d}t}|z_{\epsilon_l}(t_0)|^2 + \int_{t_0}^T \psi_l(s)\mathrm{e}^{\int_s^T \varphi(t)\mathrm{d}t}\mathrm{d}s,
$$

where

$$
\begin{aligned}
\varphi(t) &= \frac{4c_b^2}{\bar{\beta}}\|z_0(t)\|_{L^\infty}^2 - \bar{\beta}\lambda_1, \\
\psi_l(t) &= \frac{8c_b^2}{\bar{\beta}}|\Delta\bar{u}_{\epsilon_l}|^2|z_0(t)|^2.
\end{aligned}
$$

The t_0 is chosen large enough so that $\|z_0(t)\|_{L^\infty}^2 < \frac{\bar{\beta}^2\lambda_1}{4c_b^2}$ whenever $t \ge t_0$. Such t_0 exists, because by Eq. (10.19b) and the Sobolev embedding theorem (see, e.g., [3] Sect. 5.6.3 Theorem 6) for $k = 2 = p = d$, $z_0(t)$ is Hölder continuous with exponent $\gamma < 1$ for each $t \ge \tau$ and, moreover, $\sup_{t \ge \tau}\|z_0(t)\|_{C^\gamma} \le c''$. The uniform Hölder continuity of $z_0(t)$ and Eq. (10.21) implies by Lemma B.1 $z_0(t, \cdot) \longrightarrow 0$ uniformly on D as $t \to \infty$. Consequently, we have that $\sup_{t \ge t_0}\varphi(t) < 0$. Set $\bar{\varphi} > 0$

satisfying

$$2\bar{\varphi} < \min\left\{-\sup_{t\geq t_0}\varphi(t),\ 2\bar{\nu}\right\}.$$

Obviously, $\exp\left\{\int_{t_0}^{T}\varphi(t)\mathrm{d}t\right\} \leq \exp\left\{-2\bar{\varphi}(T-t_0)\right\}$. Moreover, from Eq. (10.21),

$$|\psi_l(t)| \leq \frac{8c_b^2}{\bar{\beta}}|\Delta\bar{u}_{\epsilon_l}|^2|z_0(t_0)|^2\mathrm{e}^{-2\bar{\nu}(t-t_0)} =: C_{\psi_l}\mathrm{e}^{-2\bar{\nu}(t-t_0)} \longrightarrow 0$$

decays exponentially as $t \to \infty$. Combining these results,

$$\begin{aligned}|z_{\epsilon_l}(T)|^2 &\leq \mathrm{e}^{-2\bar{\varphi}(T-t_0)}|z_{\epsilon_l}(t_0)|^2 + \int_{t_0}^{T} C_{\psi_l}\mathrm{e}^{-2\bar{\nu}(s-t_0)}\mathrm{e}^{-2\bar{\varphi}(T-s)}\mathrm{d}s \\ &\leq \mathrm{e}^{-2\bar{\varphi}(T-t_0)}|z_{\epsilon_l}(t_0)|^2 + \frac{C_{\psi_l}}{2(\bar{\nu}-\bar{\varphi})}\left(\mathrm{e}^{-2\bar{\phi}(T-t_0)}\mathrm{e}^{-2\bar{\nu}(T-t_0)}\right) \longrightarrow 0\end{aligned}$$

as $T \to \infty$. (In the first term, $|z_{\epsilon_l}(t_0)|^2$ has been shown to be finite for any finite t_0.) Since $\bar{\varphi} < \bar{\nu}$,

$$|z_{\epsilon_l}(T)|^2 \leq \left(|z_{\epsilon_l}(t_0)|^2 + \frac{C_{\psi_l}}{2(\bar{\nu}-\bar{\varphi})}\right)\mathrm{e}^{-2\bar{\varphi}(T-t_0)} =: K_{\epsilon_l}^2\mathrm{e}^{-2\bar{\varphi}(T-t_0)} \qquad (10.23)$$

for $T \geq t_0$. K_{ϵ_l} does not depend on T. For $|\alpha| \geq 2$, we prove by induction. Fix α, and assume the induction hypothesis that for each $0 < \gamma < \alpha$, for $T \geq t_0$,

$$|z_\gamma(T)| \leq K_\gamma \mathrm{e}^{-2^{1-|\gamma|}\bar{\varphi}(T-t_0)} \longrightarrow 0 \qquad (10.24)$$

as $T \to \infty$, where K_γ does not depend on T. We want to show that Eq. (10.24) also holds for α. From Eq. (10.20b) with test function $v = z_\alpha$,

$$\begin{aligned}&\frac{1}{2}\frac{\mathrm{d}}{\mathrm{d}t}|z_\alpha|^2 + \nu|\nabla z_\alpha|^2 \\ &\leq |b(z_\alpha,\bar{u}_0,z_\alpha)| + |b(z_\alpha,z_0,z_\alpha)| + |b(z_0,\bar{u}_\alpha,z_\alpha)| + |b(\bar{u}_\alpha,z_0,z_\alpha)| \\ &\quad + \sum_{0<\gamma<\alpha}\sqrt{\tbinom{\alpha}{\gamma}}\left(|b(z_\gamma,z_{\alpha-\gamma},z_\alpha)| + |b(z_\gamma,\bar{u}_{\alpha-\gamma},z_\alpha)| + |b(\bar{u}_\gamma,z_{\alpha-\gamma},z_\alpha)|\right).\end{aligned}$$

Similar to Eq. (10.22), using the ε-inequality with any $0 < \varepsilon < \bar{\beta}/2$,

$$\frac{1}{2}\frac{\mathrm{d}}{\mathrm{d}t}|z_\alpha|^2 + (\bar{\beta} - 2\varepsilon)\|z_\alpha\|_V^2 \le \frac{c_b^2}{2\varepsilon}\|z_0\|_{L^\infty}^2|z_\alpha|^2 + \frac{2c_b^2}{\varepsilon}|\Delta\bar{u}_\alpha|^2|z_0|^2$$
$$+ \frac{c_b^2}{4\varepsilon}\left(\sum_{0<\gamma<\alpha}\sqrt{\binom{\alpha}{\gamma}}\left(\|z_{\alpha-\gamma}\|_V + 2\|\bar{u}_{\alpha-\gamma}\|_V\right)|z_\gamma|_{1/2}\right)^2.$$

Using the Poincaré inequality and taking $\varepsilon = \bar{\beta}/4$,

$$\frac{\mathrm{d}}{\mathrm{d}t}|z_\alpha(t)|^2 \le \left(\frac{4c_b^2}{\bar{\beta}}\|z_0\|_{L^\infty}^2 - \lambda_1\bar{\beta}\right)|z_\alpha|^2 + \frac{16c_b^2}{\bar{\beta}}|\Delta\bar{u}_\alpha|^2|z_0|^2$$
$$+ \frac{2c_b^2}{\bar{\beta}}\left(\sum_{0<\gamma<\alpha}\sqrt{\binom{\alpha}{\gamma}}\left(\|z_{\alpha-\gamma}\|_V + 2\|\bar{u}_{\alpha-\gamma}\|_V\right)|z_\gamma|^{1/2}\|z_\gamma\|_V^{1/2}\right)^2$$
$$\le \varphi(t)|z_\alpha(t)|^2 + \psi_\alpha(t)$$

where now

$$\psi_\alpha(t) = \frac{16c_b^2}{\bar{\beta}}|\Delta\bar{u}_\alpha|^2|z_0(t)|^2$$
$$+ \frac{2c_b^2}{\bar{\beta}}\left(\sum_{0<\gamma<\alpha}\sqrt{\binom{\alpha}{\gamma}}\left(\|z_{\alpha-\gamma}(t)\|_V + 2\|\bar{u}_{\alpha-\gamma}\|_V\right)^2\|z_\gamma(t)\|_V\right)$$
$$\times\left(\sum_{0<\gamma<\alpha}\sqrt{\binom{\alpha}{\gamma}}\,|z_\gamma(t)|\right).$$

From the hypothesis (10.24),

$$|\psi_\alpha(t)| \le C_{\psi_\alpha}\mathrm{e}^{-2\bar{\nu}(t-t_0)} + \tilde{C}_{\psi_\alpha}\left(\sum_{0<\gamma<\alpha}\sqrt{\binom{\alpha}{\gamma}}\,K_\gamma\mathrm{e}^{-2^{-|\gamma|}2\bar{\varphi}(t-t_0)}\right),$$

where

$$C_{\psi_\alpha} = \frac{16c_b^2}{\bar{\beta}}\|\bar{u}_\alpha\|_{H^2}^2|z_0(t_0)|^2,$$
$$\tilde{C}_{\psi_\alpha} = \frac{2c_b^2}{\bar{\beta}}\left(\sum_{0<\gamma<\alpha}\sqrt{\binom{\alpha}{\gamma}}\left(\sup_{s\ge 0}\|z_{\alpha-\gamma}(s)\|_V + 2\|\bar{u}_{\alpha-\gamma}\|_V\right)^2\sup_{s\ge 0}\|z_\gamma(s)\|_V\right),$$

and $C_{\phi_\alpha}, \tilde{C}_{\phi_\alpha}$ do not depend on t. By Gronwall's inequality,

$$\begin{aligned}|z_\alpha(T)|^2 &\le e^{-\bar{\varphi}(T-t_0)}|z_\alpha(t_0)|^2 + \int_{t_0}^{T} \psi_\alpha(s)e^{-\bar{\varphi}(T-s)}ds\\ &\le e^{-\bar{\varphi}(T-t_0)}|z_\alpha(t_0)|^2 + \frac{C_{\psi_\alpha}}{2(\bar{\nu}-\bar{\varphi})}e^{-2\bar{\varphi}(T-t_0)}\\ &\quad + \tilde{C}_{\psi_\alpha}\sum_{0<\gamma<\alpha}\sqrt{\tbinom{\alpha}{\gamma}}\,K_\gamma\frac{e^{-2^{1-|\gamma|}\bar{\varphi}(T-t_0)}}{1-2^{-|\gamma|}}\\ &\le K_\alpha^2 e^{-2^{1-(|\alpha|-1)}\bar{\varphi}(T-t_0)},\end{aligned}$$

where K_α does not depend on T. Hence,

$$|z_\alpha(T)| \le K_\alpha e^{-2^{1-|\alpha|}\bar{\varphi}(T-t_0)} \tag{10.25}$$

for all $T \ge t_0$. It follows that Eq. (10.24) holds also for α, and the result follows. □

We proceed to deduce the long-time convergence of $u(t)$ in some Kondratiev space $\mathcal{S}_{-1,-q}(H)$. The manner of estimates in Proposition 5.1 is not directly suited for applying the Catalan numbers method. Instead, we will use a compact embedding-type argument in the following lemma to show the result.

Lemma 5.1. *For $q > 0$, let the sequence $r = (2\mathbb{N})^{-q}$. Let $u^k \in \mathcal{S}_{-1,-q}(V)$ be a sequence satisfying*

$$\sum_\alpha \frac{r^\alpha}{\alpha!}\left(\sup_k \|u_\alpha^k\|_V^2\right) < \infty,$$

that is, satisfying $\{u^k\} \in \mathcal{S}_{-1,-q}(\ell^\infty(V))$.

Then there exists a subsequence $\tilde{k}_N$ such that $u^{\tilde{k}_N}$ converges in $\mathcal{D}'(H)$ to some $\bar{u} \in \mathcal{D}'(H)$. In fact, $\bar{u} \in \mathcal{S}_{-1,-q}(V)$ and the convergence is in $\mathcal{S}_{-1,-q}(H)$.

Proof. The proof of convergence in $\mathcal{D}'(H)$ will follow a diagonalization argument and from the fact that V is compactly embedded in H. Let $\mathcal{J}_N = \{\alpha \in \mathcal{J} : |\alpha| \le N, \text{ and } \alpha_i = 0 \text{ for } i > N\}$. Since $\sup_k \|u_0^k\|_V < \infty$, there exists a subsequence $\{k_j^0\}_{j=1}^\infty$ such that $\|u_0^{k_j^0} - \bar{u}_0\|_H \to 0$ for some $\bar{u}_0 \in H$. Iteratively, for each N, there exists further subsequences $\{k_j^N\}_{j=1}^\infty \subset \{k_j^{N-1}\}_{j=1}^\infty$ such that for every $\alpha \in \mathcal{J}_N$,

$$\|u_\alpha^{k_j^N} - \bar{u}_\alpha\|_H \to 0$$

for some $\bar{u}_\alpha \in H$. In particular, for each N, we can find j_N such that

$$\|u_\alpha^{k_{j_N}^N} - \bar{u}_\alpha\|_H \le N^{-1}, \quad \text{for all } \alpha \in \mathcal{J}_N.$$

Consequently, choose the subsequence $\tilde{k}_N := k^N_{j_N}$ and we have found the limit $\bar{u} = \sum_\alpha \bar{u}_\alpha \xi_\alpha$. It follows that $u^{\tilde{k}_N} \to \bar{u}$ in $\mathcal{D}'(H)$. By Fatou's lemma,

$$\sum_\alpha \frac{r^\alpha}{\alpha!}|\bar{u}_\alpha|^2 \le \sup_k \sum_\alpha \frac{r^\alpha}{\alpha!}|u^k_\alpha|^2 \le \sup_k \sum_\alpha \frac{r^\alpha}{\alpha!}\|u^k_\alpha\|^2_V < \infty,$$

and so $\bar{u} \in \mathcal{S}_{-1,-q}(H)$.

To prove the convergence in $\mathcal{S}_{-1,-q}(H)$, let $\varepsilon > 0$ be arbitrary. For any N,

$$\|u^{\tilde{k}_N} - \bar{u}\|^2_{-1,-q;H} = \sum_{\alpha \in \mathcal{J}_N} \frac{r^\alpha}{\alpha!}\|u^{\tilde{k}_N} - \bar{u}\|^2_H + \sum_{\alpha \notin \mathcal{J}_N} \frac{r^\alpha}{\alpha!}\|u^{\tilde{k}_N} - \bar{u}\|^2_H = (I) + (II).$$

By our special choice of $\tilde{k}_N$, there exists N_I such that

$$(I) \le \sum_{\alpha \in \mathcal{J}_N} \frac{r^\alpha}{\alpha!} N^{-2} < \frac{\varepsilon}{2} \quad \text{whenever } N > N_I.$$

From the hypothesis of the lemma, there exists N_{II} such that

$$(II) \le 2 \sum_{\alpha \notin \mathcal{J}_N} \frac{r^\alpha}{\alpha!}\Big(\sup_k \|u^k\|^2_V\Big) + 2 \sum_{\alpha \notin \mathcal{J}_N} \frac{r^\alpha}{\alpha!}\|\bar{u}\|^2_H < \frac{\varepsilon}{2} \quad \text{whenever } N > N_{II}.$$

Thus, $\|u^{\tilde{k}_N} - \bar{u}\|^2_{-1,-q;H} < \varepsilon$ whenever $N > \max\{N_I, N_{II}\}$. □

The hypothesis in Lemma 5.1 is stronger than requiring $u^k \in l^\infty(\mathcal{S}_{-1,-q}(V))$; thus it is a weaker statement of what might be construed as a compact embedding result for Kondratiev spaces. It is not shown whether $\mathcal{S}_{-1,-q}(V)$ is compactly embedded in $\mathcal{S}_{-1,-q}(H)$. Nonetheless, it is sufficient for our purposes.

Corollary 5.1. *Let $d = 2$. Assume the hypotheses of Propositions 3.2 and 4.1(ii). Then, for the solutions $u(t)$ and $\bar{u}$ of Eqs.* (10.4) *and* (10.5), *we have that*

$$u(t) \longrightarrow \bar{u} \quad \text{in } \mathcal{S}_{-1,-q}(H), \text{ as } t \to \infty,$$

for $q > \max\{q_0, q_2\}$, where q_0, q_2 are the numbers from Propositions 3.2 and 4.1.

Proof. In the proof of Proposition 4.1, we have in fact shown that $u(t)$ belongs to the space $\mathcal{S}_{-1,-q}(L^\infty([0,\infty); V))$. Taking any sequence of times, $t_k \to \infty$, the sequence $\{u(t_k)\}$ satisfies the hypothesis of Lemma 5.1. So, there exists a subsequence of $u(t_k)$ converging in $\mathcal{S}_{-1,-q}(H)$ to $\bar{u}$. This is true for any sequence $\{t_k\}$; hence $u(t) \longrightarrow \bar{u}$ in $\mathcal{S}_{-1,-q}(H)$ as $t \to \infty$. □

6 Finite Approximation by Wiener Chaos Expansions

In this section, we study the accuracy of the Galerkin approximation of the solutions of the unbiased stochastic Navier–Stokes equations. The goal is to quantify the convergence rate of approximate solutions obtained from a finite truncation of the Wiener chaos expansion, where the convergence is in a suitable Kondratiev space. In relation to being a numerical approximation, quantifying the truncation error is the first step towards understanding the error from the full discretization of the unbiased stochastic Navier–Stokes equation.

In what follows, we will consider the truncation error estimates for the steady solution $\bar{u}$. Recall the estimate (10.16) for $|\Delta \bar{u}|$: for $r_\alpha^2 = \frac{(2\mathbb{N})^{-q\alpha}}{\alpha!}$, with $q > q_0$, we have

$$r_\alpha^2 |\Delta \bar{u}_\alpha|^2 \leq C_{|\alpha|-1}^2 \binom{|\alpha|}{\alpha} (2\mathbb{N})^{(1-q)\alpha} \bar{B}_0^{-2} (\bar{B}_0 \bar{K})^{2|\alpha|}.$$

This estimate arose from the method of rescaling via Catalan numbers, and will be the estimate we use for the convergence analysis. For the time-dependent equation, similar analysis can be performed using the analogous Catalan rescaled estimate, and will not be shown.

Let $\mathcal{J}_{M,P} = \{\alpha : |\alpha| \leq P, \dim(\alpha) \leq M\}$, where M, P may take value ∞. The projection of $\bar{u}$ into $\text{span}\{\xi_\alpha, \alpha \in \mathcal{J}_{M,P}\}$ is $\bar{u}^{M,P} = \sum_{\alpha \in \mathcal{J}_{M,P}} \bar{u}_\alpha \xi_\alpha$.

Then the error $e = \bar{u} - \bar{u}^{M,P}$ can be written as

$$\begin{aligned}
|\Delta e|^2 &= \sum_{\alpha \in \mathcal{J} \setminus \mathcal{J}_{M,P}} r_\alpha^2 |\Delta \bar{u}_\alpha|^2 \\
&= \sum_{|\alpha|=P+1}^{\infty} r_\alpha^2 |\Delta \bar{u}_\alpha|^2 + \sum_{\{|\alpha| \leq P,\, |\alpha_{\leq M}| < |\alpha|\}} r_\alpha^2 |\Delta \bar{u}_\alpha|^2 \\
&= \underbrace{\sum_{|\alpha|=P+1}^{\infty} r_\alpha^2 |\Delta \bar{u}_\alpha|^2}_{(IV)} + \underbrace{\sum_{|\alpha|=1}^{P} \underbrace{\sum_{i=0}^{|\alpha|-1} \underbrace{\sum_{|\alpha_{\leq M}|=i} r_\alpha^2 |\Delta \bar{u}_\alpha|^2}_{(I)}}_{(II)}}_{(III)},
\end{aligned}$$

where $\alpha_{\leq M}$ is the multi-index for which the kth entry $(\alpha_{\leq M})_k = \alpha_k$ if $k \leq M$ and $(\alpha_{\leq M})_k = 0$ for $k > M$.

We define the following values:

$$\hat{Q} := 2^{1-q}\bar{B}_0^2\bar{K}^2\sum_{i=1}^{\infty} i^{1-q},$$

$$\hat{Q}_{\leq M} := 2^{1-q}\bar{B}_0\bar{K}^2\sum_{i=1}^{M} i^{1-q}, \qquad \hat{Q}_{>M} := 2^{1-q}\bar{B}_0\bar{K}^2\sum_{i=M+1}^{\infty} i^{1-q}.$$

In particular, the term $\hat{Q}_{>M}$ decays on the order of M^{2-q}.

We proceed to estimate the terms (I)–(IV), by similar computations to Wan et al. [17]. For fixed $1 \leq p \leq P$, $|\alpha| = p$, and fixed $i < p$,

$$\begin{aligned}(I) &\leq C_{p-1}^2\bar{B}_0^{-2}\sum_{|\alpha_{\leq M}|=i,\,|\alpha_{>M}|=p-i}\binom{|\alpha|}{\alpha}(2\mathbb{N})^{(1-q)\alpha}(\bar{B}_0\bar{K})^{2p}\\ &= C_{p-1}^2\bar{B}_0^{-2}\binom{p}{i}\hat{Q}_{\leq M}^i\hat{Q}_{>M}^{p-i}.\end{aligned}$$

Then for fixed $1 \leq p \leq P$, $|\alpha| = p$,

$$\begin{aligned}(II) &= \sum_{i=0}^{p-1}(I) \leq C_{p-1}^2\bar{B}_0^{-2}\sum_{i=0}^{p-1}\binom{p}{i}\hat{Q}_{\leq M}^i\hat{Q}_{>M}^{p-i}\\ &= C_{p-1}^2\bar{B}_0^{-2}(\hat{Q}^p - \hat{Q}_{\leq M}^p).\end{aligned}$$

And finally,

$$\begin{aligned}(III) &= \sum_{|\alpha|=1}^{P}(II) \leq \sum_{p=1}^{P} C_{p-1}^2\bar{B}_0^{-2}(\hat{Q}^p - \hat{Q}_{\leq M}^p)\\ &\leq \frac{1}{\bar{B}_0^2}(\hat{Q} - \hat{Q}_{\leq M}) + \frac{1}{16\pi\bar{B}_0^2}\sum_{p=2}^{P}\frac{2^{4p}}{(p-1)^3}(\hat{Q}^p - \hat{Q}_{\leq M}^p).\end{aligned}$$

Since $\hat{Q}^p - \hat{Q}_{\leq M}^p \leq p\hat{Q}^{p-1}(\hat{Q} - \hat{Q}_{\leq M})$ by the mean value theorem for $x \mapsto x^p$,

$$\begin{aligned}(III) &\leq \frac{1}{\bar{B}_0^2}\hat{Q}_{>M} + \frac{1}{16\pi\bar{B}_0^2}\hat{Q}_{>M}\sum_{p=2}^{P}\frac{p2^{4p}\hat{Q}^{p-1}}{(p-1)^3}\\ &\leq \frac{1}{\bar{B}_0^2}\hat{Q}_{>M} + \frac{1}{\pi\bar{B}_0^2}\hat{Q}_{>M}\sum_{p=2}^{P}\frac{p(2^4\hat{Q})^{p-1}}{(p-1)^3}\end{aligned}$$

$$\le \frac{1}{\bar{B}_0^2}\hat{Q}_{>M}\sum_{p=0}^{P-1}(2^4\hat{Q})^p,$$

To estimate Term (IV),

$$(IV) \le \sum_{p=P+1}^{\infty}\sum_{|\alpha|=p} C_{p-1}^2\bar{B}_0^{-2}(2^{1-q}\bar{B}_0^2K^2)^p\binom{|\alpha|}{\alpha}(\mathbb{N})^{(1-q)\alpha}$$

$$= \bar{B}_0^{-2}\sum_{p=P+1}^{\infty} C_{p-1}^2(2^{1-q}\bar{B}_0^2\bar{K}^2)^p\Big(\sum_{i\ge 1} i^{1-q}\Big)^p$$

$$\le \bar{B}_0^{-2}\sum_{p=P+1}^{\infty}\frac{2^{4(p-1)}}{\pi(p-1)^3}\hat{Q}^p \le \frac{1}{16\pi\bar{B}_0^2}\frac{(2^4\hat{Q})^{P+1}}{1-2^4\hat{Q}}.$$

Putting the estimates together,

$$|\Delta e|^2 \le C\big((2^4\hat{Q})^{P+1} + M^{2-q}\big).$$

Notice the condition $2^4\hat{Q} < 1$ in Eq. (10.17), which ensured summability of the weighted norm of the solution, is of course a required assumption for the convergence of the error estimate.

A The Catalan Numbers Method

The Catalan numbers method was used in the preceding sections to derive estimates for the norms in Kondratiev spaces. This method was previously described in [10, 14], but we restate it here just for the record.

Lemma A.1. *Suppose L_α are a collection of positive real numbers indexed by $\alpha \in \mathcal{J}$, satisfying*

$$L_\alpha \le B\sum_{0<\gamma<\alpha} L_\gamma L_{\alpha-\gamma}.$$

Then

$$L_\alpha \le C_{|\alpha|-1}B^{|\alpha|-1}\binom{|\alpha|}{\alpha}\prod_i L_{\epsilon_i}^{\alpha_i}$$

for all α, where C_n are the Catalan numbers.

Proof. The result is clearly true for $\alpha = \epsilon_i$. By induction, let $|\alpha| \ge 2$, and suppose the result is true for all $\gamma < \alpha$. Then

$$\begin{aligned}
L_\alpha &\leq \sum_{0<\gamma<\alpha} C_{|\gamma|-1}C_{|\alpha-\gamma|-1}B^{|\alpha|-1}\binom{|\gamma|}{\gamma}\binom{|\alpha-\gamma|}{\alpha-\gamma}\Big(\prod_i L_{\epsilon_i}^{\alpha_i}\Big)\\
&= \sum_{n=1}^{|\alpha|-1}\sum_{0<\gamma<\alpha|\gamma|=n} C_{n-1}C_{|\alpha|-n-1}\frac{n!\,(|\alpha|-n)!}{\gamma!\,(\alpha-\gamma)!}B^{|\alpha|-1}\Big(\prod_i L_{\epsilon_i}^{\alpha_i}\Big)\\
&= \sum_{n=1}^{|\alpha|-1} C_{n-1}C_{|\alpha|-n-1}\underbrace{\sum_{0<\gamma<\alpha|\gamma|=n}\binom{|\alpha|}{n}^{-1}\binom{\alpha}{\gamma}}_{(*)}\frac{|\alpha|!}{\alpha!}B^{|\alpha|-1}\Big(\prod_i L_{\epsilon_i}^{\alpha_i}\Big).
\end{aligned}$$

We claim that $(*) = 1$, for any α and any $n < |\alpha|$. Indeed, let $K_\alpha = (k_1 \ldots, k_{|\alpha|})$ be the characteristic set of α. Each summand in $(*)$ is

$$\left(\frac{|\alpha|!}{\alpha!}\right)^{-1}\frac{n!\,(|\alpha|-n)!}{\gamma!\,(\alpha-\gamma)!}.$$

The term $\frac{|\alpha|!}{\alpha!}$ is the number of distinct permutations of K_α, whereas the term $\frac{n!\,(|\alpha|-n)!}{\gamma!\,(\alpha-\gamma)!}$ is the number of distinct permutations of K_α where only $K_\gamma, K_{\alpha-\gamma}$ has been permuted within themselves. On the other hand, the latter term is the number of distinct permutations of K_α corresponding to a particular γ, where the correspondence of a permutation of K_α to a $\gamma \in \{\gamma : 0 < \gamma < \alpha, |\gamma| = n\}$ can be made by taking K_γ to be the first n entries of that permutation of K_α. Thus, each summand in $(*)$ is the relative frequency of γ over all distinct permutations of K_α, and hence their sum must equal 1. To complete the proof, using the recursion property of the Catalan numbers,

$$\begin{aligned}
L_\alpha &\leq \sum_{n=1}^{|\alpha|-1} C_{n-1}C_{|\alpha|-n-1}\left(\frac{|\alpha|!}{\alpha!}\right)B^{|\alpha|-1}\prod_i L_{\epsilon_i}^{\alpha_i}\\
&= C_{|\alpha|-1}\left(\frac{|\alpha|!}{\alpha!}\right)B^{|\alpha|-1}\prod_i L_{\epsilon_i}^{\alpha_i}.
\end{aligned}$$

□

If L_α satisfies the hypothesis of Lemma A.1, and if $L_{\epsilon_i} \leq K$ for all i, then for $r = (2\mathbb{N})^{-q}$,

$$\sum_{|\alpha|=n} r^\alpha L_\alpha^2 \leq \sum_{|\alpha|=n} C_{n-1}^2 B^{2(|\alpha|-1)} K^{2|\alpha|}\binom{|\alpha|}{\alpha}(2\mathbb{N})^{(1-q)\alpha} \quad .$$

$$= B^{-2}C_{n-1}^2\left(B^2K^2 2^{1-q}\right)^n \sum_{|\alpha|=n} \binom{|\alpha|}{\alpha} \mathbb{N}^{(1-q)\alpha}$$

$$= B^{-2}C_{n-1}^2\left(B^2K^2 2^{1-q}\right)^n \left(\sum_{i=1}^{\infty} i^{(1-q)}\right)^n.$$

For large n, the Catalan numbers behave asymptotically like $C_n \sim \frac{2^{2n}}{\sqrt{\pi}n^{3/2}}$. Hence, the sum $\sum_{n=0}^{\infty}\sum_{|\alpha|=n} r^\alpha L_\alpha^2$ converges for any $q > \max\{q_0, 2\}$, where q_0 satisfies

$$B^2K^2 2^{5-q_0}\sum_{i=1}^{\infty} i^{(1-q_0)} = 1.$$

B A Lemma

Lemma B.1. *Let $z(t) \in L^2(D)$ for all $t \geq \tau$ and assume that $z(t) \longrightarrow 0$ in L^2. Also assume that $\sup_{t\geq\tau}\|z(t)\|_{C^\gamma} \geq C$ for some constant C and exponent $\gamma < 1$. Then $\sup_{t\geq\tau}\|z(t)\|_{L^\infty} \longrightarrow 0$ as $t \to \infty$.*

Proof. We prove by contradiction. Assume that there exists ϵ_0 and a sequence $t_n \to \infty$ such that $\sup_{x\in D}|z(t_n,x)| > \epsilon_0$. Since $z(t_n)$ is Hölder continuous, there exists $x_n \in D$ and a ball $B_\delta(x_n)$ of radius $\delta = \left(\frac{\epsilon_0}{2C}\right)^{1/\gamma}$ such that $|z(t_n,x_n) - z(t_n,y)| \leq C\delta^\gamma$ for $y \in B_\delta(x_n)$. Hence $|z(t_n,y)| > \frac{\epsilon_0}{2}$ for $y \in B_\delta(x_n)$. But since $z(t) \longrightarrow 0$ in L^2, $\int_D |z(t_n,y)|^2 dy < |B_\delta(x_n)|\frac{\epsilon_0^2}{4}$ for some t_n large enough. This contradicts $\int_{B_\delta(x_n)} |z(t_n,y)|^2 dy \geq |B_\delta(x_n)|\frac{\epsilon_0^2}{4}$. □

Acknowledgements B. L. Rozovskii acknowledges support from AFOSR MURI Grant 955-05-1-0613.

References

1. Bensoussan, A., Temam, R.: Équations stochastiques du type Navier-Stokes. J. Funct. Anal. **13** 195–222 (1973)
2. Cameron, R.H., Martin, W.T.: The orthogonal development of non-linear functionals in a series of Fourier-Hermite functions. Ann. of Math. **48**, 385–392 (1947)
3. Evans, L.C.: Partial Differential Equations. American Mathematical Society, Providence (1998)
4. Foiaş, C.: Statistical study of Navier-Stokes equations. I, II. Rend. Sem. Mat. Univ. Padova **48**, 219–348 (1972)

5. Foiaş, C., Rosa, R.M.S., Temam, R.: A note on statistical solutions of the three-dimensional Navier-Stokes equations: the stationary case. C. R. Math. Acad. Sci. Paris **348**(5–6), 347–353 (2010)
6. Foiaş, C., Temam, R.: Homogeneous statistical solutions of Navier-Stokes equations. Indiana Univ. Math. J.**29**(6), 913–957 (1980)
7. Flandoli, F.: Dissipativity and invariant measures for stochastic Navier-Stokes equations. NoDEA Nonlinear Differential Equations Appl. **1**(4), 403–423 (1994)
8. Flandoli, F., Gatarek, D.: Martingale and stationary solutions for stochastic Navier-Stokes equations. Probab. Theory Relat. Fields **102**(3), 367–391 (1995)
9. Guillopé, C.: Comportement à l'infini des solutions des équations de Navier-Stokes et propriété des ensembles fonctionnels invariants (ou attracteurs). Ann. Inst. Fourier (Grenoble) **32**(3) pp. ix, 1–37 (1982)
10. Kaligotla, S., Lototsky, S.V.: Wick product in stochastic Burgers equation: a curse or a cure?. Asymptotic Anal. **75**(3–4), 145–168 (2011)
11. Lototsky, S.V., Rozovskii, B.L., Seleši, D.: On generalized Malliavin calculus. Stochastic Processes and Appl. doi:10.1016/j.spa.2011.11.003 (2011)
12. Mikulevicius, R., Rozovskii, B.L.: Stochastic Navier-Stokes equations for turbulent flows. SIAM J. Math. Anal. **35**(5), 1250–1310 (2004)
13. Mikulevicius, R., Rozovskii, B.L.: Global L_2-solutions of stochastic Navier-Stokes equations. Ann. Probab. **33**(1), 137–176 (2005)
14. Mikulevicius, R., Rozovskii, B.L.: On unbiased stochastic Navier-Stokes equations. Probab. Theory Related Fields, DOI 10.1007/s00440-011-0384-1 (2011)
15. Temam, R.:Navier-Stokes equations and nonlinear functional analysis, 2nd ed., CBMS-NSF Regional Conference Series in Applied Mathematics, Society for Industrial and Applied Mathematics (SIAM), Philadelphia, PA, **66** (1995)
16. Temam, R.: Navier-Stokes Equations: Theory and Numerical Analysis. AMS Chelsea Publishing, Providence (2001)
17. Wan, X., Rozovskii, B.L., Karniadakis, G.E.: A new stochastic modeling methodology based on weighted Wiener chaos and Malliavin calculus. Proc. Natl. Acad. Sc. USA **106**34, 14189–14104 (2009)

Chapter 11
Intermittency and Chaos for a Nonlinear Stochastic Wave Equation in Dimension 1

Daniel Conus, Mathew Joseph, Davar Khoshnevisan, and Shang-Yuan Shiu

This paper is dedicated to Professor David Nualart, whose scientific innovations have influenced us greatly.

Abstract Consider a nonlinear stochastic wave equation driven by space-time white noise in dimension one. We discuss the intermittency of the solution, and then use those intermittency results in order to demonstrate that in many cases the solution is chaotic. For the most part, the novel portion of our work is about the two cases where (1) the initial conditions have compact support, where the global maximum of the solution remains bounded, and (2) the initial conditions are positive constants, where the global maximum is almost surely infinite. Bounds are also provided on the behavior of the global maximum of the solution in Case (2).

Keywords Intermittency • The stochastic wave equation • Chaos

D. Conus (✉)
Department of Mathematics/Lehigh University,
14 East Packer Avenue, Bethlehem PA 18015, USA
e-mail: daniel.conus@lehigh.edu

M. Joseph
Department of Mathematics/University of Utah,
155 South 1400 East, Salt Lake City, UT 84112, USA

School of Mathematics and Statistics/University of Sheffield,
Hicks Building, Hounsfield Road, Sheffield S3 7RH, United Kingdom
e-mail: joseph@math.utah.edu

D. Khoshnevisan
Department of Mathematics/University of Utah,
155 South 1400 East, Salt Lake City, UT 84112, USA
e-mail: davar@math.utah.edu

S.-Y. Shiu
Department of Mathematics/National Central University,
N°300 Jhongda Road, Jhongli City, Taoyuan County 32001, Taiwan
e-mail: shiu@mx.math.ncu.edu.tw

F. Viens et al. (eds.), *Malliavin Calculus and Stochastic Analysis: A Festschrift in Honor of David Nualart*, Springer Proceedings in Mathematics & Statistics 34,
DOI 10.1007/978-1-4614-5906-4_11,

MSC Subject Classication 2000: 60H15; 60H20, 60H05

Received 12/5/2011; Accepted 2/29/2012; Final 7/17/2012

1 Introduction

Let us consider the following hyperbolic stochastic PDE of the wave type

$$(\Box u)(t\,,x) = \sigma(u(t,x))\dot{W}(t,x) \qquad (t > 0,\ x \in \mathbf{R}). \tag{11.1}$$

Here, $\Box$ denotes the [massless] wave operator

$$\Box := \frac{\partial^2}{\partial t^2} - \kappa^2 \frac{\partial^2}{\partial x^2},$$

$\sigma : \mathbf{R} \to \mathbf{R}$ is a globally Lipschitz function with Lipschitz constant

$$\mathrm{Lip}_\sigma := \sup_{-\infty<x<y<\infty} \frac{|\sigma(y) - \sigma(x)|}{y - x},$$

$\dot{W}$ denotes space-time white noise, and $\kappa > 0$ is a fixed constant. The initial function and the initial velocity are denoted respectively by $u_0 : \mathbf{R} \to \mathbf{R}$ and $v_0 : \mathbf{R} \to \mathbf{R}$, and we might refer to the pair (u_0, v_0) as the "initial conditions" of the stochastic wave Eq. (11.1). [The terminology is standard in PDEs, and so we use it freely.] When the initial value $x \mapsto u_0(x)$ is assumed to be a constant, we write the constant as u_0; similar remarks apply to v_0. In those cases, we state quite clearly that u_0 and v_0 are constants in order to avoid ambiguities.

The stochastic wave Eq. (11.1) has been studied extensively by Carmona and Nualart [9] and Walsh [26]. Among other things, these references contain the theorem that the random wave Eq. (11.1) has a unique continuous solution u as long as

$$u_0 \text{ and } v_0 \text{ are bounded and measurable functions,}$$

an assumption that is made tacitly throughout this paper. All of this is about the wave equation in dimension $1 + 1$ (that is one-dimensional time and one-dimensional space). There are also some existence theorems in the more delicate dimensions $1 + d$ (that is one-dimensional time and d-dimensional space), where $d > 1$ and the 1-D wave operator $\Box$ is replaced by the d-dimensional wave operator $\partial^2_{tt} - \kappa^2\Delta$, where Δ denotes the Laplacian on $\mathbf{R}^d$; see Conus and Dalang [10], Dalang [13], Dalang and Frangos [14], and Dalang and Mueller [15].

Parabolic counterparts to the random hyperbolic Eq. (11.1) are well-studied stochastic PDEs. For example, when $\sigma(u) = u$ and the wave operator $\Box$ is replaced by the heat operator $\partial_t - \kappa^2\partial^2_{xx}$, the resulting stochastic PDE becomes a continuous *parabolic Anderson model* [8] and has connections to the study of random polymer measures and the KPZ equation [1–3, 19, 22–24] and numerous other problems of mathematical physics and theoretical chemistry [8, Introduction]. The mentioned references contain a great deal of further information about these sorts of parabolic SPDEs.

From a purely mathematical point of view, Eq. (11.1) is the hyperbolic counterpart to the stochastic heat equation, and in particular $\sigma(u) = \text{const} \cdot u$ ought to be a hyperbolic counterpart to the parabolic Anderson model. From a more pragmatic point of view, we believe that the analysis of the present hyperbolic equations might one day also lead to a better understanding of numerical analysis problems that arise when trying to solve families of chaotic hyperbolic stochastic PDEs.

It is well known, and easy to verify directly, that the Green function for the wave operator $\square$ is

$$\Gamma_t(x) := \frac{1}{2}\mathbf{1}_{[-\kappa t, \kappa t]}(x) \qquad \text{for } t > 0 \text{ and } x \in \mathbf{R}. \tag{11.2}$$

According to general theory [9, 13, 26], the stochastic wave Eq. (11.1) has an a.s.-unique continuous solution $\{u(t,x)\}_{t>0, x\in\mathbf{R}}$ which has the following mild formulation:

$$u(t,x) = U_0(t,x) + V_0(t,x) + \int_{(0,t)\times\mathbf{R}} \Gamma_{t-s}(y-x)\sigma(u(s,y))\, W(\mathrm{d}s\,\mathrm{d}y). \tag{11.3}$$

The integral is understood to be a stochastic integral in the sense of Walsh [26, Chap. 2] and

$$U_0(t,x) := \frac{u_0(x+\kappa t) + u_0(x-\kappa t)}{2}; \qquad V_0(t,x) := \frac{1}{2}\int_{x-\kappa t}^{x+\kappa t} v_0(y)\,\mathrm{d}y. \tag{11.4}$$

In the special case that u_0 and v_0 are constants, the preceding simplifies to

$$u(t,x) = u_0 + v_0\kappa t + \frac{1}{2}\int_{(0,t)\times(x-\kappa t, x+\kappa t)} \sigma(u(s,y))\, W(\mathrm{d}s\,\mathrm{d}y). \tag{11.5}$$

Recall [8, 19] that the process $\{u(t,x)\}_{t>0, x\in\mathbf{R}}$ is said to be *weakly intermittent* if the upper moment Lyapunov exponents,

$$\bar{\gamma}(p) := \limsup_{t\to\infty} \frac{1}{t} \sup_{x\in\mathbf{R}} \log \mathrm{E}\left(|u(t,x)|^p\right) \qquad (1 \le p < \infty), \tag{11.6}$$

havc the property that

$$\bar{\gamma}(2) > 0 \quad \text{and} \quad \bar{\gamma}(p) < \infty \quad \text{for every } p \in [2, l\infty). \tag{11.7}$$

Various questions from theoretical physics [24] have motivated the study of intermittency for the stochastic heat equation. A paper [19] by Foondun and Khoshnevisan introduces methods for the intermittency analysis of fully nonlinear parabolic stochastic PDEs. That paper also contains an extensive bibliography, with pointers to the large literature on the subject.

As far as we know, far less is known about the intermittent structure of the stochastic wave equation. In fact, we are aware only of two bodies of research: There is the recent work of Dalang and Mueller [16] that establishes intermittency for Eq. (11.1) in dimension $1 + 3$ (1 for time and 3 for space), where (1) $\sigma(u) = \lambda u$ (the hyperbolic Anderson model) for some $\lambda > 0$; (2) $\dot{W}$ is replaced by a generalized Gaussian field that is white in its time and has correlations in its space variable; and (3) the 1-D wave operator is replaced by the 3-D wave operator. We are aware also of a recent paper by two of the present authors [11], where the solution to Eq. (11.1) is shown to be intermittent in the case that the initial function u_0 and the initial velocity v_0 are both sufficiently smooth functions of compact support and $\dot{W}$ is a space-time white noise. The latter paper contains also detailed results on the geometry of the peaks of the solution.

The purpose of this paper is to study intermittency and chaotic properties of the fully nonlinear stochastic wave Eq. (11.1). We follow mainly the exposition style of Foondun and Khoshnevisan [19] for our results on weak intermittency: We will show that Eq. (11.7) holds provided that σ is a function of truly linear growth (Theorems 3.1 and 3.2). We will also illustrate that this condition is somehow necessary by proving that weak intermittency fails to hold when σ is bounded (Theorem 3.3).

Regarding the chaotic properties of the solution u to Eq. (11.1), we follow mainly the exposition style of Conus, Joseph, and Khoshnevisan [12] who establish precise estimates on the asymptotic behavior of $\sup_{|x|\leq R} u(t,x)$, as $R \to \infty$ for fixed $t>0$, for the parabolic counterpart to Eq. (11.1). In the present hyperbolic case, we first prove that the solution to Eq. (11.1) satisfies $\sup_{x\in\mathbf{R}} |u(t,x)| < \infty$ a.s. for all $t \geq 0$, if the initial function and the initial velocity are functions of compact support (Theorem 4.1). Then we return to the case of central importance to this paper, and prove that $\sup_{x\in\mathbf{R}} |u(t,x)| = \infty$ a.s. for all $t > 0$ when u_0 and v_0 are positive constants. Also, we obtain some quantitative estimates on the behavior of the supremum under varying assumptions on the nonlinearity σ (Theorems 7.1 and 7.2).

When considered in conjunction, the results of this paper imply that the solution to Eq. (11.1) is chaotic in the sense that slightly different initial conditions can lead to drastically different qualitative behaviors for the solution. This phenomenon is entirely due to the presence of noise in the system Eq. (11.1) and does not arise in typical deterministic wave equations.

This paper might be of interest for two main reasons: First of all, we obtain estimates on the supremum of the solution to hyperbolic stochastic PDEs and use them to show that the solution can be chaotic. We believe that these estimates might have other uses and are worthy of record in their own right. Secondly, we shall see that the analysis of the 1-D wave equation is simplified by the fact that the fundamental solution Γ of the wave operator $\Box$—see Eq. (11.2)—is a bounded function of compact support. As such, one can also view this paper, in part, as a gentle introduction to the methods of the more or less companion paper [12].

Let us conclude the Introduction with an outline of the paper. Section 3 below mainly recalls intermittency results for Eq. (11.1). These facts are mostly known in the folklore, but we document them here, in a systematic manner, for what appears to be the first time. The reader who is familiar with [19] will undoubtedly recognize some of the arguments of Sect. 3.

Section 4 is devoted to the study of the case where the initial value and velocity have compact support [and hence are *not* constants]. We will show that in such cases, $\sup_{x\in\mathbf{R}} |u(t,x)| < \infty$ a.s. for all $t > 0$. Sections 5 and 6 contain novel tail-probability estimates that depend on various forms of the nonlinearity σ. These estimates are of independent interest. Here, we use them in order to establish various localization properties. Finally, in Sect. 7, we combine our earlier estimates and use them to state and prove the main results of this paper about the asymptotic behavior of $\sup_{|x|\le R} |u(t,x)|$ as $R \to \infty$. More specifically, we prove that if u_0 is a positive constant, v_0 is a nonnegative constant, and $\inf_{z\in\mathbf{R}} |\sigma(z)| > 0$, then the peaks of the solution in $x \in [-R, R]$ grow at least as $(\kappa \log R)^{1/3}$. More precisely, we prove that there exists an almost surely finite random variable $R_0 > 0$ and a positive and finite constant a such that

$$\sup_{|x|\le R} |u(t,x)|^3 \ge a\kappa \log R \qquad \text{for all } R > R_0.$$

Furthermore, we will prove that a does not depend on κ, as long as κ is sufficiently small; this assertion measures the effect of the noise on the intermittency properties of u. If $0 < \inf \sigma \le \sup \sigma < \infty$, then we prove that the preceding can be improved to the existence of an a.s.-finite R_1 together with positive and finite constants b and c such that

$$b\kappa \log R \le \sup_{|x|\le R} |u(t,x)|^2 \le c\kappa \log R \qquad \text{for all } R > R_1.$$

2 Preliminaries

In this section we introduce some notation and preliminary results that are used throughout the paper. For a random variable Z, we denote by $\|Z\|_p := \{\mathrm{E}(|Z|^p)\}^{1/p}$ the standard norm on $L^p(\Omega)$ $(1 \le p < \infty)$.

On several occasions we apply the following form of the Burkholder–Davis–Gundy inequality [4–6] for continuous $L^2(\Omega)$ martingales: If $\{X_t\}_{t\ge 0}$ is a continuous $L^2(\Omega)$ martingale with running maximum $X_t^* := \sup_{s\in[0,t]} |X_s|$ and quadratic variation process $\langle X\rangle$, then for all $p \in [2,\infty)$ and $t \in (0,\infty)$,

$$\|X_t^*\|_p \le (4p)^{1/2} \cdot \|\langle X\rangle_t\|_{p/2}^{1/2}. \tag{11.8}$$

The multiplicative prefactor $4p$ is the asymptotically optimal bound, due to Carlen and Kree [7], for the sharp constant in the Burkholder–Davis–Gundy inequality that was discovered by Davis [18].

Given numbers $p \in [1, \infty)$ and $\beta \in (0, \infty)$ and given a space-time random field $\{Z(t,x)\}_{t>0, x\in\mathbf{R}}$, let us recall the following norm [19]:

$$\|Z\|_{p,\beta} := \left\{ \sup_{t\geq 0} \sup_{x\in\mathbf{R}} \mathrm{e}^{-\beta t} \mathrm{E}\left(|Z(t,x)|^p\right) \right\}^{1/p}. \tag{11.9}$$

We also use the following norm [12]:

$$\mathcal{N}_{p,\beta}(Z) := \left(\sup_{t\geq 0} \sup_{x\in\mathbf{R}} \mathrm{e}^{-\beta t} \|Z\|_p^2 \right)^{1/2}. \tag{11.10}$$

Clearly, the two norms are related via the elementary relations

$$\mathcal{N}_{p,\beta}(Z) = \|Z\|_{p,p\beta/2} \quad \text{and} \quad \|Z\|_{p,\beta} = \mathcal{N}_{p,2\beta/p}(Z). \tag{11.11}$$

However, the difference between the norms becomes relevant to us when we need to keep track of some constants.

Finally, we mention the following elementary formulas about the fundamental solution Γ to the wave operator $\Box$: For all $t, \beta > 0$:

$$\|\Gamma_t\|_{L^2(\mathbf{R})}^2 = \frac{\kappa t}{2}, \int_0^t \|\Gamma_s\|_{L^2(\mathbf{R})}^2 \,\mathrm{d}s = \frac{\kappa t^2}{4}, \int_0^\infty \mathrm{e}^{-\beta s} \|\Gamma_s\|_{L^2(\mathbf{R})}^2 \,\mathrm{d}s = \frac{\kappa}{2\beta^2}. \tag{11.12}$$

3 Intermittency

We are ready to state and prove the intermittency of the solution to Eq. (11.1). Our methods follow closely those of Foondun and Khoshnevisan [19], for the heat equation, and Conus and Khoshnevisan [11], for the wave equation.

In order to establish weak intermittency for the solution to Eq. (11.1) we need to obtain two different results: (1) We need to derive a finite upper bound for $\bar{\gamma}(p)$ for every $p \geq 2$; and (2) we need to establish a positive lower bound for $\bar{\gamma}(2)$. It might help to recall that the Lyapunov exponents $\bar{\gamma}(p)$ were defined in Eq. (11.6).

Theorem 3.1. *If u_0 and v_0 are both bounded and measurable functions, then*

$$\bar{\gamma}(p) \leq p^{3/2} \mathrm{Lip}_\sigma \sqrt{\kappa/2} \qquad \textit{for all } p \in [2, \infty).$$

Remark 3.1. Since the optimal constant in the Burkholder–Davis–Gundy L^2 inequality is 1, an inspection of the proof of Theorem 3.1 yields the improved bound $\bar{\gamma}(2) \leq \mathrm{Lip}_\sigma \sqrt{\kappa/2}$ in the case that $p = 2$. □

For our next result we define

$$\mathrm{L}_\sigma := \inf_{x\neq 0} |\sigma(x)/x| \,. \tag{11.13}$$

Theorem 3.2. *If u_0 and v_0 are bounded and measurable,* $\inf_{x\in\mathbf{R}} u_0(x) > 0$, $v_0 \geq 0$ *pointwise, and* $\mathrm{L}_\sigma > 0$, *then* $\bar{\gamma}(2) \geq \mathrm{L}_\sigma \sqrt{\kappa/2}$.

Theorems 3.1 and 3.2 are similar to Theorems 2.1 and 2.7 of [19] for the heat equation. Together, they prove that the solution u is weakly intermittent provided that u_0 is bounded away from 0, $v_0 \geq 0$ and σ has linear growth. Intermittency in the case where u_0 and v_0 have compact support has been proved in [11] (see also Sect. 4). Theorems 3.1 and 3.2 illustrate that the wave equation exhibits a similar qualitative behavior as the heat equation. However, the quantitative behavior is different: Here, $\bar{\gamma}(p)$ is of order $p^{3/2}$, whereas it is of order p^3 for the stochastic heat equation.

The linear growth of σ is somehow necessary for intermittency as the following result suggests.

Theorem 3.3. *If u_0, v_0, and σ are all bounded and measurable functions, then*

$$\mathrm{E}\left(|u(t,x)|^p\right) = O(t^p) \qquad \text{as } t\to\infty, \qquad \text{for all } p\in[2,\infty).$$

This estimate is sharp when $u_0(x) \geq 0$ for all $x \in \mathbf{R}$ and $\inf_{z\in\mathbf{R}} v_0(z) > 0$.

The preceding should be compared to Theorem 2.3 of [19]. There it was shown that if u were replaced by the solution to the stochastic heat equation, then there is the much smaller bound $\mathrm{E}(|u(t,x)|^p) = o(t^{p/2})$, valid under boundedness assumptions on u_0 and σ.

Analogues of the preceding three theorems above are known in the parabolic setting [11, 19]. Therefore, we will describe only outlines of their proof.

We will use a stochastic Young-type inequality for stochastic convolutions (Proposition 3.1 below), which is a ready consequence of [11, Proposition 2.5].

For a random field $\{Z(t,x)\}_{t>0,x\in\mathbf{R}}$, we denote by $\Gamma * Z\dot{W}$ the random field defined by

$$(\Gamma * Z\dot{W})(t\,,x) = \int_{(0,t)\times\mathbf{R}} \Gamma_{t-s}(y-x)Z(s,y)\,W(\mathrm{d}s\,\mathrm{d}y),$$

provided that the stochastic integral is well defined in the sense of Walsh [26].

Proposition 3.1. *For all $\beta > 0$ and $p \in (2,\infty)$,*

$$\|\Gamma * Z\dot{W}\|_{2,\beta} \leq \frac{\kappa^{1/2}}{\beta\sqrt{2}}\|Z\|_{2,\beta} \quad \text{and} \quad \|\Gamma * Z\dot{W}\|_{p,\beta} \leq \frac{p^{3/2}\kappa^{1/2}}{\beta\sqrt{2}}\|Z\|_{p,\beta}.$$

Proof. We appeal to Eq. (11.8) in order to deduce that

$$\mathrm{E}\left(|(\Gamma * Z\dot{W})(t,x)|^p\right)$$

$$\leq (4p)^{p/2}\mathrm{E}\left[\left(\int_0^t \mathrm{d}s \int_{-\infty}^{\infty} \mathrm{d}y\, \Gamma_{t-s}^2(y-x)|Z(s,y)|^2\right)^{p/2}\right]$$

$$\leq (4p)^{p/2}\left(\int_0^t \mathrm{d}s \int_{-\infty}^{\infty} \mathrm{d}y\, \Gamma_{t-s}^2(y-x)\left\{\mathrm{E}\left(|Z(s,y)|^p\right)\right\}^{2/p}\right)^{p/2}; \qquad (11.14)$$

the last inequality is justified by Minkowski's inequality. Next we raise both sides of the preceding inequality to the power $2/p$ and then multiply both sides by $\mathrm{e}^{-\beta t}$ in order to obtain

$$\begin{aligned}[\mathcal{N}_{p,\beta}(\Gamma * Z\dot{W})]^2 &\leq 4p\int_0^{\infty} \mathrm{d}s \int_{-\infty}^{\infty} \mathrm{d}y\, \mathrm{e}^{-\beta(t-s)}\Gamma_{t-s}^2(y-x)\mathrm{e}^{-\beta s}\left\{\mathrm{E}\left(|Z(s,y)|^p\right)\right\}^{2/p} \\ &\leq 4p\left[\mathcal{N}_{p,\beta}(Z)\right]^2 \int_0^{\infty} \mathrm{d}s \int_{-\infty}^{\infty} \mathrm{d}y\, \mathrm{e}^{-\beta s}[\Gamma_s(y)]^2 \\ &= \frac{2p\kappa}{\beta^2}\left[\mathcal{N}_{p,\beta}(Z)\right]^2,\end{aligned}$$

thanks to Eq. (11.12). The relation (11.11) concludes the proof in the case that $p > 2$. When $p = 2$ is handled the same way, the prefactor $(4p)^{p/2} = 8$ of Eq. (11.14) can be improved to one, owing to the $L^2(\Omega)$ isometry of Walsh integrals. □

We are now ready to prove the main results of this section.

Proof of Theorem 3.1 Since u_0 and v_0 are bounded, we clearly have

$$\sup_{x\in\mathbf{R}} |U_0(t,x) + V_0(t,x)| \leq \mathrm{const}\cdot(1+t) \qquad (t \geq 0),$$

whence

$$\|U_0 + V_0\|_{p,\beta} = \left(\sup_{t\geq 0} \mathrm{e}^{-\beta t} \sup_{x\in\mathbf{R}} |U_0(t,x) + V_0(t,x)|^p\right)^{1/p} \leq K, \qquad (11.15)$$

where $K := K_{p,\beta}$ is a positive and finite constant that depends only on p and β.

We apply Eqs. (11.3), (11.15), and Proposition 3.1, together with the fact that $|\sigma(u)| \leq |\sigma(0)| + \mathrm{Lip}_\sigma|u|$, in order to conclude that for all $\beta \in (0,\infty)$ and $p \in [2,\infty)$,

$$\|u\|_{p,\beta} \leq K + \frac{p^{3/2}\kappa^{1/2}}{\beta\sqrt{2}}\left(|\sigma(0)| + \mathrm{Lip}_\sigma\|u\|_{p,\beta}\right). \qquad (11.16)$$

This inequality implies that $\|u\|_{p,\beta} < \infty$, provided that $\beta > p^{3/2}\mathrm{Lip}_\sigma\sqrt{\kappa/2}$, and Theorem 3.1 follows. □

Proof of Theorem 3.2 We need to follow the proof of Theorem 2.7 of [19] closely, and so merely recall the necessary steps. It suffices to prove that

$$\int_0^\infty e^{-\beta t} E\left(|u(t,x)|^2\right) dt = \infty \quad \text{when } \beta \leq L_\sigma \sqrt{\kappa/2}. \tag{11.17}$$

Theorem 3.2 will follow from this. This can be seen as follows: By the very definition of $\bar{\gamma}(2)$, we know that for all fixed $\epsilon > 0$ there exists a finite constant $t_\epsilon > 1$ such that $E(|u(t,x)|^2) \leq t_\epsilon \exp((\bar{\gamma}(2)+\epsilon)t)$ whenever $t > t_\epsilon$. Consequently,

$$\int_{t_\epsilon}^\infty e^{-\beta t} E(|u(t,x)|^2)\, dt \leq t_\epsilon \int_{t_\epsilon}^\infty e^{-(\beta-\bar{\gamma}(2)-\epsilon)t}\, dt.$$

We may conclude from this and Eq. (11.17) that $\bar{\gamma}(2) \geq L_\sigma\sqrt{\kappa/2} - \epsilon$, and this completes the proof because $\epsilon > 0$ were arbitrary. It remains to verify Eq. (11.17).

A direct computation, using the L^2 isometry that defines Walsh's stochastic integrals, shows us that

$$\begin{aligned} &E\left(|u(t,x)|^2\right) \\ &= |U_0(t,x) + V_0(t,x)|^2 + \int_0^t ds \int_{-\infty}^\infty dy\, \Gamma_{t-s}^2(y-x) E\left(|\sigma(u(s,y)|^2\right) \\ &\geq \frac{C}{\beta} + L_\sigma^2 \cdot \int_0^t ds \int_{-\infty}^\infty dy\, \Gamma_{t-s}^2(y-x) E\left(|u(s,y)|^2\right), \end{aligned} \tag{11.18}$$

with $C := \inf_{z\in\mathbf{R}}[u_0(z)]^2$. Define

$$M_\beta(x) := \int_0^\infty e^{-\beta t} E\left(|u(t,x)|^2\right) dt, \quad H_\beta(x) := \int_0^\infty e^{-\beta t}[\Gamma_t(x)]^2\, dt.$$

We can rewrite Eq. (11.18) in terms of M_β and H_β as follows:

$$M_\beta(t) \geq \frac{C}{\beta} + L_\sigma^2 (M_\beta * H_\beta)(x),$$

where $*$ denotes spatial convolution. The preceding is a renewal inequation, and can be solved directly: We set

$$(\mathcal{H}f)(x) := L_\sigma^2 (H_\beta * f)(x) \qquad (x \in \mathbf{R}),$$

for every nonnegative measurable function $f : \mathbf{R} \to \mathbf{R}_+$, and deduce the functional recursion $M_\beta \geq C/\beta + (\mathcal{H} * M_\beta)$, whence

$$M_\beta(x) \geq \beta^{-1} \sum_{n=0}^\infty (\mathcal{H}^n C)(x),$$

where we have identified the constant C with the function $C(x) := C$, as usual. Now $(\mathcal{H}C)(x) = C\mathrm{L}_\sigma^2 \int_0^\infty \mathrm{e}^{-\beta t} \|\Gamma_t\|_{L^2(\mathbf{R})}^2 \, \mathrm{d}t = C[\mathrm{L}_\sigma^2 \kappa/(2\beta^2)]$; see Eq. (11.12). We can iterate this computation to see that $(\mathcal{H}^n C)(x) = C[\mathrm{L}_\sigma^2\kappa/(2\beta^2)]^n$ for all $n \geq 0$, and hence

$$M_\beta(x) \geq C\beta^{-1} \sum_{n=0}^\infty \left(\frac{\mathrm{L}_\sigma^2 \kappa}{2\beta^2}\right)^n .$$

The preceding infinite series is equal to $+\infty$ if and only if $\beta \leq \mathrm{L}_\sigma \sqrt{\kappa/2}$. This establishes Eq. (11.17) and concludes the proof of Theorem 3.2. □

Proof of Theorem 3.3 Because u_0 and v_0 are bounded, $|U_0(t,x) + V_0(t,x)| = O(t)$ as $t \to \infty$, uniformly in $x \in \mathbf{R}$. Therefore, the boundedness of σ, Eqs. (11.3), (11.8), (11.12), and (11.15) together imply that

$$\begin{aligned} \|u(t,x)\|_p &\leq O(t) + \sup_{x\in\mathbf{R}} |\sigma(x)| \left(4p \int_0^t \|\Gamma_s\|_{L^2(\mathbf{R})}^2 \, \mathrm{d}s\right)^{1/2} \\ &\leq O(t) + \sqrt{p\kappa} \sup_{x\in\mathbf{R}} |\sigma(x)| t = O(t) \qquad (t \to \infty). \end{aligned} \tag{11.19}$$

The main assertion of Theorem 3.3 follows. In order to establish the remaining claim about the sharpness of the estimator, suppose $u_0(x) \geq 0$ and $\inf_{z\in\mathbf{R}} v_0(y) > 0$, and consider $p = 2$. Thanks to Eq. (11.18), $\|u(t,x)\|_2 \geq V_0(t,x) \geq \inf_{z\in\mathbf{R}} v_0(z) \cdot \kappa t$. The claim follows from this and Jensen's inequality. □

There are many variations of the sharpness portion of Theorem 3.3. Let us conclude this section with one such variation.

Lemma 3.1. *If $\sigma(u) = \lambda$ is a constant and u_0 and v_0 are both constants, then*

$$\lim_{t\to\infty} \frac{1}{t^2} \mathrm{E}\left(|u(t,x)|^2\right) = (v_0\kappa)^2 + \frac{\lambda^2\kappa}{4} \quad \textit{for all } x \in \mathbf{R}.$$

Proof. In accord with Eq. (11.12), the second moment of $\int_{(0,t)\times\mathbf{R}} \Gamma_{t-s}(y-x) W(\mathrm{d}s\,\mathrm{d}y)$ is $\kappa t^2/4$. Therefore, Eq. (11.5) implies that

$$\mathrm{E}\left(|u(t,x)|^2\right) = (u_0 + v_0\kappa t)^2 + \frac{\lambda^2\kappa t^2}{4} = \left\{(v_0\kappa)^2 + \frac{\lambda^2\kappa}{4} + o(1)\right\} t^2,$$

as $t \to \infty$. □

4 Compact-Support Initial Data

This section is devoted to the study of the behavior of the [spatial] supremum $\sup_{x\in\mathbf{R}} |u(t,x)|$ of the solution to Eq. (11.1) when t is fixed. Throughout this section we assume the following:

$$\text{The initial function } u_0 \text{ and initial velocity } v_0 \text{ have compact support.} \tag{11.20}$$

We follow the ideas of Foondun and Khoshnevisan [20]. However, the present hyperbolic setting lends itself to significant simplifications that arise mainly because the Green function has the property that Γ_t has compact support at every fixed time $t > 0$.

Throughout this section, we assume also that

$$\sigma(0) = 0 \qquad \text{and} \qquad \mathrm{L}_\sigma > 0, \tag{11.21}$$

where L_σ was defined in Eq. (11.13). Since Eq. (11.1) has a unique solution, the preceding conditions imply that if $u_0(x) \equiv 0$, then $u_t(x) \equiv 0$ for all $t > 0$.

The idea, borrowed from [20], is to compare $\sup_{x\in\mathbf{R}} |u(t,x)|$ with the $L^2(\mathbf{R})$-norm of the infinite-dimensional stochastic process $\{u(t,\cdot)\}_{t\geq 0}$. This comparison will lead to the result, since it turns out that the compact-support property of u_0 and v_0 will lead us to show that $u(t,\cdot)$ also has compact support. This compact-support property does *not* hold for parabolic variants of Eq. (11.1); see Mueller [25].

Next is the main result of this section.

Theorem 4.1. *Suppose* $\mathrm{L}_\sigma > 0$, $\sigma(0) = 0$, *and* u_0 *is Hölder continuous with Hölder index* $\geq 1/2$. *Suppose also that* u_0 *and* v_0 *are nonnegative functions, both supported compactly in* $[-K, K]$ *for some* $K > 0$. *Then,* $u(t,\cdot) \in L^2(\mathbf{R})$ *a.s. for all* $t \geq 0$ *and*

$$\begin{aligned} \mathrm{L}_\sigma \sqrt{\frac{\kappa}{2}} &\leq \limsup_{t\to\infty} \frac{1}{t} \sup_{x\in\mathbf{R}} \log \mathrm{E}\left(|u(t,\,x)|^2\right) \\ &\leq \limsup_{t\to\infty} \frac{1}{t} \log \mathrm{E}\left(\sup_{x\in\mathbf{R}} |u(t,\,x)|^2\right) \leq \mathrm{Lip}_\sigma \sqrt{\frac{\kappa}{2}}. \end{aligned} \tag{11.22}$$

Remark 4.1. Theorem 4.1 implies that $\sup_{x\in\mathbf{R}} |u(t,x)| < \infty$ a.s. for all $t \geq 0$ provided that the initial function and the initial velocity both have compact support [and are mildly smooth]. We are going to show in Sect. 7 that $\sup_{x\in\mathbf{R}} |u(t,x)| = \infty$ a.s. if the initial function and velocity are nonzero constants, even if those constants are quite close to zero. This discrepancy suggests strongly that the stochastic wave Eq. (11.1) is chaotic [two mildly different initial conditions can lead to a drastically different solution]. This form of chaos is due entirely to the presence of the noise $\dot{W}$ in Eq. (11.1). □

Before we turn to the proof of Theorem 4.1, we will need a few intermediary results.

Proposition 4.1. *Suppose that* $\mathrm{L}_\sigma > 0$, $\sigma(0) = 0$, *and that* $u_0 \not\equiv 0$ *and* v_0 *are nonnegative functions in* $L^2(\mathbf{R})$. *Then,* $u(t,\cdot) \in L^2(\mathbf{R})$ *a.s for all* $t > 0$, *and*

$$\mathrm{L}_\sigma \sqrt{\frac{\kappa}{2}} \leq \limsup_{t\to\infty} \frac{1}{t} \log \mathrm{E}\left(\|u(t,\cdot)\|^2_{L^2(\mathbf{R})} \right) \leq \mathrm{Lip}_\sigma \sqrt{\frac{\kappa}{2}}. \tag{11.23}$$

Proof. The proof resembles that of Theorem 2.1 of [19]. The latter is valid for parabolic equations; therefore, we show how one can adapt that argument to the present hyperbolic setting.

Since $u_0 \geq 0$, it follows that

$$\frac{1}{2}\|u_0\|^2_{L^2(\mathbf{R})} \leq \|U_0(t,\cdot)\|^2_{L^2(\mathbf{R})} \leq \|u_0\|^2_{L^2(\mathbf{R})}.$$

Moreover, since $v_0 \geq 0$, we have

$$0 \leq \|V_0(t,\cdot)\|^2_{L^2(\mathbf{R})} = \int_{-\infty}^{\infty} \mathrm{d}x \left(\int_{-\kappa t}^{\kappa t} \mathrm{d}y\, v_0(y+x) \right)^2 \leq 4\kappa^2 t^2 \|v_0\|^2_{L^2(\mathbf{R})},$$

thanks to the Cauchy–Schwarz inequality.

Now, we deduce from Eq. (11.3) that

$$\begin{aligned} &\mathrm{E}\left(\|u(t,\cdot)\|^2_{L^2(\mathbf{R})} \right) \\ &\geq \|U_0(t,\cdot)\|^2_{L^2(\mathbf{R})} + \|V_0(t,\cdot)\|^2_{L^2(\mathbf{R})} + \mathrm{L}_\sigma^2 \int_0^t \mathrm{d}s\, \mathrm{E}\left(\|u(s,\cdot)\|^2_{L^2(\mathbf{R})} \right) \|\Gamma_{t-s}\|^2_{L^2(\mathbf{R})} \\ &\geq \frac{1}{2}\|u_0\|^2_{L^2(\mathbf{R})} + \mathrm{L}_\sigma^2 \int_0^t \mathrm{d}s\, \mathrm{E}\left(\|u(s,\cdot)\|^2_{L^2(\mathbf{R})} \right) \|\Gamma_{t-s}\|^2_{L^2(\mathbf{R})}. \end{aligned} \tag{11.24}$$

Define

$$U(\lambda) := \int_0^{\infty} \mathrm{e}^{-\lambda t} \mathrm{E}\left(\|u(t,\cdot)\|^2_{L^2(\mathbf{R})} \right) \mathrm{d}t. \tag{11.25}$$

In this way, we can conclude from Eqs. (11.12) and (11.24) that the nonnegative function U that was just defined satisfies the recursive inequality,

$$U(\lambda) \geq \frac{\|u_0\|^2_{L^2(\mathbf{R})}}{2\lambda} + \frac{\kappa \mathrm{L}_\sigma^2}{2\lambda^2} U(\lambda). \tag{11.26}$$

Since $u_0 \not\equiv 0$, the first term on the right-hand side of Eq. (11.26) is strictly positive, whence it follows that whenever $\lambda \leq \mathrm{L}_\sigma \sqrt{\kappa/2}$, we have $U(\lambda) = \infty$. This proves the first asserted inequality in Proposition 4.1.

As regards the other bound, we consider the Picard iteration scheme that defines u from Eq. (11.3). Namely, we set $u_0(t,x) := 0$ and then define iteratively

$$u_{n+1}(t,x) = U_0(t,x) + V_0(t,x) + \int_{(0,t)\times\mathbf{R}} \Gamma_{t-s}(y-x)\sigma(u_n(s,y))\, W(\mathrm{d}s\,\mathrm{d}y). \tag{11.27}$$

Next we may proceed, as we did for Eq. (11.24) but develop upper bounds in place of lower bounds, in order to deduce the following:

$$\begin{aligned} \mathrm{E}\left(\|u_{n+1}(t,\cdot)\|^2_{L^2(\mathbf{R})}\right) &\leq 2\|U_0(t,\cdot)\|^2_{L^2(\mathbf{R})} + 2\|V_0(t,\cdot)\|^2_{L^2(\mathbf{R})} \\ &\quad + \mathrm{Lip}_\sigma^2 \int_0^t \mathrm{d}s\, \mathrm{E}\left(\|u_n(s,\cdot)\|^2_{L^2(\mathbf{R})}\right) \|\Gamma_{t-s}\|^2_{L^2(\mathbf{R})} \\ &\leq 2\|u_0\|^2_{L^2(\mathbf{R})} + 8\kappa^2 t^2\, \|v_0\|^2_{L^2(\mathbf{R})} \\ &\quad + \mathrm{Lip}_\sigma^2 \int_0^t \mathrm{d}s\, \mathrm{E}\left(\|u_n(s,\cdot)\|^2_{L^2(\mathbf{R})}\right) \|\Gamma_{t-s}\|^2_{L^2(\mathbf{R})}. \end{aligned} \tag{11.28}$$

In order to analyze this inequality let us define

$$M_n(\lambda) := \sup_{t\geq 0}\left[\mathrm{e}^{-\lambda t}\mathrm{E}\left(\|u_n(t,\cdot)\|^2_{L^2(\mathbf{R})}\right)\right] \qquad (\lambda > 0,\ n = 1, 1, \ldots).$$

In accord with Eqs. (11.28) and (11.12), the $M_j(\lambda)$'s satisfy the recursive inequality

$$M_{n+1}(\lambda) \leq 2\|u_0\|^2_{L^2(\mathbf{R})} + \frac{8\kappa^2}{\lambda^2}\|v_0\|^2_{L^2(\mathbf{R})} + \frac{\kappa \mathrm{Lip}_\sigma^2}{2\lambda^2} M_n(\lambda).$$

It follows readily from this recursion that if $\lambda > \mathrm{Lip}_\sigma \sqrt{\kappa/2}$, then $\sup_{n\geq 0} M_n(\lambda) < \infty$. Finally, we take the limit as $n \to \infty$ in order to deduce the lower bound in Proposition 4.1. □

Among other things, Proposition 4.1 proves the first claim, made in Theorem 4.1, that $u(t,\cdot) \in L^2(\mathbf{R})$ almost surely for every $t \geq 0$.

We plan to deduce Theorem 4.1 from Proposition 4.1 by showing that $\|u(t,\cdot)\|_{L^2(\mathbf{R})}$ and $\sup_{x\in\mathbf{R}} |u(t,x)|$ are "comparable."

We start by a "compact-support property" of the solution u, which is associated strictly to the hyperbolicity of the wave operator. As such, our next result should be contrasted with Lemma 3.3 of [20], valid for parabolic stochastic partial differential equations.

Proposition 4.2. *Under the assumptions of Theorem 4.1, the random function $x \mapsto u(t,x)$ is a.s. supported in $[-K-\kappa t, K+\kappa t]$ for every $t > 0$.*

Proof. Let $u_0(t,x) := 0$ and define iteratively u_{n+1}, in terms of u_n, as Picard iterates; see (11.27). Note that Γ_{t-s} is supported in $[-\kappa(t-s), \kappa(t-s)]$ for all $s \in (0,t)$. Because $U_0(t,\cdot)$ and $V_0(t,\cdot)$ are both supported in the interval $[-K-\kappa t, K+\kappa t]$, it follows from Eq. (11.27), the fact that $\sigma(0)=0$, and induction [on $n \geq 0$] that $u_n(s,\cdot)$ is a.s. supported in $[-K-\kappa s, K+\kappa s]$ for all $s>0$ and $n \geq 0$. Now we know from Dalang's theory [13] that $\lim_{n\to\infty} u_n(t,x) = u(t,x)$ in probability. Therefore, the result follows. □

Remark 4.2. Proposition 4.2 improves some of the estimates that were obtained previously in [11]. Namely that $u(t,\cdot)$ does not have large peaks more than a distance $\kappa t + o(t)$ away from the origin as $t \to \infty$. □

In order to be able to prove Theorem 4.1, we need some continuity estimates for the solution u. The continuity of the solution itself has been known for a long time; see [13, 26] for instance. We merely state the results in the form that we need.

Lemma 4.1. *If u_0 is Hölder continuous of order $\geq 1/2$, then for all integers $p \geq 1$ and for every $\beta > \bar{\gamma}(2p)$, there exists a constant $C_{p,\beta} \in (0,\infty)$ such that, for all $t \geq 0$,*

$$\sup_{j\in\mathbf{Z}} \sup_{j\leq x<x'\leq j+1} \left\| \frac{u(t,x)-u(t,x')}{|x-x'|^{1/2}} \right\|_{2p} \leq C_{p,\beta} e^{\beta t/2p}. \tag{11.29}$$

Proof. We may observe that $|U_0(t,x)-U_0(t,x')| \leq \text{const}\cdot|x-x'|^{1/2}$ and $|V_0(t,x)-V_0(t,x')| \leq 2\sup_{z\in\mathbf{R}}|v_0(z)|\cdot|x-x'| \leq 2\sup_{z\in\mathbf{R}}|v_0(z)|\cdot|x-x'|^{1/2}$, as long as $|x-x'|\leq 1$. Therefore, we apply Eqs. (11.3) and (11.8) to deduce that uniformly for all $x, x' \in \mathbf{R}$ such that $|x-x'|\leq 1$,

$$\|u(t,x)-u(t,x')\|_{2p} \leq \text{const}\cdot|x-x'|^{1/2} + \text{Lip}_\sigma \left(4p\int_0^t \mathrm{d}s \int_{-\infty}^{\infty} \mathrm{d}y\ \|u(s,y)\|_{2p}^2 \times |\Gamma_{t-s}(y-x)-\Gamma_{t-s}(y-x')|^2\right)^{1/2}. \tag{11.30}$$

Theorem 3.1 shows that $\|u\|_{2p,\beta} < \infty$ provided $\beta > \bar{\gamma}(2p)$, and a direct calculation shows that

$$\int_{-\infty}^{\infty} \mathrm{d}y\ |\Gamma_s(y-x)-\Gamma_s(y-x')|^2 \leq 2|x-x'| \tag{11.31}$$

for all $s>0$. As a consequence,

$$\int_0^t \mathrm{d}s \int_{-\infty}^{\infty} \mathrm{d}y\ \|u(s,y)\|_{2p}^2 |\Gamma_{t-s}(y-x)-\Gamma_{t-s}(y-x')|^2 \tag{11.32}$$

$$\leq e^{\beta t/p}\int_0^t \mathrm{d}s \int_{-\infty}^{\infty} \mathrm{d}y\ e^{-\beta s/p}\|u(s,y)\|_{2p}^2\, e^{-\beta(t-s)/p}|\Gamma_{t-s}(y-x) - \Gamma_{t-s}(y-x')|^2$$

$$\leq \|u\|_{2p,\beta}^2 \, e^{\beta t/p} \int_0^\infty ds\, e^{-\beta s/p} \int_{-\infty}^\infty dy\, |\Gamma_s(y-x) - \Gamma_s(y-x')|^2$$
$$\leq \|u\|_{2p,\beta}^2 \frac{2p}{\beta} e^{\beta t/p} |x-x'|, \tag{11.33}$$

by Eq. (11.31). The theorem follows from Eqs. (11.30) and (11.33). □

By analogy with Lemmas 3.5 and 3.6 of [20], we can extend the preceding result to all real numbers $p \in (1,2)$ and to a uniform modulus of continuity estimate.

Lemma 4.2. *Suppose the conditions of Lemma 4.1 are satisfied. Then, for all $p \in (1,2)$ and $\epsilon, \delta \in (0,1)$, there exists a constant $C_{p,\epsilon,\delta} \in (0,\infty)$ such that for all $t \geq 0$,*

$$\sup_{j \in \mathbf{Z}} \left\| \sup_{j \leq x < x' \leq j+1} \frac{|u(t,x) - u(t,x')|^2}{|x-x'|^{1-\epsilon}} \right\|_p \leq C_{p,\epsilon,\delta}\, e^{(1+\delta)\lambda_p t}, \tag{11.34}$$

where $\lambda_p := (2-p)\bar{\gamma}(2) + (p-1)\bar{\gamma}(4)$.

Proof. The proof works exactly as in [20, Lemmas 3.5 and 3.6]. First, one proves that

$$\mathrm{E}\left(|u(t,x) - u(t,x')|^{2p}\right) \leq C_{p,\delta}|x-x'|^p \exp((1+\delta)\lambda(p)), \tag{11.35}$$

for all $\delta \in (0,1)$, $|x-x'| \leq 1$, and $p \in [1,2]$. This is a direct application of convexity of L^p norms and Lemma 4.1. We refer to [20, Lemma 3.5] for a detailed argument. As a second step, it is possible to use a suitable form of the Kolmogorov continuity theorem in order to obtain an estimate that holds uniformly for $j \leq x < x' \leq j+1$, as stated. We refer to [17] for a detailed proof; see in particular, the proof of Theorem 4.3 therein. □

We are ready to prove Theorem 4.1. This is similar to the proof of Theorem 1.1 in [20], but because of Proposition 4.2, some of the technical issues of [20] do not arise.

Proof of Theorem 4.1 We have already proved that $u(t,\cdot) \in L^2(\mathbf{R})$ for every $t > 0$; see Proposition 4.1. Therefore, it remains to prove Eq. (11.22).

The lower bound is a direct consequence of Propositions 4.1 and 4.2. Indeed, Proposition 4.1 implies that

$$\exp\left(\left[\mathrm{L}_\sigma\sqrt{\frac{\kappa}{2}}+o(1)\right]t\right)\leq \mathrm{E}\left(\int_{-\infty}^{\infty}|u(t,x)|^2\,\mathrm{d}x\right)$$
$$=\mathrm{E}\left(\int_{-K-\kappa t}^{K+\kappa t}|u(t,x)|^2\,\mathrm{d}x\right)$$
$$\leq 2(K+\kappa t)\sup_{x\in\mathbf{R}}\mathrm{E}\left(|u(t,x)|^2\right). \tag{11.36}$$

The first inequality in Eq. (11.22) follows.

As regards the second inequality in Eq. (11.22), we may observe that for all $p\in(1,2)$, $\epsilon\in(0,1)$, $j\in\mathbf{Z}$, and $t\geq 0$,

$$\sup_{j\leq x\leq j+1}|u(t,x)|^{2p}\leq 2^{2p-1}\left(|u(t,j)|^{2p}+\sup_{j\leq x\leq j+1}|u(t,x)-u(t,j)|^{2p}\right)$$
$$\leq 2^{2p-1}\left(|u(t,j)|^{2p}+\Omega_j^p\right),$$

where

$$\Omega_j^p:=\sup_{j\leq x\leq x'\leq j+1}\frac{|u(t,x)-u(t,x')|^2}{|x-x'|^{1-\epsilon}}. \tag{11.37}$$

Consequently,

$$\mathrm{E}\left(\sup_{j\leq x\leq j+1}|u(t,x)|^{2p}\right)\leq 2^{2p-1}\left\{\mathrm{E}\left(|u(t,j)|^{2p}\right)+\mathrm{E}\left(\Omega_j^p\right)\right\}. \tag{11.38}$$

Lemma 4.2 implies that $\mathrm{E}(\Omega_j^p)\leq C_{p,\epsilon,\delta}\,\mathrm{e}^{p(1+\delta)\lambda_p t}$. Moreover, $u(t,j)=0$ a.s. for $|j|>K+\kappa t$ [Proposition 4.2], and $\mathrm{E}(|u(t,j)|^{2p})\leq \text{const}\cdot\mathrm{e}^{(\bar{\gamma}(2p)+o(1))t}$ whenever $|j|\leq K+\kappa t$ [Theorem 3.1]. It follows that for all large t,

$$\mathrm{E}\left(\sup_{x\in\mathbf{R}}|u(t,x)|^{2p}\right)=\mathrm{E}\left(\sup_{|x|\leq\lceil K+\kappa t\rceil}|u(t,x)|^{2p}\right)$$
$$\leq \text{const}\cdot\lceil K+\kappa t\rceil\left(\mathrm{e}^{(\bar{\gamma}(2p)+o(1))t}+C_{p,\epsilon,\delta}\mathrm{e}^{p(1+\delta)\lambda_p t}\right), \tag{11.39}$$

whence

$$\limsup_{t\to\infty}\frac{1}{t}\log\mathrm{E}\left(\sup_{x\in\mathbf{R}}|u(t,x)|^{2p}\right)\leq\max\{p(1+\delta)\lambda_p\,;\,\bar{\gamma}(2p)\}. \tag{11.40}$$

We let $\delta \to 0$, then use Jensen's inequality and finally take $p \to 1$. Since $\lambda_p \to \overline{\gamma}(2)$ as $p \to 1$, this will lead us to the bounds

$$\limsup_{t\to\infty} \frac{1}{t} \log \mathrm{E}\left(\sup_{x\in\mathbf{R}} |u(t,x)|^2 \right) \le \bar{\gamma}(2) \le \mathrm{Lip}_\sigma \sqrt{\frac{\kappa}{2}}, \tag{11.41}$$

by Theorem 3.1 and Remark 3.1. The last inequality in Eq. (11.22) follows. This completes our proof. □

5 Moment and Tail-Probability Estimates

In this section, we will first present technical estimates on the L^p moments of the solution u and then use those estimates in order to establish estimates on tail probabilities of the solution. We will use the efforts of this section later in Sect. 7 in order to deduce the main results of this paper. This section contains the hyperbolic analogues of the results of [12], valid for parabolic equations. Some of the arguments of [12] can be simplified greatly, because we are in a hyperbolic setting. But in several cases, one uses arguments similar to those in [12]. Therefore, we skip some of the details.

Convention. Throughout Sect. 5, we will consider only the case that u_0 and v_0 are constants.

Without loss of much generality, we will assume that $u_0 \equiv 1$. The general case follows from this by scaling. However, we will have to keep track of the numerical value of v_0. Hence, Eq. (11.3) becomes

$$u(t,x) = 1 + v_0\kappa t + \int_{(0,t)\times\mathbf{R}} \Gamma_{t-s}(y-x)\sigma(u(s,y))\, W(\mathrm{d}s\,\mathrm{d}y), \tag{11.42}$$

for $t \ge 0$, $x \in \mathbf{R}$. In accord with the results of Dalang [13], the law of $u_t(x)$ is independent of x, since the initial velocity v_0 and position $u_0 \equiv 1$ are constants.

We start our presentation by stating a general upper bound for the moments of the solution.

Proposition 5.1. *Suppose $u_0 \equiv 1$ and v_0 is a constant. Choose and fix $T > 0$ and define $a := T\mathrm{Lip}_\sigma\sqrt{\kappa}$. Then there exists a finite constant $C > 0$ such that*

$$\sup_{0\le t\le T} \sup_{x\in\mathbf{R}} \mathrm{E}\left(|u(t,x)|^p\right) \le C^p \exp\left(ap^{3/2}\right) \qquad \textit{for all } p \in [1,\infty). \tag{11.43}$$

The preceding is a direct consequence of our proof of Theorem 3.1. Indeed, we proved there that $\|u\|_{p,\beta} < \infty$ provided that $\beta > p^{3/2}\mathrm{Lip}_\sigma\sqrt{\kappa/2}$. Proposition 5.1 follows upon unscrambling this assertion.

Let us recall the following "stretched-exponential" bound for $\log X$:

Lemma 5.1 (Lemma 3.4 of [12]). *Suppose X is a nonnegative random variable that satisfies the following: There exist finite numbers $a, C > 0$, and $b > 1$ such that* $\mathrm{E}(X^p) \le C^p \exp(ap^b)$ *for all $p \in [1, \infty)$. Then,*

$$\mathrm{E}\exp\left(\alpha\left[\log_+ X\right]^{b/(b-1)}\right) < \infty,$$

where $\log_+ u := \log(u \vee \mathrm{e})$, *provided that* $0 < \alpha < (1 - b^{-1})/(ab)^{1/(b-1)}$.

Thanks to the preceding lemma and Chebyshev's inequality, Proposition 5.1 implies readily the following upper bound on the tail of the distribution of $|u(t,x)|$.

Corollary 5.1. *For all $T \in (0, \infty)$ and $\alpha \in (0, \frac{4}{27}(T^2(\mathrm{Lip}_\sigma \vee 1)^2\kappa)^{-1})$,*

$$\sup_{0\le t\le T}\sup_{x\in\mathbf{R}} \mathrm{E}\left[\exp\left\{\alpha\left(\log_+ |u(t,x)|\right)^3\right\}\right] < \infty. \tag{11.44}$$

Consequently,

$$\limsup_{\lambda\to\infty}\frac{1}{(\log\lambda)^3}\sup_{0\le t\le T}\sup_{x\in\mathbf{R}}\log \mathrm{P}\{|u(t,x)| > \lambda\} \le -\frac{4}{27\,T^2(\mathrm{Lip}_\sigma \vee 1)^2\kappa}. \tag{11.45}$$

In plainer terms, Corollary 5.1 asserts that there is a finite constant $A := A_T > 1$ such that for all λ sufficiently large,

$$\sup_{0\le t\le T}\sup_{x\in\mathbf{R}} \mathrm{P}\left\{|u(t,x)| \ge \lambda\right\} \le A\exp\left(-A^{-1}|\log\lambda|^3\right).$$

In order to bound lower bounds on tail probabilities we need to have more specific information on the nonlinearity σ. Let us start with the case that σ is bounded uniformly away from zero.

Proposition 5.2. *If $\epsilon_0 := \inf_{z\in\mathbf{R}}\sigma(z) > 0$, then for all $t \in (0, \infty)$,*

$$\inf_{x\in\mathbf{R}} \mathrm{E}\left(|u(t,x)|^{2p}\right) \ge \left(\sqrt{2} + o(1)\right)(\mu_t p)^p \qquad \text{as } p \to \infty, \tag{11.46}$$

where the $o(1)$ term only depends on p and

$$\mu_t := \epsilon_0^2\kappa t^2/(2\mathrm{e}). \tag{11.47}$$

Proof. We follow the proof of Lemma 3.6 of [12] closely.

Since the law of $u(t,x)$ does not depend on x, the inf in Eq. (11.46) is redundant. From now on, we will consider only the case that $x = 0$.

Choose and fix a finite $t > 0$, and notice that $u(t,0) = 1 + \nu_0\kappa t + M_t$, where $(M_\tau)_{0\le\tau\le t}$ is the continuous mean-zero martingale that is defined by

$$M_\tau := \int_{(0,\tau)\times\mathbf{R}} \Gamma_{t-s}(y)\sigma(u(s,y))\,W(\mathrm{d}s\,\mathrm{d}y). \tag{11.48}$$

The quadratic variation of M is given by

$$\langle M\rangle_\tau = \int_0^\tau \mathrm{d}s \int_{-\infty}^{\infty} \mathrm{d}y\, \Gamma_{t-s}^2(y)\sigma^2(u(s,y)). \tag{11.49}$$

According to Itô's formula, if $p \in [2,\infty)$, then

$$M_t^{2p} = 2p\int_0^t M_s^{2p-1}\,\mathrm{d}M_s + p(2p-1)\int_0^t M_s^{2p-2}\mathrm{d}\langle M\rangle_s. \tag{11.50}$$

We take expectations on both sides and replace $\langle M\rangle$ using Eq. (11.49), in order to obtain the following:

$$\begin{aligned} \mathrm{E}\left(M_t^{2p}\right) &= p(2p-1)\int_0^t \mathrm{d}s \int_{-\infty}^{\infty} \mathrm{d}y\, \mathrm{E}\left(M_s^{2(p-1)}\sigma^2(u(s,y))\right)\Gamma_{t-s}^2(y) \\ &\ge p(2p-1)\epsilon_0^2\cdot \int_0^t \mathrm{d}s \int_{-\infty}^{\infty} \mathrm{d}y\, \mathrm{E}\left(M_s^{2(p-1)}\right)\Gamma_{t-s}^2(y). \end{aligned}$$

We iterate this process, using Eq. (11.49), to obtain the following lower bound for the moments of M:

$$\mathrm{E}\left(M_t^{2p}\right) \ge \sum_{k=0}^{p-1} C_k(p)\int_0^t \nu(t,\mathrm{d}s_1)\int_0^{s_1}\nu(s_1,\mathrm{d}s_2)\cdots\int_0^{s_k}\nu(s_k,\mathrm{d}s_{k+1}), \tag{11.51}$$

where

$$\nu(t,\mathrm{d}s) := \mathbf{1}_{[0,t]}(s)\,\|\Gamma_{t-s}\|_{L^2(\mathbf{R})}^2\,\mathrm{d}s = \frac{1}{2}\kappa(t-s)\mathbf{1}_{[0,t]}(s)\,\mathrm{d}s \qquad \text{[see (11.12)]},$$

and

$$C_k(p) := \epsilon_0^{2(k+1)}\cdot \prod_{j=0}^{k}\binom{2p-2j}{2}.$$

For similar moment computations, also valid for hyperbolic equations, see [10]. The right-hand side of Eq. (11.51) is the exact expression for the pth moment of u if σ were identically ϵ_0. Therefore,

$$\mathrm{E}\left(|u(t,0)|^{2p}\right) \ge \mathrm{E}\left(M_t^{2p}\right) \ge \mathrm{E}\left(N_t^{2p}\right), \tag{11.52}$$

where $N_t := \epsilon_0\cdot\int_{(0,t)\times\mathbf{R}}\Gamma_{t-s}(y)\,W(\mathrm{d}s\,\mathrm{d}y)$ is a Gaussian random variable with mean zero and variance $\mathrm{E}(N_t^2) = \epsilon_0^2\cdot\int_0^t \|\Gamma_s\|_{L^2(\mathbf{R})}^2\,\mathrm{d}s = \epsilon_0^2\kappa t^2/4$. Therefore, for every integer $p \ge 2$,

$$\mathrm{E}\left(N_t^{2p}\right) = \frac{(2p)!}{2^p\,p!}\left\{\mathrm{E}\left(N_t^2\right)\right\}^p = \frac{(2p)!}{2^p\,p!}\left(\frac{\epsilon_0^2\kappa t^2}{4}\right)^p. \tag{11.53}$$

Stirling's formula, Eqs. (11.52), and (11.53) together prove the result if $p \to \infty$ along integers. For other values of p, we use the integer case for $\lceil p \rceil$ and apply Jensen's inequality to bound the $\|u(t,0)\|_p$ by $\|u(t,0)\|_{\lceil p \rceil}$. □

The preceding moment bound yields the next probability estimate.

Proposition 5.3. *If* $\inf_{z\in\mathbf{R}} \sigma(z) = \epsilon_0 > 0$, *then there exists a constant* $C \in (0,\infty)$ *such that for all* $t > 0$,

$$\liminf_{\lambda\to\infty} \frac{1}{\lambda^3} \inf_{x\in\mathbf{R}} \log \mathrm{P}\{|u(t,x)| \geq \lambda\} \geq -C\,\frac{(\mathrm{Lip}_\sigma \vee 1)}{\epsilon_0^3 t^2 \kappa}. \tag{11.54}$$

Proof. We follow the proof of [12, Proposition 3.7].

The classical Paley–Zygmund inequality implies that

$$\begin{aligned}\mathrm{P}\left\{|u(t,x)| \geq \frac{1}{2}\|u(t,x)\|_{2p}\right\} &\geq \frac{\left\{\mathrm{E}\left(|u(t,x)|^{2p}\right)\right\}^2}{4\mathrm{E}\left(|u(t,x)|^{4p}\right)} \\ &\geq \exp\left(-8t(\mathrm{Lip}_\sigma \vee 1)\kappa^{1/2} p^{3/2}\right),\end{aligned} \tag{11.55}$$

owing to Propositions 5.1 and 5.2. Proposition 5.2 tells us that $\|u(t,x)\|_{2p}$ is bounded below by $(1+o(1))$ times $(\mu_t p)^{1/2}$ as $p \to \infty$, where μ_t is given by Eq. (11.47). Therefore,

$$\mathrm{P}\left\{|u(t,x)| \geq \frac{1}{2}(\mu_t p)^{1/2}\right\} \geq \exp\left(-8t(\mathrm{Lip}_\sigma \vee 1)\kappa^{1/2} p^{3/2}\right), \tag{11.56}$$

for all sufficiently large p. Set $\lambda := \frac{1}{2}(\mu_t p)^{1/2}$ to complete the proof. □

Let us write "$f(x) \gtrsim g(x)$ as $x \to \infty$" instead of "there exists a constant $C \in (0,\infty)$ such that $\liminf_{x\to\infty} f(x)/g(x) \geq C$." In this way, we may summarize the findings of this section, so far, as follows:

Corollary 5.2. *Suppose* $u_0 \equiv 1$ *and* $v_0 \equiv$ *a constant. If* $\inf_{z\in\mathbf{R}} \sigma(z) = \epsilon_0 > 0$, *then for all* $t > 0$,

$$-\frac{\lambda^3}{\kappa} \lesssim \log \mathrm{P}\{|u(t,x)| \geq \lambda\} \lesssim -\frac{(\log\lambda)^3}{\kappa} \qquad \text{as } \lambda \to \infty. \tag{11.57}$$

The implied constants do not depend on (x,κ).

Proposition 5.1 and Corollary 5.1 assumed that σ was a Lipschitz function. If we assume, in addition, that σ is bounded above (as well as bounded away from 0), then we can obtain a nearly optimal improvement to Corollary 5.2. In fact, the following shows that the lower bound of Proposition 5.2 is sharp in such cases.

Proposition 5.4. *If $S_0 := \sup_{z\in\mathbf{R}} \sigma(z) < \infty$, then for all $t > 0$ and all integers $p \geq 1$,*

$$\sup_{x\in\mathbf{R}} \mathrm{E}\left(|u(t,x)|^{2p}\right) \leq \left(2\sqrt{2} + o(1)\right) (\tilde{\mu}_t p)^p \qquad \text{as } p \to \infty, \tag{11.58}$$

where the $o(1)$ term only depends on p and

$$\tilde{\mu}_t := 2S_0^2 \kappa t^2/\mathrm{e}. \tag{11.59}$$

Proof. We apply an argument that is similar to the one used in the proof of Proposition 5.2. Namely, we consider the same martingale $\{M_\tau\}_{0\leq\tau\leq t}$, as we did for the proof of Proposition 5.2. We apply exactly the same argument as we did there but reverse the inequalities using the bound $\sigma(z) \leq S_0$ for all $z \in \mathbf{R}$, in order to deduce the following:

$$\begin{aligned} \mathrm{E}\left(|u(t,0)|^{2p}\right) &\leq 2^{2p}(1 + v_0\kappa t)^{2p} + 2^{2p}\mathrm{E}\left(M_t^{2p}\right) \\ &\leq 2^{2p}(1 + v_0\kappa t)^{2p} + 2^{2p}\mathrm{E}\left(N_t^{2p}\right), \end{aligned}$$

where $N_t := S_0 \cdot \int_{(0,t)\times\mathbf{R}} \Gamma_{t-s}(y)\, W(\mathrm{d}s\,\mathrm{d}y)$. Similar computations as in Proposition 5.2 prove the result. □

We can now turn this bound into Gaussian tail-probability estimates.

Proposition 5.5. *If $0 < \epsilon_0 := \inf_{z\in\mathbf{R}} \sigma(z) \leq \sup_{z\in\mathbf{R}} \sigma(z) := S_0 < \infty$, then for all $t > 0$ there exist finite constants $C > c > 0$ such that*

$$c \exp\left(-C\frac{\lambda^2}{\kappa}\right) \leq \mathrm{P}\{|u(t,x)| \geq \lambda\} \leq C \exp\left(-c\frac{\lambda^2}{\kappa}\right), \tag{11.60}$$

simultaneously for all λ large enough and $x \in \mathbf{R}$.

Proof. The lower bound is obtained in the exact same manner as in the proof of Proposition 5.3: We use the Paley–Zygmund inequality, though we now appeal to Proposition 5.4 instead of Proposition 5.1.

We establish the upper bound by first applying Proposition 5.4 in order to see that $\sup_{x\in\mathbf{R}} \mathrm{E}(|u(t,x)|^{2m}) \leq (A\kappa)^m\, m!$ for all integers $m \geq 1$, for some constant $A \in (0,\infty)$. This inequality implies that for all $0 < \xi < (Ak)^{-1}$,

$$\sup_{x\in\mathbf{R}} \mathrm{E}\left(\mathrm{e}^{\xi|u(t,x)|^2}\right) \leq \sum_{m=0}^{\infty} (\xi A\kappa)^m = \frac{1}{1 - \xi A\kappa} < \infty. \tag{11.61}$$

Therefore, Chebyshev's inequality implies that if $0 < \xi < (A\kappa)^{-1}$, then

$$\sup_{x\in\mathbf{R}} \mathrm{P}\{|u(t,x)| > \lambda\} \leq \frac{\exp(-\xi\lambda^2)}{1-\xi A\kappa} \qquad (\lambda > 0). \tag{11.62}$$

We choose $\xi = \text{const} \cdot \kappa^{-1}$, for a suitably large constant to finish. □

6 Localization

In Sect. 7 below, we will establish the chaotic behavior of the solution u to Eq. (11.1). The analysis of Sect. 7 will rest on a series of observations; one of the central ones is that the random function u is highly "localized." We will make this more clear in this section. In the mean time, let us say sketch, using only a few words, what localization means in the present context.

Essentially, localization is the property that if x_1 and x_2 are chosen "sufficiently" far apart, then $u(t,x_1)$ and $u(t,x_2)$ are "approximately independent."

As we did in Sect. 5, we will assume throughout this section that the initial conditions are identically constant and that $u_0 \equiv 1$ [Recall that the latter assumption is made without incurring any real loss in generality.]. Note, in particular, that the solution u can be written in the mild form Eq. (11.5). Equivalently,

$$u(t,x) = 1 + v_0\kappa t + \frac{1}{2}\int_{(0,t)\times(x-\kappa t,x+\kappa t)} \sigma(u(s,y))\, W(\mathrm{d}s\,\mathrm{d}y), \tag{11.63}$$

for all $t > 0$, $x \in \mathbf{R}$.

For all integers $n \geq 0$, let $\{u_n(t,x)\}_{t\geq 0, x\in\mathbf{R}}$ denote the nth step Picard approximation to u. Namely, we have $u_0 \equiv 0$ and for $n \geq 1$, $t \geq 0$, and $x \in \mathbf{R}$,

$$u_n(t,x) = 1 + v_0\kappa t + \frac{1}{2}\int_{(0,t)\times(x-\kappa t,x+\kappa t)} \sigma(u_{n-1}(s,y))\, W(\mathrm{d}s\,\mathrm{d}y). \tag{11.64}$$

Our next result estimates the order of convergence of the Picard iteration.

Proposition 6.1. *Let u denote the solution to Eq. (11.1) with constant initial velocity v_0 and constant initial function $u_0 \equiv 1$. Let u_n be defined as above. Then, for all $n \geq 0$, $t \geq 0$, and $p \in [2,\infty)$,*

$$\sup_{x\in\mathbf{R}} \mathrm{E}\left(|u(t,x) - u_n(t,x)|^p\right) \leq C^p \exp\left(ap^{3/2}t - np\right), \tag{11.65}$$

where the constants $C, a \in (0,\infty)$ do not depend on (n,t,p).

Proof. Recall the norms $\|\cdots\|_{p,\beta}$ from Eq. (11.9). In accord with Proposition 3.1 and Eq. (11.64),

$$\|u - u_n\|_{p,\beta} \leq \text{const} \cdot \frac{p^{3/2}\kappa^{1/2}\text{Lip}_\sigma^2}{4\beta\sqrt{2}} \|u - u_{n-1}\|_{p,\beta}.$$

We apply Eq. (11.12) with $\beta := \mathrm{e}2^{-5/2}\kappa^{1/2}\text{Lip}_\sigma^2 p^{3/2}$ in order to deduce, for this choice of β, the inequality $\|u - u_n\|_{p,\beta} \leq \mathrm{e}^{-1}\|u - u_{n-1}\|_{p,\beta}$, whence $\|u - u_n\|_{p,\beta} \leq \mathrm{e}^{-n}\|u - u_0\|_{p,\beta}$ by iteration. In other words, we have proved that

$$\mathrm{E}\left(|u(t,x) - u_n(t,x)|^p\right) \leq \mathrm{e}^{-np+\beta t}\|u\|_{p,\beta}^p. \tag{11.66}$$

An appeal to Proposition 5.1 concludes the proof. □

We plan to use the Picard iterates $\{u_n\}_{n=0}^\infty$ in order to establish the localization of u. The following is the next natural step in this direction.

Proposition 6.2. *Let $t > 0$ and choose and fix a positive integer n. Let $\{x_i\}_{i\geq 0}$ denote a sequence of real numbers such that $|x_i - x_j| > 2n\kappa t$ whenever $i \neq j$. Then $\{u_n(t, x_i)\}_{i\geq 0}$ is a collection of i.i.d. random variables.*

Proof. It is easy to verify, via induction, that the random variable $u_n(t,x)$ depends only on the value of the noise $\dot{W}$ evaluated on $[0,t] \times [x - n\kappa t, x + n\kappa t]$. Indeed, it follows from Eq. (11.64) that $u_1(t,x) = 1 + \nu_0\kappa t$ is deterministic, and Eq. (11.64) does the rest by induction.

With this property in mind, we now choose and fix a sequence $\{x_i\}_{i\geq 0}$ as in the statement of the proposition. Without loss of too much generality, let us consider x_1 and x_2. By the property that was proved above, $u_n(t,x_1)$ depends only on the noise on $I_1 := [0,t] \times [x_1 - n\kappa t, x_1 + n\kappa t]$, whereas $u_n(t,x_2)$ depends only on the noise on $I_2 := [0,t] \times [x_2 - n\kappa t, x_2 + n\kappa t]$. According to the defining property of the x_i's, $|x_1 - x_2| > 2n\kappa t$, and hence I_1 and I_2 are disjoint. Therefore, it follows from the independence properties of white noise that $u(t,x_1)$ and $u(t,x_2)$ are independent. Moreover, the stationarity properties of stochastic integrals imply that $u(t,x_1)$ and $u(t,x_2)$ are identically distributed as well [here we use also the assumption of constant initial data]. This proves the result for $n = 2$. The general case is proved by expanding on this case a little bit more. We omit the remaining details.

Let us conclude by mentioning that the preceding is the sketch of a complete argument. A fully rigorous proof would require us to address a few technical issues about Walsh stochastic integral. They are handled as in the proof of Lemma 4.4 in [12], and the arguments are not particularly revealing; therefore, we omit the details here as well. □

7 Chaotic Behavior

We are now ready to state and prove the two main results of this paper. The first one addresses the case that σ is bounded away uniformly from zero and shows a universal blow-up rate of $(\log R)^{1/3}$.

Theorem 7.1. *If* $u_0 > 0$, $v_0 \geq 0$, *and* $\inf_{z\in\mathbf{R}} \sigma(z) = \epsilon_0 > 0$, *then for all* $t > 0$ *there exists a constant* $c := c_t \in (0, \infty)$*—independent of* κ*—such that*

$$\liminf_{R\to\infty} \frac{1}{(\log R)^{1/3}} \sup_{x\in[-R,R]} |u(t,x)| \geq c\kappa^{1/3}.$$

Proof. The basic idea is the following: Consider a sequence of spatial points $\{x_i\}_{i\geq 0}$, as we did in Proposition 6.2, in order to obtain an i.i.d. sequence $\{u_n(t,x_i)\}_{i\geq 0}$. The tail-probability estimates of Sect. 5 will imply that every random variable $u_n(t,x_i)$ has a positive probability of being "very large." Therefore, a Borel–Cantelli argument will imply that if we have enough spatial points, then eventually one of the $u_n(t,x_i)$'s will have a "very large" value a.s. A careful quantitative analysis of this outline leads to the estimates of Theorem 7.1. Now let us add a few more details.

Fix integers $n, N > 0$ and let $\{x_i\}_{i=1}^N$ be a sequence of points as in Proposition 6.2. According to Proposition 6.2, $\{u_n(t,x_i)\}_{i=1}^N$ is a sequence of independent random variables. For every $\lambda > 0$,

$$\begin{aligned} &\mathrm{P}\left\{\max_{1\leq j\leq N} |u(t,x_j)| < \lambda\right\} \\ &\quad\leq \mathrm{P}\left\{\max_{1\leq j\leq N} |u_n(t,x_j)| < 2\lambda\right\} + \mathrm{P}\left\{\max_{1\leq j\leq N} |u(t,x_j) - u_n(t,x_j)| > \lambda\right\}. \end{aligned}$$

An inspection of the proof of Proposition 5.3 shows us that the proposition continues to hold after u is replaced by u_n. Therefore,

$$\mathrm{P}\left\{\max_{1\leq j\leq N} |u_n(t,x_j)| < 2\lambda\right\} \leq \left(1 - c_1 \mathrm{e}^{-c_2(2\lambda)^3}\right)^N, \tag{11.67}$$

for some constants c_1 and c_2. Moreover, Chebyshev's inequality and Proposition 6.1 together yield

$$\mathrm{P}\left\{\max_{1\leq j\leq N} |u(t,x_j) - u_n(t,x_j)| > \lambda\right\} \leq NC^p \mathrm{e}^{ap^{3/2}t - np} \lambda^{-p}, \tag{11.68}$$

and hence

$$\mathrm{P}\left\{\max_{1\le j\le N}|u(t,x_j)|<\lambda\right\}\le\left(1-c_1\mathrm{e}^{-c_2(2\lambda)^3}\right)^N+NC^p\mathrm{e}^{ap^{3/2}t-np}\lambda^{-p}. \quad (11.69)$$

Now, we select the various parameters with some care. Namely, we set $\lambda := p$, $N := p\exp(c_2p^3)$, and $n = \varrho p^2$ for some constant $\varrho > 8c_2$. With these parameter choices, Eq. (11.69) reduces to the following:

$$\begin{aligned}&\mathrm{P}\left\{\max_{1\le j\le N}|u(t,x_j)|<p\right\}\\&\le \mathrm{e}^{-c_1p}+\exp\left(c_2(2p)^3+\log p+atp^{3/2}-\varrho p^3-p\log p\right)\\&\le 2\mathrm{e}^{-c_1p}.\end{aligned} \quad (11.70)$$

We may consider the special case $x_i = \pm 2i\kappa tn$ in order to deduce the following:

$$\mathrm{P}\left\{\sup_{|x|\le 2N\kappa tn}|u(t,x)|<p\right\}\le 2\mathrm{e}^{-c_1p}. \quad (11.71)$$

Note that $2N\kappa tn = O(\mathrm{e}^{c_2p^3})$ as $p\to\infty$. Let us choose $R := \exp(c_2p^3)$, and equivalently $p := (\log R/c_2)^{1/3}$. Then by the Borel–Cantelli lemma,

$$\sup_{|x|<R}|u(t,x)|\ge \mathrm{const}\cdot\left(\frac{\log R}{c_2}\right)^{1/3}. \quad (11.72)$$

A monotonicity argument shows that the preceding inequality continues to hold for noninteger R [for a slightly smaller constant, possibly]. A careful examination of the content of Proposition 5.3 shows that we can at best choose $c_2 = \mathrm{const}\cdot\kappa^{-1}$. The result follows. □

The second result of this section [and the second main result of this paper] contains an analysis of the case that σ is bounded both uniformly above 0 and below ∞. In that case, we will obtain an exact order of growth for $\sup_{|x|<R}|u(t,x)|$, as $R\to\infty$. We can deduce by examining that growth order that the behavior of the solution u is similar to the case where σ is identically a constant [In the latter case, u is a Gaussian process.].

Theorem 7.2. *Assume constant initial data with $u_0 > 0$ and $v_0 \ge 0$. Suppose also that $0 < \inf_{z\in\mathbf{R}}\sigma(z)\le\sup_{z\in\mathbf{R}}\sigma(z)<\infty$. Then, for all $t > 0$, there exist finite constants $C := C_t > c := c_t > 0$ such that a.s.,*

$$c\kappa^{1/2}\le\liminf_{R\to\infty}\frac{\sup_{x\in[-R,R]}|u(t,x)|}{(\log R)^{1/2}}\le\limsup_{R\to\infty}\frac{\sup_{x\in[-R,R]}|u(t,x)|}{(\log R)^{1/2}}\le C\kappa^{1/2}.$$

Moreover, there exists a finite constant $\kappa_0 = \kappa_0(t) > 0$ *such that* c *and* C *do not depend on* κ *whenever* $\kappa \in (0, \kappa_0)$.

We first need an estimate for the quality of the spatial continuity of the solution u.

Lemma 7.1. *Suppose* $0 < \epsilon_0 := \inf_{z\in\mathbf{R}} \sigma(z) \le \sup_{z\in\mathbf{R}} \sigma(z) := S_0 < \infty$. *Then, for every* $t > 0$, *there exists a constant* $A \in (0, \infty)$ *such that*

$$\sup_{-\infty<x\neq x'<\infty} \frac{\mathrm{E}\left(|u(t,x) - u(t,x')|^{2p}\right)}{|x - x'|^p} \le (Ap)^p \qquad \text{for all } p \in [2, \infty). \quad (11.73)$$

Proof. We follow closely the proof of Lemma 6.1 of [12]. Fix $x, x' \in \mathbf{R}$ and define

$$M_\tau := \int_{(0,t)\times\mathbf{R}} (\Gamma_{t-s}(y - x) - \Gamma_{t-s}(y - x'))\sigma(u(s, y))\, W(\mathrm{d}s\, \mathrm{d}y). \qquad (11.74)$$

Then, $\{M_\tau\}_{0\le\tau\le t}$ is a mean-zero continuous $L^p(\Omega)$-martingale for every $p \in [2, \infty)$. Moreover, its quadratic variation is bounded as follows:

$$\langle M\rangle_\tau \le S_0^2 \int_0^\tau \mathrm{d}s \int_{-\infty}^\infty \mathrm{d}y\, |\Gamma_{t-s}(y - x) - \Gamma_{t-s}(y - x')|^2 \le 2\tau S_0^2 |x - x'|$$

by Eq. (11.31). Because $u(t, x) - u(t, x') = M_t$, the Burkholder–Davis–Gundy inequality Eq. (11.8) implies the result. □

Next we transform the previous lemma into an estimate of sub-Gaussian moment bounds.

Lemma 7.2. *If* $0 < \epsilon_0 := \inf_{z\in\mathbf{R}} \sigma(z) \le \sup_{z\in\mathbf{R}} \sigma(z) := S_0 < \infty$, *then for every* $t > 0$, *there exists a constant* $C = C_t \in (0, \infty)$ *such that*

$$\mathrm{E}\left[\sup_{\substack{x,x'\in I:\\ |x-x'|\le\delta}} \exp\left(\frac{|u(t, x) - u(t, x')|^2}{C\delta}\right)\right] \le \frac{2}{\delta}, \qquad (11.75)$$

uniformly for every $\delta \in (0, 1]$ *and every interval* $I \subset \mathbf{R}$ *of length at most one.*

Lemma 7.2 follows from Lemma 7.1 and a suitable form of Kolmogorov's continuity theorem. This type of technical argument appears in several places in the literature. Hence, we merely refer to the proof of [12, Lemma 6.2], where this sort of argument appears already in a different setting. Instead, we proceed with the more interesting

Proof of Theorem 7.2 We obtain lower bound by adapting the method of proof of Theorem 7.1. The only major required change is that we need to use Proposition 5.5 in place of Proposition 5.3. We also need to improve Proposition 6.1 in order to consider a moment bound that applies Proposition 5.4 instead of 5.1. After all this, Eq. (11.69) will turn into the following estimate:

$$\mathrm{P}\left\{\max_{1\le j\le N}|u(t,x_j)|<\lambda\right\}\le\left(1-c_1\mathrm{e}^{-c_2(2\lambda)^2}\right)^N+NC^p(\tilde{\mu}_t p)^p\mathrm{e}^{-np}\lambda^{-p}. \quad (11.76)$$

Next we select the parameters judiciously: We take $\lambda := p$, $N := p\mathrm{e}^{c_2p^2}$, and $n = \varrho p$ for a sufficiently large constant $\varrho \gg c_2$. In this way, Eq. (11.69) will read as follows:

$$\begin{aligned}\mathrm{P}\left\{\max_{1\le j\le N}|u(t,x_j)|<p\right\}&\le \mathrm{e}^{-c_1p}+\exp\left(c_2(2p)^2+\log(p)+p\log(\tilde{\mu}_t)-\varrho p^2\right)\\&\le 2\mathrm{e}^{-c_1p}.\end{aligned}$$

A Borel–Cantelli-type argument leads to the lower bound.

In order to establish the upper bound, let $R>0$ be an integer and $x_j := -R+j$ for $j=1,\ldots,2R$. Then, we can write

$$\begin{aligned}&\mathrm{P}\left\{\sup_{x\in[-R,R]}|u(t,x)|>2\alpha(\log R)^{1/2}\right\}\\&\le \mathrm{P}\left\{\max_{1\le j\le 2R}|u(t,x_j)|>\alpha(\log R)^{1/2}\right\}\\&\quad+\mathrm{P}\left\{\max_{1\le j\le 2R}\sup_{x\in(x_j,x_{j+1})}|u(t,x)-u(t,x_j)|>\alpha(\log R)^{1/2}\right\}. \quad (11.77)\end{aligned}$$

On one hand, Proposition 5.5 can be used to show that

$$\begin{aligned}\mathrm{P}\left\{\max_{1\le j\le 2R}|u(t,x_j)|>\alpha(\log R)^{1/2}\right\}&\le 2R\sup_{x\in\mathbf{R}}\mathrm{P}\left\{|u(t,x)|>\alpha(\log R)^{1/2}\right\}\\&\le \mathrm{const}\cdot R^{1-c\alpha^2/\kappa}.\end{aligned}$$

On the other hand, Chebyshev's inequality and Lemma 7.2 [with $\delta=1$] together imply that

$$\mathrm{P}\left\{\max_{1\le j\le 2R}\sup_{x\in(x_j,x_{j+1})}|u(t,x)-u(t,x_j)|>\alpha(\log R)^{1/2}\right\}\le \mathrm{const}\cdot R^{1-\alpha^2/C}.$$

Therefore, Eq. (11.77) has the following consequence:

$$\sum_{R=1}^{\infty}\mathrm{P}\left\{\sup_{x\in[-R,R]}|u(t,x)|>2\alpha(\log R)^{1/2}\right\}\le\sum_{R=1}^{\infty}R^{1-q\alpha^2}, \quad (11.78)$$

where

$$q := \min\left(c/\kappa\ ,\ 1/C\right).$$

The infinite sum in Eq. (11.78) converges when $\alpha > (2/q)^{1/2}$. Therefore, by an application of the Borel–Cantelli lemma,

$$\limsup_{R\to\infty:\mathbf{R}\in\mathbf{Z}} \frac{\sup_{x\in[-R,R]}|u(t,x)|}{(\log R)^{1/2}} \leq (8/q)^{1/2} \qquad \text{a.s.} \tag{11.79}$$

Clearly, $(8/q)^{1/2} \leq \kappa^{1/2}/c$ for all $\kappa > \kappa_0 := 8c^2/q$. A standard monotonicity argument can be used to replace "$\limsup_{R\to\infty:\ R\in\mathbf{Z}}$" by "$\limsup_{R\to\infty}$." This concludes the proof. □

Among other things, Theorem 7.2 implies that if σ is bounded uniformly away from 0 and infinity, then the extrema of the solution u behave as they would for the linear stochastic wave equation; i.e., they grow as $(\log R)^{1/2}$. We have shown in [12, Theorem 1.2] that the same general phenomenon holds when the stochastic wave equation is replaced by the stochastic heat equation. We may notice however that the behavior in κ is quite different in the hyperbolic setting than in the parabolic case: Here, the extrema diminish as $\kappa^{1/2}$ as $\kappa \downarrow 0$, whereas they grow as $\kappa^{-1/4}$ in the parabolic case.

Acknowledgments An anonymous referee read this paper quite carefully and made a number of critical suggestions and corrections that have improved the paper. We thank him or her wholeheartedly.

References

1. Amir G., Corwin I., Quastel J.: Probability distribution of the free energy of the continuum directed random polymer in 1 + 1 dimensions. Comm. Pure Appl. Math. **64**, 466–537 (2011)
2. Assing, S.: A rigorous equation for the Cole–Hopf solution of the conservative KPZ dynamics. Preprint (2011) available from http://arxiv.org/abs/1109.2886
3. Bertini, L., Cancrini N.: The stochastic heat equation: Feynman-Kac formula and intermittence. J. Stat. Phys. **78**(5–6), 1377–1401 (1995)
4. Burkholder, D.L.: Martingale transforms. Ann. Math. Stat. **37**, 1494–1504 (1966)
5. Burkholder, D.L., Gundy, R.F.: Extrapolation and interpolation of quasi-linear operators on martingales. Acta Math. **124**, 249–304 (1970)
6. Burkholder, D.L., Davis B.J., Gundy, R.F.: Integral inequalities for convex functions of operators on martingales. In: Proceedings of the Sixth Berkeley Symposium on Mathematical Statistics and Probability, vol. II, pp. 223–240. University of California Press, Berkeley, California (1972)
7. Carlen, E., Kree, P.: L^p estimates for multiple stochastic integrals. Ann. Probab. **19**(1), 354–368 (1991)
8. Carmona, R.A., Molchanov, S.A.: Parabolic Anderson problem and intermittency. Memoires of the American Mathematical Society, vol. 108. American Mathematical Society, Rhode Island (1994)

9. Carmona, R.A., Nualart, D.: Random nonlinear wave equations: propagation of singularities. Ann. Probab. **16**(2), 730–751 (1988)
10. Conus, D., Dalang, R.C.: The non-linear stochastic wave equation in high dimensions. Electron. J. Probab. **13**, 629–670 (2008)
11. Conus, D., Khoshnevisan, D.: On the existence and position of the farthest peaks of a family of stochastic heat and wave equations. Probab. Theory Relat. Fields. **152** n.3–4, 681–701 (2012)
12. Conus, D., Joseph, M., Khoshnevisan, D.: On the chaotic character of the stochastic heat equation, before the onset of intermittency. Ann. Probab. To appear.
13. Dalang, R.C.: Extending martingale measure stochastic integral with applications to spatially homogeneous spde's. Electron. J. Probab. **4**, 1–29 (1999)
14. Dalang, R.C., Frangos, N.E.: The stochastic wave equation in two spatial dimensions. Ann. Probab. **26**(1), 187–212 (1998)
15. Dalang, R.C., Mueller, C.: Some non-linear s.p.d.e's that are second order in time. Electron. J. Probab. **8** (2003)
16. Dalang, R.C., Mueller, C.: Intermittency properties in a hyperbolic Anderson problem. Ann. l'Institut Henri Poincaré **45**(4), 1150–1164 (2009)
17. Dalang, R.C., Khohsnevisan, D., Mueller, C., Nualart, D., Xiao, Y.: A minicourse in stochastic partial differential equations. In: Lecture Notes in Mathematics, vol. 1962. Springer, Berlin (2006)
18. Davis, B.: On the L^p norms of stochastic integrals and other martingales. Duke Math. J. **43**(4), 697–704 (1976)
19. Foondun, M., Khoshnevisan, D.: Intermittence and nonlinear parabolic stochastic partial differential equations. Electron. J. Probab. **14**(12), 548–568 (2009)
20. Foondun, M., Khoshnevisan, D.: On the global maximum of the solution to a stochastic heat equation with compact-support initial data. Ann. l'Institut Henri Poincaré **46**(4), 895–907 (2010)
21. Gärtner, J., König, W., Molchanov S.: Geometric characterization of intermittency in the parabolic Anderson model. Ann. Probab. **35**(2), 439–499 (2007)
22. Gonçalves, P., Jara, M.: Universality of KPZ equation. Preprint (2010) available from http://arxiv.org/abs/1003.4478
23. Hairer, M.: Solving the KPZ equation. To appear in Annals of Mathematics. (2012) Available from http://arxiv.org/abs/1109.6811
24. Kardar, M., Parisi, G., Zhang, Y.-C.: Dynamic scaling of growing interfaces. Phys. Rev. Lett. **56**(9), 889–892 (1985)
25. Mueller, C.: On the support of solutions to the heat equation with noise. Stochast. Stochast. Rep. **37**(4), 225–245 (1991)
26. Walsh, J.B.: An introduction to stochastic partial differential equations. In: Ecole d'Etè de Probabilitès de St-Flour, XIV, 1984. Lecture Notes in Mathematics, vol. 1180, pp. 265–439. Springer, Berlin (1986)

Chapter 12
Generalized Stochastic Heat Equations

David Márquez-Carreras

Abstract In this article, we study some properties about the solution of generalized stochastic heat equations driven by a Gaussian noise, white in time and correlated in space, and where the diffusion operator is the inverse of a Riesz potential for any positive fractional parameter. We prove the existence and uniqueness of solution and the Hölder continuity of this solution. In time, Hölder's parameter does not depend on the fractional parameter. However, in space, Hölder's parameter has a different behavior depending on the fractional parameter. Finally, we show that the law of the solution is absolutely continuous with respect to Lebesgue's measure and its density is infinitely differentiable.

Keywords Stochastic differential partial equations • Fractional derivative operators • Gaussian processes • Malliavin calculus

Mathematics Subject Classifications 2010: Primary 60H15, 60H30; Secondary 60G60, 60G15, 60H07

Received 10/10/2011; Accepted 12/24/2011; Final 3/12/2012

1 Introduction

Consider the following kind of stochastic partial differential equations (SPDEs),

$$\partial_t \eta(t,x) + (-\Delta)^{p/2}\eta(t,x) = a(\eta(t,x))\dot{W}(t,x) + b(\eta(t,x)), \qquad (12.1)$$

D. Márquez-Carreras (✉)
Facultat de Matemàtiques, Universitat de Barcelona, Gran Via 585, 08007-Barcelona, Spain
e-mail: davidmarquez@ub.edu

F. Viens et al. (eds.), *Malliavin Calculus and Stochastic Analysis: A Festschrift in Honor of David Nualart*, Springer Proceedings in Mathematics & Statistics 34,
DOI 10.1007/978-1-4614-5906-4_12,

with $t \in [0,T]$, $x \in \mathbb{R}^d$, $d \in \mathbb{N}$, and $p > 0$. We assume that $a : \mathbb{R} \to \mathbb{R}$ and $b : \mathbb{R} \to \mathbb{R}$ are Lipschitz continuous functions. The process $\dot{W}$ is a Gaussian noise, white in time, and correlated in space. We will specify later the conditions on $\dot{W}$. The initial conditions are null for the sake of simplicity. Moreover, Δ is the Laplacian operator on $\mathbb{R}^d$, and the integral operator $(-\Delta)^{p/2}$, $p > 0$, is the inverse of a Riesz potential. This last operator is widely studied in Samko et al. [23], Stein [27], and Angulo et al. [1].

This equation proposed in Eq. (12.1) is a generalization of the well-known stochastic heat equation ($p = 2$) which has been studied by many authors [10, 12, 19, 25, 28], etc. On the other hand, many researches have dealt with the following kind of SPDEs (or other similar equations):

$$\partial_t \eta(t,x) + \lambda(I - \Delta)^{q/2}(-\Delta)^{p/2}\eta(t,x) + \dot{W}(t,x), \tag{12.2}$$

with $\lambda > 0$ and where $\dot{W}(t,x)$ is a space-time white noise and the operator $(I - \Delta)^{q/2}, q > 0$, is interpreted as the inverse of the Bessel potential. This more general equation is known as generalized fractional kinetic equation or fractional diffusion equation. This kind of SPDEs has been introduced to model some physical phenomena as turbulences, diffusions in porous media, propagations of seismic waves, kinematics in viscoelastic media, ecology, hydrology, image analysis, neurophysiology, economics, and finances. The reader can find more information about these modelings in [2, 4]. This class of SPDEs (12.2) has been studied from a mathematical point of view in [1–6, 18, 22], etc. In some aspects, our Eq. (12.1) is more particular than Eq. (12.2) as a consequence of the Bessel potential. However, our Gaussian noise is more general and, moreover, we also add the functions a and b.

We would also like to mention some references: [7, 8, 13–17, 21], etc. These papers are related to the operators $(-\Delta)^{p/2}$ in Eq. (12.1) or $(I - \Delta)^{q/2}(-\Delta)^{p/2}$ in Eq. (12.2), the Gaussian noise does not appear and the study is carried out from a more deterministic point of view.

In this paper we prove the existence and uniqueness of solution and the Hölder continuity of this solution. Moreover, we show that the law of the solution is absolutely continuous with respect to Lebesgue's measure on $\mathbb{R}$ and its density is infinitely differentiable. There are some differences between this study and all the papers pointed above. Firstly, our SPDE is driven by a more general Gaussian noise (white in time and correlated in space). Secondly, thanks to the used Gaussian noise, the properties of the solution are checked for any $p > 0$ and not for a more restricted region as for instance in Boulanba et al. [6]. Moreover, these properties do not depend on the dimension of x. Finally, we generalize some results of Angulo et al. [2] to the nonlinear case. We add to the equation the functions a and b and study some new properties. Here, we deal widely with the Hölder continuity in time and in space.

The paper is organized as follows. In Sect. 2 we introduce the Gaussian noise. In Sect. 3 we describe what we understand by a solution of Eq. (12.1) and prove the existence and uniqueness of this solution. In Sect. 4 we study the Hölder continuity of the solution. Finally, adding some conditions on a and b, Sect. 5 is devoted to the proof of the existence of a smooth density. As usual, all constants will be denoted by C, independently of this value.

2 The Gaussian Noise

The noise $\dot{W}$ is the *formal* derivative of a Gaussian field, white in time and correlated in space, defined as follows: for the space of Schwartz test functions $\mathcal{D}(\mathbb{R}^{d+1})$ (see, for instance, Schwartz [26]), the noise $W = \left\{W(\phi), \phi \in \mathcal{D}(\mathbb{R}^{d+1})\right\}$ is an $L^2(\Omega, \mathcal{A}, P)$-valued Gaussian process for some probability space $(\Omega, \mathcal{A}, P)$, with mean zero and covariance functional given by

$$J(\phi, \psi) = \int_{\mathbb{R}_+} \mathrm{d}s \int_{\mathbb{R}^d} \Gamma(\mathrm{d}x) \Big[\phi(s, \bullet) * \tilde{\psi}(s, \bullet)\Big](x),$$

where $\tilde{\psi}(s, x) = \psi(s, -x)$ and Γ is a nonnegative and nonnegative definite tempered measure, therefore symmetric. There exists a nonnegative tempered measure μ which is the spectral measure of Γ such that

$$J(\phi, \psi) = \int_{\mathbb{R}_+} \mathrm{d}s \int_{\mathbb{R}^d} \mu(\mathrm{d}\xi) \mathcal{F}\phi(s, \bullet)(\xi) \overline{\mathcal{F}\psi(s, \bullet)(\xi)},$$

with $\mathcal{F}\phi$ denoting the Fourier transform of ϕ and $\bar{z}$ the complex conjugate of z. The reader interested in some examples about these two measures can find them in Boulanba et al. [6].

Since the spectral measure μ is a nontrivial tempered measure, we can ensure that there exist positive constants c_1, c_2, c_3 such that

$$c_1 < \int_{\{|\xi| \le c_3\}} \mu(\mathrm{d}\xi) < c_2. \qquad (12.3)$$

The Gaussian process W can be extended to a worthy martingale measure, in the sense given by Walsh [28],

$$F = \{F(t, A),\ t \in \mathbb{R}_+,\ A \in \mathcal{B}_b(\mathbb{R}^d)\},$$

where $\mathcal{B}_b(\mathbb{R}^d)$ are the bounded Borel subsets of $\mathbb{R}^d$.

3 Existence and Uniqueness of the Solution

A solution of Eq. (12.1) means a jointly measurable adapted process $\{\eta(t,x), (t,x) \in [0,T] \times \mathbb{R}^d\}$ such that

$$\eta(t,x) = \int_0^t \int_{\mathbb{R}^d} S_{t-s}(x-y)a(\eta(s,y))\, F(\mathrm{d}s,\mathrm{d}y) + \int_0^t \mathrm{d}s \int_{\mathbb{R}^d} \mathrm{d}y\, S_{t-s}(x-y)b(\eta(s,y)), \tag{12.4}$$

where S is the fundamental solution of Eq. (12.1) and the stochastic integral in Eq. (12.4) is defined with respect to the $\mathcal{F}_t$-martingale measure $F(t,\cdot)$.

More specifically, the fundamental solution of Eq. (12.1) is the solution of

$$\partial_t S_t(x) + (-\Delta)^{p/2} S_t(x) = 0. \tag{12.5}$$

In order to study the fundamental solution S, we need the expression of its Fourier transform. Anh and Leonenko [4] have proved that Eq. (12.5) is equivalent to

$$\partial_t \mathcal{F}S_t(\bullet)(\xi) + |\xi|^p \mathcal{F}S_t(\bullet)(\xi) = 0. \tag{12.6}$$

Using the same ideas as in Dautray and Lions [11] or in Ahn and Leonenko [4], we can prove that Eq. (12.6) has a unique solution given by

$$\mathcal{F}S_t(\bullet)(\xi) = \mathrm{e}^{-t|\xi|^p}. \tag{12.7}$$

Then, the fundamental solution of Eq. (12.5) can be written as

$$S_t(x) = \frac{1}{(2\pi)^n} \int_{\mathbb{R}^d} \mathrm{e}^{\mathrm{i}\langle x,\xi\rangle} \mathrm{e}^{-t|\xi|^p}\, \mathrm{d}\xi.$$

In [4], Ahn and Leonenko have studied widely this fundamental solution depending on the parameter p.

For more details about the stochastic integral in Eq. (12.4), we recommend the readings of Dalang [10] and also Dalang and Frangos [9]. In [10], Dalang presents an extension of Walsh's stochastic integral that requires the following integrability condition in terms of the Fourier transform of Γ:

$$\int_0^T \mathrm{d}t \int_{\mathbb{R}^d} \mu(\mathrm{d}\xi)\, |\mathcal{F}S_t(\bullet)(\xi)|^2 < \infty. \tag{12.8}$$

Assuming that Eq. (12.8) is satisfied and that a and b are Lipschitz continuous functions, we will check later that there exists the solution of Eq. (12.4).

We first prove a useful result which connects the tempered measure μ and the fundamental solution $S_t(x)$ by means of the estimate

$$\int_{\mathbb{R}^d} \frac{\mu(\mathrm{d}\xi)}{(1+|\xi|^2)^{p/2}} < \infty. \tag{12.9}$$

Lemma 3.1. *The estimates Eqs. (12.9) and (12.8) are equivalent.*

Remark. In the sequel we will use the following notation:

$$\Phi(t) = \int_0^t \mathrm{d}s \int_{\mathbb{R}^d} \mu(\mathrm{d}\xi)|\mathcal{F}S_s(\bullet)(\xi)|^2 = \int_0^t \mathrm{d}s\ \tilde{\Phi}(s), \quad \forall t \in [0,T],$$

and the following easy properties:

$$\Phi(t) = \int_0^t \mathrm{d}s\ \tilde{\Phi}(s) = \int_0^t \mathrm{d}s\ \tilde{\Phi}(t-s) = \int_0^t \mathrm{d}s \int_{\mathbb{R}^d} \mu(\mathrm{d}\xi)|\mathcal{F}S_s(x-\bullet)(\xi)|^2,$$

for any $x \in \mathbb{R}^d$.

Proof of Lemma 3.1. We first prove that Eq. (12.9) implies Eq. (12.8). From Eq. (12.7), Fubini's theorem implies that

$$\begin{aligned}\Phi(t) &= \int_0^t \mathrm{d}s \int_{\mathbb{R}^d} \mu(\mathrm{d}\xi)\mathrm{e}^{-2s|\xi|^p} = \int_{\mathbb{R}^d} \mu(\mathrm{d}\xi)\frac{1-\mathrm{e}^{-2t|\xi|^p}}{2|\xi|^p} \\ &= \int_{\{|\xi|\le K\}} \mu(\mathrm{d}\xi)\frac{1-\mathrm{e}^{-2t|\xi|^p}}{2|\xi|^p} + \int_{\{|\xi|>K\}} \mu(\mathrm{d}\xi)\frac{1-\mathrm{e}^{-2t|\xi|^p}}{2|\xi|^p} = \Phi_1(t)+\Phi_2(t),\end{aligned} \tag{12.10}$$

for some constant $K \geq 1$.

Using that $1-\mathrm{e}^{-x} \leq x$, for any $x \geq 0$, we have

$$\Phi_1(t) \leq T \int_{\{|\xi|\le K\}} \mu(\mathrm{d}\xi) < \infty, \tag{12.11}$$

since the spectral measure μ is a positive tempered measure.

As $|\xi| > K \geq 1$, applying the fact that $1+|\xi|^2 \leq 2|\xi|^2$ and Eq. (12.9), we obtain

$$\Phi_2(t) \leq \int_{\{|\xi|>K\}} \frac{\mu(\mathrm{d}\xi)}{2|\xi|^p} \leq \int_{\{|\xi|>K\}} \frac{\mu(\mathrm{d}\xi)}{(1+|\xi|^2)^{p/2}} < \infty. \tag{12.12}$$

Then, putting together Eqs. (12.10)–(12.12) we get Eq. (12.8).

The fact that Eq. (12.8) implies Eq. (12.9) can be inferred immediately after Eq. (12.10) using the inequalities

$$\frac{1}{1+x} \leq \frac{1-\mathrm{e}^{-x}}{x} \leq \frac{2}{1+x}, \qquad \forall x > 0.$$

□

We now can prove the main result of this section. The proof of this theorem could be shortened using Theorem 13 and the erratum of [9]. However, we have preferred to give the complete proof.

Theorem 3.1. *Assume Eq. (12.9) and that the functions a and b are globally Lipschitz. Then, Eq. (12.4) has a unique solution that is L^2-continuous and, for any $T > 0$ and $q \geq 1$,*

$$\sup_{0 \leq t \leq T} \sup_{x \in \mathbb{R}^d} \mathbb{E}(|\eta(t,x)|^q) < \infty. \tag{12.13}$$

Proof. We define the following Picard's approximation: $\eta_0(t,x) = 0$ and, for any $n \geq 0$,

$$\begin{aligned}\eta_{n+1}(t,x) = & \int_0^t \mathrm{d}s \int_{\mathbb{R}^d} S_{t-s}(x-y)\, a(\eta_n(s,y)F(\mathrm{d}s,\mathrm{d}y) \\ & + \int_0^t \mathrm{d}s \int_{\mathbb{R}^d} \mathrm{d}y\; S_{t-s}(x-y)\, b(\eta_n(s,y)).\end{aligned}$$

Burkholder's inequality and the Lipschitz condition on a imply that

$$\begin{aligned}&\mathbb{E}\left|\int_0^t \mathrm{d}s \int_{\mathbb{R}^d} S_{t-s}(x-y)\, a(\eta_n(s,y)F(\mathrm{d}s,\mathrm{d}y)\right|^2 \\ &\quad \leq \int_0^t \mathrm{d}s\; \tilde{\Phi}(t-s)\left[1 + \sup_{0 \leq r \leq s} \sup_{x \in \mathbb{R}^d} \mathbb{E}\eta_n^2(r,x)\right].\end{aligned} \tag{12.14}$$

Using the Cauchy–Schwartz inequality and the Lipschitz condition on b, we have that

$$\begin{aligned}&\mathbb{E}\left|\int_0^t \mathrm{d}s \int_{\mathbb{R}^d} \mathrm{d}y\; S_{t-s}(x-y)\, b(\eta(s,y))\right|^2 \\ &\quad \leq \left(\int_0^t \mathrm{d}s \int_{\mathbb{R}^d} \mathrm{d}y |S_{t-s}(x-y)|\right)\left(\int_0^t \mathrm{d}s \int_{\mathbb{R}^d} \mathrm{d}y |S_{t-s}(x-y)| \right. \\ &\qquad \left. \times \left[1 + \sup_{0 \leq r \leq s} \sup_{x \in \mathbb{R}^d} \mathbb{E}\eta_n^2(r,y)\right]\right).\end{aligned} \tag{12.15}$$

Then, Eq. (12.8) applied to Eq. (12.14) and the fact that $S_{t-s}(x-y)$ is $\mathbb{R}^d$-integrable for $t \neq s$ in Eq. (12.15) together with the induction hypothesis ensure that

$$\sup_{0\leq s\leq t} \sup_{x\in\mathbb{R}^d} \mathbb{E}\eta_{n+1}^2(t,x) < \infty,$$

and consequently $\{\eta_n(t,x), n \geq 1\}$ is well defined. Moreover, Gronwall's inequality (Lemma 15 of Dalang [10]) implies that

$$\sup_{n\geq 0} \sup_{0\leq t\leq T} \sup_{x\in\mathbb{R}^d} \mathbb{E}\eta_n^2(t,x) < \infty.$$

Secondly, we show that the process $\{\eta_n(t,x), 0 \leq t \leq T, x \in \mathbb{R}^d\}$ is L^2-continuous. We start with the time increment. For $0 \leq t \leq T$, $h > 0$ such that $t + h \leq T$ and $x \in \mathbb{R}^d$ we have

$$\mathbb{E}|\eta_{n+1}(t+h,x) - \eta_{n+1}(t,x)|^2 \leq C(A_1 + A_2 + A_3),$$

with

$$A_1 = \mathbb{E}\left|\int_0^t \int_{\mathbb{R}^d} [S_{t+h-s}(x-y) - S_{t-s}(x-y)]\, a(\eta_n(s,y))F(\mathrm{d}s,\mathrm{d}y)\right|^2,$$

$$A_2 = \mathbb{E}\left|\int_t^{t+h} \int_{\mathbb{R}^d} S_{t+h-s}(x-y)\, a(\eta_n(s,y))F(\mathrm{d}s,\mathrm{d}y)\right|^2,$$

$$A_3 = \mathbb{E}\left|\int_0^{t+h} \mathrm{d}s \int_{\mathbb{R}^d} \mathrm{d}y S_{t+h-s}(x-y) b(\eta_n(s,y)) \right.$$
$$\left. - \int_0^t \mathrm{d}s \int_{\mathbb{R}^d} \mathrm{d}y S_{t-s}(x-y) b(\eta_n(s,y))\right|^2.$$

First of all, by means of the Lipschitz condition on a, we can get that

$$A_1 \leq \left[1 + \sup_{0\leq t\leq T} \sup_{x\in\mathbb{R}^d} \mathbb{E}\eta_n^2(t,x)\right]$$
$$\times \int_0^t \mathrm{d}s \int_{\mathbb{R}^d} \mu(\mathrm{d}\xi)\, |\mathcal{F}S_{t+h-s}(\bullet)(\xi) - \mathcal{F}S_{t-s}(\bullet)(\xi))|^2,$$

where

$$|\mathcal{F}S_{t+h-s}(\bullet)(\xi) - \mathcal{F}S_{t-s}(\cdot)(\xi)|^2 = \left[\mathrm{e}^{-(t+h-s)|\xi|^p} - \mathrm{e}^{-(t-s)|\xi|^p}\right]^2$$
$$= \mathrm{e}^{-2(t-s)|\xi|^p}\left[\mathrm{e}^{-h|\xi|^p} - 1\right]^2 \leq \mathrm{e}^{-2(t-s)|\xi|^p}.$$

Then, thanks to Lemma 3.1 and the dominated convergence theorem, A_1 converges to 0 as $h \to 0$. Easier arguments show that A_2 goes to 0 as $h \to 0$. A change of rule, the Lipschitz condition on b, the Cauchy–Schwartz inequality, and the integrability of the fundamental solution imply that

$$A_3 \leq C\mathbb{E}\left|\int_0^t \mathrm{d}s \int_{\mathbb{R}^d} \mathrm{d}y\, S_{t-s}(x-y)\,[b(\eta_n(s+h,y)) - b(\eta_n(s,y))]\right|^2$$
$$+C\mathbb{E}\left|\int_0^h \mathrm{d}s \int_{\mathbb{R}^d} \mathrm{d}y S_{t+h-s}(x-y) b(\eta_n(s,y))\right|^2$$
$$\leq C\int_0^t \mathrm{d}s \sup_{x\in\mathbb{R}^d} \mathbb{E}\,|\eta_n(s+h,x) - \eta_n(s,x)|^2 + Ch.$$

The induction hypothesis and Gronwall's inequality (Lemma 15 of Dalang [10]) prove the right continuity uniformly with respect to the time variable t. The left continuity can be checked in a similar way.

We now study the spatial increment. For $t \in [0,T], x, z \in \mathbb{R}^d$, we have

$$\mathbb{E}|\eta_{n+1}(t,x) - \eta_{n+1}(t,z)|^2 \leq C(B_1 + B_2),$$

with

$$B_1 = \mathbb{E}\left|\int_0^t \int_{\mathbb{R}^d} [S_{t-s}(x-y) - S_{t-s}(z-y)]\, a(\eta_n(s,y)) F(\mathrm{d}s, \mathrm{d}y)\right|^2,$$
$$B_2 = \mathbb{E}\left|\int_0^t \mathrm{d}s \int_{\mathbb{R}^d} \mathrm{d}y\, [S_{t-s}(x-y) - S_{t-s}(z-y)]\, b(\eta_n(s,y))\right|^2.$$

Using the Lipschitz condition on a, we obtain

$$B_1 \leq \left[1 + \sup_{0\leq t\leq T} \sup_{x\in\mathbb{R}^d} \mathbb{E}\eta_n^2(t,x)\right]$$
$$\times \int_0^t \mathrm{d}s \int_{\mathbb{R}^d} \mu(\mathrm{d}\xi)\, |\mathcal{F}(S_{t-s}(x-\bullet) - S_{t-s}(z-\bullet))(\xi)|^2$$
$$\leq \left[1 + \sup_{0\leq t\leq T} \sup_{x\in\mathbb{R}^d} \mathbb{E}\eta_n^2(t,x)\right] \int_0^t \mathrm{d}s \int_{\mathbb{R}^d} \mu(\mathrm{d}\xi) |\mathrm{e}^{-\mathrm{i}\xi(x-z)} - 1|^2 |\mathcal{F}S_s(\bullet)(\xi)|^2.$$

Lemma 3.1, the fact that $|\mathcal{F}S_s(\bullet)(\xi)| \leq 1$, and the dominated convergence theorem imply again that B_1 converges to zero as $|x-z| \to 0$, uniformly in t. In order to deal with B_2, the key is the following change of rule:

$$B_2 = \mathbb{E}\left|\int_0^t \mathrm{d}s \int_{\mathbb{R}^d} \mathrm{d}y\, S_{t-s}(z-y)\, [b(\eta_n(s, y + (x-z))) - b(\eta_n(s,y))]\right|^2.$$

As before, using the Lipschitz condition on b, the Cauchy–Schwartz inequality, the integrability of the fundamental solution, the induction hypothesis, and Lemma 15 of Dalang [10], we can establish the L^2-continuity for the process $\{\eta_n(t,x), 0 \leq t \leq T, x \in \mathbb{R}^d\}$.

Not very different arguments imply that, for any $q \geq 2$,

$$\sup_{n\geq 0} \sup_{0\leq t\leq T} \sup_{x\in\mathbb{R}^d} \mathbb{E}|\eta_n(t,x)|^q < \infty.$$

Moreover, we can also prove that $\{\eta_n(t,x), n \geq 0\}$ converges uniformly in L^q to a limit denoted by $\eta(t,x)$ and that this limit satisfies Eqs. (12.4) and (12.13).

The uniqueness can be checked by a standard argument. □

4 Hölderianity of the Solution

The main result of this section is the following theorem which is a generalization of the heat equation case [24, 25].

Theorem 4.1. *Assume that the functions a and b are globally Lipschitz and the spectral measure satisfies*

$$\int_{\mathbb{R}^d} \frac{\mu(\mathrm{d}\xi)}{(1+|\xi|^2)^{\delta p/2}} < \infty, \tag{12.16}$$

for some $\delta \in (0,1)$. Then, for every $T > 0$, $q \geq 2$, $t \in [0,T]$, $h > 0$ such that $t+h \in [0,T]$, $x \in \mathbb{R}^d$, and $\gamma_1 \in (0, \frac{1-\delta}{2})$, we have

$$\mathbb{E}\,|\eta(t+h,x) - \eta(t,x)|^q \leq C h^{\gamma_1 q}; \tag{12.17}$$

and for every $T > 0$, $q \geq 2$, $t \in [0,T]$, $x \in \mathbb{R}^d$, $\vartheta \in \mathbb{R}^d$, and $\gamma_2 \in (0, 1-\delta)$, we have

$$\mathbb{E}\,|\eta(t,x+\vartheta) - \eta(t,x)|^q \leq C|\vartheta|^{\gamma_2 q}, \quad \textit{if } p \geq 2, \tag{12.18}$$

$$\mathbb{E}\,|\eta(t,x+\vartheta) - \eta(t,x)|^q \leq C|\vartheta|^{\gamma_2 q p/2}, \quad \textit{if } p \leq 2. \tag{12.19}$$

Proof. We first prove Eq. (12.17). We can decompose this expectation into three terms:

$$\mathbb{E}\,|\eta(t+h,x) - \eta(t,x)|^q \leq C(\Lambda_1 + \Lambda_2 + \Lambda_3), \tag{12.20}$$

with

$$\Lambda_1 = \mathbb{E}\left| \int_0^t \int_{\mathbb{R}^d} [S_{t+h-s}(x-y) - S_{t-s}(x-y)]\ a(\eta(s,y))\ F(\mathrm{d}s,\mathrm{d}y) \right|^q,$$

$$\Lambda_2 = \mathbb{E}\left| \int_t^{t+h} \int_{\mathbb{R}^d} S_{t+h-s}(x-y)\ a(\eta(s,y))\ F(\mathrm{d}s,\mathrm{d}y) \right|^q,$$

$$\Lambda_3 = \mathbb{E}\Bigg| \int_0^{t+h} \mathrm{d}s \int_{\mathbb{R}^d} \mathrm{d}y\ S_{t+h-s}(x-y)\ b(\eta(s,y)) - \int_0^t \mathrm{d}s \int_{\mathbb{R}^d} \mathrm{d}y\ S_{t-s}(x-y)\ b(\eta(s,y)) \Bigg|^q.$$

Applying Burkholder's inequality, the Lipschitz hypothesis on a and Eq. (12.13), we obtain

$$\Lambda_1 \leq C \left[\int_0^t \mathrm{d}s \int_{\mathbb{R}^d} \mu(\mathrm{d}\xi)\ [\mathcal{F}(S_{t+h-s}(x-\bullet) - S_{t-s}(x-\bullet))(\xi)]^2 \right]^{q/2}$$
$$\leq C \left[\int_0^t \mathrm{d}s \int_{\mathbb{R}^d} \mu(\mathrm{d}\xi)\ [\mathcal{F}(S_{t+h-s}(\bullet) - S_{t-s}(\bullet))(\xi)]^2 \right]^{q/2} \leq C\left[\Lambda_1^1 + \Lambda_1^2\right], \tag{12.21}$$

with

$$\Lambda_1^1 = \left[\int_0^t \mathrm{d}s \int_{\{|\xi|>1\}} \mu(\mathrm{d}\xi) \left[\mathrm{e}^{-(t+h-s)|\xi|^p} - \mathrm{e}^{-(t-s)|\xi|^p} \right]^2 \right]^{q/2},$$

$$\Lambda_1^2 = \left[\int_0^t \mathrm{d}s \int_{\{|\xi|\leq 1\}} \mu(\mathrm{d}\xi) \left[\mathrm{e}^{-(t+h-s)|\xi|^p} - \mathrm{e}^{-(t-s)|\xi|^p} \right]^2 \right]^{q/2}.$$

Integrating first with respect to the time and applying that $1 - \mathrm{e}^{-x} \leq 1 \wedge x$, for $x > 0$, we have that

$$\Lambda_1^1 = \left[\int_0^t \mathrm{d}s \int_{\{|\xi|>1\}} \mu(\mathrm{d}\xi)\, \mathrm{e}^{-2(t-s)|\xi|^p} \left[1 - \mathrm{e}^{-h|\xi|^p}\right]^2 \right]^{q/2}$$
$$= \left[\int_{\{|\xi|>1\}} \frac{\mu(\mathrm{d}\xi)}{2|\xi|^p} \left(1 - \mathrm{e}^{-2t|\xi|^p}\right) \left[1 - \mathrm{e}^{-h|\xi|^p}\right]^2 \right]^{q/2}$$
$$\leq \left[\int_{\{|\xi|>1\}} \frac{\mu(\mathrm{d}\xi)}{2|\xi|^p} \left[1 - \mathrm{e}^{-h|\xi|^p}\right]^{2\gamma_1} \left[1 - \mathrm{e}^{-h|\xi|^p}\right]^{2(1-\gamma_1)} \right]^{q/2}$$
$$\leq \left[\int_{\{|\xi|>1\}} \frac{h^{2\gamma_1}|\xi|^{2p\gamma_1}}{2|\xi|^p}\ \mu(\mathrm{d}\xi) \right]^{q/2} \leq h^{\gamma_1 q} \left[\int_{\{|\xi|>1\}} \frac{\mu(\mathrm{d}\xi)}{|\xi|^{p(1-2\gamma_1)}} \right]^{q/2}.$$

Now, using that $\delta p \le (1-2\gamma_1)p$, the fact that $1+|\xi|^2 \le |\xi|^2$, for $|\xi|>1$, and Eq. (12.16), we get

$$\Lambda_1^1 \le h^{\gamma_1 q}\left[\int_{\{|\xi|>1\}} \frac{\mu(\mathrm{d}\xi)}{|\xi|^{\delta p}}\right]^{q/2} \le C h^{\gamma_1 q}\left[\int_{\{|\xi|>1\}} \frac{\mu(\mathrm{d}\xi)}{(1+|\xi|^2)^{\delta p/2}}\right]^{q/2} \le C h^{\gamma_1 q}. \tag{12.22}$$

As $|\xi| \le 1$, by the mean-value theorem and Eq. (12.3) we have

$$\Lambda_1^2 \le C\left[\int_0^t \mathrm{d}s \int_{\{|\xi|\le 1\}} \mu(\mathrm{d}\xi) h^2\right]^{q/2} \le C h^q. \tag{12.23}$$

Burkholder's inequality, the Lipschitz hypothesis on a, and Eq. (12.13) imply

$$\Lambda_2 \le C\left[\int_t^{t+h} \mathrm{d}s \int_{\mathbb{R}^d} \mu(\mathrm{d}\xi) |\mathcal{F}S_{t+h-s}(x-\bullet)(\xi)|^2\right]^{q/2} \le C(\Lambda_2^1+\Lambda_2^2), \tag{12.24}$$

with

$$\Lambda_2^1 = \left[\int_t^{t+h} \mathrm{d}s \int_{\{|\xi|\le 1\}} \mu(\mathrm{d}\xi)\, \mathrm{e}^{-2(t+h-s)|\xi|^p}\right]^{q/2},$$

$$\Lambda_2^2 = \left[\int_t^{t+h} \mathrm{d}s \int_{\{|\xi|>1\}} \mu(\mathrm{d}\xi)\, \mathrm{e}^{-2(t+h-s)|\xi|^p}\right]^{q/2}.$$

Since $\mathrm{e}^{-x} \le 1$, for $x>0$, Eq. (12.3) yields

$$\Lambda_2^1 \le \left[\int_t^{t+h} \mathrm{d}s \int_{\{|\xi|\le 1\}} \mu(\mathrm{d}\xi)\right]^{q/2} \le C h^{q/2}. \tag{12.25}$$

Arguing as in Eq. (12.22)

$$\begin{aligned}
\Lambda_2^2 &= \left[\int_{\{|\xi|>1\}} \frac{\mu(\mathrm{d}\xi)}{2|\xi|^p}\left(1-\mathrm{e}^{-2h|\xi|^p}\right)\right]^{q/2} \\
&= \left[\int_{\{|\xi|>1\}} \frac{\mu(\mathrm{d}\xi)}{|\xi|^p}\left(1-\mathrm{e}^{-2h|\xi|^p}\right)^{2\gamma_1}\left(1-\mathrm{e}^{-2h|\xi|^p}\right)^{1-2\gamma_1}\right]^{q/2} \\
&\le \left[\int_{\{|\xi|>1\}} \frac{\mu(\mathrm{d}\xi)}{|\xi|^p}\left(1-\mathrm{e}^{-2h|\xi|^p}\right)^{2\gamma_1}\right]^{q/2} \le C h^{\gamma_1 q}.
\end{aligned} \tag{12.26}$$

By a change of rule

$$\Lambda_3 \le C(\Lambda_3^1+\Lambda_3^2), \tag{12.27}$$

with

$$\Lambda_3^1 = \mathbb{E}\left|\int_0^h \mathrm{d}s \int_{\mathbb{R}^d} \mathrm{d}y\ S_{t+h-s}(x-y)\, b(\eta(s,y))\right|^q,$$

$$\Lambda_3^2 = \mathbb{E}\left|\int_0^t \mathrm{d}s \int_{\mathbb{R}^d} \mathrm{d}y\ S_{t-s}(x-y)\ [b(\eta(s+h,y)) - b(\eta(s,y))]\right|^q.$$

Using the integrability of the fundamental solution and the Lipschitz condition on b and Eq. (12.13), we have

$$\Lambda_3^1 \le C h^q. \tag{12.28}$$

The integrability of S together with Hölder's inequality and the Lipschitz condition on b implies that

$$\begin{aligned}
\Lambda_3^2 &\le C\mathbb{E}\left|\int_0^t \mathrm{d}s \left[\int_{\mathbb{R}^d} \mathrm{d}y S_{t-s}(x-y)\right]^{\frac{q-1}{q}}\right. \\
&\quad \left.\times \left[\int_{\mathbb{R}^d} \mathrm{d}y S_{t-s}(x-y)\, |\eta(s+h,y) - \eta(s,y)|^q\right]^{\frac{1}{q}}\right|^q \\
&\le C\int_0^t \mathrm{d}s \left[\int_{\mathbb{R}^d} \mathrm{d}y\ S_{t-s}(x-y)\right] \sup_{x\in\mathbb{R}} \mathbb{E}\, |\eta(s+h,x) - \eta(s,x)|^q.
\end{aligned} \tag{12.29}$$

Taking into account Eqs. (12.20)–(12.29) and using Gronwall's inequality as in Lemma 15 of Dalang [10], we obtain Eq. (12.17).

We now study Eqs. (12.18) and (12.19). We have

$$\mathbb{E}\, |\eta(t, x+\vartheta) - \eta(t,x)|^q \le C(\Delta_1 + \Delta_2), \tag{12.30}$$

with

$$\Delta_1 = \mathbb{E}\left|\int_0^t \int_{\mathbb{R}^d} [S_{t-s}(x+\vartheta-y) - S_{t-s}(x-y)]\ a(\eta(s,y))\ F(\mathrm{d}s,\mathrm{d}y)\right|^q,$$

$$\Delta_2 = \mathbb{E}\left|\int_0^t \mathrm{d}s \int_{\mathbb{R}^d} \mathrm{d}y\ [S_{t-s}(x+\vartheta-y) - S_{t-s}(x-y)]\ b(\eta(s,y))\right|^q.$$

Applying Burkholder's inequality, the hypothesis on a, and Eq. (12.13), we get

$$\begin{aligned}
\Delta_1 &\le C\left[\int_0^t \mathrm{d}s \int_{\mathbb{R}^d} \mu(\mathrm{d}\xi)\ |\mathcal{F}\,[S_{t-s}(x+\vartheta-\bullet) - S_{t-s}(x-\bullet)]\,(\xi)|^2\right]^{q/2} \\
&\le C\left[\int_0^t \mathrm{d}s \int_{\mathbb{R}^d} \mu(\mathrm{d}\xi)\ \left|\mathrm{e}^{-\mathrm{i}(x+\vartheta)\xi} - \mathrm{e}^{-\mathrm{i}x\xi}\right|^2\ |\mathcal{F}S_{t-s}(\bullet)(\xi)|^2\right]^{q/2} \\
&\le C(\Delta_1^1 + \Delta_1^2),
\end{aligned} \tag{12.31}$$

where

$$\Delta_1^1 = \left[\int_0^t \mathrm{d}s \int_{\{|\xi|\le 1\}} \mu(\mathrm{d}\xi) \left| \mathrm{e}^{-\mathrm{i}(x+\vartheta)\xi} - \mathrm{e}^{-\mathrm{i}x\xi} \right|^2 |\mathcal{F}S_{t-s}(\bullet)(\xi)|^2 \right]^{q/2},$$

$$\Delta_1^2 = \left[\int_0^t \mathrm{d}s \int_{\{|\xi|> 1\}} \mu(\mathrm{d}\xi) \left| \mathrm{e}^{-\mathrm{i}(x+\vartheta)\xi} - \mathrm{e}^{-\mathrm{i}x\xi} \right|^2 |\mathcal{F}S_{t-s}(\bullet)(\xi)|^2 \right]^{q/2}.$$

The fact that the Fourier transform of S is bounded by 1, the mean-value theorem, and the property Eq. (12.3) gives that

$$\Delta_1^1 \le \left[\int_0^t \mathrm{d}s \int_{\{|\xi|\le 1\}} \mu(\mathrm{d}\xi)\, |\vartheta\xi|^2 \right]^{q/2} \le C|\vartheta|^q. \tag{12.32}$$

Now assume $p \ge 2$. Using that $\left|\mathrm{e}^{-\mathrm{i}(x+\vartheta)\xi} - \mathrm{e}^{-\mathrm{i}x\xi}\right| \le 2$ and applying the mean-value theorem, we have

$$\begin{aligned}
\Delta_1^2 &= 4\left[\int_0^t \mathrm{d}s \int_{\{|\xi|> 1\}} \mu(\mathrm{d}\xi) \left| \frac{\mathrm{e}^{-\mathrm{i}(x+\vartheta)\xi} - \mathrm{e}^{-\mathrm{i}x\xi}}{2} \right|^2 |\mathcal{F}S_{t-s}(\bullet)(\xi)|^2 \right]^{q/2} \\
&\le C\left[\int_0^t \mathrm{d}s \int_{\{|\xi|> 1\}} \mu(\mathrm{d}\xi) \left| \frac{\mathrm{e}^{-\mathrm{i}(x+\vartheta)\xi} - \mathrm{e}^{-\mathrm{i}x\xi}}{2} \right|^{2\gamma_2} \mathrm{e}^{-2(t-s)|\xi|^p} \right]^{q/2} \\
&\le C\left[\int_0^t \mathrm{d}s \int_{\{|\xi|> 1\}} \mu(\mathrm{d}\xi)\, |\vartheta|^{2\gamma_2} |\xi|^{2\gamma_2} \mathrm{e}^{-2(t-s)|\xi|^p} \right]^{q/2}.
\end{aligned} \tag{12.33}$$

The integration with respect to the time, the assumptions that $p \ge 2$ and $\gamma_2 \in (0, 1-\delta)$, the fact that $|\xi| > 1$, and the hypothesis on the spectral measure Eq. (12.16) imply that

$$\begin{aligned}
\Delta_1^2 &\le C|\vartheta|^{\gamma_2 q} \left[\int_{\{|\xi|> 1\}} \mu(\mathrm{d}\xi)\, |\xi|^{p\gamma_2} \frac{\left[1 - \mathrm{e}^{-2t|\xi|^p}\right]}{2|\xi|^p} \right]^{q/2} \\
&\le C|\vartheta|^{\gamma_2 q} \left[\int_{\{|\xi|> 1\}} \mu(\mathrm{d}\xi) \frac{1}{|\xi|^{p(1-\gamma_2)}} \right]^{q/2} \le C|\vartheta|^{\gamma_2 q} \left[\int_{\{|\xi|> 1\}} \frac{\mu(\mathrm{d}\xi)}{|\xi|^{\delta p}} \right]^{q/2} \\
&\le C|\vartheta|^{\gamma_2 q} \left[\int_{\{|\xi|> 1\}} \frac{\mu(\mathrm{d}\xi)}{(1+|\xi|^2)^{\delta p/2}} \right]^{q/2} \le C|\vartheta|^{\gamma_2 q}.
\end{aligned} \tag{12.34}$$

If we assume $p < 2$, we can ensure that

$$\Delta_1^2 \le C|\vartheta|^{\gamma_2 q p/2}. \tag{12.35}$$

Indeed, in this case, we use the following different arguments in Eq. (12.33) based on the mean-value theorem:

$$\left|\frac{e^{-i(x+\vartheta)\xi} - e^{-ix\xi}}{2}\right|^{2\gamma_2} \leq \left|\frac{e^{-i(x+\vartheta)\xi} - e^{-ix\xi}}{2}\right|^{p\gamma_2} \leq C\,|\vartheta|^{p\gamma_2}\,|\xi|^{p\gamma_2}.$$

The rest of the proof of Eq. (12.35) is similar to Eq. (12.34).

A change of rule together with the same arguments as in Eq. (12.29) allow us to obtain

$$\begin{aligned}\Delta_2 &= \mathbb{E}\left|\int_0^t \mathrm{d}s \int_{\mathbb{R}^d} \mathrm{d}y\; S_{t-s}(x-y)\; [b(\eta(s,y+\vartheta)) - b(\eta(s,y))]\right|^q \\ &\leq C \int_0^t \mathrm{d}s \left[\int_{\mathbb{R}^d} \mathrm{d}y\; S_{t-s}(x-y)\right] \sup_{+\in\mathbb{R}} \mathbb{E}\,|\eta(s,x+\vartheta) - \eta(s,x)|^q. \quad (12.36)\end{aligned}$$

The estimates Eqs. (12.30)–(12.32), (12.34) or Eqs. (12.35), and (12.36) and Gronwall's inequality as in Lemma 15 of Dalang [10] imply Eq. (12.18) or Eq. (12.19), respectively. □

5 Existence of a Smooth Density

The last result of this paper is the following theorem.

Theorem 5.1. *Assume that the spectral measure satisfies Eq. (12.16) for some* $\delta \in (0, \frac{1}{2})$. *Assume also that the functions* a *and* b *are* $\mathcal{C}^\infty$ *with bounded derivatives of any order and that there exists* $\alpha_0 > 0$ *such that* $|a(x)| > \alpha_0$*, for any* $x \in \mathbb{R}$. *Then, the law of the solution to Eq. (12.4) is absolutely continuous with respect to Lebesgue's measure on* $\mathbb{R}$ *and its density is infinitely differentiable.*

For the heat equation, the proof can be found in [20].

Proof of Theorem 5.1. In [20] we prove that the law of the solution to Eq. (12.4) will be absolutely continuous with respect to Lebesgue's measure on $\mathbb{R}$ and its density will be infinitely differentiable if the following conditions are satisfied: for fixed $t > 0$ and $x \in \mathbb{R}^d$, there exist $\varepsilon_1 > \varepsilon_2 > 0$ and $\varepsilon_3 > 0$ such that $0 < \varepsilon_1 < (2\varepsilon_2) \wedge (\varepsilon_2 + \varepsilon_3)$, positive constants C_1, C_2, and C_3, and $t_0 \in [0, t]$ such that for all $\rho \in [0, t_0]$,

$$C_1 \rho^{\varepsilon_1} \leq \int_0^\rho \mathrm{d}s \int_{\mathbb{R}^d} \mu(\mathrm{d}\xi)\; |\mathcal{F}S_s(\bullet)(\xi)|^2 \leq C_2 \rho^{\varepsilon_2} \quad (12.37)$$

and

$$\int_0^\rho \mathrm{d}s \int_{\mathbb{R}^d} \mathrm{d}y\; S_s(y) \leq C_3 \rho^{\varepsilon_3}. \quad (12.38)$$

First of all, we have that

$$\Phi(\rho) = \int_0^\rho \mathrm{d}s \int_{\mathbb{R}^d} \mu(\mathrm{d}\xi)\, \mathrm{e}^{-2s|\xi|^p} = \int_{\mathbb{R}^d} \mu(\mathrm{d}\xi)\, \frac{1-\mathrm{e}^{-2\rho|\xi|^p}}{2|\xi|^p}.$$

On the one hand, since $1-\mathrm{e}^{-x} \geq \frac{x}{1+x}$, for any $x \geq 0$, we have, for some $K > 0$,

$$\begin{aligned}\Phi(\rho) &\geq \rho \int_{\mathbb{R}^d} \frac{\mu(\mathrm{d}\xi)}{1+2\rho|\xi|^p} \geq \rho \int_{\{|\xi|<K\}} \frac{\mu(\mathrm{d}\xi)}{1+2\rho|\xi|^p} \\ &\geq \frac{\rho}{1+2TK^p} \int_{\{|\xi|<K\}} \mu(\mathrm{d}\xi) \geq C_1\rho, \end{aligned} \tag{12.39}$$

where in the last inequality we have applied Eq. (12.3) since μ is a nontrivial positive tempered measure. On the other hand, we decompose Φ into two terms:

$$\Phi(\rho) = \Phi_1(\rho) + \Phi_2(\rho), \tag{12.40}$$

with

$$\Phi_1(\rho) = \int_{\{|\xi|\geq K\}} \mu(\mathrm{d}\xi)\, \frac{1-\mathrm{e}^{-2\rho|\xi|^p}}{2|\xi|^p},$$

$$\Phi_2(\rho) = \int_{\{|\xi|<K\}} \mu(\mathrm{d}\xi)\, \frac{1-\mathrm{e}^{-2\rho|\xi|^p}}{2|\xi|^p}.$$

Taking $\gamma = 1-\delta$, using that $1-\mathrm{e}^{-x} \leq x \wedge 1$, for any $x \geq 0$, and Eq. (12.16), we obtain

$$\Phi_1(\rho) \leq \int_{\{|\xi|\geq K\}} \mu(\mathrm{d}\xi)\, \frac{\left(1-\mathrm{e}^{-2\rho|\xi|^p}\right)^\gamma}{2|\xi|^p} \leq \rho^\gamma \int_{\{|\xi|\geq K\}} \frac{\mu(\mathrm{d}\xi)}{|\xi|^{p(1-\gamma)}} \leq C_2\rho^\gamma. \tag{12.41}$$

Since μ is a nontrivial positive tempered measure (see Eq. (12.3)) and that $1-\mathrm{e}^{-x} \leq 1$, for any $x \geq 0$, we have that

$$\Phi_2(\rho) \leq \rho \int_{\{|\xi|<K\}} \mu(\mathrm{d}\xi) \leq C_2\rho. \tag{12.42}$$

Since the fundamental solution is $\mathbb{R}^d$-integrable for any $t \neq 0$, then

$$\int_0^\rho \mathrm{d}s \int_{\mathbb{R}^d} \mathrm{d}y\, S_s(y) \leq C_3 \int_0^\rho \mathrm{d}s \leq C_3\rho. \tag{12.43}$$

Then, taking $\delta \in (0, 1/2)$, Eqs. (12.39)–(12.43) imply that Eqs. (12.37) and (12.38) are satisfied for $\varepsilon_1 = \varepsilon_3 = 1$ and $\varepsilon_2 = 1-\delta$, and this finishes the proof of this theorem. □

Acknowledgements David Márquez-Carreras was partially supported by MTM2009-07203 and 2009SGR-1360.

References

1. Ahn, V.V., Angulo, J.M., Ruiz-Medina, M.D.: Possible long-range dependence in fractional randoms fields. J. Stat. Plan. Inference **80**, 95–110 (1999)
2. Angulo, J.M., Ruiz-Medina, M.D., Anh, V.V., Grecksch, W.: Fractional diffusion and fractional heat equation. Adv. Appl. Prob. **32**, 1077–1099 (2000)
3. Angulo, J.M., Anh, V.V., McVinish, R., Ruiz-Medina, M.D.: Fractional kinetic equations driven by Gaussian or infinitely divisible noise. Adv. Appl. Prob. **37**(2), 366–392 (2005)
4. Anh, V.V., Leonenko, N.N.: Spectral analysis of fractional kinetic equation with random data. J. Statist. Phys. **104**(5–6), 1349–1387 (2001)
5. Anh, V.V., Leonenko, N.N.: Renormalization and homogenization of fractional diffusion equations with random data. Prob. Theory Relat. Fields **124**(3), 381–408 (2002)
6. Boulanba, L., Eddahbi, M., Mellouk, M.: Fractional SPDEs driven by spatially correlated noise: existence of the solution and smoothness of its density. Osaka J. Math. **47**(1), 41–65 (2010)
7. Cabré, X., Sanchón, M.: Semi-stable and extremal solutions of reaction equations involving the p-Laplacian. Commun. Pure Appl. Anal. **6**(1), 43–67 (2007)
8. Caffarelli, L., Silvestre, L.: An extension problem related to the fractional Laplacian. Comm. Partial Differ. Equat. **32**(7–9), 1245–1260 (2007)
9. Dalang, R.C.: Extending the martingale measure stochastic integral with applications to spatially homogeneous spde's. Electron. J. Probab. **4**(6), 29 (electronic) (1999)
10. Dalang, R.C., Frangos, N.E.: The stochastic wave equation in two spatial dimensions. Ann. Prob. **26**(1), 187–212 (1998).
11. Dautray, R., Lions, J.L.: Mathematical Analysis and Numerical Methods for Science and Technology, vol. 2. Springer, Berlin (1995)
12. Debbi, L., Dozzi, M.: On the solutions of nonlinear stochastic fractional partial differential equations in one spatial dimension. Stochastic Proc. Appl. **115**, 1764–1781 (2005)
13. Dibenedetto, E., Gianazza, U., Vespri, V.: Intrinsic Harnack estimates for nonnegative local solutions of degenerate parabolic equations. Electron. Res. Announc. Amer. Math. Soc. **12**, 95–99 (electronic) (2006)
14. Hochberg, K.J.: A signed measure on path space related to Wiener measure. Ann. Prob. **6**(3), 433–458
15. Jourdain, B., Méléard, S., Woyczynski, W.A.: A probabilistic approach for nonlinear equations involving the fractional Laplacian and a singular operator. Potential Anal. **23**(1), 55–81 (2005)
16. Jourdain, B., Méléard, S., Woyczynski, W.A.: Probabilistic approximation and inviscid limits for one-dimensional fractional conservation laws. Bernoulli **11**(4), 689–714 (2005)
17. Krylov, V.Y.: Some properties of the distribution corresponding to the equation $\frac{\partial u}{\partial t} = (-1)^{q+1}\frac{\partial^2 q}{\partial x^{2q}}$. Soviet Math. Dokl. **1**, 760–763 (1960)
18. Márquez-Carreras, D.: Generalized fractional kinetic equations: another point of view. Adv. Appl. Prob. **41**(3), 893–910 (2009)
19. Márquez-Carreras, D., Sarrà, M.: Large deviation principle for a stochastic heat equation with spatially correlated noise. Electron. J. Probab. **8**(12), 39 (electronic) (2003)
20. Márquez-Carreras, D., Mellouk, M., Sarrà, M.: On stochastic partial differential equations with spatially correlated noise: smoothness of the law. Stoch. Process. Appl. **93**, 269–284 (2001)
21. Micu, S., Zuazua, E.: (2006) On the controllability of a fractional order parabolic equation. SIAM J. Control Optim. **44**(6), 1950–1972 (electronic)

22. Ruiz-Medina, M.D., Angulo, J.M., Anh, V.V.: Scaling limit solution of a fractional Burgers equation. Stoch. Process. Appl. **93**(2), 285–300 (2001)
23. Samko, S.G., Kilbas, A.A., Marichev, O.I.: Fractional integrals and derivatives. Gordon and Breach Science Publishers, New York (1987)
24. Sanz-Solè, M., Sarrà, M.: Path properties of a class of Gaussian processes with applications to SPDEs. CMS Proceedings **28**, 308–316 (2000)
25. Sanz-Solè, M., Sarrà, M.: Hölder continuity for the stochastic heat equation with spatially correlated noise. Seminar on Stochastic Analysis, Random Fields and Applications, III (Ascona, 1999). Progress in Probability, vol. 52, pp. 259–268. Birkhäuser, Basel (2002)
26. Schwartz, L.: Théorie des Distributions. Hermann, Paris (1966)
27. Stein, E.M.: Singular integrals and differentiating properties of Functions. Princeton University Press, Princeton (1970)
28. Walsh, J.B.: An introduction stochastic partial differential equations. Ecole d'ete de probabilites de Saint Flour XIV 1984, Lecture notes in Mathemathics, vol. 1180, pp. 265–439. Springer, Berlin (1986)

Chapter 13
Gaussian Upper Density Estimates for Spatially Homogeneous SPDEs

Lluís Quer-Sardanyons

Abstract We consider a general class of SPDEs in $\mathbb{R}^d$ driven by a Gaussian spatially homogeneous noise which is white in time. We provide sufficient conditions on the coefficients and the spectral measure associated to the noise ensuring that the density of the corresponding mild solution admits an upper estimate of Gaussian type. The proof is based on the formula for the density arising from the integration-by-parts formula of the Malliavin calculus. Our result applies to the stochastic heat equation with any space dimension and the stochastic wave equation with $d \in \{1, 2, 3\}$. In these particular cases, the condition on the spectral measure turns out to be optimal.

Keywords Stochastic partial differential equation • Spatially homogeneous Gaussian noise • Malliavin calculus

MSC Subject Classifications: 60H07, 60H15

Received 12/10/2011; Accepted 5/28/2012; Final 6/10/2012

1 Introduction

We are interested in establishing Gaussian-type upper estimates for the density of the *mild* solution of the following class of SPDEs:

$$\mathcal{L}u(t,x) = b(u(t,x)) + \sigma(u(t,x))\dot{W}(t,x), \qquad (t,x) \in [0,T] \times \mathbb{R}^d, \qquad (13.1)$$

L. Quer-Sardanyons
Departament de Matemàtiques, Universitat Autònoma de Barcelona,
08193 Bellaterra (Barcelona), Spain
e-mail: quer@mat.uab.cat

F. Viens et al. (eds.), *Malliavin Calculus and Stochastic Analysis: A Festschrift in Honor of David Nualart*, Springer Proceedings in Mathematics & Statistics 34,
DOI 10.1007/978-1-4614-5906-4_13,

where $T > 0$ is some fixed time horizon and $\mathcal{L}$ denotes a general second-order partial differential operator with constant coefficients, with appropriate initial conditions. The coefficients σ and b are real-valued functions and $\dot{W}(t,x)$ is the formal notation for a Gaussian random perturbation which is white in time and has some spatially homogeneous correlation (see Sect. 2.1 for a precise definition of this noise). The typical examples of operator $\mathcal{L}$ to which our result applies are the heat operator for any spatial dimension $d \geq 1$ and the wave operator with $d \in \{1,2,3\}$.

If $\mathcal{L}$ is first order in time, such as the heat operator $\mathcal{L} = \frac{\partial}{\partial t} - \Delta$, where Δ denotes the Laplacian operator on $\mathbb{R}^d$, then we impose initial conditions of the form

$$u(0,x) = u_0(x) \qquad x \in \mathbb{R}^d, \tag{13.2}$$

for some Borel function $u_0 : \mathbb{R}^d \to \mathbb{R}$. If $\mathcal{L}$ is second order in time, such as the wave operator $\mathcal{L} = \frac{\partial^2}{\partial t^2} - \Delta$, then we have to impose two initial conditions:

$$u(0,x) = u_0(x), \quad \frac{\partial u}{\partial t}(0,x) = v_0(x), \qquad x \in \mathbb{R}^d, \tag{13.3}$$

for some Borel functions $u_0, v_0 : \mathbb{R}^d \to \mathbb{R}$.

The above class of SPDEs has been widely studied in the last two decades. Precisely, results on existence and uniqueness of solution in such a general setting have been established in [6, 7, 10, 29], while the particular cases of heat and wave equations have been studied using several frameworks in [2, 4, 8, 15, 20, 28, 30, 35].

A fruitful line of research developed in some of the above-cited references has been to apply techniques of Malliavin calculus in order to deduce some interesting properties of the probability law of the solution at any $(t,x) \in (0,T] \times \mathbb{R}^d$. In fact, there is a whole bunch of results on existence and smoothness of the density for the stochastic heat and wave equations, for which we refer to [1, 2, 18–20, 27, 31–33]. Moreover, in the paper [23], existence and smoothness of density for the class of SPDEs (13.1) have been analyzed, unifying and improving some of the results cited so far. It is also worth mentioning that other kind of methods beyond Malliavin calculus can be used to prove the absolute continuity of the law of the solution in some particular SPDEs (see, e.g., [11]).

Once the existence (and possibly smoothness) of the density of the solution to Eq. (13.1) is established, one usually gathers at some *nice* estimates for this density, such as lower and upper Gaussian-type bounds. Exploiting again techniques of Malliavin calculus, this problem has been recently addressed by several authors. Precisely, as far as SPDEs with additive noise are concerned, using an explicit formula for the density proved in [21], the main result in [24] says the following (see Theorem 7 therein and also [25] for related results). Let μ be spectral measure associated with the spatial correlation of $\dot{W}$, and consider the following assumption, which is necessary and sufficient for the existence and uniqueness of mild solution to Eq. (13.1) (see, e.g., [7]).

Hypothesis 1.1. *Let* Γ *be the fundamental solution associated to the operator* $\mathcal{L}$. *For all* $t > 0$, $\Gamma(t)$ *defines a nonnegative distribution with rapid decrease such that*

$$\Phi(T) := \int_0^T \int_{\mathbb{R}^d} |\mathcal{F}\Gamma(t)(\xi)|^2 \, \mu(d\xi)dt < +\infty.$$

Moreover, Γ *is a non-negative measure of the form* $\Gamma(t, dx)dt$ *such that, for all* $T > 0$,

$$\sup_{0 \le t \le T} \Gamma(t, \mathbb{R}^d) < +\infty.$$

Then, under the above hypothesis, with vanishing initial data, $\sigma \equiv 1$ and assuming that $b \in \mathcal{C}^1$ has a bounded derivative, Nualart and Quer-Sardanyons [24, Theorem 7] state that, for small enough t and any $x \in \mathbb{R}^d$, the density $p_{t,x}$ of $u(t,x)$ satisfies, for almost every $z \in \mathbb{R}$,

$$\begin{aligned} &\frac{E\,|u(t,x) - m|}{C_2 \Phi(t)} \exp\left\{-\frac{(z-m)^2}{C_1 \Phi(t)}\right\} \\ &\quad \le p_{t,x}(z) \le \frac{E\,|u(t,x) - m|}{C_1 \Phi(t)} \exp\left\{-\frac{(z-m)^2}{C_2 \Phi(t)}\right\}, \end{aligned} \tag{13.4}$$

where $m = E(u(t,x))$, for some positive constants C_1, C_2. Note that the term $\Phi(t)$ is precisely the variance of the stochastic convolution in the mild form of Eq. (13.1) when $\sigma \equiv 1$. As a consequence, this result applies to the stochastic heat equation for any $d \ge 1$ and the stochastic wave equation in the case $d \in \{1, 2, 3\}$ provided that (see [7, Sect. 3])

$$\int_{\mathbb{R}^d} \frac{1}{1 + |\xi|^2} \, \mu(d\xi) < +\infty. \tag{13.5}$$

On the other hand, in the multiplicative noise setting, such kind of density estimates, particularly the lower one, becomes more difficult to obtain and Nourdin–Viens' density formula cannot be applied. This has been already illustrated by Kohatsu-Higa in [16] where, by means of Malliavin calculus techniques, a new method to obtain Gaussian lower bounds for general functionals of the Wiener sheet has been obtained. In the same paper, the author has applied this result to a stochastic heat equation in [0, 1] and driven by the space-time white noise.

In order to deal with SPDEs beyond the one-dimensional setting, in [26] Kohatsu-Higa's general method has been extended to the Gaussian space associated to the underlying spatially homogeneous noise $\dot{W}$. This allowed us to end up with the following density estimates for the stochastic heat equation in any space dimension $d \ge 1$. Assume that $b, \sigma \in \mathcal{C}^\infty$ are bounded together with all their derivatives, $|\sigma(z)| \ge c > 0$ for all $z \in \mathbb{R}$, and for some $\eta \in (0, 1)$, it holds

$$\int_{\mathbb{R}^d} \frac{1}{(1 + |\xi|^2)^\eta} \, \mu(d\xi) < +\infty. \tag{13.6}$$

Then, for all $(t, x) \in (0, T] \times \mathbb{R}^d$, the density $p_{t,x}$ of $u(t, x)$ verifies, for all $z \in \mathbb{R}$,

$$C_1 \Phi(t)^{-1/2} \exp\left\{-\frac{|z - I_0(t, x)|^2}{C_2 \Phi(t)}\right\} \leq p_{t,x}(z)$$
$$\leq c_1 \Phi(t)^{-1/2} \exp\left\{-\frac{(|z - I_0(t, x)| - c_3 T)^2}{c_2 \Phi(t)}\right\}, \quad (13.7)$$

where $I_0(t, x) = (\Gamma(t) * u_0)(x)$, u_0 being the initial data. In the case $b \equiv 0$, the constant c_3 would vanish. Note that here, in comparison to Eq. (13.4), the estimates are valid for any $T > 0$.

In fact, let us point out that the upper bound in Eq. (13.7) is much easier to obtain than the lower one, and the former comes from the expression for the density popping up from the integration-by-parts formula in the Malliavin calculus framework.

Extending the lower estimate in Eq. (13.7) to the general class of SPDEs (13.1) seems to be an open problem, for the success in the application of the general strategy of [26] is closely tied to the parabolic structure of the heat equation. However, a much more humble objective, which is the one we plan to gather in the present paper, is to tackle the upper bound. In fact, we are going to seek the minimal conditions on either the coefficients b and σ and the spectral measure μ implying that the upper estimate in Eq. (13.7) remains valid for the general class of SPDEs (13.1). In particular, we will only need b and σ to be of class $\mathcal{C}^2$ (and bounded with bounded derivatives) and, for the particular case of the heat (resp. wave) equation with any $d \geq 1$ (resp. $d \in \{1, 2, 3\}$), the condition on μ will be simply (13.5) rather than Eq. (13.6). More precisely, the main result of the paper is the following. We use the notation $I_0(t, x)$ to denote the contribution of the initial data (see Eqs. (13.12) and (13.13) for its explicit expression in the case of heat and wave equations) and suppose that the forthcoming Hypothesis 3.1 is satisfied.

Theorem 1.1. *Assume that Hypothesis 1.1 is satisfied, and that $b, \sigma \in \mathcal{C}^2$ are bounded and have bounded derivatives and $|\sigma(z)| \geq c > 0$ for all $z \in \mathbb{R}$. Moreover, suppose that, for some $\gamma > 0$, it holds*

$$C\,\tau^{\gamma} \leq \Phi(\tau) = \int_0^{\tau} \int_{\mathbb{R}^d} |\mathcal{F}\Gamma(s)(\xi)|^2 \,\mu(d\xi) ds, \quad \tau \in (0, 1]. \quad (13.8)$$

Then, for any $(t, x) \in (0, T] \times \mathbb{R}^d$, the solution $u(t, x)$ of Eq. (13.1) has a density $p_{t,x}$ which is a continuous function and satisfies, for all $z \in \mathbb{R}$,

$$p_{t,x}(z) \leq c_1 \Phi(t)^{-1/2} \exp\left\{-\frac{(|z - I_0(t, x)| - c_3 T)^2}{c_2 \Phi(t)}\right\}, \quad (13.9)$$

where the constant c_3 vanishes whenever b does.

We remark that, though the above bound does not look exactly Gaussian, it does in an *asymptotic* point of view, namely whenever T is small or z is large. On the other hand, we note that, under (13.5), condition (13.8) is satisfied for the heat and wave equations with $\gamma = 1$ and $\gamma = 3$, respectively (see, e.g., [19, Lemma 3.1] and [31, App. A]). Similarly, one can also check that the above theorem applies to the stochastic damped wave equation with any space dimension (see Example 7 in [7, Sect. 3]), where condition (13.8) is fulfilled with $\gamma = 3$.

We also point out that our result is not applicable to the case $\sigma(z) = z$ (this would be related, e.g., to the parabolic Anderson problem [3]). In fact, in such a case there are even very few results on absolute continuity of the law of solutions to SPDEs (see [27]). Nevertheless, in the recent paper [13], the authors prove existence and smoothness of the density for a stochastic heat equation with a nonlinear multiplicative noise which is white in time and with some spatial correlation (much more regular than the one considered in the present paper), and with a nondegeneracy condition on the diffusion coefficient of the form $\sigma(u_0(x_0)) \neq 0$ for some $x \in \mathbb{R}^d$. Their proof is based on a Feynman–Kac formula for the solution of the underlying equation. This technique has also been applied in [14] to study the density for a stochastic heat equation with a linear multiplicative fractional Brownian sheet.

As mentioned before, the proof of Theorem 1.1 will be based on the expression for the density arising from the application of the integration-by-parts formula (see [22, Proposition 2.1.1]). We point out that this is a well-known method that has been used in other contexts (see, e.g., [9, 12]). As far as the technical obstacles are concerned, the main two ingredients needed in the proof of Theorem 1.1 are the following:

(i) A suitable estimate, in terms of $\Phi(t)$, of the norm of the iterated Malliavin derivative in a small time interval (see Lemma 4.1 for details). This will be a consequence of a kind of analogous result for the case of the stochastic heat equation (see [26, Lemma 3.4]) and a mollifying procedure thanks to an approximation of the identity which will let us smooth the fundamental solution $\Gamma(t)$.
(ii) A precise control of the negative moments of the norm of the Malliavin derivative of the solution, again in terms of $\Phi(t)$ (see Proposition 4.1). For this, we will adapt the proof of [23, Theorem 6.2] to our setting, where the latter allowed the authors of that paper to establish that the underlying density is a smooth function (under much more regularity on the coefficients though).

The content of the paper is organized as follows. In Sect. 2, we rigorously describe the Gaussian spatially homogeneous noise $\dot{W}$ considered in Eq. (13.1) and we introduce the corresponding Gaussian setting associated to it, together with the main notations of the Malliavin calculus machinery. Section 3 will be devoted to recall the definition of mild solution to our SPDE (13.1) and summarize the main results on existence and uniqueness of solution, Malliavin differentiability, and existence and smoothness of the density. Steps (i) and (ii) detailed above will be tackled in Sect. 4. Finally, we will prove Theorem 1.1 in Sect. 5.

With a slight (but harmless) abuse of notation, as already done in this Introduction, the notation $|\cdot|$ shall denote either the modulus and norm in $\mathbb{R}^d$. Unless otherwise stated, any constant c or C appearing in our computations below is understood as a generic constant which might change from line to line without further mention.

2 Preliminaries

2.1 *Spatially Homogeneous Noise*

Let us explicitly describe here our spatially homogeneous noise (see, e.g., [7]). Precisely, on a complete probability space $(\Omega, \mathcal{F}, \mathbb{P})$, this is given by a family $W = \{W(\varphi), \varphi \in C_0^\infty(\mathbb{R}_+ \times \mathbb{R}^d)\}$ of zero mean Gaussian random variables, where $C_0^\infty(\mathbb{R}_+ \times \mathbb{R}^d)$ denotes the space of smooth functions with compact support, with the following covariance structure:

$$\mathbb{E}\big(W(\varphi)W(\psi)\big) = \int_0^\infty \int_{\mathbb{R}^d} \big(\varphi(t,\star) * \tilde{\psi}(t,\star)\big)(x)\, \Lambda(\mathrm{d}x)\mathrm{d}t. \tag{13.10}$$

In this expression, Λ denotes a nonnegative and non-negative definite tempered measure on $\mathbb{R}^d$, $*$ stands for the convolution product, the symbol $\star$ denotes the spatial variable, and $\tilde{\psi}(t,x) := \psi(t,-x)$.

In the above setting, a well-known result of harmonic analysis (see [34, Chap. VII, Théorème XVII]) implies that Λ has to be the Fourier transform of a nonnegative tempered measure μ on $\mathbb{R}^d$, where the latter is usually called the *spectral measure* of the noise W. We recall that, in particular, for some integer $m \geq 1$ it holds

$$\int_{\mathbb{R}^d} \frac{1}{(1+|\xi|^2)^m}\, \mu(\mathrm{d}\xi) < +\infty$$

and, by definition of the Fourier transform in the space $\mathcal{S}'(\mathbb{R}^d)$ of tempered distributions, $\Lambda = \mathcal{F}\mu$ means that, for all ϕ belonging to the space $\mathcal{S}(\mathbb{R}^d)$ of rapidly decreasing C^∞ functions,

$$\int_{\mathbb{R}^d} \phi(x)\Lambda(\mathrm{d}x) = \int_{\mathbb{R}^d} \mathcal{F}\phi(\xi)\mu(\mathrm{d}\xi).$$

Therefore, we have

$$\mathbb{E}\big(W(\varphi)^2\big) = \int_0^\infty \int_{\mathbb{R}^d} |\mathcal{F}\varphi(t)(\xi)|^2 \mu(\mathrm{d}\xi)\mathrm{d}t.$$

A typical example of space correlation is given by $\Lambda(dx) = f(x)\mathrm{d}x$, where f is a non-negative function which is assumed to be integrable around the origin. In this case, the covariance functional (13.10) reads

$$\int_0^\infty \int_{\mathbb{R}^d}\int_{\mathbb{R}^d} \varphi(t,x) f(x-y)\psi(t,y)\,\mathrm{d}y\mathrm{d}x\mathrm{d}t.$$

The space-time white noise would correspond to the case where f is the Dirac delta at the origin.

2.2 Gaussian Setting and Malliavin Calculus

We are going to describe the Gaussian framework which can be naturally associated to our noise W and introduce the notations involved in the Malliavin calculus techniques. To start with, let us denote by $\mathcal{H}$ the completion of the Schwartz space $\mathcal{S}(\mathbb{R}^d)$ endowed with the semi-inner product

$$\begin{aligned}\langle \phi_1, \phi_2\rangle_{\mathcal{H}} &:= \int_{\mathbb{R}^d} (\phi_1 * \tilde{\phi_2})(x)\,\Lambda(\mathrm{d}x)\\ &= \int_{\mathbb{R}^d} \mathcal{F}\phi_1(\xi)\overline{\mathcal{F}\phi_2(\xi)}\,\mu(\mathrm{d}\xi), \phi_1,\phi_2 \in \mathcal{S}(\mathbb{R}^d).\end{aligned}$$

As proved in [7, Example 6], we remind that the Hilbert space $\mathcal{H}$ may contain distributions.

Let $T > 0$ be a fixed real number and define $\mathcal{H}_T := L^2([0,T];\mathcal{H})$. Using an approximation argument, our noise W can be extended to a family of mean zero Gaussian random variables indexed by $\mathcal{H}_T$ (see, e.g., [6, Lemma 2.4]). With an innocuous abuse of notation, this family will be still denoted by $W = \{W(g), g \in \mathcal{H}_T\}$. Moreover, it holds $\mathbb{E}\big(W(g_1)W(g_2)\big) = \langle g_1, g_2\rangle_{\mathcal{H}_T}$, for all $g_1, g_2 \in \mathcal{H}_T$. Thus, this family defines an *isonormal Gaussian process* on the Hilbert space $\mathcal{H}_T$ and we shall use the differential Malliavin calculus based on it (see, e.g., [22, 33]).

As usual, we denote the Malliavin derivative operator by D. Recall that it is a closed and unbounded operator defined in $L^2(\Omega)$ and taking values in $L^2(\Omega;\mathcal{H}_T)$, whose domain is denoted by $\mathbb{D}^{1,2}$. More generally, for any integer $m \geq 1$ and any $p \geq 2$, the domain of the iterated Malliavin derivative D^m in $L^p(\Omega)$ will be denoted by $\mathbb{D}^{m,p}$, where we remind that D^m takes values in $L^p(\Omega;\mathcal{H}_T^{\otimes m})$. We also set $\mathbb{D}^\infty = \bigcap_{p\geq 1}\bigcap_{m\in\mathbb{N}} \mathbb{D}^{m,p}$. The space $\mathbb{D}^{m,p}$ can also be seen as the completion of the set of *smooth functionals* with respect to the semi-norm

$$\|F\|_{m,p} := \left\{\mathbb{E}\left(|F|^p\right) + \sum_{j=1}^m \mathbb{E}\Big(\|D^j F\|^p_{\mathcal{H}_T^{\otimes j}}\Big)\right\}^{\frac{1}{p}}.$$

For any differentiable random variable F and any $r = (r_1, \ldots, r_m) \in [0, T]^m$, $D^m F(r)$ is an element of $\mathcal{H}^{\otimes m}$ which will be denoted by $D_r^m F$.

A random variable F is said to be *smooth* if it belongs to $\mathbb{D}^\infty$, and a smooth random variable F is said to be *nondegenerate* if $\|DF\|_{\mathcal{H}_T}^{-1} \in \cap_{p\geq 1} L^p(\Omega)$. Owing to [22, Theorem 2.1.4], we know that a nondegenerate random variable has a $\mathcal{C}^\infty$ density.

For any $t \in [0, T]$, let $\mathcal{F}_t$ be the σ-field generated by the random variables $\{W_s(h), h \in \mathcal{H}, 0 \leq s \leq t\}$ and the $\mathbb{P}$-null sets, where $W_t(h) := W(1_{[0,t]}h)$.

3 Spatially Homogeneous SPDEs

We gather here a general result on existence and uniqueness of mild solution for our SPDE (13.1) and the main results on Malliavin calculus applied to it, namely Malliavin differentiability and existence and smoothness of density. As usual, we will also focus on the main examples of application that we have in mind, which are the stochastic heat and wave equations with $d \geq 1$ and $d \in \{1, 2, 3\}$, respectively.

We recall that, by definition, a mild solution of Eq. (13.1) is an $\mathcal{F}_t$-adapted random field $\{u(t, x), (t, x) \in [0, T] \times \mathbb{R}^d\}$ such that the following stochastic integral equation is satisfied:

$$
\begin{aligned}
u(t,x) = I_0(t,x) &+ \int_0^t \int_{\mathbb{R}^d} \Gamma(t-s, x-y)\sigma(u(s,y))\, W(\mathrm{d}s, \mathrm{d}y) \\
&+ \int_0^t \int_{\mathbb{R}^d} b(u(t-s, x-y))\, \Gamma(s, \mathrm{d}y)\mathrm{d}s, \quad \mathbb{P}\text{-a.s.,}
\end{aligned}
\tag{13.11}
$$

for all $(t, x) \in [0, T] \times \mathbb{R}^d$. Here, Γ denotes the fundamental solution associated to $\mathcal{L}$ and $I_0(t, x)$ is the contribution of the initial conditions, which we define below.

The (real-valued) stochastic integral on the right-hand side of Eq. (13.11) is understood with respect to the *cylindrical Wiener process* that can be naturally associated to our spatially homogeneous noise W (see [6, 23] and also [7, 35]). In particular, we will assume that Hypothesis 1.1 is satisfied. Concerning the last integral on the right-hand side of Eq. (13.11), we point out that we use the notation "$\Gamma(s, dy)$" because we will assume that $\Gamma(s)$ is a measure on $\mathbb{R}^d$.

As far as the term $I_0(t, x)$ is concerned, if $\mathcal{L}$ is a parabolic-type operator and we consider the initial condition (13.2), then

$$
I_0(t,x) = (\Gamma(t) * u_0)(x) = \int_{\mathbb{R}^d} u_0(x-y)\, \Gamma(t, \mathrm{d}y). \tag{13.12}
$$

On the other hand, in the case where $\mathcal{L}$ is second order in time with initial values (13.3),

$$
I_0(t,x) = (\Gamma(t) * v_0)(x) + \frac{\partial}{\partial t}(\Gamma(t) * u_0)(x). \tag{13.13}
$$

Example 3.1. Owing to the considerations in [7, Sect. 3] (see also [23, Examples 4.2 and 4.3]), in the case of the stochastic heat equation in any space dimension $d \geq 1$ and the stochastic wave equation in dimensions $d = 1, 2, 3$, the fundamental solutions are well known and the conditions in Hypothesis 1.1 are satisfied if and only if

$$\int_{\mathbb{R}^d} \frac{1}{1+|\xi|^2}\,\mu(\mathrm{d}\xi) < +\infty. \tag{13.14}$$

We shall consider the following assumption on the initial conditions. In the case of the stochastic heat equation in any space dimension (resp. wave equation with dimension $d = 1, 2, 3$), sufficient conditions on u_0 (resp. u_0, v_0) implying that the hypothesis below is fulfilled are provided in [6, Lemma 4.2].

Hypothesis 3.1. *$(t, x) \mapsto I_0(t, x)$ is continuous and $\sup_{(t,x)\in[0,T]\times\mathbb{R}^d} |I_0(t, x)| < +\infty$.*

The following well-posedness result, which is a quotation of [6, Theorem 4.3], is a slight extension of the results in [7].

Theorem 3.1. *Assume that Hypotheses 1.1 and 3.1 are satisfied and that σ and b are Lipschitz functions. Then there exists a unique solution $\{u(t, x),\ (t, x) \in [0, T]\times \mathbb{R}^d\}$ of equation (13.11). Moreover, for all $p \geq 1$,*

$$\sup_{(t,x)\in[0,T]\times\mathbb{R}^d} E(|u(t, x)|^p) < +\infty.$$

Let us now deal with the Malliavin differentiability of the solution $u(t, x)$ of Eq. (13.11). For this, we consider the Gaussian context described in Sect. 2.2. The following proposition summarizes a series of results in [19, 23, 32]. For the statement, we will use the following notation: for any $m \in \mathbb{N}$, set $\bar{s} := (s_1, \ldots, s_m) \in [0, T]^m$, $\bar{z} := (z_1, \ldots, z_m) \in (\mathbb{R}^d)^m$, and $\bar{s}(i) := (s_1, \ldots, s_{i-1}, s_{i+1}, \ldots, s_m)$ (resp. $\bar{z}(i)$) and, for any function f and variable X for which it makes sense, set

$$\Delta^m(f, X) := D^m f(X) - f'(X) D^m X.$$

Note that $\Delta^m(f, X) = 0$ for $m = 1$ and, if $m > 1$, it only involves iterated Malliavin derivatives up to order $m - 1$.

Proposition 3.1. *Assume that Hypothesis 1.1 is satisfied and, for some $m \in \mathbb{N} \cup \{\infty\}$, $\sigma, b \in \mathcal{C}^m(\mathbb{R})$ and their derivatives of order greater than or equal to one are bounded. Then, for all $(t, x) \in [0, T] \times \mathbb{R}^d$, the random variable $u(t, x)$ belongs to $\mathbb{D}^{j,p}$ for any $j = 1, \ldots, m$ and $p \geq 1$. Furthermore, for any $j \in \{1, \ldots, m\}$ and $p \geq 1$, the iterated Malliavin derivative $D^j u(t, x)$ satisfies the following equation in $L^p(\Omega; \mathcal{H}_T^{\otimes j})$:*

$$D^j u(t, x) = Z^j(t, x) + \int_0^t \int_{\mathbb{R}^d} \Gamma(t - s, x - y)[\Delta^j(\sigma, u(s, y))$$

$$+ D^j u(s,y)\sigma'(u(s,y))]W(\mathrm{d}s,\mathrm{d}y) + \int_0^t \int_{\mathbb{R}^d} [\Delta^j(b,u(t-s,x-y))$$
$$+ D^j u(t-s,x-y)b'(u(t-s,x-y))]\,\Gamma(s,\mathrm{d}y)\mathrm{d}s, \qquad (13.15)$$

where $Z^j(t,x)$ *is the element of* $L^p(\Omega;\mathcal{H}_T^{\otimes j})$ *given by*

$$Z^j(t,x)_{\bar{s},\bar{z}} = \sum_{i=1}^{j} \Gamma(t-s_i, x-\mathrm{d}z_i) D^{j-1}_{\bar{s}(i),\bar{z}(i)} \sigma(u(s_i,z_i)).$$

A detailed description of the construction of Hilbert-space-valued stochastic integrals as the one in Eq. (13.15) can be found in [23, Sect. 3]. Indeed, as proved in [6, Sect. 3.6], these kinds of integrals turn out to be equivalent to Hilbert-space-valued stochastic integrals *à* la Da Prato and Zabczyk [5].

Proposition 3.1 can be used to obtain the following results on existence and smoothness of the density for the solution $u(t,x)$. They are direct consequences of Theorems 5.2 and 6.2 in [23], with the only difference that the latter consider vanishing initial conditions.

Theorem 3.2. *Assume that Hypotheses 1.1 and 3.1 are satisfied,* $b,\sigma \in \mathcal{C}^1$ *have a bounded derivative, and* $|\sigma(z)| \geq c > 0$ *for all* $z \in \mathbb{R}$. *Then, for all* $(t,x) \in (0,T] \times \mathbb{R}^d$, *the random variable* $u(t,x)$ *has a law which is absolutely continuous with respect to the Lebesgue measure.*

Theorem 3.3. *Assume that Hypotheses 1.1 and 3.1 are satisfied,* $\sigma, b \in \mathcal{C}^\infty$ *and their derivatives of order greater than or equal to one are bounded, and that* $|\sigma(z)| \geq c > 0$ *for all* $z \in \mathbb{R}$. *Moreover, suppose that, for some* $\gamma > 0$,

$$C\, t^\gamma \leq \int_0^t \int_{\mathbb{R}^d} |\mathcal{F}\Gamma(s)(\xi)|^2\, \mu(\mathrm{d}\xi)\mathrm{d}s, \quad t \in (0,1). \qquad (13.16)$$

Then, for every $(t,x) \in (0,T] \times \mathbb{R}^d$, *the law of the random variable* $u(t,x)$ *has a* $\mathcal{C}^\infty$ *density.*

As commented in the Introduction, both results apply to the stochastic heat equation with $d \geq 1$ and the stochastic wave equation with $d \in \{1,2,3\}$ provided that Eq. (13.14) is satisfied, since condition (13.16) holds for these examples with $\gamma = 1$ and $\gamma = 3$, respectively.

4 Auxiliary Results

This section is devoted to prove the main two ingredients needed in the proof of Theorem 1.1. The first one establishes a suitable uniform bound for the norm of the iterated Malliavin derivative of the solution $u(t,x)$ in small time intervals. The second one deals with the negative moments of the corresponding Malliavin matrix, which here simply reduces to the norm of the Malliavin derivative of $u(t,x)$.

Lemma 4.1. *Let $0 \le a < e \le T$ and $p \ge 1$. Assume that Hypotheses 1.1 and 3.1 are satisfied and that, for some $m \in \mathbb{N}$, the coefficients b, σ belong to C^m and all their derivatives of order greater than or equal to one are bounded. Then, there exists a positive constant C, which is independent of a and e, such that, for all $\delta \in (0, e-a]$,*

$$\sup_{(\tau,y)\in[e-\delta,e]\times\mathbb{R}^d} \mathbb{E}\left(\|D^j u(\tau,y)\|^{2p}_{\mathcal{H}^{\otimes j}_{e-\delta,e}} \right) \le C\, \Phi(\delta)^{jp}, \tag{13.17}$$

for all $j \in \{1,\dots,m\}$, where we remind that, for all $t \ge 0$,

$$\Phi(t) = \int_0^t \int_{\mathbb{R}^d} |\mathcal{F}\Gamma(s)(\xi)|^2\, \mu(d\xi)ds.$$

Proof. It is similar to that of [26, Lemma 3.4], where a conditioned version of this result for the stochastic heat equation has been proved. Precisely, as already pointed out in [26, Remark 3.5], in our general setting, we need to smooth the fundamental solution Γ as follows. Let $\psi \in C_0^\infty(\mathbb{R}^d)$ be such that $\psi \ge 0$, its support is contained in the unit ball of $\mathbb{R}^d$ and $\int_{\mathbb{R}^d} \psi(x)dx = 1$. For $n \in \mathbb{N}$, set $\psi_n(x) := n^d \psi(nx)$ and, for all t, $\Gamma_n(t) := \psi_n * \Gamma(t)$. It is well known that $\Gamma_n(t)$ belongs to $\mathcal{S}(\mathbb{R}^d)$.

Let us now consider $\{u_n(t,x),\ (t,x) \in [0,T]\times\mathbb{R}^d\}$ the unique solution of

$$u_n(t,x) = I_0(t,x) + \int_0^t\int_{\mathbb{R}^d} \Gamma_n(t-s,x-y)\sigma(u_n(s,y))\, W(ds,dy)$$
$$+ \int_0^t\int_{\mathbb{R}^d} b(u_n(t-s,x-y))\,\Gamma(s,dy)ds.$$

Since $\Gamma_n(t)$ is a smooth function (such as in the case of the heat equation), we can mimic the proof of [26, Lemma 3.4], so that we end up with estimate (13.17) with u replaced by u_n. Indeed, we should remark here that the term involving the pathwise integral with respect to $\Gamma(s,dy)ds$ does not cause any problem since we only need to use that $\Gamma(t,\mathbb{R}^d)$ is uniformly bounded in t, which is part of Hypothesis 1.1.

On the other hand, a direct consequence of the proofs of [32, Theorem 1] and [23, Proposition 6.1] is that, for all $(t,x) \in [e-\delta,e]\times\mathbb{R}^d$ and $j \in \{1,\dots,m\}$,

$$D^j u(t,x) = L^2(\Omega; \mathcal{H}^{\otimes j}_{e-\delta,e}) - \lim_{n\to\infty} D^j u_n(t,x).$$

Therefore, writing down the corresponding convergence of norms and taking supremum over $[e-\delta,e]\times\mathbb{R}^d$, we conclude the proof. □

Proposition 4.1. *Assume that Hypotheses 1.1 and 3.1 are satisfied, that b, σ are C^1 functions with bounded derivatives, and that $|\sigma(z)| \ge c > 0$ for all $z \in \mathbb{R}$. Moreover,* *suppose that, for some $\gamma > 0$,*

$$C\, t^{\gamma} \leq \int_0^t \int_{\mathbb{R}^d} |\mathcal{F}\Gamma(s)(\xi)|^2 \, \mu(d\xi)ds, \quad t \in (0,1). \tag{13.18}$$

Then, for any $p > 0$, there exists a constant $C > 0$ such that, for all $(t,x) \in (0,T] \times \mathbb{R}^d$,

$$\mathbb{E}\left(\| Du(t,x) \|_{\mathcal{H}_T}^{-2p} \right) \leq C\, \Phi(t)^{-p}.$$

Proof. The proof's structure is analogous as that of the proofs of [23, Theorem 6.2] and [26, Proposition 4.3], so we will only sketch the main steps.

First, owing to [22, Lemma 2.3.1], it suffices to check that, for any $q > 2$, there exists $\varepsilon_0 = \varepsilon_0(q) > 0$ such that, for all $\varepsilon \leq \varepsilon_0$,

$$\mathbb{P}\left\{ \Phi(t)^{-1} \| Du(t,x) \|_{\mathcal{H}_T}^2 < \varepsilon \right\} \leq C \varepsilon^q. \tag{13.19}$$

Note that the Malliavin derivative $Du(t,x)$ verifies the following equation in $\mathcal{H}_T$ [take $m = 1$ in Eq. (13.15)]:

$$\begin{aligned} Du(t,x) &= \sigma(u(\cdot,\star))\Gamma(t-\cdot, x-\star) \\ &\quad + \int_0^t \int_{\mathbb{R}^d} \Gamma(t-s, x-y)\sigma'(u(s,y))Du(s,y)W(ds,dy) \\ &\quad + \int_0^t \int_{\mathbb{R}^d} b'(u(s,x-y))Du(s,x-y)\Gamma(t-s,dy)ds. \end{aligned}$$

Then, for any small $\delta > 0$, one proves that

$$\begin{aligned} \mathbb{P}\left\{ \Phi(t)^{-1} \| Du(t,x) \|_{\mathcal{H}_T}^2 < \varepsilon \right\} &\leq \mathbb{P}\left\{ \Phi(t)^{-1} I(t,x;\delta) \geq c\, \Phi(t)^{-1}\Phi(\delta) - \varepsilon \right\} \\ &\leq \left(c\, \Phi(t)^{-1}\Phi(\delta) - \varepsilon \right)^{-p} \Phi(t)^{-p}\, \mathbb{E}(|I(t,x;\delta)|^p), \end{aligned} \tag{13.20}$$

where $I(t,x;\delta) := \| R_1(t,x;\delta) \|_{\mathcal{H}_{t-\delta,t}}^2 + \| R_2(t,x;\delta) \|_{\mathcal{H}_{t-\delta,t}}^2$ and

$$R_1(t,x;\delta) := \int_{\cdot}^t \int_{\mathbb{R}^d} \Gamma(t-s, x-y)\sigma'(u(s,y))Du(s,y)W(ds,dy),$$

$$R_2(t,x;\delta) := \int_{\cdot}^t \int_{\mathbb{R}^d} b'(u(t-s,x-y))Du(t-s,x-y)\, \Gamma(s,dy)ds.$$

Using Lemma 4.1 and applying standard integral estimates, one checks that

$$\mathbb{E}(|I(t,x;\delta)|^p) \leq C\, \Phi(\delta)^p \left(\Phi(\delta)^p + \Psi(\delta)^p \right),$$

where we have set

$$\Psi(s) := \int_0^s \Gamma(r, \mathbb{R}^{\mathrm{d}})\mathrm{d}r.$$

Thus, going back to Eq. (13.20), we obtain

$$\mathbb{P}\left\{\Phi(t)^{-1}\|Du(t,x)\|^2_{\mathcal{H}_T}<\varepsilon\right\} \leq C\left(c\,\Phi(t)^{-1}\Phi(\delta)-\varepsilon\right)^{-p}\Phi(t)^{-p}\Phi(\delta)^p\left(\Phi(\delta)^p+\Psi(\delta)^p\right).$$

At this point, taking a small enough ε_0 if necessary, we can choose $\delta = \delta(\varepsilon)$ such that

$$\frac{c}{2}\,\Phi(t)^{-1}\Phi(\delta) = \varepsilon. \tag{13.21}$$

Hence, we have

$$\mathbb{P}\left\{\Phi(t)^{-1}\|Du(t,x)\|^2_{\mathcal{H}_T} < \varepsilon\right\} \leq C\left(\Phi(\delta)^p + \Psi(\delta)^p\right).$$

Note, on the one hand, that condition (13.21) implies $\Phi(\delta) \leq C\Phi(T)\varepsilon \leq C\varepsilon$. On the other hand, by Hypothesis 1.1, we have $\Psi(\delta) \leq C\delta$. Hence, the assumption (13.18) and what we have just said above let us infer that $\Psi(\delta) \leq C\varepsilon^{\frac{1}{\gamma}}$. Therefore,

$$\mathbb{P}\left\{\Phi(t)^{-1}\|Du(t,x)\|^2_{\mathcal{H}_T} < \varepsilon\right\} \leq C\left(\varepsilon^p + \varepsilon^{\frac{p}{\gamma}}\right),$$

so taking $p = q(\gamma \vee 1)$ we conclude that Eq. (13.19) is satisfied. □

5 Proof of the Main Result

In this section, we are going to prove Theorem 1.1. On the one hand, we note that Proposition 3.1 implies that, for any $(t,x) \in (0,T] \times \mathbb{R}^d$, the random variable $u(t,x)$ belongs to $\mathbb{D}^{2,p}$ for all $p \geq 1$. Moreover, an immediate consequence of Proposition 4.1 is that the Malliavin matrix associated to $u(t,x)$ has negative moments of all orders. Thus, applying a general criterion of the Malliavin calculus (see, e.g., [22, Proposition 2.1.5] or [17, Theorem 4.1]), we obtain that the law of $u(t,x)$ has a density and it is a continuous function.

On the other hand, as explained in the Introduction, the proof of Eq. (13.9) is a matter of following exactly the same arguments as in [26, Sect. 5] and invoking the results of the previous section. Let us sketch the main steps to follow.

To start with, we use the formula for the density arising from the application of the integration-by-parts formula in the Malliavin calculus context (see, e.g., [22, Proposition 2.1.1]). Precisely, denoting the density of $u(t,x)$ by $p_{t,x}$, we have

$$p_{t,x}(y) = E\left(\mathbf{1}_{\{u(t,x)>y\}}\delta\left(\frac{Du(t,x)}{\|Du(t,x)\|^2_{\mathcal{H}_T}}\right)\right), y \in \mathbb{R},$$

where here δ denotes the divergence operator or Skorohod integral, that is, the adjoint of the Malliavin derivative operator (see [22, Chap. 1]).

Next, taking into account the equation satisfied by $u(t,x)$ [i.e., Eq. (13.11)] and applying [22, Proposition 2.1.2], we obtain

$$p_{t,x}(y) \le C\ \mathbb{P}\{|M_t| > |y - I_0(t,x)| - c_3 T\}^{\frac{1}{q}}$$
$$\times \left\{ \mathbb{E}\left(\|Du(t,x)\|_{\mathcal{H}_T}^{-1}\right) + \left(\mathbb{E}\|D^2u(t,x)\|^{\alpha}_{\mathcal{H}_T^{\otimes 2}}\right)^{\frac{1}{\alpha}} \left(\mathbb{E}\|Du(t,x)\|_{\mathcal{H}_T}^{-2\beta}\right)^{\frac{1}{\beta}} \right\}, \tag{13.22}$$

where α, β, and q are any positive real numbers satisfying $\frac{1}{\alpha} + \frac{1}{\beta} + \frac{1}{q} = 1$. In the above expression, M_t denotes the *martingale part* of the solution $u(t,x)$, that is,

$$M_t = \int_0^t \int_{\mathbb{R}^d} \Gamma(t-s, x-y)\sigma(u(s,y))W(\mathrm{d}s, \mathrm{d}y),$$

and the term $c_3 T$ comes from the fact that, due to Hypothesis 1.1 and the boundedness of b, for all $(t,x) \in (0,T] \times \mathbb{R}^d$,

$$\left| \int_0^t \int_{\mathbb{R}^d} \Gamma(t-s, x-y) b(u(s,y))\, \mathrm{d}y \mathrm{d}s \right| \le c_3\, T, \mathbb{P}\text{-a.s.}$$

In order to estimate the terms in Eq. (13.22), we first apply the exponential martingale inequality in order to get a suitable exponential bound of the probability in Eq. (13.22) (using that $\langle M \rangle_t \le C\,\Phi(t)$), and then we conveniently apply Lemma 4.1 and Proposition 4.1. Thus

$$p_{t,x}(y) \le c_1\, \Phi(t)^{-1/2} \exp\left(-\frac{(|y - I_0(t,x)| - c_3 T)^2}{c_2 \Phi(t)} \right), y \in \mathbb{R},$$

where the constants c_1, c_2, and c_3 do not depend on (t,x), so we conclude the proof of Theorem 1.1. □

Acknowledgements Part of this work has been done while the author visited the Centre Interfacultaire Bernoulli at the École Polytechnique Fédérale de Lausanne, to which he would like to thank for the financial support, as well as to the National Science Foundation. Lluís Quer-Sardanyons was supported by the grant MICINN-FEDER Ref. MTM2009-08869.

References

1. Bally, V., Pardoux, E.: Malliavin calculus for white noise driven parabolic SPDEs. Potential Anal. **9**(1), 27–64 (1998)
2. Carmona, R., Nualart, D.: Random nonlinear wave equations: smoothness of the solutions. Probab. Theory Relat. Fields **79**(4), 469–508 (1988)
3. Carmona, R.A., Molchanov, S.A.: Parabolic Anderson problem and intermittency. Mem. Amer. Math. Soc. **108**(518), viii+125 (1994) MR 1185878 (94h:35080)

4. Conus, D., Dalang, R.C.: The non-linear stochastic wave equation in high dimensions. Electron. J. Probab. **13**(22), 629–670 (2008) MR 2399293 (2009c:60170)
5. Da Prato, G., Zabczyk, J.: Stochastic equations in infinite dimensions, Encyclopedia of Mathematics and its Applications, vol. 44. Cambridge University Press, Cambridge (1992)
6. Dalang, R., Quer-Sardanyons, L.: Stochastic integrals for spde's: A comparison. Expo. Math. **29**, 67–109 (2011)
7. Dalang, R.C.: Extending the martingale measure stochastic integral with applications to spatially homogeneous s.p.d.e.'s. Electron. J. Probab. **4**(6), 29 pp. (electronic) (1999)
8. Dalang, R.C., Frangos, N.E.: The stochastic wave equation in two spatial dimensions, Ann. Probab. **26**(1), 187–212 (1998)
9. Dalang, R.C., Khoshnevisan, D., Nualart, E.: Hitting probabilities for systems for non-linear stochastic heat equations with multiplicative noise, Probab. Theory Relat. Fields **144**(3–4), 371–427 (2009) MR 2496438 (2010g:60151)
10. Dalang, R.C., Mueller, C.: Some non-linear S.P.D.E.'s that are second order in time. Electron. J. Probab. **8**(1), 21 pp. (electronic) (2003)
11. Fournier, N., Printems, J.: Absolute continuity for some one-dimensional processes. Bernoulli **16**(2), 343–360 (2010)
12. Guérin, H., Méléard, S., Nualart, E.: Estimates for the density of a nonlinear Landau process. J. Funct. Anal. **238**(2), 649–677 (2006) MR 2253737 (2008e:60164)
13. Hu, Y., Nualart, D., Song, J.: A nonlinear stochastic heat equation: Hölder continuity and smoothness of the density of the solution, ArXiv Preprint arXiv:1110.4855v1
14. Hu, Y., Nualart, D., Song, J.: Feynman-Kac formula for heat equation driven by fractional white noise. Ann. Probab. **39**(1), 291–326 (2011) MR 2778803 (2012b:60208)
15. Karczewska, A., Zabczyk, J.: Stochastic PDE's with function-valued solutions, Infinite dimensional stochastic analysis (Amsterdam, 1999), Verh. Afd. Natuurkd. 1. Reeks. K. Ned. Akad. Wet., vol. 52, R. Neth. Acad. Arts Sci., Amsterdam, 197–216 (2000) MR 2002h:60132
16. Kohatsu-Higa, A.: Lower bounds for densities of uniformly elliptic random variables on Wiener space, Probab. Theory Relat. Fields **126**(3), 421–457 (2003) MR 1992500 (2004d:60141)
17. Malliavin, P.: Stochastic analysis, Grundlehren der Mathematischen Wissenschaften [Fundamental Principles of Mathematical Sciences], vol. 313. Springer, Berlin (1997) MR 1450093 (99b:60073)
18. Marinelli, C., Nualart, E., Quer-Sardanyons, L.: Existence and regularity of the density for the solution to semilinear dissipative parabolic spdes, arXiv:1202.4610
19. Mellouk, M., Márquez-Carreras, D., Sarrà, M.: On stochastic partial differential equations with spatially correlated noise: smoothness of the law. Stochastic Process. Appl. **93**(2), 269–284 (2001)
20. Millet, A., Sanz-Solé, M.: A stochastic wave equation in two space dimension: smoothness of the law. Ann. Probab. **27**(2), 803–844 (1999)
21. Nourdin, I., Viens, F.G.: Density formula and concentration inequalities with Malliavin calculus. Electron. J. Probab. **14**(78), 2287–2309 (2009) MR 2556018 (2011a:60147)
22. Nualart, D.: The Malliavin calculus and related topics, 2nd edn. Probability and its Applications (New York). Springer, Berlin (2006)
23. Nualart, D., Quer-Sardanyons, L.: Existence and smoothness of the density for spatially homogeneous SPDEs, Potential Anal. **27**(3), 281–299 (2007)
24. Nualart, D., Quer-Sardanyons, L.: Gaussian density estimates for solutions to quasi-linear stochastic partial differential equations. Stochastic Process. Appl. **119**(11), 3914–3938 (2009) MR 2552310 (2011g:60113)
25. Nualart, D., Quer-Sardanyons, L.: Optimal Gaussian density estimates for a class of stochastic equations with additive noise. Infin. Dimens. Anal. Quantum Probab. Relat. Top. **14**(1), 25–34 (2011) MR 2785746 (2012e:60155)
26. Nualart, E., Quer-Sardanyons, L.: Gaussian estimates for the density of the non-linear stochastic heat equation in any space dimension. Stochastic Process. Appl. **122**(1), 418–447 (2012) MR 2860455

27. Pardoux, É, Zhang, T.S.: Absolute continuity of the law of the solution of a parabolic SPDE. J. Funct. Anal. **112**(2), 447–458 (1993) MR MR1213146 (94k:60095)
28. Peszat, S.: The Cauchy problem for a nonlinear stochastic wave equation in any dimension. J. Evol. Equ. **2**(3), 383–394 (2002) MR 2003k:60157
29. Peszat, S., Zabczyk, J.: Stochastic evolution equations with a spatially homogeneous Wiener process. Stochastic Process. Appl. **72**(2), 187–204 (1997) MR MR1486552 (99k:60166)
30. Peszat, S., Zabczyk, J.: Nonlinear stochastic wave and heat equations. Probab. Theory Relat. Fields **116**(3), 421–443 (2000) MR 2001f:60071
31. Quer-Sardanyons, L., Sanz-Solé, M.: Absolute continuity of the law of the solution to the 3-dimensional stochastic wave equation. J. Funct. Anal. **206**(1), 1–32 (2004) MR 2 024 344
32. Quer-Sardanyons, L., Sanz-Solé, M.: A stochastic wave equation in dimension 3: smoothness of the law. Bernoulli **10**(1), 165–186 (2004) MR 2 044 597
33. Sanz-Solé, M.: Malliavin calculus, Fundamental Sciences, EPFL Press, Lausanne, 2005, With applications to stochastic partial differential equations. MR 2167213 (2006h:60005)
34. Schwartz, L.: Théorie des distributions, Publications de l'Institut de Mathématique de l'Université de Strasbourg, No. IX-X. Nouvelle édition, entiérement corrigée, refondue et augmentée, Hermann, Paris (1966) MR 35 #730
35. Walsh, J.B.: An introduction to stochastic partial differential equations, École d'été de probabilités de Saint-Flour, XIV—1984, Lecture Notes in Math., vol. 1180, pp. 265–439. Springer, Berlin (1986) MR 88a:60114

Chapter 14
Stationarity of the Solution for the Semilinear Stochastic Integral Equation on the Whole Real Line

Bijan Z. Zangeneh

I would like to dedicate this paper to Professor David Nualart for his long lasting contribution to the field of stochastic analysis.

Abstract In this article we prove the stationarity of the solution of the H-valued integral equation

$$X(t) = \int_{-\infty}^{t} U(t-s)f(X(s))\mathrm{d}s + V(t),$$

where H is a real separable Hilbert space. In this equation, $U(t)$ is a semigroup generated by a strictly negative definite, self-adjoint unbounded operator A, such that A^{-1} is compact and f is of monotone type and is bounded by a polynomial and $V(t)$ is a cadlag adapted stationary process.

Received 11/7/2011; Accepted 6/14/2012; Final 7/23/2012

1 Introduction

Let H be a real separable Hilbert space with norm $\| \ \|$ and inner product $\langle \ , \ \rangle$. Suppose $(\Omega, \mathcal{F}, \mathcal{F}_t, P)$ is a complete stochastic basis with a right continuous filtration and $\{W(t), t \geq 0\}$ is an H-valued cylindrical Brownian motion with respect to $(\Omega, \mathcal{F}, \mathcal{F}_t, P)$.

B.Z. Zangeneh (✉)
Department of Mathematical Sciences, Sharif University of Technology, Tehran, Iran
e-mail: zangeneh@sharif.edu

F. Viens et al. (eds.), *Malliavin Calculus and Stochastic Analysis: A Festschrift in Honor of David Nualart*, Springer Proceedings in Mathematics & Statistics 34,
DOI 10.1007/978-1-4614-5906-4_14,

Let g be an H-valued function defined on a set $D(g) \subset H$. Recall that g is *monotone* if for each pair $x, y \in D(g)$

$$\langle g(x) - g(y), x - y \rangle \ \geq 0,$$

and g is *semi-monotone* with *parameter* M if, for each pair $x, y \in D(g)$,

$$\langle g(x) - g(y), x - y \rangle \ \geq -M \|x - y\|^2.$$

We say g is *demicontinuous* if whenever (x_n) is a sequence in $D(g)$ which converges strongly to a point $x \in D(g)$, then $g(x_n)$ converges weakly to $g(x)$.

Consider the stochastic semilinear equation

$$\mathrm{d}X(t) = AX(t)\mathrm{d}t + f(X(t))\,\mathrm{d}t + \mathrm{d}W(t), \tag{14.1}$$

where A is a closed, self-adjoint, negative definite, unbounded operator such that A^{-1} is nuclear. A *mild solution* of Eq. (14.1) with initial condition $X(0) = X_0$ is the solution of the integral equation

$$X(t) = U(t,0)X_0 + \int_0^t U(t-s)f(X(s))\mathrm{d}s + \int_0^t U(t-s)\mathrm{d}W(s), \tag{14.2}$$

where $U(t)$ is the semigroup generated by A.

Marcus [17] has proved that when f is uniformly Lipschitz, then the solution of Eq. (14.2) is asymptotically stationary. To prove this, he studied the following integral equation:

$$X(t) = \int_{-\infty}^t U(t-s)f(X(s))\mathrm{d}s + \int_{-\infty}^t U(t-s)\mathrm{d}W(s), \tag{14.3}$$

where the parameter set of the processes is extended to the whole real line. This motivated us to study the stationarity of the solution of the more general equation

$$X(t) = \int_{-\infty}^t U(t-s)f(X(s))\mathrm{d}s + V(t), \tag{14.4}$$

where the stationary process V, the function f and the generator A of the semigroup U satisfy the following hypothesis.

Hypothesis 1.1. (a) *$U(t)$ is a semigroup generated by a strictly negative definite, self-adjoint unbounded operator A such that A^{-1} is compact. Then there is $\lambda > 0$ such that $\|U(t)\| \leq \mathrm{e}^{-\lambda t}$.*

(b) *Let $\varphi(t) = K(1+t^p)$ for some $p > 0$, $K > 0$. $-f$ is a monotone demicontinuous mapping from H to H such that $\|f(x)\| \leq \varphi(\|x\|)$ for all $x \in H$.*

(c) *Let* $r = 2p^2$. *Then,* $V(t)$ *is cadlag adapted process such that* $\sup_{t\in\mathbf{R}} E\{\|V(t)\|^r\} < \infty$.

Marcus [17] studied Eq. (14.4) when f is Lipschitz, V is an Ornstein–Uhlenbeck process, and A^{-1} is nuclear. He proved that the solution of Eq. (14.4) is a stationary process; when $f(x) = \frac{1}{2}\nabla F(x)$, he characterized its stationary measures explicitly. This result was generalized somewhat in Marcus [18] to the case where $f : B \to B^\star$, where $B \subseteq H \subseteq B^\star$ is a Gelfand triple and f satisfies

$$\langle f(x) - f(y), x - y\rangle_{B^*\times B} \leq -C\,\|x - y\|_B^p \text{ and}$$
$$\|f(x)\|_{B^*} \leq C(1 + \|x\|_B^{p-1}) \quad \text{for some } C \geq 0, \quad \text{and} \quad p \geq 1.$$

Unfortunately, we were unable to follow his proof of the stationarity of the solution of Eq. (14.4).

In this paper, we extend the above setting to a slightly more general case in which f, U and V satisfy Hypothesis 1.1 on a Hilbert space H. Our method of proof is different from that of [18]. We will give the stationary distribution of Eq. (14.5) when $\nabla F(x)$ is monotone.

In [31], we have proved the existence and the uniqueness of the solution to Eq. (14.4). In [30] we have proved that finite-dimensional Galerkin approximations converge strongly to the solution of Eq. (14.4). In this paper we prove stationarity of the solution of Eq. (14.4).

In the special case when $f(x) = -\frac{1}{2}\nabla F(x)$ is the Fréchet derivative of a potential $F(x)$ on H and $V(t)$ is the stationary Ornstein–Uhlenbeck processes $\int_{-\infty}^{t} U(t-s)\mathrm{d}W(s)$, we may consider the integral equation (14.4) as a mild solution of the infinite-dimensional Einstein–Smoluchowski equation:

$$\mathrm{d}X(t) = -AX(t)\mathrm{d}t - \frac{1}{2}\nabla F(X(t))\mathrm{d}t + \mathrm{d}W(t). \tag{14.5}$$

In finite dimensions, the solutions are diffusion processes and the stationary measures of these diffusion processes were studied by Kolmogorov [16].

Infinite-dimensional Einstein–Smoluchowski equations have been studied by many authors, for example, Marcus [17–19], Funaki [11] and Iwata [14].

The stationary measure associated with this equation has important applications in stochastic quantization (see [3, 15, 19] and Iwata [14]). In the case of Markovian stochastic evolution equations, one can study the stationarity of the solution by studying invariant measures (for comprehensive study of this subject see [7, 8] and references therein). Consider the following integral equation:

$$X(t) = U(t)X_0 + \int_0^t U(t-s)f(X(s))\mathrm{d}s + V(t), \tag{14.6}$$

where f, V and the generator A of the semigroup U satisfy Hypothesis 1.1.

Proposition 1.1. *Suppose that A, f and V satisfy Hypothesis 1.1. Then Eq. (14.6) has a unique adapted cadlag (continuous, if $V(t)$ is continuous) solution.*

Proof. See Zangeneh [32] or Hamadani and Zangeneh [12]. □

1.1 Energy Inequality

Proposition 1.2. *Let $a(.)$ be an H-valued integrable function on S. Suppose U and A satisfy Hypothesis 1.1. If*

$$X(t) = U(t,0)X_0 + \int_0^t U(t,s)a(s)\mathrm{d}s, \tag{14.7}$$

then

$$\|X(t)\|^2 \leq \mathrm{e}^{2\lambda t}\|X_0\|^2 + 2\int_0^t \mathrm{e}^{2\lambda(t-s)}\langle X(s), a(s)\mathrm{d}s\rangle, \quad t \in S. \tag{14.8}$$

Proof. See Zangeneh [28]. □

Note that this proposition works for more general semigroups and it applies to a large class of delay equations and to parabolic and hyperbolic equations.

1.2 Significant Role of Monotonicity Condition

Monotonicity condition on f plays an important role in proving different inequalities on stochastic evolution equations with monotone nonlinearity. To highlight the significance of this role, we give a proof of continuity of the solution of Eq. (14.6) with respect to V using the energy inequality.

Corollary 1.1. *Let f, V^1 and V^2 satisfy Hypothesis 1.1. Suppose A and U satisfy Hypothesis 1.1. Let $X^i(t)$, $i = 1,2$, be solutions of the integral equations:*

$$X^i(t) = \int_0^t U(t,s)f(s, X^i(s))\mathrm{d}s + V^i(t). \tag{14.9}$$

Then there is a constant C such that

$$\|X^2 - X^1\|_\infty \leq C\|V^2 - V^1\|_\infty^{\frac{1}{2}}. \tag{14.10}$$

Proof. Define $Y^i(t) = X^i(t) - V^i(t), i = 1, 2$. Then we can write Eq. (14.18) in the form

$$Y^i(t) = \int_0^t U(t.s) f(s, X^i(s)) \mathrm{d}s, \quad i = 1, 2,$$

so that

$$Y^2(t) - Y^1(t) = \int_0^t U(t, s)[f(s, X^2(s)) - f(s, X^1(s))] \mathrm{d}s.$$

Since U satisfies Hypothesis 1.1(a)–(c), then by energy inequality we have

$$\|Y^2(t) - Y^1(t)\|^2 \leq 2 \int_0^t \mathrm{e}^{2\lambda(t-s)} \langle Y^2(s) - Y^1(s), f(X^2(s)) - f(X^1(s)) \rangle \mathrm{d}s. \tag{14.11}$$

Note that because Y^i and X^i are cadlag and the f^i are bounded by φ_i, then the integrands are dominated by cadlag functions and hence are integrable. Since $Y^i = X^i - V^i$ and $-f^2$ is monotone. By the Schwarz inequality this is

$$\leq 2 \int_0^t \mathrm{e}^{2\lambda s} \|V^2(s) - V^1(s)\| \|f(X^2(s)) - f(X^1(s))\| \mathrm{d}s.$$

Since X^1 and X^2 are bounded and f is bounded then there is a constant K such that

$$\|Y^2 - Y^1\|_\infty \leq K \|V^2 - V^1\|_\infty^{\frac{1}{2}} \tag{14.12}$$

since $Y^i = X^i - V^i$ proof of theorem is complete. □

Corollary 1.2. *Let $D(S, H)$ be the set of H-valued cadlag functions on S with norm*

$$\|f\|_\infty = \sup_{t \in S} \|f(t)\|.$$

By Corollary 1.1 there is a continuous mapping $\psi : S \times D(S, H) \to D(S, H)$ such that if $X(t)$ is a solution of

$$X(t) = \int_0^t U(t, s) f(X(s)) \mathrm{d}s + V(t),$$

then $X(t) = \psi(t, V)(t)$. Moreover there is a constant C such that

$$\|\psi(., V^2) - \psi(., V^1)\|_\infty \leq C \|V^2 - V^1\|_\infty^{\frac{1}{2}},$$

so ψ is Hölder continuous with exponent 1/2.

1.3 Examples

1.3.1 A Semilinear Stochastic Evolution Equation

The existence and uniqueness of the solution of the integral equation (14.2) have been studied in Marcus [18]. He assumed that f is independent of $\omega \in \Omega$ and $t \in S$ and that there are $M > 0$ and $p \geq 1$ for which

$$\langle f(u) - f(v), u - v \rangle \leq -M \|u - v\|^p$$

and

$$\|f(u)\| \leq C(1 + \|u\|^{p-1}).$$

He proved that this integral equation has a unique solution in $L^p(\Omega, L^p(S, H))$.

As a consequence of Proposition 1.1, we can extend Marcus' result to more general f and we can show the existence of a strong solution of Eq. (14.2) which is continuous, instead of merely being in $L^p(\Omega, L^p(S, H))$.

The Ornstein–Uhlenbeck process $V(t) = \int_0^t U(t - s)\mathrm{d}W(s)$ has been well studied, for example, in [13], where they show that $V(t)$ has a continuous version. We can rewrite Eq. (14.2) as

$$X(t) = U(t)X(0) + \int_0^t U(t - s) f(X(s))\mathrm{d}s + V(t),$$

where $V(t)$ is an adapted continuous process. Then by Proposition 1.1, Eq. (14.2) has a unique continuous adapted solution.

1.3.2 A Semilinear Stochastic Partial Differential Equation

Let D be a bounded domain with a smooth boundary in $\mathbf{R}^d$. Let $-A$ be a uniformly strongly elliptic second-order differential operator with smooth coefficients on D. Let B be the operator $B = d(x)D_N + e(x)$, where D_N is the normal derivative on ∂D and d and e are in $C^\infty(\partial D)$. Let A (with the boundary condition $Bf \equiv 0$) be self-adjoint.

Consider the initial-boundary-value problem

$$\begin{cases} \frac{\partial u}{\partial t} + Au = f(u) + \dot{W} & \text{on} \quad D \times [0, \infty), \\ Bu = 0 & \text{on} \quad \partial D \times [0, \infty), \\ u(0, x) = 0 & \text{on} \quad D, \end{cases} \tag{14.13}$$

where $\dot{W} = \dot{W}(t, x)$ is a white noise in space-time (for the definition and properties of white noise, see Walsh [26]) and f is a nonlinear function that will be defined below. Let $p > \frac{d}{2}$. W can be considered as a *Brownian motion* $\tilde{W}(t)$ *on the Sobolev space* H_{-p} ; see Walsh [26], Chap. 5, p. 345. There is a complete orthonormal basis $\{e_k\}$ for H_p.

The operator A (plus boundary conditions) has eigenvalues $\{\lambda_k\}$ with respect to $\{e_k\}$, i.e., $Ae_k = \lambda_k e_k, \ \forall k$. The eigenvalues satisfy $\Sigma_j(1 + \lambda_j^{-p}) < \infty$ if $p > \frac{d}{2}$ (see Walsh [26], Chap. 5, p. 343). Then $[A^{-1}]^p$ is nuclear and $-A$ generates a contraction semigroup $U(t) \equiv \mathrm{e}^{-tA}$. This semigroup satisfies Hypothesis 1.1.

Now consider the initial-boundary-value problem (14.13) as a semilinear stochastic evolution equation

$$\mathrm{d}u(t) + Au(t)\mathrm{d}t = f(u(t))\mathrm{d}t + \mathrm{d}\tilde{W}(t), \tag{14.14}$$

with initial condition $u(0) = 0$, where $f : S \times \Omega \times H_{-p} \to H_{-p}$ satisfies Hypotheses 1.2(b) and 1.2(c) relative to the separable Hilbert space $H = H_{-p}$. We can define the mild solution of Eq. (14.14), which is also a mild solution of Eq. (14.13), to be the solution of

$$u(t) = \int_0^t U(t-s)f(u(s))\mathrm{d}s + \int_0^t U(t-s)\mathrm{d}\tilde{W}(s). \tag{14.15}$$

Since $\tilde{W}_t$ is a continuous local martingale on the separable Hilbert space H_{-p}, then $\int_0^t U(t-s)\mathrm{d}\tilde{W}(s)$ has an adapted continuous version; see, for example, Zangeneh [32]. If we define

$$V(t) := \int_0^t U(t-s)\mathrm{d}\tilde{W}(s),$$

then by Proposition 1.1, Eq. (14.15) has a unique continuous solution with values in H_{-p}.

Remark 1.1. In this section, we can replace Levy noise (see, e.g., [1, 2, 23, 24], and for study of stochastic calculus with respect to Levy noise, see [4]) or fractional noise (see, e.g., [10, 20, 25, 27], and for study of stochastic calculus with respect to fractional noise, see [5, 6, 9, 22]), with Brownian noise $\dot{W}$. In case of Levy noise $V(t)$ has a cadlag version.

1.4 Main Theorem

We close this section by introducing our main theorem. We assume V satisfies the following condition.

Hypothesis 1.2. *V is a cadlag adapted stationary processes on H, such that for some $p \geq 1$*

$$E(\|V(0)\|^r) < \infty \ \mathit{for} \ r = 2p^2. \tag{14.16}$$

Let $f : \mathbf{R} \to Y$, where Y is a topological space. Define $(\theta_s f)(t) = f(t+s)$.

Definition 1.1. A process $X = \{X(t) : t \in \mathbf{R}\}$, taking values in a topological space Y, is called *strongly stationary* if for each h and real numbers $t_1, t_2, \ldots, t_n$, the families $(X(t_1), X(t_2), \ldots, X(t_n))$ and $((\theta_h X)(t_1), \ldots, (\theta_h X)(t_n))$ have the same joint distribution.

Theorem 1.1. *If f and V satisfy Hypothesis 1.1 and if V satisfies Hypothesis 1.2, then the solution of Eq. (14.4) is a stationary process.*

2 The Semilinear Integral Equation on the Whole Real Line

Let us reduce the integral equation (14.4) to the following integral equation:

$$X(t) = \int_{-\infty}^{t} U(t-s) f(X(s) + V(s)) \mathrm{d}s. \tag{14.17}$$

The following theorem translates Proposition 1.1 to the case when the parameter set of the process is the whole real line.

Theorem 2.1. *If A, f and V satisfy Hypothesis 1.1, then the integral equation (14.17) has a unique continuous solution X such that*

$$\|X(t)\| \leq \int_{-\infty}^{t} \mathrm{e}^{-\lambda(t-s)} \varphi(\|V(s)\|) \mathrm{d}s; \tag{14.18}$$

$$E\{\|X(t)\|\} \leq \frac{1}{\lambda} \sup_{s \in R} E\{\varphi(\|V(s)\|)\} := K_1. \tag{14.19}$$

Proof. See Theorem 1 p. 272, Zangeneh [31] or Theorem 4.2, p. 57, Zangeneh [33]. □

Corollary 2.1. *If A, f and V satisfy Hypothesis 1.1, then the integral equation*

$$X(t) = \int_{-\infty}^{t} U(t-s) f(X(s)) \mathrm{d}s + V(t) \tag{14.20}$$

has a unique cadlag solution X.

Proof. Define $Y(t) = X(t) - V(t)$ and use Theorem 2.1. □

2.1 Galerkin Approximations

Let $U(t)$ be a semigroup generated by a strictly negative definite closed unbounded self-adjoint operator A such that A^{-1} is compact.

Then, there is a complete orthonormal basis (ϕ_n) and eigenvalues $0 < \lambda_0 < \lambda_1 < \lambda_2 < \cdots$ with $\lambda_n \to \infty$, such that $A\phi_n = -\lambda_n \phi_n$.

Let H_n be the subspace of H generated by $\{\phi_0, \phi_1, \ldots, \phi_{n-1}\}$ and let J_n be the projection operator on H_n.

Define

$$f_n = J_n f, \quad V_n(t) = J_n V(t), \quad U_n(t) = J_n U(t) J_n$$

and define $X_n(t)$ and $X(t)$ as solutions of

$$X_n(t) = \int_{-\infty}^{t} U_n(t-s) f_n(X_n(s)) \mathrm{d}s + V_n(t) \tag{14.21}$$

and

$$X(t) = \int_{-\infty}^{t} U(t-s) f(X(s)) \mathrm{d}s + V(t). \tag{14.22}$$

Now we can prove

Theorem 2.2. *If A, U, f and V satisfy Hypothesis 1.1, then one has*

$$E(\|X_n(t) - X(t)\|) \to 0.$$

Proof. See Zangeneh [30]. □

3 Stationarity

Let $D(\mathbf{R}, H)$ be the space of H-valued cadlag functions on $\mathbf{R}$ with the metric of uniform convergence on compacts

$$d(f, g) = \sum_{k=1}^{\infty} \frac{\|f - g\|_k}{2^k(1 + \|f - g\|_k)},$$

where

$$\|f\|_k = \sup_{-k \le t \le k} \|f(t)\|.$$

If $f \in D(\mathbf{R}, H)$, $\theta.f$ is a function from $\mathbf{R}$ to $D(\mathbf{R}, H)$.

Next, we prove the following lemma:

Lemma 3.1. *If $V = \{V(t),\ t \in \mathbf{R}\}$ is an H-valued cadlag stationary process on $\mathbf{R}$, then $\theta.V = \{\theta_s V\ s \in \mathbf{R}\}$ is a $D(\mathbf{R}, H)$-valued stationary process on $\mathbf{R}$.*

Proof. To prove this, it is enough to prove that for all real $t_1 < t_2 < \cdots < t_n$, all real $s_1 < s_2 < \cdots < s_m$ and all real h,

$$\{(\theta_{t_1} V)(s_1), (\theta_{t_2} V)(s_1), \ldots, (\theta_{t_n} V)(s_1), \ldots, (\theta_{t_1} V)(s_m), \ldots, (\theta_{t_n} V)(s_m)\}$$

and

$$\{(\theta_{t_1} + hV)(s_1), (\theta_{t_2} + hV)(s_1), \ldots, (\theta_{t_n} + hV)(s_1), \ldots, (\theta_{t_1} + hV)(s_m), \ldots, (\theta_{t_n} + hV)(s_m)\}$$

have the same joint distribution. But by definition, $(\theta_{t_i+h} V)(s_j) = V(t_i + h + s_j)$ and since V is an H-valued stationary process, then we have equality of the joint distributions, which completes the proof. □

4 The Continuity of the Solution with Respect to V_n

Let K be $D(\mathbf{R}, H_n)$, with metric

$$d_K(f,g) = \sum_{k=1}^{\infty} \frac{\|f - g\|_k}{2^k(1 + \|f - g\|_k)} + \left(\int_{-\infty}^{\infty} \mathrm{e}^{-|s|\lambda_0} \|f(s) - g(s)\|^r \mathrm{d}s \right)^{\frac{1}{r}}, \quad \lambda_0 > 0.$$

To prove that the solution $X_n(t)$ of Eq. (14.21) is a stationary process, we need to prove a result similar to Remark 3 p. 460 Zangeneh [28] for Eq. (14.21), i.e., that there is a continuous mapping

$$\psi : \mathbf{R} \times K \to D(\mathbf{R}, H_n) \text{ such that } X_n(t) = \psi(t, V_n(\cdot))(t).$$

To prove this, we first need to prove the existence of a solution of Eq. (14.21) when $V_n \in K$. Then, instead of Eq. (14.21), we consider the following integral equation:

$$Y(t) = \int_{-\infty}^{t} \mathcal{U}(t - s) f(Y(s) + g(s)) \mathrm{d}s, \tag{14.23}$$

under the following hypothesis.

Hypothesis 4.1. (a) $\mathcal{U}(t) =: U_n(t) = J_n U(t) J_n$, *and* U *satisfies Hypothesis 1.1.*
(b) $-f : H_n \to H_n$ *is a continuous monotone function such that* $\|f(x)\| \le C(1 + \|x\|^r)$, *for* $r = 2p^2$.
(c) $g \in K$.

Note that because H_n is a finite-dimensional space, the $\mathcal{U}(t)$ forms a group and $\mathcal{U}(t)$ is well defined for all $t \in \mathbf{R}$ and $\mathcal{U}(-t)\mathcal{U}(t) = I$.

Next, we prove two purely deterministic lemmas.

Lemma 4.1. *If f, $\mathcal{U}$ and g satisfy Hypothesis 4.1, then Eq. (14.23) has a unique continuous solution.*

Proof. As in Theorem 1, p. 272, Zangeneh [31], define

$$Y_k(t) = \int_{-k}^{t} \mathcal{U}(t-s) f(Y_k(s) + g(s))\mathrm{d}s. \tag{14.24}$$

Then we have

$$\|Y_k(t)\| \le C \int_{-\infty}^{t} \mathrm{e}^{-\lambda_0(t-s)}(1 + \|g(s)\|^r)\mathrm{d}s,$$

and by Hypothesis 4.1(c), there are $C(T) > 0$ and $C_1(T) > 0$ such that for all $t \in (-\infty, T]$,

$$\|Y_k(t)\| \le C(T)\mathrm{e}^{-\lambda_0 t} \le C_1(T). \tag{14.25}$$

Let $a \le t_1 \le t_2 \le T$. By Eq. (14.24) one has

$$\mathcal{U}(-t_2)\, Y_k(t_2) - \mathcal{U}(-t_1)\, Y_k(t_1) = \int_{t_1}^{t_2} \mathcal{U}(-s) f(Y_k(s) + g(s))\mathrm{d}s.$$

Now it is easy to see from Eq. (14.25) and Hypothesis 4.1(c) that there is $C(T,a) > 0$ such that

$$\|\mathcal{U}(-t_2)\, Y_k(t_2) - \mathcal{U}(-t_1)\, Y_k(t_1)\| \le C(T,a)|t_2 - t_1|.$$

Then $\mathcal{U}(-t)\, Y_k(t)$ is uniformly equicontinuous on $[a, T]$, so $Y_k(t)$ is uniformly equicontinuous on $[a, T]$. Since $Y_k(t)$ is uniformly bounded by Eq. (14.25), then by the Arzela–Ascoli theorem there is a subsequence (k_l) such that Y_{k_l} converges uniformly to a continuous function Y on $[a, T]$.

To complete the proof of the lemma we need to prove that $Y(t)$ is a solution of Eq. (14.23). As in the proof of Theorem 1 [31], we can show that $Y(t)$ is a solution of the equation

$$Y(t) = \mathcal{U}(t+T)Y(-T) + \int_{-T}^{t} \mathcal{U}(t-s) f(Y(s) + g(s))\mathrm{d}s, \; t \ge -T.$$

Then it is enough to prove that

$$Y(-T) = \int_{-\infty}^{-T} \mathcal{U}(-T-s) f(Y(s) + g(s))\mathrm{d}s. \tag{14.26}$$

But

$$\begin{aligned} Y_k(-T) &= \int_{-k}^{-T} \mathcal{U}(-T-s) f(Y_k(s)+g(s))\mathrm{d}s \\ &= \int_{-\infty}^{-T} \mathcal{U}(-T-s) f(Y_k(s)+g(s)) 1_{[-k,-T]}(s)\mathrm{d}s. \end{aligned}$$

By Hypothesis 4.1(c),

$$\|\mathcal{U}(-T-s) f(Y_k(s)+g(s)) 1_{[-k,-T]}(s)\|$$

is dominated by an integrable function. Since $Y_{k_l}(s) \rightarrow Y(s)$ and since f is continuous, then by the dominated convergence theorem we get Eq. (14.26). □

Lemma 4.2. *Suppose* $\mathcal{U}$, f *and* g_i *satisfy the conditions of Lemma 4.1. If* Y_i, $i = 1,2$ *are solutions of*

$$Y_i(t) = \int_{-\infty}^{t} \mathcal{U}(t-s) f(Y_i(s)+g_i(s))\mathrm{d}s,$$

then there is a constant $C(T) > 0$ *such that*

$$\|Y_2 - Y_1\|(T)^2 \le C(T, g_1, g_2)\left(\int_{-\infty}^{t} \mathrm{e}^{2\lambda s}\|g_2(s)-g_1(s)\|^2\mathrm{d}s\right)^2. \tag{14.27}$$

Proof. Define

$$Y_i^k(t) = \int_{-k}^{t} \mathcal{U}(t-s) f(Y_i^k(s)+g_i(s))\mathrm{d}s, \quad i = 1,2.$$

Let $Z_1^k(s) = Y_1^k(s) + g_1(s)$ and $Z_2^k(s) = Y_2^k(s) + g_2(s)$. Then

$$Y_2^k(t) - Y_1^k(t) = \int_{-k}^{t} \mathcal{U}(t-s)(f(Z_2^k(s)) - f(Z_1^k(s))\mathrm{d}s.$$

Since $\mathcal{U}$ satisfies Hypothesis 1.1, then by Proposition 1.2 (energy inequality), we have

$$\|Y_2^k(t) - Y_1^k(t)\|^2 \le 2\mathrm{e}^{-2\lambda t}\int_{-k}^{t} \mathrm{e}^{2\lambda s}\langle f(Z_2^k(s)) - f(Z_1^k(s)), Y_2^k(t) - Y_1^k(t)\rangle\mathrm{d}s.$$

The right-hand side of the above is

$$\begin{aligned} &= 2\mathrm{e}^{-2\lambda t}\int_{-k}^{t} \mathrm{e}^{2\lambda s}\langle f(Z_2^k(s)) - f(Z_1^k(s)), Z_2^k(s) - Z_1^k(s)\rangle\mathrm{d}s \\ &\quad -2\mathrm{e}^{-2\lambda t}\int_{-k}^{t} \mathrm{e}^{2\lambda s}\langle f(Z_2^k(s)) - f(Z_1^k(s)), g_2(s) - g_1(s)\rangle\mathrm{d}s, \end{aligned}$$

which by monotonicity of $-f$ is bounded by

$$\leq -2e^{-2\lambda t} \int_{-k}^{t} e^{2\lambda s} \langle f(Z_2^k(s)) - f(Z_1^k(s)), g_2^k(s) - g_1^k(s) \rangle ds$$

and by the Schwarz inequality, is

$$\leq 2e^{-2\lambda t} \int_{-k}^{t} e^{2\lambda s} \| f(Z_2^k(s)) - f(Z_1^k(s)) \| \| g_2^k(s) - g_1^k(s) \| ds.$$

Applying again the Schwarz inequality to the integral, we get

$$\leq e^{-\lambda t} I \left(\int_{-k}^{t} e^{2\lambda s} \| g_2(s) - g_1(s) \|^2 ds \right)^{\frac{1}{2}}, \tag{14.28}$$

where

$$\begin{aligned} I &= 2 \left(\int_{-k}^{t} e^{2\lambda s} \| f(Z_2^k(s)) - f(Z_1^k(s)) \|^2 ds \right)^{\frac{1}{2}} \\ &= 2 \left(\int_{-k}^{t} e^{2\lambda s} \| f(Y_2^k(s) + g_2(s)) - f(Y_1^k(s) + g_1(s)) \|^2 ds \right)^{\frac{1}{2}}. \end{aligned}$$

First, we show that I is uniformly bounded in k. Because $\| f(x) \| \leq C(1 + \|x\|^p)$ and $\int_{-\infty}^{T} e^{2\lambda s} \| g_i(s) \|^{2p^2} ds < \infty$, it is enough to show that $\int_{-k}^{T} e^{2\lambda s} \| Y_i^k(s) \|^{2p} ds$ is uniformly bounded in k. But by Eq. (14.18) of Theorem 2.1,

$$\| Y_i^k(t) \| \leq e^{-\lambda t} \int_{-\infty}^{t} e^{\lambda s} (1 + \| g_i(s) \|^p) ds, \quad i = 1, 2.$$

Using Fubini's theorem, we can show that

$$\int_{-\infty}^{T} e^{2\lambda s} \| Y_i^k(s) \|^{2p} ds \leq \int_{-\infty}^{T} e^{2p\lambda u} (1 + \| g_i(u) \|^{2p^2}) \left(\int_{u}^{T} e^{2\lambda(1-p)s} ds \right) du.$$

Then, $\int_{-k}^{T} e^{2\lambda s} \| Y_i^k(s) \|^{2p} ds$ is uniformly bounded in k, so there is $C_1(T)$ such that $I \leq C_1(T)$, and we can rewrite Eq. (14.28) as

$$\| Y_2^k(t) - Y_1^k(t) \|^2 \leq C_1(T) e^{-2\lambda t} \left[\int_{-\infty}^{t} e^{2\lambda s} \| g_2(s) - g_1(s) \|^2 ds \right]^{\frac{1}{2}}.$$

Since by the proof of Lemma 4.1, $Y_i^{k_l}(t) \to Y_i(t)$. By taking the limit over the subsequence (k_l) and taking the supremum on $[-T, T]$, we get Eq. (14.27). □

Remark 4.1. Let $\phi(g) := \int_{-\infty}^{T} e^{\lambda s}\|g(s)\|^r ds$ for $g \in K$. Then:

(i) If $\phi(g_i) \le N, \quad i = 1, 2$, there is a constant $C_N > 0$ such that

$$\|Y_2 - Y_1\|(T)^2 \le C_N \left[\int_{-\infty}^{T} e^{2\lambda s}\|g_2(s) - g_1(s)\|^2)ds\right]^{\frac{1}{2}}. \tag{14.29}$$

(ii) By Lemma 4.1, Eq. (14.21) has a unique cadlag adapted solution, and by (i) there is a constant $C_N > 0$ such that on the set where $\phi(V_i) \le N,\ i = 1, 2$,

$$\|X_2 - X_1\|(T)^2 \le C_N \left(d_K(V_2, V_1)\right)^{\frac{1}{2}}, \tag{14.30}$$

where $d_K(\cdot,\cdot)$ is a metric on K.

(iii) There is a continuous mapping $\psi_N : \mathbf{R} \times K \to D(\mathbf{R}, H_n)$ such that if $X_n(t)$ is the solution of Eq. (14.21), then $X_n(t) = \psi(t, V_n(\cdot))(t)$ on the set $\{\phi(V_n) < N\}$.

5 Proof of Theorem 1.1

Proof. To prove that the solution of Eq. (14.4) is stationary, it is enough to prove that the solution of Eq. (14.21) is stationary. Since $V(t)$ is an H-valued stationary process then $V_n(t) := J_n V(t)$ is also an H_n-valued stationary process. From Eq. (14.21), we have

$$X_n(t+h) = \int_{-\infty}^{t+h} U_n(t+h-s) f(X_n(s))ds + V_n(t+h);$$

by changing variables, we see that this is

$$\int_{-\infty}^{t} U_n(t-s) f(X_n(s+h))ds + \theta_h V_n(t).$$

Then by Remark 4.1, we have $X_n(t+h) = \psi_N(t, (0, (\theta_h V_n))(t)$ on the set $\{\phi(v) < N\}$ and in particular $X_n(h) = \psi_N(0, (\theta_h V_n))(0)$ on the set $\{\phi(v) < N\}$. But by Lemma 3.1, $\theta_h V_n$ is a $D(\mathbf{R}, H_n)$-valued stationary process; since $\varphi(f) = \psi_N(0, f)(0)$ is a continuous function from K to H_n, then $X_n(t) = \psi(\theta_t V_n) = \psi_N(0, (\theta_t V_n)(0)$ is an H_n-valued stationary process. Since $X(t)$ is the limit of $X_n(t)$ by Lemma 4.2, then $\{X(t) : t \in \mathbf{R}\}$ is also a stationary process. □

6 The Einstein–Smoluchowski Equation

Now consider Eq. (14.5) where $-\nabla F(x)$ satisfies Hypothesis 1.1. The stationary solution of Eq. (14.5) satisfies the following integral equation:

$$X(t) = \frac{-1}{2}\int_{-\infty}^{t} U(t-s)\nabla F(X(s))\mathrm{d}s + \int_{-\infty}^{t} U(t-s)\mathrm{d}W(s). \qquad (14.31)$$

By Zangeneh [29, Theorem 2], the solution of Eq. (14.31) is a limit of solutions of the finite-dimensional equations

$$X_n(t) = \frac{-1}{2}\int_{-\infty}^{t} U_n(t-s)\nabla F(X_n(s))\mathrm{d}s + \int_{-\infty}^{t} U_n(t-s)\mathrm{d}W(s). \qquad (14.32)$$

The stationary distribution of Eq. (14.32) is well known from Kolmogorov (1937) and can be given explicitly (see Marcus [17, 18]). But instead of Eq. (14.32) we are interested in a slightly different equation. Consider

$$Y_n(t) = \frac{-1}{2}\int_{-\infty}^{t} U_n(t-s)\nabla F(J_n Y_n(s))\mathrm{d}s + \int_{-\infty}^{t} U_n(t-s)\mathrm{d}W(s). \qquad (14.33)$$

It is clear that $J_n Y_n(t) = X_n(t)$. Since $Y_n(t) = J_n Y_n(t) + (Y_n(t) - J_n Y_n(t))$ and

$$Y_n(t) - J_n Y_n(t) = \int_{-\infty}^{t} (I - J_n)U(t-s)(I - J_n)\mathrm{d}W(s)$$

and $X_n(t) \to X(t)$, then we have $Y_n(t) \to X(t)$. By Theorem 1.1, $Y_n(t)$ is a stationary process. Let M be the stationary Gaussian measure of $\int_{-\infty}^{t} U(t-s)\mathrm{d}W(s)$ on H. Then we can prove the following result.

Lemma 6.1. *If U and $-\nabla F(x)$ satisfy Hypothesis 1.1, the stationary distribution of $Y_n(t)$ has a Radon–Nikodym derivative* $\exp(-F(J_n.))\int_H \exp(-F(.))\mathrm{d}M(.)$ *with respect to M on H.*

Proof. See Marcus [18, Lemma (10)].

Now we can prove □

Theorem 6.1. *If U and $-\nabla F(x)$ satisfy Hypothesis 1.1, then the distribution of the solution $X(t)$ of Eq. (14.31) has the Radon–Nikodym derivative* $\exp(-F((.))\int_H \exp(-F(.))\mathrm{d}M(.)$ *with respect to M on H.*

Proof. Since $E(\|Y_n(t) - X(t)\|) \to 0$ it is sufficient to show that

$$\lim_{n\to\infty}\int_H |\exp(-F(x)) - \exp(-F(J_n x))|\mathrm{d}M(x) = 0$$

since this implies weak convergence. Note that $\lim_{n\to\infty} F(J_n.) = F(.)$ on the set with M-measure equal to 1.

Without loss of generality, let $V(0) = 0$. Then the monotonicity of $\nabla F(x)$ ensures that F is nonnegative and $\exp(-F(.)) \leq 1$. The Lebesgue bounded convergence theorem can now be applied to show that the limit of the integral is equal to 0. □

References

1. Albeverio, S., Mandrekar, V., Rudiger, B.: Existence of mild solutions for stochastic differential equations and semilinear equations with non-Gaussian Lévy noise. Stoch. Process. Appl. **119**, 835–863 (2009)
2. Albeverio, S., Rudiger, B.: Stochastic Integrals and the Lévy-Itô decomposition theorem on separable Banach spaces. Stoch. Anal. Appl. **23**(2), 217–253 (2005)
3. Albeverio, S., Röckner, M.: Stochastic differential equations in infinite dimensions: solutions via dirichlet forms. Probab. Theor. Rel. Fields **89**, 347–386 (1991)
4. Applebaum, D.: Levy Processes and Stochastic Calculus, 2nd edn. Cambridge University Press, Cambridge (2009)
5. Alós, E., Nualart, D.: Stochastic integration with respect to the fractional Brownian motion. Stoch. Stoch. Rep. **75**(3), 129–152 (2003)
6. Caraballo, T., Garrido-Atienza, M.J., Taniguchi, T.: The existence and exponential behavior of solutions to stochastic delay evolution equations with a fractional Brownian motion. Nonlinear Anal. **74**(11), 3671–3684 (2011)
7. Da Prato, G., Zabczyk, J.: Ergodicity for infinite-dimensional systems. In: London Mathematical Society Lecture Note Series, vol. 229. Cambridge University Press, Cambridge (1996)
8. Da Prato, G., Zabcyk, J.: Stochastic equations in Infinite dimensions. In: Encyclopedia of Mathematics and its Applications, vol. 45. Cambridge University Press, Cambridge (1992)
9. Duncan, T.E., Hu, Y., Pasik-Duncan, B.: Stochastic calculus for fractional Brownian motion I Theory. SIAM J. Contr. Optim. **38**(2), 582–612 (2000)
10. Ferrante, M., Rovira, C.: Convergence of delay differential equations driven by fractional Brownian motion. J. Evol. Equat. **10**(4), 761–783 (2010)
11. Funaki, T.: Random motion of strings and related stochastic evolution equations. Nagoya Math. **89**, 129–193 (1983)
12. Hamedani, H.D., Zangeneh, B.Z.: The existence, uniqueness, and measurability of a stopped semilinear integral equation. Stoch. Anal. Appl. **25**(3), 493–518 (2007)
13. Iscoe, I., Marcus, M.B., McDonald, D., Talagrand, M., Zinn, J.: Continuity of l^2-valued Ornstein-Uhlenbeck Process. Ann. Probab. **18**(1), 68–84 (1990)
14. Iwata, K.: An infinite dimensional stochastic differential equation with state space C(R). Probab. Theor. Rel. Fields **5743**, 141–159 (1987)
15. Jona-Lasinio, P., Mitter, P.K.: On the stochastic quantization of field theory. Commun. Math. Phys. **101**, 409–436 (1985)
16. Kolmogorov, A.: Zur umker barkeit der statistischen naturgesetze. Math. Ann. **113**, 766–772 (1937)
17. Marcus, R.: Parabolic Ito equations. Trans. Am. Math. Soc. **198**, 177–190 (1974)
18. Marcus, R.: Parabolic Ito equations with monotone non-linearities. Funct. Anal. **29**, 275–286 (1978)
19. Marcus, R.: Stochastic diffusion on an unbounded domain. Pacific J. Math. **84**(1), 143–153 (1979)
20. Maslowski, B., Nualart, D.: Evolution equations driven by a fractional Brownian motion. J. Funct. Anal. **202**(1), 277–305 (2003)
21. Miyahara, Y.: Infinite dimensional Langevin equation and Fokker-Planck equation. Nagoya Math. J. **81**, 177–223 (1981)

22. Nualart, D., Rascanu, A.: Differential equations driven by fractional Brownian motion. Collect. Math. **53**(1), 55–81 (2002)
23. Peszat, S., Zabczyk, J.: Stochastic partial differential equations with Levy noise. An evolution equation approach. In: Encyclopedia of Mathematics and its Applications, vol. 113. Cambridge University Press, Cambridge, (2007)
24. Riedle, M., van Gaans, O.: Stochastic integration for Lévy processes with values in Banach spaces. Stoch. Process. Appl. **119** 1952–1974 (2009)
25. Tindel, S., Tudor, C.A., Viens, F.: Stochastic evolution equations with fractional Brownian motion. Probab. Theor. Rel. Fields **127**(2), 186–204 (2003)
26. Walsh, J.B.: An introduction to stochastic partial differential equations. Lect. Notes Math. **1180**, 266–439 (1986)
27. Zamani, S.: Reaction-diffusion equations with polynomial drifts driven by fractional Brownian motions. Stoch. Anal. Appl. **28**(6), 1020–1039 (2010)
28. Zangeneh, B.Z.: An energy-type inequality. Math. Inequal. Appl. **1**(3), 453–461 (1998)
29. Zangeneh, B.Z.: Galerkin approximations for a semilinear stochastic integral equation. Sci. Iran. **4**(1–2), 8–11 (1997)
30. Zangeneh, B.Z.: Semilinear stochastic evolution equations with monotone nonlinearities. Stoch. Stoch. Rep. **53**, 129–174 (1995)
31. Zangeneh, B.Z.: Existence and uniqueness of the solution of a semilinear stochastic evolution equation on the whole real line. In: Seminar on Stochastic Processes. Birkhãuser, Boston (1992)
32. Zangeneh, B.Z.: Measurability of the solution of a semilinear evolution equation. In: Seminar on Stochastic Processes. Birkhãuser, Boston (1990)
33. Zangeneh, B.Z.: Semilinear Stochastic Evolution Equations. Ph.D. thesis, University of British Columbia, Vancouver, B.C. Canada (1990)

Part IV
Fractional Brownian Models

Chapter 15
A Strong Approximation of Subfractional Brownian Motion by Means of Transport Processes

Johanna Garzón, Luis G. Gorostiza, and Jorge A. León

Abstract Subfractional Brownian motion is a process analogous to fractional Brownian motion but without stationary increments. In Garzón et al. (Stoch. Proc. Appl. 119:3435–3452, 2009) we proved a strong uniform approximation with a rate of convergence for fractional Brownian motion by means of transport processes. In this paper we prove a similar type of approximation for subfractional Brownian motion.

Received 10/11/2011; Accepted 1/10/2012; Final 1/27/2012

1 Introduction

Fractional Brownian motion (fBm) is well known and used in many areas of application (see [23, 26] for background and [7] for some applications). It is a centered Gaussian process $W = (W(t))_{t\geq 0}$ with covariance function

$$E(W(s)W(t)) = \frac{1}{2}(s^{2H} + t^{2H} - |s-t|^{2H}), \quad s,t \geq 0,$$

J. Garzón
Department of Mathematics, Universidad Nacıonal de Colombia, Bogota, Colombia
e-mail: mjgarzonm@unal.edu.co

L.G. Gorostiza
Department of Mathematics, CINVESTAV-IPN, Mexico city, Mexico
e-mail: lgorosti@math.cinvestav.mx

J.A. León (✉)
Department of Automatic Control, CINVESTAV-IPN, Mexico city, Mexico
e-mail: jleon@ctrl.cinvestav.mx

F. Viens et al. (eds.), *Malliavin Calculus and Stochastic Analysis: A Festschrift in Honor of David Nualart*, Springer Proceedings in Mathematics & Statistics 34,
DOI 10.1007/978-1-4614-5906-4_15,

where $H \in (0, 1/2) \cup (1/2, 1)$ (the case $H = 1/2$ corresponds to ordinary Brownian motion). H is called Hurst parameter. The main properties of fBm are that it is a continuous centered Gaussian process which is self-similar, has stationary increments with long-range dependence, and is neither a Markov process nor a semimartingale. Since it is not a semimartingale, it has been necessary to develop new theories of stochastic calculus for fBm, different from the classical Itô calculus (see, e.g., [2, 21, 23, 24] and references therein).

Subfractional Brownian motion (sfBm) is a process $S = (S(t))_{t \geq 0}$ that has the main properties of fBm except stationary increments, and its long-range dependence decays faster than that of fBm. Its covariance function is

$$E(S(s)S(t)) = s^{2H} + t^{2H} - \frac{1}{2}\left[(s+t)^{2H} + |s-t|^{2H}\right], \quad s, t \geq 0,$$

with parameter $H \in (0, 1/2) \cup (1/2, 1)$ (the case $H = 1/2$ also corresponds to ordinary Brownian motion). The main properties of sfBm were studied in [3], where it was also shown that it arises from the occupation time fluctuation limit of a branching particle system with H restricted to $(1/2, 1)$. This process appeared independently in a different context in [8].

The emergence of sfBm has motivated a series of papers where it arises in connection with several analogous but somewhat different branching particle systems, usually with $H \in (1/2, 1)$. It has been shown in [5] that it also comes out in a more natural way from a particle system without branching, and in [6] there is a different particle picture approach that yields sfBm with the full range of parameters $H \in (0, 1)$. Other long-range-dependent Gaussian processes have been obtained which are related to particle systems. A reader interested in fBm and sfBm in connection with particle systems can find some results and references in [3,4,16].

Some authors have studied further properties of sfBm for its own sake and related stochastic calculus, and possible applications of sfBm have been proposed (see [1, 9, 14, 17–20, 22, 25, 27–40]).

There are various ways of approximating fBm in distribution that can be used for simulation of paths. In [10] we obtained a strong approximation of fBm with a rate of convergence by means of the Mandelbrot-van Ness representation of fBm and a strong approximation of Brownian motion with transport processes proved in [13]. This was employed in [11] for a strong approximation of solutions of fractional stochastic differential equations with a rate of convergence, which may be used for simulation of solutions (computational efficiency was not the objective). A strong approximation of the Rosenblatt process by means of transport processes with a rate of convergence has been obtained in [12].

Since sfBm has attracted interest recently, it seems worthwhile to provide a strong approximation for it by means of transport processes with a rate of convergence, analogously as was done for fBm in [10]. This can be achieved using the same approach of [10] with some technical modifications and additional work. The aim of the present article is to prove such a strong approximation for sfBm, which moreover has the same rate of convergence as that of the transport approximation of fBm. The result is given in Corollary 2.1.

We end the Introduction by recalling the strong transport approximation of Brownian motion. For each $n = 1, 2, \ldots$, let $(Z^{(n)}(t))_{t\geq 0}$ be a process such that $Z^{(n)}(t)$ is the position on the real line at time t of a particle moving as follows. It starts from 0 with constant velocity $+n$ or $-n$, each with probability 1/2. It continues for a random time τ_1 which is exponentially distributed with parameter n^2, and at that time it switches from velocity $\pm n$ to $\mp n$ and continues that way for an additional independent random time $\tau_2 - \tau_1$, which is again exponentially distributed with parameter n^2. At time τ_2 it changes velocity as before, and so on. This process is called a (uniform) transport process.

Theorem 1.1 ([13]). *There exist versions on the transport process $(Z^{(n)}(t))_{t\geq 0}$ on the same probability space as a Brownian motion $(B(t))_{t\geq 0}$ such that for each $q > 0$,*

$$P\left(\sup_{a\leq t\leq b} |B(t) - Z^{(n)}(t)| > Cn^{-1/2}(\log n)^{5/2}\right) = o(n^{-q}) \quad \text{as} \quad n \to \infty,$$

where C is a positive constant depending on a, b, and q.

See [10, 13] for background and references.

2 Approximation

A stochastic integral representation of sfBm S with parameter H is given by

$$S(t) = C\int_{-\infty}^{\infty}\left[\left((t-s)^+\right)^{H-1/2} + ((t+s)^-)^{H-1/2} - 2((-s)^+)^{H-1/2}\right]\mathrm{d}B(s), \tag{15.1}$$

where C is a positive constant depending on H, and $B = (B(t))_{t\in\mathbb{R}}$ is Brownian motion on the whole real line (see [3]). Rewriting Eq. (15.1), we have

$$S(t) = W(t) + Y(t), \tag{15.2}$$

where W is an fBm with Hurst parameter H and Mandelbrot-van Ness representation

$$W(t) = C\left\{\int_{-\infty}^{0}\left[(t-s)^{H-1/2} - (-s)^{H-1/2}\right]\mathrm{d}B(s) + \int_0^t (t-s)^{H-1/2}\mathrm{d}B(s)\right\}, \tag{15.3}$$

and the process Y is defined by

$$Y(t) = C\left\{\int_{-\infty}^{-t}\left[(-t-s)^{H-1/2} - (-s)^{H-1/2}\right]\mathrm{d}B(s) - \int_{-t}^{0}(-s)^{H-1/2}\mathrm{d}B(s)\right\}. \tag{15.4}$$

Due to Eqs. (15.2)–(15.4), the processes S and Y have common properties in general, in particular the same Hölder continuity.

We fix $T > 0$ and $a < -T$, and we consider the following Brownian motions constructed from B:

1. $(B_1(s))_{0 \le s \le T}$, the restriction of B to the interval $[0, T]$.
2. $(B_2(s))_{a \le s \le 0}$, the restriction of B to the interval $[a, 0]$.
3. $B_3(s) = \begin{cases} sB(\frac{1}{s}) & \text{if } s \in [1/a, 0)\,, \\ 0 & \text{if } s = 0. \end{cases}$

By Theorem 1.1 there are three transport processes

$$(Z_1^{(n)}(s))_{0 \le s \le T}, \; (Z_2^{(n)}(s))_{a \le s \le 0}, \; \text{and} \; (Z_3^{(n)}(s))_{1/a \le s \le 0},$$

such that for each $q > 0$

$$P\left(\sup_{b_i \le t \le c_i} |B_i(t) - Z_i^{(n)}(t)| > C^{(i)} n^{-1/2} (\log n)^{5/2} \right) = o(n^{-q}) \quad \text{as } n \to \infty, \tag{15.5}$$

where $b_i, c_i, i = 1, 2, 3$, are the endpoints of the corresponding intervals and each $C^{(i)}$ is a positive constant depending on b_i, c_i, and q. Note that $Z_2^{(n)}$ and $Z_3^{(n)}$ are constructed going backwards in time.

We now proceed similarly as in [10]. We define the functions

$$f_t(s) = (t - s)^{H-1/2} - (-s)^{H-1/2} \quad \text{for} \quad s < 0 \le t \le T,$$

$$g_t(s) = (t - s)^{H-1/2} \quad \text{for} \quad 0 < s < t \le T,$$

and for $0 < \beta < 1/2$, we put

$$\varepsilon_n = -n^{-\beta/|H-1/2|}. \tag{15.6}$$

There are different approximations of W for $H > 1/2$ and for $H < 1/2$. We fix $0 < \beta < 1/2$. For $H > 1/2$ we define the process $W_\beta^{(n)} = \left(W_\beta^{(n)}(t)\right)_{t \in [0,T]}$ by

$$W_\beta^{(n)}(t) = C_H \left\{ \int_0^t g_t(s) \mathrm{d}Z_1^{(n)}(s) + \int_a^0 f_t(s) \mathrm{d}Z_2^{(n)}(s) + f_t(a) Z_2^{(n)}(a) \right.$$
$$\left. + \int_{1/a}^0 \left(- \int_{1/a}^{s \wedge \varepsilon_n} \partial_s f_t\left(\frac{1}{v}\right) \frac{1}{v^3} \mathrm{d}v \right) \mathrm{d}Z_3^{(n)}(s) \right\},$$

and for $H < 1/2$ we define the process $\hat{W}_\beta^{(n)} = \left(\hat{W}_\beta^{(n)}(t)\right)_{t \in [0,T]}$ by

$$\hat{W}_\beta^{(n)}(t) = C_H \left\{ \int_0^{(t+\varepsilon_n) \vee 0} g_t(s) \mathrm{d}Z_1^{(n)}(s) + \int_{(t+\varepsilon_n) \vee 0}^t g_t(\varepsilon_n + s) \mathrm{d}Z_1^{(n)}(s) \right.$$

$$+\int_a^{\varepsilon_n} f_t(s)\mathrm{d}Z_2^{(n)}(s) + f_t(a)Z_2^{(n)}(a)$$

$$+\int_{1/a}^{0}\left(-\int_{1/a}^{s} \partial_s f_t\left(\frac{1}{v}\right)\frac{1}{v^3}\mathrm{d}v\right)\mathrm{d}Z_3^{(n)}(s)\bigg\}.$$

We write $W^{(n)} = (W^{(n)}(t))_{t\in[0,T]}$, where

$$W^{(n)} = \begin{cases} W_\beta^{(n)} & \text{if } H > 1/2, \\ \hat{W}_\beta^{(n)} & \text{if } H < 1/2. \end{cases} \tag{15.7}$$

Note that $W^{(n)}$ is defined on the same probability space as the Brownian motion B in Eq. (15.3), and recall that it depends on β through Eq. (15.6).

The following theorem gives the convergence and the rate of convergence of $W^{(n)}$ to W.

Theorem 2.1 ([10]). *Let $H \neq 1/2$ and let W and $W^{(n)}$ be the processes defined by Eqs. (15.3) and (15.7), respectively. Then for each $q > 0$ and each β such that $0 < |H - 1/2| < \beta < 1/2$, there is a constant $C > 0$ such that*

$$P\left(\sup_{0\le t\le T}\left|W(t) - W^{(n)}(t)\right| > Cn^{-(1/2-\beta)}(\log n)^{5/2}\right) = o(n^{-q}) \ \text{as } n\to\infty.$$

We define another function

$$F_t(s) = (-t-s)^{H-1/2} - (-s)^{H-1/2} \quad \text{for} \quad s < -t < 0. \tag{15.8}$$

In [10] $Z_2^{(n)}(s)$ was defined for $s \in [a,0]$ and $Z_3^{(n)}(s)$ was defined for $s \in [1/a, 0]$, where $a < 0$ was arbitrary, but for the approximation of sfBm we need $a < -T$ so that $F_t(s)$ is well behaved.

Now we define approximating processes for Y and S in Eq. (15.2), again for a fixed $0 < \beta < 1/2$.

For $H > 1/2$ we define the process $Y_\beta^{(n)} = (Y_\beta^{(n)}(t))_{t\in[0,T]}$ by

$$Y_\beta^{(n)}(t) = C\bigg\{-\int_{-t}^{0}(-s)^{H-1/2}\mathrm{d}Z_2^{(n)}(s) + \int_a^{-t} F_t(s)\mathrm{d}Z_2^{(n)}(s) + F_t(a)Z_2^{(n)}(a)$$

$$+\int_{1/a}^{0}\left(-\int_{1/a}^{[\varepsilon_n\vee(1/a)]\wedge s} \partial_s F_t\left(\frac{1}{v}\right)\frac{1}{v^3}\mathrm{d}v\right)\mathrm{d}Z_3^{(n)}(s)\bigg\}, \tag{15.9}$$

and for $H < 1/2$ we define the process $\hat{Y}_\beta^{(n)} = \left(\hat{Y}_\beta^{(n)}(t)\right)_{t\in[0,T]}$ by

$$
\begin{aligned}
\hat{Y}_\beta^{(n)}(t) = C\Bigg\{ & -\int_{-t}^{\varepsilon_n \vee(-t)} (-s)^{H-1/2} \mathrm{d}Z_2^{(n)}(s) \\
& - \int_{\varepsilon_n \vee(-t)}^{0} (-s-\varepsilon_n)^{H-1/2} \mathrm{d}Z_2^{(n)}(s) + F_t(a) Z_2^{(n)}(a) \\
& - \int_{1/a}^{0} \left(- \int_{1/a}^{s} \partial_s F_t\left(\frac{1}{v}\right) \frac{1}{v^3} \mathrm{d}v \right) \mathrm{d}Z_3^{(n)}(s) \\
& + \int_{a}^{a\vee(-t+\varepsilon_n)} F_t(s) \mathrm{d}Z_2^{(n)}(s) \\
& + I_{\{-\varepsilon_n \le t\}} \int_{a\vee(-t+\varepsilon_n)}^{-t} F_{t+\varepsilon_n}(s) \mathrm{d}Z_2^{(n)}(s) \Bigg\}.
\end{aligned}
\tag{15.10}
$$

We write $Y^{(n)} = (Y^n(t))_{t\in[0,T]}$, where

$$
Y^{(n)} = \begin{cases} Y_\beta^{(n)} & \text{if } H > 1/2, \\ \hat{Y}_\beta^{(n)} & \text{if } H < 1/2 \end{cases}
\tag{15.11}
$$

(note that $Y^{(n)}$ involves only $Z_2^{(n)}$ and $Z_3^{(n)}$), and we define

$$
S^{(n)}(t) = W^{(n)}(t) + Y^{(n)}(t),
\tag{15.12}
$$

with $W^{(n)}$ as in Eq. (15.7).

The following theorem gives the convergence and the rate of convergence of $Y^{(n)}$ to Y.

Theorem 2.2. *Let $H \neq 1/2$ and let Y and $Y^{(n)}$ be the processes defined by Eqs. (15.4) and (15.11), respectively. Then for each $q > 0$ and each β such that $0 < |H - 1/2| < \beta < 1/2$, there is a constant $C > 0$ such that*

$$
P\left(\sup_{0\le t\le T} \left|Y(t) - Y^{(n)}(t)\right| > Cn^{-(1/2-\beta)} (\log n)^{5/2} \right) = o(n^{-q}) \ \text{ as } n \to \infty.
$$

From Theorems 2.1 and 2.2 we have the following result.

Corollary 2.1. *Let S and $S^{(n)}$ be the processes defined by Eqs. (15.1) and (15.12), respectively. Then for each $q > 0$ and each β such that $0 < |H - 1/2| < \beta < 1/2$, there is a constant $C > 0$ such that*

$$
P\left(\sup_{0\le t\le T} \left|S(t) - S^{(n)}(t)\right| > Cn^{-(1/2-\beta)} (\log n)^{5/2} \right) = o(n^{-q}) \ \text{ as } n \to \infty.
$$

Note that the approximation becomes better when H approaches $1/2$.

Remark 2.1. The reason that the rates of convergence for W and Y are the same is that the integral representations of W and Y, Eqs. (15.3) and (15.4), have similar kernels, and the approximations depend basically on the rate of the transport approximation for Brownian motion and on the Hölder continuity of Brownian motion. Equation (15.2) is a decomposition of sfBm S as a sum of an fBm W and a process Y, which holds everywhere on the sample space, and W and Y are dependent (but the dependence does not play a role in the proofs). In [1] (which contains an approximation of sfBm in law) and [25], for the case $H < 1/2$, sfBm has a decomposition with equality in law as the sum of an fBm and a process of the form

$$\int_0^\infty (1 - \mathrm{e}^{-rt})r^{-(1+2H)/2}\mathrm{d}B_1(r), \quad t \geq 0,$$

where B_1 is a Brownian motion. This kind of process was introduced in [15]. In that decomposition the Brownian motions B and B_1 are independent. That representation could be used for proving an approximation of sfBm with transport processes in the case $H < 1/2$, but it would require another independent set of transport processes to approximate B_1. We stress that our approximation is strong and holds for all H.

3 Proofs

The proofs are based on a series of lemmas.

Lemma 3.1. *For each fixed $t > 0$, the function F_t defined by Eq. (15.8) has the following properties:*

(1)

$$|\partial_s F_t(s)| \leq |H - 1/2| t(3/2 - H)(-t - s)^{H-5/2}, \quad s \leq -t. \tag{15.13}$$

(2)

$$\int_{-\infty}^{a} |\partial_s F_t(s)|(-s)^{1/2+\gamma}\mathrm{d}s < \infty \textit{ for each } 0 < \gamma < (1 - H) \wedge (1/2). \tag{15.14}$$

(3)

$$\lim_{b \to -\infty} F_t(b)B(b) = 0 \textit{ a.s.} \tag{15.15}$$

(4)

$$\int_{-\infty}^{a} F_t(s)\mathrm{d}B(s) = F_t(a)B_2(a) - \int_{1/a}^{0} \partial_s F_t\left(\frac{1}{v}\right)\frac{1}{v^3}B_3(v)\mathrm{d}v. \tag{15.16}$$

Proof. (1)

$$\partial_s F_t(s) = (H - 1/2)\left[(-s)^{H-3/2} - (-t - s)^{H-3/2}\right].$$

Taking $g(x) = x^{H-3/2}$, $g'(x) = (H-3/2)x^{H-5/2}$, $x \in [-t-s, -s]$. By the mean value theorem, for some $r \in [-t-s, -s]$,

$$\begin{aligned}|(-s)^{H-3/2} - (-t-s)^{H-3/2}| = |-g'(r)(-t-s+s)| &= t(3/2-H)r^{H-5/2}\\ &\leq t(3/2-H)(-t-s)^{H-5/2}.\end{aligned}$$

(2) From (1) and integration by parts we have

$$\begin{aligned}\int_{-\infty}^{a} |\partial_s F_t(s)|(-s)^{1/2+\gamma}\mathrm{d}s &\leq |H-1/2|t(3/2-H)\\ &\quad\times \int_{-\infty}^{a} (-t-s)^{H-5/2}(-s)^{1/2+\gamma}\mathrm{d}s\\ &= |H-1/2|t\left[(-s)^{\gamma+1/2}(-t-s)^{H-3/2}\Big|_{-\infty}^{a}\right.\\ &\quad\left.+\int_{-\infty}^{a}(1/2+\gamma)(-t-s)^{H-3/2}(-s)^{\gamma-1/2}\mathrm{d}s\right]. \qquad (15.17)\end{aligned}$$

Since $\gamma < (1-H) \wedge (1/2)$,

$$\lim_{s\to-\infty}(-s)^{\gamma+1/2}(-t-s)^{H-3/2} = \lim_{s\to-\infty}(-s)^{\gamma+H-1}\left(\frac{t}{s}+1\right)^{H-3/2} = 0, \qquad (15.18)$$

and

$$\begin{aligned}\int_{-\infty}^{a}(-t-s)^{H-3/2}(-s)^{\gamma-1/2}\mathrm{d}s &\leq \int_{-\infty}^{a}(-t-s)^{H+\gamma-2}\mathrm{d}s\\ &= \frac{(-t-a)^{H+\gamma-1}}{1-H-\gamma} < \infty,\end{aligned}$$

which together Eqs. (15.17) and (15.18) shows that statement (2) holds.

(3) By the pathwise Hölder continuity of B_3 on $[1/a, 0]$, taking $0 < \gamma < (1-H) \wedge (1/2)$, we have $|sB(1/s)| < Y(-s)^{1/2-\gamma}$ for each $s \in [1/a, 0]$ and a random variable Y. Then $|B(s)| < Y(-s)^{1/2+\gamma}$ for each $s \in (-\infty, a]$. Therefore,

$$\begin{aligned}|F_t(b)B(b)| &\leq \left|(-t-b)^{H-1/2} - (-b)^{H-1/2}\right| Y(-b)^{1/2+\gamma}\\ &= \left|\left(\frac{t}{b}+1\right)^{H-1/2} - 1\right| Y(-b)^{H+\gamma},\end{aligned}$$

and using l'Hôpital rule,

$$\lim_{b \to -\infty} \frac{\left| \left(\frac{t}{b}+1\right)^{H-1/2} - 1 \right|}{(-b)^{-\gamma-H}} = 0.$$

(4) Since F_t is square-integrable on $(-\infty, a)$, $\lim_{b\to-\infty} \int_b^a F_t(s)\mathrm{d}B(s) = \int_{-\infty}^a F_t(s)\mathrm{d}B(s)$. Thus, applying integration by parts,

$$\int_b^a F_t(s)\mathrm{d}B(s) = F_t(a)B(a) - F_t(b)B(b) - \int_b^a \partial_s F_t(s)B(s)\mathrm{d}s.$$

By the pathwise Hölder continuity of B (see the proof of Statement (3)) and Eq. (15.14),

$$\int_{-\infty}^a |\partial_s F_t(s)B(s)|\mathrm{d}s < \infty,$$

and using Eq. (15.15)

$$\int_{-\infty}^a F_t(s)\mathrm{d}B(s) = F_t(a)B(a) - \int_{-\infty}^a \partial_s F_t(s)B(s)\mathrm{d}s.$$

Now, with the change of variable $s = 1/v$,

$$\begin{aligned}\int_{-\infty}^a \partial_s F_t(s)B(s)\mathrm{d}s &= \int_{1/a}^0 \partial_s F_t\left(\frac{1}{v}\right)\frac{1}{v^2}B\left(\frac{1}{v}\right)\mathrm{d}v \\ &= \int_{1/a}^0 \partial_s F_t\left(\frac{1}{v}\right)\frac{1}{v^3}B_3(v)\mathrm{d}v,\end{aligned}$$

and we obtain Eq. (15.16). □

We prove Theorem 2.2 separately for $H > 1/2$ and $H < 1/2$. We denote the sup norm by $\|\ \|_\infty$, and it will always be clear from the context which interval it refers to.

3.1 *Case* H > 1/2

We fix $H - 1/2 < \beta < 1/2$ and define

$$\alpha_n = n^{-(1/2-\beta)}(\log n)^{5/2}.$$

The proof will be a consequence of the following lemmas, involving $Z_2^{(n)}$ and $Z_3^{(n)}$:

Lemma 3.2. *For each $q > 0$ there is $C > 0$ such that*

$$I_1 = P\left(\sup_{0\le t\le T}\left|F_t(a)B_2(a) - F_t(a)Z_2^{(n)}(a)\right| > C\alpha_n\right) = o(n^{-q}) \quad as\ n\to\infty.$$

Proof.

$$\begin{aligned}\left|F_t(a)B_2(a) - F_t(a)Z_2^{(n)}(a)\right| &\le \|B_2 - Z_2^{(n)}\|_\infty|(-t-a)^{H-1/2} - (-a)^{H-1/2}|\\ &\le \|B_2 - Z_2^{(n)}\|_\infty(-a)^{H-1/2},\end{aligned}$$

then, by Eq. (15.5),

$$\begin{aligned}I_1 &\le P\left(\|B_2 - Z_2^{(n)}\|_\infty(-a)^{H-1/2} > C\alpha_n\right)\\ &\le P\left(\|B_2 - Z_2^{(n)}\|_\infty > Cn^{-1/2}(\log n)^{5/2}\right) = o(n^{-q}).\end{aligned}$$

□

Lemma 3.3. *For each $q > 0$ there is $C > 0$ such that*

$$\begin{aligned}I_2 &= P\left(\sup_{0\le t\le T}\left|\int_{-t}^0(-s)^{H-1/2}\mathrm{d}B_2(s) - \int_{-t}^0(-s)^{H-1/2}\mathrm{d}Z_2^{(n)}(s)\right| > C\alpha_n\right)\\ &= o(n^{-q}) \quad as\ n\to\infty.\end{aligned}$$

Proof. By integration by parts,

$$\int_{-t}^0(-s)^{H-1/2}\mathrm{d}B_2(s) = -t^{H-1/2}B_2(-t) + (H-1/2)\int_{-t}^0(-s)^{H-3/2}B_2(s)\mathrm{d}s.$$

Analogously,

$$\int_{-t}^0(-s)^{H-1/2}\mathrm{d}Z_2^{(n)}(s) = -t^{H-1/2}Z_2^{(n)}(-t) + (H-1/2)\int_{-t}^0(-s)^{H-3/2}Z_2^{(n)}(s)\mathrm{d}s,$$

then

$$\begin{aligned}&\left|\int_{-t}^0(-s)^{H-1/2}\mathrm{d}B_2(s) - \int_{-t}^0(-s)^{H-1/2}\mathrm{d}Z_2^{(n)}(s)\right|\\ &\quad\le t^{H-1/2}|B_2(-t) - Z_2^{(n)}(-t)| + (H-1/2)\int_{-t}^0(-s)^{H-3/2}\left|B_2(s)-Z_2^{(n)}(s)\right|\mathrm{d}s\end{aligned}$$

$$\leq t^{H-1/2}\|B_2 - Z_2^{(n)}\|_\infty + (H-1/2)\|B_2 - Z_2^{(n)}\|_\infty \frac{-1}{H-1/2}(-s)^{H-1/2}\Big|_{-t}^{0}$$

$$\leq 2T^{H-1/2}\|B_2 - Z_2^{(n)}\|_\infty.$$

Consequently the result follows by Eq. (15.5). □

Lemma 3.4. *For each $q > 0$ there is $C > 0$ such that*

$$I_3 = P\left(\sup_{0\leq t\leq T}\left|\int_a^{-t} F_t(s)\mathrm{d}B_2(s) - \int_a^{-t} F_t(s)\mathrm{d}Z_2^{(n)}(s)\right| > C\alpha_n\right)$$
$$= o(n^{-q}) \quad as\ n \to \infty.$$

Proof. By integration by parts,

$$\int_a^{-t} F_t(s)\mathrm{d}B_2(s) = F_t(-t)B_2(-t) - F_t(a)B_2(a) - \int_a^{-t} \partial_s F_t(s)B_2(s)\mathrm{d}s$$

and

$$\int_a^{-t} F_t(s)\mathrm{d}Z_2^{(n)}(s) = F_t(-t)Z_2^{(n)}(-t) - F_t(a)Z_2^{(n)}(a) - \int_a^{-t} \partial_s F_t(s)Z_2^{(n)}(s)\mathrm{d}s,$$

then,

$$\left|\int_a^{-t} F_t(s)\mathrm{d}B_2(s) - \int_a^{-t} F_t(s)\mathrm{d}Z_2^{(n)}(s)\right|$$
$$\leq \|B_2 - Z_2^{(n)}\|_\infty\left[|F_t(-t)| + |F_t(a)| + \int_a^{-t} |\partial_s F_t(s)|\mathrm{d}s\right]$$
$$= \|B_2 - Z_2^{(n)}\|_\infty\Big[t^{H-1/2} + |(-t-a)^{H-1/2} - (-a)^{H-1/2}|$$
$$+ \int_a^{-t} (H-1/2)\left[(-t-s)^{H-3/2} - (-s)^{H-3/2}\right]\mathrm{d}s\Big]$$
$$\leq \|B_2 - Z_2^{(n)}\|_\infty 2T^{H-1/2}.$$

Therefore, by Eq. (15.5), the proof is complete. □

Lemma 3.5. *For each $q > 0$ there is $C > 0$ such that*

$$I_4 = P\Bigg(\sup_{0\leq t\leq T}\left|\int_{1/a}^{\varepsilon_n\vee(1/a)} \partial_s F_t\left(\frac{1}{v}\right)\frac{1}{v^3}B_3(v)\mathrm{d}v\right.$$

$$- \int_{1/a}^{0} \left(- \int_{1/a}^{[\varepsilon_n \vee (1/a)] \wedge s} \partial_s F_t \left(\frac{1}{v} \right) \frac{1}{v^3} \mathrm{d}v \right) \mathrm{d}Z_3^{(n)}(s) \right| > C\alpha_n \Big)$$
$$= o(n^{-q}) \quad as \ n \to \infty.$$

Proof. We have

$$\int_{1/a}^{\varepsilon_n \vee (1/a)} \partial_s F_t \left(\frac{1}{v} \right) \frac{1}{v^3} B_3(v) \mathrm{d}v = I_{\{\varepsilon_n > 1/a\}} \int_{1/a}^{\varepsilon_n} \partial_s F_t \left(\frac{1}{v} \right) \frac{1}{v^3} B_3(v) \mathrm{d}v.$$

Analogously, applying Fubini's theorem we have

$$\begin{aligned} &\int_{1/a}^{0} \left(- \int_{1/a}^{[\varepsilon_n \vee (1/a)] \wedge r} \partial_s F_t \left(\frac{1}{v} \right) \frac{1}{v^3} \mathrm{d}v \right) \mathrm{d}Z_3^{(n)}(r) \\ &= \int_{1/a}^{\varepsilon_n \vee (1/a)} \partial_s F_t \left(\frac{1}{v} \right) \frac{1}{v^3} Z_3^{(n)}(v) \mathrm{d}v \\ &= I_{\{\varepsilon_n \geq 1/a\}} \int_{1/a}^{\varepsilon_n} \partial_s F_t \left(\frac{1}{v} \right) \frac{1}{v^3} Z_3^{(n)}(v) \mathrm{d}v. \end{aligned}$$

Then, by Eq. (15.13),

$$\begin{aligned} &\left| \int_{1/a}^{\varepsilon_n \vee (1/a)} \partial_v f_t \left(\frac{1}{v} \right) \frac{1}{v^3} B_3(v) \mathrm{d}v - \int_{1/a}^{0} \left(- \int_{1/a}^{[\varepsilon_n \vee (1/a)] \wedge s} \partial_s F_t \left(\frac{1}{v} \right) \frac{1}{v^3} \mathrm{d}v \right) \mathrm{d}Z_3^{(n)}(s) \right| \\ &\leq \| B_3 - Z_3^{(n)} \|_\infty I_{\{\varepsilon_n > 1/a\}} \int_{1/a}^{\varepsilon_n} \left| \partial_s F_t \left(\frac{1}{s} \right) \frac{1}{s^3} \right| \mathrm{d}s \\ &\leq \| B_3 - Z_3^{(n)} \|_\infty I_{\{\varepsilon_n > 1/a\}} \int_{1/a}^{\varepsilon_n} t(3/2 - H)(H - 1/2)(tv + 1)^{H-5/2}(-v)^{-1/2-H} \mathrm{d}v \\ &\leq \| B_3 - Z_3^{(n)} \|_\infty I_{\{\varepsilon_n > 1/a\}} t(3/2 - H)(H - 1/2) \, (t/a + 1)^{H-5/2} \int_{1/a}^{\varepsilon_n} (-v)^{-1/2-H} \mathrm{d}v \\ &\leq \| B_3 - Z_3^{(n)} \|_\infty I_{\{\varepsilon_n > 1/a\}} t(3/2 - H)(1 + T/a)^{H-5/2} [(-\varepsilon_n)^{1/2-H} - (-1/a)^{1/2-H}] \\ &\leq \| B_3 - Z_3^{(n)} \|_\infty T(3/2 - H)(1 + T/a)^{H-5/2} (-\varepsilon_n)^{1/2-H}. \end{aligned}$$

Hence, since $(-\varepsilon_n)^{1/2-H} = n^\beta$, by Eq. (15.6),

$$\begin{aligned} I_4 &\leq P \left(\| B_3 - Z_3^{(n)} \|_\infty T(3/2 - H)(1 + T/a)^{H-5/2} (-\varepsilon_n)^{1/2-H} > C\alpha_n \right) \\ &\leq P \left(\| B_3 - Z_3^{(n)} \|_\infty > C n^{-\beta} \alpha_n \right) \end{aligned}$$

$$\leq P\left(\|B_3 - Z_3^{(n)}\|_\infty > Cn^{-1/2}(\log n)^{5/2}\right) = o(n^{-q}).$$

□

Lemma 3.6. *For each* $q > 0$,

$$I_5 = P\left(\sup_{0\leq t\leq T}\left|\int_{\varepsilon_n\vee(1/a)}^{0} \partial_s F_t\left(\frac{1}{v}\right)\frac{1}{v^3}B_3(v)\mathrm{d}v\right| > \alpha_n\right) = o(n^{-q}) \quad as\ n\to\infty.$$

Proof. By the pathwise Hölder continuity of B_3 with $0 < \gamma < 1 - H$, and Eq. (15.13),

$$\begin{aligned}
&\left|\int_{\varepsilon_n\vee(1/a)}^{0} \partial_s F_t\left(\frac{1}{v}\right)\frac{1}{v^3}B_3(v)\mathrm{d}v\right| \\
&\quad\leq \int_{\varepsilon_n\vee(1/a)}^{0} t(3/2-H)(H-1/2)(tv+1)^{H-5/2}(-v)^{-H-1/2}Y(-v)^{1/2-\gamma}\mathrm{d}v \\
&\quad\leq (3/2-H)(H-1/2)TY(1+T/a)^{H-5/2}\int_{\varepsilon_n\vee(1/a)}^{0}(-v)^{-H-\gamma}\mathrm{d}v \\
&\quad\leq CY(-\varepsilon_n)^{1-H-\gamma} \\
&\quad= CYn^{-\beta(1-H-\gamma)/(H-1/2)},
\end{aligned}$$

where C is a positive constant.

By Chebyshev's inequality, for $r > 0$,

$$\begin{aligned}
I_5 &\leq P\left(CYn^{-\beta(1-H-\gamma)/(H-1/2)} > \alpha_n\right) \\
&= P\left(CY > n^{\kappa}(\log n)^{5/2}\right) \\
&\leq \frac{E(|CY|^r)}{n^{r\kappa}(\log n)^{r5/2}},
\end{aligned}$$

where $\kappa = -(1/2-\beta) + \beta(1-H-\gamma)/(H-1/2)$. Taking γ close enough to 0 we have $H - 1/2 < (H-1/2)/(1-2\gamma) < \beta < 1/2$, and then $\kappa > 0$. For $q > 0$ there is $r > 0$ such that $q < r\kappa$, then

$$\lim_{n\to\infty} n^q I_5 = 0.$$

□

***Proof of Theorem 2.2 for* H > 1/2:** From Eqs. (15.4), (15.8), and (15.16) we have

$$
\begin{aligned}
Y(t) &= C\left\{-\int_{-t}^{0}(-s)^{H-1/2}\mathrm{d}B(s)+\int_{a}^{-t}F_t(s)\mathrm{d}B(s)+\int_{-\infty}^{a}F_t(s)\mathrm{d}B(s)\right\}\\
&= C\left\{-\int_{-t}^{0}(-s)^{H-1/2}\mathrm{d}B_2(s)+\int_{a}^{-t}F_t(s)\mathrm{d}B_2(s)+F_t(a)B_2(a)\right.\\
&\qquad\left.-\int_{1/a}^{0}\partial_s F_t\left(\frac{1}{v}\right)\frac{1}{v^3}B_3(v)\mathrm{d}v\right\}\\
&= C\left\{-\int_{-t}^{0}(-s)^{H-1/2}\mathrm{d}B_2(s)+\int_{a}^{-t}F_t(s)\mathrm{d}B_2(s)+F_t(a)B_2(a)\right.\\
&\qquad\left.-\int_{1/a}^{\varepsilon_n\vee(1/a)}\partial_s F_t\left(\frac{1}{v}\right)\frac{1}{v^3}B_3(v)\mathrm{d}v-\int_{\varepsilon_n\vee(1/a)}^{0}\partial_s F_t\left(\frac{1}{v}\right)\frac{1}{v^3}B_3(v)\mathrm{d}v\right\},
\end{aligned}
$$

then the definition of $Y^{(n)}$ [see Eq. (15.11)] implies

$$
\begin{aligned}
|Y(t)-Y^{(n)}(t)| &\le C\left\{\left|-\int_{-t}^{0}(-s)^{H-1/2}\mathrm{d}B_2(s)+\int_{-t}^{0}(-s)^{H-1/2}\mathrm{d}Z_2^{(n)}(s)\right|\right.\\
&+\left|\int_{a}^{-t}F_t(s)\mathrm{d}B_2(s)-\int_{a}^{-t}F_t(s)\mathrm{d}Z_2^{(n)}(s)\right|+\left|F_t(a)B_2(a)-F_t(a)Z_2^{(n)}(a)\right|\\
&+\left|\int_{1/a}^{\varepsilon_n\vee(1/a)}\partial_s F_t\left(\frac{1}{v}\right)\frac{1}{v^3}B_3(v)\mathrm{d}v\right.\\
&\left.-\int_{1/a}^{0}\left(-\int_{1/a}^{[\varepsilon_n\vee(1/a)]\wedge s}\partial_s F_t\left(\frac{1}{v}\right)\frac{1}{v^3}\mathrm{d}v\right)\mathrm{d}Z_3^{(n)}(s)\right|\\
&\left.+\left|\int_{\varepsilon_n\vee(1/a)}^{0}\partial_s F_t\left(\frac{1}{v}\right)\frac{1}{v^3}B_3(v)\mathrm{d}v\right|\right\}.
\end{aligned}
$$

Therefore, taking β such that $0 < H - 1/2 < \beta < 1/2$, by Lemmas 3.2, 3.3, 3.4, 3.5, and 3.6 we have the result. □

3.2 *Case* H < 1/2

Let $1/2 - H < \beta < 1/2$, and ε_n and α_n are as before. We proceed similarly with some lemmas.

Lemma 3.7. *For each $q > 0$ there is C such that*

$$J_1 = P\left(\sup_{0\le t\le T}\left|F_t(a)B_2(a) - F_t(a)Z_2^{(n)}(a)\right| > C\alpha_n\right) = o(n^{-q}) \quad as\ n\to\infty.$$

Proof. Similar arguments as in the proof of Lemma 3.2. □

Lemma 3.8. *For each $q > 0$ there is $C > 0$ such that*

$$\begin{aligned} J_2 = P\Bigg(\sup_{0\le t\le T}\Bigg|&\int_{-t}^{\varepsilon_n\vee(-t)}(-s)^{H-1/2}\mathrm{d}B_2(s) \\ &- \int_{-t}^{\varepsilon_n\vee(-t)}(-s)^{H-1/2}\mathrm{d}Z_2^{(n)}(s)\Bigg| > C\alpha_n\Bigg) \\ = o(n^{-q}) \quad &as\ n\to\infty. \end{aligned}$$

Proof. By integration by parts,

$$\begin{aligned} \int_{-t}^{\varepsilon_n\vee(-t)}(-s)^{H-1/2}\mathrm{d}B_2(s) &= I_{\{\varepsilon_n>-t\}}\left[\int_{-t}^{\varepsilon_n}(-s)^{H-1/2}\mathrm{d}B_2(s)\right] \\ &= I_{\{\varepsilon_n>-t\}}\Bigg[(-\varepsilon_n)^{H-1/2}B_2(\varepsilon_n) - t^{H-1/2}B_2(-t) \\ &\quad + \int_{-t}^{\varepsilon_n}(H-1/2)(-s)^{H-3/2}B_2(s)\mathrm{d}s\Bigg], \end{aligned}$$

and analogously,

$$\begin{aligned} \int_{-t}^{\varepsilon_n\vee(-t)}(-s)^{H-1/2}\mathrm{d}Z_2^{(n)}(s) = I_{\{\varepsilon_n>-t\}}\Bigg[&(-\varepsilon_n)^{H-1/2}Z_2^{(n)}(\varepsilon_n) - t^{H-1/2}Z_2^{(n)}(-t) \\ &+ \int_{-t}^{\varepsilon_n}(H-1/2)(-s)^{H-3/2}Z_2^{(n)}(s)\mathrm{d}s\Bigg]. \end{aligned}$$

We have

$$\begin{aligned} &\left|\int_{-t}^{\varepsilon_n\vee(-t)}(-s)^{H-1/2}\mathrm{d}B_2(s) - \int_{-t}^{\varepsilon_n\vee(-t)}(-s)^{H-1/2}\mathrm{d}Z_2^{(n)}(s)\right| \\ &\le I_{\{\varepsilon_n>-t\}}\|B_2 - Z_2^{(n)}\|_\infty\Bigg[(-\varepsilon_n)^{H-1/2} \\ &\quad + t^{H-1/2} + \int_{-t}^{\varepsilon_n}(1/2-H)(-s)^{H-3/2}\mathrm{d}s\Bigg] \\ &\le 2\|B_2 - Z_2^{(n)}\|_\infty n^{\beta}. \end{aligned}$$

Then,

$$J_2 \le P\left(2\|B_2 - Z_2^{(n)}\|_\infty n^\beta > C\alpha_n\right)$$
$$\le P\left(\|B_2 - Z_2^{(n)}\|_\infty > Cn^{-1/2}(\log n)^{5/2}\right) = o(n^{-q}).$$

□

Lemma 3.9. *For* $1/2 - H < \beta < 1/2$ *and each* $q > 0$,

$$J_3 = P\left(\sup_{0\le t\le T}\left|\int_{\varepsilon_n \vee(-t)}^0 [(-s)^{H-1/2} - (-s-\varepsilon_n)^{H-1/2}]\mathrm{d}B_2(s)\right| > \alpha_n\right)$$
$$= o(n^{-q}) \quad \text{as } n\to\infty.$$

Proof. By the Hölder continuity of B_2 with $0 < \gamma < H$,

$$\left|\int_{\varepsilon_n\vee(-t)}^0 [(-s)^{H-1/2} - (-s-\varepsilon_n)^{H-1/2}]\mathrm{d}B_2(s)\right|$$
$$= \left|\int_{\varepsilon_n\vee(-t)}^0 \int_{-s}^{-s-\varepsilon_n} (1/2-H)x^{H-3/2}\mathrm{d}x\mathrm{d}B_2(s)\right|$$
$$= \left|\int_0^{-\varepsilon_n-(\varepsilon_n\vee(-t))} \int_{(-t\vee\varepsilon_n)\vee(-x)}^{(-x-\varepsilon_n)\wedge 0} (1/2-H)x^{H-3/2}\mathrm{d}B_2(s)\mathrm{d}x\right|$$
$$= \left|\int_0^{-\varepsilon_n-(\varepsilon_n\vee(-t))} (1/2-H)x^{H-3/2}[B_2((-x-\varepsilon_n)\wedge 0)\right.$$
$$\left. -B_2((-t\vee\varepsilon_n)\vee(-x))]\mathrm{d}x\right|$$
$$\le (1/2-H)Y\int_0^{-\varepsilon_n-(\varepsilon_n\vee(-t))} x^{H-3/2}A_1^{1/2-\gamma}(x)\mathrm{d}x, \tag{15.19}$$

where

$$A_1(x) = |(-x-\varepsilon_n)\wedge 0 - ((-t)\vee\varepsilon_n\vee(-x))|.$$

First, if $0 \le -x-\varepsilon_n$, then $A_1(x) = |(-t)\vee(-x)|$, and if $t < x$, then

$$A_1(x) = t < x < 2x. \tag{15.20}$$

If $t \ge x$, then

$$A_1(x) = x < 2x. \tag{15.21}$$

Second, if $0 > -x-\varepsilon_n$, then

$$\begin{aligned}A_1(x) &= |-x-\varepsilon_n-((-t)\vee\varepsilon_n\vee(-x))| = |-x-\varepsilon_n-((-t)\vee\varepsilon_n)|\\ &= -x-\varepsilon_n-((-t)\vee\varepsilon_n) \le -\varepsilon_n+(t\wedge(-\varepsilon_n)).\end{aligned}$$

If $t < -\varepsilon_n$, then

$$A_1(x) \le -\varepsilon_n + t \le -2\varepsilon_n < 2x, \tag{15.22}$$

and if $t \ge -\varepsilon_n$, then

$$A_1(x) = -\varepsilon_n - \varepsilon_n \le -2\varepsilon_n < 2x. \tag{15.23}$$

From Eqs. (15.20)–(15.23) we have that $A_1(x) \le 2x$, and then from Eq. (15.19),

$$\begin{aligned}&\left|\int_{\varepsilon_n\vee(-t)}^{0}[(-s)^{H-1/2}-(-s-\varepsilon_n)^{H-1/2}]\mathrm{d}B_2(s)\right|\\ &\le (1/2-H)2^{1/2-\gamma}Y\int_0^{-\varepsilon_n-(\varepsilon_n\vee(-t))}x^{H-1-\gamma}\mathrm{d}x\\ &= \frac{(1/2-H)2^{1/2-\gamma}}{H-\gamma}Y(-\varepsilon_n-(\varepsilon_n\vee(-t)))^{H-\gamma}\\ &\le \frac{(1/2-H)2^{H-2\gamma+1/2}}{H-\gamma}Y(-\varepsilon_n)^{H-\gamma}.\end{aligned}$$

Hence

$$\begin{aligned}J_3 &\le P\left(\frac{(1/2-H)2^{H-2\gamma+1/2}}{H-\gamma}Y(-\varepsilon_n)^{H-\gamma} > \alpha_n\right)\\ &= P\left(CY > (-\varepsilon_n)^{-H+\gamma}\alpha_n\right)\\ &= P\left(CY > n^{\kappa}(\log n)^{5/2}\right)\\ &\le \frac{E(|CY|^r)}{n^{r\kappa}(\log n)^{r5/2}},\end{aligned}$$

where $\kappa = -(1/2-\beta) - \beta(H-\gamma)/(H-1/2)$. Taking γ close enough to 0 we have $0 < (1/2-H)/(1-2\gamma) < \beta < 1/2$, and then $\kappa > 0$. The result follows by analogous arguments as in proof of Lemma 3.6. □

Lemma 3.10. *For each $q > 0$ there is C such that*

$$J_4 = P\left(\sup_{0\le t\le T}\left|\int_{\varepsilon_n\vee(-t)}^{0}(-s-\varepsilon_n)^{H-1/2}\mathrm{d}B_2(s)\right.\right.$$

$$- \int_{\varepsilon_n \vee (-t)}^{0} (-s - \varepsilon_n)^{H-1/2} \mathrm{d}Z_2^{(n)}(s) \Big| > C\alpha_n \Big)$$
$$= o(n^{-q}) \quad as \ n \to \infty.$$

Proof. By integration by parts

$$\int_{\varepsilon_n \vee (-t)}^{0} (-s - \varepsilon_n)^{H-1/2} \mathrm{d}B_2(s)$$
$$= -(-(\varepsilon_n \vee (-t)) - \varepsilon_n)^{H-1/2} B_2(\varepsilon_n \vee (-t))$$
$$+ \int_{\varepsilon_n \vee (-t)}^{0} (H - 1/2)(-s - \varepsilon_n)^{H-3/2} B_2(s) \mathrm{d}s,$$

and

$$\int_{\varepsilon_n \vee (-t)}^{0} (-s - \varepsilon_n)^{H-1/2} \mathrm{d}Z_2^{(n)}(s)$$
$$= -(-(\varepsilon_n \vee (-t)) - \varepsilon_n)^{H-1/2} Z_2^{(n)}(\varepsilon_n \vee (-t))$$
$$+ \int_{\varepsilon_n \vee (-t)}^{0} (H - 1/2)(-s - \varepsilon_n)^{H-3/2} Z_2^{(n)}(s) \mathrm{d}s.$$

Then

$$\left| \int_{\varepsilon_n \vee (-t)}^{0} (-s - \varepsilon_n)^{H-1/2} \mathrm{d}B_2(s) - \int_{\varepsilon_n \vee (-t)}^{0} (-s - \varepsilon_n)^{H-1/2} \mathrm{d}Z_2^{(n)}(s) \right|$$
$$\leq \|B_2 - Z_2^{(n)}\|_\infty \Big[(-(\varepsilon_n \vee (-t)) - \varepsilon_n)^{H-1/2}$$
$$+ \int_{\varepsilon_n \vee (-t)}^{0} (1/2 - H)(-s - \varepsilon_n)^{H-3/2} \mathrm{d}s \Big]$$
$$= \|B_2 - Z_2^{(n)}\|_\infty (-\varepsilon_n)^{H-1/2}$$
$$= \|B_2 - Z_2^{(n)}\|_\infty n^{\beta}.$$

Finally,

$$J_4 \leq P\left(\|B_2 - Z_2^{(n)}\|_\infty n^{\beta} > C\alpha_n \right) = P\left(\|B_2 - Z_2^{(n)}\|_\infty > Cn^{-1/2} (\log n)^{5/2} \right)$$
$$= o(n^{-q}).$$

□

Lemma 3.11. *For each $q > 0$ there is C such that*

$$J_5 = P\left(\sup_{0\le t\le T}\left|\int_{1/a}^{0}\partial_s F_t\left(\frac{1}{v}\right)\frac{1}{v^3}B_3(v)\mathrm{d}v - \int_{1/a}^{0}\left(-\int_{1/a}^{s}\partial_s F_t\left(\frac{1}{v}\right)\frac{1}{v^3}\mathrm{d}v\right)\mathrm{d}Z_3^{(n)}(s)\right| > C\alpha_n\right) = o(n^{-q}) \quad \text{as } n\to\infty.$$

Proof. By Fubini's theorem we have

$$\int_{1/a}^{0}\left(-\int_{1/a}^{s}\partial_s F_t\left(\frac{1}{v}\right)\frac{1}{v^3}\mathrm{d}v\right)\mathrm{d}Z_3^{(n)}(s) = \int_{1/a}^{0}\partial_s F_t\left(\frac{1}{v}\right)\frac{1}{v^3}Z_3^{(n)}(v)\mathrm{d}v,$$

then, by Lemma 3.1,

$$\begin{aligned}
&\left|\int_{1/a}^{0}\partial_s F_t\left(\frac{1}{v}\right)\frac{1}{v^3}B_3(v)\mathrm{d}v - \int_{1/a}^{0}\left(-\int_{1/a}^{s}\partial_s F_t\left(\frac{1}{v}\right)\frac{1}{v^3}\mathrm{d}v\right)\mathrm{d}Z_3^{(n)}(s)\right| \\
&\le \|B_3 - Z_3^{(n)}\|_\infty \int_{1/a}^{0}\left|\partial_s F_t\left(\frac{1}{v}\right)\frac{1}{v^3}\right|\mathrm{d}v \\
&\le \|B_3 - Z_3^{(n)}\|_\infty \int_{1/a}^{0} t(3/2-H)(1/2-H)(tv+1)^{H-5/2}(-v)^{-1/2-H}\mathrm{d}v \\
&\le \|B_3 - Z_3^{(n)}\|_\infty t(3/2-H)(t/a+1)^{H-5/2}(1/2-H)\int_{1/a}^{0}(-v)^{-1/2-H}\mathrm{d}v \\
&\le \|B_3 - Z_3^{(n)}\|_\infty T(3/2-H)(T/a+1)^{H-5/2}(-1/a)^{1/2-H}.
\end{aligned}$$

Therefore,

$$J_5 \le P\left(\|B_3 - Z_3^{(n)}\|_\infty > C\alpha_n\right) = o(n^{-q}).$$

□

Lemma 3.12. *For each $q > 0$ there is C such that*

$$J_6 = P\left(\sup_{0\le t\le T}\left|\int_a^{a\vee(-t+\varepsilon_n)}F_t(s)\mathrm{d}B_2(s) - \int_a^{a\vee(-t+\varepsilon_n)}F_t(s)\mathrm{d}Z_2^{(n)}(s)\right| > C\alpha_n\right) = o(n^{-q}) \quad \text{as } n\to\infty.$$

Proof. By integration by parts,

$$\int_a^{a\vee(-t+\varepsilon_n)} F_t(s)\mathrm{d}B_2(s) = F_t(a\vee(-t+\varepsilon_n))B_2(a\vee(-t+\varepsilon_n)) - F_t(a)B_2(a) - \int_a^{a\vee(-t+\varepsilon_n)} \partial_s F_t(s)B_2(s)\mathrm{d}s,$$

and

$$\int_a^{a\vee(-t+\varepsilon_n)} F_t(s)\mathrm{d}Z_2^{(n)}(s) = F_t(a\vee(-t+\varepsilon_n))Z_2^{(n)}(a\vee(-t+\varepsilon_n)) - F_t(a)Z_2^{(n)}(a) - \int_a^{a\vee(-t+\varepsilon_n)} \partial_s F_t(s)Z_2^{(n)}(s)\mathrm{d}s.$$

Then

$$\begin{aligned}
&\left|\int_a^{a\vee(-t+\varepsilon_n)} F_t(s)\mathrm{d}B_2(s) - \int_a^{a\vee(-t+\varepsilon_n)} F_t(s)\mathrm{d}Z_2^{(n)}(s)\right| \\
&\le \|B_2 - Z_2^{(n)}\|_\infty \left[|F_t(a\vee(-t+\varepsilon_n))| + |F_t(a)| + \int_a^{a\vee(-t+\varepsilon_n)} |\partial_s F_t(s)|\mathrm{d}s\right] \\
&= \|B_2 - Z_2^{(n)}\|_\infty \Big[(-t-(a\vee(-t+\varepsilon_n)))^{H-1/2} - (-(a\vee(-t+\varepsilon_n)))^{H-1/2} \\
&\quad +(-t-a)^{H-1/2} - (-a)^{H-1/2} \\
&\quad + \int_a^{a\vee(-t+\varepsilon_n)} (1/2-H)[(-t-s)^{H-3/2} - (-s)^{H-3/2}]\mathrm{d}s\Big] \\
&= \|B_2 - Z_2^{(n)}\|_\infty 2\Big[(-t-(a\vee(-t+\varepsilon_n)))^{H-1/2} - (-(a\vee(-t+\varepsilon_n)))^{H-1/2}\Big] \\
&\le \|B_2 - Z_2^{(n)}\|_\infty 2((-\varepsilon_n)^{H-1/2} + (-T-a)^{H-1/2}).
\end{aligned}$$

Hence the result follows. □

Lemma 3.13. *For each $q > 0$ there is C such that*

$$\begin{aligned}
J_7 = P\Bigg(\sup_{0\le t\le T}\Bigg| &I_{\{-\varepsilon_n\le t\}} \int_{a\vee(-t+\varepsilon_n)}^{-t} F_{t+\varepsilon_n}(s)\mathrm{d}B_2(s) \\
&- I_{\{-\varepsilon_n\le t\}} \int_{a\vee(-t+\varepsilon_n)}^{-t} F_{t+\varepsilon_n}(s)\mathrm{d}Z_2^{(n)}(s)\Bigg| > C\alpha_n\Bigg) \\
= o(n^{-q}) \quad &\textit{as } n\to\infty.
\end{aligned}$$

Proof. By integration by parts,

$$
\begin{aligned}
& I_{\{-\varepsilon_n \le t\}} \left| \int_{a \vee (-t+\varepsilon_n)}^{-t} F_{t+\varepsilon_n}(s) \mathrm{d} B_2(s) - \int_{a \vee (-t+\varepsilon_n)}^{-t} F_{t+\varepsilon_n}(s) \mathrm{d} Z_2^{(n)}(s) \right| \\
&= I_{\{-\varepsilon_n \le t\}} \left| F_{t+\varepsilon_n}(-t)(B_2(-t) - Z_2^{(n)}(-t)) \right. \\
&\quad - F_{t+\varepsilon_n}(a \vee (-t+\varepsilon_n))(B_2(a \vee (-t+\varepsilon_n)) - Z_2^{(n)}(a \vee (-t+\varepsilon_n))) \\
&\quad \left. - \int_{a \vee (-t+\varepsilon_n)}^{-t} \partial_s F_{t+\varepsilon_n}(s)(B_2(s) - Z_2^{(n)}(s)) \mathrm{d}s \right| \\
&\le I_{\{-\varepsilon_n \le t\}} || B_2 - Z_2^{(n)} ||_\infty \Bigg(|F_{t+\varepsilon_n}(-t)| + |F_{t+\varepsilon_n}(a \vee (-t+\varepsilon_n))| \\
&\quad + \int_{a \vee (-t+\varepsilon_n)}^{-t} |\partial_s F_{t+\varepsilon_n}(s)| \mathrm{d}s \Bigg) \\
&= I_{\{-\varepsilon_n \le t\}} || B_2 - Z_2^{(n)} ||_\infty \Bigg((-\varepsilon_n)^{H-1/2} - (t)^{H-1/2} \\
&\quad + (-t - \varepsilon_n - (a \vee (-t+\varepsilon_n)))^{H-1/2} - (-(a \vee (-t+\varepsilon_n)))^{H-1/2} \\
&\quad + (1/2 - H) \int_{a \vee (-t+\varepsilon_n)}^{-t} [(-s - t - \varepsilon_n)^{H-3/2} - (-s)^{H-3/2}] \mathrm{d}s \Bigg) \\
&= I_{\{-\varepsilon_n \le t\}} || B_2 - Z_2^{(n)} ||_\infty 2 \left((-\varepsilon_n)^{H-1/2} - t^{H-1/2} \right) \\
&\le 2 \| B_2 - Z_2^{(n)} \|_\infty (-\varepsilon_n)^{H-1/2},
\end{aligned}
$$

and we have the result similarly as Lemma 3.10. □

Lemma 3.14. *For* $1/2 - H < \beta < 1/2$ *and each* $q > 0$,

$$
J_8 = P \left(\sup_{0 \le t \le T} \left| I_{\{t < -\varepsilon_n\}} \int_{a \vee (-t+\varepsilon_n)}^{-t} F_t(s) \mathrm{d} B_2(s) \right| > C \alpha_n \right) = o(n^{-q}) \quad \text{as } n \to \infty.
$$

Proof. By the Hölder continuity of B_2 with $0 < \gamma < H$,

$$
\begin{aligned}
& \left| I_{\{t < -\varepsilon_n\}} \int_{a \vee (-t+\varepsilon_n)}^{-t} F_t(s) \mathrm{d} B_2(s) \right| \\
&= \left| I_{\{t < -\varepsilon_n\}} \int_{a \vee (-t+\varepsilon_n)}^{-t} \int_{-t-s}^{-s} (1/2 - H) x^{H-3/2} \mathrm{d}x \mathrm{d} B_2(s) \right| \\
&= \left| I_{\{t < -\varepsilon_n\}} \int_0^{-(a \vee (-t+\varepsilon_n))} \int_{(-t-x) \vee a \vee (-t+\varepsilon_n)}^{(-x) \wedge (-t)} (1/2 - H) x^{H-3/2} \mathrm{d} B_2(s) \mathrm{d}x \right|
\end{aligned}
$$

$$= \left| I_{\{t<-\varepsilon_n\}} \int_0^{-(a\vee(-t+\varepsilon_n))} (1/2-H)x^{H-3/2}[B_2((-x)\wedge(-t)) \right.$$

$$\left. -B_2((-t-x)\vee a\vee(-t+\varepsilon_n))]\mathrm{d}x \right|$$

$$\leq (1/2-H)YI_{\{t<-\varepsilon_n\}} \int_0^{-(a\vee(-t+\varepsilon_n))} x^{H-3/2}(A_2(x))^{1/2-\gamma}\mathrm{d}x \tag{15.24}$$

where

$$A_2(x) = |((-x)\wedge(-t)) - ((-t-x)\vee a\vee(-t+\varepsilon_n))|.$$

First, if $-x < -t$ and $-t-x < -t+\varepsilon_n$, then

$$A_2(x) = -x + (-a\wedge(t-\varepsilon_n)) < t < x. \tag{15.25}$$

If $-x < -t$ and $-t-x \geq -t+\varepsilon_n$, then

$$A_2(x) = -x + ((t+x)\wedge(-a)) < t < x. \tag{15.26}$$

Second, if $-x \geq -t$ and $-t-x < -t+\varepsilon_n$, then

$$A_2(x) = -t + (-a\wedge(t-\varepsilon_n)) < -\varepsilon_n < x. \tag{15.27}$$

If $-x \geq -t$ and $-t-x \geq -t+\varepsilon_n$, then

$$A_2(x) = -t + ((t+x)\wedge(-a)) < x. \tag{15.28}$$

From Eqs. (15.25)–(15.28) we have that $A_2(x) \leq x$ and then by Eq. (15.24),

$$\left| I_{\{t<-\varepsilon_n\}} \int_{a\vee(-t+\varepsilon_n)}^{-t} F_t(s)\mathrm{d}B_2(s) \right|$$

$$\leq (1/2-H)YI_{\{t<-\varepsilon_n\}} \int_0^{-(a\vee(-t+\varepsilon_n))} x^{H-\gamma-1}\mathrm{d}x$$

$$\leq \frac{1/2-H}{H-\gamma} YI_{\{t<-\varepsilon_n\}}(t-\varepsilon_n)^{H-\gamma}$$

$$\leq \frac{1/2-H}{H-\gamma} 2^{H-\gamma}Y(-\varepsilon_n)^{H-\gamma}.$$

Proceeding similarly as in Lemma 3.9 we have the result. □

Lemma 3.15. *For* $1/2-H < \beta < 1/2$ *and each* $q > 0$,

$$J_9 = P\left(\sup_{0\le t\le T}\left|I_{\{-\varepsilon_n\le t\}}\int_{a\vee(-t+\varepsilon_n)}^{-t}[F_t(s)-F_{t+\varepsilon_n}(s)]\mathrm{d}B_2(s)\right| > C\alpha_n\right)$$
$$= o(n^{-q}) \quad as \;\; n\to\infty.$$

Proof. By the Hölder continuity of B_2 with $0<\gamma<H$,

$$\left|I_{\{-\varepsilon_n\le t\}}\int_{a\vee(-t+\varepsilon_n)}^{-t}[F_t(s)-F_{t+\varepsilon_n}(s)]\mathrm{d}B_2(s)\right|$$
$$= \left|I_{\{-\varepsilon_n\le t\}}\int_{a\vee(-t+\varepsilon_n)}^{-t}\int_{-t-s}^{-t-s-\varepsilon_n}(1/2-H)x^{H-3/2}\mathrm{d}x\mathrm{d}B_2(s)\right|$$
$$= \left|I_{\{-\varepsilon_n\le t\}}\int_0^{-t-\varepsilon_n+((-a)\wedge(t-\varepsilon_n))}(1/2-H)x^{H-3/2}[B_2((-t-x-\varepsilon_n)\wedge(-t))\right.$$
$$\left.-B_2((-t-x)\vee a\vee(-t+\varepsilon_n))]\mathrm{d}x\right|$$
$$\le (1/2-H)YI_{\{-\varepsilon_n\le t\}}\int_0^{-t-\varepsilon_n+((-a)\wedge(t-\varepsilon_n))}x^{H-3/2}(A_3(x))^{1/2-\gamma}\mathrm{d}x, \quad (15.29)$$

where

$$A_3(x) = |(-t-x-\varepsilon_n)\wedge(-t)-((-t-x)\vee(-t+\varepsilon_n)\vee a)| \le x.$$

Then, by Eq. (15.29),

$$\left|I_{\{-\varepsilon_n\le t\}}\int_{a\vee(-t+\varepsilon_n)}^{-t}[F_t(s)-F_{t+\varepsilon_n}(s)]\mathrm{d}B_2(s)\right|$$
$$\le (1/2-H)YI_{\{-\varepsilon_n\le t\}}\int_0^{-t-\varepsilon_n+((-a)\wedge(t-\varepsilon_n))}x^{H-1-\gamma}\mathrm{d}x$$
$$\le Y\frac{1/2-H}{H-\gamma}(-t-\varepsilon_n+((-a)\wedge(t-\varepsilon_n)))^{H-\gamma}$$
$$\le Y\frac{1/2-H}{H-\gamma}2^{H-\gamma}(-\varepsilon_n)^{H-\gamma}.$$

Proceeding similary as in Lemma 3.9 we have the result. □

***Proof of Theorem 2.2 for* H < 1/2:** From Eqs. (15.4), (15.8), and (15.16) we obtain

$$Y(t) = C\left\{-\int_{-t}^{0}(-s)^{H-1/2}dB(s)+\int_a^{-t}F_t(s)\mathrm{d}B(s)+\int_{-\infty}^{a}F_t(s)\mathrm{d}B(s)\right\}$$

$$
\begin{aligned}
&= C\left\{-\int_{-t}^{\varepsilon_n\vee(-t)}(-s)^{H-1/2}\mathrm{d}B_2(s)\right.\\
&\quad -\int_{\varepsilon_n\vee(-t)}^{0}[(-s)^{H-1/2}-(-s-\varepsilon_n)^{H-1/2}]\mathrm{d}B_2(s)\\
&\quad -\int_{\varepsilon_n\vee(-t)}^{0}(-s-\varepsilon_n)^{H-1/2}\mathrm{d}B_2(s)+\int_a^{a\vee(-t+\varepsilon_n)}F_t(s)\mathrm{d}B_2(s)\\
&\quad +I_{\{t<-\varepsilon_n\}}\int_{a\vee(-t+\varepsilon_n)}^{-t}F_t(s)\mathrm{d}B_2(s)\\
&\quad +I_{\{-\varepsilon_n\le t\}}\int_{a\vee(-t+\varepsilon_n)}^{-t}[F_t(s)-F_{t+\varepsilon_n}(s)]\mathrm{d}B_2(s)\\
&\quad +I_{\{-\varepsilon_n\le t\}}\int_{a\vee(-t+\varepsilon_n)}^{-t}F_{t+\varepsilon_n}(s)\mathrm{d}B_2(s)+F_t(a)B_2(a)\\
&\quad \left.-\int_{1/a}^{0}\partial_s F_t\left(\frac{1}{v}\right)\frac{1}{v^3}B_3(v)\mathrm{d}v\right\},
\end{aligned}
$$

and we have the result similarly as the case $H > 1/2$. □

Acknowledgements This work was done with support of CONACyT grant 98998.

References

1. Bardina, X., Bascompte, D.: Weak convergence towards two independent Gaussian processes from a unique Poisson process. Collect. Math. **61**(2), 191–204 (2010)
2. Biagini, F., Hu, Y., Øksendal, B., Zhang, T.: Stochastic Calculus for Fractional Brownian Motion and Applications. Springer, London (2008)
3. Bojdecki, T., Gorostiza, L.G., Talarczyk, A.: Sub-fractional Brownian motion and its relation to occupation times. Stat. Prob. Lett. **69**, 405–419 (2004)
4. Bojdecki, T., Gorostiza, L.G., Talarczyk, A.: Occupation times of branching systems with initial inhomogeneous Poisson states and related superprocesses. Elec. J. Probab. **14**, 1328–1371 (2009)
5. Bojdecki, T., Gorostiza, L.G., Talarczyk, A.: Particle systems with quasi-homogeneous initial states and their occupation time fluctuations. Elect. Commun. Probab. **15**, 191–202 (2010)
6. Bojdecki, T., Talarczyk, A.: Particle picture interpretation of some Gaussian processes related to fractional Brownian motion. Stoch. Proc. Appl. **122**(5), 2134–2154 (2012)
7. Doukhan, P., Oppenheim, G., Taqqu, M.S.: Theory and Applications of Long-Range Dependence. Birkhäuser, Boston (2003)
8. Dzhaparidze, K.O., van Zanten, J.H.: A series expansion of fractional Brownian motion. Probab. Theory Relat. Fields **130**, 39–55 (2004)
9. El-Nouty, C.: The lower classes of the sub-fractional Brownian motion. In: Stochastic Differential Equations and Processes. Springer Proceedings in Mathematics, vol. 7, pp. 179–196 (2012)

10. Garzón, J., Gorostiza, L.G., León, J.A.: A strong uniform approximation of fractional Brownian motion by means of transport processes. Stoch. Proc. Appl. **119**, 3435–3452 (2009)
11. Garzón, J., Gorostiza, L.G., León, J.A.: Approximations of fractional stochastic differential equations by means of transport processes. Comm. Stoch. Anal. **5**, 443–456 (2011)
12. J. Garzón, S. Torres, C.A. Tudor, A strong convergence to the Rosenblatt process. J. Math. Anal. Appl. **391**(2), 630–647 (2012)
13. Gorostiza, L.G., Griego, R.J.: Rate of convergence of uniform transport processes to Brownian motion and application to stochastic integrals. Stochastics **3**, 291–303 (1980)
14. D. Harnett, D. Nualart, Weak convergence of the Stratonovich integral with respect to a class of Gaussian processes. Stoch. Proc. Appl. **122**(10), 3460–3505 (2012)
15. Lei, P., Nualart, D.: A decomposition of the bifractional Brownian motion and some applications. Stat. Prob. Lett. **79**, 619–624 (2009)
16. Y. Li, Y. Xiao, Occupation time fluctuations of weakly degenerate branching systems. J. Theoret. Probab. **25**(14), 1119–1152 (2012)
17. Liu, J., Li, L., Yan, L.: Sub-fractional model for credit risk pricing. Int. J. Nonlinear Sci. Numerical Simulation **11**, 231–236 (2010)
18. Liu, J., Yan, L., Peng, Z., Wang, D.: Remarks on confidence intervals for self-similarity parameter of a subfractional Brownian motion. Abstract Appl. Anal. **2012**, 14 (2012). article ID 804942
19. J. Liu, L. Yan, Remarks on asymptotic behavior of weighted quadratic variation of subfractional Brownian motion. J. Korean Statist. Soc. **41**(2), 177–187 (2012)
20. Mendy, I.: On the local time of sub-fractional Brownian motion. Annales Mathématiques Blaise Pascal **17**, 357–374 (2010)
21. Mishura, Y.S.: Stochastic Calculus for Fractional Brownian Motion and Related Processes. Springer, New York (2008)
22. Norvaiša, R.: A complement to Gladyshev's theorem. Lithuanian Math. J. **51**, 26–35 (2011)
23. Nualart, D.: Stochastic integration with respect to fractional Brownian motion and applications. In: González-Barrios, J.M., León, J.A., Meda, A. (eds.) Stochastic Models. Contemporary Mathematics, vol. 336, pp. 3–39. American Mathematical Society, Providence (2003)
24. Nualart, D., Tindel, S.: A construction of the rough path above fractional Brownian motion using Volterra's representation. Ann. Probab. **39**, 1061–1096 (2011)
25. Ruiz de Chávez, J., Tudor, C.: A decomposition of sub-fractional Brownian motion. Math. Reports **11**(61,1) 67–74 (2009)
26. Samorodnitsky, G., Taqqu, M.S.: Stable Non-Gaussian Random Processes. Stochastic Models with Infinite Variance. Chapman & Hall, New York (1994)
27. Shen, G.: Necessary and sufficient condition for the smoothness of intersection local time of subfractional Brownian motions. J. Inequalities Appl. **139**, 1–16 (2011)
28. G. Shen, C. Chen, Stochastic integration with respect to the sub-fractional Brownian motion. Statist. Probab. Lett. **82**(2), 240–251 (2012)
29. Shen, G., Chen, C., Yan, L.: Remarks on sub-fractional Bessel processes. Acta Mathematica Scientia Ser. B **31**(5), 1860–1876 (2011)
30. Shen, G., Yan, L.: Remarks on an integral functional driven by sub-fractional Brownian motion. J. Korean Statist. Soc. **40**(3), 337–346 (2011)
31. Słomiński, L., Ziemkiewicz, B.: On weak approximations of integral with respect to fractional Brownian motion. Stat. Prob. Lett. **79**, 543–552 (2009)
32. Swanson, J.: Fluctuations of the empirical quantiles of independent Brownian motions. Stoch. Proc. Appl. **121**, 479–514 (2011)
33. Tudor, C.: Some aspects of stochastic calculus for the sub-fractional Brownian motion. Analele Universităţii Bucureşti, Matematicaă **LVII**, 199–230 (2008)
34. Tudor, C.: Inner product spaces of integrands associated to sub-fractional Brownian motion. Stat. Probab. Lett. **78**, 2201–2209 (2008)
35. Tudor, C.: Sub-fractional Brownian motion as a model in finance. University of Bucharest, Romania (2008)

36. Tudor, C.: Some properties of the sub-fractional Brownian motion. Stochastics **79**, 431–448 (2007)
37. Tudor, C.: On the Wiener integral with respect to a sub-fractional Brownian motion on an interval. J. Math. Anal. Appl. **351**, 456–468 (2009)
38. Tudor, C.: Berry-Esséen bounds and almost sure CLT for the quadratic variation of the sub-fractional Brownian motion. J. Math. Analysis Appl. **375**, 667–676 (2011)
39. Yan, L., Shen, G.: On the collision local time of sub-fractional Brownian motions. Stat. Prob. Lett. **80**, 296–308 (2010)
40. Yan, L., Shen, G., He, K.: Itô's formula for a sub-fractional Brownian motion. Commun. Stoch. Analysis **5**, 135–159 (2011)

Chapter 16
Malliavin Calculus for Fractional Heat Equation

Aurélien Deya and Samy Tindel

Dedicated to David Nualart on the occasion of his 60th birthday

Abstract In this article, we give some existence and smoothness results for the law of the solution to a stochastic heat equation driven by a finite dimensional fractional Brownian motion with Hurst parameter $H > 1/2$. Our results rely on recent tools of Young integration for convolutional integrals combined with stochastic analysis methods for the study of laws of random variables defined on a Wiener space.

Keywords Fractional Brownian motion • Heat equation • Malliavin calculus

MSC Subject Classifications 2000: Primary 60H35; Secondary 60H07, 60H10, 65C30.

Received 9/2/2011; Accepted 6/1/2012; Final 6/26/2012

1 Introduction

The definition and resolution of evolution type PDEs driven by general Hölder continuous signals have experienced tremendous progresses during the last past years. When the Hölder regularity of the driving noise is larger than $1/2$, this has been achieved thanks to Young integrals [9] or fractional integration [11] techniques. The more delicate issue of a Hölder exponent smaller than $1/2$ has to be handled

A. Deya • S. Tindel (✉)
Institut Élie Cartan Nancy, Université de Nancy 1, B.P. 239,
54506 Vandœuvre-lès-Nancy Cedex, France
e-mail: deya@iecn.u-nancy.fr; tindel@iecn.u-nancy.fr

F. Viens et al. (eds.), *Malliavin Calculus and Stochastic Analysis: A Festschrift in Honor of David Nualart*, Springer Proceedings in Mathematics & Statistics 34,
DOI 10.1007/978-1-4614-5906-4_16, © Springer Science+Business Media New York 2013

thanks to rough paths techniques, either by smart transformations allowing to use limiting arguments [3, 4, 8] or by an adaptation of the rough paths formalism to evolution equations [6, 7, 10]. Altogether, those contributions yield a reasonable definition of rough parabolic PDEs, driven at least by a finite dimensional signal.

With those first results in hand, a natural concern is to get a better understanding of the processes obtained as solutions to stochastic PDEs driven by rough signals. This important program includes convergence of numerical schemes (see [5] for a result in this direction), ergodic properties, and a thorough study of the law of those processes. This article makes a first step towards the last of these items.

Indeed, we shall consider here a simple case of rough evolution PDEs and see what kind of result might be obtained as far as densities of the solution are concerned. More specifically, we focus on the following mild heat equation on $(0,1)$:

$$Y_t = S_t\varphi + \int_0^t S_{t-u}(F_i(Y)_u)\,\mathrm{d}B_u^i, \qquad t \in [0,T], \tag{16.1}$$

where $T > 0$ is a finite horizon, S_t stands for the heat semigroup associated with Dirichlet boundary conditions, φ is a smooth enough initial condition, $F_i : L^2(0,1) \to L^2(0,1)$ and $B : [0,T] \to \mathbb{R}^d$ is a d-dimensional fractional Brownian motion with Hurst parameter $H > 1/2$. For this equation, we obtain the following results:

1. Existence of a density for the random variable $Y_t(\xi)$ for any $t \in (0,T]$ and $\xi \in (0,1)$, when the F_i's are rather general Nemytskii operators $F_i(\varphi)(\xi) := f_i(\varphi(\xi))$. See Theorem 3.2 for a precise statement.
2. When the F_i's are defined through some regularizing kernel (see Hypothesis 4.1), we obtain that the density of $Y_t(\xi)$ is smooth. This will be the content of Theorem 4.2.

To the best of our knowledge, these are the first density results for solutions to nonlinear PDEs driven by fractional Brownian motion. Let us point out that we could have obtained the same kind of results for a more general class of equations (operator under divergence form, general domain $D \subset \mathbb{R}^n$, drift term, Gaussian process as driving noise). We prefer however to stick to the simple case of the fBm-driven stochastic heat equation for the sake of readability and conciseness.

Our main results will obviously be based on a combination of pathwise estimates for integrals driven by rough signals and Malliavin calculus tools. In particular, a major part of our effort will be dedicated to the differentiation of the solution to Eq. (16.1) with respect to the driving noise B and to a proper estimate of the derivative. Since the equations for derivatives are always of linear type they lead to exponential type estimates, which are always a delicate issue. This is where we shall consider some regularizing vector fields F_i in Eq. (16.1), and proceed to a careful estimation procedure (see Sect. 4.1). It should also be noticed at this point that the basis of our stochastic analysis tools is contained in the celebrated book [12] by Nualart, plus the classical reference [13] as far as equations driven by fBm are concerned.

Here is how our article is structured: Section 2 is devoted to recall basic facts on both pathwise noisy evolution equations and Malliavin calculus for fractional Brownian motion. We differentiate the solution to Eq. (16.1) and obtain the existence of the density at Sect. 3. Finally, further estimates on the Malliavin derivative and smoothness of the density are derived at Sect. 4.

Throughout this article, we will use the generic notation c to refer to the constants that only depend on nonsignificant parameters. The constants which are to play a more specific role in our reasoning will be labeled $c_1, c_2, \ldots$

For any $k \in \mathbb{N}$, we will denote by $\mathcal{C}^{k,\mathbf{b}}(\mathbb{R};\mathbb{R})$ the space of functions on $\mathbb{R}$ which are k-times differentiable with bounded derivatives. For any $\gamma \in (0,1)$, $\mathcal{C}^\gamma = \mathcal{C}^\gamma([0,T];\mathbb{R}^d)$ will stand for the set of (d-dimensional) γ-Hölder paths on $[0,T]$.

2 Setting

One of the technical advantages of dealing with the simple case of a stochastic heat equation on $(0,1)$ is a simplification in the functional analysis setting based on rather elementary Fourier series considerations (notice in particular that the L^p considerations of [7] can be avoided). We shall first detail this setting, and then recall some basic facts on equations driven by noisy signals and fractional Brownian motion. Throughout the section, we assume that a (finite) horizon T has been fixed for the equation.

2.1 Fractional Sobolev Spaces

As mentioned above, we are working here with the heat equation in the Hilbert space $\mathcal{B} := L^2(0,1)$ with Dirichlet boundary conditions. The Laplace operator Δ on $\mathcal{B}$ can be diagonalized in the orthonormal basis

$$e_n(\xi) := \sqrt{2}\sin(\pi n \xi) \;\; (n \in \mathbb{N}^*), \quad \text{with eigenvalues } \lambda_n := \pi^2 n^2.$$

We shall denote by $(y^n)_n$ the (Fourier) decomposition of any function $y \in \mathcal{B}$ on this orthonormal basis.

Sobolev spaces based on $\mathcal{B}$ are then easily characterized by means of Fourier coefficients. We label their definition for further use:

Definition 2.1. For any $\alpha \geq 0$, we denote by $\mathcal{B}_\alpha$ the fractional Sobolev space of order α based on $\mathcal{B}$, defined by

$$\mathcal{B}_\alpha := \left\{ y \in L^2(0,1) \colon \sum_{n=1}^{\infty} \lambda_n^{2\alpha} (y^n)^2 < \infty \right\}. \tag{16.2}$$

This space is equipped with its natural norm $\|y\|_{\mathcal{B}_\alpha}^2 := \|\Delta^\alpha y\|_{\mathcal{B}}^2 = \sum_{n=1}^{\infty} \lambda_n^{2\alpha} (y^n)^2$. We also set $\mathcal{B}_\infty = \mathcal{C}(0,1)$.

The above-defined fractional Sobolev spaces enjoy the following classical properties (see [1, 15]):

Proposition 2.1. *Let $\mathcal{B}_\alpha, \mathcal{B}_\infty$ be the Sobolev spaces introduced at Definition 2.1. Then the following hold true:*

- Sobolev inclusions: *If $\alpha > 1/4$, then we have the continuous embedding*

$$\mathcal{B}_\alpha \subset \mathcal{B}_\infty. \tag{16.3}$$

- Algebra: *If $\alpha > 1/4$, then $\mathcal{B}_\alpha$ is a Banach algebra with respect to pointwise multiplication, or in other words*

$$\|\varphi \cdot \psi\|_{\mathcal{B}_\alpha} \leq \|\varphi\|_{\mathcal{B}_\alpha} \|\psi\|_{\mathcal{B}_\alpha}. \tag{16.4}$$

- Composition: *If $0 \leq \alpha < 1/2$, $\varphi \in \mathcal{B}_\alpha$ and $f : \mathbb{R} \to \mathbb{R}$ belongs to $C^{1,b}$, then $f(\varphi) \in \mathcal{B}_\alpha$ and*

$$\|f(\varphi)\|_{\mathcal{B}_\alpha} \leq c_f \left\{1 + \|\varphi\|_{\mathcal{B}_\alpha}\right\}. \tag{16.5}$$

Here, $f(\varphi)$ is naturally understood as $f(\varphi)(\xi) := f(\varphi(\xi))$.

Let now S_t be the heat semigroup associated with Δ, and notice that if an element $y \in L^2(0,1)$ can be decomposed as $y = \sum_{n\geq 1} y^n e_n$, then $S_t y = \sum_{n\geq 1} e^{-\lambda_n t} y^n e_n$. The general theory of fractional powers of operators provides us with sharp estimates for the semigroup S_t (see for instance [14]):

Proposition 2.2. *The heat semigroup S_t satisfies the following properties:*

- Contraction: *For all $t \geq 0$, $\alpha \geq 0$, S_t is a contraction operator on $\mathcal{B}_\alpha$.*
- Regularization: *For all $t \in (0, T]$, $\alpha \geq 0$, S_t sends $\mathcal{B}$ on $\mathcal{B}_\alpha$ and*

$$\|S_t \varphi\|_{\mathcal{B}_\alpha} \leq c_{\alpha,T}\, t^{-\alpha} \|\varphi\|_{\mathcal{B}}. \tag{16.6}$$

- Hölder regularity: *For all $t \in (0, T]$, $\varphi \in \mathcal{B}_\alpha$,*

$$\|S_t \varphi - \varphi\|_{\mathcal{B}} \leq c_{\alpha,T}\, t^{\alpha} \|\varphi\|_{\mathcal{B}_\alpha}. \tag{16.7}$$

2.2 *Young Convolutional Integrals*

The stochastic integrals involved in Eq. (16.1) will all be understood in the Young sense. In order to define them properly, let us first introduce some notation concerning Hölder type spaces in time. To begin with, for any $\alpha \geq 0$ and any subinterval $I \subset [0, T]$, set $\mathcal{C}^0(I; \mathcal{B}_\alpha)$ for the space of continuous $\mathcal{B}_\alpha$-valued functions on I, equipped with the supremum norm. Then Hölder spaces of $\mathcal{B}_\alpha$-valued functions can be defined as follows: for $\kappa \in (0, 1)$, set

$$\mathcal{C}^\kappa(I; \mathcal{B}_\alpha) := \left\{ y \in \mathcal{C}^0(I; \mathcal{B}_\alpha) : \sup_{s<t\in I} \frac{\|y_t - y_s\|_{\mathcal{B}_\alpha}}{|t-s|^\kappa} < \infty \right\}.$$

Observe now that the definition of our stochastic integrals weighted by the heat semigroup will require the introduction of a small variant of those Hölder spaces (see [7, 10] for further details): we define $\hat{\mathcal{C}}^\kappa(I;\mathcal{B}_\alpha)$ as

$$\hat{\mathcal{C}}^\kappa(I;\mathcal{B}_\alpha) := \left\{ y \in \mathcal{C}^0(I;\mathcal{B}_\alpha) : \sup_{s<t\in I} \frac{\|y_t - S_{t-s}\, y_s\|_{\mathcal{B}_\alpha}}{|t-s|^\kappa} < \infty \right\}.$$

In order to avoid confusion, the natural norms on the spaces $\mathcal{C}^\kappa(I;\mathcal{B}_\alpha)$, $\hat{\mathcal{C}}^\kappa(I;\mathcal{B}_\alpha)$ are respectively denoted by $\mathcal{N}[\cdot;\mathcal{C}^\kappa(I;\mathcal{B}_\alpha)]$, $\mathcal{N}[\cdot;\hat{\mathcal{C}}^\kappa(I;\mathcal{B}_\alpha)]$, etc. For the sake of conciseness, we shall often write $\mathcal{C}^\kappa(\mathcal{B}_\alpha)$ (resp. $\hat{\mathcal{C}}^\kappa(\mathcal{B}_\alpha)$) instead of $\mathcal{C}^\kappa([0,T];\mathcal{B}_\alpha)$ (resp. $\hat{\mathcal{C}}^\kappa([0,T];\mathcal{B}_\alpha)$). We also need to introduce a family of spaces $\hat{\mathcal{C}}^{0,\kappa}(I;\mathcal{B}_\kappa)$ in the following way:

Lemma 2.1. *For any $\kappa \in (0,1)$ and any subinterval $I \subset [0,T]$, let $\hat{\mathcal{C}}^{0,\kappa}(I;\mathcal{B}_\kappa)$ be the space associated with the norm*

$$\mathcal{N}[\cdot;\hat{\mathcal{C}}^{0,\kappa}(I;\mathcal{B}_\kappa)] := \mathcal{N}[\cdot;\mathcal{C}^0(I;\mathcal{B}_\kappa)] + \mathcal{N}[\cdot;\hat{\mathcal{C}}^\kappa(I;\mathcal{B}_\kappa)].$$

Then the following continuous embedding holds true:

$$\hat{\mathcal{C}}^{0,\kappa}(I;\mathcal{B}_\kappa) \subset \mathcal{C}^\kappa(I;\mathcal{B}). \tag{16.8}$$

More generally, for every $\lambda \geq \kappa$,

$$\mathcal{N}[y;\mathcal{C}^\kappa(I;\mathcal{B})] \leq \mathcal{N}[y;\hat{\mathcal{C}}^\kappa(I;\mathcal{B}_\lambda)] + c_\lambda\,|I|^{\lambda-\kappa}\,\mathcal{N}[y;\mathcal{C}^0(I;\mathcal{B}_\lambda)]. \tag{16.9}$$

Proof. Indeed, owing to Eq. (16.7), one has, for every $s < t \in I$,

$$\|y_t - y_s\|_{\mathcal{B}} \leq \|y_t - S_{t-s} y_s\|_{\mathcal{B}} + \|(S_{t-s} - \mathrm{Id}) y_s\|_{\mathcal{B}} \leq \|y_t - S_{t-s} y_s\|_{\mathcal{B}_\lambda} + c_\lambda\,|t-s|^\lambda\,\|y_s\|_{\mathcal{B}_\lambda}.$$

□

With those definitions in hand, the following proposition (borrowed from [7]) will be invoked in the sequel in order to give a meaning to our stochastic integrals weighted by the heat semigroup:

Proposition 2.3. *Consider a γ-Hölder real-valued function x defined on $[0,T]$. Let $I = [\ell_1, \ell_2]$ be a subinterval of $[0,T]$ and fix $\kappa \in [0,\gamma]$ such that $\gamma + \kappa > 1$. Suppose that $z \in \mathcal{C}^0(I;\mathcal{B}_\lambda) \cap \mathcal{C}^\kappa(I;\mathcal{B}_{\lambda-\alpha})$ for some parameters $\lambda \geq 0$, $0 \leq \alpha \leq \min(\kappa,\lambda)$. Then, for every $s < t \in I$, the convolutional Riemann sum*

$$\sum_{t_k \in \Pi} S_{t-t_{k+1}} z_{t_{k+1}} \left[x_{t_{k+1}} - x_{t_k} \right]$$

converges in $\mathcal{B}_\lambda$ as the mesh of the partition $\Pi := \{s = t_0 < t_1 < \ldots < t_n = t\}$ tends to 0, and we denote the limit by $\int_s^t S_{t-u} z_u \, dx_u$.

Moreover, for every $\varphi \in \mathcal{B}_\lambda$*, there exists a unique path* $y \in \hat{\mathcal{C}}^\gamma(I;\mathcal{B}_\lambda)$ *such that* $y_{\ell_1} = \varphi$ *and* $y_t - S_{t-s}y_s = \int_s^t S_{t-u}(z_u)\,\mathrm{d}x_u$ *if* $s < t \in I$*. For this function, the following estimate holds:*

$$\mathcal{N}[y;\hat{\mathcal{C}}^\gamma(I;\mathcal{B}_\lambda)] \leq c\|x\|_\gamma \left\{\mathcal{N}[z;\mathcal{C}^0(I;\mathcal{B}_\lambda)] + |I|^{\kappa-\alpha}\,\mathcal{N}[z;\mathcal{C}^\kappa(I;\mathcal{B}_{\lambda-\alpha})]\right\}, \tag{16.10}$$

for some constant c that only depends on $(\gamma,\kappa,\lambda,\alpha)$.

2.3 Malliavin Calculus Techniques

This section is devoted to present the Malliavin calculus setting which we shall work in, having in mind the differentiability properties of the solution to Eq. (16.1).

2.3.1 Wiener Space Associated to fBm

Let us first be more specific about the probabilistic setting in which we will work. For some fixed $H \in (1/2, 1)$, we consider $(\Omega, \mathcal{F}, \mathbf{P})$ the canonical probability space associated with the fractional Brownian motion with Hurst parameter H. That is, $\Omega = \mathcal{C}_0([0,T];\mathbb{R}^d)$ is the Banach space of continuous functions vanishing at 0 equipped with the supremum norm, $\mathcal{F}$ is the Borel sigma-algebra and $\mathbf{P}$ is the unique probability measure on Ω such that the canonical process $B = \{B_t,\ t \in [0,T]\}$ is a d-dimensional fractional Brownian motion with Hurst parameter H, with covariance function

$$\mathbf{E}\big[B_t^i\, B_s^j\big] = \frac{1}{2}\left(t^{2H} + s^{2H} - |t-s|^{2H}\right)\mathbf{1}_{(i=j)}, \qquad s,t \in [0,T]. \tag{16.11}$$

In particular, the paths of B are almost surely γ-Hölder continuous for all $\gamma \in (0,H)$.

Consider then a fixed parameter $H > 1/2$, and let us start by briefly describing the abstract Wiener space introduced for Malliavin calculus purposes (for a more general and complete description, we refer the reader to [13, Sect. 3]).

Let $(e_1,\ldots,e_d)$ be the canonical basis of $\mathbb{R}^d$, $\mathcal{E}$ be the set of $\mathbb{R}^d$-valued step functions on $[0,T]$ and $\mathcal{H}$ the completion of $\mathcal{E}$ with respect to the semi-inner product

$$\langle \mathbf{1}_{[0,t]}\, e_i, \mathbf{1}_{[0,s]}\, e_j\rangle_{\mathcal{H}} := R_H(s,t)\,\mathbf{1}_{(i=j)}, \qquad s,t \in [0,T].$$

Then, one constructs an isometry $K_H^* : \mathcal{H} \to L^2([0,T];\mathbb{R}^d)$ such that $K_H^*(\mathbf{1}_{[0,t]}\, e_i) = \mathbf{1}_{[0,t]}\, K_H(t,\cdot)\, e_i$, where the kernel $K = K_H$ is given by

$$K(t,s) = c_H s^{\frac{1}{2}-H} \int_s^t (u-s)^{H-\frac{3}{2}} u^{H-\frac{1}{2}}\,\mathrm{d}u$$

and verifies that $\mathbf{E}[B_s^i B_t^i] = \int_0^{s\wedge t} K(t,r)K(s,r)\,\mathrm{d}r$, for some constant c_H. Moreover, let us observe that K_H^* can be represented in the following form:

$$[K_H^*\varphi]_t = \int_t^T \varphi_r \partial_r K(r,t)\,\mathrm{d}r = d_H t^{H-1/2}\left[I_{T-}^{H-1/2}\left(u^{H-1/2}\varphi\right)\right]_t, \qquad (16.12)$$

where I_{T-}^{α} stands for the fractional integral of order α. The fractional Cameron–Martin space can be introduced in the following way: let $\mathcal{K}_H : L^2([0,T];\mathbb{R}^d) \to \mathcal{H}_H := \mathcal{K}_H(L^2([0,T];\mathbb{R}^d))$ be the operator defined by

$$[\mathcal{K}_H h](t) := \int_0^t K(t,s)\,h(s)\,\mathrm{d}s, \qquad h \in L^2([0,T];\mathbb{R}^d).$$

Then, $\mathcal{H}_H$ is the Reproducing Kernel Hilbert space associated with the fractional Brownian motion B. Observe that, in the case of the classical Brownian motion, one has that $K(t,s) = \mathbf{1}_{[0,t]}(s)$, K^* is the identity operator in $L^2([0,T];\mathbb{R}^d)$ and $\mathcal{H}_H$ is the usual Cameron–Martin space.

In order to deduce that $(\Omega, \mathcal{H}, \mathbf{P})$ defines an abstract Wiener space, we remark that $\mathcal{H}$ is continuously and densely embedded in Ω. In fact, one proves that the operator $\mathcal{R}_H : \mathcal{H} \to \mathcal{H}_H$ given by

$$\mathcal{R}_H\psi := \int_0^{\cdot} K(\cdot,s)[K^*\psi](s)\,\mathrm{d}s$$

defines a dense and continuous embedding from $\mathcal{H}$ into Ω; this is due to the fact that $\mathcal{R}_H\psi$ is H-Hölder continuous (for details, see [13, p. 400]).

Let us also recall that there exists a d-dimensional Wiener process W defined on $(\Omega, \mathcal{H}, \mathbf{P})$ such that B can be expressed as

$$B_t = \int_0^t K(t,r)\,\mathrm{d}W_r, \quad t \in [0,T]. \qquad (16.13)$$

This formula will be referred to as Volterra's representation of fBm.

2.3.2 Malliavin Calculus for B

Let us introduce now the Malliavin derivative operator on the Wiener space $(\Omega, \mathcal{H}, \mathbf{P})$. Namely, we first let $\mathcal{S}$ be the family of smooth functionals F of the form

$$F = f(B(h_1), \ldots, B(h_n)),$$

where $h_1, \ldots, h_n \in \mathcal{H}$, $n \geq 1$, and f is a smooth function having polynomial growth together with all its partial derivatives. Then, the Malliavin derivative of such a functional F is the $\mathcal{H}$-valued random variable defined by

$$\mathcal{D}F = \sum_{i=1}^{n} \frac{\partial f}{\partial x_i}(B(h_1), \ldots, B(h_n))h_i.$$

For all $p > 1$, it is known that the operator $\mathcal{D}$ is closable from $L^p(\Omega)$ into $L^p(\Omega; \mathcal{H})$ (see, e.g., [12, Chap. 1]). We will still denote by $\mathcal{D}$ the closure of this operator, whose domain is usually denoted by $\mathbb{D}^{1,p}$ and is defined as the completion of $\mathcal{S}$ with respect to the norm

$$\|F\|_{1,p} := \left(E(|F|^p) + E(\|\mathcal{D}F\|_{\mathcal{H}}^p)\right)^{\frac{1}{p}}.$$

Sobolev spaces $\mathbb{D}^{k,p}$ for any $k \in \mathbb{N}$ and $p \geq 1$ can be defined in the same way, and we denote by $\mathbb{D}^{k,p}_{\text{loc}}$ the set of random variables F for which there exists a sequence $(\Omega_n, F_n)_{n\geq 1} \subset \mathcal{F} \times \mathbb{D}^{k,p}$ such that $\Omega_n \uparrow \Omega$ a.s. and $F = F_n$ a.s. on Ω_n. We also set $\mathbb{D}^\infty = \cap_{k,p}\mathbb{D}^{k,p}$.

Remark 2.1. For $F \in \mathbb{D}^{1,2}$, one can write $\mathcal{D}F = \sum_{j=1}^d \mathcal{D}^j F\, e_j$, where $\mathcal{D}^j F$ denotes the Malliavin derivative with respect to the jth component of B.

Since we deal with pathwise equations, we shall also be able to differentiate them in a pathwise manner. The relation between almost sure and Malliavin derivatives has been established by Kusuoka, and we quote it according to [12, Proposition 4.1.3].

Proposition 2.4. *A random variable F is said to be $\mathcal{H}$-differentiable if for almost all $\omega \in \Omega$ and for any $h \in \mathcal{H}$, the map $\nu \mapsto F(\omega + \nu\mathcal{R}_H h)$ is differentiable. Those random variables belong to the space $\mathbb{D}^{1,p}_{\text{loc}}$, for any $p > 1$. Moreover, the following relation holds true:*

$$\langle \mathcal{D}F, h\rangle_{\mathcal{H}} = DF(B)(\mathcal{R}_H h), \qquad h \in \mathcal{H}, \tag{16.14}$$

where we recall that $\mathcal{D}$ stands for the Malliavin derivative and D for the pathwise differentiation operator.

Stochastic analysis techniques are widely used in order study laws of random variables defined on a Wiener space. Let us recall the main criterions we shall use in this direction:

Proposition 2.5. *Let F be a real-valued random variable defined on $(\Omega, \mathcal{F}, \mathbf{P})$. Then*

(i) *If $F \in \mathbb{D}^{1,p}_{\text{loc}}$ for $p > 1$ and $\|\mathcal{D}F\|_{\mathcal{H}} > 0$ almost surely, then the law of F admits a density p with respect to Lebesgue measure.*
(ii) *If $F \in \mathbb{D}^\infty$ and $\mathbf{E}[\|\mathcal{D}F\|_{\mathcal{H}}^{-p}]$ is finite for all $p \geq 1$, then the density p of F is infinitely differentiable.*

3 Existence of the Density in the Case of Nemytskii-Type Vector Fields

In this section, we first consider a general equation of the form

$$y_t = S_t\varphi + \int_0^t S_{t-u}(F_i(y)_u)\, \mathrm{d}x_u^i, \quad \varphi \in L^2(0,1), \; t \in [0,T], \tag{16.15}$$

driven by a d-dimensional noise $x = (x^1, \ldots, x^d)$ considered as a $\mathcal{C}^\gamma$ function with $\gamma \in (1/2, 1)$. We shall be able to handle the general case of a perturbation involving Nemytskii operators, i.e.,

$$F_i(\varphi)(\xi) := f_i(\varphi(\xi)), \quad \varphi \in \mathcal{B}, \ \xi \in (0,1),$$

for smooth enough functions $f_i : \mathbb{R} \to \mathbb{R}$, $i = 1, \ldots, d$.

Thus, Eq. (16.15) can here be written as

$$y_t = S_t \varphi + \int_0^t S_{t-u}(f_i(y_u))\, \mathrm{d}x_u^i, \quad \varphi \in \mathcal{B}, \ t \in [0,T], \tag{16.16}$$

or equivalently, in a multiparameter setting,

$$\begin{aligned} y(t,\xi) &= \int_0^1 G_t(\xi,\eta)\varphi(\eta)\, \mathrm{d}\eta \\ &\quad + \int_0^1 \int_0^t G_{t-u}(\xi,\eta) f_i(y(u,\eta))\, \mathrm{d}x_u^i \mathrm{d}\eta, \qquad t \in [0,T], \xi \in (0,1), \end{aligned}$$

where G_t stands for the heat kernel on $(0,1)$ associated with Dirichlet boundary conditions.

It is readily checked that if each f_i belongs to $\mathcal{C}^{1,\mathbf{b}}(\mathbb{R};\mathbb{R})$ and $y \in \hat{\mathcal{C}}^{0,\kappa}(\mathcal{B}_\kappa)$ for some $\kappa \in (\max(1-\gamma, 1/4), 1/2)$, then the integral in the right-hand side of Eq. (16.16) can be interpreted with Proposition 2.3. Indeed, owing to Eq. (16.5), we know that $f(y) \in \mathcal{C}^0(\mathcal{B}_\kappa)$, while, due to the embedding Eq. (16.8),

$$\mathcal{N}[f(y); \mathcal{C}^\kappa(\mathcal{B})] \leq \|f'\|_\infty \mathcal{N}[y; \mathcal{C}^\kappa(\mathcal{B})] \leq c\|f'\|_\infty \mathcal{N}[y; \hat{\mathcal{C}}^{0,\kappa}(\mathcal{B}_\kappa)] < \infty.$$

For the remainder of the section, we shall rely on the following regularity assumptions.

Hypothesis 3.1. *We consider* $\kappa \in (\max(1-\gamma, 1/4), 1/2)$ *and an initial condition* $\varphi \in \mathcal{B}_\kappa$. *The family of functions* $\{f_1, \ldots, f_d\}$ *is such that* f_i *is an element of* $\mathcal{C}^{3,\boldsymbol{b}}(\mathbb{R};\mathbb{R})$ *for* $i = 1, \ldots, d$.

In this context, the following existence and uniqueness result has been proven in [7, Theorem 3.10]:

Proposition 3.1. *Under Hypothesis 3.1, Eq. (16.16) interpreted with Proposition 2.3 admits a unique solution* $y \in \hat{\mathcal{C}}^{0,\kappa}(\mathcal{B}_\kappa)$, *where we recall that the space* $\hat{\mathcal{C}}^{0,\kappa}(\mathcal{B}_\kappa)$ *has been defined at Lemma 2.1.*

As a preliminary step towards Malliavin differentiability of the solution to Eq. (16.1), we shall study the dependence on x of the deterministic equation (16.16).

3.1 Differentiability with Respect to Driving Noise

For Eq. (16.16), consider the application

$$\Phi : \mathcal{C}^\gamma \to \hat{\mathcal{C}}^{0,\kappa}(\mathcal{B}_\kappa), \qquad x \mapsto y, \tag{16.17}$$

for a given initial condition φ. We shall elaborate on the strategy designed in [13] in order to differentiate Φ. Let us start with a lemma on linear equations:

Lemma 3.1. *Suppose that $(x, y) \in \mathcal{C}^\gamma \times \hat{\mathcal{C}}^{0,\kappa}(\mathcal{B}_\kappa)$ and fix $t_0 \in [0, T]$. Then for every $w \in \hat{\mathcal{C}}^{0,\kappa}([t_0, T]; \mathcal{B}_\kappa)$, the equation*

$$v_t = w_t + \int_{t_0}^t S_{t-u}(f_i'(y_u) \cdot v_u)\, \mathrm{d}x_u^i, \quad t \in [t_0, T], \tag{16.18}$$

admits a unique solution $v \in \hat{\mathcal{C}}^{0,\kappa}([t_0, T]; \mathcal{B}_\kappa)$, and one has

$$\mathcal{N}[v; \hat{\mathcal{C}}^{0,\kappa}([t_0, T]; \mathcal{B}_\kappa)] \le C_{x,y,T} \cdot \mathcal{N}[w; \hat{\mathcal{C}}^{0,\kappa}([t_0, T]; \mathcal{B}_\kappa)], \tag{16.19}$$

where $C_{x,y,T} := C(\|x\|_\gamma, \mathcal{N}[y; \hat{\mathcal{C}}^{0,\kappa}(\mathcal{B}_\kappa)], T)$ for some function $C : (\mathbb{R}^+)^3 \to \mathbb{R}^+$ growing with its arguments.

Proof. The existence and uniqueness of the solution stem from the same fixed-point argument as in the proof of Proposition 3.1 (see [7, Theorem 3.10]), and we only focus on the proof of Eq. (16.19).

Let $I = [\ell_1, \ell_2]$ be a subinterval of $[t_0, T]$. One has, according to Proposition 2.3,

$$\begin{aligned} &\mathcal{N}[v; \hat{\mathcal{C}}^\kappa(I; \mathcal{B}_\kappa)] \\ &\le \mathcal{N}[w; \hat{\mathcal{C}}^\kappa(\mathcal{B}_\kappa)] + c\, |I|^{\gamma-\kappa} \|x\|_\gamma \left\{ \mathcal{N}[f_i'(y) \cdot v; \mathcal{C}^0(I; \mathcal{B}_\kappa)] + \mathcal{N}[f_i'(y) \cdot v; \mathcal{C}^\kappa(I; \mathcal{B})] \right\}. \end{aligned} \tag{16.20}$$

Now, by using successively Eqs. (16.4) and (16.5), we get

$$\begin{aligned} \mathcal{N}[f_i'(y) \cdot v; \mathcal{C}^0(I; \mathcal{B}_\kappa)] &\le c \mathcal{N}[f_i'(y); \mathcal{C}^0(\mathcal{B}_\kappa)] \mathcal{N}[v; \mathcal{C}^0(I; \mathcal{B}_\kappa)] \\ &\le c \left\{ 1 + \mathcal{N}[y; \mathcal{C}^0(\mathcal{B}_\kappa)] \right\} \mathcal{N}[v; \mathcal{C}^0(I; \mathcal{B}_\kappa)], \end{aligned}$$

while, owing to Eqs. (16.8) and (16.3),

$$\begin{aligned} &\mathcal{N}[f_i'(y) \cdot v; \mathcal{C}^\kappa(I; \mathcal{B})] \\ &\le \mathcal{N}[f_i'(y); \mathcal{C}^\kappa(I; \mathcal{B})] \mathcal{N}[v; \mathcal{C}^0(I; \mathcal{B}_\infty)] + \mathcal{N}[f_i'(y); \mathcal{C}^0(I; \mathcal{B}_\infty)] \mathcal{N}[v; \mathcal{C}^\kappa(I; \mathcal{B})] \\ &\le c \left\{ 1 + \mathcal{N}[y; \hat{\mathcal{C}}^{0,\kappa}(\mathcal{B}_\kappa)] \right\} \mathcal{N}[v; \hat{\mathcal{C}}^{0,\kappa}(I; \mathcal{B}_\kappa)]. \end{aligned}$$

Going back to Eq. (16.20), these estimates lead to

$$\mathcal{N}[v;\hat{\mathcal{C}}^{\kappa}(I;\mathcal{B}_\kappa)] \leq \mathcal{N}[w;\hat{\mathcal{C}}^{\kappa}(\mathcal{B}_\kappa)] + c_{x,y}\,|I|^{\gamma-\kappa}\,\mathcal{N}[v;\hat{\mathcal{C}}^{0,\kappa}(I;\mathcal{B}_\kappa)],$$

and hence

$$\mathcal{N}[v;\hat{\mathcal{C}}^{0,\kappa}(I;\mathcal{B}_\kappa)] \leq \|v_{\ell_1}\|_{\mathcal{B}_\kappa} + c\mathcal{N}[w;\hat{\mathcal{C}}^{\kappa}(\mathcal{B}_\kappa)] + c_{x,y}\,|I|^{\gamma-\kappa}\,\mathcal{N}[v;\hat{\mathcal{C}}^{0,\kappa}(I;\mathcal{B}_\kappa)].$$

Control (16.19) is now easily deduced with a standard patching argument. □

The following lemma on flow-type linear equations will also be technically important for our computations below.

Lemma 3.2. *Fix $(x,y) \in \mathcal{C}^\gamma \times \hat{\mathcal{C}}^{0,\kappa}(\mathcal{B}_\kappa)$ and for every $u \in [0,T]$, consider the system of equations*

$$\begin{aligned}\Psi^i_{t,u} &= S_{t-u}(f_i(y_u)) \\ &\quad + \int_u^t S_{t-w}(f_j'(y_w)\cdot \Psi^i_{w,u})\,\mathrm{d}x^j_w\,, \quad t\in[u,T]\,,\ i\in\{1,\ldots,m\}. \qquad (16.21)\end{aligned}$$

Then, for every $i \in \{1,\ldots,m\}$ and $t \in [0,T]$, the mapping $u \mapsto \Psi^i_{t,u}$ is continuous from $[0,t]$ to $\mathcal{B}_\kappa$. In particular, for every $\xi \in (0,1)$, $u \mapsto \Psi^i_{t,u}(\xi)$ is a continuous function on $[0,t]$.

Proof. Let us fix $i \in \{1,\ldots,m\}$, $t \in [0,T]$. For any $0 \leq u < v \leq t$, set

$$\Gamma^i_{v,u}(s) := \Psi^i_{s,v} - \Psi^i_{s,u}\,, \quad s \in [v,T].$$

It is easy to check that $\Gamma^i_{v,u}$ is solution of the equation on $[v,T]$

$$\Gamma^i_{v,u}(s) = S_{s-v}(\Psi^i_{v,v} - \Psi^i_{v,u}) + \int_v^s S_{s-w}(f_j'(y_w)\cdot\Gamma^i_{v,u}(w))\,\mathrm{d}x^j_w.$$

Therefore, according to the estimate (16.19),

$$\|\Psi^i_{t,v} - \Psi^i_{t,u}\|_{\mathcal{B}_\kappa} = \|\Gamma^i_{v,u}(t)\|_{\mathcal{B}_\kappa} \leq \mathcal{N}[\Gamma^i_{v,u};\mathcal{C}^0([v,T];\mathcal{B}_\kappa)] \leq c_{x,y,T}\,\|\Psi^i_{v,v} - \Psi^i_{v,u}\|_{\mathcal{B}_\kappa}.$$

Now, observe that

$$\Psi^i_{v,v} - \Psi^i_{v,u} = f_i(y_v) - S_{v-u}(f_i(y_u)) + \int_u^v S_{v-w}(f_j'(y_w)\cdot\Psi^i_{w,u})\,\mathrm{d}x^j_w,$$

and since $y \in \hat{\mathcal{C}}^{0,\kappa}(\mathcal{B}_\kappa)$, it becomes clear that $\|\Psi^i_{v,v} - \Psi^i_{v,u}\|_{\mathcal{B}_\kappa} \xrightarrow{v\to u} 0$. □

We now show how to differentiate a function which is closely related to Eq. (16.16).

Lemma 3.3. *The application* $F : \mathcal{C}^\gamma \times \hat{\mathcal{C}}^{0,\kappa}(\mathcal{B}_\kappa) \to \hat{\mathcal{C}}^{0,\kappa}(\mathcal{B}_\kappa)$ *defined by*

$$F(x,y)_t := y_t - S_t\varphi - \int_0^t S_{t-u}(f_i(y_u))\,dx_u^i,$$

is differentiable in the Fréchet sense and denoting by D_1F *(resp.* D_2F*) the derivative of* F *with respect to* x *(resp.* y*), we obtain*

$$D_1F(x,y)(h)_t = -\int_0^t S_{t-u}(f_i(y_u))\,dh_u^i, \tag{16.22}$$

$$D_2F(x,y)(v)_t = v_t - \int_0^t S_{t-u}(f_i'(y_u)\cdot v_u)\,dx_u^i. \tag{16.23}$$

Besides, for any $x \in \mathcal{C}^\gamma$*, the mapping* $D_2F(x,\Phi(x))$ *is a homeomorphism of* $\hat{\mathcal{C}}^{0,\kappa}(\mathcal{B}_\kappa)$.

Proof. One has, for every $h \in \mathcal{C}^\gamma, v \in \hat{\mathcal{C}}^{0,\kappa}(\mathcal{B}_\kappa)$,

$$\begin{aligned} F(x+h,y+v)_t - F(x,y)_t &= v_t - \int_0^t S_{t-u}(f_i'(y_u)\cdot v_u)\,dx_u^i \\ &\quad - \int_0^t S_{t-u}(f_i(y_u))\,dh_u^i - \left[R_t^1(v) + R_t^2(h,v)\right], \end{aligned} \tag{16.24}$$

with

$$R_t^1(v) := \int_0^t S_{t-s}z_s^i\,dx_s^i, \quad z_s^i := \int_0^1 dr \int_0^1 dr'\, r\, f_i''(y_s + rr'v_s)\cdot v_s^2,$$

$$R_t^2(v,h) := \int_0^t S_{t-s}\tilde{z}_s^i\,dh_s^i, \quad \tilde{z}_s^i := \int_0^1 dr\, f_i'(y_s + rv_s)\cdot v_s,$$

and we now have to show that

$$\mathcal{N}[R^1_\cdot(v) + R^2_\cdot(h,v); \hat{\mathcal{C}}^{0,\kappa}(\mathcal{B}_\kappa)] = o\left(\left[\|h\|_\gamma^2 + \mathcal{N}[v;\hat{\mathcal{C}}^{0,\kappa}(\mathcal{B}_\kappa)]^2\right]^{1/2}\right).$$

Observe first that $\mathcal{N}[R^1_\cdot(v);\hat{\mathcal{C}}^{0,\kappa}(\mathcal{B}_\kappa)] \le c\mathcal{N}[R^1_\cdot(v);\hat{\mathcal{C}}^{\kappa}(\mathcal{B}_\kappa)]$. Thanks to Eqs. (16.4) and (16.5), we get

$$\begin{aligned} \|z_s\|_{\mathcal{B}_\kappa} &\le c \iint_{[0,1]^2} drdr'\, \|f_i''(y_s + rr'v_s)\|_{\mathcal{B}_\kappa} \|v_s\|_{\mathcal{B}_\kappa}^2 \\ &\le c\,\{1 + \|y_s\|_{\mathcal{B}_\kappa} + \|v_s\|_{\mathcal{B}_\kappa}\}\, \|v_s\|_{\mathcal{B}_\kappa}^2. \end{aligned}$$

Besides, owing to Eqs. (16.3) and (16.8) and setting $M_{ts}(r,r') = f_i''(y_t + rr'v_t) - f_i''(y_s + rr'v_s)$ we end up with

$$
\begin{aligned}
\|z_t - z_s\|_{\mathcal{B}} \le & \iint_{[0,1]^2} \mathrm{d}r\mathrm{d}r' \, \|M_{ts}(r,r')\|_{\mathcal{B}} \, \|v_s\|_{\mathcal{B}_\infty}^2 \\
& + c\|v_t - v_s\|_{\mathcal{B}} \left\{\|v_t\|_{\mathcal{B}_\infty} + \|v_s\|_{\mathcal{B}_\infty}\right\} \\
\le & \, c\left\{\|y_t - y_s\|_{\mathcal{B}} + \|v_t - v_s\|_{\mathcal{B}}\right\} \|v_s\|_{\mathcal{B}_\kappa}^2 \\
& + c\|v_t - v_s\|_{\mathcal{B}} \left\{\|v_t\|_{\mathcal{B}_\kappa} + \|v_s\|_{\mathcal{B}_\kappa}\right\}. \\
\le & \, c\,|t-s|^\kappa \Big\{ \left(\mathcal{N}[y;\hat{\mathcal{C}}^{0,\kappa}(\mathcal{B}_\kappa)] + \mathcal{N}[v;\hat{\mathcal{C}}^{0,\kappa}(\mathcal{B}_\kappa)]\right) \mathcal{N}[v;\mathcal{C}^0(\mathcal{B}_\kappa)]^2 \\
& \qquad + \mathcal{N}[v;\hat{\mathcal{C}}^{0,\kappa}(\mathcal{B}_\kappa)] \mathcal{N}[v;\mathcal{C}^0(\mathcal{B}_\kappa)] \Big\}.
\end{aligned}
$$

The estimate (16.10) for the Young convolutional integral now provides us with the expected control $\mathcal{N}[R^1(v);\hat{\mathcal{C}}^{0,\kappa}(\mathcal{B}_\kappa)] = O(\mathcal{N}[v;\hat{\mathcal{C}}^{0,\kappa}(\mathcal{B}_\kappa)]^2)$. In the same way, one can show that $\mathcal{N}[R^2(h,v);\hat{\mathcal{C}}^{0,\kappa}(\mathcal{B}_\kappa)] = O(\|h\|_\gamma \cdot \mathcal{N}[v;\hat{\mathcal{C}}^{0,\kappa}(\mathcal{B}_\kappa)])$, and the differentiability of F is thus proved.

Of course, the two expressions (16.22) and (16.23) for the partial derivatives are now easy to derive from Eq. (16.24). As for the bijectivity of $D_2F(x,\Phi(x))$, it is a consequence of Lemma 3.1. □

We are now ready to differentiate the application Φ defined by Eq. (16.17):

Proposition 3.2. *The map $\Phi : \mathcal{C}^\gamma \to \hat{\mathcal{C}}^{0,\kappa}(\mathcal{B}_\kappa)$ is differentiable in the Fréchet sense. Moreover, for every $x \in \mathcal{C}^\gamma$ and $h \in \mathcal{C}^\infty$, the following representation holds: if $t \in [0,T], \xi \in (0,1)$,*

$$
D\Phi(x)(h)_t(\xi) = \int_0^t \Psi^i_{t,u}(\xi)\,\mathrm{d}h^i_u, \tag{16.25}
$$

where $\Psi^i_{t,\cdot} \in \mathcal{C}([0,t];\mathcal{B}_\kappa)$ is defined through the equation

$$
\Psi^i_{t,u} = S_{t-u}(f_i(\Phi(x)_u)) + \int_u^t S_{t-w}(f'_j(\Phi(x)_w)\cdot\Psi^i_{w,u})\,\mathrm{d}x^j_w. \tag{16.26}
$$

Proof. Thanks to Lemma 3.3, the differentiability of Φ is a consequence of the implicit function theorem, which gives in addition

$$
D\Phi(x) = -D_2F(x,\Phi(x))^{-1} \circ D_1F(x,\Phi(x)), \quad x \in \mathcal{C}^\gamma.
$$

In particular, for every $x, h \in \mathcal{C}^\gamma$, $z := D\Phi(x)(h)$ is the (unique) solution of the equation

$$
z_t = \int_0^t S_{t-u}(f_i(\Phi(x)_u))\,\mathrm{d}h^i_u + \int_0^t S_{t-u}(f'_i(\Phi(x)_u)\cdot z_u)\,\mathrm{d}x^i_u, \quad t \in [0,T]. \tag{16.27}
$$

If $x \in \mathcal{C}^\gamma$ and $h \in \mathcal{C}^\infty$, an application of the Fubini theorem shows (as in the proof of [13, Proposition 4]) that the path $\tilde{z}_t := \int_0^t \Psi^i_{t,u}\, dh^i_u$ (which is well-defined thanks to Lemma 3.2) is also solution of Eq. (16.27), and this provides us with the identification (16.25). □

As the reader might expect, one can obtain derivatives of any order for the solution when the coefficients of Eq. (16.16) are smooth:

Proposition 3.3. *Suppose that $f_i \in C^{\infty,b}(\mathbb{R};\mathbb{R})$ for every $i \in \{1,\dots,m\}$. Then the function $\Phi : \mathcal{C}^\gamma \to \hat{\mathcal{C}}^{0,\kappa}(\mathcal{B}_\kappa)$ defined by Eq. (16.17) is infinitely differentiable in the Fréchet sense. Moreover, for every $n \in \mathbb{N}^*$ and every $x, h_1, \dots, h_n \in \mathcal{C}^\gamma$, the path $z_t := D^n\Phi(x)(h_1,\dots,h_n)_t$ satisfies a linear equation of the form*

$$z_t = w_t + \int_0^t S_{t-u}(f_i'(\Phi(x)_u)\cdot z_u)\, \mathrm{d}x^i_u, \quad t \in [0,T], \tag{16.28}$$

where $w \in \hat{\mathcal{C}}^{0,\kappa}(\mathcal{B}_\kappa)$ only depends on $x, h_1, \dots, h_n$.

Proof. The details of this proof are omitted for the sake of conciseness, since they simply mimic the formulae contained in the proof of [13, Proposition 5]. As an example, let us just observe that for $x, h, k \in \mathcal{C}^\gamma$, the path $z_t := D^2\Phi(x)(h,k)_t$ is the unique solution of Eq. (16.28) with

$$\begin{aligned} w_t := & \int_0^t S_{t-u}(f_i'(\Phi(x)_u)\cdot D\Phi(x)(h)_u)\, \mathrm{d}k^i_u \\ & + \int_0^t S_{t-u}(f_i'(\Phi(x)_u)\cdot D\Phi(x)(k)_u)\, \mathrm{d}h^i_u \\ & + \int_0^t S_{t-u}(f_i''(\Phi(x)_u)\cdot D\Phi(x)(h)_u\cdot D\Phi(x)(k)_u)\, \mathrm{d}x^i_u. \end{aligned}$$

□

3.2 Existence of the Density

We will now apply the results of the previous section to an evolution equation driven by a fractional Brownian motion $B = (B^1,\dots,B^d)$ with Hurst parameter $H > 1/2$. Namely, we fix $\kappa \in (\max(1/4, 1-\gamma), 1/2)$ and an initial condition $\varphi \in \mathcal{B}_\kappa$. We also assume that $f_i \in C^{3,\mathbf{b}}(\mathbb{R};\mathbb{R})$ for $i = 1,\dots,m$. We denote by $Y = \Phi(B)$ the solution of

$$Y_t = S_t\varphi + \int_0^t S_{t-u}(f_i(Y_u))\, \mathrm{d}B^i_u, \quad t \in [0,T]. \tag{16.29}$$

Notice that since $H > 1/2$ the paths of B are almost surely γ-Hölder continuous with Hölder exponent greater than $1/2$. Thus, Eq. (16.29) can be solved by a direct application of Proposition 3.1. Moreover, one can invoke Proposition 3.2 in order to obtain the Malliavin differentiability of $Y_t(\xi)$:

Lemma 3.4. *For every $t \in [0,T], \xi \in (0,1)$, $Y_t(\xi) \in \mathbb{D}^{1,2}_{\text{loc}}$ and one has, for any $h \in \mathcal{H}$,*

$$\langle \mathcal{D}(Y_t(\xi)), h\rangle_{\mathcal{H}} = D\Phi(B)(\mathcal{R}_H h)_t(\xi). \tag{16.30}$$

Proof. According to Eq. (16.14), we have that

$$\langle \mathcal{D}(Y_t(\xi)), h\rangle_{\mathcal{H}} = D(Y_t(\xi))(\mathcal{R}_H h) = \frac{\mathrm{d}}{\mathrm{d}\varepsilon}_{|\varepsilon=0} \Phi(B + \varepsilon \mathcal{R}_H h)_t(\xi).$$

Furthermore, Proposition 3.2 asserts that $\Phi : \mathcal{C}^\gamma \to \hat{\mathcal{C}}^{0,\kappa}(\mathcal{B}_\kappa)$ is differentiable. Therefore

$$\frac{1}{\varepsilon}[\Phi(x + \varepsilon \mathcal{R}_H h)_t(\xi) - \Phi(x)_t(\xi)] = D\Phi(x)(\mathcal{R}_H h)_t(\xi) + \frac{1}{\varepsilon} R(\varepsilon \mathcal{R}_H h)_t(\xi),$$

with

$$|R(\varepsilon \mathcal{R}_H h)_t(\xi)| \le \mathcal{N}[R(\varepsilon \mathcal{R}_H h); \mathcal{C}^0(\mathcal{B}_\infty)] \le c\mathcal{N}[R(\varepsilon \mathcal{R}_H h); \mathcal{C}^0(\mathcal{B}_\kappa)] = o(\varepsilon),$$

and hence $\frac{\mathrm{d}}{\mathrm{d}\varepsilon}_{|\varepsilon=0}\Phi(B + \varepsilon \mathcal{R}_H h)_t(\xi) = D\Phi(B)(\mathcal{R}_H h)_t(\xi)$, which trivially yields both the inclusion $Y_t(\xi) \in \mathbb{D}^{1,2}_{\text{loc}}$ and expression (16.30). □

With this differentiation result in hand plus some non degeneracy assumptions, we now obtain the existence of a density for the random variable $Y_t(\xi)$:

Theorem 3.1. *Suppose that for all $\lambda \in \mathbb{R}$, there exists $i \in \{1, \ldots, d\}$ such that $f_i(\lambda) \neq 0$. Then, for all $t \in (0,1]$ and $\xi \in (0,1)$, the law of $Y_t(\xi)$ is absolutely continuous with respect to Lebesgue measure.*

Proof. We apply Proposition 2.5 part (i), and we will thus prove that $\|\mathcal{D}(Y_t(\xi))\|_{\mathcal{H}} > 0$ almost surely. Assume then that $\|\mathcal{D}(Y_t(\xi))\|_{\mathcal{H}} = 0$. In this case, owing to Eq. (16.30), we have $D\Phi(B)(\mathcal{R}_H h)_t(\xi) = 0$ for every $h \in \mathcal{H}$. In particular, due to Eq. (16.25), one has $\int_0^t \Psi^i_{t,u}(\xi)\, dh^i_u = 0$ for every $h \in \mathcal{C}^\infty$. As $u \mapsto \Psi^i_{t,u}(\xi)$ is known to be continuous, it is easily deduced that $\Psi^i_{t,u}(\xi) = 0$ for every $u \in [0,t]$ and every $i \in \{1, \ldots, d\}$, and so $0 = \Psi^i_{t,t}(\xi) = f_i(Y_t(\xi))$ for every $i \in \{1, \ldots d\}$, which contradicts our nonvanishing hypothesis. □

4 Smoothness of the Density in the Case of Regularizing Vector Fields

Up to now, we have been able to differentiate the solution to Eq. (16.16) when the coefficients are fairly general Nemytskii operators. However, we have only obtained the inclusion $Y_t(\xi) \in \mathbb{D}^{1,2}_{\text{loc}}$. Additional problems arise when one tries to prove

$Y_t(\xi) \in \mathbb{D}^{1,2}$, due to bad behavior of linear equations driven by rough signals in terms of moment estimates. This is why we shall change our setting here, and consider an equation of the following type

$$y_t = S_t\varphi + \int_0^t S_{t-u}(L(f_i(y_u)))\, dx_u^i, \quad t \in [0,T],\ \varphi \in \mathcal{B}, \tag{16.31}$$

where $x \in \mathcal{C}^\gamma([0,T];\mathbb{R}^d)$ with $\gamma > 1/2$, each $f_i : \mathbb{R} \to \mathbb{R}$ ($i \in \{1,\dots,d\}$) is seen as a Nemytskii operator (see the beginning of Sect. 3), and L stands for a regularizing linear operator of $\mathcal{B}$. Let us be more specific about the assumptions in this section:

Hypothesis 4.1. *We assume that for every $i \in \{1,\dots,d\}$, f_i is infinitely differentiable with bounded derivatives. Moreover, the operator $L : \mathcal{B} \to \mathcal{B}$ is taken of the form*

$$L(\phi)(\xi) := \int_0^1 d\eta\, U(\xi,\eta)\phi(\eta),$$

for some positive kernel U such that: (i) U is regularizing, i.e., L is continuous from $\mathcal{B}$ to $\mathcal{B}_\lambda$ for every $\lambda \geq 0$, and (ii) one has $c_U := \min_{\xi\in(0,1)} \int_0^1 d\eta\, U(\xi,\eta) > 0$.

In other words, we are now concerned with the following equation on $[0,T] \times (0,1)$:

$$y(t,\xi) = \int_0^1 G_t(\xi,\eta)\varphi(\eta)\, d\eta + \int_0^1 \int_0^1 \int_0^t G_{t-u}(\xi,\eta)U(\eta,\mu) f_i(y(u,\mu))\, dx_u^i d\mu d\eta,$$

with U satisfying the above conditions (i)–(ii).

This setting covers for instance the case of an (additional) heat kernel $U = G_\varepsilon$ on $(0,1)$ for any fixed $\varepsilon > 0$. The following existence and uniqueness result then holds true:

Proposition 4.1. *Under Hypothesis 4.1, for any $\lambda \geq \gamma$ and any initial condition $\varphi \in \mathcal{B}_\lambda$, Eq. (16.31) interpreted with Proposition 2.3 admits a unique solution in $\hat{\mathcal{C}}^\gamma(\mathcal{B}_\lambda)$.*

Proof. As in the proof of Proposition 3.1, the result can be obtained with a fixed-point argument. Observe indeed that if $y \in \hat{\mathcal{C}}^\gamma(I;\mathcal{B}_\lambda)$ ($I := [\ell_1,\ell_2] \subset [0,1]$) and z is the path defined by $z_{\ell_1} = y_{\ell_1}$, $z_t - S_{t-s}z_s = \int_s^t S_{t-u}(L(f_i(y_u)))\, dx_u^i$ ($s < t \in I$), then, according to Proposition 2.3, $z \in \hat{\mathcal{C}}^\gamma(I;\mathcal{B}_\lambda)$ and one has

$$\begin{aligned}\mathcal{N}[z;\hat{\mathcal{C}}^\gamma(I;\mathcal{B}_\lambda)] \leq c\|x\|_\gamma \{&\mathcal{N}[L(f(y));\mathcal{C}^0(I;\mathcal{B}_\lambda^m)] \\ &+ |I|^\gamma \mathcal{N}[L(f(y));\mathcal{C}^\gamma(I;\mathcal{B}_\lambda^m)]\}.\end{aligned} \tag{16.32}$$

Now, owing to the regularizing effect of L, it follows that $\mathcal{N}[L(f(y));\mathcal{C}^0(I;\mathcal{B}_\lambda^m)] \leq \|L\|_{\mathcal{L}(\mathcal{B},\mathcal{B}_\lambda)}\|f\|_\infty$ and

$$\mathcal{N}[L(f(y));\mathcal{C}^\gamma(I;\mathcal{B}_\lambda^m)] \leq \|L\|_{\mathcal{L}(\mathcal{B},\mathcal{B}_\lambda)}\|f'\|_\infty \mathcal{N}[y;\mathcal{C}^\gamma(I;\mathcal{B})] \leq c\mathcal{N}[y;\hat{\mathcal{C}}^{0,\gamma}(I;\mathcal{B}_\lambda)],$$

which, together with Eq. (16.32), allows to settle the fixed-point argument. □

For the sake of clarity, we henceforth assume that $T = 1$. The generalization to any (fixed) horizon $T > 0$ easily follows from slight modifications of our estimates.

Moreover, for some technical reasons that will arise in the proofs of Propositions 4.2 and 4.3, we will focus on the case $\lambda = 2 + \gamma$ in the statement of Proposition 4.1. In other words, from now on, *we fix the initial condition φ in the space $\mathcal{B}_{2+\gamma}$*.

4.1 Estimates on the Solution

Under our new setting, let us find an appropriate polynomial control on the solution to Eq. (16.31) in terms of x.

Proposition 4.2. *Suppose that y is the solution of Eq. (16.31) in $\hat{\mathcal{C}}^\gamma(\mathcal{B}_{2+\gamma})$ with initial condition φ. Then there exists a constant $C_{\gamma,f,L}$ such that*

$$\mathcal{N}[y;\hat{\mathcal{C}}^\gamma([0,1];\mathcal{B}_{2+\gamma})] \leq C_{\gamma,f,L}\left(1+\|x\|_\gamma\right)\left(\max\left(\|x\|_\gamma^{1/\gamma},\|\varphi\|_{\mathcal{B}_{2+\gamma}}^{1/2}\right)\right)^{1-\gamma}. \tag{16.33}$$

Proof. For any $N \in \mathbb{N}^*$, let us introduce the two sequences

$$\varepsilon_k = \varepsilon_{N,k} := \frac{1}{N+k}, \quad \ell_0 := 0\,,\ \ell_{k+1} = \ell_{k+1}^N := \ell_k^N + \varepsilon_{N,k}.$$

The first step of the proof consists in showing that we can pick N such that for every k,

$$\varepsilon_k^2 \|y_{\ell_k}\|_{\mathcal{B}_{2+\gamma}} \leq 1. \tag{16.34}$$

For the latter control to hold at time 0 (i.e., for $k = 0$), we must first assume that $N \geq \|\varphi\|_{\mathcal{B}_{2+\gamma}}^{1/2}$. Now, observe that for any k, one has, owing to Eq. (16.10),

$$\begin{aligned}
&\mathcal{N}[y;\hat{\mathcal{C}}^\gamma([\ell_k,\ell_{k+1}];\mathcal{B}_{2+\gamma})]\\
&\leq c\|x\|_\gamma\Big\{\mathcal{N}[L(f(y));\mathcal{C}^0([\ell_k,\ell_{k+1}];\mathcal{B}_{2+\gamma}^m)]\\
&\quad+\varepsilon_k^\gamma\mathcal{N}[L(f(y));\mathcal{C}^\gamma([\ell_k,\ell_{k+1}];\mathcal{B}_{2+\gamma}^m)]\Big\}\\
&\leq c\|x\|_\gamma\|L\|_{\mathcal{L}(\mathcal{B},\mathcal{B}_{2+\gamma})}\left\{1+\varepsilon_k^\gamma\mathcal{N}[y;\mathcal{C}^\gamma([\ell_k,\ell_{k+1}];\mathcal{B})]\right\}\\
&\leq c\|x\|_\gamma\Big\{1+\varepsilon_k^\gamma\mathcal{N}[y;\hat{\mathcal{C}}^\gamma([\ell_k,\ell_{k+1}];\mathcal{B}_{2+\gamma})]+\varepsilon_k^{\gamma+2}\mathcal{N}[y;\mathcal{C}^0([\ell_k,\ell_{k+1}];\mathcal{B}_{2+\gamma})]\Big\}\\
&\leq c_1\|x\|_\gamma\Big\{1+\varepsilon_k^\gamma\mathcal{N}[y;\hat{\mathcal{C}}^\gamma([\ell_k,\ell_{k+1}];\mathcal{B}_{2+\gamma})]+\varepsilon_k^{\gamma+2}\|y_{\ell_k}\|_{\mathcal{B}_{2+\gamma}}\Big\},
\end{aligned} \tag{16.35}$$

where we have used Eq. (16.9) to get the third inequality. Consequently, if we take N such that $2c_1 N^{-\gamma}\|x\|_\gamma \leq 1$ (i.e., $N \geq (2c_1\|x\|_\gamma)^{1/\gamma}$), we retrieve

$$\mathcal{N}[y;\hat{\mathcal{C}}^\gamma([\ell_k,\ell_{k+1}];\mathcal{B}_{2+\gamma})] \leq 2c_1\|x\|_\gamma + \varepsilon_k^2\|y_{\ell_k}\|_{\mathcal{B}_{2+\gamma}} \tag{16.36}$$

and hence

$$\|y_{\ell_{k+1}}\|_{\mathcal{B}_{2+\gamma}} \leq 1 + (1 + \varepsilon_k^{2+\gamma})\|y_{\ell_k}\|_{\mathcal{B}_{2+\gamma}}.$$

From this estimate, if we assume that $\varepsilon_k^2 \|y_{\ell_k}\|_{\mathcal{B}_{2+\gamma}} \leq 1$, then

$$\begin{aligned}\varepsilon_{k+1}^2 \|y_{\ell_{k+1}}\|_{\mathcal{B}_{2+\gamma}} &\leq \varepsilon_{k+1}^2 + \varepsilon_{k+1}^2 \left\{\|y_{\ell_k}\|_{\mathcal{B}_{2+\gamma}} + \varepsilon_k^{\gamma}\right\} \\ &\leq 2\varepsilon_{k+1}^2 + \frac{\varepsilon_{k+1}^2}{\varepsilon_k^2} = \frac{2 + (N+k)^2}{(N+k+1)^2} \leq 1\end{aligned}$$

and Eq. (16.34) is thus proved by induction. Going back to Eq. (16.36), we get, for every k,

$$\mathcal{N}[y; \hat{\mathcal{C}}^{\gamma}([\ell_k, \ell_{k+1}]; \mathcal{B}_{2+\gamma})] \leq 2c_1 \|x\|_{\gamma} + 1. \tag{16.37}$$

By a standard patching argument, this estimate yields

$$\mathcal{N}[y; \hat{\mathcal{C}}^{\gamma}([0,1]; \mathcal{B}_{2+\gamma})] \leq \left\{2c_1 \|x\|_{\gamma} + 1\right\} K^{1-\gamma},$$

where K stands for the smallest integer such that $\sum_{k=0}^{K} \varepsilon_k \geq 1$.

Finally, observe that $2 \geq \sum_{k=0}^{K} \varepsilon_k = \sum_{k=N}^{N+K} \frac{1}{k}$, and thus one can check that $K \leq (\mathrm{e}^2 - 1)N \leq 7N$. To achieve the proof of Eq. (16.33), it now suffices to notice that N can be picked proportional to $\max(\|x\|_{\gamma}^{1/\gamma}, \|\varphi\|_{\mathcal{B}_{2+\gamma}}^{1/2})$. □

We now consider a linear equation, which is equivalent to Eq. (16.18) in our regularized context:

$$z_t = w_t + \int_0^t S_{t-u}(L(f_i'(y_u) \cdot z_u))\, \mathrm{d}x_u^i, \quad t \in [0,1], \tag{16.38}$$

where $w \in \hat{\mathcal{C}}^{\gamma}(\mathcal{B}_{2+\gamma})$ and y stands for the solution of Eq. (16.31) with initial condition $\varphi \in \mathcal{B}_{2+\gamma}$.

The existence and uniqueness of a solution for Eq. (16.38) can be proved along the same lines as Proposition 4.1, that is to say via a fixed-point argument. We shall get a suitable exponential control for this solution.

Proposition 4.3. *There exists constants C_1, C_2 which only depends on f, L and γ such that*

$$\mathcal{N}[z; \mathcal{C}^0([0,1]; \mathcal{B}_{2+\gamma})] \leq C_1 \mathcal{N}[w; \hat{\mathcal{C}}^{0,\gamma}(\mathcal{B}_{2+\gamma})] \exp\left(C_2 \max\left(\|\varphi\|_{\mathcal{B}_{2+\gamma}}^{1/2}, \|x\|_{\gamma}^{1/\gamma}\right)\right). \tag{16.39}$$

Moreover, if $w_t = S_t \psi$ for some function $\psi \in \mathcal{B}_{2+\gamma}$, there exists an additional constant C_3 which only depends on f, L and γ such that

$$\begin{aligned}&\mathcal{N}[z; \hat{\mathcal{C}}^{\gamma}([0,1]; \mathcal{B}_{2+\gamma})] \\ &\quad \leq C_3 \|\psi\|_{\mathcal{B}_{2+\gamma}} \max\left(\|\varphi\|_{\mathcal{B}_{2+\gamma}}^{1/2}, \|x\|_{\gamma}^{1/\gamma}\right) \exp\left(C_2 \max\left(\|\varphi\|_{\mathcal{B}_{2+\gamma}}^{1/2}, \|x\|_{\gamma}^{1/\gamma}\right)\right)\end{aligned} \tag{16.40}$$

Proof. We go back to the notation $\varepsilon_{N,k}$, ℓ_k^N of the proof of Proposition 4.2, and set, for every $c \geq 0$,

$$N(c) := \max\left(\|\varphi\|_{\mathcal{B}_{2+\gamma}}^{1/2}, (2c\|x\|_\gamma)^{1/\gamma}\right).$$

We have seen in the proof of Proposition 4.2 that there exists a constant c_1 such that for every $N \geq N(c_1)$ and every k, one has simultaneously

$$\varepsilon_{N,k}^2 \|y_{l_k^N}\|_{\mathcal{B}_{2+\gamma}} \leq 1\,, \quad \mathcal{N}[y; \hat{\mathcal{C}}^\gamma([\ell_k^N, \ell_{k+1}^N]; \mathcal{B}_{2+\gamma})] \leq 2c_1\|x\|_\gamma + 1. \tag{16.41}$$

Suppose that $N \geq N(c_1)$ and set $\mathcal{N}_w := \mathcal{N}[w; \hat{\mathcal{C}}^\gamma(\mathcal{B}_{2+\gamma})]$. One has, similarly to Eq. (16.35),

$$\begin{aligned}
&\mathcal{N}[z; \hat{\mathcal{C}}^\gamma([\ell_k^N, \ell_{k+1}^N]; \mathcal{B}_{2+\gamma})] \\
&\leq \mathcal{N}_w + c\|x\|_\gamma \left\{\mathcal{N}[f'(y)\cdot z; \mathcal{C}^0([\ell_k^N, \ell_{k+1}^N]; \mathcal{B}^m)] + \varepsilon_{N,k}^\gamma \mathcal{N}[f'(y)\cdot z; \mathcal{C}^\gamma([\ell_k^N, \ell_{k+1}^N]; \mathcal{B}^m)]\right\} \\
&\leq \mathcal{N}_w + c\|x\|_\gamma \{\mathcal{N}[z; \mathcal{C}^0([\ell_k^N, \ell_{k+1}^N]; \mathcal{B})] + \varepsilon_{N,k}^\gamma \mathcal{N}[z; \mathcal{C}^\gamma([\ell_k^N, \ell_{k+1}^N]; \mathcal{B})] \\
&\quad + \varepsilon_{N,k}^\gamma \mathcal{N}[y; \mathcal{C}^\gamma([\ell_k^N, \ell_{k+1}^N]; \mathcal{B})] \mathcal{N}[z; \mathcal{C}^0([\ell_k^N, \ell_{k+1}^N]; \mathcal{B}_\infty)]\} \\
&\leq \mathcal{N}_w + c_2\|x\|_\gamma \mathcal{N}[z; \mathcal{C}^0([\ell_k^N, \ell_{k+1}^N]; \mathcal{B}_{2+\gamma})] \left\{1 + \varepsilon_{N,k}^\gamma \mathcal{N}[y; \mathcal{C}^\gamma([\ell_k^N, \ell_{k+1}^N]; \mathcal{B})]\right\} \\
&\quad + c_2\|x\|_\gamma \varepsilon_{N,k}^\gamma \mathcal{N}[z; \hat{\mathcal{C}}^\gamma([\ell_k^N, \ell_{k+1}^N]; \mathcal{B}_{2+\gamma})],
\end{aligned}$$

where we have used Eqs. (16.3) and (16.9) to derive the last inequality. Therefore, if we choose $N_2 \geq \max\left(N(c_1), (2c_2\|x\|_\gamma)^{1/\gamma}\right)$, one has, for any $N \geq N_2$ and any k,

$$\begin{aligned}
&\mathcal{N}[z; \hat{\mathcal{C}}^\gamma([\ell_k^N, \ell_{k+1}^N]; \mathcal{B}_{2+\gamma})] \\
&\quad \leq 2\mathcal{N}_w + 2c_2\|x\|_\gamma \mathcal{N}[z; \mathcal{C}^0([\ell_k^N, \ell_{k+1}^N]; \mathcal{B}_{2+\gamma})] \left\{1 + \varepsilon_{N,k}^\gamma \mathcal{N}[y; \mathcal{C}^\gamma([\ell_k^N, \ell_{k+1}^N]; \mathcal{B})]\right\}.
\end{aligned}$$

Thanks to Eqs. (16.9) and (16.41), we know that

$$\begin{aligned}
\mathcal{N}[y; \mathcal{C}^\gamma([\ell_k^N, \ell_{k+1}^N]; \mathcal{B})] &\leq c\left\{\mathcal{N}[y; \hat{\mathcal{C}}^\gamma([\ell_k^N, \ell_{k+1}^N]; \mathcal{B}_{2+\gamma})] + \varepsilon_{N,k}^2 \|y_{\ell_n^N}\|_{\mathcal{B}_{2+\gamma}}\right\} \\
&\leq c\left\{2c_1\|x\|_\gamma + 2\right\}.
\end{aligned}$$

As a consequence, there exists c_3 such that for any $N \geq N_2$,

$$\begin{aligned}
&\mathcal{N}\left[z; \hat{\mathcal{C}}^\gamma\left([\ell_k^N, \ell_{k+1}^N]; \mathcal{B}_{2+\gamma}\right)\right] \\
&\quad \leq 2\mathcal{N}_w + c_3\|x\|_\gamma \mathcal{N}\left[z; \mathcal{C}^0\left([\ell_k^N, \ell_{k+1}^N]; \mathcal{B}_{2+\gamma}\right)\right] \left\{1 + \varepsilon_{N,k}^\gamma \|x\|_\gamma\right\}. \qquad (16.42)
\end{aligned}$$

Then, for any $N \geq N_2$,

$$\mathcal{N}[z; \mathcal{C}^0([\ell_k^N, \ell_{k+1}^N]; \mathcal{B}_{2+\gamma})]$$
$$\leq 2\mathcal{N}_w + \|z_{\ell_k^N}\|_{\mathcal{B}_{2+\gamma}} + c_3\|x\|_\gamma \varepsilon_{N,k}^\gamma \mathcal{N}[z; \mathcal{C}^0([\ell_k^N, \ell_{k+1}^N]; \mathcal{B}_{2+\gamma})] \left\{1 + \varepsilon_{N,k}^\gamma \|x\|_\gamma\right\}.$$

Pick now an integer $N_3 \geq N_2$ such that

$$c_3 N_3^{-\gamma} \|x\|_\gamma \left\{1 + N_3^{-\gamma}\|x\|_\gamma\right\} \leq \frac{1}{2},$$

and we get, for any k,

$$\mathcal{N}[z; \mathcal{C}^0([\ell_k^{N_3}, \ell_{k+1}^{N_3}]; \mathcal{B}_{2+\gamma})] \leq 2\|z_{\ell_k^{N_3}}\|_{\mathcal{B}_{2+\gamma}} + 4\mathcal{N}_w,$$

so $\mathcal{N}[z; \mathcal{C}^0([0,1]; \mathcal{B}_{2+\gamma})] \leq 2^{K(N_3)}\|w_0\|_{\mathcal{B}_{2+\gamma}} + 2^{K(N_3)+2}\mathcal{N}_w$, where $K(N_3)$ stands for the smallest integer such that $\sum_{k=0}^{K(N_3)} \varepsilon_{N_3,k} \geq 1$. As in the proof of Proposition 4.2, one can check that $K(N_3) \leq cN_3$. In order to get Eq. (16.39), it suffices to observe that there exists a constant c_4 such that any integer $N_3 \geq c_4 \max(\|\varphi\|_{\mathcal{B}_{2+\gamma}}^{1/2}, \|x\|_\gamma^{1/\gamma})$ meets the above requirements.

Suppose now that $w_t = S_t\psi$. In particular, $\mathcal{N}_w = 0$. Then we go back to Eq. (16.42) to obtain, thanks to Eq. (16.39),

$$\mathcal{N}[z; \hat{\mathcal{C}}^\gamma([\ell_k^{N_3}, \ell_{k+1}^{N_3}]; \mathcal{B}_{2+\gamma})] \leq C_1\|\psi\|_{\mathcal{B}_{2+\gamma}} N_3^\gamma \exp\left(C_2 \max\left(\|\varphi\|_{\mathcal{B}_{2+\gamma}}^{1/2}, \|x\|_\gamma^{1/\gamma}\right)\right),$$

which entails

$$\mathcal{N}[z; \hat{\mathcal{C}}^\gamma([0,1]; \mathcal{B}_{2+\gamma})]$$
$$\leq C_1\|\psi\|_{\mathcal{B}_{2+\gamma}} N_3^\gamma K(N_3)^{1-\gamma} \exp\left(C_2 \max\left(\|\varphi\|_{\mathcal{B}_{2+\gamma}}^{1/2}, \|x\|_\gamma^{1/\gamma}\right)\right)$$
$$\leq C_3\|\psi\|_{\mathcal{B}_{2+\gamma}} N_3 \exp\left(C_2 \max\left(\|\varphi\|_{\mathcal{B}_{2+\gamma}}^{1/2}, \|x\|_\gamma^{1/\gamma}\right)\right),$$

and Eq. (16.40) is thus proved. □

Remark 4.1. For any $t_0 \in [0,1]$, the proof of Proposition 4.3 can be easily adapted to the equation starting at time t_0

$$z_t = w_{t,t_0} + \int_{t_0}^t S_{t-u}(L(f_i'(y_u) \cdot z_u))\, dx_u^i, \quad w_{.,t_0} \in \hat{\mathcal{C}}^\gamma([t_0,1]; \mathcal{B}_{2+\gamma}),\ t \in [t_0, 1],$$

and both estimates (16.39) and (16.40) remain of course true in this situation.

4.2 Smoothness of the Density

Let us now go back to the fractional Brownian situation

$$Y_t = S_t\varphi + \int_0^t S_{t-u}(L(f_i(Y_u)))\,\mathrm{d}B_u^i\,,\quad t \in [0,1]\,,\ \varphi \in \mathcal{B}_{2+\gamma}, \tag{16.43}$$

where $\gamma \in (\frac{1}{2}, H)$ is a fixed parameter. We suppose, for the rest of the section, that the initial condition φ is fixed in $\mathcal{B}_{2+\gamma}$ and that Hypothesis 4.1 is satisfied. We denote by Y the solution of Eq. (16.43) in $\hat{\mathcal{C}}^\gamma(\mathcal{B}_{2+\gamma})$ given by Proposition 4.1.

As in Sect. 3.2, we wish to study the law of $Y_t(\xi)$ for $t \in [0,1]$ and $\xi \in (0,1)$. Without loss of generality, we focus more exactly on the law of $Y_1(\xi)$, for $\xi \in (0,1)$.

The first thing to notice here is that the whole reasoning of Sect. 3 can be transposed without any difficulty to Eq. (16.43), which is more easy to handle due to the regularizing effect of L. Together with the estimates (16.33) and (16.39), this observation leads us to the following statement:

Proposition 4.4. *For every $\xi \in (0,1)$, $Y_1(\xi) \in \mathbb{D}^\infty$ and the law of $Y_1(\xi)$ is absolutely continuous with respect to the Lebesgue measure.*

Proof. The absolute continuity of the law of $Y_1(\xi)$ can be obtained by following the lines of Sect. 3, which gives $Y_1(\xi) \in \mathbb{D}^\infty_{\mathrm{loc}}$ as well. Then, like in Proposition 3.3, observe that nth (Fréchet) derivatives Z^n of the flow associated with Eq. (16.43) satisfy a linear equation of the form

$$Z_t^n = W_t^n + \int_0^t S_{t-u}\left(L(f_i'(Y_u)\cdot Z_u^n)\right)\mathrm{d}B_u^i,\quad t\in[0,1].$$

The explicit expression for W^n ($n \geq 1$) can be derived from the formulae contained in [13, Proposition 5], and it is easy to realize that due to Eq. (16.33), one has $\mathcal{N}[W^n;\hat{\mathcal{C}}^{0,\gamma}(\mathcal{B}_{2+\gamma})] \in L^p(\Omega)$ for any n and any p. Then, thanks to Eq. (16.39), we deduce that $\mathcal{N}[Z^n;\mathcal{C}^0(\mathcal{B}_{2+\gamma})]$ is a square-integrable random variable, which allows us to conclude that $Y_1(\xi) \in \mathbb{D}^\infty$ (see [12, Lemma 4.1.2]). □

The following proposition, which can be seen as an improvement of Lemma 3.2 (in this regularized situation), provides us with the key-estimate to prove the smoothness of the density:

Proposition 4.5. *For every $s \in [0,1]$, consider the system of equations*

$$\begin{aligned}\Psi_{t,s}^i &= S_{t-s}(L(f_i(Y_s)))\\ &\quad + \int_s^t S_{t-u}(L(f_j'(Y_u)\cdot\Psi_{u,s}^i))\,\mathrm{d}B_u^j,\quad t\in[s,1],\ i\in\{1,\dots,m\}.\end{aligned} \tag{16.44}$$

Then, for every $i \in \{1,\dots,m\}$ and every $t \in [0,1]$, $\Psi_{t,.}^i \in \mathcal{C}^\gamma([0,t];\mathcal{B}_{2+\gamma})$. In particular, for any $\xi \in (0,1)$, $\Psi_{t,.}^i(\xi) \in \mathcal{C}^\gamma([0,t])$.

Moreover, one has the following estimate

$$\mathcal{N}[\Psi^i_{t,\cdot};\mathcal{C}^\gamma([0,t];\mathcal{B}_{2+\gamma})]$$
$$\leq Q(\|\varphi\|_{\mathcal{B}_{2+\gamma}},\|B\|_\gamma)\cdot\exp\left(c\max\left(\|\varphi\|^{1/2}_{\mathcal{B}_{2+\gamma}},\|B\|^{1/\gamma}_\gamma\right)\right), \tag{16.45}$$

for some polynomial expression Q.

Proof. As in the proof of Lemma 3.2, we introduce the path

$$\Gamma^i_{v,u}(s) := \Psi^i_{s,v} - \Psi^i_{s,u}, \quad s\in[v,1]\,,\ 0\leq u<v\leq t,$$

and it is readily checked that $\Gamma^i_{v,u}$ solves the equation on $[v,1]$

$$\Gamma^i_{v,u}(s) = S_{s-v}(\Psi^i_{v,v}-\Psi^i_{v,u}) + \int_v^s S_{s-w}(L(f'_j(Y_w)\cdot\Gamma^i_{v,u}(w)))\,\mathrm{d}B^j_w.$$

Therefore, thanks to the estimate (16.39), we get

$$\begin{aligned}\|\Psi^i_{t,v}-\Psi^i_{t,u}\|_{\mathcal{B}_{2+\gamma}} &= \|\Gamma^i_{v,u}(t)\|_{\mathcal{B}_{2+\gamma}} \leq \mathcal{N}[\Gamma^i_{v,u};\mathcal{C}^0([v,1];\mathcal{B}_{2+\gamma})]\\ &\leq c\|\Psi^i_{v,v}-\Psi^i_{v,u}\|_{\mathcal{B}_{2+\gamma}}\exp\left(c\max\left(\|\varphi\|^{1/2}_{\mathcal{B}_{2+\gamma}},\|B\|^{1/\gamma}_\gamma\right)\right).\end{aligned} \tag{16.46}$$

Then, by writing

$$\begin{aligned}\Psi^i_{v,v}-\Psi^i_{v,u} &= L(f_i(Y_v)-f_i(Y_u)) - [S_{v-u}-\mathrm{Id}]\,(L(f_i(Y_u)))\\ &\quad -\int_u^v S_{v-w}(L(f'_j(Y_w)\cdot\Psi^i_{w,u}))\,\mathrm{d}B^j_w,\end{aligned}$$

we deduce that

$$\begin{aligned}\|\Psi^i_{v,v}-\Psi^i_{v,u}\|_{\mathcal{B}_{2+\gamma}} &\leq c\,|v-u|^\gamma\left\{\|L\|_{\mathcal{L}(\mathcal{B},\mathcal{B}_{2+\gamma})}\mathcal{N}[Y;\mathcal{C}^\gamma(\mathcal{B})]+\|L\|_{\mathcal{L}(\mathcal{B},\mathcal{B}_{2+2\gamma})}\right.\\ &\quad +\|L\|_{\mathcal{L}(\mathcal{B},\mathcal{B}_{2+\gamma})}\|B\|_\gamma\big(\mathcal{N}[f'(Y)\cdot\Psi^i_{\cdot,u};\mathcal{C}^0(\mathcal{B}^m)]\\ &\quad \left.+\mathcal{N}[f'(Y)\cdot\Psi^i_{\cdot,u};\mathcal{C}^\gamma(\mathcal{B}^m)]\big)\right\}\\ &\leq c|v-u|^\gamma\left\{1+\|B\|_\gamma\right\}\{1+\mathcal{N}[Y;\mathcal{C}^\gamma(\mathcal{B})]\}\\ &\quad\times\left\{1+\mathcal{N}[\Psi^i_{\cdot,u};\mathcal{C}^0(\mathcal{B}_{2+\gamma})]+\mathcal{N}[\Psi^i_{\cdot,u};\mathcal{C}^\gamma(\mathcal{B})]\right\}.\end{aligned}$$

Going back to Eq. (16.46), the result now easily follows from the embedding $\hat{\mathcal{C}}^{0,\gamma}(\mathcal{B}_{2+\gamma})\subset\mathcal{C}^\gamma(\mathcal{B})$ and the three controls (16.33), (16.39) and (16.40). □

Proposition 4.5 implies in particular that the Young integral $\int_0^t \Psi^i_{t,u}(\xi)\,dh^i_u$ is well-defined for every $h\in\mathcal{C}^\gamma$, $t\in[0,1]$ and $\xi\in(0,1)$. We are thus in a position to apply the Fubini-type argument of [13, Propositions 4 and 7] so as to retrieve the following convenient expression for the Malliavin derivative:

Corollary 4.1. *For every $\xi \in (0,1)$, the Malliavin derivative of $Y_1(\xi)$ is given by*

$$\mathcal{D}_s^i(Y_1(\xi)) = \Psi_{1,s}^i(\xi), \quad s \in [0,1], \ i \in \{1,\dots,m\}, \tag{16.47}$$

where $\Psi^i_{\cdot,s}$ stands for the solution of Eq. (16.44) on $[s,1]$.

Theorem 4.2. *Suppose that there exists $\lambda_0 > 0$ such that for every $i \in \{1,\dots,m\}$ and every $\eta \in \mathbb{R}$, $f_i(\eta) \geq \lambda_0$. Then, for every $\xi \in (0,1)$, the density of $Y_1(\xi)$ with respect to the Lebesgue measure is infinitely differentiable.*

Proof. We shall apply here the criterion stated at Proposition 2.5 item (ii). Notice that we already know that $Y_1(\xi) \in \mathbb{D}^\infty$, so it remains to show that for every $p \geq 2$, there exists $\varepsilon_0(p) > 0$ such that if $\varepsilon < \varepsilon_0(p)$, then $P\left(\|\mathcal{D}_\cdot(Y_1(\xi))\|_{\mathcal{H}} < \varepsilon\right) \leq \varepsilon^p$.

To this end, we resort to the following practical estimate, borrowed from [2, Corollary 4.5]: for every $\beta > H - 1/2$, there exist $\alpha > 0$ such that

$$\mathbf{P}\left(\|\mathcal{D}_\cdot(Y_1(\xi))\|_{\mathcal{H}} < \varepsilon\right) \leq \mathbf{P}\left(\|\mathcal{D}_\cdot(Y_1(\xi))\|_\infty < \varepsilon^\alpha\right) + \mathbf{P}\left(\|\mathcal{D}_\cdot(Y_1(\xi))\|_\beta > \varepsilon^{-\alpha}\right). \tag{16.48}$$

The first term in the right-hand side of Eq. (16.48) is easy to handle. Indeed, owing to the expression (16.47) for the Malliavin derivative of $Y_1(\xi)$, one has

$$\begin{aligned}\|\mathcal{D}_\cdot(Y_1(\xi))\|_\infty &\geq \inf_{i=1,\dots,m} |\Psi_{1,1}^i(\xi)| = \inf_{i=1,\dots,m} |L(f_i(Y_1))(\xi)| \\ &= \inf_{i=1,\dots,m} \left| \int_0^1 d\eta\, U(\xi,\eta) f_i(Y_1(\eta)) \right| \geq c_U \lambda_0 > 0\end{aligned}$$

(remember that U and c_U have been defined in Hypothesis 4.1), so that $\mathbf{P}(\|\mathcal{D}_\cdot(Y_1(\xi))\|_\infty < \varepsilon^\alpha) = 0$ for ε small enough.

Then, in order to cope with $\mathbf{P}\left(\|\mathcal{D}_\cdot(Y_1(\xi))\|_\beta > \varepsilon^{-\alpha}\right)$, one can simply rely on the Markov inequality, since, according to Eq. (16.45),

$$\begin{aligned}\|\mathcal{D}_\cdot(Y_1(\xi))\|_\beta &= \|\Psi_{1,\cdot}(\xi)\|_\beta \leq c \sup_{i \in \{1,\dots,m\}} \mathcal{N}[\Psi^i_{1,\cdot}; \mathcal{C}^\gamma([0,1]; \mathcal{B}_{2+\gamma})] \\ &\leq c\, Q\left(\|\varphi\|_{\mathcal{B}_{2+\gamma}}, \|B\|_\gamma\right) \cdot \exp\left(c \max\left(\|\varphi\|_{\mathcal{B}_{2+\gamma}}, \|B\|_\gamma^{1/\gamma}\right)\right),\end{aligned}$$

which proves that $\|\mathcal{D}_\cdot(Y_1(\xi))\|_\beta \in L^q(\Omega)$ for every $q \geq 1$. □

Acknowledgements S. Tindel is partially supported by the (French) ANR grant ECRU.

References

1. Adams, R.A.: Sobolev spaces. Pure and Applied Mathematics, vol. 65. Academic [A subsidiary of Harcourt Brace Jovanovich, Publishers], New York (1975)
2. Baudoin, F., Hairer, M.: A version of Hörmander's theorem for the fractional Brownian motion. Probab. Theory Relat. Fields **139**(3–4), 373–395 (2007)
3. Caruana, M., Friz, P.: Partial differential equations driven by rough paths. J. Differ. Equations **247**(1), 140–173 (2009)
4. Caruana, M., Friz, P.K., Oberhauser, H.: A (rough) pathwise approach to a class of non-linear stochastic partial differential equations. Ann. Inst. H. Poincaré Anal. Non Linéaire **28**(1), 27–46 (2011)
5. Deya, A.: Numerical schemes for rough parabolic equations. Appl. Math. Optim. **65**(2), 253–292 (2012)
6. Deya, A.: A discrete approach to Rough Parabolic Equations. Electron. J. Probab. **16**, 1489–1518 (2011)
7. Deya, A., Gubinelli, M., Tindel, S.: Non-linear rough heat equations. Probab. Theory Relat. Fields **153**(1–2), 97–147 (2012)
8. Friz, P., Oberhauser, H.: Rough path stability of SPDEs arising in non-linear filtering. ArXiv preprint (2010)
9. Gubinelli, M., Lejay, A., Tindel, S.: Young integrals and SPDEs. Potential Anal. **25**(4), 307–326 (2006)
10. Gubinelli, M., Tindel, S.: Rough evolution equations. Ann. Probab. **38**(1), 1–75 (2010)
11. Maslowski, B., Nualart, D.: Evolution equations driven by a fractional Brownian motion. J. Funct. Anal. **202**(1), 277–305 (2003)
12. Nualart, D.: The Malliavin calculus and related topics. Probability and its Applications (New York), 2nd edn. Springer, Berlin (2006)
13. Nualart, D., Saussereau, B.: Malliavin calculus for stochastic differential equations driven by a fractional Brownian motion. Stoch. Process. Appl. **119**(2), 391–409 (2009)
14. Pazy, A.: Semigroups of linear operators and applications to partial differential equations. Applied Mathematical Sciences, vol. 44. Springer, New York (1983)
15. Runst, T., Sickel, W.: Sobolev spaces of fractional order, Nemytskij operators, and nonlinear partial differential equations. de Gruyter Series in Nonlinear Analysis and Applications, vol. 3. Walter de Gruyter, Berlin (1996)

Chapter 17
Parameter Estimation for α-Fractional Bridges

Khalifa Es-Sebaiy and Ivan Nourdin

It is a great pleasure for us to dedicate this paper to our friend David Nualart, in celebration of his 60th birthday and with all our admiration.

Abstract Let $\alpha, T > 0$. We study the asymptotic properties of a least squares estimator for the parameter α of a fractional bridge defined as $\mathrm{d}X_t = -\alpha \frac{X_t}{T-t}\,\mathrm{d}t + \mathrm{d}B_t$, $0 \leq t < T$, where B is a fractional Brownian motion of Hurst parameter $H > \frac{1}{2}$. Depending on the value of α, we prove that we may have strong consistency or not as $t \to T$. When we have consistency, we obtain the rate of this convergence as well. Also, we compare our results to the (known) case where B is replaced by a standard Brownian motion W.

Received 2/6/2011; Accepted 9/9/2011; Final 1/25/2012

1 Introduction

Let W be a standard Brownian motion and let α be a nonnegative real parameter. In recent years, the study of various problems related to the (so-called) α-Wiener bridge, that is, to the solution X to

$$X_0 = 0; \quad \mathrm{d}X_t = -\alpha \frac{X_t}{T-t}\,\mathrm{d}t + \mathrm{d}W_t, \quad 0 \leq t < T, \tag{17.1}$$

K. Es-Sebaiy
Institut de Mathématiques de Bourgogne, Université de Bourgogne, Dijon, France
e-mail: khalifasbai@gmail.com

I. Nourdin (✉)
Institut Élie Cartan, Université de Lorraine, BP 70239, 54506 Vandoeuvre-lès-Nancy, France
e-mail: inourdin@gmail.com

F. Viens et al. (eds.), *Malliavin Calculus and Stochastic Analysis: A Festschrift in Honor of David Nualart*, Springer Proceedings in Mathematics & Statistics 34,
DOI 10.1007/978-1-4614-5906-4_17,

has attracted interest. For a motivation and further references, we refer the reader to Barczy and Pap [2, 3], as well as Mansuy [6]. Because Eq. (17.1) is linear, it is immediate to solve it explicitly; one then gets the following formula:

$$X_t = (T-t)^\alpha \int_0^t (T-s)^{-\alpha} \mathrm{d}W_s, \quad t \in [0,T),$$

the integral with respect to W being a Wiener integral.

An example of interesting problem related to X is the statistical estimation of α when one observes the whole trajectory of X. A natural candidate is the maximum likelihood estimator (MLE), which can be easily computed for this model, due to the specific form of Eq. (17.1): one gets

$$\hat{\alpha}_t = -\left(\int_0^t \frac{X_u}{T-u}\mathrm{d}X_u\right) \Big/ \left(\int_0^t \frac{X_u^2}{(T-u)^2}\mathrm{d}u\right), \quad t < T. \tag{17.2}$$

In Eq. (17.2), the integral with respect to X must of course be understood in the Itô sense. On the other hand, at this stage it is worth noticing that $\hat{\alpha}_t$ coincides with a least squares estimator (LSE) as well; indeed, $\widehat{\alpha}_t$ (formally) minimizes

$$\alpha \mapsto \int_0^t \left|\dot{X}_u + \alpha \frac{X_u}{T-u}\right|^2 du.$$

Also, it is worth bearing in mind an alternative formula for $\widehat{\alpha}_t$, which is more easily amenable to analysis and which is immediately shown thanks to Eq. (17.1):

$$\alpha - \widehat{\alpha}_t = \left(\int_0^t \frac{X_u}{T-u}\mathrm{d}W_u\right) \Big/ \left(\int_0^t \frac{X_u^2}{(T-u)^2}\mathrm{d}u\right). \tag{17.3}$$

When dealing with Eq. (17.3) by means of a semimartingale approach, it is not very difficult to check that $\widehat{\alpha}_t$ is indeed a strongly consistent estimator of α. The next step generally consists in studying the second-order approximation. Let us describe what is known about this problem: as $t \to T$,

- If $0 < \alpha < \frac{1}{2}$ then

$$(T-t)^{\alpha-\frac{1}{2}}\left(\alpha - \widehat{\alpha}_t\right) \xrightarrow{\text{law}} T^{\alpha-\frac{1}{2}}(1-2\alpha) \times \mathcal{C}(1), \tag{17.4}$$

  with $\mathcal{C}(1)$ the standard Cauchy distribution, see [4, Theorem 2.8].
- If $\alpha = \frac{1}{2}$ then

$$|\log(T-t)|\left(\alpha - \widehat{\alpha}_t\right) \xrightarrow{\text{law}} \frac{\int_0^T W_s \mathrm{d}W_s}{\int_0^T W_s^2 \mathrm{d}s}, \tag{17.5}$$

  see [4, Theorem 2.5];
- If $\alpha > \frac{1}{2}$ then

$$\sqrt{|\log(T-t)|}\left(\alpha - \widehat{\alpha}_t\right) \xrightarrow{\text{law}} \mathcal{N}(0, 2\alpha - 1), \tag{17.6}$$

  see [4, Theorem 2.11].

Thus, we have the full picture for the asymptotic behavior of the MLE/LSE associated to α-Wiener bridges.

In this paper, our goal is to investigate what happens when, in Eq. (17.1), the standard Brownian motion W is replaced by a fractional Brownian motion B. More precisely, suppose from now on that $X = \{X_t\}_{t\in[0,T)}$ is the solution to

$$X_0 = 0; \quad \mathrm{d}X_t = -\alpha \frac{X_t}{T-t}\,\mathrm{d}t + \mathrm{d}B_t, \quad 0 \le t < T, \tag{17.7}$$

where B is a fractional Brownian motion with known parameter H, whereas $\alpha > 0$ is considered as an unknown parameter. Although X could have been defined for all H in $(0, 1)$, for technical reasons and in order to keep the length of our paper within bounds we restrict ourself to the case $H \in (\frac{1}{2}, 1)$ in the sequel.

In order to estimate the unknown parameter α when the whole trajectory of X is observed, we continue to consider the estimator $\widehat{\alpha}_t$ given by Eq. (17.2) (It is no longer the MLE, but it is still an LSE.). Nevertheless, there is a major difference with respect to the standard Brownian motion case. Indeed, the process X being no longer a semimartingale, in Eq. (17.2) one cannot utilize the Itô integral to integrate with respect to it. One may however choose, instead, the young integral [see Sect. 2.3 for the main properties of this integral, notably its chain rule (17.17) and how Eq. (17.18) relies it to Skowrohod integral]. This is because X has[1] r-Hölder continuous paths on [0,t] for all $r\epsilon(\frac{1}{2}, H)$ and all $t\epsilon[0, T]$.

Let us now describe the results we prove in this paper. First, in Theorem 3.1 we show that the (strong) consistency of $\widehat{\alpha}_t$ as $t \to T$ holds true if and only if $\alpha \le \frac{1}{2}$. Then, depending on the precise value of $\alpha \in (0, \frac{1}{2}]$, we derive the asymptotic behavior of the error $\widehat{\alpha}_t - \alpha$. It turns out that, once adequately renormalized, this error converges either in law or almost surely to a limit that we are able to compute explicitly. More specifically, we show in Theorem 3.2 the following convergences [below and throughout the paper, $\mathcal{C}(1)$ always stands for the standard Cauchy distribution and $\beta(a,b) = \int_0^1 x^{a-1}(1-x)^{b-1}\mathrm{d}x$ for the usual Beta function]: as $t \to T$,

- If $0 < \alpha < 1 - H$ then

$$(T-t)^{\alpha-H}\left(\alpha-\widehat{\alpha}_t\right) \xrightarrow{\text{law}} T^{\alpha-H}(1-2\alpha)\sqrt{\frac{(H-\alpha)\beta(2-2H-\alpha,2H-1)}{(1-H-\alpha)\beta(1-\alpha,2H-1)}}\times\mathcal{C}(1). \tag{17.8}$$

- If $\alpha = 1 - H$ then

$$\frac{(T-t)^{1-2H}}{\sqrt{|\log(T-t)|}}\left(\alpha-\widehat{\alpha}_t\right) \xrightarrow{\text{law}} T^{1-2H}(2H-1)^{\frac{3}{2}}\sqrt{\frac{2\,\beta(1-H,2H-1)}{\beta(H,2H-1)}}\times\mathcal{C}(1). \tag{17.9}$$

- If $1 - H < \alpha < \frac{1}{2}$ then

[1]More precisely, we assume throughout the paper that we work with a suitable γ-Hölder continuous version of X, which is easily shown to exist by the Kolmogorov–Centsov theorem.

$$(T-t)^{2\alpha-1}\big(\alpha-\widehat{\alpha}_t\big) \overset{\text{a.s.}}{\longrightarrow} (1-2\alpha)\int_0^T \frac{\mathrm{d}B_u}{(T-u)^{1-\alpha}} \times \int_0^u \frac{\mathrm{d}B_s}{(T-s)^{\alpha}} \Big/ \left(\int_0^T \frac{\mathrm{d}B_s}{(T-s)^{\alpha}}\right)^2. \quad (17.10)$$

- If $\alpha = \frac{1}{2}$ then

$$|\log(T-t)|\big(\alpha-\widehat{\alpha}_t\big) \overset{\text{a.s.}}{\longrightarrow} \frac{1}{2}. \quad (17.11)$$

When comparing the convergences (17.8) to (17.11) with those arising in the standard Brownian motion case [i.e., Eq. (17.4) to Eq. (17.6)], we observe a new and interesting phenomenon when the parameter α ranges from $1-H$ to $\frac{1}{2}$ (of course, this case is immaterial in the standard Brownian motion case).

We hope our proofs of Eqs. (17.8)–(17.11) to be elementary. Indeed, except maybe the link (17.18) between Young and Skorohod integrals, they only involve soft arguments, often based on the mere derivation of suitable equivalent for some integrals. In particular, unlike the classical approach (as used, e.g., in [4]) we stress that, here, we use no tool coming from the semimartingale realm.

Before concluding this introduction, we would like to mention the recent paper [5] by Hu and Nualart, which has been a valuable source of inspiration. More specifically, the authors of [5] study the estimation of the parameter $\alpha > 0$ arising in the fractional Ornstein–Uhlenbeck model, defined as $\mathrm{d}X_t = -\alpha X_t \mathrm{d}t + \mathrm{d}B_t$, $t \geq 0$, where B is a fractional Brownian motion of (known) index $H \in (\frac{1}{2}, \frac{3}{4})$. They show the strong consistency of a LSE $\widehat{\alpha}_t$ as $t \to \infty$ (with, however, a major difference with respect to us: they are forced to use Skorohod integral rather than Young integral to define $\widehat{\alpha}_t$, otherwise $\widehat{\alpha}_t \not\to \alpha$ as $t \to \infty$; unfortunately, this leads to an impossible-to-simulate estimator, and this is why they introduce an alternative estimator for α). They then derive the associated rate of convergence as well, by exhibiting a central limit theorem. Their calculations are of completely different nature than ours because, to achieve their goal, the authors of [5] make use of the fourth moment theorem of Nualart and Peccati [8].

The rest of our paper is organized as follows. In Sect. 2 we introduce the needed material for our study, whereas Sect. 3 contains the precise statements and proofs of our results.

2 Basic Notions for Fractional Brownian Motion

In this section, we briefly recall some basic facts concerning stochastic calculus with respect to fractional Brownian motion; we refer to [7] for further details. Let $B = \{B_t\}_{t\in[0,T]}$ be a fractional Brownian motion with Hurst parameter $H \in (0,1)$, defined on some probability space $(\Omega, \mathcal{F}, P)$ (Here, and throughout the text, we do assume that $\mathcal{F}$ is the sigma-field generated by B.). This means that B is a centered Gaussian process with the covariance function $E[B_s B_t] = R_H(s,t)$, where

$$R_H(s,t) = \frac{1}{2}\left(t^{2H} + s^{2H} - |t-s|^{2H}\right). \qquad (17.12)$$

If $H = \frac{1}{2}$, then B is a Brownian motion. From Eq. (17.12), one can easily see that $E\left[|B_t - B_s|^2\right] = |t-s|^{2H}$, so B has γ–Hölder continuous paths for any $\gamma \in (0, H)$ thanks to the Kolmogorov–Centsov theorem.

2.1 Space of Deterministic Integrands

We denote by $\mathcal{E}$ the set of step $\mathbb{R}$-valued functions on $[0,T]$. Let $\mathcal{H}$ be the Hilbert space defined as the closure of $\mathcal{E}$ with respect to the scalar product

$$\left\langle \mathbf{1}_{[0,t]}, \mathbf{1}_{[0,s]} \right\rangle_{\mathcal{H}} = R_H(t,s).$$

We denote by $|\cdot|_{\mathcal{H}}$ the associated norm. The mapping $\mathbf{1}_{[0,t]} \mapsto B_t$ can be extended to an isometry between $\mathcal{H}$ and the Gaussian space associated with B. We denote this isometry by

$$\varphi \mapsto B(\varphi) = \int_0^T \varphi(s) \mathrm{d}B_s. \qquad (17.13)$$

When $H \in (\frac{1}{2}, 1)$, it follows from [9] that the elements of $\mathcal{H}$ may not be functions but distributions of negative order. It will be more convenient to work with a subspace of $\mathcal{H}$ which contains only functions. Such a space is the set $|\mathcal{H}|$ of all measurable functions φ on $[0,T]$ such that

$$|\varphi|^2_{|\mathcal{H}|} := H(2H-1)\int_0^T \int_0^T |\varphi(u)||\varphi(v)||u-v|^{2H-2}\mathrm{d}u\mathrm{d}v < \infty.$$

If $\varphi, \psi \in |\mathcal{H}|$ then

$$E\left[B(\varphi)B(\psi)\right] = H(2H-1)\int_0^T \int_0^T \varphi(u)\psi(v)|u-v|^{2H-2}\mathrm{d}u\mathrm{d}v. \qquad (17.14)$$

We know that $(|\mathcal{H}|, \langle\cdot,\cdot\rangle_{|\mathcal{H}|})$ is a Banach space, but that $(|\mathcal{H}|, \langle\cdot,\cdot\rangle_{\mathcal{H}})$ is not complete (see, e.g., [9]). We have the dense inclusions $L^2([0,T]) \subset L^{\frac{1}{H}}([0,T]) \subset |\mathcal{H}| \subset \mathcal{H}$.

2.2 Malliavin Derivative and Skorohod Integral

Let $\mathcal{S}$ be the set of all smooth cylindrical random variables, which can be expressed as $F = f(B(\phi_1), \ldots, B(\phi_n))$ where $n \geq 1$, $f : \mathbb{R}^n \to \mathbb{R}$ is a C^∞-function such that f and all its derivatives have at most polynomial growth, and $\phi_i \in \mathcal{H}$, $i =$

$1,\ldots,n$. The Malliavin derivative of F with respect to B is the element of $L^2(\Omega,\mathcal{H})$ defined by

$$D_sF = \sum_{i=1}^{n} \frac{\partial f}{\partial x_i}(B(\phi_1),\ldots,B(\phi_n))\phi_i(s), \quad s \in [0,T].$$

In particular $D_sB_t = \mathbf{1}_{[0,t]}(s)$. As usual, $\mathbb{D}^{1,2}$ denotes the closure of the set of smooth random variables with respect to the norm

$$\|F\|_{1,2}^2 = E[F^2] + E\big[|DF|_{\mathcal{H}}^2\big].$$

The Malliavin derivative D verifies the chain rule: if $\varphi : \mathbb{R}^n \to \mathbb{R}$ is C_b^1 and if $(F_i)_{i=1,\ldots,n}$ is a sequence of elements of $\mathbb{D}^{1,2}$, then $\varphi(F_1,\ldots,F_n) \in \mathbb{D}^{1,2}$ and we have, for any $s \in [0,T]$,

$$D_s\varphi(F_1,\ldots,F_n) = \sum_{i=1}^{n} \frac{\partial \varphi}{\partial x_i}(F_1,\ldots,F_n)D_sF_i.$$

The Skorohod integral δ is the adjoint of the derivative operator D. If a random variable $u \in L^2(\Omega,\mathcal{H})$ belongs to the domain of the Skorohod integral (denoted by domδ), that is, if it verifies

$$|E\langle DF,u\rangle_{\mathcal{H}}| \le c_u\sqrt{E[F^2]} \quad \text{for any } F \in \mathcal{S},$$

then $\delta(u)$ is defined by the duality relationship

$$E[F\delta(u)] = E\big[\langle DF,u\rangle_{\mathcal{H}}\big],$$

for every $F \in \mathbb{D}^{1,2}$. In the sequel, when $t \in [0,T]$ and $u \in \text{dom}\delta$, we shall sometimes write $\int_0^t u_s\delta B_s$ instead of $\delta(u\mathbf{1}_{[0,t]})$. If $h \in \mathcal{H}$, notice moreover that $\int_0^T h_s\delta B_s = \delta(h) = B(h)$.

For every $q \ge 1$, let $\mathcal{H}_q$ be the qth Wiener chaos of B, that is, the closed linear subspace of $L^2(\Omega)$ generated by the random variables $\{H_q(B(h)), h \in \mathcal{H}, \|h\|_{\mathcal{H}} = 1\}$, where H_q is the qth Hermite polynomial. The mapping $I_q(h^{\otimes q}) = H_q(B(h))$ provides a linear isometry between the symmetric tensor product $\mathcal{H}^{\odot q}$ (equipped with the modified norm $\|\cdot\|_{\mathcal{H}^{\odot q}} = \frac{1}{\sqrt{q!}}\|\cdot\|_{\mathcal{H}^{\otimes q}}$) and $\mathcal{H}_q$. Specifically, for all $f,g \in \mathcal{H}^{\odot q}$ and $q \ge 1$, one has

$$E\big[I_q(f)I_q(g)\big] = q!\langle f,g\rangle_{\mathcal{H}^{\otimes q}}. \tag{17.15}$$

On the other hand, it is well known that any random variable Z belonging to $L^2(\Omega)$ admits the following chaotic expansion:

$$Z = E[Z] + \sum_{q=1}^{\infty} I_q(f_q), \tag{17.16}$$

where the series converges in $L^2(\Omega)$ and the kernels f_q, belonging to $\mathcal{H}^{\odot q}$, are uniquely determined by Z.

2.3 Young Integral

For any $\gamma \in [0,1]$, we denote by $\mathscr{C}^\gamma([0,T])$ the set of γ-Hölder continuous functions, that is, the set of functions $f : [0,T] \to \mathbb{R}$ such that

$$|f|_\gamma := \sup_{0 \le s < t \le T} \frac{|f(t) - f(s)|}{(t-s)^\gamma} < \infty.$$

(Notice the calligraphic difference between a space $\mathscr{C}$ of Hölder continuous functions and a space C of continuously differentiable functions!) We also set $|f|_\infty = \sup_{t \in [0,T]} |f(t)|$, and we equip $\mathscr{C}^\gamma([0,T])$ with the norm

$$\|f\|_\gamma := |f|_\gamma + |f|_\infty.$$

Let $f \in \mathscr{C}^\gamma([0,T])$, and consider the operator $T_f : C^1([0,T]) \to C^0([0,T])$ defined as

$$T_f(g)(t) = \int_0^t f(u) g'(u) \mathrm{d}u, \quad t \in [0,T].$$

It can be shown (see, e.g., [10, Sect. 2.2]) that, for any $\beta \in (1-\gamma, 1)$, there exists a constant $C_{\gamma,\beta,T} > 0$ depending only on γ, β, and T such that, for any $g \in \mathscr{C}^\beta([0,T])$,

$$\left\| \int_0^\cdot f(u) g'(u) \mathrm{d}u \right\|_\beta \le C_{\gamma,\beta,T} \|f\|_\gamma \|g\|_\beta.$$

We deduce that, for any $\gamma \in (0,1)$, any $f \in \mathscr{C}^\gamma([0,T])$ and any $\beta \in (1-\gamma,1)$, the linear operator $T_f : C^1([0,T]) \subset \mathscr{C}^\beta([0,T]) \to \mathscr{C}^\beta([0,T])$, defined as $T_f(g) = \int_0^\cdot f(u) g'(u) \mathrm{d}u$, is continuous with respect to the norm $\| \cdot \|_\beta$. By density, it extends (in a unique way) to an operator defined on $\mathscr{C}^\beta$. As a consequence, if $f \in \mathscr{C}^\gamma([0,T])$, if $g \in \mathscr{C}^\beta([0,T])$, and if $\gamma + \beta > 1$, then the (so-called) Young integral $\int_0^\cdot f(u) \mathrm{d}g(u)$ is (well) defined as being $T_f(g)$.

The Young integral obeys the following chain rule. Let $\phi : \mathbb{R}^2 \to \mathbb{R}$ be a C^2 function, and let $f, g \in \mathscr{C}^\gamma([0,T])$ with $\gamma > \frac{1}{2}$. Then $\int_0^\cdot \frac{\partial \phi}{\partial f}(f(u), g(u)) \mathrm{d}f(u)$ and $\int_0^\cdot \frac{\partial \phi}{\partial g}(f(u), g(u)) \mathrm{d}g(u)$ are well defined as Young integrals. Moreover, for all $t \in [0,T]$,

$$\phi(f(t), g(t)) = \phi(f(0), g(0)) + \int_0^t \frac{\partial \phi}{\partial f}(f(u), g(u)) \mathrm{d}f(u) + \int_0^t \frac{\partial \phi}{\partial g}(f(u), g(u)) \mathrm{d}g(u). \tag{17.17}$$

2.4 Link Between Young and Skorohod Integrals

Assume $H > \frac{1}{2}$, and let $u = (u_t)_{t \in [0,T]}$ be a process with paths in $\mathscr{C}^\gamma([0,T])$ for some fixed $\gamma > 1 - H$. Then, according to the previous section, the integral $\int_0^T u_s \mathrm{d}B_s$ exists pathwise in the Young sense. Suppose moreover that u_t belongs to $\mathbb{D}^{1,2}$ for all $t \in [0,T]$, and that u satisfies

$$P\left(\int_0^T\int_0^T |D_s u_t||t-s|^{2H-2}\mathrm{d}s\mathrm{d}t < \infty\right) = 1.$$

Then $u \in \mathrm{dom}\delta$, and we have [1], for all $t \in [0, T]$,

$$\int_0^t u_s \mathrm{d}B_s = \int_0^t u_s \delta B_s + H(2H-1)\int_0^t\int_0^t D_s u_x |x-s|^{2H-2}\mathrm{d}s\mathrm{d}x. \quad (17.18)$$

In particular, notice that

$$\int_0^T \varphi_s \mathrm{d}B_s = \int_0^T \varphi_s \delta B_s = B(\varphi) \quad (17.19)$$

when φ is non-random.

3 Statement and Proofs of Our Main Results

In all this section, we fix a fractional Brownian motion B of Hurst index $H \in (\frac{1}{2}, 1)$, as well as a parameter $\alpha > 0$. Let us consider the solution X to Eq. (17.7). It is readily checked that we have the following explicit expression for X_t:

$$X_t = (T-t)^\alpha \int_0^t (T-s)^{-\alpha}\mathrm{d}B_s, \quad t \in [0, T), \quad (17.20)$$

where the integral can be understood either in the Young sense or in the Skorohod sense, see indeed Eq. (17.19).

For convenience, and because it will play an important role in the forthcoming computations, we introduce the following two processes related to X: for $t \in [0, T]$,

$$\xi_t = \int_0^t (T-s)^{-\alpha}\mathrm{d}B_s; \quad (17.21)$$

$$\eta_t = \int_0^t \mathrm{d}B_u (T-u)^{\alpha-1}\int_0^u \mathrm{d}B_s (T-s)^{-\alpha} = \int_0^t (T-u)^{\alpha-1}\xi_u \mathrm{d}B_u. \quad (17.22)$$

In particular, we observe that

$$X_t = (T-t)^\alpha \xi_t \quad \text{and} \quad \int_0^t \frac{X_u}{T-u}\mathrm{d}B_u = \eta_t \quad \text{for } t \in [0, T). \quad (17.23)$$

When α is between 0 and H (resp. $1-H$ and H), in Lemma 3.2 (resp. Lemma 3.3), we shall actually show that the process ξ (resp. η) is well defined on the whole interval $[0, T]$ (notice that we could have had a problem at $t = T$) and that it admits a continuous modification. This is why we may and will assume in the sequel, without loss of generality, that ξ (resp. η) is continuous when $0 < \alpha < H$ (resp. $1-H < \alpha < H$).

Recall the definition (17.2) of $\widehat{\alpha}_t$. By using Eq. (17.7) and then Eq. (17.23), as well as the definitions (17.21) and (17.22), we arrive to the following formula:

$$\alpha - \widehat{\alpha}_t = \frac{\int_0^t X_u (T-u)^{-1} \mathrm{d}B_u}{\int_0^t X_u^2 (T-u)^{-2} \mathrm{d}s} = \frac{\eta_t}{\int_0^t (T-u)^{2\alpha-2} \xi_u^2 \mathrm{d}u}.$$

Thus, in order to prove the convergences (17.8)–(17.11) of the introduction (i.e., our main result!), we are left to study the (joint) asymptotic behaviors of η_t and $\int_0^t (T-u)^{2\alpha-2} \xi_u^2 \mathrm{d}u$ as $t \to T$. The asymptotic behavior of $\int_0^t (T-u)^{2\alpha-2} \xi_u^2 \mathrm{d}u$ is rather easy to derive (see Lemma 3.7), because it looks like a convergence *à la* Cesàro when $\alpha \le \frac{1}{2}$. In contrast, the asymptotic behavior of η_t is more difficult to obtain and will depend on the relative position of α with respect to $1-H$. It is actually the combination of Lemmas 3.1, 3.3, 3.4, 3.5, 3.6 that will allow to derive it for the full range of values of α.

We are now in position to prove our two main results that we restate here as theorems for convenience.

Theorem 3.1. *We have* $\widehat{\alpha}_t \overset{\text{prob.}}{\longrightarrow} \alpha \wedge \frac{1}{2}$ *as* $t \to T$. *When* $\alpha < H$ *we have almost sure convergence as well.*

As a corollary, we find that $\widehat{\alpha}_t$ is a strong consistent estimator of α if and only if $\alpha \le \frac{1}{2}$. The next result precises the associated rate of convergence in this case.

Theorem 3.2. *Let* $G \sim \mathcal{N}(0,1)$ *be independent of* B, *let* $\mathcal{C}(1)$ *stand for the standard Cauchy distribution, and let* $\beta(a,b) = \int_0^1 x^{a-1}(1-x)^{b-1}\mathrm{d}x$ *denote the usual Beta function:*

1. Assume $\alpha \in (0, 1-H)$. *Then, as* $t \to T$,

$$(T-t)^{\alpha-H}\left(\alpha - \widehat{\alpha}_t\right) \overset{\text{law}}{\longrightarrow} (1-2\alpha)\sqrt{H(2H-1)\frac{\beta(2-\alpha-2H, 2H-1)}{1-H-\alpha}} \times \frac{G}{\xi_T}$$

$$\overset{\text{law}}{=} T^{\alpha-H}(1-2\alpha)\sqrt{\frac{(H-\alpha)\beta(2-2H-\alpha, 2H-1)}{(1-H-\alpha)\beta(1-\alpha, 2H-1)}} \times \mathcal{C}(1).$$

2. Assume $\alpha = 1-H$. *Then, as* $t \to T$,

$$\frac{(T-t)^{1-2H}}{\sqrt{|\log(T-t)|}}(\alpha - \widehat{\alpha}_t) \overset{\text{law}}{\longrightarrow} (2H-1)^{\frac{3}{2}}\sqrt{2H\,\beta(1-H, 2H-1)} \times \frac{G}{\xi_T}$$

$$\overset{\text{law}}{=} T^{1-2H}(2H-1)^{\frac{3}{2}}\sqrt{\frac{2\,\beta(1-H, 2H-1)}{\beta(H, 2H-1)}} \times \mathcal{C}(1).$$

3. *Assume* $\alpha \in \left(1-H, \frac{1}{2}\right)$. *Then, as* $t \to T$,

$$(T-t)^{2\alpha-1}\big(\alpha - \widehat{\alpha}_t\big) \overset{\text{a.s.}}{\longrightarrow} \frac{(1-2\alpha)\,\eta_T}{(\xi_T)^2}.$$

4. *Assume* $\alpha = \frac{1}{2}$. *Then, as* $t \to T$,

$$|\log(T-t)|\big(\alpha - \widehat{\alpha}_t\big) \overset{\text{a.s.}}{\longrightarrow} \frac{1}{2}.$$

The rest of this section is devoted to the proofs of Theorems 3.1 and 3.2. Before to be in position to do so, we need to state and prove some auxiliary lemmas. In what follows we use the same symbol c for all constants whose precise value is not important for our consideration.

Lemma 3.1. *Let* $\alpha, \beta \in (0,1)$ *be such that* $\alpha + \beta < 2H$. *Then, for all* $T > 0$,

$$\int_0^T \mathrm{d}s\,(T-s)^{-\beta}\int_0^T \mathrm{d}r\,(T-r)^{-\alpha}|s-r|^{2H-2} = \int_0^T \mathrm{d}s\,s^{-\beta}\int_0^T \mathrm{d}r\,r^{-\alpha}|s-r|^{2H-2} < \infty.$$

Proof. By homogeneity, we first notice that

$$\int_0^T \mathrm{d}s\,s^{-\beta}\int_0^T \mathrm{d}r\,r^{-\alpha}|s-r|^{2H-2} = T^{2H-\alpha-\beta}\int_0^1 \mathrm{d}s\,s^{-\beta}\int_0^1 \mathrm{d}r\,r^{-\alpha}|s-r|^{2H-2}$$

so that it is not a loss of generality to assume in the proof that $T=1$. If $\alpha+1<2H$, then $\int_0^{1/s} r^{-\alpha}|1-r|^{2H-2}\mathrm{d}r \le c s^{-2H+1+\alpha}$, implying in turn

$$\int_0^1 \mathrm{d}s\,s^{-\beta}\int_0^1 \mathrm{d}r\,r^{-\alpha}|s-r|^{2H-2} = \int_0^1 \mathrm{d}s\,s^{2H-\alpha-\beta-1}\int_0^{1/s} \mathrm{d}r\,r^{-\alpha}|1-r|^{2H-2}$$
$$\le c\int_0^1 s^{-\beta}\mathrm{d}s < \infty.$$

If $\alpha+1=2H$, then $\int_0^{1/s} r^{1-2H}|1-r|^{2H-2}\mathrm{d}r \le c(1+|\log s|)$, implying in turn

$$\int_0^1 \mathrm{d}s\,s^{-\beta}\int_0^1 \mathrm{d}r\,r^{-\alpha}|s-r|^{2H-2} = \int_0^1 \mathrm{d}s\,s^{-\beta}\int_0^1 \mathrm{d}r\,r^{1-2H}|s-r|^{2H-2}$$
$$= \int_0^1 \mathrm{d}s\,s^{-\beta}\int_0^{1/s} \mathrm{d}r\,r^{1-2H}|1-r|^{2H-2} \le c\int_0^1 s^{-\beta}\big(1+|\log s|\big)\mathrm{d}s < \infty.$$

Finally, if $\alpha+1>2H$, then

$$\int_0^1 \mathrm{d}s\,s^{-\beta}\int_0^1 \mathrm{d}r\,r^{-\alpha}|s-r|^{2H-2} = \int_0^1 \mathrm{d}s\,s^{2H-\alpha-\beta-1}\int_0^{1/s} \mathrm{d}r\,r^{-\alpha}|1-r|^{2H-2}$$
$$\le \int_0^1 s^{2H-\alpha-\beta-1}\mathrm{d}s \times \int_0^\infty r^{-\alpha}|1-r|^{2H-2}\mathrm{d}r < \infty.$$

□

Lemma 3.2. *Assume $\alpha \in (0, H)$. Recall the definition (17.21) of ξ_t. Then $\xi_T := \lim_{t\to T} \xi_t$ exists in L^2. Moreover, for all $\varepsilon \in (0, H - \alpha)$, the process $\{\xi_t\}_{t\in[0,T]}$ admits a modification with $(H - \alpha - \varepsilon)$-Hölder continuous paths, still denoted ξ in the sequel. In particular, $\xi_t \to \xi_T$ almost surely as $t \to T$.*

Proof. Because $\alpha < H$, by Lemma 3.1 we have that $\int_0^T \mathrm{d}s\, s^{-\alpha} \int_0^T \mathrm{d}u\, u^{-\alpha}|s - u|^{2H-2} < \infty$. For all $s \le t < T$, we thus have, using Eq. (17.14) to get the first equality,

$$
\begin{aligned}
&E\left[(\xi_t - \xi_s)^2\right] \\
&= H(2H-1)\int_s^t \mathrm{d}u (T-u)^{-\alpha}\int_s^t \mathrm{d}v (T-v)^{-\alpha}|v-u|^{2H-2} \\
&= H(2H-1)\int_{T-t}^{T-s} \mathrm{d}u\, u^{-\alpha}\int_{T-t}^{T-s} \mathrm{d}v\, v^{-\alpha}|v-u|^{2H-2} \\
&= H(2H-1)\int_0^{t-s} \mathrm{d}u\,(u+T-t)^{-\alpha}\int_0^{t-s} \mathrm{d}v\,(v+T-t)^{-\alpha}|v-u|^{2H-2} \\
&\le H(2H-1)\int_0^{t-s} \mathrm{d}u\, u^{-\alpha}\int_0^{t-s} \mathrm{d}v\, v^{-\alpha}|v-u|^{2H-2} \\
&= H(2H-1)(t-s)^{2H-2\alpha}\int_0^{1} \mathrm{d}u\, u^{-\alpha}\int_0^{1} \mathrm{d}v\, v^{-\alpha}|v-u|^{2H-2} \\
&= c(t-s)^{2H-2\alpha}.
\end{aligned}
$$

By the Cauchy criterion, we deduce that $\xi_T := \lim_{t\to T} \xi_t$ exists in L^2. Moreover, because the process ξ is centered and Gaussian, the Kolmogorov–Centsov theorem applies as well, thus leading to the desired conclusion. □

Lemma 3.3. *Assume $\alpha \in (1 - H, H)$. Recall the definition (17.22) of η_t. Then $\eta_T := \lim_{t\to T} \eta_t$ exists in L^2. Moreover, there exist $\gamma > 0$ such that $\{\eta_t\}_{t\in[0,T]}$ admits a modification with γ-Hölder continuous paths, still denoted η in the sequel. In particular, $\eta_t \to \eta_T$ almost surely as $t \to T$.*

Proof. As a first step, fix $\beta_1, \beta_2 \in (1 - H, H)$ and let us show that there exists $\varepsilon = \varepsilon(\beta_1, \beta_2, H) > 0$ and $c = c(\beta_1, \beta_2, H) > 0$ such that, for all $0 \le s \le t \le T$,

$$
\int_{[0,t]\times[s,t]} (T-u)^{-\beta_1}(T-v)^{-\beta_2}|u-v|^{2H-2}\mathrm{d}u\mathrm{d}v \le c(t-s)^{\varepsilon}. \tag{17.24}
$$

Indeed, we have

$$
\begin{aligned}
&\int_{[0,t]\times[s,t]} (T-u)^{-\beta_1}(T-v)^{-\beta_2}|u-v|^{2H-2}\mathrm{d}u\mathrm{d}v \\
&\quad = \int_{T-t}^{T} \mathrm{d}u\, u^{-\beta_1} \int_{T-t}^{T-s} \mathrm{d}v\, v^{-\beta_2}|u-v|^{2H-2} \\
&\quad = \int_0^t \mathrm{d}u(u+T-t)^{-\beta_1} \int_0^{t-s} \mathrm{d}v(v+T-t)^{-\beta_2}|u-v|^{2H-2} \\
&\quad \le \int_0^t \mathrm{d}u\, u^{-\beta_1} \int_0^{t-s} \mathrm{d}v\, v^{-\beta_2}|u-v|^{2H-2} \\
&\quad = \int_0^{t-s} \mathrm{d}u\, u^{-\beta_1} \int_0^{t-s} \mathrm{d}v\, v^{-\beta_2}|u-v|^{2H-2} \\
&\qquad + \int_{t-s}^{t} \mathrm{d}u\, u^{-\beta_1} \int_0^{t-s} \mathrm{d}v\, v^{-\beta_2}(u-v)^{2H-2} \\
&\quad = (t-s)^{2H-\beta_1-\beta_2} \int_0^1 \mathrm{d}u\, u^{-\beta_1} \int_0^1 \mathrm{d}v\, v^{-\beta_2}|u-v|^{2H-2} \\
&\qquad + \int_{t-s}^{t} \mathrm{d}u\, u^{-\beta_1-\beta_2+2H-1} \int_0^{(t-s)/u} \mathrm{d}v\, v^{-\beta_2}(1-v)^{2H-2} \\
&\quad \le c(t-s)^{2H-\beta_1-\beta_2} + c(t-s)^{1-\beta_2} \\
&\qquad \times \int_{t-s}^{T} \mathrm{d}u\, u^{-\beta_1+2H-2} \quad \text{(see Lemma 3.1 for the first integral and} \\
&\qquad\qquad \text{use } 1-v \le 1 \text{ for the second one)} \\
&\quad = c(t-s)^{2H-\beta_1-\beta_2} + c(t-s)^{1-\beta_2} \\
&\qquad \times \begin{cases} 1 & \text{if } \beta_1 < 2H-1 \\ 1+|\log(t-s)| & \text{if } \beta_1 = 2H-1 \\ (t-s)^{2H-1-\beta_1} & \text{if } \beta_1 > 2H-1 \end{cases} \\
&\quad \le c(t-s)^{\varepsilon} \quad \text{for some } \varepsilon \in (0, 1\wedge(2H-\beta_1)-\beta_2);
\end{aligned}
$$

hence Eq. (17.24) is shown.

Now, let $t < T$. Using Eq. (17.18), we can write

$$
\begin{aligned}
\eta_t = &\int_0^t \xi_u (T-u)^{\alpha-1}\delta B_u + H(2H-1)\int_0^t \mathrm{d}u(T-u)^{\alpha-1} \\
&\times \int_0^u \mathrm{d}v(T-v)^{-\alpha}(u-v)^{2H-2}. \qquad (17.25)
\end{aligned}
$$

To have the right to write Eq. (17.25), according to Sect. 2.4 we must check that: (i) $u \to (T-u)^{\alpha-1}\xi_u$ belongs almost surely to $\mathscr{C}^\gamma([0,t])$ for some $\gamma > 1-H$; (ii) $\xi_u \in \mathbb{D}^{1,2}$ for all $u \in [0,t]$, and (iii) $\int_{[0,t]^2}(T-u)^{\alpha-1}|D_v\xi_u|\,|u-v|^{2H-2}\mathrm{d}u\mathrm{d}v < \infty$ almost surely. To keep the length of this paper within bounds, we will do it completely here, and this will serve as a basis for the proof of the other instances where a similar verification should have been made as well. The main reason why (i) to (iii) are easy to check is because we are integrating on the compact interval $[0,t]$ with t *strictly less than* T.

Proof of (i). Firstly, $u \to (T-u)^{\alpha-1}$ is $\mathcal{C}^\infty$ and bounded on $[0,t]$. Secondly, for $u, v \in [0,t]$ with, say, $u < v$, we have

$$E[(\xi_u - \xi_v)^2] = H(2H-1)\int_u^v \mathrm{d}x(T-x)^{-\alpha}\int_u^v \mathrm{d}y(T-y)^{-\alpha}|y-x|^{2H-2}$$

$$\leq (T-t)^{-2\alpha}H(2H-1)\int_u^v \mathrm{d}x\int_u^v \mathrm{d}y|y-x|^{2H-2} = (T-t)^{-2\alpha}|v-u|^{2H}.$$

Hence, by combining the Kolmogorov–Centsov theorem with the fact that ξ is Gaussian, we get that (almost) all the sample paths of ξ are θ-Hölderian on $[0,t]$ for any $\theta \in (0,H)$. Consequently, by choosing $\gamma \in (1-H, H)$ (which is possible since $H > 1/2$), the proof of (i) is concluded. □

Proof of (ii). This is evident, using the representation (17.21) of ξ as well as the fact that $s \to (T-s)^{-\alpha}\mathbf{1}_{[0,t]}(s) \in |\mathcal{H}|$, see Sect. 2.1. □

Proof of (iii). Here again, it is easy: indeed, we have $D_v\xi_u = (T-v)^{-\alpha}\mathbf{1}_{[0,u]}(v)$, so

$$\int_{[0,t]^2}(T-u)^{\alpha-1}|D_v\xi_u|\,|u-v|^{2H-2}\mathrm{d}u\mathrm{d}v$$

$$= \int_{[0,t]^2}(T-u)^{\alpha-1}(T-v)^{-\alpha}\,|u-v|^{2H-2}\mathrm{d}u\mathrm{d}v < \infty.\Big)$$

Let us go back to the proof. We deduce from Eq. (17.25), after setting

$$\varphi_t(u,v) = \frac{1}{2}(T-u\vee v)^{\alpha-1}(T-u\wedge v)^{-\alpha}\,\mathbf{1}_{[0,t]^2}(u,v),$$

that

$$\eta_t = I_2(\varphi_t) + H(2H-1)\int_0^t \mathrm{d}u(T-u)^{\alpha-1}\int_0^u \mathrm{d}v(T-v)^{-\alpha}(u-v)^{2H-2}.$$

Hence, because of Eq. (17.15),

$$E\left[(\eta_t - \eta_s)^2\right] = 2\|\varphi_t - \varphi_s\|^2_{\mathcal{H}^{\otimes 2}} + H^2(2H-1)^2$$

$$\times\left(\int_s^t \mathrm{d}u(T-u)^{\alpha-1}\int_0^u \mathrm{d}v(T-v)^{-\alpha}(u-v)^{2H-2}\right)^2. \quad (17.26)$$

We have, by observing that $\varphi_t - \varphi_s \in |\mathcal{H}|^{\odot 2}$,

$$\begin{aligned}\|\varphi_t - \varphi_s\|^2_{\mathcal{H}^{\otimes 2}} &= H^2(2H-1)^2 \int_{[0,T]^4} \big[\varphi_t(u,v) - \varphi_s(u,v)\big] \\ &\quad \times \big[\varphi_t(x,y) - \varphi_s(x,y)\big] |u-x|^{2H-2} |v-y|^{2H-2} \mathrm{d}u\mathrm{d}v\mathrm{d}x\mathrm{d}y \\ &= \frac{1}{4} H^2(2H-1)^2 \int_{([0,t]^2 \setminus [0,s]^2)^2} (T - u\vee v)^{\alpha-1} (T - x \vee y)^{\alpha-1} \\ &\quad \times (T - u\wedge v)^{-\alpha} (T - x\wedge y)^{-\alpha} |u-x|^{2H-2} |v-y|^{2H-2} \mathrm{d}u\mathrm{d}v\mathrm{d}x\mathrm{d}y.\end{aligned}$$

Taking into account the form of the domain in the previous integral and using that $\varphi_t - \varphi_s$ is symmetric, we easily show that $\|\varphi_t - \varphi_s\|^2_{\mathcal{H}^{\otimes 2}}$ is upper bounded (up to constant and without seeking for sharpness) by a sum of integrals of the type

$$\begin{aligned}&\int_{[0,t]\times[s,t]\times[0,T]^2} (T-u)^{-\beta_1} (T-v)^{-\beta_2} (T-x)^{-\beta_3} \\ &\quad \times (T-y)^{-\beta_4} |u-x|^{2H-2} |v-y|^{2H-2} \mathrm{d}u\mathrm{d}v\mathrm{d}x\mathrm{d}y,\end{aligned}$$

with $\beta_1, \beta_2, \beta_3, \beta_4 \in \{\alpha, 1-\alpha\}$. Hence, combining Lemma 3.1 with Eq. (17.24), we deduce that there exists $\varepsilon > 0$ small enough and $c > 0$ such that, for all $s, t \in [0, T]$,

$$\|\varphi_t - \varphi_s\|^2_{\mathcal{H}^{\otimes 2}} \le c|t-s|^{\varepsilon}. \tag{17.27}$$

On the other hand, we can write, for all $s \le t < T$,

$$\begin{aligned}&\int_s^t \mathrm{d}u (T-u)^{\alpha-1} \int_0^u \mathrm{d}v (T-v)^{-\alpha} (u-v)^{2H-2} \\ &= \int_{T-t}^{T-s} \mathrm{d}u\, u^{\alpha-1} \int_u^T \mathrm{d}v\, v^{-\alpha} (v-u)^{2H-2} \\ &= \int_0^{t-s} \mathrm{d}u (u+T-t)^{\alpha-1} \int_u^t \mathrm{d}v (v+T-t)^{-\alpha} (v-u)^{2H-2} \\ &\le \int_0^{t-s} \mathrm{d}u\, u^{\alpha-1} \int_u^T \mathrm{d}v\, v^{-\alpha} (v-u)^{2H-2} \\ &= (t-s)^{2H-1} \int_0^1 \mathrm{d}u\, u^{\alpha-1} \int_u^{\frac{T}{t-s}} \mathrm{d}v\, v^{-\alpha} (v-u)^{2H-2} \\ &= (t-s)^{2H-1} \int_0^1 \mathrm{d}u\, u^{2H-2} \int_1^{\frac{T}{(t-s)u}} \mathrm{d}v\, v^{-\alpha} (v-1)^{2H-2}.\end{aligned} \tag{17.28}$$

Let us consider three cases. Assume first that $\alpha > 2H-1$: in this case,

$$\int_1^{\frac{T}{(t-s)u}} v^{-\alpha} (v-1)^{2H-2} \mathrm{d}v \le \int_1^{\infty} v^{-\alpha} (v-1)^{2H-2} \mathrm{d}v < \infty;$$

leading, thanks to Eq. (17.28), to

$$\int_s^t du(T-u)^{\alpha-1}\int_0^u dv(T-v)^{-\alpha}(u-v)^{2H-2} \le c(t-s)^{2H-1}.$$

The second case is when $\alpha = 2H-1$: we then have

$$\int_1^{\frac{T}{(t-s)u}} v^{-\alpha}(v-1)^{2H-2}dv \le c\big(1+|\log(t-s)|+|\log u|\big)$$

so that, by Eq. (17.28),

$$\int_s^t du(T-u)^{\alpha-1}\int_0^u dv(T-v)^{-\alpha}(u-v)^{2H-2} \le c(t-s)^{2H-1}\big(1+|\log(t-s)|\big).$$

Finally, the third case is when $\alpha < 2H-1$: in this case,

$$\int_1^{\frac{T}{(t-s)u}} v^{-\alpha}(v-1)^{2H-2}dv \le c(t-s)^{\alpha-2H+1}u^{\alpha-2H+1};$$

so that, by Eq. (17.28),

$$\int_s^t du(T-u)^{\alpha-1}\int_0^u dv(T-v)^{-\alpha}(u-v)^{2H-2} \le c(t-s)^{\alpha}.$$

To summarize, we have shown that there exists $c>0$ such that, for all $s,t\in[0,T]$,

$$\int_s^t du(T-u)^{\alpha-1}\int_0^u dv(T-v)^{-\alpha}(u-v)^{2H-2}$$
$$\le c\big(1+|\log(|t-s|)|\mathbf{1}_{\{\alpha=2H-1\}}\big)|t-s|^{(2H-1)\wedge\alpha}. \tag{17.29}$$

By inserting Eqs. (17.27) and (17.29) into Eq. (17.26), we finally get that there exists $\varepsilon>0$ small enough and $c>0$ such that, for all $s,t\in[0,T]$,

$$E\left[(\eta_t-\eta_s)^2\right] \le c|t-s|^{\varepsilon}.$$

By the Cauchy criterion, we deduce that $\eta_T := \lim_{t\to T}\eta_t$ exists in L^2. Moreover, because $\eta_t-\eta_s-E[\eta_t]+E[\eta_s]$ belongs to the second Wiener chaos of B (where all the L^p norms are equivalent), the Kolmogorov–Centsov theorem applies as well, thus leading to the desired conclusion. □

Lemma 3.4. *Recall the definition (17.22) of η_t. For any $t\in[0,T)$, we have*

$$\eta_t = \int_0^t (T-u)^{\alpha-1}dB_u \times \int_0^t (T-s)^{-\alpha}dB_s - \int_0^t \delta B_s\,(T-s)^{-\alpha}\int_0^s \delta B_u\,(T-u)^{\alpha-1}$$
$$-H(2H-1)\int_0^t ds\,(T-s)^{-\alpha}\int_0^s du\,(T-u)^{\alpha-1}(s-u)^{2H-2}.$$

Proof. Fix $t \in [0,T)$. Applying the change of variable formula (17.17) to the right-hand side of the first equality in Eq. (17.22) leads to

$$\eta_t = \int_0^t (T-u)^{\alpha-1}\mathrm{d}B_u \times \int_0^t (T-s)^{-\alpha}\mathrm{d}B_s - \int_0^t \mathrm{d}B_s\,(T-s)^{-\alpha}\int_0^s \mathrm{d}B_u\,(T-u)^{\alpha-1}. \tag{17.30}$$

On the other hand, by Eq. (17.18) we have that

$$\begin{aligned}\int_0^t \mathrm{d}B_s\,(T-s)^{-\alpha}\int_0^s \mathrm{d}B_u\,(T-u)^{\alpha-1} &= \int_0^t \delta B_s\,(T-s)^{-\alpha}\int_0^s \delta B_u\,(T-u)^{\alpha-1} + H(2H-1)\\ &\times \int_0^t \mathrm{d}s(T-s)^{-\alpha}\int_0^s \mathrm{d}u(T-u)^{\alpha-1}(s-u)^{2H-2}.\end{aligned} \tag{17.31}$$

The desired conclusion follows. (We omit the justification of Eqs. (17.30) and (17.31) because it suffices to proceed as in the proof (17.25).) □

Lemma 3.5. *Let $\beta(a,b) = \int_0^1 x^{a-1}(1-x)^{b-1}\mathrm{d}x$ denote the usual Beta function, let Z be any $\sigma\{B\}$-measurable random variable satisfying $P(Z<\infty)=1$, and let $G \sim \mathcal{N}(0,1)$ be independent of B:*

1. Assume $\alpha \in (0, 1-H)$. Then, as $t \to T$,

$$\left(Z, (T-t)^{1-H-\alpha}\int_0^t (T-u)^{\alpha-1}\mathrm{d}B_u\right) \xrightarrow{\text{law}} \left(Z, \sqrt{H(2H-1)\frac{\beta(2-\alpha-2H, 2H-1)}{1-H-\alpha}}\,G\right). \tag{17.32}$$

2. Assume $\alpha = 1-H$. Then, as $t \to T$,

$$\left(Z, \frac{1}{\sqrt{|\log(T-t)|}}\int_0^t (T-u)^{-H}\mathrm{d}B_u\right) \xrightarrow{\text{law}} \left(Z, \sqrt{2H(2H-1)\beta(1-H, 2H-1)}\,G\right). \tag{17.33}$$

Proof. By a standard approximation procedure, we first notice that it is not a loss of generality to assume that Z belongs to $L^2(\Omega)$ (using, e.g., that $Z\,\mathbf{1}_{\{|Z|\le n\}} \xrightarrow{\text{a.s.}} Z$ as $n \to \infty$):

1. Set $N = \sqrt{H(2H-1)\frac{\beta(2-\alpha-2H,2H-1)}{1-H-\alpha}}\,G$. For any $d \ge 1$ and any $s_1, \ldots, s_d \in [0,T)$, we shall prove that

$$\left(B_{s_1},\ldots,B_{s_d},(T-t)^{1-H-\alpha}\int_0^t(T-u)^{\alpha-1}\mathrm{d}B_u\right)\xrightarrow{\text{law}}\left(B_{s_1},\ldots,B_{s_d},N\right)$$
$$\text{as } t\to T. \tag{17.34}$$

Suppose for a moment that Eq. (17.34) has been shown, and let us proceed with the proof of Eq. (17.32). By the very construction of $\mathcal{H}$ and by reasoning by approximation, we deduce that, for any $l \geq 1$ and any $h_1,\ldots,h_l \in \mathcal{H}$ with unit norms,

$$\left(B(h_1),\ldots,B(h_l),(T-t)^{1-H-\alpha}\int_0^t(T-u)^{\alpha-1}\mathrm{d}B_u\right)$$
$$\xrightarrow{\text{law}}\left(B(h_1),\ldots,B(h_l),N\right)\quad\text{as } t\to T.$$

This implies that, for any $l \geq 1$, any $h_1,\ldots,h_l \in \mathcal{H}$ with unit norms and any integers $q_1,\ldots,q_l \geq 0$,

$$\left(H_{q_1}(B(h_1)),\ldots,H_{q_l}(B(h_l)),(T-t)^{1-H-\alpha}\int_0^t(T-u)^{\alpha-1}\mathrm{d}B_u\right)$$
$$\xrightarrow{\text{law}}\left(H_{q_1}(B(h_1)),\ldots,H_{q_l}(B(h_l)),N\right)\quad\text{as } t\to T,$$

with H_q the qth Hermite polynomial. Using now the very definition of the Wiener chaoses and by reasoning by approximation once again, we deduce that, for any $l \geq 1$, any integers $q_1,\ldots,q_l \geq 0$ and any $f_1 \in \mathcal{H}^{\odot q_1},\ldots,f_l \in \mathcal{H}^{\odot q_l}$,

$$\left(I_{q_1}(f_1),\ldots,I_{q_l}(f_l),(T-t)^{1-H-\alpha}\int_0^t(T-u)^{\alpha-1}\mathrm{d}B_u\right)$$
$$\xrightarrow{\text{law}}\left(I_{q_1}(f_1),\ldots,I_{q_l}(f_l),N\right)\quad\text{as } t\to T.$$

Thus, for any random variable $F \in L^2(\Omega)$ with a *finite* chaotic decomposition, we have

$$\left(F,(T-t)^{1-H-\alpha}\int_0^t(T-u)^{\alpha-1}\mathrm{d}B_u\right)\xrightarrow{\text{law}}(F,N)\quad\text{as } t\to T. \tag{17.35}$$

To conclude, let us consider the chaotic decomposition (17.16) of Z. By applying Eq. (17.35) to $F = E[Z] + \sum_{q=1}^{n} I_q(f_q)$ and then letting $n \to \infty$, we finally deduce that Eq. (17.32) holds true.

Now, let us proceed with the proof of Eq. (17.34). Because the left-hand side of Eq. (17.34) is a Gaussian vector, to get Eq. (17.34) it is sufficient to check the convergence of covariance matrices. Let us first compute the limiting variance of $(T-t)^{1-H-\alpha}\int_0^t(T-u)^{\alpha-1}\mathrm{d}B_u$ as $t \to T$. By Eq. (17.14), for any $t \in [0,T)$ we have

$$
\begin{aligned}
&E\left[\left((T-t)^{1-H-\alpha}\int_0^t (T-u)^{\alpha-1}\mathrm{d}B_u\right)^2\right]\\
&= H(2H-1)(T-t)^{2-2H-2\alpha}\int_0^t \mathrm{d}s(T-s)^{\alpha-1}\int_0^t \mathrm{d}u(T-u)^{\alpha-1}|s-u|^{2H-2}\\
&= H(2H-1)(T-t)^{2-2H-2\alpha}\int_{T-t}^T \mathrm{d}s\, s^{\alpha-1}\int_{T-t}^T \mathrm{d}u\, u^{\alpha-1}|s-u|^{2H-2}\\
&= H(2H-1)\int_1^{\frac{T}{T-t}} \mathrm{d}s\, s^{\alpha-1}\int_1^{\frac{T}{T-t}} \mathrm{d}u\, u^{\alpha-1}|s-u|^{2H-2}\\
&\to H(2H-1)\int_1^{\infty} \mathrm{d}s\, s^{\alpha-1}\int_1^{\infty} \mathrm{d}u\, u^{\alpha-1}|s-u|^{2H-2} \quad \text{as } t\to T,
\end{aligned}
$$

with

$$
\begin{aligned}
&\int_1^{\infty} \mathrm{d}s\, s^{\alpha-1}\int_1^{\infty} \mathrm{d}u\, u^{\alpha-1}|s-u|^{2H-2}\\
&\quad= \int_1^{\infty} \mathrm{d}s\, s^{2\alpha+2H-3}\int_{1/s}^{\infty} \mathrm{d}u\, u^{\alpha-1}|1-u|^{2H-2}\\
&\quad= \int_1^{\infty} s^{2\alpha+2H-3}\mathrm{d}s\int_1^{\infty} u^{\alpha-1}(u-1)^{2H-2}\mathrm{d}u\\
&\qquad+\int_1^{\infty} \mathrm{d}s\, s^{2\alpha+2H-3}\int_{1/s}^{1} \mathrm{d}u\, u^{\alpha-1}(1-u)^{2H-2}\\
&\quad= \frac{\beta(2-\alpha-2H,2H-1)}{2(1-H-\alpha)}+\int_0^1 \mathrm{d}u\, u^{\alpha-1}(1-u)^{2H-2}\int_{1/u}^{\infty} \mathrm{d}s\, s^{2\alpha+2H-3}\\
&\quad= \frac{\beta(2-\alpha-2H,2H-1)}{1-H-\alpha}.
\end{aligned}
$$

Thus,

$$
\begin{aligned}
&\lim_{t\to T} E\left[\left((T-t)^{1-H-\alpha}\int_0^t (T-u)^{\alpha-1}\mathrm{d}B_u\right)^2\right]\\
&\quad= \frac{H(2H-1)}{1-H-\alpha}\beta(2-\alpha-2H,2H-1).
\end{aligned}
$$

On the other hand, by Eq. (17.14) we have, for any $v<t<T$,

$$
\begin{aligned}
&E\left[B_v \times (T-t)^{1-H-\alpha} \int_0^t (T-u)^{\alpha-1} \mathrm{d}B_u\right] \\
&= H(2H-1)(T-t)^{1-H-\alpha} \int_0^t \mathrm{d}u\,(T-u)^{\alpha-1} \int_0^v \mathrm{d}s\,|u-s|^{2H-2} \\
&= H(T-t)^{1-H-\alpha} \int_0^t (T-u)^{\alpha-1}\big(u^{2H-1} + \mathrm{sign}(v-u) \times |v-u|^{2H-1}\big)\mathrm{d}u \\
&\to 0 \quad \text{as } t \to T,
\end{aligned}
$$

because $\int_0^T (T-u)^{\alpha-1}\big(u^{2H-1} + \mathrm{sign}(v-u) \times |v-u|^{2H-1}\big)\mathrm{d}u < \infty$. Convergence (17.34) is then shown, and Eq. (17.32) follows.

2. By Eq. (17.14), for any $t \in [0 \vee (T-1), T)$, we have

$$
\begin{aligned}
&E\left[\left(\frac{1}{\sqrt{|\log(T-t)|}} \int_0^t (T-u)^{-H} \mathrm{d}B_u\right)^2\right] \\
&= \frac{H(2H-1)}{|\log(T-t)|} \int_0^t \mathrm{d}s(T-s)^{-H} \int_0^t \mathrm{d}u(T-u)^{-H}|s-u|^{2H-2} \\
&= \frac{H(2H-1)}{|\log(T-t)|} \int_{T-t}^T \mathrm{d}s\, s^{-H} \int_{T-t}^T \mathrm{d}u\, u^{-H}|s-u|^{2H-2} \\
&= \frac{2H(2H-1)}{|\log(T-t)|} \int_{T-t}^T \mathrm{d}s\, s^{-H} \int_{T-t}^s \mathrm{d}u\, u^{-H}(s-u)^{2H-2} \\
&= \frac{2H(2H-1)}{|\log(T-t)|} \int_{T-t}^T \frac{\mathrm{d}s}{s} \int_{\frac{T-t}{s}}^1 \mathrm{d}u\, u^{-H}(1-u)^{2H-2} \\
&= \frac{2H(2H-1)}{|\log(T-t)|} \int_{\frac{T-t}{T}}^1 \mathrm{d}u\, u^{-H}(1-u)^{2H-2} \int_{\frac{T-t}{u}}^T \frac{\mathrm{d}s}{s} \\
&= 2H(2H-1) \int_{\frac{T-t}{T}}^1 \mathrm{d}u\, u^{-H}(1-u)^{2H-2}\left(1 + \frac{\log(Tu)}{|\log(T-t)|}\right).
\end{aligned}
$$

Because $\int_0^1 |\log(Tu)|u^{-H}(1-u)^{2H-2}\mathrm{d}u < \infty$, we get that

$$
\begin{aligned}
&E\left[\left(\frac{1}{\sqrt{|\log(T-t)|}} \int_0^t (T-s)^{-H} \mathrm{d}B_s\right)^2\right] \\
&\to 2H(2H-1)\beta(1-H, 2H-1) \quad \text{as } t \to T.
\end{aligned}
$$

On the other hand, fix $v \in [0, T)$. For all $t \in [0 \vee (T-1), T)$, using Eq. (17.14) we can write

$$
\begin{aligned}
&E\left[B_v \times \frac{1}{\sqrt{|\log(T-t)|}} \int_0^t (T-u)^{-H} \mathrm{d}B_u\right] \\
&= \frac{H(2H-1)}{\sqrt{|\log(T-t)|}} \int_0^t \mathrm{d}u\,(T-u)^{-H} \int_0^v \mathrm{d}s\,|u-s|^{2H-2} \\
&= \frac{H}{\sqrt{|\log(T-t)|}} \int_0^T (T-u)^{\alpha-1}\big(u^{2H-1} + \mathrm{sign}(v-u) \times |v-u|^{2H-1}\big)\mathrm{d}u \\
&\longrightarrow 0 \quad \text{as } t \to T
\end{aligned}
$$

because $\int_0^T (T-u)^{\alpha-1}\big(u^{2H-1} + \mathrm{sign}(v-u) \times |v-u|^{2H-1}\big)\mathrm{d}u < \infty$. Thus, we have shown that, for any $d \geq 1$ and any $s_1, \ldots, s_d \in [0, T)$,

$$
\begin{aligned}
&\left(B_{s_1}, \ldots, B_{s_d}, (T-t)^{1-H-\alpha} \int_0^t (T-u)^{\alpha-1} \mathrm{d}B_u\right) \\
&\qquad \xrightarrow{\text{law}} \left(B_{s_1}, \ldots, B_{s_d}, \sqrt{2H(2H-1)\beta(1-H, 2H-1)}\, G\right) \qquad (17.36)
\end{aligned}
$$

as $t \to T$. Finally, the same reasoning as in point 1 above allows to go from Eqs. (17.36) to (17.33). The proof of the lemma is concluded. □

Lemma 3.6. *Assume $\alpha \in (0, 1-H]$. Then, as $t \to T$,*

$$
\limsup_{t \to T} E\left[\left(\int_0^t \delta B_u\,(T-u)^{-\alpha} \int_0^s \delta B_v\,(T-v)^{\alpha-1}\right)^2\right] < \infty.
$$

Proof. Set $\phi_t(u,v) = \frac{1}{2}(T-u\vee v)^{-\alpha}(T-u\wedge v)^{\alpha-1}\mathbf{1}_{[0,t]^2}(u,v)$. We have $\phi_t \in |\mathcal{H}|^{\odot 2}$ and $\int_0^t \delta B_u\,(T-u)^{-\alpha} \int_0^u \delta B_v\,(T-v)^{\alpha-1} = I_2(\phi_t)$ so that

$$
\begin{aligned}
&\limsup_{t \to T} E\left[\left(\int_0^t \delta B_u\,(T-u)^{-\alpha} \int_0^u \delta B_v\,(T-v)^{\alpha-1}\right)^2\right] = 2 \limsup_{t \to T} \|\phi_t\|^2_{\mathcal{H}^{\otimes 2}} \\
&= 2H^2(2H-1)^2 \limsup_{t \to T} \int_{[0,T]^4} \phi_t(u,v)\phi_t(x,y)|u-x|^{2H-2} \\
&\quad \times |v-y|^{2H-2} \mathrm{d}u\mathrm{d}v\mathrm{d}x\mathrm{d}y \\
&= \frac{1}{2} H^2(2H-1)^2 \int_{[0,T]^4} (T-u\vee v)^{-\alpha}(T-u\wedge v)^{\alpha-1}(T-x\vee y)^{-\alpha} \\
&\quad \times (T-x\wedge y)^{\alpha-1}|u-x|^{2H-2}|v-y|^{2H-2}\mathrm{d}u\mathrm{d}v\mathrm{d}x\mathrm{d}y \\
&= 2H^2(2H-1)^2 \int_0^T \mathrm{d}u\,(T-u)^{-\alpha} \int_0^T \mathrm{d}x\,(T-x)^{-\alpha}|u-x|^{2H-2} \\
&\quad \times \int_0^u \mathrm{d}v\,(T-v)^{\alpha-1} \int_0^x \mathrm{d}v\,(T-y)^{\alpha-1}|v-y|^{2H-2}
\end{aligned}
$$

$$
\begin{aligned}
&= 2H^2(2H-1)^2 \int_0^T du\, u^{-\alpha} \int_0^T dx\, x^{-\alpha} |u-x|^{2H-2} \int_u^T dv\, v^{\alpha-1} \\
&\quad \times \int_x^T dy\, y^{\alpha-1} |v-y|^{2H-2} \\
&= 2H^2(2H-1)^2 \int_0^T du\, u^{-\alpha} \int_0^T dx\, x^{-\alpha} |u-x|^{2H-2} \int_u^T dv\, v^{2H+2\alpha-3} \\
&\quad \times \int_{x/v}^{T/v} dy\, y^{\alpha-1} |1-y|^{2H-2}.
\end{aligned}
$$

Because $\alpha \leq 1 - H$ and $H < 1$, we have $\alpha < 2 - 2H$, so that

$$
\int_{x/v}^{T/v} y^{\alpha-1}|1-y|^{2H-2} dy \leq \int_0^\infty y^{\alpha-1}|1-y|^{2H-2} dy < \infty.
$$

Moreover, because $2H + 2\alpha - 3 \leq -1$ due to our assumption on α, we have

$$
\int_u^T dv\, v^{2H+2\alpha-3} \leq c \begin{cases} u^{2H+2\alpha-2} & \text{if } \alpha < 1 - H \\ 1 + |\log u| & \text{if } \alpha = 1 - H \end{cases}.
$$

Consequently, if $\alpha = 1 - H$, then

$$
\begin{aligned}
&\int_0^T du\, u^{-\alpha} \int_0^T dx\, x^{-\alpha} |u-x|^{2H-2} \int_u^T dv\, v^{2H+2\alpha-3} \int_{x/v}^{T/v} dy\, y^{\alpha-1} |1-y|^{2H-2} \\
&\leq c \int_0^T du\, u^{H-1} \big(1 + |\log u|\big) \int_0^T dx\, x^{H-1} |u-x|^{2H-2} \\
&= c \int_0^T du\, u^{4H-3} \big(1 + |\log u|\big) \int_0^{T/u} dx\, x^{H-1} |1-x|^{2H-2} \\
&\leq c \int_0^T du\, u^{4H-3} \big(1 + |\log u|\big) \times \begin{cases} 1 & \text{if } H < \frac{2}{3} \\ 1 + |\log u| & \text{if } H = \frac{2}{3} \\ u^{2-3H} & \text{if } H > \frac{2}{3} \end{cases} \\
&< \infty,
\end{aligned}
$$

and the proof is concluded in this case. Assume now that $\alpha < 1 - H$. Then

$$\begin{aligned}
&\int_0^T du\, u^{-\alpha} \int_0^T dx\, x^{-\alpha} |u-x|^{2H-2} \\
&\qquad \times \int_u^T dv\, v^{2H+2\alpha-3} \int_{x/v}^{T/v} dy\, y^{\alpha-1} |1-y|^{2H-2} \\
&\leq c \int_0^T du\, u^{2H+\alpha-2} \int_0^T dx\, x^{-\alpha} |u-x|^{2H-2} \\
&= c \int_0^T du\, u^{4H-3} \int_0^{T/u} dx\, x^{-\alpha} |1-x|^{2H-2}.
\end{aligned}$$

Let us distinguish three different cases. First, if $\alpha < 2H - 1$ then

$$\int_0^T du\, u^{4H-3} \int_0^{T/u} dx\, x^{-\alpha} |1-x|^{2H-2} \leq c \int_0^T u^{2H-2+\alpha} du < \infty.$$

Second, if $\alpha = 2H - 1$, then

$$\begin{aligned}
\int_0^T du\, u^{4H-3} \int_0^{T/u} dx\, x^{-\alpha} |1-x|^{2H-2} &= \int_0^T du\, u^{4H-3} \int_0^{T/u} dx\, x^{1-2H} |1-x|^{2H-2} \\
&\leq c \int_0^T u^{4H-3} (1 + |\log u|) du < \infty.
\end{aligned}$$

Third, if $\alpha > 2H - 1$, then

$$\begin{aligned}
&\int_0^T du\, u^{4H-3} \int_0^{T/u} dx\, x^{-\alpha} |1-x|^{2H-2} \\
&\leq \int_0^T u^{4H-3} du \int_0^\infty x^{-\alpha} |1-x|^{2H-2} dx < \infty.
\end{aligned}$$

Thus, in all the possible cases we see that $\limsup_{t\to T} E[(\int_0^t \delta B_u (T-u)^{-\alpha} \int_0^u \delta B_v (T-v)^{\alpha-1})^2]$ is finite, and the proof of the lemma is done. □

Lemma 3.7. *Assume $\alpha \in (0, H)$, and recall the definition (17.21) of ξ_t. Then, as $t \to T$:*

1. If $0 < \alpha < \frac{1}{2}$, then

$$(T-t)^{1-2\alpha} \int_0^t \xi_s^2 (T-s)^{2\alpha-2} ds \overset{\text{a.s.}}{\to} \frac{\xi_T^2}{1-2\alpha}.$$

2. *If* $\alpha = \frac{1}{2}$, *then*

$$\frac{1}{|\log(T-t)|}\int_0^t \frac{\xi_s^2}{T-s}\,\mathrm{d}s \overset{\text{a.s.}}{\to} \xi_T^2.$$

3. *If* $\frac{1}{2} < \alpha < H$, *then*

$$\int_0^t \xi_s^2 (T-s)^{2\alpha-2}\,\mathrm{d}s \overset{\text{a.s.}}{\to} \int_0^T \xi_s^2 (T-s)^{2\alpha-2}\,\mathrm{d}s < \infty.$$

Proof. 1. Using the $(\frac{H}{2}-\frac{\alpha}{2})$-Hölderianity of ξ (Lemma 3.2), we can write

$$\begin{aligned}
&\left|(T-t)^{1-2\alpha}\int_0^t \xi_s^2 (T-s)^{2\alpha-2}\,\mathrm{d}s - \frac{\xi_T^2}{1-2\alpha}\right| \\
&\le (T-t)^{1-2\alpha}\int_0^t \left|\xi_s^2-\xi_T^2\right|(T-s)^{2\alpha-2}\,\mathrm{d}s + (T-t)^{1-2\alpha}\frac{T^{2\alpha-1}}{1-2\alpha}\xi_T^2 \\
&\le c|\xi|_\infty (T-t)^{1-2\alpha}\int_0^t (T-s)^{\frac{H}{2}+\frac{3\alpha}{2}-2}\,\mathrm{d}s + (T-t)^{1-2\alpha}\frac{T^{2\alpha-1}}{1-2\alpha}\xi_T^2 \\
&\le c|\xi|_\infty\left((T-t)^{\frac{H}{2}-\frac{\alpha}{2}} + (T-t)^{1-2\alpha}T^{\frac{H}{2}+\frac{3\alpha}{2}-1}\right) + (T-t)^{1-2\alpha}\frac{T^{2\alpha-1}}{1-2\alpha}\xi_T^2 \\
&\to 0 \quad \text{almost surely as } t \to T.
\end{aligned}$$

2. Using the $(\frac{H}{2}-\frac{1}{4})$-Hölderianity of ξ (Lemma 3.2), we can write

$$\begin{aligned}
&\left|\frac{1}{|\log(T-t)|}\int_0^t \frac{\xi_s^2}{T-s}\mathrm{d}s - \xi_T^2\right| \\
&\le \frac{1}{|\log(T-t)|}\int_0^t \frac{|\xi_s^2-\xi_T^2|}{T-s}\mathrm{d}s + \frac{\log(T)}{|\log(T-t)|}\xi_T^2 \\
&\le \frac{c|\xi|_\infty}{|\log(T-t)|}\int_0^t (T-s)^{\frac{H}{2}-\frac{5}{4}}\mathrm{d}s + \frac{\log(T)}{|\log(T-t)|}\xi_T^2 \\
&\le \frac{c|\xi|_\infty}{|\log(T-t)|}\left(T^{\frac{H}{2}-\frac{1}{4}} + (T-t)^{\frac{H}{2}-\frac{1}{4}}\right) + \frac{\log(T)}{|\log(T-t)|}\xi_T^2 \\
&\to 0 \quad \text{almost surely as } t \to T.
\end{aligned}$$

3. By Lemma 3.2, the process ξ is continuous on $[0, T]$, hence integrable. Moreover, $s \mapsto (T-s)^{2\alpha-2}$ is integrable at $s = T$ because $\alpha > \frac{1}{2}$. The convergence in point 3 is then clear, with a finite limit.

□

We are now ready to prove Theorems 3.1 and 3.2.

Proof of Theorem 3.1. Fix $\alpha > 0$. Thanks to the change of variable formula (17.17) (which can be well applied here, as is easily shown by proceeding as in the proof (17.25)), we can write, for any $t \in [0, T)$,

$$\frac{1}{2}(T-t)^{2\alpha-1}\xi_t^2 = \frac{1-2\alpha}{2}\int_0^t (T-u)^{2\alpha-2}\xi_u^2 \mathrm{d}u + \int_0^t (T-u)^{2\alpha-1}\xi_u d\xi_u$$
$$= \frac{1-2\alpha}{2}\int_0^t (T-u)^{2\alpha-2}\xi_u^2 \mathrm{d}u + \eta_t,$$

so that

$$\alpha - \widehat{\alpha}_t = \frac{\xi_t^2}{2(T-t)^{1-2\alpha}\int_0^t \xi_u^2 (T-u)^{2\alpha-2}\mathrm{d}u} + \alpha - \frac{1}{2}. \tag{17.37}$$

When $\alpha \in (0, \frac{1}{2})$, we have $(T-t)^{1-2\alpha}\int_0^t \xi_u^2(T-u)^{2\alpha-2}\mathrm{d}u \overset{\text{a.s.}}{\to} \frac{\xi_T^2}{1-2\alpha}$ (resp. $\xi_t^2 \overset{\text{a.s.}}{\to} \xi_T^2$) as $t \to T$ by Lemma 3.7 (resp. Lemma 3.2); hence, as desired one gets that $\alpha - \widehat{\alpha}_t \overset{\text{a.s.}}{\to} 0$ as $t \to T$.

When $\alpha = \frac{1}{2}$, the identity (17.37) becomes

$$\alpha - \widehat{\alpha}_t = \frac{\xi_t^2}{2\int_0^t \xi_u^2 (T-u)^{-1}\mathrm{d}u}; \tag{17.38}$$

as $t \to T$, we have $\int_0^t \xi_u^2(T-u)^{-1}\mathrm{d}u \overset{\text{a.s.}}{\sim} |\log(T-t)|\xi_T^2$ (resp. $\xi_t^2 \overset{\text{a.s.}}{\to} \xi_T^2$) by Lemma 3.7 (resp. Lemma 3.2). Hence, here again we have $\alpha - \widehat{\alpha}_t \overset{\text{a.s.}}{\to} 0$ as $t \to T$.

Suppose now that $\alpha \in (\frac{1}{2}, H)$. As $t \to T$, we have $\int_0^t \xi_u^2(T-u)^{2\alpha-2}\mathrm{d}u \overset{\text{a.s.}}{\to} \int_0^T \xi_u^2(T-u)^{2\alpha-2}\mathrm{d}u$ (resp. $\xi_t^2 \overset{\text{a.s.}}{\to} \xi_T^2$) by Lemma 3.7 (resp. Lemma 3.2). Hence Eq. (17.37) yields this time that $\alpha - \widehat{\alpha}_t \overset{\text{a.s.}}{\to} \alpha - \frac{1}{2}$ as $t \to T$, that is, $\widehat{\alpha}_t \overset{\text{a.s.}}{\to} \frac{1}{2}$.

Assume finally that $\alpha \geq H$. By Eq. (17.15), we have

$$\begin{aligned}
&E\big[(T-t)^{2\alpha-1}\xi_t^2\big] \\
&= (T-t)^{2\alpha-1}\int_0^t \mathrm{d}u(T-u)^{-\alpha}\int_0^t \mathrm{d}v(T-v)^{-\alpha}|v-u|^{2H-2} \\
&= (T-t)^{2\alpha-1}\int_{T-t}^T \mathrm{d}u\,u^{-\alpha}\int_{T-t}^T \mathrm{d}v\,v^{-\alpha}|v-u|^{2H-2} \\
&= (T-t)^{2H-1}\int_1^{\frac{T}{T-t}} \mathrm{d}u\,u^{-\alpha}\int_1^{\frac{T}{T-t}} \mathrm{d}v\,v^{-\alpha}|v-u|^{2H-2} \\
&= (T-t)^{2H-1}\int_{\frac{T-t}{T}}^1 \mathrm{d}u\,u^{\alpha-2H}\int_{\frac{T-t}{T}}^1 \mathrm{d}v\,v^{\alpha-2H}|v-u|^{2H-2}
\end{aligned}$$

$$
\begin{aligned}
&\leq (T-t)^{2H-1} \int_{\frac{T-t}{T}}^{1} \mathrm{d}u\, u^{2\alpha-2H-1} \int_0^{\frac{1}{u}} \mathrm{d}v\, v^{\alpha-2H} |v-1|^{2H-2} \\
&\leq c(T-t)^{2H-1} \int_{\frac{T-t}{T}}^{1} \mathrm{d}u\, u^{2\alpha-2H-1} \times \begin{cases} 1 & \text{if } \alpha < 1 \\ 1 + |\log u| & \text{if } \alpha = 1 \\ u^{1-\alpha} & \text{if } \alpha > 1 \end{cases} \\
&\leq c(T-t)^{2H-1} \times \begin{cases} |\log(T-t)| & \text{if } \alpha = H \\ 1 & \text{if } \alpha > H \end{cases} \\
&\longrightarrow 0 \text{ as } t \to T.
\end{aligned}
$$

Hence, having a look at Eq. (17.37) and because $\int_0^t \xi_u^2 (T-u)^{2\alpha-2} \mathrm{d}u \overset{\text{a.s.}}{\to} \int_0^T \xi_u^2 (T-u)^{2\alpha-2} \mathrm{d}u \in (0,\infty]$ as $t \to T$, we deduce that $\alpha - \widehat{\alpha}_t \overset{\text{prob.}}{\to} \alpha - \frac{1}{2}$ as $t \to T$, that is, $\widehat{\alpha}_t \overset{\text{prob.}}{\to} \frac{1}{2}$.

The proof of Theorem 3.1 is done. □

Proof of Theorem 3.2.

1. Assume that α belongs to $(0, 1-H)$. We have, by using Lemma 3.4 to go from the first to the second line,

$$
\begin{aligned}
(T-t)^{\alpha-H}\left(\alpha - \widehat{\alpha}_t\right) &= \frac{(T-t)^{1-H-\alpha} \eta_t}{(T-t)^{1-2\alpha} \int_0^t \xi_s^2 (T-s)^{2\alpha-2} \mathrm{d}s} \\
&= \frac{(T-t)^{1-H-\alpha} \int_0^t (T-u)^{\alpha-1} \mathrm{d}B_u \int_0^t (T-s)^{-\alpha} \mathrm{d}B_s}{(T-t)^{1-2\alpha} \int_0^t \xi_s^2 (T-s)^{2\alpha-2} \mathrm{d}s} \\
&\quad - \frac{(T-t)^{1-H-\alpha} \int_0^t \delta B_s (T-s)^{-\alpha} \int_0^s \delta B_u (T-u)^{\alpha-1}}{(T-t)^{1-2\alpha} \int_0^t \xi_s^2 (T-s)^{2\alpha-2} \mathrm{d}s} \\
&\quad - H(2H-1) \frac{(T-t)^{1-H-\alpha} \int_0^t \mathrm{d}s\, (T-s)^{-\alpha} \int_0^s \mathrm{d}u\, (T-u)^{\alpha-1} (s-u)^{2H-2}}{(T-t)^{1-2\alpha} \int_0^t \xi_s^2 (T-s)^{2\alpha-2} \mathrm{d}s} \\
&= \frac{1-2\alpha}{\xi_T} (T-t)^{1-H-\alpha} \int_0^t (T-u)^{\alpha-1} \mathrm{d}B_u \\
&\quad \times \frac{\int_0^t (T-s)^{-\alpha} \mathrm{d}B_s}{\xi_T} \times \frac{\xi_T^2}{(1-2\alpha)(T-t)^{1-2\alpha} \int_0^t \xi_s^2 (T-s)^{2\alpha-2} \mathrm{d}s}
\end{aligned}
$$

$$
\begin{aligned}
&-\frac{(T-t)^{1-H-\alpha}\int_0^t \delta B_s(T-s)^{-\alpha}\int_0^s \delta B_u(T-u)^{\alpha-1}}{(T-t)^{1-2\alpha}\int_0^t \xi_s^2(T-s)^{2\alpha-2}\mathrm{d}s} \\
&-H(2H-1)\frac{(T-t)^{1-H-\alpha}\int_0^t \mathrm{d}s\,(T-s)^{-\alpha}\int_0^s \mathrm{d}u\,(T-u)^{\alpha-1}(s-u)^{2H-2}}{(T-t)^{1-2\alpha}\int_0^t \xi_s^2(T-s)^{2\alpha-2}\mathrm{d}s} \\
&= a_t \times b_t \times c_t - d_t - e_t, \qquad (17.39)
\end{aligned}
$$

with clear definitions for a_t to e_t. Lemma 3.5 yields

$$
a_t \overset{\text{law}}{\to} (1-2\alpha)\sqrt{H(2H-1)\frac{\beta(2-\alpha-2H,2H-1)}{1-H-\alpha}} \times \frac{G}{\xi_T} \quad \text{as } t \to T,
$$

where $G \sim \mathcal{N}(0,1)$ is independent of B, whereas Lemma 3.2 (resp. Lemma 3.7) implies that $b_t \overset{\text{a.s.}}{\to} 1$ (resp. $c_t \overset{\text{a.s.}}{\to} 1$) as $t \to T$. On the other hand, by combining Lemma 3.7 with Lemma 3.6 (resp. Lemma 3.1), we deduce that $d_t \overset{\text{prob.}}{\to} 0$ (resp. $e_t \overset{\text{prob.}}{\to} 0$) as $t \to T$. By plugging all these convergences together, we get that, as $t \to T$,

$$
(T-t)^{\alpha-H}\left(\widehat{\alpha}_t - \alpha\right) \overset{\text{law}}{\to} (1-2\alpha)\sqrt{H(2H-1)\frac{\beta(2-\alpha-2H,2H-1)}{1-H-\alpha}} \times \frac{G}{\xi_T}.
$$

Because it is well known that the ratio of two independent $\mathcal{N}(0,1)$-random variables is $\mathcal{C}(1)$-distributed, to conclude it remains to compute the variance σ^2 of $\xi_T \sim \mathcal{N}(0,\sigma^2)$. By Eq. (17.15), we have

$$
\begin{aligned}
E[\xi_T^2] &= H(2H-1)\int_0^T \mathrm{d}u(T-u)^{-\alpha}\int_0^T \mathrm{d}v(T-v)^{-\alpha}|v-u|^{2H-2} \\
&= H(2H-1)\int_0^T \mathrm{d}u\, u^{-\alpha}\int_0^T \mathrm{d}v\, v^{-\alpha}|v-u|^{2H-2} \\
&= 2H(2H-1)\int_0^T \mathrm{d}u\, u^{-\alpha}\int_0^u \mathrm{d}v\, v^{-\alpha}(u-v)^{2H-2} \\
&= 2H(2H-1)\int_0^T u^{2H-2\alpha-1}\mathrm{d}u\int_0^1 v^{-\alpha}(1-v)^{2H-2}\mathrm{d}v \\
&= \frac{H(2H-1)}{H-\alpha}T^{2H-2\alpha}\beta(1-\alpha,2H-1), \qquad (17.40)
\end{aligned}
$$

and the proof of the first part of Theorem 3.2 is done.

2. Assume that $\alpha = 1-H$. The proof follows the same lines as in point 1 above. The counterpart of decomposition (17.39) is here:

$$
\begin{aligned}
&\frac{(T-t)^{1-2H}}{\sqrt{|\log(T-t)|}}\left(\alpha-\widehat{\alpha}_t\right)\\
&= \frac{2H-1}{\xi_T\sqrt{|\log(T-t)|}}\int_0^t (T-s)^{-H}\mathrm{d}B_s\\
&\quad\times\frac{\int_0^t (T-u)^{H-1}\mathrm{d}B_u}{\xi_T}\times\frac{\xi_T^2}{(2H-1)(T-t)^{2H-1}\int_0^t \xi_s^2 (T-s)^{-2H}\mathrm{d}s}\\
&\quad-\frac{\int_0^t \delta B_s (T-s)^{H-1}\int_0^s \delta B_u (T-u)^{-H}}{\sqrt{|\log(T-t)|}(T-t)^{2H-1}\int_0^t \xi_s^2 (T-s)^{-2H}\mathrm{d}s}\\
&\quad-H(2H-1)\frac{\int_0^t \mathrm{d}s\,(T-s)^{H-1}\int_0^s \mathrm{d}u\,(T-u)^{-H}(s-u)^{2H-2}}{\sqrt{|\log(T-t)|}(T-t)^{2H-1}\int_0^t \xi_s^2(T-s)^{-2H}\mathrm{d}s}\\
&=\widetilde{a}_t\times\widetilde{b}_t\times\widetilde{c}_t-\widetilde{d}_t-\widetilde{e}_t.
\end{aligned}
$$

Lemma 3.5 yields

$$
\widetilde{a}_t \overset{\text{law}}{\to} (2H-1)^{\frac{3}{2}}\sqrt{2H\,\beta(1-H,2H-1)}\times\frac{G}{\xi_T}\quad\text{as } t\to T,
$$

where $G\sim\mathcal{N}(0,1)$ is independent of B, whereas Lemma 3.2 (resp. Lemma 3.7) implies that $\widetilde{b}_t\overset{\text{a.s.}}{\to}1$ (resp. $\widetilde{c}_t\overset{\text{a.s.}}{\to}1$) as $t\to T$. On the other hand, by combining Lemma 3.7 with Lemma 3.6 (resp. Lemma 3.1), we deduce that $\widetilde{d}_t\overset{\text{prob.}}{\to}0$ (resp. $\widetilde{e}_t\overset{\text{prob.}}{\to}0$) as $t\to T$. By plugging all these convergences together we get that, as $t\to T$,

$$
\frac{(T-t)^{1-2H}}{\sqrt{|\log(T-t)|}}\left(\widehat{\alpha}_t-\alpha\right)\overset{\text{law}}{\to}(2H-1)^{\frac{3}{2}}\sqrt{2H\,\beta(1-H,2H-1)}\times\frac{G}{\xi_T}.
$$

Moreover, by Eq. (17.40) we have that $\xi_T\sim\mathcal{N}\big(0,HT^{4H-2}\beta(H,2H-1)\big)$. Thus,

$$
\begin{aligned}
&(2H-1)^{\frac{3}{2}}\sqrt{2H\,\beta(1-H,2H-1)}\times\frac{G}{\xi_T}\\
&\overset{\text{law}}{=}T^{1-2H}(2H-1)^{\frac{3}{2}}\sqrt{\frac{2\,\beta(1-H,2H-1)}{\beta(H,2H-1)}}\times\mathcal{C}(1),
\end{aligned}
$$

and the convergence in point 2 is shown.

3. Assume that α belongs to $(1-H,\frac{1}{2})$. Using the decomposition

$$(T-t)^{2\alpha-1}\left(\alpha-\widehat{\alpha}_t\right)=\frac{\eta_t}{(T-t)^{1-2\alpha}\int_0^t \xi_u^2(T-u)^{2\alpha-2}\mathrm{d}u},$$

we immediately see that the second part of Theorem 3.2 is an obvious consequence of Lemmas 3.3 and 3.7.

4. Assume that $\alpha=\frac{1}{2}$. Recall the identity (17.38) for this particular value of α:

$$\alpha-\widehat{\alpha}_t=\frac{\xi_t^2}{2\int_0^t \xi_u^2(T-u)^{-1}\mathrm{d}u}.$$

As $t\to T$, we have $\xi_t^2 \overset{\text{a.s.}}{\to} \xi_T^2$ by Lemma 3.2, whereas $\int_0^t \xi_u^2(T-u)^{-1}\mathrm{d}u \overset{\text{a.s.}}{\sim} |\log(T-t)|\xi_T^2$ by Lemma 3.7. Therefore, we deduce as announced that $|\log(T-t)|\left(\alpha-\widehat{\alpha}_t\right) \overset{\text{a.s.}}{\to} \frac{1}{2}$ as $t\to T$. □

Acknowledgments We thank an anonymous referee for his/her careful reading of the manuscript and for his/her valuable suggestions and remarks. Ivan Nourdin was supported in part by the (French) ANR grant 'Exploration des Chemins Rugueux'

References

1. Alòs, E., Nualart, D.: Stochastic integration with respect to the fractional Brownian motion. Stoch. Stoch. Reports **75**(3), 129–152 (2003)
2. Barczy, M., Pap, G.: Explicit formulas for Laplace transforms of certain functionals of some time inhomogeneous diffusions. J. Math. Anal. Appl. **380**(2), 405–424 (2011)
3. Barczy, M., Pap, G.: α-Wiener bridges: singularity of induced measures and sample path properties. Stoch. Anal. Appl. **28**(3), 447–466 (2010)
4. Barczy, M., Pap, G.: Asymptotic behavior of maximum likelihood estimator for time inhomogeneous diffusion processes. J. Statist. Plan. Infer. **140**(6), 1576–1593 (2010)
5. Hu, Y., Nualart, D.: Parameter estimation for fractional Ornstein-Uhlenbeck processes. Statist. Probab. Lett. **80**, 1030–1038 (2010)
6. Mansuy, R.: On a one-parameter generalization of the Brownian bridge and associated quadratic functionals. J. Theoret. Probab. **17**(4), 1021–1029 (2004)
7. Nualart, D.: The Malliavin Calculus and Related Topics, 2nd edn. Springer, Berlin (2006)
8. Nualart, D., Peccati, G.: Central limit theorems for sequences of multiple stochastic integrals. Ann. Probab. **33**(1), 177–193 (2005)
9. Pipiras, V., Taqqu, M.S.: Integration questions related to fractional Brownian motion. Probab. Theory Rel. Fields **118**(2), 251–291 (2000)
10. Russo, F., Vallois, P.: Elements of stochastic calculus via regularization. Séminaire de Probabilités XL. In: Lecture Notes in Math, vol. 1899, pp. 147–185. Springer, Berlin (2007)

Chapter 18
Gradient Bounds for Solutions of Stochastic Differential Equations Driven by Fractional Brownian Motions

Fabrice Baudoin and Cheng Ouyang

To Pr. David Nualart's 60th birthday

Abstract We study some functional inequalities satisfied by the distribution of the solution of a stochastic differential equation driven by fractional Brownian motions. Such functional inequalities are obtained through new integration by parts formulas on the path space of a fractional Brownian motion.

Received 2/22/2011; Accepted 5/28/2012; Final 6/26/2012

1 Introduction

Let $(X_t^x)_{t\geq 0}$ be the solution of a stochastic differential equation

$$X_t^x = x + \sum_{i=1}^{n} \int_0^t V_i(X_s^x) \mathrm{d}B_s^i,$$

where $(B_t)_{t\geq 0}$ is an n-dimensional fractional Brownian motion with Hurst parameter $H > \frac{1}{2}$. Under ellipticity assumptions and classical boundedness conditions [2, 10], the random variable X_t^x, $t > 0$ admits a smooth density with respect to the Lebesgue measure of $\mathbb{R}^n$ and the functional operator

$$P_t f(x) = \mathbb{E}\left(f(X_t^x)\right)$$

F. Baudoin (✉) • C. Ouyang
Department of Mathematics, Purdue University,
West Lafayette, IN 47907, USA
e-mail: fbaudoin@math.purdue.edu; couyang@math.purdue.edu

F. Viens et al. (eds.), *Malliavin Calculus and Stochastic Analysis: A Festschrift in Honor of David Nualart*, Springer Proceedings in Mathematics & Statistics 34,
DOI 10.1007/978-1-4614-5906-4_18,

is regularizing in the sense that it transforms a bounded Borel function f into a smooth function $P_t f$ for $t > 0$. In this note we aim to quantify precisely this regularization property and prove that, under the above assumptions, bounds of the type

$$|V_{i_1} \cdots V_{i_k} P_t f(x)| \leq C_{i_1 \cdots i_k}(t,x) \|f\|_\infty, \quad t > 0, x \in \mathbb{R}^n,$$

are satisfied. We are moreover able to get an explicit blow up rate when $t \to 0$: for a fixed $x \in \mathbb{R}^n$, when $t \to 0$,

$$C_{i_1 \cdots i_k}(t,x) = O\left(\frac{1}{t^{kH}}\right).$$

Our strategy to prove such bounds is the following. If f is a C^∞ bounded function on $\mathbb{R}^n$, we first prove (see Lemma 4.1) that the following commutation holds

$$V_i P_t f(x) = \mathbb{E}\left(\sum_{k=1}^{n} \alpha_k^i(t,x) V_k f(X_t^x)\right),$$

where the $\alpha(t,x)$'s solve an explicit system of stochastic differential equations. Then, using an integration by parts formula in the path space of the underlying fractional Brownian motion (see Theorem 3.1), we may rewrite the expectation of the right-hand side of the above inequality as $\mathbb{E}\left(\Phi_i(t,x) f(X_t^x)\right)$ where $\Phi_i(t,x)$ is shown to be bounded in L^p, $1 \leq p < +\infty$ with a blow up rate that may be controlled when $t \to 0$. It yields a bounds on $|V_i P_t f(x)|$. Bounds on higher order derivatives are obtained in a similar way, by iterating the procedure just described. Let us mention here that the bounds we obtain depend on L^p bounds for the inverse of the Malliavin matrix of X_t^x. As of today, to the knowledge of the authors, such bounds have not yet been obtained in the rough case $H < \frac{1}{2}$. The extension of our results to the case $H < \frac{1}{2}$ is thus not straightforward.

We close the paper by an interesting geometric situation where we may prove an optimal and global gradient bound with a constant that is independent from the starting point x. In the situation where the equation is driven by a Brownian motion such global gradient bound is usually related to lower bounds on the Ricci curvature of the Riemannian geometry given by the vector fields V_i's, which makes interesting the fact that the bound also holds with fractional Brownian motions.

2 Stochastic Differential Equations Driven by Fractional Brownian Motions

We consider the Wiener space of continuous paths:

$$\mathbb{W}^n = (\mathcal{C}([0,1], \mathbb{R}^n), (\mathcal{B}_t)_{0 \leq t \leq 1}, \mathbb{P})$$

where:

1. $\mathcal{C}([0,1],\mathbb{R}^n)$ is the space of continuous functions $[0,1] \to \mathbb{R}^n$.
2. $(\beta_t)_{t\geq 0}$ is the coordinate process defined by $\beta_t(f) = f(t)$, $f \in \mathcal{C}([0,1],\mathbb{R}^n)$.
3. $\mathbb{P}$ is the Wiener measure.
4. $(\mathcal{B}_t)_{0\leq t\leq 1}$ is the ($\mathbb{P}$-completed) natural filtration of $(\beta_t)_{0\leq t\leq 1}$.

An n-dimensional fractional Brownian motion with Hurst parameter $H \in (0,1)$ is a Gaussian process:

$$B_t = (B_t^1, \dots, B_t^n), t \geq 0,$$

where $B^1, \dots, B^n$ are n independent centered Gaussian processes with covariance function

$$R(t,s) = \frac{1}{2}\left(s^{2H} + t^{2H} - |t-s|^{2H}\right).$$

It can be shown that such a process admits a continuous version whose paths are Hölder γ continuous, $\gamma < H$. Throughout this paper we will always consider the "regular" case, $H > 1/2$. In this case the fractional Brownian motion can be constructed on the Wiener space by a Volterra type representation [4]. Namely, under the Wiener measure, the process

$$B_t = \int_0^t K_H(t,s)\mathrm{d}\beta_s, t \geq 0 \tag{18.1}$$

is a fractional Brownian motion with Hurst parameter H, where

$$K_H(t,s) = c_H s^{\frac{1}{2}-H} \int_s^t (u-s)^{H-\frac{3}{2}} u^{H-\frac{1}{2}} \mathrm{d}u\,, \qquad t > s,$$

and c_H is a suitable constant.

Denote by $\mathcal{E}$ the set of step functions on $[0,1]$. Let $\mathcal{H}$ be the Hilbert space defined as the closure of $\mathcal{E}$ with respect to the scalar product

$$\langle \mathbf{1}_{[0,t]}, \mathbf{1}_{[0,s]} \rangle_{\mathcal{H}} = R_H(t,s).$$

The isometry K_H^* from $\mathcal{H}$ to $L^2([0,1])$ is given by

$$(K_H^*\varphi)(s) = \int_s^1 \varphi(t)\frac{\partial K_H}{\partial t}(t,s)\mathrm{d}t.$$

Moreover, for any $\varphi \in L^2([0,1])$ we have

$$\int_0^1 \varphi(s)\mathrm{d}B_s = \int_0^1 (K_H^*\varphi)(s)\mathrm{d}\beta_s.$$

Let us consider for $x \in \mathbb{R}^n$ the solution $(X_t^x)_{t\geq 0}$ of the stochastic differential equation:

$$X_t^x = x + \sum_{i=1}^n \int_0^t V_i(X_s^x)\mathrm{d}B_s^i, \tag{18.2}$$

where the V_i's are C^∞-bounded vector fields in $\mathbb{R}^n$. Existence and uniqueness of solutions for such equations have widely been studied and are known to hold in this framework (see for instance [9]). Moreover, the following bounds were proved by Hu and Nualart as an application of fractional calculus methods.

Lemma 2.1 (Hu-Nualart [6]). *Consider the stochastic differential equation (18.2). If the derivatives of V_i's are bounded and Hölder continuous of order $\gamma > 1/H - 1$, then*

$$\mathbb{E}\left(\sup_{0\leq t\leq T}|X_t|^p\right) < \infty$$

for all $p \geq 1$. If furthermore V_i's are bounded and $\mathbb{E}(\exp(\lambda|X_0|^q)) < \infty$ for any $\lambda > 0$ and $q < 2H$, then

$$\mathbb{E}\left(\exp\lambda\left(\sup_{0\leq t\leq T}|X_t|^q\right)\right) < \infty$$

for any $\lambda > 0$ and $q < 2H$.

Throughout our discussion, we assume that the following assumption is in force:

Hypothesis 2.1. *(1) $V_i(x)$'s are bounded smooth vector fields on $\mathbb{R}^n$ with bounded derivatives at any order.*
(2) For every $x \in \mathbb{R}^n$, $(V_1(x), \cdots, V_n(x))$ is a basis of $\mathbb{R}^n$.

Therefore, in this framework, we can find functions ω_{ij}^k such that

$$[V_i, V_j] = \sum_{k=1}^{n}\omega_{ij}^k V_k, \tag{18.3}$$

where the ω_{ij}^k's are bounded smooth functions on $\mathbb{R}^n$ with bounded derivatives at any order.

3 Integration by Parts Formulas

We first introduce notations and basic relations for the purpose of our discussion. Consider the diffeomorphism $\Phi(t,x) = X_t^x : \mathbb{R}^n \to \mathbb{R}^n$. Denote by $\mathbf{J_t} = \frac{\partial X_t^x}{\partial x}$ the Jacobian of $\Phi(t,\cdot)$. It is standard (see [10] for details) that

$$d\mathbf{J}_t = \sum_{i=1}^{n}\partial V_i(X_t^x)\mathbf{J}_t dB_t^i, \text{ with } \mathbf{J}_0 = \mathbf{I}, \tag{18.4}$$

and

$$d\mathbf{J}_t^{-1} = -\sum_{i=1}^{n}\mathbf{J}_t^{-1}\partial V_i(X_t^x)dB_t^i, \text{ with } \mathbf{J}_0^{-1} = \mathbf{I}. \tag{18.5}$$

For any C_b^∞ vector field W on $\mathbb{R}^n$, we have that

$$(\Phi_{t*}W)(X_t^x) = \mathbf{J}_t W(x), \quad \text{and} \quad (\Phi_{t*}^{-1}W)(x) = \mathbf{J}_t^{-1} W(X_t^x).$$

Here Φ_{t*} is the push-forward operator with respect to the diffeomorphism $\Phi(t,x) : \mathbb{R}^n \to \mathbb{R}^n$. Introduce the nondegenerate $n \times n$ matrix value process

$$\alpha(t,x) = (\alpha_j^i(t,x))_{i,j=1}^n \tag{18.6}$$

by

$$(\Phi_{t*}V_i)(X_t^x) = \mathbf{J}_t(V_i(x)) = \sum_{k=1}^n \alpha_k^i(t,x) V_k(X_t^x) \quad i = 1, 2, \ldots, n.$$

Note that $\alpha(t,x)$ is nondegenerate since we assume V_i's form a basis at each point $x \in \mathbb{R}^n$. Denote by

$$\beta(t,x) = \alpha^{-1}(t,x). \tag{18.7}$$

Clearly we have

$$(\Phi_{t*}^{-1}V_i(X_t^x))(x) = \mathbf{J}_t^{-1}V_i(X_t^x) = \sum_{k=1}^n \beta_k^i(t,x) V_k(x) \quad i = 1, 2, \ldots, n. \tag{18.8}$$

Lemma 3.1. *Let $\alpha(t,x)$ and $\beta(t,x)$ be as above, we have*

$$d\alpha_j^i(t,x) = -\sum_{k,l=1}^n \alpha_k^i(t,x)\omega_{lk}^j(X_t^x)dB_t^l, \qquad \text{with} \quad \alpha_j^i(0,x) = \delta_j^i; \tag{18.9}$$

and

$$d\beta_j^i(t,x) = \sum_{k,l=1}^n \omega_{li}^k(X_t^x)\beta_j^k(t,x)dB_t^l, \qquad \text{with} \qquad \beta_j^i(0,x) = \delta_j^i. \tag{18.10}$$

Proof. The initial values are apparent by the definition of α and β. We show how to derive Eq. (18.10). Once the equation for $\beta(t,x)$ is obtained, it is standard to obtain that $\alpha(t,x) = \beta^{-1}(t,x)$.

Consider the $n \times n$ matrix $V = (V_1, V_2, \ldots, V_n) = (V_j^i)$ obtained from the vector fields V. Let W be the inverse matrix of V. It is not hard to see we have

$$\beta_j^i(t,x) = \sum_{k=1}^n W_k^j(x)(\mathbf{J}_t^{-1}V_i(X_t^x))^k(x).$$

By the equation for X_t^x, relation (18.3), Eq. (18.5), and Itô's formula, we obtain

$$d(\mathbf{J}_t^{-1}V_i(X_t^x))(x) = \sum_{k=1}^{n}(\mathbf{J}_t^{-1}[V_k, V_i](X_t^x))(x)dB_t^k$$
$$= \sum_{k,l=1}^{n} \omega_{ki}^l(X_t^x)(\mathbf{J}_t^{-1}V_l(X_t^x))(x)dB_t^k.$$

Hence

$$d\beta_j^i(t,x) = \sum_{k,l=1}^{n} \omega_{ki}^l(X_t^x)\beta_j^l(t,x)dB_t^k.$$

This completes our proof. □

Define now $h_i(t,x) : [0,1] \times \mathbb{R}^n \to \mathcal{H}$ by

$$h_i(t,x) = (\beta_i^k(s,x)\mathbb{I}_{[0,t]}(s))_{k=1,\ldots,n}, \qquad i = 1,\ldots,n. \tag{18.11}$$

Introduce $M_{i,j}(t,x)$ given by

$$M_{i,j}(t,x) = \frac{1}{t^{2H}}\langle h_i(t,x), h_j(t,x)\rangle_{\mathcal{H}}. \tag{18.12}$$

For each $t \in [0,1]$, consider the semi-norms

$$\|f\|_{\gamma,t} := \sup_{0\le v<u\le t} \frac{|f(u)-f(v)|}{(u-v)^\gamma}.$$

The semi-norm $\|f\|_{\gamma,1}$ will simply be denoted by $\|f\|_\gamma$.
We have the following two important estimates.

Lemma 3.2. *Let $\alpha(t,x)$, $\beta(t,x)$, and $h_i(t,x)$ be as above. We have:*

(1) For any multi-index ν, integers $k,p \ge 1$, there exists a constant $C_{k,p}(x) > 0$ depending on k, p, and x such that for all $x \in \mathbb{R}^n$

$$\sup_{0\le t\le 1}\left\|\frac{\partial^{|\nu|}}{\partial x^\nu}\alpha(t,x)\right\|_{k,p} < C_{k,p}(x), \quad \sup_{0\le t\le 1}\left\|\frac{\partial^{|\nu|}}{\partial x^\nu}\beta(t,x)\right\|_{k,p} < C_{k,p}(x).$$

(2) For all integers $k,p \ge 1$, $\delta h_i(t,x) \in \mathbb{D}^{k,p}$. Moreover, there exists a constant $C_{k,p}(x)$ depending on k, p and x such that

$$\|\delta h_i(t,x)\|_{k,p} < C_{k,p}(x)t^H, \qquad t \in [0,1].$$

In the above δ is the adjoint operator of $\mathbf{D}$.

Proof. The result in (1) follows from Eqs. (18.9), (18.10), and Lemma 2.1. In what follows, we show (2). Note that we have (c.f. Nualart[8])

$$\begin{aligned}\delta h_i(t,x) &= \int_0^1 h_i(t,x)_u \mathrm{d}B_u - \alpha_H \int_0^1 \int_0^1 \mathbf{D}_u h(t,x)_v |u-v|^{2H-2} \mathrm{d}u\mathrm{d}v \\ &= \int_0^t \beta_i(u,x)\mathrm{d}B_u - \alpha_H \int_0^t \int_0^t \mathbf{D}_u \beta_i(v,x)|u-v|^{2H-2}\mathrm{d}u\mathrm{d}v.\end{aligned}$$

Here $\alpha_H = H(2H-1)$. From the above representation of δh_i and what we have just proved in the first statement of the lemma, it follows immediately that $\delta h_i(t,x) \in \mathbb{D}^{k,p}$ for all integers $k, p \geq 1$. To show

$$\|\delta h_i(t,x)\|_{k,p} < C_{k,p}(x)t^H \qquad \text{for all } t \in [0,1],$$

it suffices to prove

$$\left|\int_0^t \beta_i(u,x)\mathrm{d}B_u\right| \leq C(x)t^H \qquad t \in [0,1].$$

Here $C(x)$ is a random variable in $L^p(\mathbb{P})$. Indeed, by standard estimate, we have

$$\left\|\int_0^{\cdot} (\beta_i(u,x) - \beta_i(0,x))\mathrm{d}B_u\right\|_{\gamma,t} \leq C\|\beta(\cdot,x)\|_{\tau,t}\|B\|_{\gamma,t}, \qquad t \in [0,1].$$

In the above $\frac{1}{2} < \tau, \gamma < H$ and $\tau + \gamma > 1$, and $C > 0$ is a constant only depending on γ. Therefore

$$\left|\int_0^t \beta_i(u,x)\mathrm{d}B_u\right| \leq C\|\beta(\cdot,x)\|_{\tau,t}\|B\|_{\gamma,t}t^\gamma + |\beta(0,x)||B_t|, \qquad t \in [0,1].$$

Together with the fact that for any $\tau < H$, there exists a random variable $G_\tau(x)$ in $L^p(\mathbb{P})$ for all $p > 1$ such that

$$|\beta(t,x) - \beta(s,x)| < G_\tau(x)|t-s|^\tau,$$

the proof is now completed. □

Lemma 3.3. *Let $M(t,x) = (M_{i,j}(t,x))$ be given in Eq. (18.12). We have for all $p \geq 1$,*

$$\sup_{t\in[0,1]} \mathbb{E}\left[\det(M(t,x))^{-p}\right] < \infty.$$

Proof. Denote the Malliavin matrix of X_t^x by $\Gamma(t,x)$. By definition

$$\Gamma_{i,j}(t,x) = \langle \mathbf{D}_s X_t^i, \mathbf{D}_s X_t^j\rangle_{\mathcal{H}} = \alpha_H \int_0^t \int_0^t \mathbf{D}_u X_t^i \mathbf{D}_v X_t^j |u-v|^{2H-2}\mathrm{d}u\mathrm{d}v.$$

It can be shown that for all $p > 1$ (cf. Baudoin-Hairer [2], Hu-Nualart [6], and Nualart-Saussereau [10])

$$\sup_{t\in[0,1]} \mathbb{E}\left(\det \frac{\Gamma(t,x)}{t^{2H}}\right)^{-p} < \infty. \tag{18.13}$$

Introduce γ by

$$\gamma_{i,j}(t,x) = \alpha_H \int_0^t \int_0^t \sum_{k=1}^n \left(\mathbf{J}_u^{-1} V_k(X_u)\right)^i \left(\mathbf{J}_v^{-1} V_k(X_v)\right)^j |u-v|^{2H-2} dudv.$$

Since $\mathbf{D}_s^k X_t = \mathbf{J}_t \mathbf{J}_s^{-1} V_k(X_s)$, we obtain

$$\Gamma(t,x) = \mathbf{J}_t \gamma(t,x) \mathbf{J}_t^T. \tag{18.14}$$

Recall

$$M_{i,j}(t,x) = \frac{1}{t^{2H}} \langle h_i(t,x), h_j(t,x) \rangle_{\mathcal{H}},$$

where

$$h_i(t,x) = (\beta_i^k(s,x) \mathbb{I}_{[0,t]}(s))_{k=1,\ldots,n}, \qquad i = 1,\ldots,n.$$

By Eqs. (18.8) and (18.14), we have

$$V(x)M(t,x)V(x)^T = \frac{1}{t^{2H}} \gamma(t,x) = \mathbf{J}^{-1} \frac{\Gamma(t,x)}{t^{2H}} (\mathbf{J}^{-1})^T. \tag{18.15}$$

Finally, by Eq. (18.4), Lemma 2.1 and estimate (18.13) we have for all $p \geq 1$

$$\sup_{t \in [0,1]} \mathbb{E}\left[\det(M_{i,j})^{-p}\right] < \infty,$$

which is the desired result. □

The following definition is inspired by Kusuoka [7].

Definition 3.1. Let H be a separable real Hilbert space and $r \in \mathbb{R}$ be any real number. Introduce $\mathcal{K}_r(H)$ the set of mappings $\Phi(t,x) : (0,1] \times \mathbb{R}^n \to \mathbb{D}^\infty(H)$ satisfying:

(1) $\Phi(t,x)$ is smooth in x and $\frac{\partial^\nu \Phi}{\partial x^\nu}(t,x)$ is continues in $(t,x) \in (0,1] \times \mathbb{R}^n$ with probability one for any multi-index ν.
(2) For any $n, p > 1$ we have

$$\sup_{0<t\leq 1} t^{-rH} \left\| \frac{\partial^\nu \Phi}{\partial^\nu x}(t,x) \right\|_{\mathbb{D}^{k,p}(H)} < \infty.$$

We denote $\mathcal{K}_r(\mathbb{R})$ by $\mathcal{K}_r$.

Lemma 3.4. *With probability one, we have*

(1) $\alpha(t,x), \beta(t,x) \in \mathcal{K}_0$.
(2) $\delta h_i(t,x) \in \mathcal{K}_1$.

(3) *Let* $(M_{i,j}^{-1})$ *be the inverse matrix of* $(M_{i,j})$. *Then* $M_{i,j}^{-1} \in \mathcal{K}_0$ *for all* $i, j = 1, \ldots, n$.

Proof. The first two statements are immediate consequences of Lemma 3.2. The third statement follows by writing $M^{-1} = \frac{\mathrm{adj} M}{\det M}$, estimates in Lemma 3.2 (1) and Lemma 3.3. □

Now we can state one of our main results in this note.

Theorem 3.1. *Let* f *be any* C^∞*-bounded function and* $\Phi(t,x) : \Omega \to \mathcal{K}_r$ *we have*

$$\mathbb{E}\left(\Phi(t,x) V_i f(X_t^x)\right) = \mathbb{E}\left((T_{V_i}^* \Phi(t,x)) f(X_t^x)\right),$$

where $T_{V_i}^* \Phi(t,x)$ *is an element in* $\mathcal{K}_{r-1}$ *with probability one.*

Proof. This is primarily integration by parts together with the estimates obtained before. First note

$$\begin{aligned}
\mathbf{D}_s^j f(X_t) &= \langle \nabla f(X_t), \mathbf{D}_s^j X_t \rangle \\
&= \langle \nabla f(X_t), \mathbf{J}_t \mathbf{J}_s^{-1} V_j(X_s) \rangle \\
&= \sum_{k,l=1}^{n} h_k^j(t) \alpha_l^k(t) (V_l f)(X_t).
\end{aligned}$$

Hence

$$V_i f(X_t) = \frac{1}{t^{2H}} \sum_{j,l=1}^{n} \beta_j^i(t) M_{jl}^{-1} \langle \mathbf{D} f(X_t), h_l(t) \rangle_{\mathcal{H}}. \tag{18.16}$$

Therefore, we have

$$\begin{aligned}
&\mathbb{E}\left(\Phi(t,x) V_i f(X_t)\right) \\
&= \frac{1}{t^{2H}} \sum_{k,l=1}^{n} \mathbb{E}\left(\langle \mathbf{D} f(X_t), \Phi(t,x) \beta_k^i(t) M_{kl}^{-1}(t) h_l(t) \rangle_{\mathcal{H}}\right) \\
&= \frac{1}{t^{2H}} \sum_{k,l=1}^{n} \mathbb{E}\left(\left[\delta\left(\Phi(t,x) \beta_k^i(t) M_{kl}^{-1}(t) h_l(t)\right)\right] f(X_t)\right) \\
&= \sum_{k,l=1}^{n} \mathbb{E}\Bigg(\bigg[\frac{1}{t^{2H}} \Phi(t,x) \beta_k^i(t) M_{kl}^{-1}(t) \delta h_l(t) \\
&\qquad - \frac{1}{t^{2H}} \langle \mathbf{D}(\Phi(t,x) \beta_k^i(t) M_{kl}^{-1}(t)), h_l(t) \rangle_{\mathcal{H}}\bigg] f(X_t)\Bigg).
\end{aligned}$$

By Lemma 3.4, the first term in the brackets above is in $\mathcal{K}_{r-1}$ and the second term is in $\mathcal{K}_r$. Finally, denote

$$T^*_{V_i}\Phi(t,x) = \sum_{k,l=1}^{n} \left[\frac{1}{t^{2H}} \Phi(t,x)\beta^i_k(t) M^{-1}_{kl}(t)\delta h_l(t) \right.$$
$$\left. - \frac{1}{t^{2H}} \langle \mathbf{D}(\Phi(t,x)\beta^i_k(t)M^{-1}_{kl}(t)), h_l(t)\rangle_{\mathcal{H}} \right].$$

It is clear that $T^*_{V_i}\Phi(t,x) \in \mathcal{K}_{r-1}$. The proof is completed. □

4 Gradient Bounds

With the integration by parts formula of Theorem 3.1 in hand we can now prove our gradient bounds. We start with the following basic commutation formula:

Lemma 4.1. *For $i = 1, 2, \ldots, n$, we have the commutation*

$$V_i P_t f(x) = \mathbb{E}\left(((\mathbf{J}_t V_i) f)(X^x_t) \right) = \mathbb{E}\left(\sum_{k=1}^{n} \alpha^i_k(t,x) V_k f(X^x_t) \right),$$

where the $\alpha(t,x)$ solve the system of stochastic differential equations (18.9).

Proof. For any C^∞_b-vector field W on $\mathbb{R}^n$ we have

$$W P_t f(x) = \mathbb{E}\left(((\mathbf{J}_t W) f)(X^x_t) \right).$$

The remainder of the proof is then clear from the computations in the previous section. □

Finally we have the following gradient bounds.

Theorem 4.1. *Let $p > 1$. For $i_1, \ldots, i_k \in \{1, \ldots, n\}$, and $x \in, \mathbb{R}^n$, we have*

$$|V_{i_1} \ldots V_{i_k} P_t f(x)| \leq C(t,x)(P_t f^p(x))^{\frac{1}{p}} \quad t \in [0,1]$$

with $C(t,p,x) = O\left(\frac{1}{t^{Hk}}\right)$ when $t \to 0$.

Proof. By Theorem 3.1 and Lemma 4.1, for each $k \geq 1$ there exists a $\Phi^{(-k)}(t,x) \in \mathcal{K}_{-k}$ such that

$$V_{i_1} \ldots V_{i_k} P_t f(x) = \mathbb{E}\left(\Phi^{(-k)}(t,x) f(X_t) \right).$$

Now an application of Hölder's inequality gives us the desired result. □

Remark 4.1. Here let us emphasize a simple but important consequence of the above theorem that suppose f is uniformly bounded, then

$$|V_{i_1} \ldots V_{i_k} P_t f(x)| \leq C(t,x)\|f\|_\infty \quad t \in [0,1]$$

where $C(t,x) = O\left(\frac{1}{t^{Hk}}\right)$ as $t \to 0$.

Another direct corollary of Theorem 4.1 is the following inverse Poincaré inequality.

Corollary 4.1. *For* $i_1, \ldots, i_k \in \{1, \ldots, n\}$, *and* $x \in \mathbb{R}^n$,

$$|V_{i_1} \ldots V_{i_k} P_t f(x)|^2 \leq C(t,x)(P_t f^2(x) - (P_t f)^2(x)) \quad t \in [0,1]$$

with $C(t,x) = O\left(\frac{1}{t^{2Hk}}\right)$ *when* $t \to 0$.

Proof. By Theorem 4.1, for any constant $C \in \mathbb{R}$ we have

$$\begin{aligned} |V_{i_1} \ldots V_{i_k} P_t f(x)|^2 &= |V_{i_1} \ldots V_{i_k} P_t (f - C)(x)|^2 \\ &\leq C(t,x)(P_t (f - C)^2(x)) \quad t \in [0,1] \end{aligned}$$

with $C(t,x) = O\left(\frac{1}{t^{2Hk}}\right)$ when $t \to 0$. Now minimizing $C \in \mathbb{R}$ gives us the desired result. □

Remark 4.2. For each smooth function $f : \mathbb{R}^n \to \mathbb{R}$, denote

$$\Gamma(f) = \sum_{i=1}^{n} (V_i f)^2.$$

We also have, for $i_1, \ldots, i_k \in \{1, \ldots, n\}$, and $x \in \mathbb{R}^n$,

$$|V_{i_1} \ldots V_{i_k} P_t f(x)|^2 \leq C(t,x) P_t \Gamma(f)(x), \quad t \in [0,1]$$

with $C(t,x) = O\left(\frac{1}{t^{2H(k-1)}}\right)$ when $t \to 0$. Indeed, by Theorem 3.1 and Lemma 4.1, we know that for each $k \geq 1$, there exists $\Phi_j^{(1-k)}(t,x) \in \mathcal{K}_{1-k}, j = 1, 2, \ldots, n$ such that

$$V_{i_1} \ldots V_{i_k} P_t f(x) = \mathbb{E}\left(\Phi_j^{(1-k)}(t,x)(V_j f)(X_t)\right).$$

The sequel of the argument is then clear.

5 A Global Gradient Bound

Throughout our discussion in this section, we show that under some additional conditions on the vector fields $V_i, \ldots, V_n$, we are able to obtain

$$\sqrt{\Gamma(P_t f)} \leq P_t(\sqrt{\Gamma(f)}),$$

uniformly in x, where we denoted as above

$$\Gamma(f) = \sum_{i=1}^{n} (V_i f)^2.$$

For this purpose, we need the following additional structure equation imposed on vector fields $V_i, \dots, V_d$.

Hypothesis 5.1. *In addition to Hypothesis 2.1, we assume the smooth and bounded functions ω_{ij}^k satisfy*

$$\omega_{ij}^k = -\omega_{ik}^j, \qquad 1 \le i, j, k \le d.$$

Interestingly, such an assumption already appeared in a previous work of the authors [3] where they proved an asymptotic expansion of the density of X_t when $t \to 0$. In a Lie group structure, that is if the ω_{ij}^k's are constant, then this assumption is equivalent to the fact that the Lie algebra is of compact type. So, in particular, this assumption is satisfied on any compact Lie group.

Remark 5.1. In the case of a stochastic differential equation driven by a Brownian motion, the functional operator P_t is a diffusion semigroup with infinitesimal generator $L = \frac{1}{2}\left(\sum_{i=1}^n V_i^2\right)$. The gradient subcommutation,

$$\sqrt{\Gamma(P_t f)} \le P_t(\sqrt{\Gamma f}),$$

is then known to be equivalent to the fact that the Ricci curvature of the Riemannian geometry given by the vector fields V_i's is nonnegative (see for instance [1]).

The following approximation result, which can be found for instance in [5], will also be used in the sequel:

Proposition 5.1. *For $m \ge 1$, let $B^m = \{B_t^m; t \in [0,1]\}$ be the sequence of linear interpolations of B along the dyadic subdivision of $[0,1]$ of mesh m; that is, if $t_i^m = i2^{-m}$ for $i = 0, \dots, 2^m$, then for $t \in (t_i^m, t_{i+1}^m]$*

$$B_t^m = B_{t_{im}} + \frac{t - t_{im}}{t_{i+1}^m - t_i^m}(B_{t_{i+1}^m} - B_{t_i^m}).$$

Consider X^m the solution to Eq. (18.2) restricted to $[0,1]$, where B has been replaced by B^m. Then almost surely, for any $\gamma < H$ and $t \in [0,1]$, the following holds true:

$$\lim_{m \to \infty} \|X^x - X^m\|_\gamma = 0. \tag{18.17}$$

Theorem 5.1. *Recall the definition of $\alpha(t,x)$ in Eq. (18.6) and*

$$d\alpha_j^i(t,x) = -\sum_{k,l=1}^n \alpha_k^i(t,x)\omega_{lk}^j(X_t^x)dB_t^l, \qquad \text{with} \quad \alpha_j^i(0,x) = \delta_j^i. \tag{18.18}$$

Under Assumption 5.1, $\alpha(t,x)$ is an orthogonal matrix almost surely for all t and x. In particular, we have

$$\sum_{j=1}^n \alpha_j^i(t,x)^2 = 1; \qquad \text{and} \quad \sum_{i=1}^n \alpha_j^i(t,x)^2 = 1, \tag{18.19}$$

almost surely.

Proof. Let $\omega_i = (\omega^k_{i,j})_{1\le j.k\le d}$ be the $d\times d$ matrix formed by $\omega^k_{i,j}$. From Eq. (18.18) and Hypothesis 5.1, we have

$$\begin{aligned} d(\alpha(t,x)\alpha^T(t,x)) &= (d\alpha(t,x))\,\alpha^T(t,x) + \alpha(t,x)\left(d\alpha^T(t,x)\right) \\ &= -\alpha(t,x)\left(\sum_{i=1}^d \left(\omega_i(X_t^x) + \omega_i^T(X_t^x)\right) dB_t^i\right)\alpha^T(t,x) \\ &= 0 \end{aligned}$$

In the above, $\alpha^T(t,x)$ is the transposed matrix of $\alpha(t,x)$. Taking into account the initial condition $\alpha(0,x) = I$, we conclude that $\alpha(t,x)$ is an orthogonal matrix for all t and x almost surely. This concludes our proof. □

As a direct consequence of Lemma 4.1 and Theorem 5.1, we have the main result of this section.

Theorem 5.2. *Under Assumption 5.1, we have uniformly in x*

$$\sqrt{\Gamma(P_t f)} \le P_t(\sqrt{\Gamma(f)}).$$

Proof. By applying Lemma 4.1, Cauchy–Schwarz inequality, and then Theorem 5.1, we have for any vector $a = (a_i) \in \mathbb{R}^n$

$$\begin{aligned} \sum_{i=1}^n a_i V_i P_t f(x) &= \mathbb{E}\left(\sum_{i,k=1}^n a_i \alpha_k^i(t,x) V_k f(X_t^x)\right) \\ &\le \mathbb{E}\left[\left(\sum_{k=1}^n\left(\sum_{i=1}^n a_i\alpha_k^i(t,x)\right)^2\right)^{\frac{1}{2}}\left(\sum_{k=1}^n (V_k f(X_t^x))^2\right)^{\frac{1}{2}}\right] \\ &\le \|a\|\mathbb{E}\left(\sum_{k=1}^n V_k f(X_t^x)^2\right)^{\frac{1}{2}}. \end{aligned}$$

By choosing

$$a_i = \frac{(V_i P_t f)(x)}{\sqrt{\sum_{i=1}^n (V_i P_t f)^2(x)}},$$

we obtain

$$\sqrt{\Gamma(P_t f)} = \sqrt{\sum_{i=1}^n (V_i P_t f)^2} \le P_t(\sqrt{\Gamma(f)}).$$

The proof is completed. □

Remark 5.2. Since P_t comes from probability measure, we observe from Jensen inequality that

$$\sqrt{\Gamma(P_t f)} \leq P_t(\sqrt{\Gamma(f)})$$

implies

$$\Gamma(P_t f) \leq P_t(\Gamma f).$$

Acknowledgements First author supported in part by NSF Grant DMS 0907326.

References

1. Bakry, D.: L'hypercontractivité et son utilisation en théorie des semigroupes, Ecole d'Eté de Probabilites de St-Flour, Lecture Notes in Math (1994)
2. Baudoin, F., Hairer, M.: A version of Hörmander's theorem for the fractional Brownian motion. Prob. The. Rel. Fields. **139**, 373–395 (2007)
3. Baudoin, F., Ouyang, C.: Small-time kernel expansion for solutions of stochastic differential equations driven by fractional Brownian motions. Stoch. Process. Appl. **121**(4), 759–792 (2011)
4. Decreusefond, L., Üstünel, A.S.: Stochastic analysis of the fractional Brownian motion. Potential Anal. **10**, 177–214 (1998)
5. Friz, P., Victoir, N.: Multidimensional Dimensional Processes Seen as Rough Paths. Cambridge University Press, Cambridge (2010)
6. Hu, Y., Nualart, D.: Differential equations driven by Holder continuous functions of order greater than 1/2. In: Stochastic Analysis and Applications, pp. 399–413, Abel Symp., 2, Springer, Berlin (2007)
7. Kusuoka, S.: Malliavin calculus revisited. J. Math. Sci. Univ. Tokyo **10**, 261–277 (2003)
8. Nualart, D.: The Malliavin calculus and related topics. In: Probability and Its Applications, 2nd edn. Springer, Berlin (2006)
9. Nualart, D., Rascanu, A.: Differential equations driven by fractional Brownian motion. Collect. Math. **53**(1), 55–81 (2002)
10. Nualart, D., Saussereau, B.: Malliavin calculus for stochastic differential equations driven by a fractional Brownian motion. Stoch. Process. Appl. **119**(2), 391–409 (2009)

Chapter 19
Parameter Estimation for Fractional Ornstein–Uhlenbeck Processes with Discrete Observations

Yaozhong Hu and Jian Song

Abstract Consider an Ornstein–Uhlenbeck process, $dX_t = -\theta X_t dt + \sigma dB_t^H$, driven by fractional Brownian motion B^H with known Hurst parameter $H \geq \frac{1}{2}$ and known variance σ. But the parameter $\theta > 0$ is unknown. Assume that the process is observed at discrete time instants $t = h, 2h, \ldots, nh$. We construct an estimator $\hat{\theta}_n$ of θ which is strongly consistent, namely, $\hat{\theta}_n$ converges to θ almost surely as $n \to \infty$. We also obtain a central limit type theorem and a Berry–Esseen type theorem for this estimator $\hat{\theta}_n$ when $1/2 \leq H < 3/4$. The tool we use is some recent results on central limit theorems for multiple Wiener integrals through Malliavin calculus. It should be pointed out that no condition on the step size h is required, contrary to the existing conventional assumptions.

Received 9/28/2011; Accepted 2/21/2012; Final 2/29/2012

1 Introduction

The Ornstein–Uhlenbeck process X_t driven by a certain type of noise Z_t is described by the following Langevin equation:

$$dX_t = -\theta X_t dt + \sigma dZ_t. \tag{19.1}$$

If the parameter θ is unknown and if the process $(X_t, 0 \leq t \leq T)$ can be observed continuously, then an important problem is to estimate the parameter θ based on the (single-path) observation $(X_t, 0 \leq t \leq T)$. See [6] and the references therein

Y. Hu (✉)
Department of Mathematics, University of Kansas, Lawrence, KS 66045, USA
e-mail: hu@math.ku.edu

J. Song
Department of Mathematics, Rutgers University, Piscataway, NJ 08854-8019, USA
e-mail: jsong2@math.rutgers.edu

F. Viens et al. (eds.), *Malliavin Calculus and Stochastic Analysis: A Festschrift in Honor of David Nualart*, Springer Proceedings in Mathematics & Statistics 34, DOI 10.1007/978-1-4614-5906-4_19, © Springer Science+Business Media New York 2013

for a short account of the research works relevant to this problem. In this paper, we consider the case Z_t is a fractional Brownian motion with Hurst parameter H. Namely, we consider the following stochastic Langevin equation:

$$dX_t = -\theta X_t dt + \sigma dB_t^H, \quad X_0 = x, \tag{19.2}$$

where θ is an unknown parameter. We assume $\theta > 0$ throughout the paper so that the process is ergodic (when $\theta < 0$ the solution to Eq. (19.2) will diverge as T goes to infinity). If the process $(X_t, 0 \le t \le T)$ can be observed continuously, then the least square estimator $\tilde{\theta}_T$, defined by

$$\tilde{\theta}_T = -\frac{\int_0^T X_t dX_t}{\int_0^T X_t^2 dt} \tag{19.3}$$

was studied in [8], where it is proved that $\tilde{\theta}_T \to \theta$ almost surely as $T \to \infty$ and that $\sqrt{T}\left(\tilde{\theta}_T - \theta\right)$ converges in law to a mean zero normal random variable. The variance of this normal is also calculated. As a consequence it is also proved in [8] that the following estimator

$$\bar{\theta}_T := \left(\frac{1}{\sigma^2 H \Gamma(2H) T} \int_0^T X_t^2 dt\right)^{-\frac{1}{2H}} \tag{19.4}$$

is also strongly consistent and $\sqrt{T}\left(\bar{\theta}_T - \theta\right)$ converges in law to a mean zero normal with explicit variance given by $\frac{\theta \sigma_H^2}{(2H)^2}$.

In applications usually the process cannot be observed continuously. Only discrete-time observations are available. To simplify presentation of the paper, we assume that the process X_t is observed at discrete-time instants $t_k = kh$, $k = 1, 2, \ldots, n$, for some fixed $h > 0$. We seek to estimate θ based on $X_h, X_{2h}, \ldots, X_{nh}$.

Motivated by the estimator Eq. (19.4), we propose to use a function of $\frac{1}{n}\sum_{k=1}^n |X_{kh}|^p$ as a statistic to estimate θ. More precisely, we define

$$\hat{\theta}_n = \left(\frac{1}{n\sigma^2 H \Gamma(2H)} \sum_{k=1}^n X_{kh}^2\right)^{-\frac{1}{2H}}.$$

We shall show that $\hat{\theta}_n$ converges to θ almost surely as n tends to ∞. We shall also show that $\sqrt{n}\left(\hat{\theta}_n - \theta\right)$ converges in law to mean zero normal random variable with variance $\frac{\theta^2}{2H^2}$ as $n \to \infty$. The following Berry–Esseen type theorem will also be shown:

$$\sup_{z \in \mathbb{R}} \left| P\left(\sqrt{\frac{2H^2 n}{\theta^2}}\left(\hat{\theta}_n - \theta\right) \le z\right) - \Psi(z)\right| \le C n^{4H-3},$$

where $\Psi(z) = \frac{1}{\sqrt{2\pi}} \int_{-\infty}^z e^{-\frac{u^2}{2}} du$ is the error function.

Usually, to obtain consistency result for discrete-time observations, one has to assume that the length h of the time interval between two consecutive observations depends on n (namely $h = h_n$) and h_n converges to 0 as $n \to \infty$ and h_n and n must satisfy some other conditions (see [2, 6, 7, 12], and references therein). Surprisingly enough, for our simple model Eq. (19.2) and for our estimator defined above we do not need to assume h depends on n. In fact, we do not have any condition on h! Let us also point out that throughout the paper, we assume that the observation times are uniform: $t_k = kh, k = 1, \ldots, n$. General deterministic observation times t_k can be also considered in a similar way.

The paper is organized as follows. In Sect. 2, some known results that we will use are recalled. The strong consistency of a slight more general estimator is proved in Sect. 3. Sect. 4 deals with the central limit type theorem and Sect. 5 concerns with the Berry–Esseen type theorem.

Along the paper, we denote by C a generic constant possibly depending on θ and/or h which may be different from line to line.

2 Preliminaries

In this section we first introduce some basic facts on the Malliavin calculus for the fractional Brownian motion and recall the main results in [10, 11] concerning the central limit theorem and Berry–Esseen type results for multiple stochastic integrals.

We are working on some complete probability space $(\Omega, \mathcal{F}, P)$. The expectation on this probability space is denoted by $\mathbb{E}$. The fractional Brownian motion with Hurst parameter $H \in (0, 1)$, $(B_t^H, t \in \mathbb{R})$ is a zero mean Gaussian process with the following covariance structure:

$$\mathbb{E}(B_t^H B_s^H) = R_H(t, s) = \frac{1}{2}\left(|t|^{2H} + |s|^{2H} - |t - s|^{2H}\right). \tag{19.5}$$

Fix a time interval $[0, T]$. Denote by $\mathcal{E}$ the set of real-valued step functions on $[0, T]$ and let $\mathcal{H}$ be the Hilbert space defined as the closure of $\mathcal{E}$ with respect to the inner product:

$$\langle \mathbf{1}_{[0,t]}, \mathbf{1}_{[0,s]} \rangle_{\mathcal{H}} = R_H(t, s),$$

where R_H is the covariance function of the fBm, given in Eq. (19.5). The mapping $\mathbf{1}_{[0,t]} \longmapsto B_t^H$ can be extended to a linear isometry between $\mathcal{H}$ and the Gaussian space $\mathcal{H}_1$ spanned by B^H (see also [9]). We denote this isometry by $\varphi \longmapsto B^H(\varphi)$, which can be also considered as the stochastic integral of φ with respect to B^H (denoted by $B^H(\varphi) = \int_0^T \varphi(t) \mathrm{d}B_t^H$). For $H = \frac{1}{2}$ we have $\mathcal{H} = L^2([0, T])$, whereas for $H > \frac{1}{2}$ we have $L^{\frac{1}{H}}([0, T]) \subset \mathcal{H}$ and for $\varphi, \psi \in L^{\frac{1}{H}}([0, T])$ we have

$$\mathbb{E}\left(B^H(\varphi)B^H(\psi)\right) = \mathbb{E}\left(\int_0^T \varphi(t)\mathrm{d}B_t^H \int_0^T \psi(t)\mathrm{d}B_t^H\right)$$
$$= \langle \varphi, \psi \rangle_{\mathcal{H}} = \int_0^T \int_0^T \varphi_s \psi_t \phi(t-s)\mathrm{d}s\mathrm{d}t, \tag{19.6}$$

where

$$\phi(u) = H(2H-1)|u|^{2H-2}. \tag{19.7}$$

Let $\mathcal{S}$ be the space of smooth and cylindrical random variables of the form

$$F = f(B^H(\varphi_1), \ldots, B^H(\varphi_n)), \quad \varphi_1, \ldots, \varphi_n \in L^{\frac{1}{H}}([0,T]) \subseteq \mathcal{H}, \tag{19.8}$$

where $f \in C_b^\infty(\mathbb{R}^n)$ (f and all its partial derivatives are bounded). For a random variable F of the form Eq. (19.8) we define its Malliavin derivative as the $\mathcal{H}$-valued random element:

$$DF = \sum_{i=1}^n \frac{\partial f}{\partial x_i}(B^H(\varphi_1), \ldots, B^H(\varphi_n))\varphi_i.$$

By iteration, one can define the mth derivative $D^m F$, which is an element of $L^2(\Omega; \mathcal{H}^{\otimes m})$, for every $m \geq 2$. For $m \geq 1$, $\mathbb{D}^{m,2}$ denotes the closure of $\mathcal{S}$ with respect to the norm $\|\cdot\|_{m,2}$, defined by the relation

$$\|F\|_{m,2}^2 = \mathbb{E}\left[|F|^2\right] + \sum_{i=1}^m \mathbb{E}\left(\|D^i F\|_{\mathcal{H}^{\otimes i}}^2\right).$$

Let δ be the adjoint of the operator D, also called the *divergence operator*. A random element $u \in L^2(\Omega, \mathcal{H})$ belongs to the domain of δ, denoted by $\mathrm{Dom}(\delta)$, if and only if it verifies

$$|\mathbb{E}\langle DF, u\rangle_{\mathcal{H}}| \leq c_u \|F\|_{L^2},$$

for any $F \in \mathbb{D}^{1,2}$, where c_u is a constant depending only on u. If $u \in \mathrm{Dom}(\delta)$, then the random variable $\delta(u)$ is defined by the duality relationship

$$\mathbb{E}(F\delta(u)) = \mathbb{E}\langle DF, u\rangle_{\mathcal{H}}, \tag{19.9}$$

which holds for every $F \in \mathbb{D}^{1,2}$. The divergence operator δ is also called the Skorohod integral because in the case of the Brownian motion it coincides with the anticipating stochastic integral introduced by Skorohod in [13]. We will make use of the notation $\delta(u) = \int_0^T u_t \mathrm{d}B_t^H$.

For every $n \geq 1$, let $\mathcal{H}_n$ be the nth Wiener chaos of B^H, that is, the closed linear subspace of $L^2(\Omega, \mathcal{F}, P)$ generated by the random variables: $\{H_n(B^H(h)), h \in \mathcal{H}, \|h\|_{\mathcal{H}} = 1\}$, where H_n is the nth Hermite polynomial. The mapping $h^{\otimes n} \in \mathcal{H}^{\odot n} \to I_n(h^{\otimes n}) \in \mathcal{H}_n$, defined by $I_n(h^{\otimes n}) = H_n(B^H(h))$, provides a linear isometry between the symmetric tensor product $\mathcal{H}^{\odot n}$ and $\mathcal{H}_n$. For $H = \frac{1}{2}$, I_n coincides with the multiple Itô stochastic integral. On the other hand, $I_n(h^{\otimes n})$ coincides with the iterated divergence $\delta^n(h^{\otimes n})$ and coincides with the multiple Itô type stochastic integral introduced in [1, 3, 5].

We will make use of the following central limit theorem for multiple stochastic integrals (see [11]).

Proposition 2.1. *Let $\{F_n, n \geq 1\}$ be a sequence of random variables in the pth Wiener chaos, $p \geq 2$, such that $\lim_{n\to\infty} \mathbb{E}(F_n^2) = \sigma^2$. Then, the following conditions are equivalent:*

(i) F_n converges in law to $N(0, \sigma^2)$ as n tends to infinity.
(ii) $\|DF_n\|_{\mathcal{H}}^2$ converges in L^2 to a constant as n tends to infinity.

Remark 2.1. In [11] it is proved that (i) is equivalent to the fact that $\|DF_n\|_{\mathcal{H}}^2$ converges in L^2 to $p\sigma^2$ as n tends to infinity. If we assume (ii), the limit of $\|DF_n\|_{\mathcal{H}}^2$ must be equal to $p\sigma^2$ because

$$\mathbb{E}(\|DF_n\|_{\mathcal{H}}^2) = p\mathbb{E}(F_n^2).$$

To obtain Berry–Esseen type estimate, we shall use a result from [10], which we shall state in our fractional Brownian motion framework. The validity is straightforward.

Assume that $F = \int_0^T \int_0^T f(s,t) \mathrm{d}B_s^H \mathrm{d}B_t^H$ is an element in the second chaos, where f is symmetric functions of two variables. Then with this kernel f we can define a Hilbert–Schmidt operator H_f from $\mathcal{H}$ to $\mathcal{H}$ by

$$H_f g(t) = \langle f(t,\cdot), g(\cdot)\rangle_{\mathcal{H}}.$$

If g is a continuous function on $[0, T]$, then

$$H_f g(t) = \int_0^T \int_0^T f(t,u) g(v) \phi(u-v) \mathrm{d}u \mathrm{d}v,$$

where ϕ is defined by Eq. (19.7). For $p \geq 2$, the pth cumulant of F is well known (see, e.g., [4] for a proof).

$$\begin{aligned}\kappa_p(F) &= 2^{p-1}(p-1)!\mathrm{Tr}(H_f^p) \\ &= 2^{p-1}(p-1)! \int_{[0,T]^{2p}} f(s_1,s_2) f(s_3,s_4) \cdots f(s_{2p-1},s_{2p}) \phi(s_2,s_3) \cdots \\ &\quad \phi(s_{2p-2},s_{2p-1}) \phi(s_{2p},s_1) \mathrm{d}s_1 \cdots \mathrm{d}s_{2p}.\end{aligned}$$

Let

$$F_n = I_2(f_n) = \int_0^T \int_0^T f_n(s,t) \mathrm{d}B_s^H \mathrm{d}B_t^H$$

be a sequence of random variables in the second chaos. We shall use the following result from [10], Proposition 3.8.

Proposition 2.2. *If* $\kappa_2(F_n) = \mathbb{E}\,(F_n^2) \to 1$ *and* $\kappa_4(F_n) \to 0$*, then*

$$\sup_{z\in\mathbb{R}} |P(F_n \le z) - \Psi(z)| \le \sqrt{\frac{\kappa_4(F_n)}{6} + (\kappa_2(F_n) - 1)^2},$$

where $\Psi(z) = \frac{1}{\sqrt{2\pi}} \int_{-\infty}^{z} \mathrm{e}^{-\frac{u^2}{2}} \mathrm{d}u$ *is the error function.*

3 Construction and Strong Consistency of the Estimator

As in [8], we can assume that $X_0 = 0$, and

$$X_t = \sigma \int_0^t \mathrm{e}^{-\theta(t-s)} \mathrm{d}B_s^H .$$

[We can express $X_t = Y_t - \mathrm{e}^{-\theta t}\xi$, where $Y_t = \sigma \int_{-\infty}^t \mathrm{e}^{-\theta(t-s)} \mathrm{d}B_s^H$ is stationary and $\xi = \sigma \int_{-\infty}^0 \mathrm{e}^{\theta s} \mathrm{d}B_s^H$ has the limiting (normal) distribution of X_t.]

Let $p > 0$ be a positive number and denote

$$\eta_{p,n} = \frac{1}{n} \sum_{k=1}^{n} |X_{kh}|^p. \tag{19.10}$$

It is easy to argue that

$$\lim_{n\to\infty} \eta_{p,n} = \lim_{n\to\infty} \frac{1}{n} \sum_{k=1}^{n} |Y_{kh}|^p.$$

Thus by the ergodic theorem we see that $\eta_{p,n}$ converges almost surely to

$$\begin{aligned} \lim_{n\to\infty} \eta_{p,n} &= \mathbb{E}\,(|Y_h|^p) = \lim_{n\to\infty} \mathbb{E}\,(|X_{nh}|^p) \\ &= c_p \lim_{n\to\infty} (\mathrm{Var}(X_{nh}))^{p/2} \\ &= c_p \sigma^p \theta^{-Hp} (H\Gamma(2H))^{p/2}, \end{aligned}$$

where

$$c_p = \frac{1}{\sqrt{2\pi}\sigma} \int_{-\infty}^{\infty} |x|^p \mathrm{e}^{-\frac{x^2}{2\sigma^2}} \mathrm{d}x = \pi^{-1/2} \Gamma\left(\frac{p+1}{2}\right).$$

Thus we obtain

Proposition 3.1. *Let $p > 0$, and $h > 0$. Define*

$$\hat{\theta}_{p,n} = \left(\frac{1}{c_p \sigma^p (H\Gamma(2H))^{p/2}} \eta_{p,n}\right)^{-\frac{1}{pH}}$$
$$= \left(\frac{1}{n\pi^{-1/2}\Gamma\left(\frac{p+1}{2}\right)\sigma^p (H\Gamma(2H))^{p/2}} \sum_{k=1}^{n} |X_{kh}|^p\right)^{-\frac{1}{pH}}. \quad (19.11)$$

Then $\hat{\theta}_{p,n} \to \theta$ almost surely as $n \to \infty$.

4 Central Limit Theorem

In this section we shall show that $\sqrt{n}\left(\hat{\theta}_{p,n} - \theta\right)$ converges in law to a mean zero normal and we shall also compute the limiting variance. But we shall study the case $p = 2$. More general case may be discussed with the same approach, but it will be much more sophisticated. When $p = 2$, we denote $\hat{\theta}_n = \hat{\theta}_{2,n}$. Namely,

$$\hat{\theta}_n = \left(\frac{1}{n\sigma^2 H\Gamma(2H)} \sum_{k=1}^{n} X_{kh}^2\right)^{-\frac{1}{2H}}. \quad (19.12)$$

Denote

$$\xi_n = \frac{1}{n} \sum_{k=1}^{n} X_{kh}^2 \quad (19.13)$$

and $\rho = \sigma^2 H\Gamma(2H)$. Then $\hat{\theta}_n = \left(\frac{\xi_n}{\rho}\right)^{-\frac{1}{2H}}$. From the last section, we see

$$\lim_{n\to\infty} \xi_n = \lim_{n\to\infty} \mathbb{E}(\xi_n) = \sigma^2\theta^{-2H} H\Gamma(2H) = \rho\theta^{-2H}.$$

First we want to show that

$$F_n := \sqrt{n}\left(\xi_n - \mathbb{E}(\xi_n)\right) \quad (19.14)$$

converges in law. We shall use Proposition 2.1.

Lemma 4.1. *When $H \in [\frac{1}{2}, \frac{3}{4})$, we have*

$$\lim_{n\to\infty} \mathbb{E}(F_n^2) = 2\rho^2\theta^{-4H} \quad (19.15)$$

and

$$\left|\mathbb{E}\,(F_n^2) - 2\rho^2\theta^{-4H}\right| \le C n^{4H-3},$$

where and in what follows $C > 0$ denotes a generic constant independent of n (but it may depend on θ, H).

Proof. From the definition of F_n we see

$$\begin{aligned}
\mathbb{E}\,(F_n^2) =& \frac{1}{n}\left[\sum_{k,k'=1}^{n} \mathbb{E}\,(X_{kh}^2 X_{k'h}^2) - \sum_{k,k'=1}^{n} \mathbb{E}\,(X_{kh}^2)\mathbb{E}\,(X_{k'h}^2)\right] \\
=& \frac{2}{n}\sum_{k,k'=1}^{n} [\mathbb{E}\,(X_{kh}X_{k'h})]^2 \\
=& \frac{2}{n}\sum_{k\neq k';k,k'=1}^{n} [\mathbb{E}\,(X_{kh}X_{k'h})]^2 + \frac{2}{n}\sum_{k=1}^{n}\left[\mathbb{E}\,(X_{kh}^2)\right]^2 \\
=& A_n + B_n .
\end{aligned}$$

We shall prove that $\lim_{n\to\infty} A_n = 0$ and $\lim_{n\to\infty} B_n = 2\rho^2\theta^{-4H}$. By Lemma 5.4 in [8], we have

$$\begin{aligned}
A_n \le& C\frac{1}{n}\sum_{k\neq k',k,k'=1}^{n} |k-k'|^{4H-4} \\
\le& C\frac{1}{n}\sum_{i=1}^{n}\sum_{j=i+1}^{n} (j-i)^{4H-4} \\
\le& C\frac{1}{n}\sum_{i=1}^{n}(n-i)^{4H-3} \\
\le& C n^{4H-3}
\end{aligned}$$

which implies that $\lim_{n\to\infty} A_n = 0$ when $H < \frac{3}{4}$. On the other hand,

$$\lim_{n\to\infty} B_n = 2\lim_{n\to\infty}(\mathbb{E}\,X_{nh}^2)^2 = 2H^2\Gamma^2(2H)\sigma^4\theta^{-4H} = 2\rho^2\theta^{-4H}.$$

To prove the second inequality, it suffices to show that

$$\left|\frac{1}{n}\sum_{k=1}^{n}\left[\mathbb{E}\,(X_{kh}^2)\right]^2 - \rho^2\theta^{-4H}\right| \le C n^{4H-3}.$$

In fact,

$$\begin{aligned}
&\left|\frac{1}{n}\sum_{k=1}^{n}\left[\mathbb{E}\,(X_{kh}^2)\right]^2-\rho^2\theta^{-4H}\right| \\
&\leq \frac{1}{n}\sum_{k=1}^{n}\left|\mathbb{E}\,(X_{kh}^2)-\rho\theta^{-2H}\right|\left(\mathbb{E}\,(X_{kh}^2)+\rho\theta^{-2H}\right) \\
&\leq C\frac{1}{n}\sum_{k=1}^{n}\left|\mathbb{E}\,(X_{kh}^2)-\rho\theta^{-2H}\right|.
\end{aligned}$$

However, we have

$$\begin{aligned}
\left|\mathbb{E}\,(X_{kh}^2)-\rho\theta^{-2H}\right| &= C\Bigg(\int_0^{\infty}\int_0^{s}\mathrm{e}^{-\theta(u+s)}|s-u|^{2H-2}\mathrm{d}u\mathrm{d}s \\
&\quad -\int_0^{kh}\int_0^{s}\mathrm{e}^{-\theta(u+s)}|s-u|^{2H-2}\mathrm{d}u\mathrm{d}s\Bigg). \\
&= C\int_{kh}^{\infty}\int_0^{s}\mathrm{e}^{-\theta(u+s)}|s-u|^{2H-2}\mathrm{d}u\mathrm{d}s \\
&= C\int_{kh}^{\infty}\int_0^{s}\mathrm{e}^{\theta(x-2s)}x^{2H-2}\mathrm{d}x\mathrm{d}s \\
&\leq C\int_{kh}^{\infty}\mathrm{e}^{\theta(-s)}s^{2H-1}\mathrm{d}s \\
&\leq C\int_{kh}^{\infty}\mathrm{e}^{\theta(-s/2)}\mathrm{d}s \\
&\leq C\mathrm{e}^{-kh/2}.
\end{aligned}$$

Hence, we have

$$\left|\frac{1}{n}\sum_{k=1}^{n}\left[\mathbb{E}\,(X_{kh}^2)\right]^2-\rho^2\theta^{-4H}\right| \leq Cn^{-1}\leq Cn^{4H-3}$$

which completes the proof. □

Now we have

$$DF_n=\frac{2}{\sqrt{n}}\sum_{k=1}^{n}X_{kh}DX_{kh}.$$

Thus

$$G_n:=\langle DF_n,DF_n\rangle_{\mathcal{H}}=\frac{4}{n}\sum_{k,k'=1}^{n}X_{kh}X_{k'h}\langle DX_{kh},DX_{k'h}\rangle_{\mathcal{H}}.$$

Since X_{kh} is normal random variable, it is easy to see that

$$\langle DX_{kh}, DX_{k'h}\rangle_{\mathcal{H}} = \mathbb{E}\ (X_{kh}X_{k'h}).$$

Thus

$$G_n = \frac{4}{n}\sum_{k,k'=1}^{n} X_{kh}X_{k'h}\mathbb{E}\ (X_{kh}X_{k'h}).$$

It is easy to check

$$\mathbb{E}\ (G_n) = 2\mathbb{E}\ (F_n^2)$$

which converges to $4\rho^2\theta^{-4H}$ as $n \to \infty$ by Lemma 4.1. Thus to verify (ii) of Proposition 2.1, it suffices to show that

$$\lim_{n\to\infty} \mathbb{E}\ [G_n - \mathbb{E}\ (G_n)]^2 = 0. \tag{19.16}$$

However,

$$\begin{aligned}\mathbb{E}\ [G_n - \mathbb{E}\ (G_n)]^2 &= \mathbb{E}\ \left(G_n^2\right) - [\mathbb{E}\ (G_n)]^2 \\ &= \frac{1}{n^2}\sum_{k,k';j,j'=1}^{n} \Big\{\mathbb{E}\ \left[X_{kh}X_{k'h}X_{jh}X_{j'h}\right]\mathbb{E}\left[X_{kh}X_{k'h}\right]\mathbb{E}\left[X_{jh}X_{j'h}\right] \\ &\quad - \left(\mathbb{E}\ [X_{kh}X_{k'h}]\,\mathbb{E}\ \left[X_{jh}X_{j'h}\right]\right)^2\Big\}.\end{aligned}$$

The expectation $\mathbb{E}\ (X_1X_2\cdots X_p)$ can be computed by the well-known Feynman diagram. In the case $p = 4$, we have

$$\mathbb{E}\ (X_1X_2X_3X_4) = \mathbb{E}\ (X_1X_2)\mathbb{E}\ (X_3X_4) + \mathbb{E}\ (X_1X_3)\mathbb{E}\ (X_2X_4) + \mathbb{E}\ (X_1X_4)\mathbb{E}\ (X_2X_3).$$

Thus

$$\begin{aligned}&\mathbb{E}\ [G_n - \mathbb{E}\ (G_n)]^2 \\ &= \frac{32}{n^2}\sum_{k,k',j,j'=1}^{n} \mathbb{E}\ \left[X_{kh}X_{jh}\right]\mathbb{E}\ \left[X_{k'h}X_{j'h}\right]\mathbb{E}\ [X_{kh}X_{k'h}]\,\mathbb{E}\ \left[X_{jh}X_{j'h}\right].\end{aligned}$$

From Lemma 5.4 (Eq. (5.7)) of [8], we have

$$|\mathbb{E}\ [X_{kh}X_{k'h}]| \le \sigma^2 C_{\theta,h,H}|k-k'|^{2H-2}.$$

Therefore,

$$\mathbb{E}\ [G_n - \mathbb{E}\ (G_n)]^2$$

$$\begin{aligned}
&\leq \frac{C}{n^2} \sum_{k,k',j,j'=1}^{n} |k-j|^{2H-2}|k'-j'|^{2H-2}|k-k'|^{2H-2}|j-j'|^{2H-2} \\
&\leq \frac{C}{n^2} \int_{[0,n]^4} |u-v|^{2H-2}|u'-v'|^{2H-2}|u-v'|^{2H-2}|v-v'|^{2H-2}\mathrm{d}u\mathrm{d}v\mathrm{d}u'\mathrm{d}v' \\
&= Cn^{4(2H-2)+4-2} \int_{[0,1]^4} |u-v|^{2H-2}|u'-v'|^{2H-2} \\
&\quad \times |u-v'|^{2H-2}|v-v'|^{2H-2}\mathrm{d}u\mathrm{d}v\mathrm{d}u'\mathrm{d}v' \\
&\leq Cn^{8H-6}, \qquad (19.17)
\end{aligned}$$

which converges to 0 as $n \to \infty$ if $H < 3/4$.

Summarizing the above, we can state

Theorem 4.1. *Let X_t be the Ornstein–Uhlenbeck process defined by Eq. (19.2) and let ξ_n be defined by Eq. (19.13). If $1/2 \leq H < 3/4$, then*

$$\sqrt{n}\,(\xi_n - \mathbb{E}\,(\xi_n)) \to N(0,\Sigma), \qquad (19.18)$$

where

$$\Sigma = \lim_{n\to\infty} \mathbb{E}\,(F_n^2) = 2\rho^2\theta^{-4H}. \qquad (19.19)$$

To study the weak convergence of $\sqrt{n}\left(\hat{\theta}_n - \theta\right)$, we need the following lemma.

Lemma 4.2. *Let $H \geq 1/2$. Then*

$$\sqrt{n}\left|\mathbb{E}\,(\xi_n) - \rho\theta^{-2H}\right| \leq Cn^{-\frac{1}{2}},$$

and hence

$$\lim_{n\to\infty} \sqrt{n}(\mathbb{E}\,(\xi_n) - \rho\theta^{-2H}) = 0.$$

Proof. From the definition of ξ_n, we have

$$\begin{aligned}
&\sqrt{n}\left|\mathbb{E}\,(\xi_n) - \rho\theta^{-2H}\right| \\
&= \frac{C}{\sqrt{n}} \sum_{k=1}^{n} \left| \int_0^{kh}\int_0^{kh} \mathrm{e}^{-\theta(u+s)}|u-s|^{2H-2}\mathrm{d}s\mathrm{d}u \right. \\
&\quad \left. - \int_0^{\infty}\int_0^{\infty} \mathrm{e}^{-\theta(u+s)}|u-s|^{2H-2}\mathrm{d}s\mathrm{d}u \right| \\
&= \frac{C}{\sqrt{n}} \sum_{k=1}^{n} \left| \int_0^{\theta kh}\int_0^{u} \mathrm{e}^{-(u+s)}|u-s|^{2H-2}\mathrm{d}s\mathrm{d}u \right.
\end{aligned}$$

$$- \int_0^\infty \int_0^u \mathrm{e}^{-(u+s)} |u-s|^{2H-2} \mathrm{d}s\mathrm{d}u \Bigg|$$

$$= \frac{C}{\sqrt{n}} \sum_{k=1}^n \int_{\theta kh}^\infty \int_0^u \mathrm{e}^{-(u+s)} |u-s|^{2H-2} \mathrm{d}s\mathrm{d}u$$

$$= \frac{C}{\sqrt{n}} \sum_{k=1}^n \int_{\theta kh}^\infty \int_0^u \mathrm{e}^{-2u+x} x^{2H-2} \mathrm{d}x\mathrm{d}u \le \frac{C}{\sqrt{n}} \sum_{k=1}^n \int_{\theta kh}^\infty \mathrm{e}^{-2u} \mathrm{e}^u u^{2H-1} \mathrm{d}u$$

$$\le \frac{C}{\sqrt{n}} \sum_{k=1}^n \int_{\theta kh}^\infty \mathrm{e}^{-\frac{1}{2}u} \mathrm{d}u \le \frac{C}{\sqrt{n}} \sum_{k=1}^n \mathrm{e}^{-\frac{\theta kh}{2}} \le \frac{C}{\sqrt{n}}.$$

This proves the lemma. □

Let us recall that

$$\hat{\theta}_n = \left(\frac{\xi_n}{\rho} \right)^{-1/(2H)}.$$

Therefore

$$\sqrt{n} \left(\hat{\theta}_n - \theta \right) = -\frac{1}{2H} \tilde{\xi}_n^{-1/(2H)-1} \sqrt{n} \left(\frac{\xi_n}{\rho} - \theta^{-2H} \right),$$

where $\tilde{\xi}_n$ is between θ^{-2H} and $\frac{\xi_n}{\rho}$. Since $\tilde{\xi}_n \to \theta^{-2H}$ almost surely and since $\sqrt{n} \left(\xi_n - \rho\theta^{-2H} \right)$ converges to $N(0, \Sigma)$ in law by Theorem 4.1 and Lemma 4.2, we see that $\sqrt{n} \left(\hat{\theta}_n - \theta \right)$ converges in law to

$$N\left(0, \frac{\theta^{4H+2}}{4H^2\rho^2} \Sigma \right) = N\left(0, \frac{\theta^2}{2H^2}\right).$$

Thus we arrive at our main theorem of this section.

Theorem 4.2. *Let* $1/2 \le H < 3/4$. *Then*

$$\sqrt{n} \left(\hat{\theta}_n - \theta \right) \to N\left(0, \frac{\theta^2}{2H^2}\right) \quad \textit{in law} \quad \textit{as} \quad n \to \infty. \tag{19.20}$$

5 Berry–Esseen Asymptotics

Theorem 4.2 shows that when $n \to \infty$, $Q_n := \sqrt{\frac{2H^2 n}{\theta^2}} \left(\hat{\theta}_n - \theta \right)$ converges to $N(0, 1)$ in law. In this section we shall obtain a rate of this convergence. We shall use Proposition 2.2. To this end we need to compute the fourth cumulant $\kappa_4(F_n)$.

Let us develop a general approach to estimate $\kappa_4(Q_n)$ which is particularly useful for our situation. To simplify notation we omit the explicit dependence on n. It is clear that if $Z_k = \int_0^T f_k(s)\mathrm{d}B_s^H$ for some (deterministic) $f_k \in \mathcal{H}$, then

$$V = \sum_{k=1}^{N} (Z_k^2 - \mathbb{E}\,(Z_k^2)) = \sum_{k=1}^{N} I_2(f_k^{\otimes 2}). \tag{19.21}$$

Thus

$$f = \sum_{k=1}^{N} f_k \otimes f_k$$

and

$$H_f^4 = \sum_{k_1,k_2,k_3,k_4=1}^{N} f_{k_1} \otimes f_{k_4} \langle f_{k_1}, f_{k_2}\rangle_{\mathcal{H}} \langle f_{k_2}, f_{k_3}\rangle_{\mathcal{H}} \langle f_{k_3}, f_{k_4}\rangle_{\mathcal{H}},$$

which is a map from $\mathcal{H}$ to $\mathcal{H}$ such that for $g \in \mathcal{H}$,

$$H_f^4(g)(t) = \sum_{k_1,k_2,k_3,k_4=1}^{N} \langle f_{k_1}, f_{k_2}\rangle_{\mathcal{H}} \langle f_{k_2}, f_{k_3}\rangle_{\mathcal{H}} \langle f_{k_3}, f_{k_4}\rangle_{\mathcal{H}} \langle f_{k_4}, g\rangle_{\mathcal{H}} f_{k_1}(t).$$

If V is given by Eq. (19.21), then the fourth cumulant of V is

$$\begin{aligned}\kappa_4(V) &= \mathrm{Tr}(H_f^4) \qquad (19.22)\\ &= \sum_{k_1,k_2,k_3,k_4=1}^{N} \langle f_{k_1}, f_{k_2}\rangle_{\mathcal{H}} \langle f_{k_2}, f_{k_3}\rangle_{\mathcal{H}} \langle f_{k_3}, f_{k_4}\rangle_{\mathcal{H}} \langle f_{k_4}, f_{k_1}\rangle_{\mathcal{H}}\\ &= \sum_{k_1,k_2,k_3,k_4=1}^{N} \mathbb{E}\,(Z_{k_1}Z_{k_2})\mathbb{E}\,(Z_{k_2}Z_{k_3})\mathbb{E}\,(Z_{k_3}Z_{k_4})\mathbb{E}\,(Z_{k_4}Z_{k_1}).\end{aligned} \tag{19.23}$$

If we apply this computation Eq. (19.23) to F_n defined in Sect. 4, then we see that $\kappa_4(F_n)$ is the same as $\mathbb{E}\,(G_n - \mathbb{E}\,(G_n))^2$ studied in Sect. 4. Thus we have from Eq. (19.17)

$$\kappa_4(F_n) \le Cn^{8H-6}.$$

By Lemma 4.1, we have

$$|\kappa_2(F_n) - \Sigma| = \left|E(F_n^2) - \Sigma\right| \le Cn^{4H-3}.$$

Therefore, by Proposition 2.2, we have

Lemma 5.1.

$$\sup_{z\in\mathbb{R}}\left|P\left(-\frac{F_n}{\sqrt{\Sigma}}\le z\right)-\Psi(z)\right|\le Cn^{4H-3},$$

where $\Psi(z)=\frac{1}{\sqrt{2\pi}}\int_{-\infty}^{z}\mathrm{e}^{-\frac{x^2}{2}}\mathrm{d}x$ *is the error function.*

We also have the following lemma.

Lemma 5.2. *Let* $1/2\le H<3/4$*. There is a constant* C *such that*

$$\sup_{y\in\mathbb{R}}\left|P\left(\sqrt{\frac{n}{2}}\theta^{2H}\left(\theta^{-2H}-\hat{\theta}_n^{-2H}\right)\le y\right)-\Psi(y)\right|\le Cn^{(4H-3)\vee(-\frac{1}{2})}. \tag{19.24}$$

Proof. Recall that $F_n=\sqrt{n}\,(\xi_n-\mathbb{E}\,(\xi_n))$.
Let $\tilde{F}_n=\sqrt{n}\left(\xi_n-\rho\theta^{-2H}\right)$, and $a_n=\tilde{F}_n-F_n=\sqrt{n}(\mathbb{E}\,(\xi_n)-\rho\theta^{-2H})$, then $|a_n|\le Cn^{-\frac{1}{2}}$ by Lemma 4.2.

$$\begin{aligned}&\left|P\left(-\frac{\tilde{F}_n}{\sqrt{\Sigma}}\le z\right)-\Psi(z)\right|\\&\quad=\left|P\left(-\frac{F_n+a_n}{\sqrt{\Sigma}}\le z\right)-\Psi(z)\right|\\&\quad\le\left|P\left(-\frac{F_n}{\sqrt{\Sigma}}\le z+\frac{a_n}{\sqrt{\Sigma}}\right)-\Psi(z+\frac{a_n}{\sqrt{\Sigma}})\right|+\left|\Psi(z+\frac{a_n}{\sqrt{\Sigma}})-\Psi(z)\right|\\&\quad\le C(n^{4H-3}+n^{-\frac{1}{2}}).\end{aligned}$$

The inequality (19.24) is obtained since $\xi_n=\rho\hat{\theta}_n^{-2H}$ and $\Sigma=2\rho^2\theta^{-4H}$. □

Now we can prove our main theorem.

Theorem 5.1. *Let* $1/2\le H<3/4$*. For any* $K>0$*, there exist a constant* C_K *depending on* K *and* H *and a constant* $N_K>0$ *depending on* K*, such that when* $n>N_K$*,*

$$\sup_{|z|\le K}\left|P\left(\frac{\sqrt{2n}H}{\theta}\left(\hat{\theta}_n-\theta\right)\le z\right)-\Psi(z)\right|\le C_Kn^{(4H-3)\vee(-\frac{1}{2})}. \tag{19.25}$$

Proof. Now we have

$$\begin{aligned}&P\left(\sqrt{\frac{n}{2}}\theta^{2H}\left(\theta^{-2H}-\hat{\theta}_n^{-2H}\right)\le y\right)\\&\quad=P\left(\hat{\theta}_n\le\theta\left(1-\sqrt{\frac{2}{n}}y\right)^{-\frac{1}{2H}}\right)\end{aligned}$$

$$= P\left(\frac{\sqrt{2nH}}{\theta}\left(\hat{\theta}_n - \theta\right) \leq \sqrt{2nH}\left[\left(1 - \sqrt{\frac{2}{n}}y\right)^{-\frac{1}{2H}} - 1\right]\right).$$

Choose $y_{n,z}$ so that

$$\sqrt{2nH}\left[\left(1 - \sqrt{\frac{2}{n}}y_{n,z}\right)^{-\frac{1}{2H}} - 1\right] = z,$$

namely,

$$y_{n,z} = \sqrt{\frac{n}{2}}\left[1 - \left(1 + \frac{z}{\sqrt{2nH}}\right)^{-2H}\right].$$

Then

$$\left|P\left(\frac{\sqrt{2nH}}{\theta}\left(\hat{\theta}_n - \theta\right) \leq z\right) - \Psi(z)\right|$$

$$= \left|P\left(\sqrt{\frac{n}{2}}\theta^{2H}\left(\theta^{-2H} - \hat{\theta}_n^{-2H}\right) \leq y_{n,z}\right) - \Psi(z)\right|$$

$$\leq \left|P\left(\sqrt{\frac{n}{2}}\theta^{2H}\left(\theta^{-2H} - \hat{\theta}_n^{-2H}\right) \leq y_{n,z}\right) - \Psi(y_{n,z})\right| + \left|\Psi(y_{n,z}) - \Psi(z)\right|.$$

The inequality Eq. (19.24) implies that the above first term is bounded by $Cn^{(4H-3)\vee(-\frac{1}{2})}$. It is easy to check that there exists a constant C_K depending on K and H and a number N_K depending on K, such that when $n > N_K$, $|\Psi(y_{n,z}) - \Psi(z)| \leq |y_{n,z} - z| \leq C_K n^{-1/2}$ for all $|z| \leq K$. □

Remark 5.1. Throughout this paper we did not discuss the case $H = 1/2$ in detail, which is easy.

Acknowledgements We appreciate Chihoon Lee and the referee's careful reading of this paper. Yaozhong Hu is partially supported by a grant from the Simons Foundation #209206.

References

1. Biagini, F., Hu, Y., Øksendal, B., Zhang, T.: Stochastic Calculus for Fractional Brownian Motion and Applications. Springer, London (2008)
2. Dacunha-Castelle, D., Florens-Zmirou, D.: Estimation of the coefficients of a diffusion from discrete observations. Stochastics **19**(4), 263–284 (1986)
3. Duncan, T.E., Hu, Y., Pasik-Duncan, B.: Stochastic calculus for fractional Brownian motion. I. Theory. SIAM J. Control Optim. **38**, 582–612 (2000)

4. Fox, R., Taqqu, M.S.: Central limit theorems for quadratic forms in random variables having long-range dependence. Probab. Theory Relat. Fields **74**, 213–240 (1987)
5. Hu, Y.: Integral transformations and anticipative calculus for fractional Brownian motions. Mem. Amer. Math. Soc. **175**, 825 (2005)
6. Hu, Y., Long, H.: Least squares estimator for Ornstein-Uhlenbeck processes driven by α-stable motions. Stoch. Process Appl. **119**, 2465–2480 (2009)
7. Hu, Y., Long, H.: Least squares estimator for Ornstein-Uhlenbeck processes driven by α-stable motions. Stoch. Process Appl. **119**(8), 2465–2480 (2009)
8. Hu, Y., Nualart, D.: Parameter estimation for fractional Ornstein-Uhlenbeck processes. Statist. Probab. Lett. **80**(11–12), 1030–1038 (2010)
9. Neveu, J.: Processus aléatoires gaussiens. Les Presses de l'Universit de Montral, Montreal, Que (1968)
10. Nourdin, I., Peccati, G.: Stein's method and exact Berry-Esseen asymptotics for functionals of Gaussian fields. Ann. Probab. **37**, 2231–2261 (2009)
11. Nualart, D., Ortiz-Latorre, S.: Central limit theorems for multiple stochastic integrals and Malliavin calculus. Stoch. Process. Appl. **118**, 614–628 (2008)
12. Prakasa Rao, B.L.S.: Statistical inference for diffusion type processes. In: Kendall's Library of Statistics, vol. 8. Oxford University Press, Oxford (1999)
13. Skorohod, A.V.: On a generalization of a stochastic integral. Theory Probab. Appl. **20**, 219–233 (1975)

Part V
Applications of Stochastic Analysis

Chapter 20
The Effect of Competition on the Height and Length of the Forest of Genealogical Trees of a Large Population

Mamadou Ba and Etienne Pardoux

Abstract We consider a population generating a forest of genealogical trees in continuous time, with m roots (the number of ancestors). In order to model competition within the population, we superimpose to the traditional Galton–Watson dynamics (births at constant rate μ, deaths at constant rate λ) a death rate which is γ times the size of the population alive at time t raised to some power $\alpha > 0$ ($\alpha = 1$ is a case without competition). If we take the number of ancestors at time 0 to be equal to $[xN]$, weight each individual by the factor $1/N$ and choose adequately μ, λ and γ as functions of N, then the population process converges as N goes to infinity to a Feller SDE with a negative polynomial drift. The genealogy in the continuous limit is described by a real tree [in the sense of Aldous (Ann Probab 19:1–28, 1991)]. In both the discrete and the continuous case, we study the height and the length of the genealogical tree as an (increasing) function of the initial population. We show that the expectation of the height of the tree remains bounded as the size of the initial population tends to infinity iff $\alpha > 1$, while the expectation of the length of the tree remains bounded as the size of the initial population tends to infinity iff $\alpha > 2$.

Keywords Galton–Watson processes • Feller diffusion

AMS Subject Classification 2000: (Primary) 60J80, 60F17 (Secondary) 92D25.

Received 11/27/2011; Accepted 4/15/2012; Final 6/1/2012

M. Ba • E. Pardoux (✉)
LATP-UMR 6632, CMI, Université de Provence, 39 rue F. Joliot Curie, Marseille cedex 13, France
e-mail: azobak@yahoo.fr; pardoux@cmi.univ-mrs.fr

F. Viens et al. (eds.), *Malliavin Calculus and Stochastic Analysis: A Festschrift in Honor of David Nualart*, Springer Proceedings in Mathematics & Statistics 34, DOI 10.1007/978-1-4614-5906-4_20, © Springer Science+Business Media New York 2013

1 Introduction

Consider a Galton–Watson binary branching process in continuous time with m ancestors at time $t = 0$, in which each individual gives birth to children at a constant rate μ and dies after an exponential time with parameter λ. Suppose we superimpose deaths due to competition. For instance, we might decide to add to each individual a death rate equal to γ times the number of presently alive individuals in the population, which amounts to add a global death rate equal to $\gamma(X_t^m)^2$, if X_t^m denotes the total number of alive individuals at time t. It is rather clear that the process which describes the evolution of the total population, which is not a branching process (due to the interactions between branches, created by the competition term), goes extinct in finite time a.s.

If we consider this population with $m = [Nx]$ ancestors at time $t = 0$, weight each individual with the factor $1/N$ and choose $\mu_N = 2N + \theta$, $\lambda_N = 2N$ and $\gamma_N = \gamma/N$, then it is shown in Le, Pardoux and Wakolbinger [4] that the "total population mass process" converges weakly to the solution of the Feller SDE with logistic drift

$$dZ_t^x = \left[\theta Z_t^x - \gamma(Z_t^x)^2\right]dt + 2\sqrt{Z_t^x}dW_t,\ Z_0^x = x.$$

This equation has been studied in Lambert [3], who shows in particular that the population goes extinct in finite time a.s.

There is a natural way of describing the genealogical tree of the discrete population. The notion of genealogical tree is discussed for this limiting continuous population as well in [4, 6], in terms of continuous random trees in the sense of Aldous [1]. Clearly that forest of trees is finite a.s., and one can define the height H^m and the length L^m of the discrete forest of genealogical trees, as well as the height of the continuous "forest of genealogical trees", equal to the lifetime T^x of the process Z^x and the length of the same forest of trees, given as $S^x := \int_0^{T^x} Z_t^x dt$.

Let us now generalize the above models, both in the discrete and in the continuous case, replacing in the first case the death rate $\gamma(X_t^m)^2$ by $\gamma(X_t^m)^\alpha$ and in the second case the drift term $-\gamma(Z_t^x)^2$ by $-\gamma(Z_t^x)^\alpha$, for some $\alpha > 0$. In the case $\alpha = 1$, there is no competition; we are back to branching processes, both discrete and continuous. The case $0 < \alpha < 1$ corresponds to a situation where an increase of the population size reduces the per capita death rate, by allowing for an improvement of the living conditions (one can argue that this a reasonable model, at least for moderate population size compared to the available resources). The case $\alpha > 1$ is the case of competition, where an increase of the population size increases the per capita death rate, because for instance of the limitation of available resources.

The main result of this paper is the following:

Theorem 1.1. *Both* $\mathbb{E}[\sup_m H^m] < \infty$ *and* $\mathbb{E}[\sup_x T^x] < \infty$ *if* $\alpha > 1$, *while* $H^m \to \infty$ *as* $m \to \infty$ *and* $T^x \to \infty$ *as* $x \to \infty$ *a.s. if* $\alpha \le 1$. *Both* $\mathbb{E}[\sup_m L^m] < \infty$ *and* $\mathbb{E}[\sup_x S^x] < \infty$ *if* $\alpha > 2$, *while* $L^m \to \infty$ *as* $m \to \infty$ *and* $S^x \to \infty$ *as* $x \to \infty$ *a.s. if* $\alpha \le 2$.

Note that the monotonicity in α is not a surprise, $\sup_m H^m = \infty$ a.s. when $\alpha = 1$ follows rather easily from the branching property, $\mathbb{E}[\sup_m H^m] < \infty$ and $\mathbb{E}[\sup_m L^m] = \infty$ in case $\alpha = 2$ follow from results in [5], and again in the case $\alpha = 2$, $\sup_x S^x = \infty$ has been established in [4]. The main novelty of our results concerns the case $\alpha > 2$, which we discovered while trying to generalize the quadratic competition term.

Our theorem necessitates to define in a consistent way the population processes jointly for all initial population sizes, that is, we will need to define the two-parameter processes $\{X_t^m,\ t \geq 0, m \geq 1\}$ and $\{Z_t^x,\ t \geq 0, x > 0\}$. One of the objectives of this paper is also to prove that the renormalized discrete two-parameter processes converge weakly, for an appropriate topology, towards $\{Z_t^x,\ t \geq 0, x > 0\}$.

The paper is organized as follows. In the first section we present the discrete model, which provides the coupling for different values n of the initial size of the population. We describe in Sect. 3 the renormalized model for large population sizes. We then construct in Sect. 4 a random field indexed by t and x in the case of the continuous model, for which we precise the laws. After that we establish the convergence of the renormalized discrete random field to the continuous random field model in Sect. 5. We finally study the finiteness of the supremum over the initial population size of the height and of the length of the forest of genealogical trees in the discrete case in Sect. 6 and in the continuous case in Sect. 7.

2 The Discrete Model

We first present the discrete model. As declared in the introduction, we consider a continuous time $\mathbb{Z}_+$-valued population process $\{X_t^m,\ m \geq 1\}$, which starts at time zero from the initial condition $X_0^m = m$, that is, m is the number of ancestors of the whole population. The process X_t^m evolves as follows. Each individual, independently of the others, spawns a child at a constant rate μ and dies either "from natural death" at constant rate λ or from the competition pressure, which results in a total additional death rate equal at time t to $\gamma(X_t^m)^\alpha$ (in fact it will be a quantity close to that one, see below). This description is valid for one initial condition m. But it is not sufficiently precise to describe the joint evolution of $\{(X_t^m, X_t^n),\ t \geq 0\}$, with say $1 \leq m < n$. We must precise the effect of the competition upon the death rate of each individual. In order to be consistent, we need to introduce a non-symmetric picture of the effect of the competition term, exactly as it was first introduced in [4] in the case $\alpha = 2$, in order to describe the exploration process of the genealogical tree. The idea is that the progeny X_t^m of the m "first" ancestors should not feel the competition due to the progeny $X_t^n - X_t^m$ of the $n - m$ "additional" ancestors which is present in the population X_t^n. One way to do so is to model the effect of the competition in the following asymmetric way. We order the ancestors from left to right, this order being passed to their progeny. This means that the forest of genealogical trees of the population is a planar forest of trees, where the ancestor of the population X_t^1 is placed on the far left, the ancestor of $X_t^2 - X_t^1$ immediately on his right, etc. Moreover, we draw the genealogical trees in such a way that distinct branches never cross. This defines in a non-ambiguous way an order from left to right within the population alive at each time t. Now we model the competition as each individual being "under attack" from his contemporaries located on his left in the planar tree. Let $\mathcal{L}_i(t)$ denote the number of alive individuals at time t, who are located at his left on the planar tree. At any time t, the individual which is such that $\mathcal{L}_i(t) = 0$ does not feel the competition. For any $i \geq 1$ and $t \geq 0$ such that $\mathcal{L}_i(t) \geq 1$, the individual i is subject to a "competition death rate" equal to $\gamma[\mathcal{L}_i(t)^\alpha - (\mathcal{L}_i(t) - 1)^\alpha]$. Note that this rate, as a function of $\mathcal{L}_i(t)$, is decreasing if $0 < \alpha < 1$, constant if $\alpha = 1$ and increasing if $\alpha > 1$. Of course, conditionally upon $\mathcal{L}_i(\cdot)$, the occurrence of a "competition death event" for individual i is independent of the other birth/death events and of what happens to the other individuals.

The resulting total death rate endured by the population X_t^m at time t is then

$$\gamma \sum_{k=2}^{X_t^m} [(k-1)^\alpha - (k-2)^\alpha] = \gamma(X_t^m - 1)^\alpha,$$

which is a reasonable approximation of $\gamma(X_t^m)^\alpha$.

As a result, $\{X_t^m,\ t \geq 0\}$ is a continuous time $\mathbb{Z}_+$-valued Markov process, which evolves as follows. If $X_t^m = 0$, then $X_s^m = 0$ for all $s \geq t$. While at state $k \geq 1$, the process

$$X_t^m \text{ jumps to } \begin{cases} k+1, & \text{at rate } \mu k, \\ k-1, & \text{at rate } \lambda k + \gamma(k-1)^\alpha. \end{cases}$$

The above description specifies the joint evolution of all $\{X_t^m,\ t \geq 0\}_{m \geq 0}$, or in other words of the two-parameter process $\{X_t^m,\ t \geq 0, m \geq 0\}$. Let us rephrase it in more mathematical terms.

In the case $\alpha = 1$, for each fixed $t > 0$, $\{X_t^m,\ m \geq 1\}$ is an independent increments process. In the case $\alpha \neq 1$, $\{X_t^m,\ m \geq 1\}$ is not a Markov chain for fixed t. That is to say, the conditional law of X_t^{n+1} given X_t^n differs from its conditional law given $(X_t^1, X_t^2, \ldots, X_t^n)$. The intuitive reason for that is that the additional information carried by $(X_t^1, X_t^2, \ldots, X_t^{n-1})$ gives us a clue as to the level of competition which the progeny of the $n+1$st ancestor had to suffer, between time 0 and time t.

However, $\{X_\cdot^m,\ m \geq 0\}$ is a Markov chain with values in the space $D([0,\infty); \mathbb{Z}_+)$ of càdlàg functions from $[0,\infty)$ into $\mathbb{Z}_+$, which starts from 0 at $m = 0$. Consequently, in order to describe the law of the whole process, that is of the two-parameter process $\{X_t^m,\ t \geq 0, m \geq 0\}$, it suffices to describe the conditional law of $X_\cdot^n$, given $\{X_\cdot^{n-1}\}$. We now describe that conditional law for arbitrary $0 \leq m < n$. Let $V_t^{m,n} := X_t^n - X_t^m$, $t \geq 0$. Conditionally upon $\{X_\cdot^\ell,\ \ell \leq m\}$ and given that $X_t^m = x(t)$, $t \geq 0$, $\{V_t^{m,n},\ t \geq 0\}$ is a $\mathbb{Z}_+$-valued time inhomogeneous Markov process starting from $V_0^{m,n} = n - m$, whose time-dependent infinitesimal generator $\{Q_{k,\ell}(t),\ k,\ell \in \mathbb{Z}_+\}$ is such that its off-diagonal terms are given by

$$\begin{aligned} Q_{0,\ell}(t) &= 0, \quad \forall \ell \geq 1, \quad \text{and for any } k \geq 1, \\ Q_{k,k+1}(t) &= \mu k, \\ Q_{k,k-1}(t) &= \lambda k + \gamma\left[(x(t)+k-1)^\alpha - (x(t)-1)^\alpha\right], \\ Q_{k,\ell}(t) &= 0, \quad \forall \ell \notin \{k-1,k,k+1\}. \end{aligned}$$

The reader can easily convince himself that this description of the conditional law of $\{X_t^n - X_t^m,\ t \geq 0\}$ given $X_\cdot^m$ is prescribed by what we have said above, and that $\{X_\cdot^m,\ m \geq 0\}$ is indeed a Markov chain.

3 Renormalized Discrete Model

We consider a family of models like in the previous section, indexed by $N \in \mathbb{N}$. We choose the number of ancestors to be $m = \lfloor Nx \rfloor$, for some fixed $x > 0$, the birth rate to be $\mu_N = 2N + \theta$, for some $\theta > 0$, the "natural death rate" to be $\lambda_N = 2N$, and the competition death parameter to be $\gamma_N = \gamma/N^{\alpha-1}$. We now weight each individual by a factor N^{-1}, which means that we want to study the limit, as $N \to \infty$, of the "reweighted total mass population" process $Z_t^{N,x} := X_t^{\lfloor Nx \rfloor}/N$. The process $\{Z_t^{N,x},\ t \geq 0\}$ is a $\mathbb{Z}_+/N$-valued continuous time Markov process which starts from $Z_0^{N,x} = \lfloor Nx \rfloor/N$ such that if $Z_t^{N,x} = 0$, then $Z_s^{N,x} = 0$, for all $s \geq t$, and while at state k/N, $k \geq 1$,

$$Z^{N,x} \text{ jumps to } \begin{cases} (k+1)/N, & \text{at rate } 2Nk + k\theta; \\ (k-1)/N, & \text{at rate } 2Nk + \gamma N \left(\frac{k-1}{N}\right)^\alpha. \end{cases}$$

Clearly there exist three mutually independent standard Poisson processes P_1, P_2 and P_3 such that

$$\begin{aligned} X_t^{\lfloor Nx \rfloor} = \lfloor Nx \rfloor &+ P_1\left(\int_0^t (2N+\theta) X_r^{\lfloor Nx \rfloor} \mathrm{d}r\right) - P_2\left(2N \int_0^t X_r^{\lfloor Nx \rfloor} \mathrm{d}r\right) \\ &- P_3\left(\gamma N \int_0^t \left[\frac{X_r^{\lfloor Nx \rfloor} - 1}{N}\right]^\alpha \mathrm{d}r\right). \end{aligned}$$

Consequently there exists a martingale $M^{N,x}$ such that

$$Z_t^{N,x} = Z_0^{N,x} + \int_0^t \{\theta Z_r^{N,x} - \gamma (Z_r^{N,x} - 1/N)^\alpha\}\,\mathrm{d}r + M_t^{N,x}, \quad \text{with}$$

$$\langle M^{N,x} \rangle_t = \int_0^t \left\{4 Z_r^{N,x} + \frac{\theta}{N} Z_r^{N,x} + \frac{\gamma}{N}(Z_r^{N,x} - 1/N)^\alpha\right\} \mathrm{d}r. \tag{20.1}$$

Now for $0 < x < y$, let $V_t^{N,x,y} := Z_t^{N,y} - Z_t^{N,x}$. It is not too hard to show that there exist three further standard Poisson processes P_4, P_5 and P_6, such that the six Poisson processes P_1, P_2, P_3, P_4, P_5 and P_6 are mutually independent, and moreover

$$V_t^{N,x,y} = V_0^{N,x,y} + \frac{1}{N} P_4\Big(\int_0^t N(2N+\theta)V_r^{N,x,y}\mathrm{d}r\Big) - \frac{1}{N} P_5\Big(2N^2 \int_0^t V_r^{N,x,y}\mathrm{d}r\Big)$$
$$- N^{-1} P_6\left(\gamma N \int_0^t [(Z_r^{N,x} + V_r^{N,x,y} - 1/N)^\alpha - (Z_r^{N,x} - 1/N)^\alpha]\mathrm{d}r\right),$$

from which we deduce that there exists a martingale $M^{N,x,y}$ such that

$$V_t^{N,x,y} = V_0^{N,x,y} + \int_0^t \Big\{\theta V_r^{N,x,y} - \gamma\big[(Z_r^{N,x} + V_r^{N,x,y} - 1/N)^\alpha$$
$$-(Z_r^{N,x} - 1/N)^\alpha\big]\Big\}\mathrm{d}r + M_t^{N,x,y},$$

$$\langle M^{N,x,y}\rangle_t = \int_0^t \Big\{4V_r^{N,x,y} + \frac{\theta}{N}V_r^{N,x,y}$$
$$+ \frac{\gamma}{N}\big[(Z_r^{N,x} + V_r^{N,x,y} - 1/N)^\alpha - (Z_r^{N,x} - 1/N)^\alpha\big]\Big\}\mathrm{d}r \qquad (20.2)$$

and moreover

$$\langle M^{N,x,y}, M^{N,x}\rangle_t \equiv 0. \qquad (20.3)$$

The formulas for $\langle M^{N,x}\rangle_t$ and $\langle M^{N,x,y}\rangle_t$, as well as Eq. (20.3), rely on the following lemma, for the statement of which we need to introduce some notations. Let

$$M_t^{N,1} = N^{-1}P_1\left((2N^2+\theta N)\int_0^t Z_r^{N,x}\mathrm{d}r\right) - \int_0^t (2N+\theta)Z_r^{N,x}\mathrm{d}r,$$

$$M_t^{N,2} = N^{-1}P_2\left(2N^2\int_0^t Z_r^{N,x}\mathrm{d}r\right) - 2N\int_0^t Z_r^{N,x}\mathrm{d}r,$$

$$M_t^{N,3} = N^{-1}P_3\left(\gamma N\int_0^t \big[Z_r^{N,x} - 1/N\big]^\alpha\mathrm{d}r\right) - \gamma\int_0^t \big[Z_r^{N,x} - 1/N\big]^\alpha\mathrm{d}r,$$

$$M_t^{N,4} = N^{-1}P_4\left(\int_0^t (2N^2+\theta N)V_r^{N,x,y}\mathrm{d}r\right) - \int_0^t (2N+\theta)V_r^{N,x,y}\mathrm{d}r,$$

$$M_t^{N,5} = N^{-1}P_5\left(2N^2\int_0^t V_r^{N,x,y}\mathrm{d}r\right) - 2N\int_0^t V_r^{N,x,y}\mathrm{d}r,$$

$$M_t^{N,6} = N^{-1}P_6\left(\gamma N\int_0^t \big[(Z_r^{N,x} + V_r^{N,x,y} - 1/N)^\alpha - (Z_r^{N,x} - 1/N)^\alpha\big]\mathrm{d}r\right)$$
$$-\gamma\int_0^t \big[(Z_r^{N,x} + V_r^{N,x,y} - 1/N)^\alpha - (Z_r^{N,x} - 1/N)^\alpha\big]\mathrm{d}r.$$

Lemma 3.1. *For any* $1 \leq i \neq j \leq 6$, *the martingales* $M^{N,i}$ *and* $M^{N,j}$ *are orthogonal, in the sense that*

$$\langle M^{N,i}, M^{N,j} \rangle \equiv 0.$$

Proof. All we have to show is that $M^{N,i}$ and $M^{N,j}$ have a.s. no common jump time. In other words we need to show that

$$P_i\left(\int_0^t \varphi_i(r)\mathrm{d}r\right) \quad \text{and} \quad P_j\left(\int_0^t \varphi_j(r)\mathrm{d}r\right)$$

have no common jump time, where

$$\varphi_i(r) = f_i(Z_r^{N,x}, V_r^{N,x,y}) \quad \text{and } \varphi_j(r) = f_j(Z_r^{N,x}, V_r^{N,x,y}),$$

for some functions f_i and f_j from $(\mathbb{Z}_+/N)^2$ into $\mathbb{R}_+$.

Let

$$A_i(t) = \int_0^t \varphi_i(r)\mathrm{d}r, \quad \eta_i(t) = \inf\{s > 0,\ A_i(s) > t\},$$

$$A_j(t) = \int_0^t \varphi_j(r)\mathrm{d}r, \quad \eta_j(t) = \inf\{s > 0,\ A_j(s) > t\}.$$

Suppose the lemma is not true, that is, for some jump time T_k^i of P_i and some jump time T_ℓ^j of P_j, $\eta_i(T_k^i) = \eta_j(T_\ell^j)$. Let $S = \eta_i(T_{k-1}^i) \vee \eta_j(T_{\ell-1}^j)$. On the interval $[A^i(S), T_k^i)$, $\varphi_i(r)$ depends upon the jump times $T_1^i, \dots T_{k-1}^i$ of P_i, the jump times $T_1^j, \dots, T_{\ell-1}^j$ of P_j, plus upon some of the jump times of the other four Poisson processes, which are independent of (P_i, P_j). The same is true for $\varphi_j(r)$ on the interval $[A^j(S), T_\ell^j)$. It is now easy to show that conditionally upon those values of φ_i and φ_j, the two random variables $\eta_i(T_k^i) - S$ and $\eta_j(T_\ell^j) - S$ are independent, and their laws are absolutely continuous. Consequently $\mathbb{P}(\eta_i(T_k^i) = \eta_j(T_\ell^j)) = 0$. □

4 The Continuous Model

We now define an $\mathbb{R}_+$-valued two-parameter stochastic process $\{Z_t^x,\ t \geq 0, x \geq 0\}$ which such that for each fixed $x > 0$, $\{Z_t^x, t \geq 0\}$ is continuous process, solution of the SDE

$$\mathrm{d}Z_t^x = \left[\theta Z_t^x - \gamma(Z_t^x)^\alpha\right]\mathrm{d}t + 2\sqrt{Z_t^x}\mathrm{d}W_t,\ Z_0^x = x, \tag{20.4}$$

where $\theta \in \mathbb{R}$, $\gamma > 0$, $\alpha > 0$ and $\{W_t,\ t \geq 0\}$ is a standard scalar Brownian motion. Similarly as in the discrete case, the process $\{Z^x_\cdot,\ x \geq 0\}$ is a Markov process with values in $C([0,\infty),\mathbb{R}_+)$, the space of continuous functions from $[0,\infty)$ into $\mathbb{R}_+$, starting from 0 at $x = 0$. The transition probabilities of this Markov process are specified as follows. For any $0 < x < y$, $\{V^{x,y}_t := Z^y_t - Z^x_t,\ t \geq 0\}$ solves the SDE

$$dV^{x,y}_t = \left[\theta V^{x,y}_t - \gamma \left\{(Z^x_t + V^{x,y}_t)^\alpha - (Z^x_t)^\alpha\right\}\right] dt + 2\sqrt{V^{x,y}_t}\, dW'_t,\ V^{x,y}_0 = y - x, \tag{20.5}$$

where the standard Brownian motion $\{W'_t,\ t \geq 0\}$ is independent from the Brownian motion W which drives the SDE (20.4) for Z^x_t. It is an easy exercise to show that $Z^y_t = Z^x_t + V^{x,y}_t$ solves the same equation as Z^x_t, with the initial condition $Z^y_0 = y$ and a different driving standard Brownian motion. Moreover we have that whenever $0 \leq x < y$, $Z^x_t \leq Z^y_t$ for all $t \geq 0$, a.s., and in the case $\alpha = 1$, the increment of the mapping $x \to Z^x_t$ are independent, for each $t > 0$. Moreover, the conditional law of $Z^y_\cdot$, given that $Z^x_t = z(t)$, $t \geq 0$, is the law of the sum of z plus the solution of Eq. (20.5) with Z^x_t replaced by $z(t)$.

5 Convergence as $N \to \infty$

The aim of this section is to prove the convergence in law as $N \to \infty$ of the two-parameter process $\{Z^{N,x}_t,\ t \geq 0, x \geq 0\}$ defined in Sect. 3 towards the process $\{Z^x_t,\ t \geq 0, x \geq 0\}$ defined in Sect. 4. We need to make precise the topology for which this convergence will hold. We note that the process $Z^{N,x}_t$ (resp. Z^x_t) is a Markov processes indexed by x, with values in the space of càdlàg (resp. continuous) functions of t $D(([0,\infty);\mathbb{R}_+)$ (resp. $C(([0,\infty);\mathbb{R}_+))$ (note that the trajectories have compact support—the population process goes extinct in finite time—except in the cases $\alpha < 1, \theta > 0$ and $\alpha = 1, \theta > \gamma$). So it will be natural to consider a topology of functions of x, with values in functions of t.

The second point to notice is that for each fixed x, the process $t \to Z^{N,x}_t$ is càdlàg, constant between its jumps, with jumps of size $\pm N^{-1}$, while the limit process $t \to Z^x_t$ is continuous. On the other hand, both $Z^{N,x}_t$ and Z^x_t are discontinuous as functions of x. $x \to Z^x_\cdot$ has countably many jumps on any compact interval, but the mapping $x \to \{Z^x_t,\ t \geq \varepsilon\}$, where $\varepsilon > 0$ is arbitrary, has finitely many jumps on any compact interval, and it is constant between its jumps. Recall that $D([0,\infty);\mathbb{R}_+)$ equipped with the distance d^0_∞ defined by Eq. (16.4) in [2] is separable and complete; see Theorem 16.3 in [2]. We have the following statement:

Theorem 5.1. *As $N \to \infty$,*

$$\left\{Z^{N,x}_t,\ t \geq 0, x \geq 0\right\} \Rightarrow \{Z^x_t,\ t \geq 0, x \geq 0\}$$

in $D([0,\infty); D([0,\infty);\mathbb{R}_+))$, equipped with the Skorokhod topology of the space of càdlàg functions of x, with values in the separable complete metric space $(D([0,\infty);\mathbb{R}_+), d_\infty^0)$.

We first establish tightness for fixed x.

5.1 Tightness of $Z^{N,x}$, x Fixed

Let us prove the tightness of the sequence $\{Z^{N,x},\ N \geq 0\}$. For this end, we first establish some a priori estimates.

Lemma 5.1. $\forall\, T > 0$, *there exists a constant $C_1 > 0$ such that*

$$\sup_{N\geq 1}\sup_{0\leq t\leq T}\mathbb{E}\left\{Z_t^{N,x} + \int_0^t (Z_r^{N,x})^\alpha \mathrm{d}r\right\} \leq C_1.$$

It follows from this and the expression for $\langle M^{N,x}\rangle$ that the local martingale $M^{N,x}$ is in fact a square integrable martingale. We then have:

Lemma 5.2. $\forall\, T > 0$, *there exists a constant $C_2 > 0$ such that*

$$\sup_{N\geq 1}\sup_{0\leq t\leq T}\mathbb{E}\left\{\left(Z_t^{N,x}\right)^2 + \int_0^t \left(Z^{N,x}\right)^{\alpha+1} \mathrm{d}r\right\} \leq C_2.$$

The proof of those two lemmas is obtained easily using Eq. (20.1), elementary stochastic calculus and Gronwall and Fatou's lemmas.
We want to check the tightness of the sequence $\{Z^{N,x},\ N \geq 0\}$ using the Aldous criterion. Let $\{\tau_N,\ N \geq 1\}$ be a sequence of stopping time with values in $[0,T]$. We deduce from Lemma 5.1.

Proposition 5.1. *For any $T > 0$ and $\eta, \epsilon > 0$, there exists $\delta > 0$ such that*

$$\sup_{N\geq 1}\sup_{0\leq a\leq \delta}\mathbb{P}\left(\int_{\tau_N}^{(\tau_N+a)\wedge T} Z_r^{N,x}\mathrm{d}r \geq \eta\right) \leq \epsilon.$$

Proof. We have that

$$\begin{aligned}\sup_{0\leq a\leq \delta}\mathbb{P}\left(\int_{\tau_N}^{(\tau_N+a)\wedge T} Z_r^{N,x}\mathrm{d}r \geq \eta\right) &\leq \sup_{0\leq a\leq\delta}\frac{1}{\eta}\mathbb{E}\int_{\tau_N}^{(\tau_N+a)\wedge T} Z_r^{N,x}\mathrm{d}r \\ &\leq \frac{\delta}{\eta}\sup_{0\leq t\leq T}\mathbb{E}(Z_t^{N,x}) \\ &\leq C_1\frac{\delta}{\eta}.\end{aligned}$$

Hence the result with $\delta = \epsilon\eta/C_1$. □

We also deduce from Lemma 5.2:

Proposition 5.2. *For any $T > 0$ and η, $\epsilon > 0$, there exists $\delta > 0$ such that*

$$\sup_{N\geq 1}\sup_{0\leq a\leq\delta}\mathbb{P}\left(\int_{\tau_N}^{(\tau_N+a)\wedge T}(Z_r^{N,x})^\alpha \mathrm{d}r \geq \eta\right)\leq \epsilon.$$

Proof. For any $M > 0$, we have

$$\int_{\tau_N}^{(\tau_N+a)\wedge T}(Z_r^{N,x})^\alpha \mathrm{d}r \leq M^\alpha a + M^{-1}\int_0^T (Z_r^{N,x})^{\alpha+1}\mathrm{d}r.$$

This implies that

$$\begin{aligned}
&\sup_{N\geq 1}\sup_{0\leq a\leq\delta}\mathbb{P}\left(\int_{\tau_N}^{(\tau_N+a)\wedge T}(Z_r^{N,x})^\alpha \mathrm{d}r \geq \eta\right)\\
&\leq \sup_{N\geq 1}\sup_{0\leq a\leq\delta}\eta^{-1}\mathbb{E}\left(\int_{\tau_N}^{(\tau_N+a)\wedge T}(Z_r^{N,x})^\alpha \mathrm{d}r\right)\\
&\leq \frac{M^\alpha\delta}{\eta}+\frac{C_2}{M\eta}.
\end{aligned}$$

The result follows by choosing first $M = 2C_2/\epsilon\eta$, and then $\delta = \epsilon\eta/2M^\alpha$. □

From Eq. (20.1), Propositions 5.1 and 5.2, we deduce:

Proposition 5.3. *For each fixed $x > 0$, the sequence of processes $\{Z^{N,x},\ N\geq 1\}$ is tight in $D([0,\infty))$.*

5.2 *Proof of Theorem 5.1*

From Theorem 13.5 in [2], Theorem 5.1 follows from the two next propositions.

Proposition 5.4. *For any $n\in\mathbb{N}$, $0\leq x_1 < x_2 < \cdots < x_n$,*

$$\left(Z^{N,x_1}, Z^{N,x_2},\cdots, Z^{N,x_n}\right)\Rightarrow (Z^{x_1}, Z^{x_2},\cdots, Z^{x_n})$$

as $N\to\infty$, for the topology of locally uniform convergence in t.

Proof. We prove the statement in the case $n = 2$ only. The general statement can be proved in a very similar way. For $0\leq x_1 < x_2$, we consider the process $(Z^{N,x_1}, V^{N,x_1,x_2})$, using the notations from Sect. 3. The process $(Z^{N,x_1}, V^{N,x_1,x_2})$

is tight, as a consequence of Proposition 5.3, and thanks to Eqs. (20.1)–(20.3), any weak limit (Z^{x_1}, V^{x_1,x_2}) of a subsequence of $\{(Z^{N,x_1}, V^{N,x_1,x_2}), N \geq 1\}$ is the unique weak solution of the pair of coupled SDEs (20.4) and (20.5). □

Proposition 5.5. *There exists a constant C, which depends only upon θ and T, such that for any $0 \leq x < y < z$, which are such that $y - x \leq 1$, $z - y \leq 1$,*

$$\mathbb{E}\left[\sup_{0\leq t\leq T} |Z_t^{N,y} - Z_t^{N,x}|^2 \times \sup_{0\leq t\leq T} |Z_t^{N,z} - Z_t^{N,y}|^2\right] \leq C|z-x|^2.$$

Proof. For any $0 \leq x < y < z$, we have

$$\sup_{0\leq t\leq T} |Z_t^{N,y} - Z_t^{N,x}|^2 = \sup_{0\leq t\leq T} (V_t^{N,x,y})^2 \leq \sup_{0\leq t\leq T} (U_t^{N,y,x})^2,$$

$$\sup_{0\leq t\leq T} |Z_t^{N,z} - Z_t^{N,y}|^2 = \sup_{0\leq t\leq T} (V_t^{N,z,y})^2 \leq \sup_{0\leq t\leq T} (U_t^{N,z,y})^2,$$

where $U_t^{N,x,y}$ and $U_t^{N,z,y}$ are mutually independent branching processes, with in particular

$$U_t^{N,x,y} = y - x + \theta \int_0^t U_r^{N,x,y} \mathrm{d}r + \tilde{M}_t^{N,x,y},$$

with $\tilde{M}^{N,x,y}$ a local martingale such that $\langle \tilde{M}^{N,x,y}\rangle_t = (4 + \frac{\theta}{N}) \int_0^t U_r^{N,x,y} \mathrm{d}r$. Consequently

$$\mathbb{E}\left[\sup_{0\leq s\leq t} (U_t^{N,y,x})^2\right] \leq 3|y-x|^2 + 3\theta^2 t \int_0^t \mathbb{E}\left[\sup_{0\leq r\leq s} (U_r^{N,y,x})^2\right] \mathrm{d}s$$
$$+ 3\mathbb{E}\left[\sup_{0\leq s\leq t} (\tilde{M}_s^{N,x,y})^2\right]$$
$$\leq 3|y-x|^2 + 3\theta^2 t \int_0^t \mathbb{E}\left[\sup_{0\leq r\leq s} (U_r^{N,y,x})^2\right] \mathrm{d}s + 3C\mathbb{E}\langle \tilde{M}^{N,x,y}\rangle_t.$$

But clearly

$$\mathbb{E}[U_s^{N,y,x}] = |x-y| \exp(\theta s);$$

hence

$$\mathbb{E}\langle \tilde{M}^{N,x,y}\rangle_T \leq C(\theta, T)|x-y|.$$

Note that since $|x-y| \leq 1$, $|x-y|^2 \leq |x-y|$. The above computations, combined with Gronwall's lemma, lead to

$$\mathbb{E}\left[\sup_{0\leq t\leq T} \left(U_t^{N,y,x}\right)^2\right] \leq C'(\theta, T)|x-y|.$$

We obtain similarly

$$\mathbb{E}\left[\sup_{0\le t\le T}\left(U_s^{N,z,y}\right)^2\right] \le C'(\theta,T)|z-y|.$$

Since moreover the two random processes $U_t^{N,y,x}$ and $U_t^{N,z,y}$ are independent, the proposition follows from the above computations. □

Proof of Theorem 5.1. We will show that for any $T > 0$,

$$\{Z_t^{N,x},\ 0\le t\le T,\ x\ge 0\} \Rightarrow \{Z_t^x,\ 0\le t\le T,\ x\ge 0\}$$

in $D([0,\infty); D([0,T],\mathbb{R}_+))$. From Theorems 13.1 and 16.8 in [2], since from Proposition 5.4, for all $n \ge 1$, $0 < x_1 < \cdots < x_n$,

$$(Z_\cdot^{N,x_1},\ldots,Z_\cdot^{N,x_n}) \Rightarrow (Z_\cdot^{x_1},\ldots,Z_\cdot^{x_n})$$

in $D([0,T];\mathbb{R}^n)$, it suffices to show that for all $\bar{x} > 0$, $\varepsilon, \eta > 0$, there exists $N_0 \ge 1$ and $\delta > 0$ such that for all $N \ge N_0$,

$$\mathbb{P}(w_{\bar{x},\delta}(Z^N) \ge \varepsilon) \le \eta, \tag{20.6}$$

where for a function $(x,t) \to z(x,t)$

$$w_{\bar{x},\delta}(z) = \sup_{0\le x_1\le x\le x_2\le \bar{x},x_2-x_1\le\delta} \inf\left\{\|z(x,\cdot)-z(x_1,\cdot)\|,\|z(x_2,\cdot)-z(x,\cdot)\|\right\},$$

with the notation $\|z(x,\cdot)\| = \sup_{0\le t\le T}|z(x,t)|$. But from the proof of Theorem 13.5 in [2], Eq. (20.6) for Z^N follows from Proposition 5.5. □

6 Height and Length of the Genealogical Tree in the Discrete Case

6.1 Height of the Discrete Tree

We consider the two-parameter $\mathbb{Z}_+$-valued stochastic process $\{X_t^m,\ t\ge 0, m\ge 1\}$ defined in Sect. 2 and define the height and length of its genealogical tree by

$$H^m = \inf\{t > 0,\ X_t^m = 0\},\quad L^m = \int_0^{H^m} X_t^m dt,\quad \text{for } m \ge 1.$$

We shall occasionally write $X_t^{\alpha,m}$ when we want to emphasize the dependence upon the value of α. We first prove the

Proposition 6.1. *If* $0 < \alpha \leq 1$, *then*

$$\sup_{m \geq 1} H^m = +\infty \quad a.s.$$

Proof. Since for any $0 < \alpha < 1$, $j \geq 2$, $(j-1)^\alpha - (j-2)^\alpha < 1$, it is not hard to couple the two-parameter processes $\{X_t^{\alpha,m},\ t \geq 0, m \geq 1\}$ and $\{X_t^{1,m},\ t \geq 0, m \geq 1\}$ in such a way that $X_t^{\alpha,m} \geq X_t^{1,m}$, for all $m \geq 1$, $t \geq 0$, a.s. Consequently it suffices to prove the proposition in the case $\alpha = 1$.

But in that case $\{X_t^m,\ t \geq 0\}$ is the sum of m mutually independent copies of $\{X_t^1,\ t \geq 0\}$. Hence H^m is the sup of m independent copies of H^1, and the result follows from the fact that $\mathbb{P}(H^1 > t) > 0$, for all $t > 0$. □

We now prove the

Theorem 6.1. *If* $\alpha > 1$, *then*

$$\mathbb{E}\left[\sup_{m \geq 1} H^m\right] < \infty.$$

Proof. Since $m \to H^m$ is a.s. increasing, it suffices to prove that there exists a constant $C > 0$ such that

$$\mathbb{E}[H^m] \leq C, \quad \text{for any } m \geq 1.$$

We first show that $\lim_{m\to\infty} \mathbb{E}[H_1^m] < \infty$, where

$$H_1^m = \inf\{s \geq 0; X_s^m = 1\}.$$

It suffices to prove this result in the case $\lambda = 0$, which implies the result in the case $\lambda > 0$.

Proposition 6.2. *For* $\alpha > 1$, $\lambda = 0$, $\forall\, m \geq 1$, $\mathbb{E}\left(H_1^m\right)$ *is given by*

$$\mathbb{E}\left(H_1^m\right) = \sum_{k=2}^{m} \frac{1}{\gamma(k-1)^\alpha} \sum_{n=0}^{\infty} \frac{\mu^n}{\gamma^n} \frac{1}{[k(k+1)\cdots(k+n-1)]^{\alpha-1}}.$$

Proof. Define $u_m = \mathbb{E}\left(H_1^m\right)$. It is clear that $u_1 = 0$. The waiting time of X^m at state k is an exponential variable with mean $\frac{1}{\mu k+\gamma(k-1)^\alpha}$, and either X^m jumps from k to $k-1$ with probability $\frac{\gamma(k-1)^\alpha}{\mu k+\gamma(k-1)^\alpha}$ or either from k to $k+1$ with probability $\frac{\mu k}{\mu k+\gamma(k-1)^\alpha}$. We then have the recursive formula for u_m for any $m \geq 1$.

$$u_m = \frac{1}{\mu m + \gamma(m-1)^\alpha} + \frac{\gamma(m-1)^\alpha}{\mu m + \gamma(m-1)^\alpha} u_{m-1} + \frac{\mu m}{\mu m + \gamma(m-1)^\alpha} u_{m+1}.$$

If we define $w_m = u_m - u_{m-1}$, we obtain, for any $n \geq 0$, the following relation.

$$w_m = \frac{[(m-1)!]^{\alpha-1}}{\gamma(m-1)^\alpha}\left(\sum_{k=0}^{n-1}\frac{\mu^k}{\gamma^k}\frac{1}{[(m+k-1)!]^{\alpha-1}}+\frac{\mu^n}{\gamma^{n-1}}\frac{m+n-1}{[(m+n-2)!]^{\alpha-1}}w_{m+n}\right)$$

Define the random variable τ_{m+n} by

$$\tau_{m+n} = \inf\left\{t \geq 0;\ X_t^{m+n} = m+n-1\right\}.$$

We have that $w_{m+n} = \mathbb{E}(\tau_{m+n})$. Let R_{m+n} be the number of births which occur before Z^{m+n} reaches the value $m+n-1$, starting from $m+n$. For any $k \geq 0$ we have

$$\mathbb{P}\left(R_{m+n} = k\right) \leq a_k\left(\frac{\mu(m+n)}{\gamma(m+n-1)^\alpha}\right)^k,$$

where a_k is the cardinal of the set of binary trees with $k+1$ leaves. It is called a Catalan number and is given by

$$a_k = \frac{1}{k+1}\binom{2k}{k}, \quad a_k \sim \frac{4^k}{k^{3/2}\sqrt{\pi}}.$$

Moreover we have that

$$\mathbb{E}\left(\tau_{m+n}|R_{m+n} = k\right) \leq \frac{2k+1}{\gamma(m+n-1)^\alpha}.$$

Finally, with $c = \sup_{k\geq 1}(2k+1)4^{-k}a_k/\gamma$,

$$\mathbb{E}\left(\tau_{m+n}\right) \leq \frac{c}{(m+n-1)^\alpha}\sum_{k=1}^{\infty}\left(\frac{\mu}{\gamma}\frac{4(m+n)}{(m+n-1)^\alpha}\right)^k.$$

For large n, we have $\frac{\mu}{\gamma}\frac{4(m+n)}{(m+n-1)^\alpha} < \frac{1}{2}$. This implies that there exists another constant C such that $\mathbb{E}\left(\tau_{m+n}\right) \leq C(m+n-1)^{-\alpha}$ and $\lim_{n\to\infty} w_{m+n} = 0$. We then deduce that

$$w_m = \frac{[(m-1)!]^{\alpha-1}}{\gamma(m-1)^\alpha}\left(\sum_{k=0}^{\infty}\frac{\mu^k}{\gamma^k}\frac{1}{[(m+k-1)!]^{\alpha-1}}\right).$$

Consequently, for $\alpha > 1$, we have

$$u_m = \sum_{k=1}^{m} w_k = \sum_{k=2}^{m}\frac{1}{\gamma(k-1)^\alpha}\sum_{n=0}^{\infty}\left(\frac{\mu}{\gamma}\right)^n\frac{1}{[k(k+1)\cdots(k+n-1)]^{\alpha-1}}.$$

□

End of the proof of Theorem 6.1

Furthermore, for $0 \le \ell \le n-1$, we have

$$\frac{1}{[k(k+1)\dots(k+n-1)]^{\alpha-1}} \le \frac{(k+\ell)^{\ell(\alpha-1)}}{[k(k+1)\cdots(k+\ell-1)]^{\alpha-1}} \frac{1}{(k+\ell)^{n(\alpha-1)}}.$$

Let $K = \lfloor (2\mu/\gamma)^{1/(\alpha-1)} \rfloor$. We conclude that

$$\text{if } K \ge 3, \quad u_m \le \frac{2}{\gamma}\left(\sum_{k=2}^{K-1}\left(\frac{K^K}{k-1}\right)^{\alpha-1} + \frac{1}{(\alpha-1)(K-2)^{\alpha-1}}\right),$$

$$\text{if } K \le 2, \quad u_m \le \frac{2}{\gamma}\frac{\alpha}{\alpha-1}.$$

In all cases, $\sup_{m\ge1} \mathbb{E}[H_1^m] < \infty$.

Finally starting from 1 at time H_1^m, the probability p that X_t^m hits zero before hitting 2 is $\frac{\lambda}{\mu+\lambda}$. Let G be a random variable defined as follows. Let X_t^1 start from 1 at time 0. If X_t^1 hits zero before hitting 2, then $G = 1$. If not, we wait until X_t^1 goes back to 1. This time is less than $T_1 + H_1^2$, where T_1 is an exponential random variable with mean $1/(\lambda+\mu)$, which is independent of G. If starting again from 1 at that time, if X_t^1 reaches 0 before 2, we stop and $G = 2$. If not, we continue and so on. The random variable G is geometric with parameter p and independent of H_1^m. Clearly we have that

$$H^m \le H_1^m + GH_1^2 + \sum_{i=1}^{G} T_i,$$

We conclude that $\sup_{m\ge1} \mathbb{E}[H^m] < \infty$. □

We have proved in particular that (in the terminology used in coalescent theory) the population process comes down from infinity if $\alpha > 1$. This means that if the population starts with an infinite number of individuals at time $t = 0$, instantaneously the population becomes finite, that is, for all $t > 0$, $X_t^\infty < \infty$.

6.2 *Length of the Discrete Tree*

Define now

$$A_t^m := \int_0^t X_r^m \mathrm{d}r, \quad \eta_t^m = \inf\{s > 0;\ A_s^m > t\}.$$

We consider the process $U^m := X^m \circ \eta^m$. Let S^m be the stopping time defined by

$$S^m = \inf\{r > 0; U_r^m = 0\}.$$

Note that $S^m = L^m$, the length of the genealogical forest of trees of the population X^m, since we have $S^m = \int_0^{H^m} X_r^m dr$. The process X^m can be expressed using two mutually independent standard Poisson processes, as

$$X_t^m = m + P_1\left(\int_0^t \mu X_r^m \mathrm{d}r\right) - P_2\left(\int_0^t \left[\lambda X_r^m + \gamma(X_r^m - 1)^\alpha\right] \mathrm{d}r\right).$$

Consequently the process $U^m = X^m \circ \eta^m$ satisfies

$$U_t^m = m + P_1(\mu t) - P_2\left(\int_0^t \left[\lambda + \gamma(U_r^m)^{-1}(U_r^m - 1)^\alpha\right] \mathrm{d}r\right).$$

On the interval $[0, S^m)$, $U_t^m \geq 1$, and consequently we have the two inequalities

$$m - P_2\left(\int_0^t \left[\lambda U_r^m + \gamma(U_r^m - 1)^{\alpha-1}\right] \mathrm{d}r\right) \leq U_t^m$$
$$\leq m + P_1\left(\int_0^t \mu U_r^m \mathrm{d}r\right) - P_2\left(\int_0^t \left[\frac{\gamma}{2}(U_r^m - 1)^{\alpha-1}\right] \mathrm{d}r\right).$$

The following result is now a consequence of Proposition 6.1 and Theorem 6.1.

Theorem 6.2. *If $\alpha \leq 2$, then*

$$\sup_{m\geq 0} L^m = \infty \quad a.s.$$

If $\alpha > 2$, then

$$\mathbb{E}\left[\sup_{m\geq 0} L^m\right] < \infty.$$

7 Height and Length of the Continuous Tree

Now we study the same quantities in the continuous model. We first need to establish some preliminary results on SDEs with infinite initial condition.

7.1 SDE with Infinite Initial Condition

Let $f : \mathbb{R}_+ \to \mathbb{R}$ be locally Lipschitz and such that

$$\lim_{x\to\infty} \frac{|f(x)|}{x^\alpha} = 0. \tag{20.7}$$

Theorem 7.1. *Let $\alpha > 1$, $\gamma > 0$ and f satisfy the assumption* (20.7). *Then there exists a minimal $X \in C\left((0,+\infty);\mathbb{R}\right)$ which solves*

$$\begin{cases} \mathrm{d}X_t = [f(X_t) - \gamma(X_t)^\alpha]\mathbf{1}_{\{X_t \geq 0\}}\mathrm{d}t + \mathrm{d}W_t, \\ X_t \to \infty, \text{ as } t \to 0. \end{cases} \tag{20.8}$$

Moreover, if $T_0 := \inf\{t > 0,\ X_t = 0\}$, then $\mathbb{E}[T_0] < \infty$.

Proof (A Priori Estimate). Setting $V_t = X_t - W_t$, the result is equivalent to the existence of a minimal $V \in C\left((0,+\infty);\mathbb{R}\right)$ solution of the ODE

$$\begin{cases} \dfrac{\mathrm{d}V_t}{\mathrm{d}t} = f(V_t + W_t) - \gamma(V_t + W_t)^\alpha, \\ V_t \to \infty, \text{ as } t \to 0. \end{cases} \tag{20.9}$$

Let first

$$M = \inf\{x > 0;\ |f(x)| \leq \gamma x^\alpha/2\},$$
$$\tau = \inf\{t > 0,\ W_t \notin [-M, 2M]\}.$$

Suppose there exists a solution $\{V_t,\ t \geq 0\}$ to the ODE (20.9). Then the following random time is positive a.s.

$$S := \inf\{t > 0,\ V_t < 2M\}.$$

Now on the time interval $[0, \tau \wedge S]$,

$$\begin{aligned}
-\frac{V_t}{2} &\leq W_t \leq V_t, \\
M \leq \frac{V_t}{2} &\leq V_t + W_t \leq 2V_t, \\
-\frac{3\gamma}{2}(V_t + W_t)^\alpha &\leq f(V_t + W_t) - \gamma(V_t + W_t)^\alpha \leq -\frac{1}{2}\gamma(V_t + W_t)^\alpha, \\
-3\gamma 2^{\alpha-1} V_t^\alpha &\leq \frac{\mathrm{d}V_t}{\mathrm{d}t} \leq -\frac{\gamma}{2^{\alpha+1}} V_t^\alpha, \\
\frac{\gamma(\alpha-1)}{2^{\alpha+1}} &\leq \frac{\mathrm{d}}{\mathrm{d}t}\left[(V_t)^{-(\alpha-1)}\right] \leq 3\gamma(\alpha-1)2^{\alpha-1}, \\
\frac{\gamma(\alpha-1)}{2^{\alpha+1}}t &\leq \frac{1}{(V_t)^{\alpha-1}} \leq 3\gamma(\alpha-1)2^{\alpha-1}t, \\
\frac{c_{\alpha,\gamma}}{t^{1/(\alpha-1)}} &\leq V_t \leq \frac{C_{\alpha,\gamma}}{t^{1/(\alpha-1)}},
\end{aligned}$$

where

$$c_{\alpha,\gamma} = \frac{1}{2}[3\gamma(\alpha-1)]^{-1/(\alpha-1)}, \quad C_{\alpha,\gamma} = \frac{2^{\frac{\alpha+1}{\alpha-1}}}{[\gamma(\alpha-1)]^{1/(\alpha-1)}}.$$

Proof of Existence To each $x > 0$, we associate the unique solution X^x of Eq. (20.8), but with the initial condition $X_0^x = x$. Clearly $x \leq y$ implies that $X_t^x \leq X_t^y$ for all $t \geq 0$ a.s. Consider an increasing sequence $x_n \to \infty$, the corresponding increasing sequence of processes $\{X_t^{x_n},\ t \geq 0\}_{n\geq 1}$, and define $V_t^n := X_t^{x_n} - W_t$, $S_n = \inf\{t > 0,\ V_t^n < 2M\}$. Note that S_n is increasing. A minor modification of the computations in the first part of this proof shows that for $0 \leq t \leq S_n$,

$$\frac{1}{(c_{\alpha,\gamma}^{\alpha-1}t + x_n^{-(\alpha-1)})^{1/(\alpha-1)}} \leq V_t^n \leq \frac{1}{(C_{\alpha,\gamma}^{\alpha-1}t + x_n^{-(\alpha-1)})^{1/(\alpha-1)}}.$$

It readily follows that $V_t := \lim_{n\to\infty} V_t^n$ solves Eq. (20.9), while $X_t := V_t + W_t$ solves Eq. (20.8). Those solutions do not depend upon the choice of a particular sequence $x_n \to \infty$, and the thus constructed solution is clearly the minimal solution of Eq. (20.8).

Proof of $\mathbb{E}[T_0] < \infty$; *Step1* We first show that $S \wedge \tau$ is bounded, and $V_{S\wedge\tau}$ is integrable.

For that sake, start noting that $V_{S\wedge\tau} \geq 2M$. It then follows from one of the inequalities obtained in the first part of the proof that

$$S \wedge \tau \leq \left(\frac{C_{\alpha,\gamma}}{2M}\right)^{\alpha-1}.$$

On the set $S < \tau$, $V_{S\wedge\tau} = 2M$. Consequently

$$\begin{aligned} V_{S\wedge\tau} &= 2M\mathbf{1}_{\{S<\tau\}} + V_\tau \mathbf{1}_{\{\tau\leq S\}} \\ &\leq 2M + C_{\alpha,\gamma}\tau^{-1/(\alpha-1)}. \end{aligned}$$

We now show that

$$\mathbb{E}[V_{S\wedge\tau}] < \infty. \tag{20.10}$$

From the last inequality,

$$\mathbb{E}[V_{S\wedge\tau}] \leq 2M + C_{\alpha,\gamma}\mathbb{E}\left[\tau^{-1/(\alpha-1)}\right].$$

We need to compute (below $W_t^* = \sup_{0\leq s\leq t}|W_s|$)

$$\begin{aligned}
\mathbb{E}\left[\tau^{-1/(\alpha-1)}\right] &= \int_0^\infty \mathbb{P}\left(\tau^{-1/(\alpha-1)} > t\right) \mathrm{d}t \\
&= \int_0^\infty \mathbb{P}\left(\tau < t^{-(\alpha-1)}\right) \mathrm{d}t \\
&\le \int_0^\infty \mathbb{P}\left(W^*_{t^{-(\alpha-1)}} \ge 2M\right) \mathrm{d}t \\
&\le 4\int_0^\infty \mathbb{P}\left(W_{t^{-(\alpha-1)}} \ge 2M\right) \mathrm{d}t \\
&\le 4\int_0^\infty \mathbb{P}\left(W_1 \ge 2Mt^{(\alpha-1)/2}\right) \mathrm{d}t \\
&\le 4\sqrt{e}\int_0^\infty \exp\left(-2Mt^{(\alpha-1)/2}\right) \mathrm{d}t \\
&< \infty,
\end{aligned}$$

where we have used for the fourth inequality the following Tchebysheff inequality:

$$\mathbb{P}(W_1 > A) = \mathbb{P}\left(e^{W_1 - 1/2} > e^{A-1/2}\right) \le \frac{\sqrt{e}}{e^A}.$$

Equation (20.10) follows.

Proof of $\mathbb{E}[T_0] < \infty$; *Step 2* Now we turn back to the X equation, and we show in this step that X comes down to level M in time which has finite expectation. Since $V_{S\wedge\tau} \ge 2M$ and $-M \le W_{S\wedge\tau} \le 2M$,

$$M \le X_{S\wedge\tau} \le V_{S\wedge\tau} + 2M.$$

In order to simplify notations, let us write $\xi := X_{S\wedge\tau}$. Until X_t reaches the level M,

$$f(X_t) - \gamma(X_t)^\alpha \le -\frac{\gamma}{2}(X_t)^\alpha \le -1,$$

from the choice of M. Consequently, if $T_M = \inf\{t > 0, X_t \le M\}$, $X_{S\wedge\tau+r} \le Y_r$ for all $0 \le r \le T_M - S\wedge\tau$ a.s., where Y solves

$$\mathrm{d}Y_r = -\mathrm{d}r + \mathrm{d}W_r, \quad Y_0 = \xi,$$

where W is a standard Brownian motion independent of ξ. Let $R_M := \inf\{r > 0,\ Y_r \le M\}$. Clearly $T_M \le S\wedge\tau + R_M$. So since $S\wedge\tau$ is bounded, Step 2 will follow if we show that $\mathbb{E}(R_M) < \infty$. But the time taken by Y to descend a given level is linear in that level. So, given that $\mathbb{E}(\xi) < \infty$, it suffices to show that the time needed for Y to descend a distance one is integrable, which is easy, since if $\tau := \inf\{t > 0,\ W_t - t \le -1\}$, for $t \ge 2$,

$$\mathbb{P}(\tau > t) \le \mathbb{P}(W_t > t - 1)$$

$$\begin{aligned}
&\leq \mathbb{P}(W_t > t/2) \\
&\leq \mathbb{P}(W_1 > \sqrt{t}/2) \\
&\leq \sqrt{e}\exp(-\sqrt{t}/2).
\end{aligned}$$

Proof of $\mathbb{E}[T_0] < \infty$*; Step 3* For proving that $\mathbb{E}[T_0] < \infty$, it remains to show that the time taken by X to descend from M to 0 is integrable, which we now establish. Given any fixed $T > 0$, let p denote the probability that starting from M at time $t = 0$, X hit zero before time T. Clearly $p > 0$. Let α be a geometric random variable with success probability p, which is defined as follows. Let X start from M at time 0. If X hits zero before time T, then $\alpha = 1$. If not, we look at the position X_T of X at time T. If $X_T > M$, we want until X goes back to M. The time needed is bounded by the integrable random variable η, which is the time needed for X to descend to M, when starting from $+\infty$. If however $X_T \leq M$, we start afresh from there, since the probability to reach zero in less than T is greater than or equal to p, for all starting points in the interval $(0, M]$. So either at time T, or at time $T + \eta$, we start again from a level which is less than or equal to M. If zero is reached during the next interval of length T, then $\alpha = 2$. Repeating this procedure, we see that the time needed to reach 0, starting from M, is bounded by

$$\alpha T + \sum_{i=1}^{\alpha} \eta_i,$$

where the r.v.'s η_i are i.i.d., with the same law as η, globally independent of α. Now the total time needed to descend from $+\infty$ to 0 is bounded by

$$\alpha T + \sum_{i=0}^{\alpha} \eta_i,$$

whose expectation is $T/p + (1 + 1/p)\mathbb{E}(\eta) < \infty$. □

7.2 *Height of the Continuous Tree*

We consider again the process $\{Z_t^x,\ t \geq 0\}$ solution of Eq. (20.4):

$$dZ_t^x = \left[\theta Z_t^x - \gamma (Z_t^x)^\alpha\right] dt + 2\sqrt{Z_t^x} dW_t,\ Z_0^x = x,$$

with $\theta \geq 0, \gamma > 0, \alpha > 0$, and define $T^x = \inf\{t > 0,\ Z_t^x = 0\}$.

We first prove

Theorem 7.2. *If* $0 < \alpha < 1$, $0 < \mathbb{P}(T^x = \infty) < 1$ *if* $\theta > 0$, *while* $T^x < \infty$ *a.s. if* $\theta = 0$.

If $\alpha = 1$, $T^x < \infty$ *a.s. if* $\gamma \geq \theta$, *while* $0 < \mathbb{P}(T^x = \infty) < 1$ *if* $\gamma < \theta$.
If $\alpha > 1$, $T^x < \infty$ *a. s.*

Proof. Clearly, if $\theta > 0$ and $\alpha < 1$, then for large values of Z_t^x, the nonlinear term $-\gamma(Z_t^x)^\alpha$ is negligible with respect to the linear term θZ_t^x; hence the process behaves as in the supercritical branching case: both extinction in finite time and infinite time survival happen with positive probability. If however $\theta = 0$, then the process goes extinct in finite time a.s., since on the interval $[1, \infty)$ the process is bounded from above by the Brownian motion with constant negative drift (equal to $-\gamma$), which comes back to 1 as many times as necessary, until it hits 0, hence $T^x < \infty$ a.s.

In case $\alpha = 1$ we have a continuous branching process, whose behaviour is well known.

In case $\alpha > 1$, the nonlinear term $-\gamma(Z_t^x)^\alpha$ dominates for large values of Z_t^x; hence the process comes back to 1 as many times as necessary, until it hits 0, hence $T^x < \infty$ a.s. □

We now establish the large x behaviour of T^x.

Theorem 7.3. *If* $\alpha \leq 1$, *then* $T^x \to \infty$ *a.s., as* $x \to \infty$.

Proof. The result is equivalent to the fact that the time to reach 1, starting from x, goes to ∞ as $x \to \infty$. But when $Z_t^x \geq 1$, a comparison of SDEs for various values of α shows that it suffices to consider the case $\alpha = 1$. In that case, T^n is the maximum of the extinction times of n mutually independent copies of Z_t^1, hence the result. □

Theorem 7.4. *If* $\alpha > 1$, *then* $\mathbb{E}\left[\sup_{x>0} T^x\right] < \infty$.

Proof. It follows from the Itô formula that the process $Y_t^x := \sqrt{Z_t^x}$ solves the SDE

$$dY_t^x = \left[\frac{\theta}{2} Y_t^x - \frac{\gamma}{2}(Y_t^x)^{2\alpha-1} - \frac{1}{8Y_t^x}\right] dt + dW_t, \; Y_0^x = \sqrt{x}.$$

By a well known comparison theorem, $Y_t^x \leq U_t^x$, where U_t^x solves

$$dU_t^x = \left[\frac{\theta}{2} U_t^x - \frac{\gamma}{2}(U_t^x)^{2\alpha-1}\right] dt + dW_t, \; U_0^x = \sqrt{x}.$$

The result now follows readily from Theorem 7.1, since $\alpha > 1$ implies that $2\alpha - 1 > 1$. □

7.3 Length of the Continuous Tree

Recall that in the continuous case, the length of the genealogical tree is given as

$$S^x = \int_0^{T^x} Z_t^x dt.$$

For fixed values of x, S^x is finite iff T^x is finite (remind that $T^x = \infty$ requires that $Z_t^x \to \infty$ as $t \to \infty$), hence the result of Theorem 7.2 translates immediately into a result for S^x. We next consider the limit of S^x as $x \to \infty$. Consider the additive functional

$$A_t = \int_0^t Z_s^x \mathrm{d}s, \ t \geq 0,$$

and the associated time change

$$\eta(t) = \inf\{s > 0, \ A_s > t\}.$$

We now define $U_t^x = Z^x \circ \eta(t)$, $t \geq 0$. It is easily seen that the process U^x solves the SDE

$$\mathrm{d}U_t^x = \left[\theta - \gamma(U_t^x)^{\alpha-1}\right] \mathrm{d}t + 2\mathrm{d}W_t, \ U_0^x = x. \tag{20.11}$$

Let $\tau^x := \inf\{t > 0, \ U_t^x = 0\}$. It follows from the above that $\eta(\tau^x) = T^x$, hence $S^x = \tau^x$.

We have the following results.

Theorem 7.5. *If $\alpha \leq 2$, then $S^x \to \infty$ a.s. as $x \to \infty$.*

Proof. $\alpha \leq 2$ means $\alpha - 1 \leq 1$. The same argument as in Theorem 7.3 implies that it suffices to consider the case $\alpha = 2$. But in that case Eq. (20.11) has the explicit solution

$$U_t^x = e^{-\gamma t} x + \int_0^t e^{-\gamma(t-s)}[\theta \mathrm{d}s + 2\mathrm{d}W_s];$$

hence

$$S^x = \inf\left\{t > 0, \ \int_0^t e^{\gamma s}(\theta \mathrm{d}s + 2\mathrm{d}W_s) \leq -x\right\},$$

which clearly goes to infinity, as $x \to \infty$. □

Theorem 7.6. *If $\alpha > 2$, then $\mathbb{E}\left[\sup_{x>0} S^x\right] < \infty$.*

Proof. This theorem follows readily from Theorem 7.1 applied to the U^x-Eq. (20.11), since $\alpha > 2$ means $\alpha - 1 > 1$. □

References

1. Aldous, D.: The continuum random tree I. Ann. Probab. **19**, 1–28 (1991)
2. Billingsley, P.: Convergence of Probability Measures, 2nd edn. Wiley, New York (1999)
3. Lambert, A.: The branching process with logistic growth. Ann. Probab. **15**, 1506–1535 (2005)
4. Le, V., Pardoux, E., Wakolbinger, A.: Trees under attack: a Ray-Knight representation of Feller's branching diffusion with logistic growth. Probab. Theory Relat. Fields (2012)

5. Pardoux, E., Salamat, M.: On the height and length of the ancestral recombination graph. J. Appl. Probab. **46**, 979–998 (2009)
6. Pardoux, E., Wakolbinger, A.: From Brownian motion with a local time drift to Feller's branching diffusion with logistic growth. Elec. Comm. Probab. **16**, 720–731 (2011)

Chapter 21
Linking Progressive and Initial Filtration Expansions

Younes Kchia, Martin Larsson, and Philip Protter

Abstract In this article, we study progressive filtration expansions with random times. We show how semimartingale decompositions in the expanded filtration can be obtained using a natural link between progressive and initial expansions. The link is, on an intuitive level, that the two coincide after the random time. We make this idea precise and use it to establish known and new results in the case of expansion with a single random time. The methods are then extended to the multiple time case, without any restrictions on the ordering of the individual times. Finally we study the link between the expanded filtrations from the point of view of filtration shrinkage. As the main analysis progresses, we indicate how the techniques can be generalized to other types of expansions.

Received 9/25/2011; Accepted 8/15/2012; Final 8/19/2012

1 Introduction

Expansion of filtrations is a well-studied topic that has been investigated both in theoretical and applied contexts; see for instance [8–10]. There are two main types of filtration expansion: initial expansion and progressive expansion. The initial

Y. Kchia
ANZ Bank, Singapore
e-mail: younes.kchia@anz.com

M. Larsson
Swiss Finance Institute and Ecole Polytechnique Fédérale de Lausanne, Lausanne, Switzerland
e-mail: martin.larsson@epfl.ch

P. Protter (✉)
Statistics Department, Columbia University, New York, NY 10027, USA
e-mail: pep2117@columbia.edu

F. Viens et al. (eds.), *Malliavin Calculus and Stochastic Analysis: A Festschrift in Honor of David Nualart*, Springer Proceedings in Mathematics & Statistics 34,
DOI 10.1007/978-1-4614-5906-4_21,

expansion of a filtration $\mathbb{F} = (\mathcal{F}_t)_{t \geq 0}$ with a random variable τ is the filtration $\mathbb{H}$ obtained as the right-continuous modification of $(\mathcal{F}_t \vee \sigma(\tau))_{t \geq 0}$. A priori there is no particular interpretation attached to this random variable. The progressive expansion $\mathbb{G}$ is obtained as any right-continuous filtration containing $\mathbb{F}$ and making τ a stopping time. In this case τ should of course be nonnegative since it has the interpretation of a random time. We often use the smallest such filtration in this paper. On a complete probability space $(\Omega, \mathcal{F}, P)$ a filtration $\mathbb{F}$ is said to satisfy the "usual hypotheses" if it is right-continuous and if $\mathcal{F}_0$ contains all the P null sets of $\mathcal{F}$. Therefore $\mathcal{F}_t$ contains all the P null sets of $\mathcal{F}$ as well, for any $t \geq 0$. We will assume throughout this paper that all filtrations satisfy the usual hypotheses.

Yor [15] noted that the decompositions in the initially and progressively expanded filtrations are related and showed how one can obtain the decompositions of some $\mathbb{F}$ local martingales in the progressively expanded filtration under representability assumptions of some crucial $\mathbb{F}$ martingales related to the random time τ. We refer the reader to [15] for more details. However, these two filtrations have typically been viewed and studied in the literature independently from one another. The purpose of the present paper is to demonstrate that there is a very natural connection between the initial and progressive expansions. The reason is, on an intuitive level, that the filtrations $\mathbb{G}$ and $\mathbb{H}$ coincide after time τ. We make this idea precise for filtrations $\mathbb{H}$ that are not necessarily obtained as initial expansions. This, in combination with a classical theorem by Jeulin and Yor, allows us to show how the semimartingale decomposition of an $\mathbb{F}$ local martingale, when viewed in the progressively expanded filtration $\mathbb{G}$, can be obtained on all of $[0, \infty)$, provided that its decomposition in the filtration $\mathbb{H}$ is known. One well-known situation where this is the case is when *Jacod's criterion* is satisfied. This is, however, not the only case, and we give an example using techniques based on Malliavin calculus developed by Imkeller et al. [7]. These developments, which all concern expansion with a single random time, are treated in Sect. 2. The technique is, however, applicable in more general situations than expansion with a single random time. As an indication of this, we perform en passant the same analysis for what we call the (τ, X)-progressive expansion of $\mathbb{F}$, denoted $\mathbb{G}^{(\tau,X)}$. Here τ becomes a stopping time in the larger filtration, and the random variable X becomes $\mathcal{G}_\tau^{(\tau,X)}$-measurable.

In Sect. 3 we extend these ideas in order to deal with the case where the base filtration $\mathbb{F}$ is expanded progressively with a whole vector $\boldsymbol{\tau} = (\tau_1, \ldots, \tau_n)$ of random times. Unlike previous work in the literature (see for instance [9]), we do not impose any conditions on the ordering of the individual times. After establishing a general semimartingale decomposition result we treat the special case where Jacod's criterion is satisfied for the whole vector $\boldsymbol{\tau}$, and we show how the decompositions may be expressed in terms of $\mathbb{F}$ conditional densities of $\boldsymbol{\tau}$ with respect to its law.

Finally, in Sect. 4 we take a different point of view and study the link between the filtrations $\mathbb{G}$ and $\mathbb{H}$ from the perspective of filtration shrinkage. In the case of a random time that avoids all $\mathbb{F}$ stopping times and whose $\mathbb{F}$ conditional probabilities are equivalent to some deterministic measure, Callegaro, Jeanblanc, and Zargari already used the idea of projecting down the $\mathbb{H}$ decomposition of an $\mathbb{F}$ local martingale to obtain its decomposition in $\mathbb{G}$; see [1]. Some of their results follow from ours, and our technique allows us to avoid the heavy manual computations involving dual predictable projections that arise naturally in this problem.

2 Expansion with One Random Time

Assume that a filtered probability space $(\Omega, \mathcal{F}, \mathbb{F}, P)$ is given, and let τ be a *random time*, i.e., a nonnegative random variable. Typically τ is not a stopping time with respect to $\mathbb{F}$. Consider now the larger filtrations $\mathbb{G}^\tau = (\mathcal{G}_t^\tau)_{t\geq 0}$ and $\tilde{\mathbb{G}} = (\tilde{\mathcal{G}}_t)_{t\geq 0}$ given by

$$\mathcal{G}_t^\tau = \bigcap_{u>t} \mathcal{G}_u^{0,\tau} \qquad \text{where} \qquad \mathcal{G}_t^{0,\tau} = \mathcal{F}_t \vee \sigma(\tau \wedge t)$$

and

$$\tilde{\mathcal{G}}_t = \left\{A \in \mathcal{F}, \exists A_t \in \mathcal{F}_t \mid A \cap \{\tau > t\} = A_t \cap \{\tau > t\}\right\}.$$

One normally refers to $\mathbb{G}^\tau$ as the *progressive expansion* of $\mathbb{F}$ with τ, and it can be characterized as the smallest right-continuous filtration that contains $\mathbb{F}$ and makes τ a stopping time.

2.1 Initial and Progressive Enlargements When They Are Related

Throughout this section $\mathbb{G}$ will denote any right-continuous filtration containing $\mathbb{F}$, making τ a stopping time and satisfying $\mathcal{G}_t \cap \{\tau > t\} = \mathcal{F}_t \cap \{\tau > t\}$ for all $t \geq 0$. In [5], the authors study the compensator of τ in $\mathbb{G}$ and notice that these assumptions on the filtration are sufficient to obtain Jeulin–Yor type results. Our goal in this section is to analyze how $\mathbb{F}$ semimartingales behave in the progressively expanded filtration $\mathbb{G}$. In particular, in case they remain semimartingales in $\mathbb{G}$, we are interested in their canonical decompositions. Under a well-known and well-studied hypothesis due to Jacod [8], this has been done by Jeanblanc and Le Cam [9] in the filtration $\mathbb{G}^\tau$. As one consequence of our approach, we are able to provide a short proof of their main result. Moreover, our technique also works for a larger class of progressively expanded filtrations and under other conditions than Jacod's criterion.

The $\mathbb{G}$ decomposition before time τ of an $\mathbb{F}$ local martingale M follows from a classical and very general theorem by Jeulin and Yor [10], which we now recall.

Theorem 2.1. *Fix an $\mathbb{F}$ local martingale M and define $Z_t = P(\tau > t \mid \mathcal{F}_t)$ as the optional projection of $\mathbf{1}_{[\![\tau,\infty[\![}$ onto $\mathbb{F}$, let μ be the martingale part of its Doob–Meyer decomposition, and let J be the dual predictable projection of $\Delta M_\tau \mathbf{1}_{[\![\tau,\infty[\![}$ onto $\mathbb{F}$. Then*

$$M_{t\wedge\tau} - \int_0^{t\wedge\tau} \frac{\mathrm{d}\langle M, \mu\rangle_s + \mathrm{d}J_s}{Z_{s-}}$$

is a local martingale in both $\mathbb{G}^\tau$ *and* $\tilde{\mathbb{G}}$.

The $\mathbb{G}$ decomposition before time τ follows as a straightforward corollary of Theorem 2.1 using the following shrinkage result of Föllmer and Protter; see [4].

Lemma 2.1. *Let* $\mathbb{E} \subset \mathbb{F} \subset \mathbb{G}$ *be three filtrations. Let* X *be a* $\mathbb{G}$ *local martingale. If the optional projection of* X *onto* $\mathbb{E}$ *is an* $\mathbb{E}$ *local martingale, then the optional projection of* X *onto* $\mathbb{F}$ *is an* $\mathbb{F}$ *local martingale.*

Putting Theorem 2.1 and Lemma 2.1 together provides the $\mathbb{G}$ decomposition before time τ of an $\mathbb{F}$ local martingale M as given in the following theorem.

Theorem 2.2. *Fix an* $\mathbb{F}$ *local martingale* M. *Then*

$$M_{t\wedge\tau} - \int_0^{t\wedge\tau} \frac{\mathrm{d}\langle M, \mu\rangle_s + \mathrm{d}J_s}{Z_{s-}}$$

is a $\mathbb{G}$ *local martingale. Here, the quantities* Z, μ, *and* J *are defined as in Theorem 2.1.*

Finding the decomposition after τ is more complicated, but it can be obtained provided that it is known with respect to a suitable auxiliary filtration $\mathbb{H}$. More precisely, one needs that $\mathbb{H}$ and $\mathbb{G}$ coincide after τ, in a certain sense. One such filtration $\mathbb{H}$ when $\mathbb{G}$ is taken to be $\mathbb{G}^\tau$ is, as we will see later, the initial expansion of $\mathbb{F}$ with τ. We now make precise what it means for two filtrations to coincide after τ.

Definition 2.1. Let $\mathbb{G}$ and $\mathbb{H}$ be two filtrations such that $\mathbb{G} \subset \mathbb{H}$, and let τ be an $\mathbb{H}$ stopping time. Then $\mathbb{G}$ and $\mathbb{H}$ are said to *coincide after* τ if for every $\mathbb{H}$ optional process X, the process

$$\mathbf{1}_{[\![\tau,\infty[\![}(X - X_\tau)$$

is $\mathbb{G}$ adapted.

The following lemma establishes some basic properties of filtrations that coincide after τ. Recall that all filtrations are assumed right-continuous.

Lemma 2.2. *Assume that* $\mathbb{G}$ *and* $\mathbb{H}$ *coincide after* τ. *Then:*

(i) For every $\mathbb{H}$ *stopping time* T, $T \vee \tau$ *is* $\mathbb{G}$ *stopping time. In particular,* τ *itself is a* $\mathbb{G}$ *stopping time.*
(ii) For every $\mathbb{H}$ *optional (predictable) process* X, *the process* $\mathbf{1}_{[\![\tau,\infty[\![}(X - X_\tau)$ *is* $\mathbb{G}$ *optional (predictable).*

Proof. For (i), let T be an $\mathbb{H}$ stopping time. Then $T \vee \tau$ is again an $\mathbb{H}$ stopping time, so $X = \mathbf{1}_{[\![0,T\vee\tau]\!]}$ is $\mathbb{H}$ optional. Thus

$$\mathbf{1}_{[\![\tau,\infty[\![}(r)(X_r - X_\tau) = -\mathbf{1}_{\{T\vee\tau<r\}}$$

is $\mathcal{G}_r$-measurable by the assumption that $\mathbb{G}$ and $\mathbb{H}$ coincide after τ. This holds for every $r \geq 0$, so $T \vee \tau$ is a $\mathbb{G}$ stopping time since $\mathbb{G}$ is right-continuous.

For (ii), let X be of the form $X = h\mathbf{1}_{[s,t[}$ for an $\mathcal{H}_s$-measurable random variable h and fixed $0 \leq s < t$. Then

$$Y_r = \mathbf{1}_{[\tau,\infty[}(r)(X_r - X_\tau) = \mathbf{1}_{\{\tau \leq r\}}\big(h\mathbf{1}_{\{s \leq r < t\}} - h\mathbf{1}_{\{s \leq \tau < t\}}\big)$$

is $\mathcal{G}_r$-measurable by assumption and defines a càdlàg process. Hence Y is $\mathbb{G}$ optional, and the monotone class theorem implies the claim for $\mathbb{H}$ optional processes. The predictable case is similar. □

Before giving the first result on the $\mathbb{G}$ semimartingale decomposition of an $\mathbb{F}$ local martingale, under the general assumption that such a decomposition is available in some filtration $\mathbb{H}$ that coincides with $\mathbb{G}$ after τ, we provide examples of such pairs of filtrations $(\mathbb{G}, \mathbb{H})$. For the progressively expanded filtration $\mathbb{G}^\tau$, it turns out that the filtration $\mathbb{H}$ given as the initial expansion of $\mathbb{F}$ with τ coincides with $\mathbb{G}^\tau$ after τ. Recall that the initial expansion is defined as follows:

$$\mathcal{H}_t = \bigcap_{u>t} \mathcal{H}_u^0 \qquad \text{where} \qquad \mathcal{H}_t^0 = \mathcal{F}_t \vee \sigma(\tau).$$

Lemma 2.3. *Let $\mathbb{H}$ be the initial expansion of $\mathbb{F}$ with τ. Then $\mathbb{G}^\tau$ and $\mathbb{H}$ coincide after τ.*

Proof. Let $X = h\mathbf{1}_{[s,t[}$ for an $\mathcal{H}_s$-measurable random variable h and fixed $0 \leq s < t$. Then

$$\begin{aligned} Y_r = \mathbf{1}_{[\tau,\infty[}(r)(X_r - X_\tau) &= \mathbf{1}_{\{\tau \leq r\}}\big(h\mathbf{1}_{\{s \leq r < t\}} - h\mathbf{1}_{\{s \leq \tau < t\}}\big) \\ &= h\mathbf{1}_{\{s \leq r < t\}}\mathbf{1}_{\{\tau \leq r\}} - h\mathbf{1}_{\{s \leq \tau < t\}}\mathbf{1}_{\{t \leq r\}} \\ &\quad - h\mathbf{1}_{\{s \leq \tau \leq r\}}\mathbf{1}_{\{r < t\}}. \end{aligned}$$

It is enough to prove that the three terms on the right side are $\mathcal{G}_r^\tau$-measurable. Let us consider the first term, the other ones being similar. First let h be of the form $fk(\tau)$ for some $\mathcal{F}_s$-measurable f and Borel function k. Then

$$h\mathbf{1}_{\{s \leq r < t\}}\mathbf{1}_{\{\tau \leq r\}} = fk(\tau)\mathbf{1}_{\{s \leq r < t\}}\mathbf{1}_{\{\tau \leq r\}} = fk(r \wedge \tau)\mathbf{1}_{\{s \leq r < t\}}\mathbf{1}_{\{\tau \leq r\}},$$

which is $\mathcal{G}_r^\tau$-measurable. Using the monotone class theorem the result follows for every $\mathcal{H}_s^0$-measurable h and finally for every $\mathcal{H}_s$-measurable h by a standard argument. □

As a corollary we may give an alternative characterization of the progressively enlarged filtration $\mathbb{G}^\tau$ as the smallest right-continuous filtration that contains $\mathbb{F}$ and coincides with $\mathbb{H}$ after τ, where $\mathbb{H}$ is the initial expansion of $\mathbb{F}$.

Corollary 2.1. *Let $\mathbb{H}$ be the initial expansion of $\mathbb{F}$ with τ. Then*

$$\mathbb{G}^\tau = \bigcap \{\tilde{\mathbb{G}} : \mathbb{F} \subset \tilde{\mathbb{G}} \subset \mathbb{H},\ \tilde{\mathbb{G}} \text{ is right-continuous, and } \tilde{\mathbb{G}} \text{ and } \mathbb{H} \text{ coincide after } \tau\}$$

Proof. To show "$\subset$," let $\tilde{\mathbb{G}}$ be an arbitrary element in the class over which we take the intersection. By Lemma 2.2, $\tilde{\mathbb{G}}$ is a right-continuous filtration that makes τ a stopping time and contains $\mathbb{F}$. Hence $\mathbb{G}^\tau$ is included in each $\tilde{\mathbb{G}}$ and thus in their intersection. For "$\supset$," Lemma 2.3 implies that $\mathbb{G}^\tau$ coincides with $\mathbb{H}$ after τ. It is right-continuous, so it is one of the filtrations we are intersecting. □

Let X be a random variable. Consider now the filtration $\mathbb{G}^{(\tau,X)}$, which we call the (τ, X)-*expansion of* $\mathbb{F}$, given by

$$\mathcal{G}_t^{(\tau,X)} = \bigcap_{u>t} \mathcal{G}_u^{0,(\tau,X)} \qquad \text{where} \qquad \mathcal{G}_t^{0,(\tau,X)} = \mathcal{F}_t \vee \sigma(\tau \wedge t) \vee \sigma(X 1_{\{\tau \le t\}}).$$

The filtration $\mathbb{G}^{(\tau,X)}$ is the smallest right-continuous filtration containing $\mathbb{F}$, which makes τ a stopping time and such that X is $\mathcal{G}_\tau^{(\tau,X)}$-measurable. See Dellacherie and Meyer [2] for a related discussion. As in Lemma 2.3 it is easy to prove the following result.

Lemma 2.4. *Let $\mathbb{H}$ be the initial expansion of $\mathbb{F}$ with (τ, X). Then $\mathbb{G}^{(\tau,X)}$ coincides with $\mathbb{H}$ after τ.*

Also, $\mathbb{G}^{(\tau,X)}$ satisfies the crucial condition $\mathcal{G}_t^{(\tau,X)} \cap \{\tau > t\} = \mathcal{F}_t \cap \{\tau > t\}$ for all $t \ge 0$.

Lemma 2.5. *The progressively expanded filtration $\mathbb{G}^{(\tau,X)}$ satisfies*

$$\mathcal{G}_t^{(\tau,X)} \cap \{\tau > t\} = \mathcal{F}_t \cap \{\tau > t\}$$

for all $t \ge 0$.

Proof. It is well known that $\mathcal{G}_t^\tau \cap \{\tau > t\} = \mathcal{F}_t \cap \{\tau > t\}$. Let $H_t = Y_t h(X 1_{\{\tau \le t\}})$, where Y_t is $\mathcal{F}_t$-measurable and bounded and h is a bounded Borel function. Then $H_t 1_{\{\tau > t\}} = Y_t h(0) 1_{\{\tau > t\}}$ which is measurable with respect to $\{\tau > t\} \cap \mathcal{G}_t^\tau = \{\tau > t\} \cap \mathcal{F}_t$. The monotone class theorem now proves that $\mathcal{G}_t^{(\tau,X)} \cap \{\tau > t\} \subset \mathcal{F}_t \cap \{\tau > t\}$. The reverse inclusion is clear. □

We are now ready to give the first result on the $\mathbb{G}$ semimartingale decomposition of an $\mathbb{F}$ local martingale, under the general assumption that such a decomposition is available in some filtration $\mathbb{H}$ that coincides with $\mathbb{G}$ after τ.

Theorem 2.3. *Let M be an $\mathbb{F}$ local martingale. Let $\mathbb{G}$ be any progressive expansion of $\mathbb{F}$ with τ satisfying $\mathcal{G}_t \cap \{\tau > t\} = \mathcal{F}_t \cap \{\tau > t\}$ for all $t \ge 0$ and let $\mathbb{H}$ be a filtration that coincides with $\mathbb{G}$ after τ. Suppose there exists an $\mathbb{H}$ predictable finite variation process A such that $M - A$ is an $\mathbb{H}$ local martingale. Then M is a $\mathbb{G}$ semimartingale, and*

$$M_t - \int_0^{t\wedge\tau} \frac{\mathrm{d}\langle M,\mu\rangle_s + \mathrm{d}J_s}{Z_{s-}} - \int_{t\wedge\tau}^t \mathrm{d}A_s$$

is the local martingale part of its $\mathbb{G}$ decomposition. Here Z, μ, and J are defined as in Theorem 2.1.

Proof. The process $M_t^{\mathbb{G}} = M_{t\wedge\tau} - \int_0^{t\wedge\tau} \frac{\mathrm{d}\langle M,\mu\rangle_s + \mathrm{d}J_s}{Z_{s-}}$ is a $\mathbb{G}$ local martingale by the Jeulin–Yor theorem (Theorem 2.2). Next, define

$$M^{\mathbb{H}} = \mathbf{1}_{[\![\tau,\infty[\![}(\tilde{M} - \tilde{M}_\tau),$$

where $\tilde{M}_t = M_t - A_t$. Since $\tilde{M}$ is an $\mathbb{H}$ local martingale, $M^{\mathbb{H}}$ is also. Moreover, if $(T_n)_{n\geq 1}$ is a sequence of $\mathbb{H}$ stopping times that reduce $\tilde{M}$, then $T_n' = T_n \vee \tau$ yields a reducing sequence for $M^{\mathbb{H}}$. Lemma 2.2(i) shows that the T_n' are in fact $\mathbb{G}$ stopping times, and since $\mathbb{G}$ and $\mathbb{H}$ coincide after τ, $M^{\mathbb{H}}$ is $\mathbb{G}$ adapted. This implies that $M^{\mathbb{H}}_{\cdot\wedge T_n'}$ is an $\mathbb{H}$ martingale that is $\mathbb{G}$ adapted and is therefore a $\mathbb{G}$ martingale. It follows that $M^{\mathbb{H}}$ is a $\mathbb{G}$ local martingale. It now only remains to observe that

$$M_t^{\mathbb{G}} + M_t^{\mathbb{H}} = M_t - \int_0^{t\wedge\tau} \frac{\mathrm{d}\langle M,\mu\rangle_s + \mathrm{d}J_s}{Z_{s-}} - \int_{t\wedge\tau}^t \mathrm{d}A_s,$$

which thus is a $\mathbb{G}$ local martingale. Finally, by Lemma 2.2(ii), the last term is $\mathbb{G}$ predictable, so we obtain indeed the $\mathbb{G}$ semimartingale decomposition. □

Part of the proof of Theorem 2.3 can be viewed as a statement about filtration shrinkage. According to a result by Föllmer and Protter [4], if $\mathbb{G} \subset \mathbb{H}$ are two nested filtrations and L is an $\mathbb{H}$ local martingale that can be reduced using $\mathbb{G}$ stopping times, then its optional projection onto $\mathbb{G}$ is again a local martingale. In our case L corresponds to $M^{\mathbb{H}}$, which is $\mathbb{G}$ adapted and hence coincides with its optional projection.

We now proceed to examine two particular situations where $\mathbb{G}$ and $\mathbb{H}$ coincide after τ and where the $\mathbb{H}$ decomposition $M - A$ is available. First we make an absolute continuity assumption on the $\mathcal{F}_t$ conditional laws of τ or (τ, X), known as Jacod's criterion. We then assume that $\mathbb{F}$ is a Wiener filtration and impose a condition related to the Malliavin derivatives of the process of $\mathcal{F}_t$ conditional distributions. This is based on theory developed by Imkeller, Pontier, and Weisz [7] and Imkeller [6].

2.2 Jacod's Criterion

In this section we study the case where τ or (τ, X) satisfy *Jacod's criterion*, which we now recall. Let ξ be a random variable. We state Jacod's criterion for the case where ξ takes values in $\mathbb{R}^d$. In this subsection, $\mathbb{H}$ will always denote the initial expansion of $\mathbb{F}$ with ξ. Our results will be obtained with ξ being either τ or (τ, X) and in both cases $\mathbb{H}$ indeed coincides with $\mathbb{G}^\xi$ after τ by Lemma 2.3 and Lemma 2.4.

Assumption 2.1 (Jacod's criterion). *There exists a σ-finite measure η on $\mathcal{B}(\mathbb{R}^d)$ such that $P(\xi \in \cdot \mid \mathcal{F}_t)(\omega) \ll \eta(\cdot)$ a.s.*

Without loss of generality, η may be chosen as the law of ξ. Under Assumption 2.1, the $\mathcal{F}_t$ conditional density

$$p_t(u;\omega) = \frac{P(\xi \in du \mid \mathcal{F}_t)(\omega)}{\eta(du)}$$

exists and can be chosen so that $(u,\omega,t) \mapsto p_t(u;\omega)$ is càdlàg in t and measurable for the optional σ-field associated with the filtration $\widehat{\mathbb{F}}$ given by $\widehat{\mathcal{F}}_t = \cap_{u>t}\mathcal{B}(\mathbb{R}^d) \otimes \mathcal{F}_u$. See Lemma 1.8 in [8].

Theorem 2.4. *Let M be an $\mathbb{F}$ local martingale.*

(i) If τ satisfies Jacod's criterion (Assumption 2.1), then M is a $\mathbb{G}^\tau$ semimartingale.
(ii) Let X be a random variable such that (τ, X) satisfies Jacod's criterion (Assumption 2.1); then M is a $\mathbb{G}^{(\tau,X)}$ semimartingale.

Proof. We prove (i). Let $\mathbb{H}^\tau$ be the initial expansion of F with τ. It follows from Jacod's theorem (see Theorems VI.10 and VI.11 in [7]) that M is an $\mathbb{H}^\tau$ semimartingale, which is $\mathbb{G}^\tau$ adapted. It is also a $\mathbb{G}^\tau$ semimartingale by Stricker's theorem. The proof of (ii) is similar. □

We provide the explicit decompositions using the following classical result by Jacod; see [8], Theorem 2.5.

Theorem 2.5. *Let M be an $\mathbb{F}$ local martingale, and assume Assumption 2.1 is satisfied. Then there exists set $B \in \mathcal{B}(\mathbb{R}^d)$, with $\eta(B) = 0$, such that:*

(i) $\langle p(u), M\rangle$ exists on $\{(\omega,t) : p_{t-}(u;\omega) > 0\}$ for every $u \notin B$.
(ii) There is an increasing predictable process A and an $\widehat{\mathbb{F}}$ predictable function $k_t(u;\omega)$ such that for every $u \notin B$, $\langle p(u), M\rangle_t = \int_0^t k_s(u)p_{s-}(u)\mathrm{d}A_s$ on $\{(\omega,t) : p_{t-}(u;\omega) > 0\}$.
(iii) $\int_0^t |k_s(\xi)|\mathrm{d}A_s < \infty$ a.s. for every $t \geq 0$ and $M_t - \int_0^t k_s(\xi)\mathrm{d}A_s$ is an $\mathbb{H}$ local martingale.

Immediate consequences of Theorem 2.3 and Theorem 2.5 are the following corollaries.

Corollary 2.2. *Let M be an $\mathbb{F}$ local martingale, and assume that τ satisfies Assumption 2.1. Then M is a $\mathbb{G}^\tau$ semimartingale, and*

$$M_t - \int_0^{t\wedge\tau} \frac{\mathrm{d}\langle M,\mu\rangle_s + \mathrm{d}J_s}{Z_{s-}} - \int_{t\wedge\tau}^t k_s(\tau)\mathrm{d}A_s$$

is the local martingale part of its $\mathbb{G}^\tau$ decomposition. Here Z, μ and J are defined as in Theorem 2.1, and k, and A as in Theorem 2.5 with $d = 1$ and $\xi = \tau$.

Notice that this recovers the main result in [9] (Theorem 3.1), since by Theorem 2.5(ii) we may write

$$\int_{t\wedge\tau}^{t} k_s(\tau)\mathrm{d}A_s = \int_{t\wedge\tau}^{t} \frac{\mathrm{d}\langle p(u), M\rangle_s}{p_{s-}(u)}\Bigg|_{u=\tau},$$

whenever the right side makes sense. See [9] for a detailed discussion.

Corollary 2.3. *Let M be an $\mathbb{F}$ local martingale. Let X be a random variable and assume that (τ, X) satisfies Assumption 2.1. Then M is a $\mathbb{G}^{(\tau,X)}$ semimartingale, and*

$$M_t - \int_0^{t\wedge\tau} \frac{\mathrm{d}\langle M,\mu\rangle_s + \mathrm{d}J_s}{Z_{s-}} - \int_{t\wedge\tau}^{t} k_s(\tau, X)\mathrm{d}A_s$$

is the local martingale part of its $\mathbb{G}^{(\tau,X)}$ decomposition. Here Z, μ, and J are defined as in Theorem 2.1, and k and A as in Theorem 2.5 with $d = 2$ and $\xi = (\tau, X)$.

2.3 Absolute Continuity of the Malliavin Trace

In two papers on models for insider trading in mathematical finance, Imkeller et al. [7] and Imkeller [6] introduced an extension of Jacod's criterion for initial expansions, based on the Malliavin calculus. Given a measure-valued random variable $F(du;\omega)$ defined on Wiener space with coordinate process $(W_t)_{0\le t\le 1}$, they introduce a Malliavin derivative $D_t F(du;\omega)$, defined for all F satisfying certain regularity conditions. The full details are outside the scope of the present paper, and we refer the interested reader to [7] and [6]. Filtration expansion using the Malliavin calculus has also been developed elsewhere; see for instance Kohatsu-Higa [11] and Sulem et al. [14]. We continue to let $\mathbb{H}$ be the initial expansion of $\mathbb{F}$.

The extension of Jacod's criterion is the following. Let $P_t(du,\omega) = P(\tau \in du \mid \mathcal{F}_t)(\omega)$, and assume that $D_t P_t(du,\omega)$ exists and satisfies

$$\sup_{f\in C_b(\mathbb{R}),\ \|f\|\le 1} E\left(\int_0^1 \langle D_s P_s(du), f\rangle^2 \mathrm{d}s\right) < \infty.$$

Here $C_b(\mathbb{R})$ is the space of bounded and continuous functions on $\mathbb{R}$, $\|\cdot\|$ is the supremum norm, and $\langle F(du), f\rangle = \int_{\mathbb{R}_+} f(u)F(\mathrm{d}u)$ for a random measure F and $f \in C_b(\mathbb{R})$. Assume also that

$$D_t P_t(du,\omega) \ll P_t(du,\omega) \quad a.s. \text{ for all } t \in [0,1],$$

and let $g_t(u; \omega)$ be a suitably measurable version of the corresponding density. Then they prove the following result.

Theorem 2.6. *Under the above conditions, if* $\int_0^1 |g_t(\tau)|\mathrm{d}t < \infty$ *a.s., then*

$$W_t - \int_0^t g_s(\tau)\mathrm{d}s$$

is a Brownian motion in the initially expanded filtration $\mathbb{H}$.

One example where this holds but Jacod's criterion fails is $\tau = \sup_{0 \le t \le 1} W_t$. In this case $g_t(\tau)$ can be computed explicitly and the $\mathbb{H}$ decomposition of W obtained. Due to the martingale representation theorem in $\mathbb{F}$, this allows one to obtain the $\mathbb{H}$ decomposition for every $\mathbb{F}$ local martingale. Using Theorem 2.3, the decomposition in the progressively expanded filtration $\mathbb{G}$ can then also be obtained.

Corollary 2.4. *Under the assumptions of Theorem 2.6,*

$$W_t - \int_0^{t\wedge\tau} \frac{\mathrm{d}\langle W, Z\rangle_s}{Z_{s-}} - \int_{t\wedge\tau}^t g_s(\tau)\mathrm{d}s$$

is a $\mathbb{G}$ *Brownian motion.*

3 Expansion with Multiple Random Times

We now move on to progressive expansions with multiple random times. We start again with a filtered probability space $(\Omega, \mathcal{F}, \mathbb{F}, P)$, but instead of a single random time we consider a vector of random times

$$\boldsymbol{\tau} = (\tau_1, \ldots, \tau_n).$$

We emphasize that there are no restrictions on the ordering of the individual times. This is a significant departure from previous work in the field, where the times are customarily assumed to be ordered. The progressive expansion of $\mathbb{F}$ with $\boldsymbol{\tau}$ is

$$\mathcal{G}_t = \bigcap_{u>t} \mathcal{G}_u^0 \qquad \text{where} \qquad \mathcal{G}_t^0 = \mathcal{F}_t \vee \sigma(\tau_i \wedge t; i = 1, \ldots, n),$$

and we are interested in the semimartingale decompositions of $\mathbb{F}$ local martingales in the $\mathbb{G}$ filtration. Several other filtrations will also appear, and we now introduce notation that will be in place for the remainder of this section, except in Theorem 3.2 and its corollary. Let $I \subset \{1, \ldots, n\}$ be an index set:

- $\sigma_I = \max_{i\in I} \tau_i$ and $\rho_I = \min_{j\notin I} \tau_j$.
- $\mathbb{G}^I$ denotes the initial expansion of $\mathbb{F}$ with the random vector $\boldsymbol{\tau}_I = (\tau_i)_{i\in I}$.
- $\mathbb{H}^I$ denotes the progressive expansion of $\mathbb{G}^I$ with the random time ρ_I.

If $I = \emptyset$, then $\mathbb{G}^I = \mathbb{F}$ and $\mathbb{H}^I$ is the progressive expansion of $\mathbb{F}$ with $\rho_\emptyset = \min_{i=1,\dots,n} \tau_i$. If on the other hand $I = \{1,\dots,n\}$, then $\mathbb{G}^I = \mathbb{H}^I$ and coincides with the initial expansion of $\mathbb{F}$ with $\boldsymbol{\tau}_I = \boldsymbol{\tau}$. Notice also that we always have $\mathbb{G}^I \subset \mathbb{H}^I$.

The idea from Sect. 2 can be modified to work in the present context. The intuition is that the filtrations $\mathbb{G}$ and $\mathbb{H}^I$ coincide on $[\![\sigma_I, \rho_I[\![$. The $\mathbb{G}$ decomposition on $[\![\sigma_I, \rho_I[\![$ of an $\mathbb{F}$ local martingale M can then be obtained by computing its decomposition in $\mathbb{H}^I$. This is done in two steps. First it is obtained in $\mathbb{G}^I$ using, for instance, Jacod's theorem (Theorem 2.5), and then in $\mathbb{H}^I$ up to time ρ_I using the Jeulin–Yor theorem (Theorem 2.1).

The following results collect some properties of the relationship between $\mathbb{H}^I$ and $\mathbb{G}$, thereby clarifying in which sense they coincide on $[\![\sigma_I, \rho_I[\![$. We take the index set I to be given and fixed.

Lemma 3.1. *Let X be an $\mathcal{H}^I_t$-measurable random variable. Then the quantity $X\mathbf{1}_{\{\sigma_I \le t\}}$ is $\mathcal{G}_t$-measurable. As a consequence, if H is an $\mathbb{H}^I$ optional (predictable) process, then $\mathbf{1}_{[\![\sigma_I,\rho_I[\![}(H - H_{\sigma_I})$ is $\mathbb{G}$ optional (predictable).*

Proof. Let X be of the form $X = f k(\rho_I \wedge t) \prod_{i\in I} h_i(\tau_i)$ for some $\mathcal{F}_t$-measurable random variable f and Borel functions k, h_i. Then

$$X\mathbf{1}_{\{\sigma_I \le t\}} = f k(\rho_I \wedge t) \prod_{i\in I} h_i(\tau_i \wedge t)\mathbf{1}_{\{\sigma_I \le t\}},$$

which is $\mathcal{G}_t$-measurable, since σ_I and ρ_I are $\mathbb{G}$ stopping times and $\tau_i \wedge t$ is $\mathcal{G}_t$-measurable by construction. The monotone class theorem shows that the statement holds for every X that is measurable for $\mathcal{F}_t \vee \sigma(\tau_i : i \in I) \vee \sigma(\rho_I \wedge t)$. $\mathbb{H}^I$ is the right-continuous version of this filtration, so the result follows.

Now consider an $\mathbb{H}^I$ predictable process of the form $H = h\mathbf{1}_{]\!]s,t]\!]}$ with $s \le t$ and h an $\mathcal{H}^I_s$-measurable random variable. Then

$$H_r \mathbf{1}_{[\![\sigma_I,\rho_I[\![}(r) = h\mathbf{1}_{\{s<r\le t\}}\mathbf{1}_{\{\sigma_I \le r<\rho_I\}} - h\mathbf{1}_{\{s<\sigma_I\le r<\rho_I\}},$$

which defines a $\mathbb{G}$ predictable process using the first part of the lemma. An application of the monotone class theorem yields the desired result in the predictable case. The optional case is similar. □

Lemma 3.2. *Let T_n be an $\mathbb{H}^I$ stopping time and define $T_n' = (\sigma_I \vee T_n) \wedge (\rho_I \vee n)$. Then T_n' is a $\mathbb{G}$ stopping time.*

Proof. Note that $\{T_n' \le t\} = \{\rho_I \vee n \le t\} \cup \{\sigma_I \vee T_n \le t\}$. Since ρ_I is a $\mathbb{G}$ stopping time and the filtration is right-continuous, $\{\rho_I \vee n \le t\} \in \mathcal{G}_t$. Next, the complement of $\{\sigma_I \vee T_n \le t\}$ is $\mathcal{G}_t$-measurable. Indeed,

$$\{\sigma_I \vee T_n > t\} = \{\sigma_I > t\} \cup \{\sigma_I \le t,\ T_n > t\},$$

and $\mathbf{1}_{\{\sigma_I \le t\}}\mathbf{1}_{\{T_n > t\}}$ is $\mathcal{G}_t$-measurable by Lemma 3.1. We conclude that $\{T_n' \le t\} \in \mathcal{G}_t$ for every $t \ge 0$. □

The next theorem generalizes Theorem 2.3 to the case of progressive enlargement with multiple, not necessarily ranked times.

Theorem 3.1. *Let M be an $\mathbb{F}$ local martingale such that $M_0 = 0$. For each $I \subset \{1,\dots,n\}$, suppose there exists a $\mathbb{G}^I$ predictable finite variation process A^I such that $M - A^I$ is a $\mathbb{G}^I$ local martingale. Moreover, define*

$$Z_t^I = P(\rho_I > t \mid \mathcal{G}_t^I),$$

and let μ^I and J^I be as in Theorem 2.1. Then M is a $\mathbb{G}$ semimartingale with local martingale part $M_t - A_t$, where

$$A_t = \sum_I \mathbf{1}_{\{\sigma_I \le \rho_I\}} \left(\int_{t\wedge\sigma_I}^{t\wedge\rho_I} \mathrm{d}A_s^I + \int_{t\wedge\sigma_I}^{t\wedge\rho_I} \frac{\mathrm{d}\langle M, \mu^I\rangle_s + \mathrm{d}J_s^I}{Z_{s-}^I} \right).$$

Here the sum is taken over all $I \subset \{1,\dots,n\}$.

Notice that when $I = \emptyset$, then $\sigma_I = 0$ and $\mathbb{G}^I = \mathbb{F}$, so that $A^I = 0$ and does not contribute to A_t. Similarly, when $I = \{1,\dots,n\}$, $\rho_I = \infty$ and we have $Z_t^I = 1$, causing both $\langle M, \mu^I\rangle$ and J to vanish.

Before proving Theorem 3.1 we need the two following technical lemmas.

Lemma 3.3. *Let M be a process. Then*

$$\sum_{I\subset\{1,\dots,n\}} \mathbf{1}_{\{\sigma_I \le \rho_I\}}(M_{t\wedge\rho_I} - M_{t\wedge\sigma_I}) = M_t - M_0.$$

Proof. Fix $\omega \in \Omega$ and let $0 \le \tau_{\eta(1)} \le \tau_{\eta(2)} \le \cdots \le \tau_{\eta(n)}$ be the ordered times. For I of the form $I = \{\eta(1),\dots,\eta(k)\}$, $0 \le k \le n$, we have

$$\mathbf{1}_{\{\sigma_I \le \rho_I\}}(M_{t\wedge\rho_I} - M_{t\wedge\sigma_I}) = M_{t\wedge\tau_{\eta(k+1)}} - M_{t\wedge\tau_{\eta(k)}},$$

with the convention $\tau_{\eta(0)} = 0$ and $\tau_{\eta(n+1)} = \infty$. For any other I, the left side vanishes since $\sigma_I \ge \rho_I$. It follows that

$$\sum_{I\subset\{1,\dots,n\}} \mathbf{1}_{\{\sigma_I \le \rho_I\}}(M_{t\wedge\rho_I} - M_{t\wedge\sigma_I}) = \sum_{k=0}^{n} \left(M_{t\wedge\tau_{\eta(k+1)}} - M_{t\wedge\tau_{\eta(k)}}\right) = M_t - M_0,$$

as desired. □

Lemma 3.4. *Let L be a local martingale in some filtration $\mathbb{F}$, suppose σ and ρ are two stopping times, and define a process $N_t = \mathbf{1}_{\{\sigma \le \rho\}}(L_{t\wedge\rho} - L_{t\wedge\sigma})$:*

(i) N is again a local martingale.
(ii) Let T be a stopping time and define $T' = (\sigma \vee T) \wedge (\rho \vee n)$ for a fixed n. Then $N_{t\wedge T} = N_{t\wedge T'}$.

Proof.

Part (i): The claim follows from the observation that $N_t = L_{t\wedge\rho} - L_{t\wedge\rho\wedge\sigma}$.

Part (ii): The proof consists of a careful analysis of the interplay between the indicators involved in the definition of N and T'. First note that

$$t \wedge T' \wedge \rho = t \wedge (\sigma \vee T) \wedge (\rho \vee n) \wedge \rho = t \wedge (\sigma \vee T) \wedge \rho$$

and thus also

$$t \wedge T' \wedge \rho \wedge \sigma = t \wedge (\sigma \vee T) \wedge \rho \wedge \sigma = t \wedge \rho \wedge \sigma.$$

Therefore

$$N_{t\wedge T'} = L_{t\wedge(\sigma\vee T)\wedge\rho} - L_{t\wedge\sigma\wedge\rho}.$$

On $\{\sigma > \rho\}$, $N_{t\wedge T'}$ is zero, as is $N_{t\wedge T}$. On the set $\{\sigma \le \rho\} \cap \{T \le \sigma\}$, we have $N_{t\wedge T'} = L_{t\wedge\sigma} - L_{t\wedge\sigma} = 0 = N_{t\wedge T}$. Finally, on $\{\sigma \le \rho\} \cap \{T > \sigma\}$, we have $(\sigma \vee T)\wedge\rho = T\wedge\rho$ and $\sigma\wedge\rho = T\wedge\sigma\wedge\rho$, so that $N_{t\wedge T'} = L_{t\wedge T\wedge\rho} - L_{t\wedge T\wedge\rho\wedge\sigma} = N_{t\wedge T}$. □

Proof of Theorem 3.1 For each fixed index set I, the process $M_t - \int_0^t \mathrm{d}A_s^I$ is a local martingale in the initially expanded filtration $\mathbb{G}^I$ by assumption. Now, $\mathbb{H}^I$ is obtained from $\mathbb{G}^I$ by a progressive expansion with ρ_I, so Theorem 2.1 yields that

$$M_t^I = M_{t\wedge\rho_I} - \int_0^{t\wedge\rho_I} \mathrm{d}A_s^I - \int_0^{t\wedge\rho_I} \frac{\mathrm{d}\langle M, \mu^I\rangle_s + \mathrm{d}J_s^I}{Z_{s-}^I}$$

is an $\mathbb{H}^I$ local martingale. Define the process

$$N_t^I = \mathbf{1}_{\{\sigma_I \le \rho_I\}} \left(M_{t\wedge\rho_I}^I - M_{t\wedge\sigma_I}^I \right).$$

Our goal is to prove that N^I is a $\mathbb{G}$ local martingale. This will imply the statement of the theorem, since summing the N^I over all index sets I and using Lemma 3.3 yield precisely $M - A$.

By part (i) of Lemma 3.4, N^I is a local martingale in $\mathbb{H}^I$. Write

$$N_t^I = \mathbf{1}_{\{\sigma_I \le t < \rho_I\}} \left(M_t^I - M_{\sigma_I}^I \right) + \mathbf{1}_{\{\sigma_I \le \rho_I < t\}} \left(M_{\rho_I}^I - M_{\sigma_I}^I \right) = Y_1 + Y_2$$

and apply Lemma 3.1 with $X = Y_1$ for the first term and $X = Y_2$ for the second to see that N^I is in fact $\mathbb{G}$ adapted. The use of Lemma 3.1 is valid because both Y_1 and Y_2 are $\mathcal{H}_t^I$-measurable.

Next, let $(T_n)_{n\ge1}$ be a reducing sequence for N^I in $\mathbb{H}^I$. By Lemma 3.2 we know that $T_n' = (\sigma_I \vee T_n) \wedge (\rho_I \vee n)$ are $\mathbb{G}$ stopping times, and since $T_n' \ge T_n \wedge n$ we have $T_n' \uparrow \infty$ a.s. Moreover, part (ii) of Lemma 3.4 implies that $N_{t\wedge T_n}^I = N_{t\wedge T_n'}^I$. Hence $(N_{t\wedge T_n'}^I)_{t\ge0}$ is an $\mathbb{H}^I$ martingale that is $\mathbb{G}$ adapted and therefore even a $\mathbb{G}$ martingale. We deduce from the above that N^I is a $\mathbb{G}$ local martingale. A final application of Lemma 3.1 shows that A is $\mathbb{G}$ predictable, so we obtain indeed the $\mathbb{G}$ semimartingale decomposition of M. This completes the proof of Theorem 3.1. □

We now proceed to study the special case where the vector $\boldsymbol{\tau}$ of random times satisfies Jacod's criterion, meaning that Assumption 2.1 holds, now with state space $E = \mathbb{R}_+^n$. Again there is no loss of generality to let η be the law of $\boldsymbol{\tau}$. We will further assume that η is absolutely continuous w.r.t. to Lebesgue measure, so that $\eta(d\boldsymbol{u}) = h(\boldsymbol{u})d\boldsymbol{u}$. Provided the law of $\boldsymbol{\tau}$ does not have atoms, this restriction is not essential—everything that follows goes through without it—but it simplifies the already quite cumbersome notation.

The joint $\mathcal{F}_t$ conditional density corresponding to this choice of η is denoted $p_t(\boldsymbol{u};\omega)$. That is,

$$P(\boldsymbol{\tau} \in d\boldsymbol{u} \mid \mathcal{F}_t) = p_t(\boldsymbol{u})du_1 \cdots du_n,$$

where we suppressed the dependence on ω. Now, for an index set $I \subset \{1,\ldots,n\}$ with $|I| = m$, we have, for $\boldsymbol{u}_I \in \mathbb{R}_+^m$,

$$P(\boldsymbol{\tau}_I \leq \boldsymbol{u}_I \mid \mathcal{F}_t) = \int_{\boldsymbol{v}_I \leq \boldsymbol{u}_I} \int_{\boldsymbol{v}_{-I} \geq 0} p_t(\boldsymbol{v}_I;\boldsymbol{v}_{-I})\mathrm{d}\boldsymbol{v}_{-I}\mathrm{d}\boldsymbol{v}_I,$$

where $\boldsymbol{\tau}_I$ is the subvector of $\boldsymbol{\tau}$ whose components have indices in I and where $\boldsymbol{v}_I$ and $\boldsymbol{v}_{-I}$ are the subvectors of $\boldsymbol{v}$ with indices in I, respectively, not in I. Inequalities should be interpreted componentwise. The above shows that $\boldsymbol{\tau}_I$ also satisfies Assumption 2.1, so that there is an appropriately measurable function $p_t^I(\boldsymbol{u}_I;\omega)$ such that

$$P(\boldsymbol{\tau}_I \in d\boldsymbol{u}_I \mid \mathcal{F}_t) = p_t^I(\boldsymbol{u}_I)d\boldsymbol{u}_I.$$

Moreover, this conditional density p^I is given by

$$p_t^I(\boldsymbol{u}_I) = \int_{\mathbb{R}_+^{n-m}} p_t(\boldsymbol{u}_I;\boldsymbol{u}_{-I})\mathrm{d}\boldsymbol{u}_{-I}.$$

Define

$$p_t(\boldsymbol{u}_{-I} \mid \boldsymbol{\tau}_I) = \frac{p_t(\boldsymbol{\tau}_I;\boldsymbol{u}_{-I})}{\int_0^\infty \cdots \int_0^\infty p_t(\boldsymbol{\tau}_I;\boldsymbol{v}_{-I})\mathrm{d}\boldsymbol{v}_{-I}}.$$

This is the conditional density of $\boldsymbol{\tau}_{-I}$ given $\mathcal{F}_t \vee \sigma(\tau_I)$, as one can verify using standard arguments. Defining

$$Z_t^I = \int_t^\infty \cdots \int_t^\infty p_t(\boldsymbol{u}_{-I} \mid \boldsymbol{\tau}_I)\mathrm{d}\boldsymbol{u}_{-I}, \tag{21.1}$$

we have $Z_t^I = P(\rho_I > t \mid \mathcal{F}_t \vee \sigma(\tau_I))$. One then readily checks that we also have

$$Z_t^I = P(\rho_I > t \mid \mathcal{G}_t^I).$$

We can now state the decomposition theorem for continuous $\mathbb{F}$ local martingales in the progressively expanded filtration $\mathbb{G}$, when Jacod's criterion is satisfied. Recall that $\sigma_I = \max_{i \in I} \tau_i$ and $\rho_I = \min_{j \notin I} \tau_j$.

Corollary 3.1. *Let M be an $\mathbb{F}$ local martingale, and assume Assumption 2.1 is satisfied for $\boldsymbol{\tau} = (\tau_1, \ldots, \tau_n)$. Then M is a $\mathbb{G}$ semimartingale. Furthermore, assume that M is continuous. For each $I \subset \{1, \ldots, n\}$, let Z^I be given by Eq. (21.1) and let μ^I and J^I be as in Theorem 2.1. Then $M_t - A_t$ is a $\mathbb{G}$ local martingale, where*

$$A_t = \sum_I \mathbf{1}_{\{\sigma_I \leq \rho_I\}} \left(\int_{t \wedge \sigma_I}^{t \wedge \rho_I} \frac{\mathrm{d}\langle p^I(\boldsymbol{u}_I), M \rangle_s}{p_{s-}^I(\boldsymbol{u}_I)} \bigg|_{\boldsymbol{u}_I = \boldsymbol{\tau}_I} + \int_{t \wedge \sigma_I}^{t \wedge \rho_I} \frac{\mathrm{d}\langle M, \mu^I \rangle_s + \mathrm{d}J_s^I}{Z_{s-}^I} \right).$$

Here the sum is taken over all $I \subset \{1, \ldots, n\}$.

Proof. We apply Theorem 3.1 and notice that it follows from Theorem 2.5 that

$$A_t^I = \int_0^t k_s^I(\boldsymbol{\tau}_I) \mathrm{d}\langle M, M \rangle_s = \int_0^t \frac{\mathrm{d}\langle p^I(\boldsymbol{u}_I), M \rangle_s}{p_{s-}^I(\boldsymbol{u}_I)} \bigg|_{\boldsymbol{u}_I = \boldsymbol{\tau}_I}.$$

This completes the proof. □

We end this section by pointing out that the filtration $\mathbb{G}^{(\boldsymbol{\tau}, \boldsymbol{X})}$ introduced in Sect. 2 can be generalized to the multiple time case. To state the precise result, we first suppose that each random time τ_i is accompanied by a random variable X_i and let $\boldsymbol{X} = (X_1, \ldots, X_n)$. Define the filtration $\mathbb{G}^{(\boldsymbol{\tau}, \boldsymbol{X})}$ by

$$\mathcal{G}_t^{(\boldsymbol{\tau}, \boldsymbol{X})} = \bigcap_{u > t} \mathcal{G}_u^{0, (\boldsymbol{\tau}, \boldsymbol{X})},$$

where

$$\mathcal{G}_t^{0, (\boldsymbol{\tau}, \boldsymbol{X})} = \mathcal{F}_t \vee \sigma(\tau_i \wedge t : i = 1, \ldots, n) \vee \sigma(X_i 1_{\{\tau_i \leq t\}} : i = 1, \ldots, n).$$

Let $I \subset \{1, \ldots, n\}$ be an index set. Assume for simplicity that $P(\tau_i = \tau_j) = 0$ for $i \neq j$. We may then define the following:

- $\boldsymbol{X}_I = (X_i)_{i \in I}$.
- $Y_I = X_{i^*}$, where $i^* \in I$ is the index for which $\rho_I = \tau_{i^*}$.

For the statement and proof of Theorem 3.2, we redefine the objects $\mathbb{G}^I$ and $\mathbb{H}^I$ as follows. For an index set $I \subset \{1, \ldots, n\}$:

- $\mathbb{G}^I$ denotes the initial expansion of $\mathbb{F}$ with the random vector $(\boldsymbol{\tau}_I, \boldsymbol{X}_I) = (\tau_i, X_i)_{i \in I}$.

- $\mathbb{H}^I$ denotes the (ρ_I, Y_I)-expansion of $\mathbb{G}^I$.

Theorem 3.2. *Let M be an $\mathbb{F}$ local martingale such that $M_0 = 0$. For each $I \subset \{1, \ldots, n\}$, suppose there exists a $\mathbb{G}^I$ predictable finite variation process A^I such that $M - A^I$ is a $\mathbb{G}^I$ local martingale. Moreover, define*

$$Z_t^I = P(\rho_I > t \mid \mathcal{G}_t^I),$$

and let μ^I and J^I be as in Theorem 2.1. Then M is a $\mathbb{G}^{(\tau,X)}$ semimartingale with local martingale part $M_t - A_t$, where

$$A_t = \sum_I \mathbf{1}_{\{\sigma_I \le \rho_I\}} \left(\int_{t\wedge\sigma_I}^{t\wedge\rho_I} \mathrm{d}A_s^I + \int_{t\wedge\sigma_I}^{t\wedge\rho_I} \frac{\mathrm{d}\langle M, \mu^I\rangle_s + \mathrm{d}J_s^I}{Z_{s-}^I} \right).$$

Here the sum is taken over all $I \subset \{1, \ldots, n\}$.

Proof. The proof is the same as that of Theorem 3.1, except for the following points: instead of Theorem 2.1, we invoke Theorem 2.2, which is justified by Lemma 2.5. Moreover, it must be verified that Lemma 3.1 remains valid for the redefined $\mathbb{H}^I$ and $\mathbb{G} = \mathbb{G}^{(\tau,X)}$. This is easily done: in the proof of Lemma 3.1, simply replace $X = fk(\rho_I \wedge t)\prod_{i\in I} h_i(\tau_i)$ by

$$X = fk(\rho_I \wedge t)\ell(Y_I \mathbf{1}_{\{\rho_I \le t\}}) \prod_{i\in I} h_i(\tau_i) g_i(X_i \mathbf{1}_{\{\tau_i \le t\}}),$$

where $\ell(\cdot)$ and $g_i(\cdot)$ are Borel functions. □

Define the process

$$N_t^n = \sum_{i=1}^n X_i 1_{\{\tau_i \le t\}}.$$

Let $\mathbb{N}^n$ be the smallest right-continuous filtration containing $\mathbb{F}$ and to which the process N^n is adapted. Under the same assumption as in Theorem 3.2, $\mathbb{F}$ semimartingales remain $\mathbb{N}^n$ semimartingales.

Corollary 3.2. *Let M be an $\mathbb{F}$ local martingale such that $M_0 = 0$. For each $I \subset \{1, \ldots, n\}$, suppose there exists a $\mathbb{G}^I$ predictable finite variation process A^I such that $M - A^I$ is a $\mathbb{G}^I$ local martingale. Then M is a $\mathbb{N}^n$ semimartingale.*

Proof. Applying Theorem 3.2, M is a $\mathbb{G}^{(\tau,X)}$ semimartingale. Since $\mathbb{N}^n \subset \mathbb{G}^{(\tau,X)}$ and M is adapted to $\mathbb{N}^n$, it follows from Stricker's theorem [13] (see also [12], Theorem II.4, for a short proof) that M is a $\mathbb{N}^n$ semimartingale. □

4 Connection to Filtration Shrinkage

It has been observed that the optional projection of a local martingale M onto a filtration to which it is not adapted may lose the local martingale property; see Föllmer and Protter [4]. In that paper, the following general condition was given that guarantees that the local martingale property is preserved (see [4], Theorem 3.7).

Lemma 4.1. *Consider two filtrations $\mathbb{F} \subset \mathbb{G}$ and a $\mathbb{G}$ local martingale N. If there exists a sequence of $\mathbb{F}$ stopping times that reduce N, then its optional projection onto $\mathbb{F}$ is again a local martingale.*

Using this result we establish that certain local martingales that arise in the context of filtration expansion in fact retain their local martingale property when projected to various subfiltrations. In particular, if M is an $\mathbb{F}$ local martingale and a $\mathbb{G}$ semimartingale, then ${}^{o}M^{\mathbb{G}}$, the optional projection onto $\mathbb{F}$ of the local martingale part in the $\mathbb{G}$ semimartingale decomposition of M, always remains a local martingale.

Theorem 4.1. *Consider three filtrations $\mathbb{E} \subset \mathbb{F} \subset \mathbb{G}$, and let M be an $\mathbb{E}$ local martingale. Suppose M is also a $\mathbb{G}$ semimartingale with canonical decomposition $M = N + A$. Then the optional projection of N onto $\mathbb{F}$, when it exists, is again a local martingale.*

Remark. Note that M remains a special semimartingale in $\mathbb{G}$, given that it remains a semimartingale.

Proof. By Lemma 4.1 it suffices to show that N can be reduced using $\mathbb{E}$, and hence $\mathbb{F}$, stopping times. Now, since M is an $\mathbb{E}$ local martingale, it is locally in $\mathcal{H}^1_{loc}(\mathbb{E})$. Therefore it can be reduced with a sequence $(T_n)_{n\geq 1}$ of $\mathbb{E}$ stopping times such that for each n, $M^{T_n} \in \mathcal{H}^1(\mathbb{E})$. Also, M^{T_n} is a $\mathbb{G}$ semimartingale with canonical decomposition:

$$M^{T_n} = N^{T_n} + A^{T_n}.$$

A result by Yor ([16], Théorème 5) then yields

$$\|N^{T_n}\|_{\mathcal{H}^1(\mathbb{G})} \leq c\|M^{T_n}\|_{\mathcal{H}^1(\mathbb{E})}$$

for some absolute constant c. Since M^{T_n} is in $\mathcal{H}^1(\mathbb{E})$, the right side is finite and it readily follows that N^{T_n} is a uniformly integrable $\mathbb{G}$ martingale for each n, which was what we set out to prove. □

At this point, it is worth mentioning that the result above helps illustrate the role of our assumptions and shows how they fail to hold for the explicit counterexample given in [4]. There the authors consider a standard three-dimensional Brownian motion $(B_t)_{t\geq 0} = (B_t^1, B_t^2, B_t^3)_{t\geq 0}$ starting at $x_0 = (1, 0, 0)$. Let $\mathbb{H}$ be its natural filtration. It is well known that $(\|B_t\|)_{t\geq 0}$ is a Bessel(3) process with initial value 1 and that its reciprocal $N_t = \|B_t\|^{-1}$ is an $\mathbb{H}$ strict local martingale. Let $\mathbb{F}$ be the natural filtration of B^1. It is proved in [4] that the optional projection of N onto $\mathbb{F}$ is not a local martingale. Theorem 4.1 therefore implies that N can never appear as the local martingale part in the $\mathbb{H}$ decomposition $M = N + A$ of any $\mathbb{F}$ local martingale M that is also an $\mathbb{H}$ semimartingale with A nontrivial, since then the projection of N would not be an $\mathbb{F}$ local martingale.

As an application of Theorem 4.1 we obtain the following. The filtrations $\mathbb{F}$, $\mathbb{G}$, and $\mathbb{H}$ are now as in Theorem 2.3.

Corollary 4.1. *Let M be an $\mathbb{F}$ local martingale and let $\mathbb{H}$ be a filtration that coincides with $\mathbb{G}$ after τ. Suppose M is an $\mathbb{H}$ semimartingale with decomposition*

$$M_t = M_t^{\mathbb{H}} + A_t^{\mathbb{H}},$$

and suppose $A^{\mathbb{H}}$ has a $\mathbb{G}$ optional projection ${}^oA^{\mathbb{H}}$ which is locally integrable in $\mathbb{G}$.[1]

Then M is a $\mathbb{G}$ semimartingale with decomposition $M = M^{\mathbb{G}} + A^{\mathbb{G}}$, and $A^{\mathbb{G}} = (A^{\mathbb{H}})^p$. In particular, it follows that

$$(A^{\mathbb{H}})_t^p = \int_0^{t\wedge\tau} \frac{\mathrm{d}\langle M, \mu\rangle_s + \mathrm{d}J_s}{Z_{s-}} + \int_{t\wedge\tau}^t \mathrm{d}A_s^{\mathbb{H}}.$$

Here, the quantities Z, μ, and J are defined as in Theorem 2.1 and $(A^{\mathbb{H}})^p$ is the dual predictable projection of $A^{\mathbb{H}}$ onto $\mathbb{G}$.

Proof. M is an $\mathbb{H}$ semimartingale, hence a $\mathbb{G}$ semimartingale by Stricker's theorem [13]. Since ${}^oA^{\mathbb{H}}$ exists and M is $\mathbb{G}$ adapted, ${}^oM^{\mathbb{H}}$ exists. Taking optional projections of the relation $M = M^{\mathbb{H}} + A^{\mathbb{H}}$ and adding and subtracting $(A^{\mathbb{H}})^p$ yield

$$M = \left[{}^oM^{\mathbb{H}} + {}^oA^{\mathbb{H}} - (A^{\mathbb{H}})^p\right] + (A^{\mathbb{H}})^p.$$

By Theorem 4.1, ${}^oM^{\mathbb{H}}$ is a $\mathbb{G}$ local martingale, and therefore the quantity in brackets is a $\mathbb{G}$ local martingale. We deduce that $A^{\mathbb{G}} = (A^{\mathbb{H}})^p$. The expression for $(A^{\mathbb{H}})^p$ now follows from Theorem 2.3. □

As a particular case, we recover Proposition 3.3 in [1]. There $\mathbb{G}$ and $\mathbb{H}$ are the progressive, respectively initial, expansions of $\mathbb{F}$ with a random time τ that avoids all $\mathbb{F}$ stopping times, and whose $\mathbb{F}$ conditional probabilities are equivalent to some deterministic measure. Under these assumptions, the authors compute explicitly the $\mathbb{G}$ dual predictable projection of $A_t^{\mathbb{H}} = \int_0^t k_s(\tau)\mathrm{d}s$, where k is as in Theorem 2.5, and prove that

$$(A^{\mathbb{H}})_t^p = \int_0^{t\wedge\tau} \frac{\mathrm{d}\langle M, Z\rangle_s}{Z_{s-}} + \int_{t\wedge\tau}^t k_s(\tau)\mathrm{d}s.$$

This follows from Corollary 4.1 since $J = 0$ when τ avoids all $\mathbb{F}$ stopping times.

---

[1]This assumption guarantees that the dual predictable projection $(A^{\mathbb{H}})^p$ exists, and that ${}^oA^{\mathbb{H}} - (A^{\mathbb{H}})^p$ is a $\mathbb{G}$ local martingale; see Theorem VI.80 in [3]. That $(A^{\mathbb{H}})^p$ is $\mathbb{G}$ locally integrable means that there exists a sequence of $\mathbb{G}$ stopping times $(T_n)_{n\geq 1}$ increasing to infinity and such that $(A^{\mathbb{H}})^p_{\cdot\wedge T_n}$ is integrable.

Acknowledgements Philip Protter was supported in part by NSF grant DMS-0906995.

References

1. Callegaro, G., Jeanblanc, M., Zargari, B.: Carthaginian enlargement of filtrations. ESAIM: Probability and Statistics, doi:10.1051/ps/2011162
2. Dellacherie, C., Meyer, P.A.: A propos du travail de Yor sur le grossissement des tribus. Séminaire de Probabilités **XII**, 70–77 (1976)
3. Dellacherie, C., Meyer, P.A.: Probabilities and Potential B, Theory of Martingales, North-Holland Publishing Company, Amsterdam (1982)
4. Föllmer, H., Protter, P.: Local martingales and filtration shrinkage. ESAIM: Probability and Statistics, Available on CJO 2006 doi:10.1051/ps/2010023 (2011)
5. Guo, X., Zeng, Y.: Intensity process and compensator: a new filtration expansion approach and the Jeulin-Yor Theorem. The Ann. Appl. Probab. **18**(1), 120–142 (2008)
6. Imkeller, P.: Malliavin's calculus in insider models: additional utility and free lunches. Math. Finance **13**, 153–169 (2003)
7. Imkeller, P., Pontier, M., and Weisz, F.: Free lunch and arbitrage possibilities in a financial market model with an insider. Stoch. Proc. Appl. **92**, 103–130 (2001)
8. Jacod, J.: Grossissement initial, hypothèse (H') et théorème de Girsanov. In: Lecture Notes in Mathematics, vol. 1118, pp. 15–35. Springer, Berlin (1987)
9. Jeanblanc, M., Le Cam, Y.: Progressive enlargement of filtrations with initial times. Stoch. Proc. Appl. **119**, 2523–2543 (2009)
10. Jeulin, T., Yor, M. (eds.): Grossissement de filtrations: examples et applications. In: Lecture Notes in Mathematics, vol. 1118. Springer, Berlin (1985)
11. Kohatsu-Higa, A.: Models for insider trading with finite utility. In: Paris-Princeton Lectures on Mathematical Finance, Lecture Notes in Mathematics. Springer, Berlin (2004)
12. Protter, P.: Stochastic Integration and Differential Equations, 2nd edn. Springer, Heidelberg (2005)
13. Stricker, C.: Quasimartingales, martingales locales, semimartingales et filtration naturelle. Z. Wahrscheinlishkeitstheorie verw. Gebiete **39**, 55–63 (1977)
14. Sulem, A., Kohatsu-Higa, A., Øksendal, B., Proske, F., Di Nunno, G.: Anticipative stochastic control for Lévy processes with application to insider trading. In: Mathematical Modeling and Numerical Methods in Finance, Elsevier (2008)
15. Yor, M.: Grossissement de filtrations et absolue continuité de noyaux. In: Jeulin, T., Yor, M. (eds.) Grossissements de filtrations: exemples et applications, Springer (1985)
16. Yor, M.: Inégalités entre processus minces et applications. C.R. Acad. Sci. Paris, Sér. A **286**, 799–801 (1978)

Chapter 22
A Malliavin Calculus Approach to General Stochastic Differential Games with Partial Information

An Ta Thi Kieu, Bernt Øksendal, and Yeliz Yolcu Okur

Abstract In this article, we consider stochastic differential game where the state process is governed by a controlled Itô–Lévy process and the information available to the controllers is possibly less than the general information. All the system coefficients and the objective performance functional are assumed to be random. We use Malliavin calculus to derive a maximum principle for the optimal control of such problem. The results are applied to solve a worst-case scenario portfolio problem in finance.

Keywords Malliavin calculus • Stochastic differential game • Stochastic control, Jump diffusion • Partial information • Optimal worst-case scenario portfolio

Received 2/18/2011; Accepted 5/23/2012; Final 5/29/2012

1 Introduction

Suppose the dynamics of a state process $X(t) = X^{(u_0,u_1)}(t,\omega); t \geq 0, \omega \in \Omega$, is a controlled Itô–Lévy process in $\mathbb{R}$ of the form

A.T.T. Kieu • B. Øksendal (✉)
Center of Mathematics for Applications (CMA), University of Oslo,
Box 1053 Blindern, N-0316 Oslo, Norway
e-mail: a.t.k.ta@mn.uio.no; oksendal@math.uio.no

Y.Y. Okur
Institute of Applied Mathematics, Middle East Technical Unoversity (METU),
Ankara 06800, Turkey
e-mail: yyolcu@metu.edu.tr

F. Viens et al. (eds.), *Malliavin Calculus and Stochastic Analysis: A Festschrift in Honor of David Nualart*, Springer Proceedings in Mathematics & Statistics 34,
DOI 10.1007/978-1-4614-5906-4_22,

$$\begin{cases} dX(t) = b(t, X(t), u_0(t), \omega)dt + \sigma(t, X(t), u_0(t), \omega)dB(t) \\ \qquad\qquad + \int_{\mathbb{R}_0} \gamma(t, X(t^-), u_0(t^-), u_1(t^-, z), z, \omega)\tilde{N}(dt, dz); \\ X(0) \ = x \in \mathbb{R}, \end{cases} \tag{22.1}$$

where the coefficients $b : [0, T] \times \mathbb{R} \times U \times \Omega \to \mathbb{R}$, $\sigma : [0, T] \times \mathbb{R} \times U \times \Omega \to \mathbb{R}$ and $\gamma : [0, T] \times \mathbb{R} \times U \times K \times \mathbb{R}_0 \times \Omega$ are all continuously differentiable (C^1) with respect to $x \in \mathbb{R}$ and $u_0 \in U$, $u_1 \in K$ for each $t \in [0, T]$ and a.a. $\omega \in \Omega$; U, K are given open convex subsets of $\mathbb{R}^2$ and $\mathbb{R} \times \mathbb{R}_0$, respectively. Here $\mathbb{R}_0 = \mathbb{R} - \{0\}$, $B(t) = B(t, \omega)$ and $\eta(t) = \eta(t, \omega)$, given by

$$\eta(t) = \int_0^t \int_{\mathbb{R}_0} z\tilde{N}(ds, dz); \ t \geq 0, \ \omega \in \Omega, \tag{22.2}$$

are a one-dimensional Brownian motion and an independent pure jump Lévy martingale, respectively, on a given filtered probability space $(\Omega, \mathcal{F}, \{\mathcal{F}_t\}_{t\geq 0}, P)$. Thus

$$\tilde{N}(dt, dz) := N(dt, dz) - \nu(dz)dt \tag{22.3}$$

is the *compensated Poisson jump measure* of $\eta(\cdot)$, where $N(dt, dz)$ is the *Poisson jump measure* and $\nu(dz)$ is the *Lévy measure* of the pure jump Lévy process $\eta(\cdot)$. For simplicity, we assume that

$$\int_{\mathbb{R}_0} z^2\nu(dz) < \infty. \tag{22.4}$$

The processes $u_0(t)$ and $u_1(t, z)$ are the control processes and have values in a given open convex set U and K, respectively, for a.a. $t \in [0, T]$, $z \in \mathbb{R}_0$ for a given fixed $T > 0$. Also, $u_0(\cdot)$ and $u_1(\cdot)$ are càdlàg and adapted to a given filtration $\{\mathcal{E}_t\}_{t\geq 0}$, where

$$\mathcal{E}_t \subseteq \mathcal{F}_t, \qquad t \in [0, T].$$

$\{\mathcal{E}_t\}_{t\geq 0}$ represents the information available to the controller at time t. For example, we could have

$$\mathcal{E}_t = \mathcal{F}_{(t-\delta)^+}; \quad t \in [0, T], \ \delta > 0 \text{ is a constant},$$

meaning that the controller gets a delayed information compared to $\mathcal{F}_t$. We refer to [15, 12] for more information about stochastic control of Itô diffusions and jump diffusions, respectively, and to [2], [4], [8], [9], [14] for other papers dealing with optimal control under partial information/observation.

Let $f : [0, T] \times \mathbb{R} \times U \times K \times \Omega \to \mathbb{R}$ and $g : \mathbb{R} \times \Omega \to \mathbb{R}$ are given continuously differentiable (C^1) with respect to $x \in \mathbb{R}$ and $u_0 \in U$, $u_1 \in K$. Suppose there are two players in the stochastic differential game and the given performance functionals for players are as follows:

$$J_i(u_0, u_1) = \mathbb{E}^x \left[\int_0^T \int_{\mathbb{R}_0} f_i(t, X(t), u_0(t), u_1(t, z), z, \omega)\mu(dz)dt + g_i(X(T), \omega) \right],$$

$$i = 1, 2,$$

where μ is a measure on the given measurable space $(\Omega, \mathcal{F})$ and $\mathbb{E}^x = \mathbb{E}_P^x$ denotes the expectation with respect to P given that $X(0) = x$. Suppose that the controls $u_0(t)$ and $u_1(t, z)$ have the form

$$u_0(t) = (\pi_0(t), \theta_0(t)); \quad t \in [0, T]; \tag{22.5}$$

$$u_1(t, z) = (\pi_1(t, z), \theta_1(t, z)); \quad (t, z) \in [0, T] \times \mathbb{R}_0. \tag{22.6}$$

Let $\mathscr{A}_\Pi$ and $\mathscr{A}_\Theta$ denote the given family of controls $\pi = (\pi_0, \pi_1)$ and $\theta = (\theta_0, \theta_1)$ such that they are contained in the set of càdlàg $\mathcal{E}_t$-adapted controls, Eq. (22.1) has a unique strong solution up to time T and

$$\mathbb{E}^x \left[\int_0^T \int_{\mathbb{R}_0} |f_i(t, X(t), \pi_0(t), \pi_1(t, z), \theta_0(t), \theta_1(t, z), z, \omega)| \mu(\mathrm{d}z)\mathrm{d}t + |g_i(X(T), \omega)| \right]$$
$$< \infty, \quad i = 1, 2.$$

The partial information non-zero-sum stochastic differential game problem we consider is the following:

Problem 1.1. Find $(\pi^*, \theta^*) \in \mathscr{A}_\Pi \times \mathscr{A}_\Theta$ (if it exists) such that

(i) $J_1(\pi, \theta^*) \leq J_1(\pi^*, \theta^*)$ for all $\pi \in \mathscr{A}_\Pi$,
(ii) $J_2(\pi^*, \theta) \leq J_2(\pi^*, \theta^*)$ for all $\theta \in \mathscr{A}_\Theta$.

Such a control (π^*, θ^*) is called a *Nash equilibrium* (if it exists). The intuitive idea is that there are two players, players I and II. While player I controls π, player II controls θ. Given that each player knows the equilibrium strategy chosen by the other player, none of the players has anything to gain by changing only his or her own strategy only (i.e., by changing unilaterally). Note that since we allow b, σ, γ, f and g to be stochastic processes and also because our controls are required to be $\mathcal{E}_t$-adapted, this problem is not of Markovian type and hence cannot be solved by dynamic programming. Our paper is related to the recent paper [1, 10], where a maximum principle for stochastic differential games with partial information and a mean-field maximum principle are dealt with, respectively. However, the approach in [1] needs the solution of the backward stochastic differential equation (BSDE) for the adjoint processes. This is often a difficult point, particularly in the partial information case. In the current paper, we use Malliavin calculus techniques to obtain a maximum principle for this general non-Markovian stochastic differential game with partial information, without the use of BSDEs.

2 The General Maximum Principle for the Stochastic Differential Games

In this section we base on Malliavin calculus to solve Problem 1.1. We assume the following:

(A1) For all $s, r, t \in (0,T)$, $t \leq r$, and all bounded $\mathscr{E}_t$-measurable random variables $\alpha = \alpha(\omega)$, $\xi = \xi(\omega)$ the controls $\beta_\alpha(s) := (0, \beta^i_\alpha(s))$ and $\eta_\xi(s) := (0, \eta^i_\xi)$, $i = 1, 2$, with

$$\beta^i_\alpha(s) = \alpha^i(\omega)\chi_{[t,r]}(s) \text{ and } \eta^i_\xi(s) = \xi^i(\omega)\chi_{[t,r]}(s); \quad s \in [0,T],$$

belong to $\mathscr{A}_\Pi$ and $\mathscr{A}_\Theta$, respectively. Also, we will denote the transposed of the vectors β and η by β^*, η^*, respectively.

(A2) For all $\pi, \beta \in \mathscr{A}_\Pi$; $\theta, \eta \in \mathscr{A}_\Theta$ with β and η are bounded, there exists $\delta > 0$ such that the controls $\pi(t) + y\beta(t)$ and $\theta(t) + \upsilon\eta(t)$, $t \in [0,T]$, belong to $\mathscr{A}_\Pi$ and $\mathscr{A}_\Theta$, respectively, for all $\upsilon \in (-\delta, \delta)$, and such that the families

$$\Big\{\frac{\partial f_1}{\partial x}(t, X^{(\pi+y\beta,\theta)}(t), \pi + y\beta, \theta, z)\frac{\mathrm{d}}{\mathrm{d}y}X^{(\pi+y\beta,\theta)}(t) + \nabla_\pi f_1(t, X^{(\pi+y\beta,\theta)}(t), \pi + y\beta, \theta, z)\beta^*(t)\Big\}_{y\in(-\delta,\delta)},$$

$$\Big\{\frac{\partial f_2}{\partial x}(t, X^{(\pi,\theta+\upsilon\eta)}(t), \pi, \theta + \upsilon\eta, z)\frac{\mathrm{d}}{\mathrm{d}y}X^{(\pi,\theta+\upsilon\eta)}(t) + \nabla_\theta f_2(t, X^{(\pi,\theta+\upsilon\eta)}(t), \pi(t), \theta + \upsilon\eta, z)\eta^*(t)\Big\}_{\upsilon\in(-\delta,\delta)}$$

are $\lambda \times \nu \times P$-uniformly integrable and the families

$$\Big\{g_1'(X^{\pi+y\beta}(T))\frac{\mathrm{d}}{\mathrm{d}y}X^{(\pi+y\beta,\theta)}(T)\Big\}_{y\in(-\delta,\delta)},$$

$$\Big\{g_2'(X^{\pi,\theta+\upsilon\eta}(T))\frac{\mathrm{d}}{\mathrm{d}y}X^{(\pi,\theta+\upsilon\eta)}(T)\Big\}_{\upsilon\in(-\delta,\delta)}$$

are P-uniformly integrable.

In the following, $D_t F$ denotes the *Malliavin derivative with respect to* $B(\cdot)$ *(at* t*)* of a given (Malliavin differentiable) random variable $F = F(\omega)$; $\omega \in \Omega$. Similarly, $D_{t,z}F$ denotes the *Malliavin derivative with respect to* $\widetilde{N}(\cdot,\cdot)$ *(at* t, z*)* of F. We let $\mathbb{D}_{1,2}$ denote the set of all random variables which are Malliavin differentiable with respect to both $B(\cdot)$ and $N(\cdot,\cdot)$. We will use the following *duality formula* for Malliavin derivatives:

$$\mathbb{E}\left[F\int_0^T \varphi(t)\mathrm{d}B(t)\right] = \mathbb{E}\left[\int_0^T \varphi(t)D_t F\mathrm{d}t\right], \tag{22.7}$$

$$\mathbb{E}\left[F\int_0^T \int_{\mathbb{R}_0} \psi(t,z)\tilde{N}(\mathrm{d}t,\mathrm{d}z)\right] = \mathbb{E}\left[\int_0^T \int_{\mathbb{R}_0} \psi(t,z)D_{t,z}F\nu(\mathrm{d}z)\mathrm{d}t\right] \tag{22.8}$$

valid for all Malliavin differentiable F- and all $\mathcal{F}_t$-predictable processes φ and ψ such that the integrals on the right converge absolutely. We also need the following basic properties of Malliavin derivatives:

If $F \in \mathbb{D}_{1,2}$ is $\mathcal{F}_s$-measurable, then

$$D_t F = D_{t,z}F = 0, \text{ for all } t > s. \tag{22.9}$$

(Fundamental theorem)

$$D_t\left(\int_0^T \varphi(s)\delta B(s)\right) = \int_0^T D_t\varphi(s)\delta B(s) + \varphi(t) \qquad \text{for a.a. } (t,\omega), \tag{22.10}$$

where $\int_0^T u(s)\delta B(s)$ denotes Skorohod integral of u with respect to $B(\cdot)$. (See [11], p. 35–38 for a definition of Skorohod integrals and for more details.)

$$D_{t,z}\left(\int_0^T \int_{\mathbb{R}} \psi(s,y)\tilde{N}(\mathrm{d}s,\mathrm{d}y)\right) = \int_0^T \int_{\mathbb{R}} D_{t,z}\psi(s,y)\tilde{N}(\mathrm{d}s,\mathrm{d}y) + \psi(t,z), \tag{22.11}$$

provided that all terms involved are well defined. We refer to [3], [5], [6], [7], [10] and [11] for more information about the Malliavin calculus for Lévy processes and its applications.

(A3) For all $(\pi,\theta) \in \mathscr{A}_\Pi \times \mathscr{A}_\Theta$, we assume the following processes, $i = 1,2$:

$$K_i(t) = g_i'(X(T)) + \int_t^T \int_{\mathbb{R}_0} \frac{\partial f_i}{\partial x}(s,X(s),\pi,\theta,z_1)\mu(\mathrm{d}z_1)\mathrm{d}s, \tag{22.12}$$

$$\begin{aligned} H_i^0(s,x,\pi,\theta) =& K_i(s)b(s,x,\pi_0,\theta_0) + D_sK_i(s)\sigma(s,x,\pi_0,\theta_0) \\ &+ \int_{\mathbb{R}_0} D_{s,z}K_i(s)\gamma(s,x,\pi,\theta,z)\nu(\mathrm{d}z), \end{aligned} \tag{22.13}$$

$$\begin{aligned} G(t,s) := \exp\Bigg(\int_t^s \Big\{&\frac{\partial b}{\partial x}(r,X(r),\pi_0(r),\theta_0(r)) \\ &-\frac{1}{2}\left(\frac{\partial \sigma}{\partial x}\right)^2(r,X(r),\pi_0(r),\theta_0(r))\Big\}\mathrm{d}r \end{aligned}$$

$$+\int_t^s \frac{\partial \sigma}{\partial x}(r, X(r), \pi_0(r), \theta_0(r))\mathrm{d}B(r)$$

$$+\int_t^s \int_{\mathbb{R}_0} \ln\left(1+\frac{\partial \gamma}{\partial x}(r, X(r^-), \pi(r^-, z), \theta(r^-, z), z)\right) \tilde{N}(\mathrm{d}r, \mathrm{d}z)$$

$$+\int_t^s \int_{\mathbb{R}_0} \left\{\ln\left(1+\frac{\partial \gamma}{\partial x}(r, X(r), \pi, \theta, z)\right)\right.$$

$$\left.-\frac{\partial \gamma}{\partial x}(r, X(r), \pi, \theta, z)\right\}\nu(\mathrm{d}z)\mathrm{d}r\Big), \tag{22.14}$$

$$F_i(t,s) := \frac{\partial H_i^0}{\partial x}(s)G(t,s), \tag{22.15}$$

$$p_i(t) = K_i(t) + \int_t^T \frac{\partial H_i^0}{\partial x}(s, X(s), \pi_0(s), \pi_1(s,z), \theta_0(s), \theta_1(s,z))G(t,s)\mathrm{d}s, \tag{22.16}$$

$$q_i(t) = D_t p_i(t), \tag{22.17}$$

$$r_i(t,z) = D_{t,z} p_i(t), \tag{22.18}$$

all exist for $0 \le t \le s$, $z \in \mathbb{R}_0$.

We now define the *Hamiltonians* for this general problem as follows:

Definition 2.1 (The General Stochastic Hamiltonian). The general stochastic Hamiltonians for the stochastic differential game in Problem 1.1 are the functions

$$H_i(t, x, \pi, \theta, \omega) : [0,T] \times \mathbb{R} \times U \times K \times \Omega \to \mathbb{R}, \quad i = 1, 2,$$

defined by

$$H_i(t, x, \pi, \theta, \omega)$$

$$= \int_{\mathbb{R}_0} f_i(t, x, \pi, \theta, z, \omega)\mu(\mathrm{d}z) + p_i(t)b(t, x, \pi_0, \theta_0, \omega) + q_i(t)\sigma(t, x, \pi_0, \theta_0, \omega)$$

$$+\int_{\mathbb{R}_0} r_i(t,z)\gamma(t, x, \pi, \theta, z, \omega)\nu(\mathrm{d}z), \quad i = 1, 2, \tag{22.19}$$

where $\pi = (\pi_0, \pi_1)$ and $\theta = (\theta_0, \theta_1)$.

Remark 2.1. In the classical case, the Hamiltonian $H_i^* : [0,T] \times \mathbb{R} \times U \times K \times \mathbb{R} \times \mathbb{R} \times \mathcal{R} \to \mathbb{R}$ is defined by

$$H_i^*(t,x,\pi,\theta,p,q,r) = \int_{\mathbb{R}_0} f_i(t,x,\pi,\theta)\mu(dz) + p_i\, b(t,x,\pi_0,\theta_0) + q_i\, \sigma(t,x,\pi_0,\theta_0) + \int_{\mathbb{R}_0} r_i(t,z)\gamma(t,x,\pi,\theta,z)\nu(dz), \qquad (22.20)$$

where $\mathcal{R}$ is the set of functions $r_i : \mathbb{R} \times \mathbb{R}_0 \to \mathbb{R}$; $i = 1,2$; see [12]. Thus the relation between H_i^* and H_i is that

$$H_i(t,x,\pi,\theta,\omega) = H_i^*(t,x,\pi,\theta,p(t),q(t),r(t,\cdot)), \quad i = 1,2, \qquad (22.21)$$

where $p(\cdot), q(\cdot)$ and $r(\cdot,\cdot)$ are given by Eqs. (22.16)–(22.18).

Theorem 2.1 (Maximum principle for non-zero-sum games).

(i) Let $(\hat{\pi},\hat{\theta}) \in \mathcal{A}_\Pi \times \mathcal{A}_\Theta$ be a Nash equilibrium with corresponding state process $\hat{X}(t) = X^{(\hat{\pi},\hat{\theta})}(t)$, i.e.,

$$J_1(\pi,\hat{\theta}) \le J_1(\hat{\pi},\hat{\theta}), \qquad \textit{for all } \pi \in \mathcal{A}_\Pi,$$
$$J_2(\hat{\pi},\theta) \le J_2(\hat{\pi},\hat{\theta}), \qquad \textit{for all } \theta \in \mathcal{A}_\Theta.$$

Assume that the random variables $\frac{\partial f_i}{\partial x}$ and $F_i(t,s)$, $i = 1,2$, belong to $\mathbb{D}_{1,2}$. Then

$$\mathbb{E}^x[\nabla_\pi \hat{H}_1(t, X^{(\pi,\hat{\theta})}(t), \pi, \hat{\theta}, \omega)|_{\pi=\hat{\pi}} \,|\mathcal{E}_t] = 0, \qquad (22.22)$$
$$\mathbb{E}^x[\nabla_\theta \hat{H}_2(t, X^{(\hat{\pi},\theta)}(t), \hat{\pi}, \theta, \omega)|_{\theta=\hat{\theta}} \,|\mathcal{E}_t] = 0, \qquad (22.23)$$

for a.a. t,ω.

(ii) Conversely, suppose that there exists $(\hat{\pi},\hat{\theta}) \in \mathcal{A}_\Pi \times \mathcal{A}_\Theta$ such that Eqs. (22.22) and (22.23) hold. Then

$$\frac{\partial}{\partial y} J_1(\hat{\pi} + y\beta, \hat{\theta})\Big|_{y=0} = 0 \quad \textit{for all } \beta,$$
$$\frac{\partial}{\partial \upsilon} J_2(\hat{\pi}, \hat{\theta} + \upsilon\eta)\Big|_{\upsilon=0} = 0 \quad \textit{for all } \eta.$$

In particular, if

$$\pi \to J_1(\pi,\hat{\theta}) \qquad \textit{and} \qquad \theta \to J_2(\hat{\pi},\theta), \qquad (22.24)$$

are concave, then $(\hat{\pi},\hat{\theta})$ is a Nash equilibrium.

Proof. (i) Suppose $(\hat{\pi},\hat{\theta}) \in \mathcal{A}_\Pi \times \mathcal{A}_\Theta$ is a Nash equilibrium. Since (i) and (ii) hold for all π and θ, $(\hat{\pi},\hat{\theta})$ is a directional critical point for $J_i(\pi,\theta)$, $i = 1,2$, in the sense that for all bounded $\beta \in \mathscr{A}_\Pi$ and $\eta \in \mathscr{A}_\Theta$, there exists $\delta > 0$ such that $\hat{\pi} + y\beta \in \mathscr{A}_\Pi$, $\hat{\theta} + \upsilon\eta \in \mathscr{A}_\Theta$ for all $y, \upsilon \in (-\delta, \delta)$. Then we have

$$
\begin{aligned}
&= \frac{\partial}{\partial y} J_1\left(\hat{\pi} + y\beta, \hat{\theta}\right)\Big|_{y=0} \\
&= \mathbb{E}^x\Bigg[\int_0^T\int_{\mathbb{R}_0}\Big\{\frac{\partial f_1}{\partial x}(t, \hat{X}(t), \hat{\pi}_0(t), \hat{\pi}_1(t,z), \hat{\theta}_0(t), \hat{\theta}_1(t,z), z)\frac{\mathrm{d}}{\mathrm{d}y}X^{(\hat{\pi}+y\beta,\hat{\theta})}(t)\Big|_{y=0} \\
&\quad +\nabla_\pi f_1(t, X^{(\pi,\hat{\theta})}(t), \pi_0(t), \pi_1(t,z), \hat{\theta}_0(t), \hat{\theta}_1(t,z), z)\Big|_{\pi=\hat{\pi}}\beta^*(t)\Big\}\mu(\mathrm{d}z)\mathrm{d}t \\
&\quad +g_1'(\hat{X}(T))\frac{\mathrm{d}}{\mathrm{d}y}X^{(\hat{\pi}+y\beta,\hat{\theta})}(T)\Big|_{y=0}\Bigg] \\
&= \mathbb{E}^x\Big[\int_0^T\int_{\mathbb{R}_0}\Big\{\frac{\partial f_1}{\partial x}(t, \hat{X}(t), \hat{\pi}_0(t), \hat{\pi}_1(t,z), \hat{\theta}_0(t), \hat{\theta}_1(t,z), z)Y(t) \\
&\quad +\nabla_\pi f_1(t, X^{(\pi,\hat{\theta})}(t), \pi_0(t), \pi_1(t,z), \hat{\theta}_0(t), \hat{\theta}_1(t,z), z)\Big|_{\pi=\hat{\pi}}\beta^*(t)\Big\} \\
&\quad \times\mu(\mathrm{d}z)\mathrm{d}t + g_1'(\hat{X}(T))Y(T)\Big],
\end{aligned} \tag{22.25}
$$

where

$$
\begin{aligned}
Y(t) =& Y^{(\beta)}(t) = \frac{\mathrm{d}}{\mathrm{d}y}X^{(\hat{\pi}+y\beta,\hat{\theta})}(t)|_{y=0} \\
=& \int_0^t\Big\{\frac{\partial b}{\partial x}(s, \hat{X}(s), \hat{\pi}_0(s), \hat{\theta}_0(s))Y(s) \\
& + \nabla_\pi b(s, X^{(\pi,\hat{\theta})}(s), \pi_0(s), \hat{\theta}_0(s))\Big|_{\pi=\hat{\pi}}\beta^*(s)\Big\}\mathrm{d}s \\
& + \int_0^t\Big\{\frac{\partial \sigma}{\partial x}(s, \hat{X}(s), \hat{\pi}_0(s), \hat{\theta}_0(s))Y(s) \\
& + \nabla_\pi \sigma(s, X^{(\pi,\hat{\theta})}(s), \pi_0(s), \hat{\theta}_0(s))\Big|_{\pi=\hat{\pi}}\beta^*(s)\Big\}\mathrm{d}B(s) \\
& + \int_0^t\int_{\mathbb{R}_0}\Big\{\frac{\partial \gamma}{\partial x}(s, \hat{X}(s^-), \hat{\pi}(s^-), \hat{\theta}(s^-), z)Y(s) \\
& + \nabla_\pi \gamma(s, X^{(\pi,\hat{\theta})}(s^-), \pi(s^-), \hat{\theta}(s^-), z)\Big|_{\pi=\hat{\pi}}\beta^*(s)\Big\}\tilde{N}(\mathrm{d}s, \mathrm{d}z).
\end{aligned} \tag{22.26}
$$

If we use the shorthand notation

$$
\frac{\partial f_1}{\partial x}(t, \hat{X}(t), \hat{\pi}, \hat{\theta}, z) = \frac{\partial f_1}{\partial x}(t,z), \quad \nabla_\pi f_1(t, X^{(\pi,\hat{\theta})}(t), \pi, \hat{\theta}, z)|_{\pi=\hat{\pi}} = \nabla_\pi f_1(t,z),
$$

and similarly for $\frac{\partial b}{\partial x}$, $\nabla_\pi b$, $\frac{\partial \sigma}{\partial x}$, $\nabla_\pi \sigma$, $\frac{\partial \gamma}{\partial x}$ and $\nabla_\pi \gamma$, we can write

$$\begin{cases} \mathrm{d}Y(t) = (\frac{\partial b}{\partial x}(t)Y(t)+\nabla_\pi b(t)\beta^*(t))\mathrm{d}t+(\frac{\partial \sigma}{\partial x}(t)Y(t) + \nabla_\pi \sigma(t)\beta^*(t))\mathrm{d}B(t) \\ \qquad\quad + \int_{\mathbb{R}_0}(\frac{\partial \gamma}{\partial x}(t)Y(t) + \nabla_\pi \gamma(t,z)\beta^*(t))\tilde{N}(\mathrm{d}t,\mathrm{d}z); \\ Y(0) \qquad = 0. \end{cases} \tag{22.27}$$

By the duality formulas (22.7) and (22.8) and the Fubini theorem, we get

$$\begin{aligned} &\mathbb{E}^x\Big[\int_0^T \int_{\mathbb{R}_0} \frac{\partial f_1}{\partial x}(t,z)Y(t)\mu(\mathrm{d}z)\mathrm{d}t\Big] \\ &\quad = \mathbb{E}^x\Big[\int_0^T \int_{\mathbb{R}_0} \Big\{\int_0^t \Big(\frac{\partial f_1}{\partial x}(t,z)\Big[\frac{\partial b}{\partial x}(s)Y(s) + \nabla_\pi b(s)\beta^*(s)\Big] \\ &\quad + D_s\frac{\partial f_1}{\partial x}(t,z)\Big[\frac{\partial \sigma}{\partial x}(s)Y(s) + \nabla_\pi \sigma(s)\beta^*(s)\Big] \\ &\quad + \int_{\mathbb{R}_0} D_{s,z_1}\frac{\partial f_1}{\partial x}(t,z)\Big[\frac{\partial \gamma}{\partial x}(s,z_1)Y(s) \\ &\quad + \nabla_\pi \gamma(s,z_1)\beta^*(s)\Big]\nu(\mathrm{d}z_1)\Big)\mathrm{d}s\Big\}\mu(\mathrm{d}z)\mathrm{d}t\Big] \\ &\quad = \mathbb{E}^x\Big[\int_0^T \Big\{\Big(\int_s^T \int_{\mathbb{R}_0} \frac{\partial f_1}{\partial x}(t,z)\mu(\mathrm{d}z)\mathrm{d}t\Big)\Big[\frac{\partial b}{\partial x}Y(s) + \nabla_\pi b(s)\beta^*(s)\Big] \\ &\quad + \Big(\int_s^T \int_{\mathbb{R}_0} D_s\frac{\partial f_1}{\partial x}(t,z)\mu(\mathrm{d}z)\mathrm{d}t\Big)\Big[\frac{\partial \sigma}{\partial x}Y(s) + \nabla_\pi \sigma\beta^*(s)\Big] \\ &\quad + \int_{\mathbb{R}_0}\Big(\int_s^T \int_{\mathbb{R}_0} D_{s,z_1}\frac{\partial f_1}{\partial x}(t,z)\mu(\mathrm{d}z)\mathrm{d}t\Big) \\ &\quad \times\Big[\frac{\partial \gamma}{\partial x}Y(s) + \nabla_\pi \gamma\beta^*(s)\Big]\nu(\mathrm{d}z_1)\Big\}\mathrm{d}s\Big]. \end{aligned} \tag{22.28}$$

Changing notation $s \to t$ and $z_1 \to z$ this becomes

$$\begin{aligned} &\mathbb{E}^x\Big[\int_0^T \int_{\mathbb{R}_0} \frac{\partial f_1}{\partial x}(t,z)Y(t)\mu(\mathrm{d}z)\mathrm{d}t\Big] \\ &\quad = \mathbb{E}^x\Big[\int_0^T \Big\{\Big(\int_t^T \int_{\mathbb{R}_0} \frac{\partial f_1}{\partial x}(s,z_1)\mu(\mathrm{d}z_1)\mathrm{d}s\Big)\Big[\frac{\partial b}{\partial x}(t)Y(t) + \nabla_\pi b(t)\beta^*(t)\Big] \\ &\quad + \Big(\int_t^T \int_{\mathbb{R}_0} D_t\frac{\partial f_1}{\partial x}(s,z_1)\mu(\mathrm{d}z_1)\mathrm{d}s\Big)\Big[\frac{\partial \sigma}{\partial x}(t)Y(t) + \nabla_\pi \sigma(t)\beta^*(t)\Big] \end{aligned}$$

$$+\int_{\mathbb{R}_0}\Big(\int_t^T\int_{\mathbb{R}_0} D_{t,z}\frac{\partial f_1}{\partial x}(s,z_1)\mu(\mathrm{d}z_1)\mathrm{d}s\Big)$$
$$\times\Big[\frac{\partial\gamma}{\partial x}(t,z)Y(t)+\nabla_\pi\gamma(t,z)\beta^*(t)\Big]\nu(\mathrm{d}z)\Big\}\mathrm{d}t\Big]. \tag{22.29}$$

On the other hand, by the duality formulas (22.7) and (22.8), we get

$$\begin{aligned}\mathbb{E}^x\Big[g_1'(\hat{X}(T))Y(T)\Big] &= \mathbb{E}^x\Big[g_1'(\hat{X}(T))\Big(\textstyle\int_0^T\{\frac{\partial b}{\partial x}(t)Y(t)+\nabla_\pi b(t)\beta^*(t)\}\,\mathrm{d}t\\
&\quad+\textstyle\int_0^T\{\frac{\partial\sigma}{\partial x}(t)Y(t)+\nabla_\pi\sigma(t)\beta^*(t)\}\,\mathrm{d}B(t)\\
&\quad+\textstyle\int_0^T\int_{\mathbb{R}_0}\Big\{\frac{\partial\gamma}{\partial x}(t,z)Y(t)+\nabla_\pi\gamma(t,z)\beta(t)\Big\}\,\tilde{N}(\mathrm{d}t,\mathrm{d}z)\Big)\Big]\\
&= \mathbb{E}^x\Big[\textstyle\int_0^T\Big\{g_1'(\hat{X}(T))\frac{\partial b}{\partial x}(t)Y(t)+g_1'(\hat{X}(T))\nabla_\pi b(t)\beta^*(t)\\
&\quad+D_t(g_1'(\hat{X}(T)))\frac{\partial\sigma}{\partial x}(t)Y(t)+D_t(g_1'(\hat{X}(T)))\nabla_\pi\sigma(t)\beta^*(t)\\
&\quad+\textstyle\int_{\mathbb{R}_0}[D_{t,z}(g_1'(\hat{X}(T)))\frac{\partial\gamma}{\partial x}(t,z)Y(t)\\
&\quad+D_{t,z}(g_1'(\hat{X}(T)))\nabla_\pi\gamma(t,z)\beta^*(t)]\nu(\mathrm{d}z)\Big\}\mathrm{d}t\Big].\end{aligned}$$

We recall that

$$\hat{K}_1(t):=g_1'(\hat{X}(T))+\int_t^T\int_{\mathbb{R}_0}\frac{\partial f_1}{\partial x}(s,z_1)\mu(\mathrm{d}z_1)\,\mathrm{d}s,$$

and combining Eqs. (22.27)–(22.29), we get

$$\begin{aligned}\mathbb{E}^x\Big[\int_0^T&\Big\{\hat{K}_1(t)\Big(\frac{\partial b}{\partial x}(t)Y(t)+\nabla_\pi b(t)\beta^*(t)\Big)\\
&+D_t\hat{K}_1(t)\Big(\frac{\partial\sigma}{\partial x}(t)Y(t)+\nabla_\pi\sigma(t)\beta^*(t)\Big)\\
&+\int_{\mathbb{R}_0}D_{t,z}\hat{K}_1(t)\Big(\frac{\partial\gamma}{\partial x}(t,z)Y(t)+\nabla_\pi\gamma(t,z)\beta^*(t)\Big)\nu(\mathrm{d}z)\\
&+\int_{\mathbb{R}_0}\nabla_\pi f_1(t,z)\beta^*(t)\mu(\mathrm{d}z)\Big\}\mathrm{d}t\Big]=0.\end{aligned} \tag{22.30}$$

Now apply this to $\beta=\beta_\alpha\in\mathscr{A}_\Pi$ of the form $\beta_\alpha(s)=\alpha\chi_{[t,t+h]}(s)$, for some $t,h\in(0,T), t+h\le T$, where $\alpha=\alpha(\omega)$ is bounded and $\mathscr{E}_t$-measurable. Then $Y^{(\beta_\alpha)}(s)=0$ for $0\le s\le t$. Hence Eq. (22.30) becomes

$$A_1+A_2=0, \tag{22.31}$$

where

$$A_1 = \mathbb{E}^x\Big[\int_t^T \Big\{\hat{K}_1(s)\frac{\partial b}{\partial x}(s) + D_s\hat{K}_1(s)\frac{\partial \sigma}{\partial x}(s) + \int_{\mathbb{R}_0} D_{s,z}\hat{K}_1(s)\frac{\partial \gamma}{\partial x}(s)\nu(\mathrm{d}z)\Big\}Y^{(\beta_\alpha)}(s)\mathrm{d}s\Big],$$

$$A_2 = \mathbb{E}^x\Big[\Big\{\int_t^{t+h}\Big(\hat{K}_1(s)\nabla_\pi b(s) + D_s\hat{K}_1(s)\nabla_\pi\sigma(s) + \int_{\mathbb{R}_0} D_{s,z}\hat{K}_1(s)\nabla_\pi\gamma(s,z)\nu(\mathrm{d}z) + \int_{\mathbb{R}_0}\nabla_\pi f_1(s,z)\mu(\mathrm{d}z)\Big)\mathrm{d}s\Big\}\alpha\Big].$$

Note that, by Eq. (22.26), with $Y(s) = Y^{(\beta_\alpha)}(s)$ and $s \geq t + h$,

$$\mathrm{d}Y(s) = Y(s^-)\Big\{\frac{\partial b}{\partial x}(s)\mathrm{d}s + \frac{\partial \sigma}{\partial x}(s)\mathrm{d}B(s) + \int_{\mathbb{R}_0}\frac{\partial \gamma}{\partial x}(s^-,z)\tilde{N}(\mathrm{d}s,\mathrm{d}z)\Big\},$$

for $s \geq t + h$. Hence, by the Itô formula,

$$Y(s) = Y(t+h)G(t+h,s); \qquad s \geq t+h, \tag{22.32}$$

where, in general, for $s \geq t$,

$$G(t,s) = \exp\Big(\int_t^s\Big\{\frac{\partial b}{\partial x}(r) - \tfrac{1}{2}\Big(\frac{\partial \sigma}{\partial x}\Big)^2(r)\Big\}\mathrm{d}r + \int_t^s\frac{\partial \sigma}{\partial x}(r)\mathrm{d}B(r) + \int_t^s\int_{\mathbb{R}_0}\ln\Big(1+\frac{\partial \gamma}{\partial x}(r^-,z)\Big)\tilde{N}(\mathrm{d}r,\mathrm{d}z) + \int_t^s\int_{\mathbb{R}_0}\Big\{\ln\Big(1+\frac{\partial \gamma}{\partial x}(r,z)\Big) - \frac{\partial \gamma}{\partial x}(r,z)\Big\}\nu(\mathrm{d}z)\mathrm{d}r\Big). \tag{22.33}$$

Note that $G(t,s)$ does not depend on h. Put

$$H_1^0(s,x,\pi,\theta) = K_1(s)b(s,x,\pi_0,\theta_0) + D_sK_1(s)\sigma(s,x,\pi_0,\theta_0) + \int_{\mathbb{R}_0} D_{s,z}K_1(s)\gamma(s,x,\pi,\theta,z)\nu(\mathrm{d}z), \tag{22.34}$$

and $\hat{H}_1^0(s) = H_1^0(s,\hat{X}(s),\hat{\pi},\hat{\theta})$. Then

$$A_1 = \mathbb{E}^x\Big[\int_t^T\frac{\partial \hat{H}_1^0}{\partial x}(s)Y(s)\mathrm{d}s\Big].$$

Differentiating with respect to h at $h = 0$ we get

$$\frac{\mathrm{d}}{\mathrm{d}h}A_1\big|_{h=0} = \frac{\mathrm{d}}{\mathrm{d}h}\mathbb{E}^x\Big[\int_t^{t+h}\frac{\partial \hat{H}_1^0}{\partial x}(s)Y(s)\mathrm{d}s\Big]_{h=0} + \frac{\mathrm{d}}{\mathrm{d}h}\mathbb{E}^x\Big[\int_{t+h}^{T}\frac{\partial \hat{H}_1^0}{\partial x}(s)Y(s)\mathrm{d}s\Big]_{h=0}. \tag{22.35}$$

Since $Y(t) = 0$ we see that

$$\frac{\mathrm{d}}{\mathrm{d}h}\mathbb{E}^x\Big[\int_t^{t+h}\frac{\partial \hat{H}_1^0}{\partial x}(s)Y(s)\mathrm{d}s\Big]_{h=0} = 0. \tag{22.36}$$

Therefore, by Eq. (22.31),

$$\begin{aligned}\frac{\mathrm{d}}{\mathrm{d}h}A_1\big|_{h=0} &= \frac{\mathrm{d}}{\mathrm{d}h}\mathbb{E}^x\Big[\int_{t+h}^{T}\frac{\partial \hat{H}_1^0}{\partial x}(s)Y(t+h)G(t+h,s)\mathrm{d}s\Big]_{h=0}\\ &= \int_t^T\frac{\mathrm{d}}{\mathrm{d}h}\mathbb{E}^x\Big[\frac{\partial \hat{H}_1^0}{\partial x}(s)Y(t+h)G(t+h,s)\Big]_{h=0}\mathrm{d}s\\ &= \int_t^T\frac{\mathrm{d}}{\mathrm{d}h}\mathbb{E}^x\Big[\frac{\partial \hat{H}_1^0}{\partial x}(s)G(t,s)Y(t+h)\Big]_{h=0}\mathrm{d}s.\end{aligned} \tag{22.37}$$

On the other hand, Eq. (22.26) gives

$$\begin{aligned}Y(t+h) =&\alpha\int_t^{t+h}\Big\{\nabla_\pi b(r)\mathrm{d}r + \nabla_\pi\sigma\mathrm{d}B(r) + \int_{\mathbb{R}_0}\nabla_\pi\gamma(r^-,z)\tilde{N}(\mathrm{d}r,\mathrm{d}z)\Big\}\\ &+\int_t^{t+h}Y(r^-)\Big\{\frac{\partial b}{\partial x}(r)\mathrm{d}r + \frac{\partial\sigma}{\partial x}(r)\mathrm{d}B(r) + \int_{\mathbb{R}_0}\frac{\partial\gamma}{\partial x}(r^-,z)\tilde{N}(\mathrm{d}r,\mathrm{d}z)\Big\}.\end{aligned} \tag{22.38}$$

Combining this with Eqs. (22.36) and (22.37), we have

$$\frac{\mathrm{d}}{\mathrm{d}h}A_1\big|_{h=0} = \Lambda_1 + \Lambda_2, \tag{22.39}$$

where

$$\begin{aligned}\Lambda_1 = \int_t^T\frac{\mathrm{d}}{\mathrm{d}h}\mathbb{E}^x\Big[\frac{\partial \hat{H}_1^0}{\partial x}(s)G(t,s)\alpha\int_t^{t+h}\Big\{\nabla_\pi b(r)\mathrm{d}r + \nabla_\pi\sigma(r)\mathrm{d}B(r)\\ + \int_{\mathbb{R}_0}\nabla_\pi\gamma(r^-,z)\tilde{N}(\mathrm{d}r,\mathrm{d}z)\Big\}\Big]_{h=0}\mathrm{d}s\end{aligned} \tag{22.40}$$

and

$$\Lambda_2 = \int_t^T \frac{\mathrm{d}}{\mathrm{d}h}\mathbb{E}^x\Big[\frac{\partial \hat{H}_1^0}{\partial x}(s)G(t,s)\int_t^{t+h} Y(r^-)\Big\{\frac{\partial b}{\partial x}(r)\mathrm{d}r + \frac{\partial \sigma}{\partial x}(r)\mathrm{d}B(r)$$
$$+ \int_{\mathbb{R}_0} \frac{\partial \gamma}{\partial x}(r^-,z)\tilde{N}(\mathrm{d}r,\mathrm{d}z)\Big\}\Big]_{h=0}\mathrm{d}s. \tag{22.41}$$

By the duality formulae (22.7) and (22.8), we have

$$\Lambda_1 = \int_t^T \frac{\mathrm{d}}{\mathrm{d}h}\mathbb{E}^x\Big[\alpha\int_t^{t+h}\Big\{\nabla_\pi b(r)F_1(t,s) + \nabla_\pi \sigma(r)D_r F_1(t,s)$$
$$+ \int_{\mathbb{R}_0}\nabla_\pi \gamma(r,z)D_{r,z}F_1(t,s)\nu(\mathrm{d}z)\Big\}\mathrm{d}r\Big]_{h=0}\mathrm{d}s$$
$$= \int_t^T \mathbb{E}^x\Big[\alpha\Big\{\nabla_\pi b(t)F_1(t,s) + \nabla_\pi \sigma(t)D_t F_1(t,s)$$
$$+ \int_{\mathbb{R}_0}\nabla_\pi \gamma(t,z)D_{t,z}F_1(t,s)\nu(\mathrm{d}z)\Big\}\Big]\mathrm{d}s. \tag{22.42}$$

Since $Y(t) = 0$ we see that

$$\Lambda_2 = 0. \tag{22.43}$$

We conclude that

$$\frac{\mathrm{d}}{\mathrm{d}h}A_1\Big|_{h=0} = \Lambda_1$$
$$= \int_t^T \mathbb{E}^x\Big[\alpha\Big\{F_1(t,s)\nabla_\pi b(t) + D_t F_1(t,s)\nabla_\pi \sigma(t)$$
$$+ \int_{\mathbb{R}_0} D_{t,z}F_1(t,s)\nabla_\pi \gamma(t,z)\nu(\mathrm{d}z)\Big\}\Big]\mathrm{d}s. \tag{22.44}$$

Moreover, we see directly that

$$\frac{\mathrm{d}}{\mathrm{d}h}A_2\Big|_{h=0} = \mathbb{E}^x\Big[\alpha\Big\{\hat{K}_1(t)\nabla_\pi b(t) + D_t\hat{K}_1(t)\nabla_\pi \sigma(t)$$
$$+ \int_{\mathbb{R}_0}\{D_{t,z}\hat{K}_1(t)\nabla_\pi \gamma(t,z) + \nabla_\pi f_1(t,z)\}\nu(\mathrm{d}z)\Big\}\Big]. \tag{22.45}$$

Therefore, differentiating Eq. (22.30) with respect to h at $h = 0$ gives the equation

$$\mathbb{E}^x\Big[\alpha\Big\{\Big(\hat{K}_1(t) + \int_t^T F_1(t,s)\mathrm{d}s\Big)\nabla_\pi b(t) + D_t\Big(\hat{K}_1(t) + \int_t^T F_1(t,s)\mathrm{d}s\Big)\nabla_\pi \sigma(t)$$
$$+ \int_{\mathbb{R}_0} D_{t,z}\Big(\hat{K}_1(t) + \int_t^T F_1(t,s)\mathrm{d}s\Big)\nabla_\pi \gamma(t,z) + \nabla_\pi f_1(t,z)\nu(\mathrm{d}z)\Big\}\Big] = 0. \tag{22.46}$$

We can reformulate this as follows: If we define, as in Eq. (22.16),

$$\hat{p}_1(t) = \hat{K}_1(t) + \int_t^T F_1(t,s)\mathrm{d}s = \hat{K}_1(t) + \int_t^T \frac{\partial \hat{H}_1^0}{\partial x}(s)G(t,s)\mathrm{d}s, \quad (22.47)$$

then Eq. (22.44) can be written:

$$\mathbb{E}^x\Big[\nabla_\pi\Big\{\int_{\mathbb{R}_0} f_1(t,\hat{X}(t),\pi,\hat{\theta},z)\mu(\mathrm{d}z) + \hat{p}_1(t)b(t,\hat{X}(t),\pi_0,\hat{\theta}_0)$$
$$+ D_t\hat{p}_1(t)\sigma(t,\hat{X}(t),\pi_0,\hat{\theta}_0)$$
$$+ \int_{\mathbb{R}_0} D_{t,z}\hat{p}_1(t)\gamma(t,\hat{X}(t),\pi,\hat{\theta},z)\nu(\mathrm{d}z)\Big\}_{\pi=(\pi_0(t),\pi_1(t,z))}\alpha\Big] = 0.$$

Since this holds for all bounded $\mathcal{E}_t$-measurable random variable α, we conclude that

$$\mathbb{E}^x\Big[\nabla_\pi \hat{H}_1(t, X^{(\pi,\hat{\theta})}(t),\pi,\hat{\theta}) \mid_{\pi=\hat{\pi}(t)} \mid \mathcal{E}_t\Big] = 0.$$

Similarly, we have

$$0 = \frac{\partial}{\partial \upsilon} J_2(\hat{\pi},\hat{\theta}+\upsilon\eta)\Big|_{\upsilon=0}$$
$$= \mathbb{E}^x\Big[\int_0^T \int_{\mathbb{R}_0} \Big\{\frac{\partial f_2}{\partial x}(t, X^{(\hat{\pi},\theta)}(t),\hat{\pi}(t,z),\hat{\theta}(t,z),z)D(t)$$
$$+\nabla_\theta f_2(t,X^{(\hat{\pi},\theta)}(t),\hat{\pi}(t,z),\theta(t,z),z)\Big|_{\theta=\hat{\theta}}\eta(t)\Big\}\mu(\mathrm{d}z)\mathrm{d}t + g_2'(\hat{X}(T))D(T)\Big], \quad (22.48)$$

where

$$D(t) = D^{(\eta)}(t) = \frac{\mathrm{d}}{\mathrm{d}\upsilon} X^{(\hat{\pi},\hat{\theta}+\upsilon\eta)}(t)\Big|_{\upsilon=0}$$
$$= \int_0^t \Big\{\frac{\partial b}{\partial x}(s,\hat{X}(s),\hat{\pi}_0(s),\hat{\theta}_0(s))D(s)$$
$$+ \nabla_\theta b(s, X^{(\hat{\pi}_0,\theta_0)}(s),\hat{\pi}_0(s),\theta_0(s))\Big|_{\theta=\hat{\theta}}\eta^*(s)\Big\}\mathrm{d}s$$
$$+ \int_0^t \Big\{\frac{\partial \sigma}{\partial x}(s,\hat{X}(s),\hat{\pi}_0(s),\hat{\theta}_0(s))Y(s)$$
$$+ \nabla_\theta \sigma(s, X^{(\hat{\pi},\theta)}(s),\hat{\pi}_0(s),\theta_0(s))\Big|_{\theta=\hat{\theta}}\eta^*(s)\Big\}\mathrm{d}B(s)$$
$$+ \int_0^t \int_{\mathbb{R}_0} \Big\{\frac{\partial \gamma}{\partial x}(s,\hat{X}(s^-),\hat{\pi}(s^-,z),\hat{\theta}(s^-),z)D(s)$$
$$+ \nabla_\theta \gamma(s, X^{(\hat{\pi},\theta)}(s^-),\hat{\pi}(s^-,z),\theta(s^-,z),z)\Big|_{\theta=\hat{\theta}}\eta^*(s)\Big\}\tilde{N}(\mathrm{d}s,\mathrm{d}z). \quad (22.49)$$

Define

$$D(s) = D(t+h)G(t+h,s); \quad s \geq t+h,$$

where $G(t,s)$ is defined as in Eq. (22.32). By using similar arguments as above, we get

$$\mathbb{E}^x\Big[\nabla_\theta \hat{H}_2(t, X^{(\hat{\pi},\theta)}(t), \hat{\pi}, \theta) \mid_{\theta=\hat{\theta}(t)} \mid \mathscr{E}_t\Big] = 0.$$

This completes the proof of (i).

(ii) Conversely, suppose that there exists $(\hat{\pi}, \hat{\theta}) \in \mathscr{A}_\Pi \times \mathscr{A}_\Theta$ such that Eqs. (22.22) and (22.23) hold. Then by reversing the above arguments, we obtain that Eq. (22.31) holds for all $\beta_\alpha(s,\omega) = \alpha(\omega)\chi_{(t,t+h]}(s) \in \mathscr{A}_\Pi$, where

$$\begin{aligned} A_1 =& \mathbb{E}^x\Big[\int_t^T \Big\{\hat{K}_1(s)\frac{\partial b}{\partial x}(s) + D_s\hat{K}_1(s)\frac{\partial \sigma}{\partial x}(s) \\ &+ \int_{\mathbb{R}_0} D_{s,z}\hat{K}_1(s)\frac{\partial \gamma}{\partial x}(s)\nu(\mathrm{d}z)\Big\}Y^{(\beta_\alpha)}(s)\mathrm{d}s\Big], \end{aligned}$$

$$\begin{aligned} A_2 =& \mathbb{E}^x\Big[\Big\{\int_t^{t+h} \Big(\hat{K}_1(s)\nabla_\pi b(s) + D_s\hat{K}_1(s)\nabla_\pi\sigma(s) \\ &+ \int_{\mathbb{R}_0} D_{s,z}\hat{K}_1(s)\nabla_\pi\gamma(s,z)\nu(\mathrm{d}z) + \int_{\mathbb{R}_0} \nabla_\pi f_1(s,z)\mu(\mathrm{d}z)\Big)\mathrm{d}s\Big\}\alpha\Big], \end{aligned}$$

for some $t, h \in [0,T]$ with $t+h \leq T$ and some bounded $\mathcal{E}_t$-measurable α. Similarly,

$$A_3 + A_4 = 0 \tag{22.50}$$

for all $\eta_\xi(s,\omega) = \xi(\omega)\chi_{(t,t+h]}(s) \in \mathscr{A}_\Theta$, where

$$\begin{aligned} A_3 =& \mathbb{E}^x\Big[\int_t^T \Big\{\hat{K}_2(s)\frac{\partial b}{\partial x}(s) + D_s\hat{K}_2(s)\frac{\partial \sigma}{\partial x}(s) \\ &+ \int_{\mathbb{R}_0} D_{s,z}\hat{K}_2(s)\frac{\partial \gamma}{\partial x}(s)\nu(\mathrm{d}z)\Big\}Y^{(\eta_\xi)}(s)\mathrm{d}s\Big], \end{aligned}$$

$$\begin{aligned} A_4 =& \mathbb{E}^x\Big[\Big\{\int_t^{t+h} \Big(\hat{K}_2(s)\nabla_\theta b(s) + D_s\hat{K}_2(s)\nabla_\theta\sigma(s) \\ &+ \int_{\mathbb{R}_0} D_{s,z}\hat{K}_2(s)\nabla_\theta\gamma(s,z)\nu(\mathrm{d}z) + \int_{\mathbb{R}_0} \nabla_\theta f_2(s,z)\mu(\mathrm{d}z)\Big)\mathrm{d}s\Big\}\alpha\Big], \end{aligned}$$

for some $t, h \in [0,T]$ with $t+h \leq T$ and some bounded $\mathcal{E}_t$-measurable ξ. Hence, these equalities hold for all linear combinations of β_α and η_ξ. Since all bounded $\beta \in \mathscr{A}_\Pi$ and $\eta \in \mathscr{A}_\Theta$ can be approximated pointwise boundary in

(t, ω) by such linear combinations, it follows that Eqs. (22.31) and (22.50) hold for all bounded $(\beta, \eta) \in \mathscr{A}_\Pi \times \mathscr{A}_\Theta$. Hence, by reversing the remaining part of the proof above, we conclude that

$$\frac{\partial}{\partial y} J_1(\hat{\pi} + y\beta, \hat{\theta})\Big|_{y=0} = 0,$$

$$\frac{\partial}{\partial \upsilon} J_2(\hat{\pi}, \hat{\theta} + \upsilon\eta)\Big|_{\upsilon=0} = 0,$$

for all β and η.

□

3 Zero-Sum Games

Suppose that the given performance functional of player I is the negative of the player II, i.e.,

$$\begin{aligned} J_1(u_0, u_1) &= \mathbb{E}^x\Big[\int_0^T \int_{\mathbb{R}_0} f(t, X(t), u_0(t), u_1(t,z), z, \omega)\mu(\mathrm{d}z)\mathrm{d}t + g(X(T), \omega)\Big] \\ &= -J_2(u_0, u_1), \end{aligned} \tag{22.51}$$

where $\mathbb{E}^x = \mathbb{E}_P^x$ denotes the expectation with respect to P given that $X(0) = x$. Suppose that the controls $u_0(t)$ and $u_1(t,z)$ have the form as in Eqs. (22.5) and (22.6). Let $\mathscr{A}_\Pi$ and $\mathscr{A}_\Theta$ denote the given family of controls $\pi = (\pi_0, \pi_1)$ and $\theta = (\theta_0, \theta_1)$ such that they are contained in the set of càdlàg $\mathscr{E}_t$-adapted controls, Eq. (22.1) has a unique strong solution up to time T and

$$\begin{aligned} \mathbb{E}^x\Big[\int_0^T \int_{\mathbb{R}_0} &|f(t, X(t), \pi_0(t), \pi_1(t,z), \theta_0(t), \theta_1(t,z), z, \omega)|\mu(\mathrm{d}z)\mathrm{d}t \\ &+ |g(X(T), \omega)|\Big] < \infty. \end{aligned} \tag{22.52}$$

Then the partial information zero-sum stochastic differential game problem is the following:

Problem 3.1. Find $\Phi_{\mathscr{E}} \in \mathbb{R}$, $\pi^* \in \mathscr{A}_\Pi$ and $\theta^* \in \mathscr{A}_\Theta$ (if it exists) such that

$$\Phi_{\mathscr{E}} = \inf_{\theta \in \mathscr{A}_\Theta} \big(\sup_{\pi \in \mathscr{A}_\Pi} J(\pi, \theta)\big) = J(\pi^*, \theta^*) = \sup_{\pi \in \mathscr{A}_\Pi} \big(\inf_{\theta \in \mathscr{A}_\Theta} J(\pi, \theta)\big). \tag{22.53}$$

Such a control (π^*, θ^*) is called an *optimal control* (if it exists). The intuitive idea is that while player I controls π, player II controls θ. The actions of the players are antagonistic, which means that between players I and II, there is a payoff $J(\pi, \theta)$

which is a reward for player I and a cost for Player II. Note that since we allow b, σ, γ, f and g to be stochastic processes and also because our controls are $\mathscr{E}_t$-adapted, this problem is not of Markovian type and hence cannot be solved by dynamic programming.

Theorem 3.1 (Maximum principle for zero-sum games).

(i) Suppose $(\hat{\pi}, \hat{\theta}) \in \mathscr{A}_\Pi \times \mathscr{A}_\Theta$ *is a directional critical point for* $J(\pi, \theta)$ *in the sense that for all bounded* $\beta \in \mathscr{A}_\Pi$ *and* $\eta \in \mathscr{A}_\Theta$*, there exists* $\delta > 0$ *such that* $\hat{\pi} + y\beta \in \mathscr{A}_\Pi$, $\hat{\theta} + \upsilon\eta \in \mathscr{A}_\Theta$ *for all* $y, \upsilon \in (-\delta, \delta)$ *and*

$$c(y, \upsilon) := J(\hat{\pi} + y\beta, \hat{\theta} + \upsilon\eta), \quad y, \upsilon \in (-\delta, \delta)$$

has a critical point at 0*, i.e.,*

$$\frac{\partial c}{\partial y}(0,0) = \frac{\partial c}{\partial \upsilon}(0,0) = 0. \tag{22.54}$$

Then

$$\mathbb{E}^x[\nabla_\pi \hat{H}(t, X^{(\pi,\hat{\theta})}(t), \pi, \hat{\theta}, \omega)|\mathscr{E}_t]_{\pi=\hat{\pi}} = 0, \tag{22.55}$$

$$\mathbb{E}^x[\nabla_\theta \hat{H}(t, X^{(\hat{\pi},\theta)}(t), \hat{\pi}, \theta, \omega)|\mathscr{E}_t]_{\theta=\hat{\theta}} = 0 \quad \text{for a.a. } t, \omega, \tag{22.56}$$

where

$$\hat{X}(t) = X^{(\hat{\pi},\hat{\theta})}(t),$$

$$\begin{aligned}\hat{H}(t, \hat{X}(t), \pi, \theta) = &\int_{\mathbb{R}_0} f(t, \hat{X}(t), \pi, \theta, z)\mu(\mathrm{d}z) + \hat{p}(t)b(t, \hat{X}(t), \pi_0, \theta_0)\\ &+ \hat{q}(t)\sigma(t, \hat{X}(t), \pi_0, \theta_0) + \int_{\mathbb{R}_0} \hat{r}(t,z)\gamma(t, \hat{X}(t^-), \pi, \theta, z)\nu(\mathrm{d}z),\end{aligned} \tag{22.57}$$

with

$$\hat{p}(t) = \hat{K}(t) + \int_t^T \frac{\partial \hat{H}^0}{\partial x}(s, \hat{X}(s), \hat{\pi}(s), \hat{\theta}(s))\,\hat{G}(t,s)\mathrm{d}s, \tag{22.58}$$

$$\hat{K}(t) = K^{(\hat{\pi},\hat{\theta})}(t) = g'(\hat{X}(T)) + \int_t^T \int_{\mathbb{R}_0} \frac{\partial f}{\partial x}(s, \hat{X}(s), \hat{\pi}(s,z), \hat{\theta}(s,z), z)\mu(\mathrm{d}z)\mathrm{d}s, \tag{22.59}$$

$$\begin{aligned}\hat{H}^0(s, \hat{X}, \hat{\pi}, \hat{\theta}) = &\hat{K}(s)b(s, \hat{X}, \hat{\pi}_0, \hat{\theta}_0) + D_s\hat{K}(s)\sigma(s, \hat{X}, \hat{\pi}_0, \hat{\theta}_0)\\ &+ \int_{\mathbb{R}_0} D_{s,z}\hat{K}(s)\gamma(s, \hat{X}, \hat{\pi}, \hat{\theta}, z)\nu(\mathrm{d}z),\end{aligned} \tag{22.60}$$

$$\begin{aligned}\hat{G}(t,s) := \exp\Big(\int_t^s \Big\{ \frac{\partial b}{\partial x}(r, \hat{X}(r), \hat{\pi}_0(r), \hat{\theta}_0(r)) \\ - \frac{1}{2}\Big(\frac{\partial \sigma}{\partial x}\Big)^2 (r, \hat{X}(r), \hat{\pi}_0(r), \hat{\theta}_0(r))\Big\} \mathrm{d}r \\ + \int_t^s \frac{\partial \sigma}{\partial x}(r, \hat{X}(r), \hat{\pi}_0(r), \hat{\theta}_0(r)) \mathrm{d}B(r) \\ + \int_t^s \int_{\mathbb{R}_0} \ln\Big(1 + \frac{\partial \gamma}{\partial x}(r, \hat{X}(r^-), \hat{\pi}(r^-, z), \hat{\theta}(r^-, z), z)\Big) \tilde{N}(\mathrm{d}r, \mathrm{d}z) \\ + \int_t^s \int_{\mathbb{R}_0} \Big\{ \ln\Big(1 + \frac{\partial \gamma}{\partial x}(r, \hat{X}(r), \hat{\pi}, \hat{\theta}, z)\Big) \\ - \frac{\partial \gamma}{\partial x}(r, \hat{X}(r), \hat{\pi}, \hat{\theta}, z)\Big\} \nu(\mathrm{d}z)\mathrm{d}r \Big);\end{aligned}$$

$$\hat{q}(t) := D_t \hat{p}(t),$$

and

$$\hat{r}(t,z) := D_{t,z}\hat{p}(t). \qquad (22.61)$$

(*ii*) *Conversely, suppose that there exists* $(\hat{\pi}, \hat{\theta}) \in \mathscr{A}_\Pi \times \mathscr{A}_\Theta$ *such that Eqs. (22.55) and (22.56) hold. Furthermore, suppose that* g *is an affine function,* H *is concave in* π *and convex in* θ*. Then* $(\hat{\pi}, \hat{\theta})$ *satisfies Eq. (22.54).*

4 Application: Worst-Case Scenario Optimal Portfolio Under Partial Information

We illustrate the results in the previous section by looking at an application to robust portfolio choice in finance:

Consider a financial market with the following two investment possibilities:

1. A *risk free asset*, where the unit price $S_0(t)$ at time t is

$$\mathrm{d}S_0(t) = r(t)S_0(t)\mathrm{d}t; \quad S_0(0) = 1; \quad 0 \le t \le T,$$

where $T > 0$ is a given constant.

2. A *risky asset*, where the unit price $S_1(t)$ at time t is given by

$$\begin{cases} \mathrm{d}S_1(t) = S_1(t^-)[\theta(t)\mathrm{d}t + \sigma_0(t)\mathrm{d}B(t) + \int_{\mathbb{R}_0} \gamma_0(t,z)\tilde{N}(\mathrm{d}t, \mathrm{d}z)], \\ S_1(0) > 0, \end{cases} \qquad (22.62)$$

where r, θ, σ_0 and γ_0 are predictable processes such that

$$\int_0^T \{| \theta(s) | + \sigma_0^2(s) + \int_{\mathbb{R}_0} \gamma_0^2(s,z)\nu(dz)\}ds < \infty \quad \text{a.s.}$$

We assume that θ is adapted to a given subfiltration $\mathcal{E}_t$ and that

$$\gamma_0(t,z,\omega) \geq -1 + \delta \qquad \text{for all } t,z,\omega \in [0,T] \times \mathbb{R}_0 \times \Omega,$$

for some constant $\delta > 0$.

Let $\pi(t) = \pi(t,\omega)$ be a *portfolio*, representing the amount invested in the risky asset at time t. We require that π be càdlàg and $\mathcal{E}_t$-adapted and self–financing and hence that the corresponding wealth $X(t) = X^{(\pi,\theta)}(t)$ at time t is given by

$$\begin{cases} dX(t) = [X(t) - \pi(t)]r(t)dt + \pi(t^-)[\theta(t)dt + \sigma_0(t)dB(t) \\ \qquad\qquad + \int_{\mathbb{R}_0} \gamma_0(t,z)\tilde{N}(dt,dz)] \\ X(0) = \ x > 0. \end{cases} \tag{22.63}$$

Let us assume that the mean relative growth rate $\theta(t)$ of the risky asset is not known to the trader, but subject to uncertainty. We may regard θ as a *market scenario* or a *stochastic control* of the market, which is playing against the trader. Let $\mathcal{A}_\Pi^\varepsilon$ and $\mathcal{A}_\Theta^\varepsilon$ denote the set of admissible controls π, θ, respectively. *The worst-case partial information scenario optimal problem* for the trader is to find $\pi^* \in \mathcal{A}_\Pi^\varepsilon$ and $\theta^* \in \mathcal{A}_\Theta^\varepsilon$ and $\Phi \in \mathbb{R}$ such that

$$\begin{aligned} \Phi &= \inf_{\theta \in \mathcal{A}_\Theta^\varepsilon} (\sup_{\pi \in \mathcal{A}_\Pi^\varepsilon} \mathbb{E}[U(X^{(\pi,\theta)}(T))]) \\ &= \mathbb{E}[U(X^{(\pi^*,\theta^*)(T)})], \end{aligned} \tag{22.64}$$

where $U : [0,\infty) \to \mathbb{R}$ is a given utility function, assumed to be concave, strictly increasing and C^1 on $(0,\infty)$. We want to study this problem by using Theorem 3.1. In this case we have

$$b(t,x,\pi,\theta) = \pi(\theta - r(t)) + xr(t), \quad K(t) = U^{'}(X^{(\pi,\theta)}(T)), \tag{22.65}$$

$$\begin{aligned} H_0(t,x,\pi,\theta) =& U^{'}(X^{(\pi,\theta)}(T))[\pi(\theta - r(t)) + xr(t)] \\ &+ D_t(U^{'}(X^{(\pi,\theta)}(T)))\pi\sigma_0(t) \\ &+ \int_{\mathbb{R}_0} D_{t,z}(U^{'}(X^{(\pi,\theta)}(T)))\pi\gamma_0(t,z)\nu(dz), \end{aligned} \tag{22.66}$$

and

$$p(t) = U^{'}(X^{(\pi,\theta)}(T)) \left[1 + \int_t^T r(s)G(t,s)ds \right],$$

where

$$G(t,s) = \exp\left(\int_t^s r(v)\mathrm{d}v\right).$$

Hence,

$$\int_t^T r(s)G(t,s)\mathrm{d}s = \Big|_t^T \exp\left(\int_t^s r(v)\mathrm{d}v\right) = \exp\left(\int_t^T r(v)\mathrm{d}v\right) - 1$$

and

$$p(t) = U'(X^{(\pi,\theta)}(T))\exp\left(\int_t^T r(s)\mathrm{d}s\right). \tag{22.67}$$

With this value for $p(t)$ we have

$$\begin{aligned} H(t, X^{(\pi,\theta)}(t), \pi, \theta) =& p(t)[\pi(\theta - r(t)) + r(t)X(t)] \\ &+ D_t p(t)\pi\sigma_0(t) + \int_{\mathbb{R}_0} D_{t,z} p(t)\pi\gamma_0(t,z)\nu(\mathrm{d}z). \end{aligned} \tag{22.68}$$

Hence Eq. (22.55) becomes

$$\begin{aligned} \mathbb{E}\left[\frac{\partial H}{\partial \pi}(t, \hat{X}(t), \pi, \hat{\theta})\Big|\mathcal{E}_t\right]_{\pi=\hat{\pi}(t)} &= \mathbb{E}\Big[p(t)(\hat{\theta} - r(t)) + D_t p(t)\sigma_0(t) \\ &+ \int_{\mathbb{R}_0} D_{t,z} p(t)\gamma_0(t,z)\nu(\mathrm{d}z)\Big|\mathcal{E}_t\Big] = 0 \end{aligned} \tag{22.69}$$

and Eq. (22.56) becomes

$$\mathbb{E}\left[\frac{\partial H}{\partial \theta}(t, \hat{X}(t), \hat{\pi}, \theta)\Big|\mathcal{E}_t\right]_{\theta=\hat{\theta}(t)} = \mathbb{E}[p(t)\hat{\pi}(t) \mid \mathcal{E}_t] = \mathbb{E}[p(t) \mid \mathcal{E}_t]\hat{\pi}(t) = 0. \tag{22.70}$$

Since $p(t) > 0$ we conclude that

$$\hat{\pi}(t) = 0. \tag{22.71}$$

This implies that

$$\hat{X}(t) = x\exp\left(\int_0^t r(s)\mathrm{d}s\right) \tag{22.72}$$

and

$$\hat{p}(t) = U'\left(x\exp\left(\int_0^T r(s)\mathrm{d}s\right)\right)\exp\left(\int_t^T r(s)\mathrm{d}s\right); \quad t \in [0,T]. \tag{22.73}$$

Substituting this into Eq. (22.69), we get

$$\hat{\theta}(t) = \frac{\mathbb{E}\left[\hat{p}(t)r(t) - D_t\hat{p}(t)\sigma_0(t) - \int_{\mathbb{R}_0} D_{t,z}\hat{p}(t)\gamma_0(t,z)\nu(dz)\Big|\mathscr{E}_t\right]}{\mathbb{E}[\hat{p}(t)|\mathscr{E}_t]}. \tag{22.74}$$

We have proved the following theorem:

Theorem 4.1 (Worst-case scenario optimal portfolio under partial information). *Suppose there exists a solution $(\pi^*, \theta^*) \in (\mathscr{A}_\Pi^\varepsilon, \mathscr{A}_\Theta^\varepsilon)$ of the stochastic differential game Eq. (22.64). Then*

$$\pi^* = \hat{\pi} = 0, \tag{22.75}$$

and

$$\theta^* = \hat{\theta} \text{ is given by } (22.74). \tag{22.76}$$

In particular, if $r(s)$ is deterministic, then

$$\pi^* = 0 \quad \text{and} \quad \theta^*(t) = r(t). \tag{22.77}$$

Remark 4.1. (i) If $r(s)$ is deterministic, then Eq. (22.77) states that the worst-case scenario is when $\hat{\theta}(t) = r(t)$, for all $t \in [0, T]$, i.e., when the normalized risky asset price

$$e^{-\int_0^t r(s)ds} S_1(t)$$

is a martingale. In such a situation the trader might as well put all her money in the risk free asset, i.e., choose $\pi(t) = \hat{\pi}(t) = 0$. This trading strategy remains optimal if $r(s)$ is not deterministic, but now the worst-case scenario $\hat{\theta}(t)$ is given by the more complicated expression (22.74).

(ii) This is a new approach to, and a partial extension of, Theorem 2.2 in [13] and Theorem 4.1 in the subsequent paper [1]. Both of these papers consider the case with deterministic $r(t)$ only. On the other hand, in these papers the scenario is represented by a probability measure and not by the drift.

Acknowledgements The research leading to these results has received funding from the European Research Council under the European Community's Seventh Framework Programme (FP7/2007–2013)/ ERC grant agreement no [228087].

References

1. An, T.T.K., Øksendal, B.: A maximum principle for stochastic differential games with partial information. J. Optim. Theor. Appl. **139**, 463–483 (2008)
2. Bensoussan, A.: Stochastic Control of Partially Observable Systems. Cambridge University Press, Cambridge (1992)

3. Benth, F.E., Di Nunno, G., Løkka, A., Øksendal, B., Proske, F.: Explicit representation of the minimal variance portfolio in markets driven by Lévy processes. Math. Financ. **13**, 55–72 (2003)
4. Baghery, F., Øksendal, B.: A maximum principle for stochastic control with partial information. Stoch. Anal. Appl. **25**, 705–717 (2007)
5. Di Nunno, G., Meyer-Brandis, T., Øksendal, B., Proske, F.: Malliavin calculus and anticipative Itô formulae for Lévy processes. Inf. Dim. Anal. Quant. Probab. **8**, 235–258 (2005)
6. Di Nunno, G., Øksendal, B., Proske, F.: Malliavin Calculus for Lévy Processes and Applications to Finance. Springer, Berlin (2009)
7. Karatzas, I., Ocone, D.: A generalized Clark representation formula, with application to optimal portfolios. Stoch. Stoch. Rep. **34**, 187–220 (1991)
8. Karatzas, I., Xue, X.: A note on utility maximization under partial observations. Math. Financ. **1**, 57–70 (1991)
9. Lakner, P.: Optimal trading strategy for an investor: the case of partial information. Stoch. Process. Appl. **76**, 77–97 (1998)
10. Meyer-Brandis, T., Øksendal, B., Zhou, X.Y.: A mean-field stochastic maximum principle via Malliavin calculus. Stoch. **84**, 643–666 (2012)
11. Nualart, D.: Malliavin Calculus and Related Topics, 2nd edn. Springer, Berlin (2006)
12. Øksendal, B., Sulem, A.: Applied Stochastic Control of Jump Diffusions, 2nd edn. Springer, Berlin (2007)
13. Øksendal, B., Sulem, A.: A game theoretic approach to martingale measures in incomplete markets. Surv. Appl. Ind. Math. **15**, 18–24 (2008)
14. Pham, H., Quenez M.-C.: Optimal portfolio in partially observed stochastic volatility models. Ann. Appl. Probab. **11**, 210–238 (2001)
15. Yong, J., Zhou, X.Y.: Stochastic Controls: Hamiltonian Systems and HJB Equations. Springer, New York (1999)

Chapter 23
Asymptotics for the Length of the Longest Increasing Subsequence of a Binary Markov Random Word

Christian Houdré and Trevis J. Litherland

Abstract Let $(X_n)_{n\geq 0}$ be an irreducible, aperiodic, and homogeneous binary Markov chain and let LI_n be the length of the longest (weakly) increasing subsequence of $(X_k)_{1\leq k\leq n}$. Using combinatorial constructions and weak invariance principles, we present elementary arguments leading to a new proof that (after proper centering and scaling) the limiting law of LI_n is the maximal eigenvalue of a 2×2 Gaussian random matrix. In fact, the limiting shape of the RSK Young diagrams associated with the binary Markov random word is the spectrum of this random matrix.

Keywords Longest increasing subsequence • Markov chains • Functional central limit theorem • Random matrices • Young diagrams

Received 9/29/2011; Accepted 7/1/2012; Final 7/16/2012

1 Introduction

The identification of the limiting distribution of the length of the longest increasing subsequence of a random permutation or of a random word has attracted a lot of interest in the past decade, in particular in light of its connections with random matrices (see [1–4, 6, 8, 12–15, 17, 18]). For random words, both the iid uniform and nonuniform settings are understood, leading respectively to the maximal eigenvalue of a traceless (or generalized traceless) element of the Gaussian unitary ensemble (GUE) as limiting laws of LI_n. In a dependent framework, Kuperberg [16] conjectured that if the word is generated by an irreducible, doubly stochastic, cyclic,

C. Houdré (✉) • T.J. Litherland
School of Mathematics, Georgia Institute of Technology, Atlanta, GA 30332-0160, USA
e-mail: houdre@math.gatech.edu; tlitherland@scoringsolutions.com

F. Viens et al. (eds.), *Malliavin Calculus and Stochastic Analysis: A Festschrift in Honor of David Nualart*, Springer Proceedings in Mathematics & Statistics 34,
DOI 10.1007/978-1-4614-5906-4_23,

Markov chain with state space an ordered m-letter alphabet, then the limiting distribution of the length LI_n is still that of the maximal eigenvalue of a traceless $m \times m$ element of the GUE. More generally, the conjecture asserts that the shape of the Robinson–Schensted–Knuth (RSK) Young diagrams associated with the Markovian random word is that of the joint distribution of the eigenvalues of a traceless $m \times m$ element of the GUE. For $m = 2$, Chistyakov and Götze [7] positively answered this conjecture, and in the present paper this result is rederived in an elementary way.

The precise class of homogeneous Markov chains with which Kuperberg's conjecture is concerned is more specific than the ones we shall study. The irreducibility of the chain is a basic property we certainly must demand: each letter has to occur at some point following the occurrence of any given letter. The cyclic (also called *circulant*) criterion, i.e., the Markov transition matrix P has entries satisfying $p_{i,j} = p_{i+1,j+1}$, for $1 \le i, j \le m$ (where $m + 1 = 1$), ensures a uniform stationary distribution.

Let us also note that Kuperberg implicitly assumes the Markov chain to also be aperiodic. Indeed, the simple 2-state Markov chain for the letters α_1 and α_2 described by $\mathbb{P}(X_{n+1} = \alpha_i | X_n = \alpha_j) = 1$, for $i \neq j$, produces a sequence of alternating letters, so that LI_n is always either $n/2$ or $n/2 + 1$, for n even, and $(n + 1)/2$, for n odd, and so has a degenerate limiting distribution. Even though this Markov chain is irreducible and cyclic, it is periodic.

By the end of this introduction, the reader might certainly have wondered how the binary results do get modified for ordered alphabets of arbitrary fixed size m. As shown in [10], for $m = 3$, Kuperberg's conjecture is indeed true. However, for $m \ge 4$, this is no longer the case; and some, but not all, cyclic Markov chains lead to a limiting law as in the iid uniform case.

2 Combinatorics

As in [9], one can express LI_n in a combinatorial manner. For convenience, this short section recapitulates that development.

Let $(X_n)_{n \ge 1}$ consist of a sequence of values taken from an m-letter ordered alphabet, $\mathcal{A}_m = \{\alpha_1 < \alpha_2 < \cdots < \alpha_m\}$. Let a_k^r be the number of occurrences of α_r among $(X_i)_{1 \le i \le k}$. Each increasing subsequence of $(X_i)_{1 \le i \le k}$ consists simply of consecutive identical values, with these values forming an increasing subsequence of α_r. Moreover, the number of occurrences of $\alpha_r \in \{\alpha_1, \ldots, \alpha_m\}$ among $(X_i)_{k+1 \le i \le \ell}$, where $1 \le k < \ell \le n$, is simply $a_\ell^r - a_k^r$. The length of the longest increasing subsequence of $X_1, X_2, \ldots, X_n$ is thus given by

$$LI_n = \max_{\substack{0 \le k_1 \le \cdots \\ \le k_{m-1} \le n}} \left[(a_{k_1}^1 - a_0^1) + (a_{k_2}^2 - a_{k_1}^2) + \cdots + (a_n^m - a_{k_{m-1}}^m) \right], \tag{23.1}$$

i.e.,

$$LI_n = \max_{\substack{0 \le k_1 \le \cdots \\ \le k_{m-1} \le n}} [(a_{k_1}^1 - a_{k_1}^2) + (a_{k_2}^2 - a_{k_2}^3) + \cdots + (a_{k_{m-1}}^{m-1} - a_{k_{m-1}}^m) + a_n^m], \quad (23.2)$$

where $a_0^r = 0$.

For $i = 1, \ldots, n$ and $r = 1, \ldots, m-1$, let

$$Z_i^r = \begin{cases} 1, & \text{if } X_i = \alpha_r, \\ -1, & \text{if } X_i = \alpha_{r+1}, \\ 0, & \text{otherwise}, \end{cases} \quad (23.3)$$

and let $S_k^r = \sum_{i=1}^k Z_i^r$, $k = 1, \ldots, n$, with also $S_0^r = 0$. Then clearly $S_k^r = a_k^r - a_k^{r+1}$. Hence,

$$LI_n = \max_{\substack{0 \le k_1 \le \cdots \\ \le k_{m-1} \le n}} \left\{ S_{k_1}^1 + S_{k_2}^2 + \cdots + S_{k_{m-1}}^{m-1} + a_n^m \right\}. \quad (23.4)$$

By the telescoping nature of the sum $\sum_{k=r}^{m-1} S_n^k = \sum_{k=r}^{m-1} (a_n^k - a_n^{k+1})$, we find that, for each $1 \le r \le m-1$, $a_n^r = a_n^m + \sum_{k=r}^{m-1} S_n^k$. Since $a_k^1, \ldots, a_k^m$ must evidently sum up to k, we have

$$\begin{aligned} n &= \sum_{r=1}^{m} a_n^r \\ &= \sum_{r=1}^{m-1} \left(a_n^m + \sum_{k=r}^{m-1} S_n^k \right) + a_n^m \\ &= \sum_{r=1}^{m-1} r S_n^r + m a_n^m. \end{aligned}$$

Solving for a_n^m gives us

$$a_n^m = \frac{n}{m} - \frac{1}{m} \sum_{r=1}^{m-1} r S_n^r.$$

Substituting into Eq. (23.4), we finally obtain

$$LI_n = \frac{n}{m} - \frac{1}{m} \sum_{r=1}^{m-1} r S_n^r + \max_{\substack{0 \le k_1 \le \cdots \\ \le k_{m-1} \le n}} \left\{ S_{k_1}^1 + S_{k_2}^2 + \cdots + S_{k_{m-1}}^{m-1} \right\}. \quad (23.5)$$

As emphasized in [9], Eq. (23.5) is of a *purely combinatorial nature or, in more probabilistic terms, is of a pathwise nature*. We now proceed to analyze Eq. (23.5) for a binary Markovian sequence.

3 Binary Markovian Alphabet

In the context of binary Markovian alphabets, $(X_n)_{n\geq 0}$ is described by the following transition probabilities between the two states (which we identify with the two letters α_1 and α_2): $\mathbb{P}(X_{n+1} = \alpha_2 | X_n = \alpha_1) = a$ and $\mathbb{P}(X_{n+1} = \alpha_1 | X_n = \alpha_2) = b$, where $0 < a + b < 2$. We later examine the degenerate cases $a = b = 0$ and $a = b = 1$. In keeping with the common usage within the Markov chain literature, we begin our sequence at $n = 0$, although our focus will be on $n \geq 1$. Denoting by (p_n^1, p_n^2) the vector describing the probability distribution on $\{\alpha_1, \alpha_2\}$ at time n, we have

$$\left(p_{n+1}^1, p_{n+1}^2\right) = \left(p_n^1, p_n^2\right) \begin{pmatrix} 1-a & a \\ b & 1-b \end{pmatrix}. \tag{23.6}$$

The eigenvalues of the matrix in Eq. (23.6) are $\lambda_1 = 1$ and $-1 < \lambda_2 = 1-a-b < 1$, with respective left eigenvectors $(\pi_1, \pi_2) = (b/(a+b), a/(a+b))$ and $(1, -1)$. Moreover, (π_1, π_2) is also the stationary distribution. Given any initial distribution (p_0^1, p_0^2), we find that

$$\left(p_n^1, p_n^2\right) = \left(\pi_1, \pi_2\right) + \lambda_2^n \frac{ap_0^1 - bp_0^2}{a+b} (1, -1) \to \left(\pi_1, \pi_2\right), \tag{23.7}$$

as $n \to \infty$, since $|\lambda_2| < 1$.

Our goal is now to use these probabilistic expressions to describe the random variables Z_k^1 and S_k^1 defined in the previous section. (We retain the redundant superscript "1" in Z_k^1 and S_k^1 in the interest of uniformity.)

Setting $\beta = ap_0^1 - bp_0^2$, we easily find that

$$\begin{aligned} \mathbb{E}Z_k^1 &= (+1)\left(\pi_1 + \frac{\beta}{a+b}\lambda_2^k\right) + (-1)\left(\pi_2 - \frac{\beta}{a+b}\lambda_2^k\right) \\ &= \frac{b-a}{a+b} + 2\frac{\beta}{a+b}\lambda_2^k, \end{aligned} \tag{23.8}$$

for each $1 \leq k \leq n$. Thus,

$$\mathbb{E}S_k^1 = \frac{b-a}{a+b}k + 2\left(\frac{\beta\lambda_2}{a+b}\right)\left(\frac{1-\lambda_2^k}{1-\lambda_2}\right), \tag{23.9}$$

and so $\mathbb{E}S_k^1/k \to (b-a)/(a+b)$, as $k \to \infty$.

Turning to the second moments of Z_k^1 and S_k^1, first note that $\mathbb{E}(Z_k^1)^2 = 1$, since $(Z_k^1)^2 = 1$ a.s. Next, we consider $\mathbb{E}Z_k^1 Z_\ell^1$, for $k < \ell$. Using the Markovian structure of $(X_n)_{n\geq 0}$, it quickly follows that

$$\mathbb{P}((X_k, X_\ell) = (x_k, x_\ell))$$
$$= \begin{cases} \left(\pi_1 + \lambda_2^{\ell-k}\frac{a}{a+b}\right)\left(\pi_1 + \lambda_2^k \frac{\beta}{a+b}\right), & \text{if } (x_k, x_\ell) = (\alpha_1, \alpha_1), \\ \left(\pi_1 - \lambda_2^{\ell-k}\frac{b}{a+b}\right)\left(\pi_2 - \lambda_2^k \frac{\beta}{a+b}\right), & \text{if } (x_k, x_\ell) = (\alpha_1, \alpha_2), \\ \left(\pi_2 - \lambda_2^{\ell-k}\frac{a}{a+b}\right)\left(\pi_1 + \lambda_2^k \frac{\beta}{a+b}\right), & \text{if } (x_k, x_\ell) = (\alpha_2, \alpha_1), \\ \left(\pi_2 + \lambda_2^{\ell-k}\frac{b}{a+b}\right)\left(\pi_2 - \lambda_2^k \frac{\beta}{a+b}\right), & \text{if } (x_k, x_\ell) = (\alpha_2, \alpha_2). \end{cases} \tag{23.10}$$

For simplicity, we will henceforth assume that our initial distribution is the stationary one, i.e., $(p_0^1, p_0^2) = (\pi_1, \pi_2)$. (This assumption can be dropped as explained in the Concluding Remarks of [10].) Under this assumption, $\beta = 0$, $\mathbb{E}S_k^1 = k\mu$, where $\mu = \mathbb{E}Z_k^1 = (b-a)/(a+b)$, and Eq. (23.10) simplifies to

$$\mathbb{P}((X_k, X_\ell) = (x_k, x_\ell))$$
$$= \begin{cases} \left(\pi_1 + \lambda_2^{\ell-k}\frac{a}{a+b}\right)\pi_1, & \text{if } (x_k, x_\ell) = (\alpha_1, \alpha_1), \\ \left(\pi_1 - \lambda_2^{\ell-k}\frac{b}{a+b}\right)\pi_2, & \text{if } (x_k, x_\ell) = (\alpha_1, \alpha_2), \\ \left(\pi_2 - \lambda_2^{\ell-k}\frac{a}{a+b}\right)\pi_1, & \text{if } (x_k, x_\ell) = (\alpha_2, \alpha_1), \\ \left(\pi_2 + \lambda_2^{\ell-k}\frac{b}{a+b}\right)\pi_2, & \text{if } (x_k, x_\ell) = (\alpha_2, \alpha_2). \end{cases} \tag{23.11}$$

We can now compute $\mathbb{E}Z_k^1 Z_\ell^1$:

$$\begin{aligned} \mathbb{E}Z_k^1 Z_\ell^1 &= \mathbb{P}(Z_k^1 Z_\ell^1 = +1) - \mathbb{P}(Z_k^1 Z_\ell^1 = -1) \\ &= \mathbb{P}((X_k, X_\ell) \in \{(\alpha_1, \alpha_1), (\alpha_2, \alpha_2)\}) \\ &\quad - \mathbb{P}((X_k, X_\ell) \in \{(\alpha_1, \alpha_2), (\alpha_2, \alpha_1)\}) \\ &= \left(\pi_1^2 + \lambda_2^{\ell-k}\frac{a}{a+b}\pi_1 + \pi_2^2 + \lambda_2^{\ell-k}\frac{b}{a+b}\pi_2\right) \\ &\quad - \left(\pi_1\pi_2 - \lambda_2^{\ell-k}\frac{b}{a+b}\pi_2 + \pi_1\pi_2 - \lambda_2^{\ell-k}\frac{a}{a+b}\pi_1\right) \\ &= \left(\pi_1^2 + \pi_2^2 + \frac{2ab}{(a+b)^2}\lambda_2^{\ell-k}\right) - \left(2\pi_1\pi_2 - \frac{2ab}{(a+b)^2}\lambda_2^{\ell-k}\right) \\ &= \frac{(b-a)^2}{(a+b)^2} + \frac{4ab}{(a+b)^2}\lambda_2^{\ell-k}. \end{aligned} \tag{23.12}$$

Hence, recalling that $\beta = 0$,

$$\sigma^2 := \mathrm{Var} Z_k^1 = 1 - \left(\frac{b-a}{a+b}\right)^2$$
$$= \frac{4ab}{(a+b)^2}, \tag{23.13}$$

for all $k \geq 1$ and, for $k < \ell$, the covariance of Z_k^1 and Z_ℓ^1 is

$$\mathrm{Cov}(Z_k^1, Z_\ell^1) = \frac{(b-a)^2}{(a+b)^2} + \sigma^2 \lambda_2^{\ell-k} - \left(\frac{b-a}{a+b}\right)^2 = \sigma^2 \lambda_2^{\ell-k}. \tag{23.14}$$

Proceeding to the covariance structure of S_k^1, we first find that

$$\mathrm{Var} S_k^1 = \sum_{j=1}^{k} \mathrm{Var} Z_j^1 + 2 \sum_{j<\ell} \mathrm{Cov}(Z_j^1, Z_l^1)$$
$$= \sigma^2 k + 2\sigma^2 \sum_{j<\ell} \lambda_2^{\ell-j}$$
$$= \sigma^2 k + 2\sigma^2 \left(\frac{\lambda_2^{k+1} - k\lambda_2^2 + (k-1)\lambda_2}{(1-\lambda_2)^2}\right)$$
$$= \sigma^2 \left(\frac{1+\lambda_2}{1-\lambda_2}\right) k + 2\sigma^2 \left(\frac{\lambda_2(\lambda_2^k - 1)}{(1-\lambda_2)^2}\right). \tag{23.15}$$

Next, for $k < \ell$, and using Eqs. (23.14) and (23.15), the covariance of S_k^1 and S_ℓ^1 is given by

$$\mathrm{Cov}(S_k^1, S_\ell^1) = \sum_{i=1}^{k} \sum_{j=1}^{\ell} \mathrm{Cov}(Z_i^1, Z_j^1)$$
$$= \sum_{i=1}^{k} \mathrm{Var} Z_i^1 + 2 \sum_{i<j<k} \mathrm{Cov}(Z_i^1, Z_j^1) + \sum_{i=1}^{k} \sum_{j=k+1}^{\ell} \mathrm{Cov}(Z_i^1, Z_j^1)$$
$$= \mathrm{Var} S_k^1 + \sum_{i=1}^{k} \sum_{j=k+1}^{\ell} \mathrm{Cov}(Z_i^1, Z_j^1)$$
$$= \mathrm{Var} S_k^1 + \sigma^2 \left(\frac{\lambda_2(1-\lambda_2^k)(1-\lambda_2^{\ell-k})}{(1-\lambda_2)^2}\right)$$
$$= \sigma^2 \left(\left(\frac{1+\lambda_2}{1-\lambda_2}\right) k - \frac{\lambda_2(1-\lambda_2^k)(1+\lambda_2^{\ell-k})}{(1-\lambda_2)^2}\right). \tag{23.16}$$

From Eqs. (23.15) and (23.16) we see that, as $k \to \infty$,

$$\frac{\mathrm{Var} S_k^1}{k} \to \sigma^2 \left(\frac{1+\lambda_2}{1-\lambda_2} \right), \tag{23.17}$$

and, moreover, as $k \wedge \ell \to \infty$,

$$\frac{\mathrm{Cov}(S_k^1, S_\ell^1)}{(k \wedge \ell)} \to \sigma^2 \left(\frac{1+\lambda_2}{1-\lambda_2} \right). \tag{23.18}$$

When $a = b$, $\mathbb{E} S_k^1 = 0$, and in Eq. (23.17) the asymptotic variance becomes

$$\begin{aligned} \frac{\mathrm{Var} S_k^1}{k} &\to \frac{4a^2}{(2a)^2} \left(\frac{1+(1-2a)}{1-(1-2a)} \right) \\ &= \frac{1}{a} - 1. \end{aligned}$$

For a small, we have a "lazy" Markov chain, i.e., a Markov chain which tends to remain in a given state for long periods of time. In this regime, the random variable S_k^1 has long periods of increase followed by long periods of decrease. In this way, linear asymptotics of the variance with large constants occur. If, on the other hand, a is close to 1, the Markov chain rapidly shifts back and forth between α_1 and α_2, and so the constant associated with the linearly increasing variance of S_k^1 is small.

As in [9], Brownian functionals play a central rôle in describing the limiting distribution of LI_n. To move towards a Brownian functional expression for the limiting law of LI_n, define the polygonal function

$$\hat{B}_n(t) = \frac{S_{[nt]}^1 - [nt]\mu}{\sigma\sqrt{n(1+\lambda_2)/(1-\lambda_2)}} + \frac{(nt - [nt])(Z_{[nt]+1}^1 - \mu)}{\sigma\sqrt{n(1+\lambda_2)/(1-\lambda_2)}}, \tag{23.19}$$

for $0 \le t \le 1$. In our finite-state, irreducible, aperiodic, stationary Markov chain setting, we may conclude that $\hat{B}_n \Rightarrow B$, as desired. (See, e.g., the more general settings for Gordin's martingale approach to dependent invariance principles and the stationary ergodic invariance principle found in Theorem 19.1 of [5].)

Turning now to LI_n, we see that for the present 2-letter situation, Eq. (23.5) simply becomes

$$LI_n = \frac{n}{2} - \frac{1}{2} S_n^1 + \max_{1 \le k \le n} S_k^1.$$

To find the limiting distribution of LI_n from this expression, recall that $\pi_1 = b/(a+b)$, $\pi_2 = a/(a+b)$, $\mu = \pi_1 - \pi_2 = (b-a)/(a+b)$, $\sigma^2 = 4ab/(a+b)^2$, and that $\lambda_2 = 1-a-b$. Define $\pi_{\max} = \max\{\pi_1, \pi_2\}$ and $\tilde{\sigma}^2 = \sigma^2(1+\lambda_2)/(1-\lambda_2)$. Rewriting Eq. (23.19) as

$$\hat{B}_n(t) = \frac{S_{[nt]}^1 - [nt]\mu}{\tilde{\sigma}\sqrt{n}} + \frac{(nt - [nt])(Z_{[nt]+1}^1 - \mu)}{\tilde{\sigma}\sqrt{n}},$$

LI_n becomes

$$
\begin{aligned}
LI_n &= \frac{n}{2} - \frac{1}{2}\left(\tilde{\sigma}\sqrt{n}\hat{B}_n(1) + \mu n\right) + \max_{0 \le t \le 1}\left(\tilde{\sigma}\sqrt{n}\hat{B}_n(t) + \mu n t\right) \\
&= n\pi_2 - \frac{1}{2}\left(\tilde{\sigma}\sqrt{n}\hat{B}_n(1)\right) + \max_{0 \le t \le 1}\left(\tilde{\sigma}\sqrt{n}\hat{B}_n(t) + (\pi_1 - \pi_2)nt\right) \\
&= n\pi_{\max} - \frac{1}{2}\left(\tilde{\sigma}\sqrt{n}\hat{B}_n(1)\right) \\
&\quad + \max_{0 \le t \le 1}\left(\tilde{\sigma}\sqrt{n}\hat{B}_n(t) + (\pi_1 - \pi_2)nt - (\pi_{\max} - \pi_2)n\right). \qquad (23.20)
\end{aligned}
$$

This immediately gives

$$
\begin{aligned}
\frac{LI_n - n\pi_{\max}}{\tilde{\sigma}\sqrt{n}} &= -\frac{1}{2}\hat{B}_n(1) \\
&\quad + \max_{0 \le t \le 1}\left(\hat{B}_n(t) + \frac{\sqrt{n}}{\tilde{\sigma}}((\pi_1 - \pi_2)t - (\pi_{\max} - \pi_2))\right). \qquad (23.21)
\end{aligned}
$$

Let us examine Eq. (23.21) on a case-by-case basis. First, if $\pi_{\max} = \pi_1 = \pi_2 = 1/2$, i.e., if $a = b$, then $\sigma = 1$ and $\tilde{\sigma} = (1-a)/a$, and so Eq. (23.21) becomes

$$
\frac{LI_n - n/2}{\sqrt{(1-a)n/a}} = -\frac{1}{2}\hat{B}_n(1) + \max_{0 \le t \le 1}\hat{B}_n(t). \qquad (23.22)
$$

Then, by the invariance principle and the continuous mapping theorem,

$$
\frac{LI_n - n/2}{\sqrt{(1-a)n/a}} \Rightarrow -\frac{1}{2}B(1) + \max_{0 \le t \le 1}B(t). \qquad (23.23)
$$

Next, if $\pi_{\max} = \pi_2 > \pi_1$, Eq. (23.21) becomes

$$
\begin{aligned}
\frac{LI_n - n\pi_{\max}}{\tilde{\sigma}\sqrt{n}} &= -\frac{1}{2}\hat{B}_n(1) \\
&\quad + \max_{0 \le t \le 1}\left(\hat{B}_n(t) - \frac{\sqrt{n}}{\tilde{\sigma}}(\pi_{\max} - \pi_1)t\right). \qquad (23.24)
\end{aligned}
$$

On the other hand, if $\pi_{\max} = \pi_1 > \pi_2$, Eq. (23.21) becomes

$$
\begin{aligned}
\frac{LI_n - n\pi_{\max}}{\tilde{\sigma}\sqrt{n}} &= -\frac{1}{2}\hat{B}_n(1) \\
&\quad + \max_{0 \le t \le 1}\left(\hat{B}_n(t) - \frac{\sqrt{n}}{\tilde{\sigma}}(\pi_{\max} - \pi_2)(1-t)\right)
\end{aligned}
$$

$$= \frac{1}{2}\hat{B}_n(1)$$

$$+ \max_{0\le t\le 1}\left(\hat{B}_n(t) - \hat{B}_n(1) - \frac{\sqrt{n}}{\tilde{\sigma}}(\pi_{\max} - \pi_2)(1-t)\right). \quad (23.25)$$

In both Eqs. (23.24) and (23.25) we have a term in our maximal functional which is linear in t or $1-t$, with a negative slope. We now show, in an elementary fashion, that in both cases, as $n \to \infty$, the maximal functional goes to zero in probability.

Consider first Eq. (23.24). Let $c_n = \sqrt{n}(\pi_{\max} - \pi_1)/\tilde{\sigma} > 0$, and for any $c > 0$, let $M_c = \max_{0\le t\le 1}(B(t) - ct)$, where $B(t)$ is a standard Brownian motion. Now for n large enough,

$$\hat{B}_n(t) - ct \ge \hat{B}_n(t) - c_n t$$

a.s., for all $0 \le t \le 1$. Then for any $z > 0$ and n large enough,

$$\mathbb{P}(\max_{0\le t\le 1}(\hat{B}_n(t) - c_n t) > z) \le \mathbb{P}(\max_{0\le t\le 1}(\hat{B}_n(t) - ct) > z), \quad (23.26)$$

and so by the Invariance Principle and the Continuous Mapping Theorem,

$$\limsup_{n\to\infty}\mathbb{P}(\max_{0\le t\le 1}(\hat{B}_n(t) - c_n t) > z) \le \lim_{n\to\infty}\mathbb{P}(\max_{0\le t\le 1}(\hat{B}_n(t) - ct) > z)$$
$$= \mathbb{P}(M_c > z). \quad (23.27)$$

Now, as is well known, $\mathbb{P}(M_c > z) \to 0$ as $c \to \infty$. One can confirm this intuitive fact with the following simple argument. For $z > 0$, $c > 0$, and $0 < \varepsilon < 1$, we have that

$$\mathbb{P}(M_c > z) \le \mathbb{P}(\max_{0\le t\le \varepsilon}(B(t) - ct) > z) + \mathbb{P}(\max_{\varepsilon< t\le 1}(B(t) - ct) > z)$$
$$\le \mathbb{P}(\max_{0\le t\le \varepsilon} B(t) > z) + \mathbb{P}(\max_{\varepsilon< t\le 1}(B(t) - c\varepsilon) > z)$$
$$\le \mathbb{P}(\max_{0\le t\le \varepsilon} B(t) > z) + \mathbb{P}(\max_{0< t\le 1} B(t) > c\varepsilon + z)$$
$$= 2\left(1 - \Phi\left(\frac{z}{\sqrt{\varepsilon}}\right)\right) + 2\left(1 - \Phi(c\varepsilon + z)\right). \quad (23.28)$$

But, as c and ε are arbitrary, we can first take the limsup of Eq. (23.28) as $c \to \infty$, and then let $\varepsilon \to 0$, proving the claim.

We have thus shown that

$$\limsup_{n\to\infty} \mathbb{P}(\max_{0\le t\le 1}(\hat{B}_n(t) - c_n t) > z) \le 0,$$

and since the functional clearly is equal to zero when $t = 0$, we have

$$\max_{0\le t\le 1}(\hat{B}_n(t) - c_n t) \xrightarrow{\mathbb{P}} 0, \tag{23.29}$$

as $n \to \infty$. Thus, by the continuous mapping theorem and the converging together lemma, we obtain the weak convergence result

$$\frac{LI_n - n\pi_{\max}}{\tilde{\sigma}\sqrt{n}} \Rightarrow -\frac{1}{2}B(1). \tag{23.30}$$

Lastly, consider Eq. (23.25). Here we need simply note the following equality in law, which follows from the stationary and Markovian nature of the underlying sequence $(X_n)_{n\ge 0}$:

$$\begin{aligned}\hat{B}_n(t) - \hat{B}_n(1) - \frac{\sqrt{n}}{\tilde{\sigma}}(\pi_{\max} - \pi_2)(1-t)&\\ \stackrel{\mathcal{L}}{=} -\hat{B}_n(1-t) - \frac{\sqrt{n}}{\tilde{\sigma}}(\pi_{\max} - \pi_2)(1-t),&\end{aligned} \tag{23.31}$$

for $t = 0, 1/n, \ldots, (n-1)/n, 1$. With a change of variables ($u = 1-t$) and noting that $B(t)$ and $-B(t)$ are equal in law, our previous convergence result Eq. (23.29) implies that

$$\max_{0\le t\le 1}(\hat{B}_n(t) - \hat{B}_n(1) - c_n(1-t)) \stackrel{\mathcal{L}}{=} \max_{0\le u\le 1}(-\hat{B}_n(u) - c_n u) \xrightarrow{\mathbb{P}} 0, \tag{23.32}$$

as $n \to \infty$. Our limiting functional is thus of the form

$$\frac{LI_n - n\pi_{\max}}{\tilde{\sigma}\sqrt{n}} \Rightarrow \frac{1}{2}B(1). \tag{23.33}$$

Since $B(1)$ is simply a standard normal random variable, the different signs in Eqs. (23.30) and (23.33) are inconsequential.

Finally, consider the degenerate cases. If either $a = 0$ or $b = 0$, then the sequence $(X_n)_{n\ge 0}$ will be a.s. constant, regardless of the starting state, and so $LI_n \sim n$. On the other hand, if $a = b = 1$, then the sequence oscillates back and forth between α_1 and α_2, so that $LI_n \sim n/2$. Combining these trivial cases with the previous development gives

Theorem 3.1. *Let $(X_n)_{n\geq 0}$ be a 2-state Markov chain, with $\mathbb{P}(X_{n+1} = \alpha_2 | X_n = \alpha_1) = a$ and $\mathbb{P}(X_{n+1} = \alpha_1 | X_n = \alpha_2) = b$. Let the law of X_0 be the invariant distribution $(\pi_1, \pi_2) = (b/(a+b), a/(a+b))$, for $0 < a + b \leq 2$, and be $(\pi_1, \pi_2) = (1, 0)$, for $a = b = 0$. Then, for $a = b > 0$,*

$$\frac{LI_n - n/2}{\sqrt{n}} \Rightarrow \sqrt{\frac{1-a}{a}} \left(-\frac{1}{2} B(1) + \max_{0\leq t\leq 1} B(t) \right), \tag{23.34}$$

where $(B(t))_{0\leq t\leq 1}$ is a standard Brownian motion. For $a \neq b$ or $a = b = 0$,

$$\frac{LI_n - n\pi_{\max}}{\sqrt{n}} \Rightarrow N(0, \tilde{\sigma}^2/4), \tag{23.35}$$

with $\pi_{\max} = \max\{\pi_1, \pi_2\}$ and where $N(0, \tilde{\sigma}^2/4)$ is a centered normal random variable with variance $\tilde{\sigma}^2/4 = ab(2-a-b)/(a+b)^3$, for $a \neq b$, and $\tilde{\sigma}^2 = 0$, for $a = b = 0$. (If $a = b = 1$ or $\tilde{\sigma}^2 = 0$, then the distributions in Eqs. (23.34) *and* (23.35)*, respectively, are understood to be degenerate at the origin.)*

To extend this result to the entire RSK Young diagrams, let us introduce the following notation. By

$$(Y_n^{(1)}, Y_n^{(2)}, \ldots, Y_n^{(k)}) \Rightarrow (Y_\infty^{(1)}, Y_\infty^{(2)}, \ldots, Y_\infty^{(k)}), \tag{23.36}$$

we shall indicate the weak convergence of the *joint* law of the k-vector $(Y_n^{(1)}, Y_n^{(2)}, \ldots, Y_n^{(k)})$ to that of $(Y_\infty^{(1)}, Y_\infty^{(2)}, \ldots, Y_\infty^{(k)})$, as $n \to \infty$. Since LI_n is the length of the top row of the associated Young diagrams, the length of the second row is simply $n - LI_n$. Denoting the length of the ith row by R_n^i, Eq. (23.36), together with an application of the Cramér–Wold theorem, recovers the result of Chistyakov and Götze [7] as part of the following easy corollary, which is in fact equivalent to Theorem 3.1:

Corollary 3.1. *For the sequence in Theorem 3.1, if $a = b > 0$, then*

$$\left(\frac{R_n^1 - n/2}{\sqrt{n}}, \frac{R_n^2 - n/2}{\sqrt{n}} \right) \Rightarrow Y_\infty := (R_\infty^1, R_\infty^2), \tag{23.37}$$

where the law of Y_∞ is supported on the second main diagonal of $\mathbb{R}^2$ and with

$$R_\infty^1 \stackrel{\mathcal{L}}{=} \sqrt{\frac{1-a}{a}} \left(-\frac{1}{2} B(1) + \max_{0\leq t\leq 1} B(t) \right).$$

If $a \neq b$ or $a = b = 0$, then setting $\pi_{\min} = \min\{\pi_1, \pi_2\}$, we have

$$\left(\frac{R_n^1 - n\pi_{\max}}{\sqrt{n}}, \frac{R_n^2 - n\pi_{\min}}{\sqrt{n}} \right) \Rightarrow N((0,0), \tilde{\Sigma}), \tag{23.38}$$

where $\tilde{\Sigma}$ is the covariance matrix

$$(\tilde{\sigma}^2/4)\begin{pmatrix} 1 & -1 \\ -1 & 1 \end{pmatrix},$$

where $\tilde{\sigma}^2 = 4ab(2-a-b)/(a+b)^3$, *for* $a \neq b$, *and* $\tilde{\sigma}^2 = 0$, *for* $a = b = 0$.

Remark 3.1. The joint distributions in Eqs. (23.37) and (23.38) are of course degenerate, in that the sum of the two components is a.s. identically zero in each case. In Eq. (23.37), the density of the first component of R_∞ is easy to find and is given by (e.g., see [11])

$$f(y) = \frac{16}{\sqrt{2\pi}}\left(\frac{a}{1-a}\right)^{3/2} y^2 \mathrm{e}^{-2ay^2/(1-a)}, \qquad y \geq 0. \tag{23.39}$$

As in Chistyakov and Götze [7], Eq. (23.37) can then be stated as follows: for any bounded, continuous function $g : \mathbb{R}^2 \to \mathbb{R}$,

$$\lim_{n\to\infty}\left(g\left(\frac{R_n^1 - n/2}{\sqrt{(1-a)n/a}}, \frac{R_n^2 - n/2}{\sqrt{(1-a)n/a}} \right)\right)$$
$$= 2\sqrt{2\pi}\int_0^\infty g(x,-x)\phi_{\mathrm{GUE},2}(x,-x)\mathrm{d}x,$$

where $\phi_{\mathrm{GUE},2}$ is the density of the eigenvalues of the 2×2 GUE and is given by

$$\phi_{\mathrm{GUE},2}(x_1, x_2) = \frac{1}{\pi}(x_1 - x_2)^2 \mathrm{e}^{-(x_1^2 + x_2^2)}.$$

To see the GUE connection more explicitly, consider the 2×2 traceless GUE matrix

$$M_0 = \begin{pmatrix} X_1 & Y + iZ \\ Y - iZ & X_2 \end{pmatrix},$$

where X_1, X_2, Y, and Z are centered, normal random variables. Since $\mathrm{Corr}\,(X_1, X_2) = -1$, the largest eigenvalue of M_0 is

$$\lambda_{1,0} = \sqrt{X_1^2 + Y^2 + Z^2},$$

almost surely, so that $\lambda_{1,0}^2 \sim \chi_3^2$ if $\mathrm{Var}\, X_1 = \mathrm{Var}\, Y = \mathrm{Var}\, Z = 1$. Hence, up to a scaling factor, the density of $\lambda_{1,0}$ is given by Eq. (23.39). Next, let us perturb M_0 to

$$M = \alpha G I + \beta M_0,$$

where α and β are constants, G is a standard normal random variable independent of M_0, and I is the identity matrix. The covariance of the diagonal elements of M is then computed to be $\rho := \alpha^2 - \beta^2$. Hence, to obtain a desired value of ρ, we may take $\alpha = \sqrt{(1+\rho)/2}$ and $\beta = \sqrt{(1-\rho)/2}$. Clearly, the largest eigenvalue of M can then be expressed as

$$\lambda_1 = \sqrt{\frac{1+\rho}{2}} G + \sqrt{\frac{1-\rho}{2}} \lambda_{1,0}. \tag{23.40}$$

At one extreme, $\rho = -1$, we recover $\lambda_1 = \lambda_{1,0}$. At the other extreme, $\rho = 1$, we obtain $\lambda_1 = Z$. Midway between these two extremes, at $\rho = 0$, we have a standard GUE matrix, so that

$$\lambda_1 = \sqrt{\frac{1}{2}} (G + \lambda_{1,0}) .$$

Acknowledgements Research supported in part by the NSA Grant H98230-09-1-0017.

References

1. Baik, J., Deift, P., Johansson, K.: On the distribution of the length of the longest increasing subsequence of random permutations. J. Am. Math. Soc. **12**(4), 1119–1178 (1999)
2. Baik, J. Deift, P., Johansson, K.: On the distribution of the length of the second row of a Young diagram under Plancherel measure. Geom. Funct. Anal. **10**(4), 702–731 (2000)
3. Baik, J. Deift, P., Johansson, K.: Addendum to: on the distribution of the length of the second row of a Young diagram under Plancherel measure. Geom. Funct. Anal. **10**(6), 1606–1607 (2000)
4. Baryshnikov, Y.: GUEs and queues. Probab. Theory Related Fields **119**(2), 256–274 (2001)
5. Billingsley, P.: Convergence of probability measures. In: Wiley Series in Probability and Statistics: Probability and Statistics, 2nd edn. Wiley, New York (1999) (A Wiley-Interscience Publication)
6. Borodin, A., Okounkov, A., Olshanski, G.: Asymptotics of Plancherel measures for symmetric groups. J. Am. Math. Soc. **13**(3), 481–515 (electronic) (2000)
7. Chistyakov, G.P., Götze, F.: Distribution of the shape of Markovian random words. Probab. Theory Related Fields **129**(1), 18–36 (2004)
8. Gravner, J., Tracy, C., Widom, H.: Limit theorems for height fluctuations in a class of discrete space and time growth models. J. Statist. Phys. **102**(5–6), 1085–1132 (2001)
9. Houdré, C., Litherland, T.: On the longest increasing subsequence for finite and countable alphabets. In: High Dimensional Probability V: The Luminy Volume, pp. 185–212. Institute of Mathematical Statistics, Beachwood (2009)
10. Houdré C. and Litherland, T., On the limiting shape of Young diagrams associated with Markov random words. ArXiv #math.Pr/1110.4570 (2011)
11. Houdré C., Lember, J., Matzinger, H.: On the longest common increasing binary subsequence. C. R. Acad. Sci. Paris **343**(9), 589–594 (2006)
12. Houdré C., Xu, H.: On the limiting shape of Young diagrams associated with inhomogeneous random words, HDP VI: The Banff Volume (Progress in Probability), Birkhaüser (2013)
13. Its, A.R., Tracy, C.A., Widom, H.: Random words, Toeplitz determinants, and integrable systems. I. In: Random Matrix Models and their Applications, vol. 40 of Mathematical Sciences Research Institute Publications, pp. 245–258. Cambridge University Press, Cambridge (2001)
14. Its, A.R., Tracy, C.A., Widom, H.: Random words, Toeplitz determinants, and integrable systems. II. Advances in Nonlinear Mathematics and Science, Phys. D **152/153**, 199–224 (2001)

15. Johansson, K.: Discrete orthogonal polynomial ensembles and the Plancherel measure. Ann. Math. **153**(2), 259–296 (2001)
16. Kuperberg, G.: Random words, quantum statistics, central limits, random matrices. Methods. Appl. Anal. **9**(1), 99–118 (2002)
17. Okounkov, A.: Random matrices and random permutations. Int. Math. Res. Notices **2000**(20), 1043–1095 (2000)
18. Tracy, C.A., Widom, H.: On the distributions of the lengths of the longest monotone subsequences in random words. Probab. Theory Related Fields **119**(3), 350–380 (2001)

Chapter 24
A Short Rate Model Using Ambit Processes

José Manuel Corcuera, Gergely Farkas, Wim Schoutens, and Esko Valkeila

Abstract In this article, we study a bond market where short rates evolve as

$$r_t = \int_{-\infty}^{t} g(t-s)\sigma_s W(\mathrm{d}s)$$

where $g : (0, \infty) \to \mathbb{R}$ is deterministic, $\sigma \geq 0$ is also deterministic, and W is the stochastic Wiener measure. Processes of this type are also called Brownian semistationary processes and they are particular cases of ambit processes. These processes are, in general, not of the semimartingale kind. We also study a fractional version of the Cox–Ingersoll–Ross model. Some calibration and simulations are also done.

Keywords Bond market • Gaussian processes • Nonsemimartingales • Short rates • Volatility • Cox–Ingersoll–Ross model

Received 12/1/2011; Accepted 2/23/2012; Final 4/3/2012

J.M. Corcuera (✉)
Universitat de Barcelona, Gran Via de les Corts Catalanes 585, 08007 Barcelona, Spain
e-mail: jmcorcuera@ub.edu.

G. Farkas
Universitat de Barcelona, Barcelona, Spain
e-mail: farkasge@gmail.com.

W. Schoutens
K.U. Leuven, Leuven, Belgium
e-mail: wim.schoutens@wis.kuleuven.be.

E. Valkeila
Department of Mathematics and Systems Analysis, Aalto University,
P.O. Box 11100, 00076 Aalto, Helsinki, Finland
e-mail: esko.valkeila@aalto.fi

F. Viens et al. (eds.), *Malliavin Calculus and Stochastic Analysis: A Festschrift in Honor of David Nualart*, Springer Proceedings in Mathematics & Statistics 34,
DOI 10.1007/978-1-4614-5906-4_24,

1 Introduction

In this paper we study a bond market where short rates evolve as

$$r_t = \int_{-\infty}^{t} g(t-s)\sigma_s W(\mathrm{d}s),$$

where $g : (0,\infty) \to \mathbb{R}$ is deterministic, $\sigma \geq 0$ is also deterministic, and W is the stochastic Wiener measure. Processes of this type are particular cases of ambit processes. These processes are, in general, not of the semimartingale kind. Our purpose is to see if these new models can capture the features of the bond market by extending popular models like the Vasicek model. Affine models are quite popular as short rate models (see for instance [5]) but they imply a perfect correlation between bond prices and short rates, something unobservable in real markets. Moreover, the long-range dependence in the short interest rates (see [7]) and also in the intensity of default in credit risk models (see [3, 8]) is not captured by these affine models.

We model the short rates under the risk neutral probability and we obtain formulas for bond prices and options on bonds. We also consider defaultable bonds where the short and intensity rates show long-range dependence. We also try to establish the dynamics corresponding to this ad hoc or statistical modelling. This leads us to study the stochastic calculus associated with certain ambit processes. The paper is structured as follows: in the next section we introduce the short rate model. In the second section we calculate the bond and option prices as well as the hedging strategies. In the third section we look for a dynamic version of the model that lead us to a stochastic calculus in a nonsemimartingale setting. In the fourth we discuss a credit risk model with long-range dependence and finally, in the fifth section, we discuss the analogous of the Cox–Ingersoll–Ross (CIR) model in this context and we do some calibration and simulations to see, as a first step, how these models can work in practice.

2 The Model of Short Rates

Let $(\Omega, \mathcal{F}, \mathbb{F}, P)$ be a filtered, complete probability space with $\mathbb{F} = (\mathcal{F}_t)_{t\in\mathbb{R}_+}$. Assume that, in this probability space

$$r_t = \int_{-\infty}^{t} g(t-s)\sigma_s W(\mathrm{d}s) + \mu_t, \tag{24.1}$$

where W is the stochastic Wiener measure under the risk neutral probability, $P^* \sim P$, g is a deterministic function on $\mathbb{R}_+$, $g \in L^2((0,\infty))$, and $\sigma \geq 0$ and μ are also deterministic. Notice that the process r is not a semimartingale if $g' \notin L^2((0,\infty))$. Furthermore, we also assume that

$$\int_{-\infty}^{t} g^2(t-s)\sigma_s^2 \mathrm{d}s < \infty \tag{24.2}$$

which ensures that $r_t < \infty$ almost surely. By an $(\mathcal{F}_t)$-stochastic Wiener measure we understand an L^2-valued measure such that, for any Borelian set A with $E(W(A)^2) < \infty$

$$W(A) \sim N(0, m(A)),$$

where m is the Lebesgue measure and if $A \subseteq [t, +\infty)$ then $W(A)$ is independent of $\mathcal{F}_t$. Note that for $a \in \mathbb{R}$ the process $\left\{B_t := \int_a^{t+a} W(\mathrm{d}s), t \geq 0\right\}$ is a standard Brownian motion.

3 Pricing and Hedging

3.1 Bond Prices

Set

$$P(t,T) = E_{P^*}\left(\exp\left(-\int_t^T r_s \mathrm{d}s\right)\middle| \mathcal{F}_t\right)$$

for the price at t of the zero-coupon bond with maturity time T. We assume that $\exp\left(-\int_0^T r_s \mathrm{d}s\right) \in L^1(P^*)$ in such a way that the discounted prices $\tilde{P}(t,T) := P(t,T)\exp\left\{-\int_0^t r_s \mathrm{d}s\right\}$ are P^*-martingales. Then we have

$$\begin{aligned}
\int_t^T r_s \mathrm{d}s &= \int_t^T \left(\int_{-\infty}^s g(s-u)\sigma_u W(\mathrm{d}u)\right)\mathrm{d}s + \int_t^T \mu_s \mathrm{d}s \\
&= \int_{-\infty}^t \sigma_u \left(\int_t^T g(s-u)\mathrm{d}s\right) W(\mathrm{d}u) \\
&\quad + \int_t^T \sigma_u \left(\int_u^T g(s-u)\mathrm{d}s\right) W(\mathrm{d}u) + \int_t^T \mu_s \mathrm{d}s \\
&= \int_{-\infty}^t \sigma_u c(u;t,T) W(\mathrm{d}u) \\
&\quad + \int_t^T \sigma_u c(u;u,T) W(\mathrm{d}u) + \int_t^T \mu_s \mathrm{d}s,
\end{aligned}$$

where

$$c(u;t,T) := \int_t^T g(s-u)\mathrm{d}s, t \geq u$$

and where we use the stochastic Fubini theorem. Its use is guaranteed by (24.2). Then

$$P(t,T) = \exp\left(A(t,T) - \int_{-\infty}^{t} \sigma_u c(u;t,T)W(\mathrm{d}u)\right),$$

where

$$\begin{aligned}A(t,T) &= \log E_{P*}\left(\exp\left(-\int_t^T \sigma_u c(u;u,T)W(\mathrm{d}u) - \int_t^T \mu_s \mathrm{d}s\right)\Big|\mathcal{F}_t\right)\\ &= \frac{1}{2}\int_t^T \sigma_u^2 c^2(u;u,T)\mathrm{d}u - \int_t^T \mu_s \mathrm{d}s\end{aligned}$$

and the variance of the yield $-\frac{1}{T-t}\log P(t,T)$ is given by

$$\operatorname{var}\left(-\frac{1}{T-t}\log P(t,T)\right) = \frac{1}{(T-t)^2}\int_{-\infty}^{t} \sigma_u^2 c^2(u;t,T)\mathrm{d}u.$$

The corresponding forward rates are given by

$$\begin{aligned}f(t,T) &= -\partial_T \log P(t,T)\\ &= -\partial_T\left(\frac{1}{2}\int_t^T \sigma_u^2 c^2(u;u,T)\mathrm{d}u\right) + \partial_T\left(\int_{-\infty}^{t} \sigma_u c(u;t,T)W(\mathrm{d}u)\right)\\ &\quad +\partial_T\left(\int_t^T \mu_s \mathrm{d}s\right)\\ &= -\int_t^T \sigma_u^2 g(T-u)c(u;u,T)\mathrm{d}u + \int_{-\infty}^{t} \sigma_u g(T-u)W(\mathrm{d}u) + \mu_T\end{aligned}$$

and

$$\operatorname{var}(f(t,T)) = \int_{-\infty}^{t} \sigma_u^2 g^2(T-u)\mathrm{d}u.$$

Note that

$$\begin{aligned}\mathrm{d}_t f(t,T) &= \sigma_t^2 g(T-t)c(t;t,T)\mathrm{d}t + \sigma_t g(T-t)W(\mathrm{d}t)\\ &= \alpha(t,T)\mathrm{d}t + \sigma(t,T)W(\mathrm{d}t),\end{aligned}$$

with

$$\begin{aligned}\sigma(t,T) &= \sigma_t g(T-t),\\ \alpha(t,T) &= \sigma_t^2 g(T-t)c(t;t,T).\end{aligned}$$

Obviously it satisfies the HJM condition (see Chap. 18 in [5]) of absence of arbitrage:

$$\begin{aligned}\alpha(t,T) &= \sigma(t,T)\int_t^T \sigma(t,s)\mathrm{d}s\\ &= \sigma_t g(T-t)\int_t^T \sigma_t g(t-s)\mathrm{d}s\\ &= \sigma_t^2 g(T-t)c(t;t,T).\end{aligned}$$

3.2 Completeness of the Market

It is easy to see that

$$\begin{aligned}\tilde{P}(t,T) &:= \frac{P(t,T)}{\exp\left\{\int_0^t r_s \mathrm{d}s\right\}}\\ &= P(0,T)\exp\left(-\int_0^t \sigma_u c(u;u,T)W(\mathrm{d}u) - \frac{1}{2}\int_0^t \sigma_u^2 c(u;u,T)^2\mathrm{d}u\right).\end{aligned}$$

In fact

$$\begin{aligned}A(0,T) &= \frac{1}{2}\int_0^T \sigma_u^2 c(u;u,T)^2\mathrm{d}u - \int_0^T \mu_s\mathrm{d}s\\ &= A(t,T) - \frac{1}{2}\int_0^t \sigma_u^2 c(u;u,T)^2\mathrm{d}u - \int_0^t \mu_s\mathrm{d}s,\end{aligned}$$

so

$$\begin{aligned}P(t,T) &= \exp\left(A(t,T) - \int_{-\infty}^t \sigma_u c(u;t,T)W(\mathrm{d}u)\right)\\ &= \exp\left(A(0,T) - \int_{-\infty}^0 \sigma_u c(u;0,T)W(\mathrm{d}u)\right)\\ &\quad\times\exp\left(-\frac{1}{2}\int_0^t \sigma_u^2 c^2(u;u,T)\mathrm{d}u + \int_0^t \mu_s\mathrm{d}s\right)\\ &\quad\times\exp\left(\int_{-\infty}^0 \sigma_u\left(c(u;0,T)-c(u;t,T)\right)W(\mathrm{d}u)\right)\\ &\quad\times\exp\left(-\int_0^t \sigma_u c(u;t,T)W(\mathrm{d}u)\right),\end{aligned}$$

consequently

$$P(t,T) = P(0,T)\exp\left(-\frac{1}{2}\int_0^t \sigma_u^2 c^2(u;u,T)\mathrm{d}u + \int_0^t \mu_s \mathrm{d}s\right)$$
$$\times\exp\left(\int_{-\infty}^0 \sigma_u c(u;0,t)W(\mathrm{d}u) - \int_0^t \sigma_u c(u;t,T)W(\mathrm{d}u)\right),$$

$$\exp\left\{\int_0^t r_s \mathrm{d}s\right\} = \exp\left\{\int_0^t \left(\int_{-\infty}^s \sigma_u g(s-u)W(\mathrm{d}u)\right)\mathrm{d}s + \int_0^t \mu_s \mathrm{d}s\right\}$$
$$= \exp\left\{\int_{-\infty}^0 \sigma_u c(u;0,t)W(\mathrm{d}u) + \int_0^t \sigma_u c(u;u,t)W(\mathrm{d}u) + \int_0^t \mu_s \mathrm{d}s\right\},$$

$$\tilde{P}(t,T) = P(0,T)\exp\left(-\frac{1}{2}\int_0^t \sigma_u^2 c^2(u;u,T)\mathrm{d}u\right)$$
$$\times\exp\left(-\int_0^t \sigma_u(c(u;t,T) + c(u;u,t))W(\mathrm{d}u)\right)$$
$$= P(0,T)\exp\left(-\int_0^t \sigma_u c(u;u,T)W(\mathrm{d}u) - \frac{1}{2}\int_0^t \sigma_u^2 c^2(u;u,T)\mathrm{d}u\right).$$

Therefore,

$$\mathrm{d}\tilde{P}(t,T) = -\tilde{P}(t,T)\sigma_t c(t;t,T)W(\mathrm{d}t), t \geq 0.$$

Let X be a P^*-square integrable, $\mathcal{F}_T$-measurable payoff. Consider the $(\mathcal{F}_t)$-martingale

$$M_t := E_{P^*}(X|\mathcal{F}_t), t \geq 0,$$

then by an extension of Brownian martingale representation theorem we can write

$$\mathrm{d}M_t = H_t W(\mathrm{d}t),$$

where H is an adapted square integrable process. The proof of this extension follows the same steps as the proof of the classical result (for more details, see [14], pp. 198–200). But we need a wider set of functions $\mathcal{E} = \left\{\exp\left(\int_{-\infty}^T f(s)\, W(\mathrm{d}s)\right) : f \in \mathcal{S}\right\}$ as total set in $L^2(\mathcal{F}_T, P^*)$, where $\mathcal{S}$ is the set of step functions with compact support on $(-\infty, T]$.

Let (ϕ_t^0, ϕ_t^1) be a self-financing portfolio built with a bank account and a T-bond; its value process is given by

$$V_t = \phi_t^0 \mathrm{e}^{\int_0^t r_s \mathrm{d}s} + \phi_t^1 P(t,T),$$

and, by the self-financing condition, the discounted value process $\tilde{V}$. satisfies

$$d\tilde{V}_t = \phi_t^1 d\tilde{P}(t,T).$$

So, if we take

$$\phi_t^1 = -\frac{H_t}{\tilde{P}(t,T)\sigma_t c(t;t,T)}$$

we can replicate X. In particular the bond with maturity T^* can be replicated by taking

$$\frac{P(t,T^*)c(t;t,T^*)}{P(t,T)c(t;t,T)}$$

bonds with maturity time $T \geq T^*$.

3.3 Option Prices

Consider a bond with maturity $\bar{T} > T$, where T is the maturity time of a call option for this bond with strike K. Its price is given by (see [5], Chap. 19)

$$\begin{aligned}\Pi(t;T) &= P(t,\bar{T})P^{\bar{T}}(P(T,\bar{T}) \geq K|\mathcal{F}_t) - KP(t,T)P^T(P(T,\bar{T}) \geq K|\mathcal{F}_t)\\ &= P(t,\bar{T})P^{\bar{T}}\left(\frac{P(T,T)}{P(T,\bar{T})} \leq \frac{1}{K}\middle| \mathcal{F}_t\right) - KP(t,T)P^T\left(\frac{P(T,\bar{T})}{P(T,T)} \geq K\middle| \mathcal{F}_t\right),\end{aligned}$$

where P^T is the T-forward measure and analogously for $P^{\bar{T}}$. Define

$$U(t,T,\bar{T}) := \frac{P(t,T)}{P(t,\bar{T})}.$$

Then

$$U(t;T,\bar{T}) = \exp\left\{-A(t,\bar{T}) + A(t,T) - \int_{-\infty}^t \sigma_u \left(c(u;t,T) - c(u;t,\bar{T})\right) W(du)\right\}.$$

If we take the $\bar{T}$-forward measure $P^{\bar{T}}$, we will have that

$$W(du) = W^{\bar{T}}(du) - a(u)du,$$

where $W^{\bar{T}}(du)$ is a random Wiener measure in $\mathbb{R}$ again. Then, since $U(t,T,\bar{T})$ has to be a martingale with respect to $P^{\bar{T}}$, $a(u)$ is deterministic and we also have that

$$U(t;T,\bar{T}) = \exp\left\{-\int_{-\infty}^{t} \sigma_u \left(c(u;t,T) - c(u;t,\bar{T})\right) W^{\bar{T}}(\mathrm{d}u) \right.$$
$$\left. -\frac{1}{2}\int_{-\infty}^{t} \sigma_u^2 \left(c(u;t,T) - c(u;t,\bar{T})\right)^2 \mathrm{d}u\right\},$$

so

$$U(T) := U(T;T,\bar{T}) = U(t;T,\bar{T}) \exp\left\{ \int_{t}^{T} \sigma_u c(u;T,\bar{T}) W^{\bar{T}}(\mathrm{d}u) \right.$$
$$\left. -\frac{1}{2}\int_{t}^{T} \sigma_u^2 c(u;T,\bar{T})^2 \mathrm{d}u\right\}$$

and analogously

$$U(T)^{-1} = U(T;\bar{T},T) = U^{-1}(t;T,\bar{T}) \exp\left\{ -\int_{t}^{T} \sigma_u c(u;T,\bar{T}) W^{T}(\mathrm{d}u) \right.$$
$$\left. -\frac{1}{2}\int_{t}^{T} \sigma_u^2 c(u;T,\bar{T})^2 \mathrm{d}u\right\}.$$

Therefore

$$\begin{aligned}
\Pi(t;T) &= P(t,\bar{T})P^{\bar{T}}(U(T) \leq \frac{1}{K}|\mathcal{F}_t) - KP(t,T)P^{T}(U^{-1}(T) \geq K|\mathcal{F}_t) \\
&= P(t,\bar{T})P^{\bar{T}}(\log U(T) \leq -\log K|\mathcal{F}_t) - KP(t,T)P^{T}(\log U^{-1}(T) \\
&\geq \log K|\mathcal{F}_t) \\
&= P(t,\bar{T})\Phi(d_+) - KP(t,T)\Phi(d_-),
\end{aligned}$$

where

$$d\pm = \frac{\log \frac{P(t,\bar{T})}{KP(t,T)} \pm \frac{1}{2}\Sigma^2_{t,T,\bar{T}}}{\Sigma_{t,T,\bar{T}}}$$

and

$$\Sigma^2_{t,T,\bar{T}} := \int_{t}^{T} \sigma_u^2 c(u;T,\bar{T})^2 \mathrm{d}u.$$

3.4 Examples

Example 3.1. If

$$g(t) = \mathrm{e}^{-bt}, \quad \sigma_u = \sigma, \text{ and } \mu = a,$$

we have

$$\begin{aligned} r_t &= a + \mathrm{e}^{-bt} \int_{-\infty}^{0} \mathrm{e}^{bs} \sigma W(\mathrm{d}s) + \mathrm{e}^{-bt} \int_{0}^{t} \mathrm{e}^{bs} \sigma W(\mathrm{d}s) \\ &= r_0 \mathrm{e}^{-bt} + a(1 - \mathrm{e}^{-bt}) + \mathrm{e}^{-bt} \int_{0}^{t} \mathrm{e}^{bs} \sigma W(\mathrm{d}s), \end{aligned}$$

that is the Vasicek model, and

$$\begin{aligned} P(t,T) &= \exp\left(A(t,T) - \int_t^T \left(\int_{-\infty}^{t} \sigma g(s-u) W(\mathrm{d}u) \right) \mathrm{d}s \right) \\ &= \exp\left(A(t,T) - \int_t^T \left(\int_{-\infty}^{t} \frac{g(s-u)}{g(t-u)} \sigma g(t-u) W(\mathrm{d}u) \right) \mathrm{d}s \right) \\ &= \exp\left(A(t,T) - \int_t^T \mathrm{e}^{-b(s-t)} \left(\int_{-\infty}^{t} \sigma \mathrm{e}^{-b(t-u)} W(\mathrm{d}u) \right) \mathrm{d}s \right) \\ &= \exp\left(A(t,T) - (r_t - a) \int_t^T \mathrm{e}^{-b(s-t)} \mathrm{d}s \right) \\ &= \exp\left(A(t,T) + aB(t,T) - r_t B(t,T) \right), \end{aligned}$$

with

$$B(t,T) = \frac{1}{b}(1 - \mathrm{e}^{-b(T-t)})$$

and

$$\begin{aligned} A(t,T) &= \frac{\sigma^2}{2} \int_t^T \left(\int_u^T g(s-u) \mathrm{d}s \right)^2 \mathrm{d}u - a(T-t) \\ &= \frac{\sigma^2}{2} \int_t^T B(u,T)^2 \mathrm{d}u - a(T-t). \end{aligned}$$

Here

$$c(u;t,T) = \frac{1}{b} \left(\mathrm{e}^{-b(t-u)} - \mathrm{e}^{-b(T-u)} \right), u \le t \le T,$$

so

$$\begin{aligned} \operatorname{var}\left(-\frac{1}{T-t} \log P(t,T) \right) &= \frac{1}{(T-t)^2} \int_{-\infty}^{t} \sigma_u^2 c^2(u;t,T) \mathrm{d}u \\ &= \frac{\sigma^2}{2b^3} \frac{(1 - \mathrm{e}^{-b(T-t)})^2}{(T-t)^2} \sim T^{-2}, \end{aligned}$$

when $T \to \infty$. The corresponding instantaneous forward rates are given by

$$f(t,T) = -\frac{\sigma^2}{2b^2}\left(1 - e^{-b(T-t)}\right)^2 + \sigma e^{-b(T-t)}(r_t - a) + a,$$

$$\begin{aligned}\operatorname{var}(f(t,T)) &= \int_{-\infty}^{t} \sigma_u^2 g^2(T-u)\mathrm{d}u \\ &= \sigma^2 \int_{-\infty}^{t} e^{-2b(T-u)}\mathrm{d}u = \frac{\sigma^2}{2b} e^{-2b(T-t)} \sim e^{-2bT},\end{aligned}$$

when $T \to \infty$. Moreover the volatility of the forward rates is given by $\sigma(t,T) = \sigma e^{-b(T-t)}$ and this is not too realistic.

Example 3.2. Assume that $\sigma_t = \sigma \mathbf{1}_{\{t \geq 0\}}$ and

$$g(t-u) = e^{-b(t-u)} \int_0^{t-u} e^{bs} \beta s^{\beta-1} \mathrm{d}s,$$

for $\beta \in (0, 1/2)$. We have that

$$c(u;t,T) := \int_t^T g(s-u)\mathrm{d}s = c(0;0,T-u) - c(0;0,t-u),$$

with

$$c(0;0,x) = e^{-bx} \int_0^x e^{bs} s^{\beta} \mathrm{d}s.$$

Then

$$\begin{aligned}\operatorname{var}\left(-\frac{1}{T-t}\log P(t,T)\right) &= \frac{1}{(T-t)^2}\int_{-\infty}^{t} \sigma_u^2 c^2(u;t,T)\mathrm{d}u \\ &= \frac{\sigma^2}{2}\frac{1}{(T-t)^2}\int_0^t (c(0;0,T-u) - c(0;0,t-u))^2 \mathrm{d}u \\ &\sim \frac{1}{T^2}\int_0^t c(0;0,T-u)^2 \mathrm{d}u \sim T^{2\beta-2},\end{aligned}$$

when $T \to \infty$. In fact

$$c(0;0,x) = e^{-bx} \int_0^x e^{bs} s^{\beta} \mathrm{d}s = x^{\beta} \int_0^x e^{-bs}\left(1 - \frac{s}{x}\right)^{\beta} \mathrm{d}s,$$

and by the monotone convergence theorem

$$\lim_{x \to \infty} \int_0^x e^{-bs}(1 - \frac{s}{x})^{\beta} \mathrm{d}s = \int_0^{\infty} e^{-bs} \mathrm{d}s = \frac{1}{b}.$$

Moreover

$$\mathrm{var}\,(f(t,T)) = \int_{-\infty}^{t} \sigma_u^2 g^2(T-u)\mathrm{d}u \sim T^{2\beta-2}.$$

Since for $x \geq 0$

$$g(x) = \mathrm{e}^{-bx}\int_0^x \mathrm{e}^{bs}\beta s^{\beta-1}\mathrm{d}s = \beta x^{\beta-1}\int_0^x \mathrm{e}^{-bs}(1-\frac{s}{x})^{\beta-1}\mathrm{d}s$$
$$= \beta x^{\beta-1}\left(\int_0^{x/2} \mathrm{e}^{-bs}(1-\frac{s}{x})^{\beta-1}\mathrm{d}s + \int_{x/2}^{x} \mathrm{e}^{-bs}(1-\frac{s}{x})^{\beta-1}\mathrm{d}s\right),$$

and

$$\lim_{x\to\infty}\int_0^{x/2} \mathrm{e}^{-bs}(1-\frac{s}{x})^{\beta-1}\mathrm{d}s = \int_0^{\infty} \mathrm{e}^{-bs}\mathrm{d}s = \frac{1}{b},$$

$$\int_{x/2}^{x} \mathrm{e}^{-bs}(1-\frac{s}{x})^{\beta-1}\mathrm{d}s \leq \mathrm{e}^{-bx/2}\int_{x/2}^{x}(1-\frac{s}{x})^{\beta-1}\mathrm{d}s$$
$$= x\mathrm{e}^{-bx/2}\int_0^{1/2} v^{\beta-1}\mathrm{d}v = \frac{x\mathrm{e}^{-bx/2}}{\beta 2^{\beta}} \to 0,$$

when $x \to \infty$. Also observe that the volatility of the forward rates $\sigma(t,T) = \sigma^2 g(T-t) \sim T^{\beta-1}$, when $T \to \infty$, that is more realistic (see Sect. 4.1 in [7] and also [2]) than the exponential decay in the Vasicek model. For $\beta \in (-1/2, 0)$ consider the memory function

$$g(x) = \mathrm{e}^{-bx}x^{\beta} + \beta\int_0^x (\mathrm{e}^{-b(x-u)} - \mathrm{e}^{-bx})u^{\beta-1}\mathrm{d}u,$$

and then

$$g(x) \sim x^{\beta-1}$$

when $x \to \infty$. In such a way that we obtain analogous asymptotic results to the previous case.

4 An SDE Approach

We have postulated that

$$r_t = \int_{-\infty}^{t} g(t-s)\sigma_s W(\mathrm{d}s) + \mu_t,$$

and the question is if this process $(r_t)_{t\in\mathbb{R}}$ can be seen as the solution of such a stochastic differential equation. For instance, assume that

$$\mathrm{d}r_t = b(a - r_t)\mathrm{d}t + \sigma W(\mathrm{d}t),$$

then we have

$$r_t = r_0\mathrm{e}^{-bt} + a(1 - \mathrm{e}^{-bt}) + \mathrm{e}^{-bt}\int_0^t \mathrm{e}^{bs}\sigma W(\mathrm{d}s),$$

and if we take

$$r_0 = \int_{-\infty}^0 \mathrm{e}^{bs}\sigma W(\mathrm{d}s) + a,$$

we obtain that

$$r_t = a + \int_{-\infty}^t \mathrm{e}^{-b(t-s)}\sigma W(\mathrm{d}s).$$

So, it corresponds to $g(t) = \mathrm{e}^{-bt}$, $\sigma_s = \sigma$, and $\mu_t = a$.

4.1 Ambit Processes as Noises of SDE

Consider the processes W^g given by

$$W_t^g := \int_{-\infty}^t g(s,t)W(\mathrm{d}s),$$

where $g : \mathbb{R}^2 \to \mathbb{R}$ deterministic, continuously differentiable with respect to the second variable, $g(s,t) = 0$ if $s > t$ and $\int_{-\infty}^t g^2(s,t)\mathrm{d}s < \infty$. In this section we explain how a stochastic calculus can be developed with respect to these processes. Here we follow [1, 7, 13]. First, formally,

$$W_t^g(\mathrm{d}t) = g(t,t)W(\mathrm{d}t) + \left(\int_{-\infty}^t \partial_t g(s,t)W(\mathrm{d}s)\right)\mathrm{d}t,$$

and for a deterministic function $f(\cdot,\cdot)$, we can define

$$\begin{aligned}
&\int_{-\infty}^t f(u,t)W_t^g(\mathrm{d}u) \\
&= \int_{-\infty}^t f(u,t)\left(g(u,u)W(\mathrm{d}u) + \left(\int_{-\infty}^u \partial_u g(s,u)W(\mathrm{d}s)\right)\mathrm{d}u\right) \\
&= \int_{-\infty}^t \left(\int_{-\infty}^u (f(u,t) - f(s,t))\,\partial_u g(s,u)W(\mathrm{d}s)\right)\mathrm{d}u
\end{aligned}$$

$$
\begin{aligned}
&+\int_{-\infty}^{t}\left(\int_{s}^{t} f(s,t)\partial_u g(s,u)\mathrm{d}u\right) W(\mathrm{d}s)\\
&+\int_{-\infty}^{t} f(u,t)g(u,u)W(\mathrm{d}u)\\
=&\int_{-\infty}^{t}\left(\int_{s}^{t} \left(f(u,t)-f(s,t)\right)\partial_u g(s,u)\mathrm{d}u\right) W(\mathrm{d}s)\\
&+\int_{-\infty}^{t} f(s,t)g(s,t)W(\mathrm{d}s)\\
=&\int_{-\infty}^{t}\left(\int_{s}^{t} \left(f(u,t)-f(s,t)\right)\partial_u g(s,u)\mathrm{d}u + f(s,t)g(s,t)\right) W(\mathrm{d}s).
\end{aligned}
$$

Then, the latest integral is well defined in an L^2 sense, provided that

$$
\int_{-\infty}^{t}\left(\int_{s}^{t} \left(f(u,t)-f(s,t)\right)\partial_u g(s,u)\mathrm{d}u + f(s,t)g(s,t)\right)^2 \mathrm{d}s < \infty.
$$

Now, if we construct the operator

$$
K_t^g(f)(s,t) := \int_{s}^{t} \left(f(u,t)-f(s,t)\right)\partial_u g(s,u)\mathrm{d}u + f(s,t)g(s,t),
$$

it is natural to define

$$
\int_{-\infty}^{t} f(s,t)W_t^g(\mathrm{d}s) := \int_{-\infty}^{t} K_t^g(f)(s,t)W(\mathrm{d}s),
$$

provided that $f(\cdot,t) \in \left(K_t^g\right)^{-1}(L^2(-\infty,t])$.

Note that if $g(s,s)=0$, then we can write

$$
K_t^g(f)(s,t) := \int_{s}^{t} f(u,t)\partial_u g(s,u)\mathrm{d}u, \tag{24.3}
$$

and in the particular case that $\Delta f = 0$, we have

$$
\begin{aligned}
K_t^g(f)(s,t) &= \partial_t \int_{s}^{t} f(u,t)g(s,u)\mathrm{d}u\\
&= \partial_t\left(f * g\right)(s,t),
\end{aligned}
$$

and

$$
\begin{aligned}
\int_{-\infty}^{t} f(s,t)W_t^g(\mathrm{d}s) &= \int_{-\infty}^{t}\left(\partial_t \int_{s}^{t} f(u,t)g(s,u)\mathrm{d}u\right) W(\mathrm{d}s)\\
&= \frac{\mathrm{d}}{\mathrm{d}t}\int_{-\infty}^{t}\int_{s}^{t} f(u,t)g(s,u)\mathrm{d}uW(\mathrm{d}s)
\end{aligned}
$$

$$= \frac{\mathrm{d}}{\mathrm{d}t}\int_{-\infty}^{t} f(u,t)\left(\int_{-\infty}^{u} g(s,u)W(\mathrm{d}s)\right)\mathrm{d}u$$
$$= \frac{\mathrm{d}}{\mathrm{d}t}\int_{-\infty}^{t} f(u,t)W_u^g\mathrm{d}u.$$

Consider now

$$r_t = b\int_0^t (a - r_s)\mathrm{d}s + \sigma\int_0^t (t-s)^\beta W(\mathrm{d}s),$$

with $\beta \in (-1/2, 0) \cup (0, 1/2)$, then if we define

$$W_t^\beta := \int_0^t (t-s)^\beta W(\mathrm{d}s),$$

$$r_t = b\int_0^t (a - r_s)\mathrm{d}s + \sigma W^\beta(t).$$

In such a way that (r_t) is an Ornstein–Uhlenbeck process driven by W^β.

We obtain

$$r_t = r_0\mathrm{e}^{-bt} + a(1-\mathrm{e}^{-bt}) + \mathrm{e}^{-bt}\int_0^t \mathrm{e}^{bs}\sigma W^\beta(\mathrm{d}s)$$
$$= r_0\mathrm{e}^{-bt} + a(1-\mathrm{e}^{-bt}) + \int_0^t \sigma g(t-s)W(\mathrm{d}u).$$

Then, if $\beta \in (0, 1/2)$, by (24.3) we have

$$\int_0^t \mathrm{e}^{-b(t-s)}W^\beta(\mathrm{d}s) = \int_0^t \left(\int_u^t \mathrm{e}^{-b(t-s)}\beta(s-u)^{\beta-1}\mathrm{d}s\right)W(\mathrm{d}u)$$
$$= \int_0^t \left(\int_0^{t-u} \mathrm{e}^{-b(t-s-u)}\beta s^{\beta-1}\mathrm{d}s\right)W(\mathrm{d}u).$$
$$= \int_0^t \mathrm{e}^{-b(t-u)}\left(\int_0^{t-u} \mathrm{e}^{bs}\beta s^{\beta-1}\mathrm{d}s\right)W(\mathrm{d}u).$$

In such a way that

$$g(t-s) = \mathrm{e}^{-b(t-s)}\left(\int_0^{t-s} \mathrm{e}^{bu}\beta u^{\beta-1}\mathrm{d}u\right),$$

and if $\beta \in (-1/2, 0)$

$$g(t-s) = \mathrm{e}^{-b(t-s)}(t-s)^\beta + \beta\mathrm{e}^{-b(t-s)}\int_0^{t-s} (\mathrm{e}^{bu}-1)v^{\beta-1}\mathrm{d}u.$$

5 A Defaultable Zero-Coupon Bond

The purpose in this section is to price a zero-coupon bond with possibility of default. The payoff of this contract at the maturity time is $1_{\{\tau>T\}}$, where τ is the default time. Then, an arbitrage free price at time t is given by

$$D(t,T) = 1_{\{\tau>t\}} E\left(1_{\{\tau>T\}} e^{-\int_t^T r_s ds} \middle| \mathcal{G}_t\right), 0 \le t \le T,$$

where the expectation is taken with respect to a risk neutral probability, P^*, and where the filtration $\mathbb{G} = (\mathcal{G}_t)_{t\ge 0}$ represents the information available to the market. Here we follow the hazard process approach (for more details, see Sect. 8.2 in [4]). In this approach we consider two filtrations, one is the default-free filtration $\mathbb{F} = (\mathcal{F}_t)_{t\ge 0}$ that typically incorporates the history of the short rates. The default time is modelled by a random variable τ that is not necessarily an $\mathbb{F}$-stopping time, then the other filtration is $\mathbb{G} = (\mathcal{G}_t)_{t\ge 0}$, where

$$\mathcal{G}_t = \mathcal{F}_t \vee \sigma(\tau \wedge t),$$

in such a way that τ is a $\mathbb{G}$-stopping time. Now, if we assume that there exists an $\mathbb{F}$-adapted process $(\lambda_t)_{t\ge 0}$, the so-called hazard process, such that

$$P^*(\tau > t|\mathcal{F}_t) = e^{-\int_0^t \lambda_s ds},$$

it can be shown (see [12], Chap. 8) that

$$D(t,T) = 1_{\{\tau>t\}} E\left(1_{\{\tau>T\}} e^{-\int_t^T r_s ds} \middle| \mathcal{G}_t\right) = 1_{\{\tau>t\}} E\left(e^{-\int_t^T (r_s+\lambda_s) ds} \middle| \mathcal{F}_t\right).$$

Then we need a model for $(r_t)_{t\ge 0}$ and $(\lambda_t)_{t\ge 0}$. A classical model is a Vasicek model for both processes

$$\begin{aligned} dr_t &= b(a - r_t)dt + \sigma dW(t), \\ d\lambda_t &= \breve{b}(\breve{a} - \lambda_t)dt + \breve{\sigma} d\breve{W}(t), \end{aligned}$$

where W and $\breve{W}$ are correlated Brownian motions and here $\mathcal{F}_t = \sigma(W_s, \breve{W}_s, 0 \le s \le t)$. The idea is to extend this model by considering ambit processes as noises in the stochastic differential equations. For instance we can have

$$\begin{aligned} r_t &= \int_{-\infty}^t \sigma_s g(t-s) W(ds) + \mu_t, \\ \lambda_t &= \int_{-\infty}^t \breve{\sigma}_s \breve{g}(t-s) \breve{W}(ds) + \breve{\mu}_t. \end{aligned}$$

See [3] for a similar modelling. Then, the price of a defaultable zero-coupon bond at time t will be given by

$$D(t,T) = 1_{\{\tau>t\}} \exp\left(A(t,T) - \int_{-\infty}^{t} (\sigma_u c(u;t,T) W(\mathrm{d}u) + \breve{\sigma}_u \breve{c}(u;t,T)) \breve{W}(\mathrm{d}u) \right),$$

where

$$A(t,T) = \frac{1}{2} \int_t^T \left(\sigma_u^2 c^2(u;t,T) + \breve{\sigma}_u^2 \breve{c}^2(u;t,T) + 2\rho \sigma_u \breve{\sigma}_u c(u;t,T) \breve{c}(u;t,T)\right) \mathrm{d}u$$
$$- \int_t^T (\mu_u + \breve{\mu}_u) \,\mathrm{d}u$$

and ρ is the correlation coefficient between W and $\breve{W}$. Interesting cases are $\sigma_u = \sigma 1_{\{u \geq 0\}}$, $\sigma_u = \breve{\sigma} 1_{\{u \geq 0\}}$, $\mu_u = \mu$, $\breve{\mu}_u = \breve{\mu}$,

$$g(t-s) = \mathrm{e}^{-b(t-s)} \int_0^{t-s} \mathrm{e}^{bu} \beta u^{\beta-1} \mathrm{d}u,$$

$$\breve{g}(t-s) = \mathrm{e}^{-\breve{b}(t-s)} \int_0^{t-s} \mathrm{e}^{\breve{b}u} \breve{\beta} u^{\breve{\beta}-1} \mathrm{d}u,$$

$\beta, \breve{\beta} \in (-1/2, 0) \cup (0, 1/2)$. Note that

$$\operatorname{var}\left(-\frac{1}{T-t} \log D(t,T) \right) \sim T^{2(\beta \vee \breve{\beta})-2}.$$

6 The Analogue of a CIR Model

One of the drawbacks of the previous model is that it allows for negative short rates. An obvious way of avoiding this is to take

$$r_t = \sum_{i=1}^{d} \left(\int_0^t g(t-s) \sigma_s \mathrm{d}W_i(s) \right)^2 + r_0, \quad t \geq 0, r_0 > 0,$$

where $(W_i))_{1 \leq i \leq d}$ is a Brownian motion in $\mathbb{R}^d$.

6.1 Bond Prices

$$r_t = \sum_{i=1}^{d} \int_0^t \int_0^t g(t-u) g(t-v) \sigma_s \sigma_u \mathrm{d}W_i(u) \mathrm{d}W_i(v),$$

where by simplicity we take $r_0 = 0$, then

$$
\begin{aligned}
\int_t^T r_s \mathrm{d}s &= \sum_{i=1}^d \int_t^T \left(\int_0^s g(s-u)g(s-v)\sigma_u\sigma_v \mathrm{d}W_i(u)\mathrm{d}W_i(v) \right) \mathrm{d}s \\
&= \sum_{i=1}^d \int_0^t \int_0^t \sigma_u\sigma_v \left(\int_t^T g(s-u)g(s-v)\mathrm{d}s \right) \mathrm{d}W_i(u)\mathrm{d}W_i(v) \\
&\quad +2\sum_{i=1}^d \int_0^t \int_t^T \sigma_u\sigma_v \left(\int_u^T g(s-u)g(s-v)\mathrm{d}s \right) W_i(\mathrm{d}u)W_i(\mathrm{d}v) \\
&\quad +\sum_{i=1}^d \int_t^T \int_t^T \sigma_u\sigma_v \left(\int_{u\vee v}^T g(s-u)g(s-v)\mathrm{d}s \right) W_i(\mathrm{d}u)W_i(\mathrm{d}v) \\
&= \sum_{i=1}^d \int_0^t \int_0^t \sigma_u\sigma_v c_2(u,v;t,T)\mathrm{d}W_i(u)\mathrm{d}W_i(v) \\
&\quad +2\sum_{i=1}^d \int_0^t \int_t^T \sigma_u\sigma_v c_2(u,v;u,T)\mathrm{d}W_i(u)\mathrm{d}W_i(v) \\
&\quad +\sum_{i=1}^d \int_t^T \int_t^T \sigma_u\sigma_v c_2(u,v;u\vee v,T)\mathrm{d}W_i(u)\mathrm{d}W_i(v),
\end{aligned}
$$

with $c_2(u,v;t,T) := \int_t^T g(s-u)g(s-v)\mathrm{d}s$.

$$
\begin{aligned}
P(0,T) &= E\left(\exp\left\{ -\int_0^T r_s \mathrm{d}s \right\} \right) \\
&= E\left(\exp\left\{ -\sum_{i=1}^d \int_0^T \int_0^T \sigma_u\sigma_v c_2(u,v;u\vee v,T)\mathrm{d}W_i(u)\mathrm{d}W_i(v) \right\} \right) \\
&= \prod_{i=1}^d E\left(\exp\left\{ -T \int_0^1 \int_0^1 \sigma_{Tu}\sigma_{Tv} c_2(Tu,Tv;T(u\vee v),T)\mathrm{d}W_i(u)\mathrm{d}W_i(v) \right\} \right) \\
&= \left(1 + \sum_{n=1}^\infty \frac{(2T)^n}{n!} \int_0^1 \cdots \int_0^1 \begin{vmatrix} R(s_1,s_1) & \cdots & R(s_1,s_n) \\ \vdots & & \vdots \\ R(s_n,s_1) & \cdots & R(s_n,s_n) \end{vmatrix} \mathrm{d}s_1 \cdots \mathrm{d}s_n \right)^{-d/2},
\end{aligned}
$$

where

$$
R(u,v) = \sigma_{Tu}\sigma_{Tv} c_2(Tu,Tv;T(u\vee v),T).
$$

In the second equality we use the scaling property of the Brownian motion and in the third Corollary 4 in [15].

Example 6.1. Assume that $g(t) = \mathbf{1}_{\{t\geq 0\}}$ and $\sigma_t = \sigma$. Then r_t is a squared Bessel process of dimension d (see for instance [10]) and

$$R(u,v) = \sigma^2 T(1-(u \vee v)),$$

consequently

$$P(0,T) = (\cosh(\sqrt{2}\sigma T)^{-\frac{d}{2}} = \frac{2^{\frac{d}{2}}}{\left(\mathrm{e}^{\sqrt{2}\sigma T} + \mathrm{e}^{-\sqrt{2}\sigma T}\right)^{\frac{d}{2}}}$$

(see [15] for the calculations of the Fredholm determinant),

$$d(\lambda) := \left(1 + \sum_{n=1}^{\infty} \frac{\lambda^n}{n!} \int_0^1 \cdots \int_0^1 \begin{vmatrix} R(s_1,s_1) & \cdots & R(s_1,s_n) \\ \vdots & & \vdots \\ R(s_n,s_1) & \cdots & R(s_n,s_n) \end{vmatrix} \mathrm{d}s_1 \cdots \mathrm{d}s_n \right).$$

Another procedure to calculate the Fredholm determinants is given in [11], where it is shown that provided the kernel $R(u,v)$ is of the form

$$R(u,v) = M(u \vee v) N(u \wedge v)$$

we have that

$$d(\lambda) = B_\lambda(1),$$

and therefore

$$P(0,T) = (B_{2T}(1))^{-\frac{d}{2}},$$

where $B_\lambda(t)$is defined by the linear differential equation system

$$\begin{pmatrix} \dot{A}_\lambda(t) \\ \dot{B}_\lambda(t) \end{pmatrix} = \lambda \begin{pmatrix} -N(t)M(t) & N^2(t) \\ -M^2(t) & N(t)M(t) \end{pmatrix} \begin{pmatrix} A_\lambda(t) \\ B_\lambda(t) \end{pmatrix},$$

$$\begin{pmatrix} A_\lambda(0) \\ B_\lambda(0) \end{pmatrix} = \begin{pmatrix} 0 \\ 1 \end{pmatrix}.$$

In our case $M(t) = \sigma^2 T(1-t)$ and $N(t) = 1$ and by straightforward calculations we obtain

$$B_\lambda(t) = \sigma^2 T^2 \left((1-t) \frac{\mathrm{e}^{\sigma\sqrt{\lambda T} t} - \mathrm{e}^{-\sigma\sqrt{\lambda T} t}}{\sigma\sqrt{\lambda T}} + \frac{\mathrm{e}^{\sigma\sqrt{\lambda T} t} + \mathrm{e}^{-\sigma\sqrt{\lambda T} t}}{\left(\sigma\sqrt{\lambda T}\right)^2} \right).$$

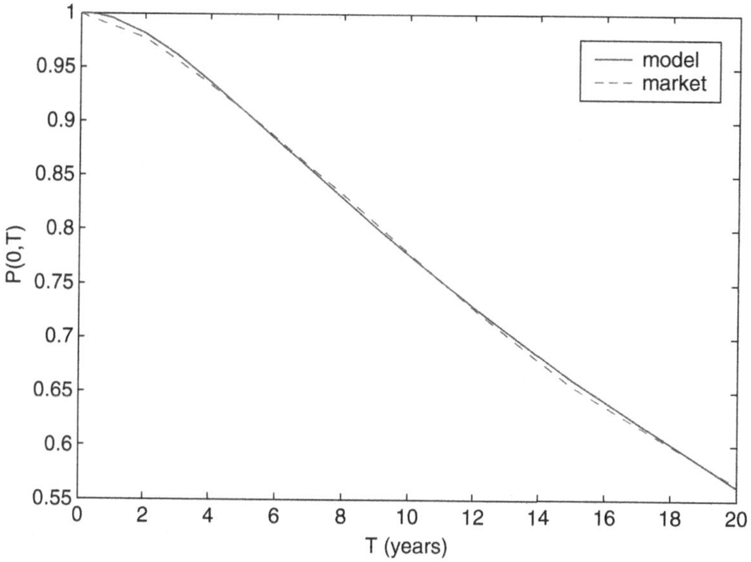

Fig. 24.1 EUR – Discount curve 04/11/2011 : $\sigma = 21.90\%$ and d $= 0.2093$

Note that we can consider squared Bessel processes of dimension $d \geq 0$, where d is not necessarily integer (see [10] and Corollary 6.2.5.5 therein). A calibration of this model is given in Fig. 24.1. We have performed a calibration of the model on the market discount curve of the 4th of November 2011. More precisely, we have on that date calibrated the d and σ parameters on the EUR market implied discount curve up to 20 years of maturity. The optimal parameters were obtained using a least-squared-error minimization employing a Nelder–Mead search algorithm. The calibrating is performed very fast and the optimal parameters are obtained in less than a second, due to the fact that discount values under the model are available in close form. Even though this model is not mean reverting the fit to real data is quite good.

Example 6.2. Another interesting example is the classical CIR model. In such a case

$$\begin{aligned} R(u,v) &= \sigma^2 \int_{T(u\vee v)}^{T} \mathrm{e}^{-b(s-u)}\mathrm{e}^{-b(s-v)}\mathrm{d}s = \frac{\sigma^2}{2b}\mathrm{e}^{bT((u\wedge v)-1)}(\mathrm{e}^{-bT((u\vee v)-1)} \\ &\quad -\mathrm{e}^{bT((u\vee v)-1)}) \\ &= M(u\vee v)N(u\wedge v), \end{aligned}$$

where

$$M(t) = \frac{\sigma}{\sqrt{2b}}\left(\mathrm{e}^{-bT(t-1)} - \mathrm{e}^{bT(t-1)}\right),$$
$$N(t) = \frac{\sigma}{\sqrt{2b}}\mathrm{e}^{bT(t-1)}.$$

Then we have the system

$$\begin{pmatrix} \dot{A}_\lambda(t) \\ \dot{B}_\lambda(t) \end{pmatrix} = \frac{\lambda\sigma^2}{2b}\begin{pmatrix} \mathrm{e}^{2bT(t-1)}\left(1-\mathrm{e}^{-2bT(t-1)}\right) & \mathrm{e}^{2bT(t-1)} \\ -\left(\mathrm{e}^{2bT(t-1)}-1\right)\left(1-\mathrm{e}^{-2bT(t-1)}\right) & -\left(\mathrm{e}^{2bT(t-1)}-1\right) \end{pmatrix} \times \begin{pmatrix} A_\lambda(t) \\ B_\lambda(t) \end{pmatrix},$$
$$\begin{pmatrix} A_\lambda(0) \\ B_\lambda(0) \end{pmatrix} = \begin{pmatrix} 0 \\ 1 \end{pmatrix}.$$

So,

$$\dot{B}_\lambda(t) = \left(\mathrm{e}^{-2bT(t-1)} - 1\right)\dot{A}_\lambda(t)$$

and

$$\ddot{A}_\lambda(t) = 2bT\dot{A}_\lambda(t) + \lambda\sigma^2 T A_\lambda(t);$$

from here we obtain that

$$A_{2T}(t) = C\left(\mathrm{e}^{T(b+\sqrt{b^2+2\sigma^2})t} - \mathrm{e}^{T(b-\sqrt{b^2+2\sigma^2})t}\right)$$

and that

$$B_{2T}(t) = C\left(\mathrm{e}^{-2bT(t-1)} - 1\right)\left(\mathrm{e}^{T\left(b+\sqrt{b^2+2\sigma^2}\right)t} - \mathrm{e}^{T(b-\sqrt{b^2+2\sigma^2})t}\right)$$
$$+C(-2b)T\mathrm{e}^{2bT}\left(\frac{\mathrm{e}^{T\left(-b+\sqrt{b^2+2\sigma^2}\right)t}}{-b+\sqrt{b^2+2\sigma^2}} - \frac{\mathrm{e}^{T\left(-b-\sqrt{b^2+2\sigma^2}\right)t}}{-b-\sqrt{b^2+2\sigma^2}}\right),$$

where $C = -\frac{\sigma^2\mathrm{e}^{-2bT}}{2bT\sqrt{b^2+2\sigma^2}}$. Therefore

$$B_{2T}(1) = \frac{1}{2\sqrt{b^2+2\sigma^2}}\left((b+\sqrt{b^2+2\sigma^2})\mathrm{e}^{T(-b+\sqrt{b^2+2\sigma^2})}\right.$$
$$\left.+(-b+\sqrt{b^2+2\sigma^2})\mathrm{e}^{-T(b+\sqrt{b^2+2\sigma^2})}\right).$$

6.2 Numerical Methods for Pricing

In case that the Fredholm determinant appearing in the price formula cannot be calculated analytically, efficient numerical methods are known [6]. The idea of the approximation is the following: first let denote

$$d^R(\lambda) = \sum_{n=0}^{\infty} \frac{\lambda^n}{n!} \int_0^1 \cdots \int_0^1 \begin{vmatrix} R(s_1, s_1) & \cdots & R(s_1, s_n) \\ \vdots & & \vdots \\ R(s_n, s_1) & \cdots & R(s_n, s_n) \end{vmatrix} ds_1 \cdots ds_n,$$

the price we are looking for equals $\left[d^R(2T)\right]^{-d/2}$; then, for a given quadrature formula

$$Q_m(f) = \sum_{j=1}^{m} w_j f(x_j) \approx \int_0^1 f(x)\, dx,$$

we consider the *Nyström-type* approximation of $d(\lambda)$:

$$d^R_{Q_m}(\lambda) = \det\left[\delta_{ij} + \lambda w_i R(x_i, x_j)\right]_{i,j=1}^{m}. \tag{24.4}$$

By the von Koch formula (see [6]), we can write

$$d^R_{Q_m}(\lambda) = 1 + \sum_{n=1}^{\infty} \frac{\lambda^n}{n!} Q^n_m(R_n)$$

where, for functions f on $\mathbb{R}^n$,

$$Q^n_m(f) := \sum_{j_1,\ldots,j_n=1}^{m} w_{j_1} \ldots w_{j_n} f(x_{j_1}, \ldots x_{j_n})$$

and $R_n(s_1, \ldots, s_n) := \det\left[R(s_i, s_j)\right]_{i,j=1}^{n}$. Note that the previous series terminates in fact at $n = m$. Nevertheless, the error is given by the exponentially generating function of the quadrature errors for the functions R_n

$$d^R_Q(\lambda) - d(\lambda) = \sum_{n=1}^{\infty} \frac{\lambda^n}{n!} \left[Q^n_m(R_n) - \int_{[0,1]^n} R_n(t_1, \ldots, t_n)\, dt_1 \cdots dt_n\right].$$

So, this method approximates the Fredholm determinant by the determinant of an $m \times m$ matrix applied in (24.4). If the weights are positive (which is always a better choice), its equivalent symmetric variant is

$$d^R_{Q_m}(\lambda) = \det\left[\delta_{ij} + \lambda w_i^{1/2} R(x_i, x_j)\, w_j^{1/2}\right]_{i,j=1}^{m}.$$

Using Gauss–Legendre quadrature rule, the computation cost is of order $O\left(m^3\right)$ and simple codes for Matlab and Mathematica can be found on page in [6]. Also, Theorem 6.1 in [6] shows that if a family Q_m of quadrature rules converges for continuous functions, when m goes to infinity, then the corresponding Nyström-type approximation of the Fredholm determinant converges to $d\left(\lambda\right)$, uniformly for bounded λ. Moreover Theorem 6.2 in [6] shows that if $R \in C^{k-1,1}\left([0,1]^2\right)$, then for each quadrature rule Q of order $v \geq k$ with positive weights there holds the error estimate

$$\left|d^R_{Q_m}\left(\lambda\right) - d^R\left(\lambda\right)\right| \leq c_k 2^k \left(b-a\right) v^{-k} \Phi\left(|z|\left(b-a\right)\|R\|_k\right),$$

where c_k is a constant depending only on k:

$$\|R\|_k = \max_{i+j\leq k}\left\|\partial_1^i \partial_2^j R\right\|_{L^\infty}$$

and

$$\Phi\left(z\right) = \sum_{n=1}^{\infty} \frac{n^{(n+2)/2}}{n!} z^n$$

is an entire function on $\mathbb{C}$.

Figure 24.2 shows the relative error

$$R\left(T\right) = \left|\frac{P\left(0,T\right) - d^R_{Q_{100}}\left(2T\right)}{P\left(0,T\right)}\right|$$

in the classical CIR model as presented in Example 6.2 (with $m = 100$).

Now, we can apply this method to evaluate numerically Fredholm determinants and consequently prices for bonds in the CIR models. With the notation used above, we have the following proposition:

Proposition 6.1. *Assume* $\sigma_t = \mathbf{1}_{\{t\geq 0\}}$, $g(s) = s^\alpha$, *for* $\alpha \in (-1/2, 1/2)$, *let*

$$\tilde{R}\left(u,v\right) = \left[\frac{2(1-u)(1-v)}{2-u-v}\right]^{2\alpha+1} - \frac{1}{2}\left(\frac{|u-v|}{2}\right)^{2\alpha+1} \times\left[B\left(\frac{1}{2}-\alpha,\alpha+1\right) - B^\gamma\left(\frac{1}{2}-\alpha,\alpha+1\right)\right]$$

for $\gamma = \left(\frac{u-v}{2-(u+v)}\right)^2$, *and where* B *and* B^γ *are the beta and the incomplete beta functions, respectively. Then, the price of a zero-coupon bond, for the corresponding CIR model, is given by*

$$P\left(0,T\right) = \left[d^R\left(2T\right)\right]^{-d/2} = \left[d^{\tilde{R}}\left(\frac{2T^{2\alpha+2}}{1+2\alpha}\right)\right]^{-d/2} \approx \left[d^{\tilde{R}}_{Q_m}\left(\frac{2T^{2\alpha+2}}{1+2\alpha}\right)\right]^{-d/2}.$$

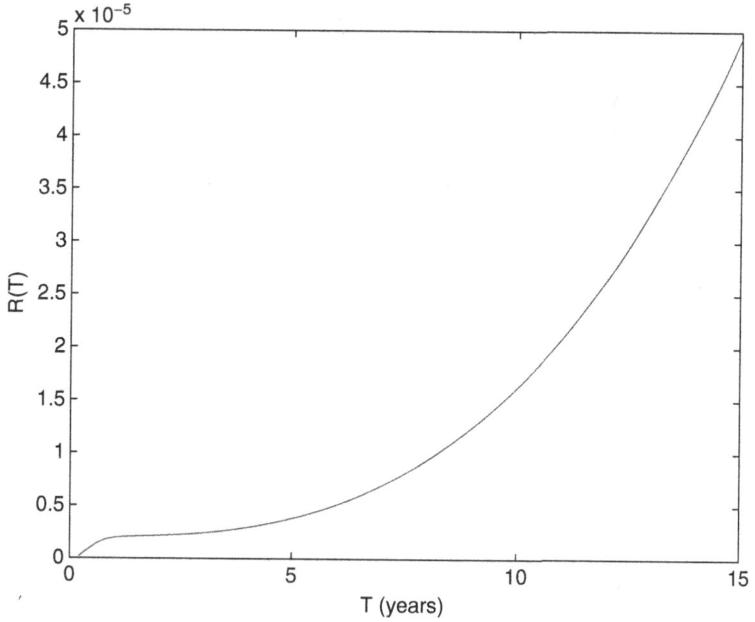

Fig. 24.2 R(T), Relative error, classical CIR model, d = 2, sigma = 0.2, m = 100

Proof. Assume that $0 \leq v \leq u \leq 1$, then

$$
\begin{aligned}
c_2(Tu, Tv; Tu, T) &= \int_{Tu}^{T} g(s - Tu) g(s - Tv) \mathrm{d}s = \int_{Tu}^{T} (s - Tu)^{\alpha} (s - Tv)^{\alpha} \, \mathrm{d}s \\
&= T^{2\alpha+1} \int_{u}^{1} (s - u)^{\alpha} (s - v)^{\alpha} \, \mathrm{d}s = T^{2\alpha+1} c_2(u, v; u, 1).
\end{aligned}
\tag{24.5}
$$

Now, for $u \neq v$, we have

$$
\int_u^1 (s-u)^{\alpha} (s-v)^{\alpha} \, \mathrm{d}s = \left(\frac{u-v}{2}\right)^{2\alpha} \int_u^1 \left[\left(s \frac{2}{u-v} - \frac{u+v}{u-v} \right)^2 - 1 \right]^{\alpha} \mathrm{d}s,
$$

and we obtain

$$
c_2(u, v; u, 1) = \left(\frac{u-v}{2}\right)^{2\alpha+1} \int_1^b \left(x^2 - 1\right)^{\alpha} \mathrm{d}x,
$$

where $b = \frac{2-(u+v)}{u-v}$. Now, by writing $1/b^2 = \gamma$, we have

$$\int_1^b \left(x^2-1\right)^\alpha \mathrm{d}x$$

$$= \frac{1}{2}\int_\gamma^1 (1-x)^\alpha x^{-\frac{3}{2}-\alpha}\mathrm{d}x = \int_\gamma^1 (1-x)^{\alpha+1} x^{-\frac{3}{2}-\alpha}\mathrm{d}x$$

$$+ \int_\gamma^1 (1-x)^\alpha x^{-\frac{1}{2}-\alpha}\mathrm{d}x$$

$$= \frac{1}{2}\left\{\left[(1-x)^{\alpha+1}\frac{x^{-\frac{1}{2}-\alpha}}{-\frac{1}{2}-\alpha}\right]_\gamma^1 - \frac{1}{1+2\alpha}\int_\gamma^1 (1-x)^\alpha x^{-\frac{1}{2}-\alpha}\mathrm{d}x\right\}$$

$$= \frac{1}{1+2\alpha}\left\{-\left[(1-x)^{\alpha+1} x^{-\frac{1}{2}-\alpha}\right]_\gamma^1 - \int_\gamma^1 (1-x)^\alpha x^{-\frac{1}{2}-\alpha}\mathrm{d}x\right\}$$

$$= \frac{1}{1+2\alpha}\left\{2\gamma^{-\frac{1}{2}-\alpha}(1-\gamma)^{\alpha+1} - \int_\gamma^1 (1-x)^{\alpha+1-1} x^{\left(\frac{1}{2}-\alpha\right)-1}\mathrm{d}x\right\}.$$

Then, since $\alpha \in \left(-\frac{1}{2},\frac{1}{2}\right)$, $\frac{1}{2}-\alpha > 0$, and $\alpha+1 > 0$, and we can write

$$B\left(\frac{1}{2}-\alpha,\alpha+1\right) = \int_0^1 (1-x)^{\alpha+1-1} x^{\left(\frac{1}{2}-\alpha\right)-1}\mathrm{d}x,$$

where $B(\cdot,\cdot)$ is the beta function. If we denote the incomplete beta function by $B^z(\cdot,\cdot)$

$$B^z(\alpha,\beta) = \int_0^z x^{\alpha-1}(1-x)^{\beta-1}\,\mathrm{d}x, \qquad \alpha,\beta > 0,$$

we can also write, for $v < u \le 1$,

$$c_2(u,v;u,1)$$

$$= \frac{1}{1+2\alpha}\left(\frac{u-v}{2}\right)^{2\alpha+1}$$

$$\times\left\{\gamma^{-\frac{1}{2}-\alpha}(1-\gamma)^{\alpha+1} - \frac{1}{2}\left(B\left(\frac{1}{2}-\alpha,\alpha+1\right) - B^\gamma\left(\frac{1}{2}-\alpha,\alpha+1\right)\right)\right\}$$

$$= \frac{1}{1+2\alpha}\left\{\left(\frac{2(1-u)(1-v)}{2-u-v}\right)^{2\alpha+1}\right.$$

$$\left.-\frac{1}{2}\left(\frac{u-v}{2}\right)^{2\alpha+1}\left[\left(B\left(\frac{1}{2}-\alpha,\alpha+1\right) - B^\gamma\left(\frac{1}{2}-\alpha,\alpha+1\right)\right)\right]\right\}. \tag{24.6}$$

In case of $v = u \leq 1$,

$$c_2(u, u; u, 1) = \int_u^1 (s-u)^{2\alpha} \, ds = \left[\frac{(s-u)^{2\alpha+1}}{2\alpha+1} \right]_u^1 = \frac{(1-u)^{2\alpha+1}}{2\alpha+1}$$

Then, by (24.5) and (24.6), we have

$$R(u, v) = \frac{T^{2\alpha+1}}{1+2\alpha} \left\{ \left[\frac{2(1-u)(1-v)}{2-u-v} \right]^{2\alpha+1} - \frac{1}{2} \left(\frac{u-v}{2} \right)^{2\alpha+1} \right.$$
$$\left. \times \left[\left(B\left(\frac{1}{2} - \alpha, \alpha + 1 \right) - B^{\gamma}\left(\frac{1}{2} - \alpha, \alpha + 1 \right) \right) \right] \right\}.$$

Therefore

$$\begin{aligned} d^R(\lambda) &= \sum_{n=0}^{\infty} \frac{\lambda^n}{n!} \int_0^1 \cdots \int_0^1 \begin{vmatrix} R(s_1, s_1) & \cdots & R(s_1, s_n) \\ \vdots & & \vdots \\ R(s_n, s_1) & \cdots & R(s_n, s_n) \end{vmatrix} ds_1 \cdots ds_n \\ &= \sum_{n=0}^{\infty} \frac{\left(\frac{\lambda T^{2\alpha+1}}{1+2\alpha} \right)^n}{n!} \int_0^1 \cdots \int_0^1 \begin{vmatrix} \tilde{R}(s_1, s_1) & \cdots & \tilde{R}(s_1, s_n) \\ \vdots & & \vdots \\ \tilde{R}(s_n, s_1) & \cdots & \tilde{R}(s_n, s_n) \end{vmatrix} ds_1 \cdots ds_n \\ &= d^{\tilde{R}} \left(\frac{\lambda T^{2\alpha+1}}{1+2\alpha} \right), \end{aligned}$$

and the price is given by

$$P(0, T) = \left[d^R(2T) \right]^{-d/2} = \left[d^{\tilde{R}} \left(\frac{2T^{2\alpha+2}}{1+2\alpha} \right) \right]^{-d/2} \approx \left[d^{\tilde{R}}_{Q_m} \left(\frac{2T^{2\alpha+2}}{1+2\alpha} \right) \right]^{-d/2}.$$

Remark 6.1. In order to include the case of the volatility not being constant, one only has to substitute $\sigma_{Tu}\sigma_{Tv}R(u, v)$ for $R(u, v)$ or $\sigma_{Tu}\sigma_{Tv}\tilde{R}(u, v)$ for $\tilde{R}(u, v)$.

Remark 6.2. The incomplete beta function ratio defined by

$$I_x(\alpha, \beta) = \frac{1}{B(\alpha, \beta)} \int_0^x t^{\alpha-1} (1-t)^{\beta-1}$$

can be obtained by using the function `betainc(x,`α, β`)` in matlab, so we can compute $B^b(\alpha, \beta)$ easily.

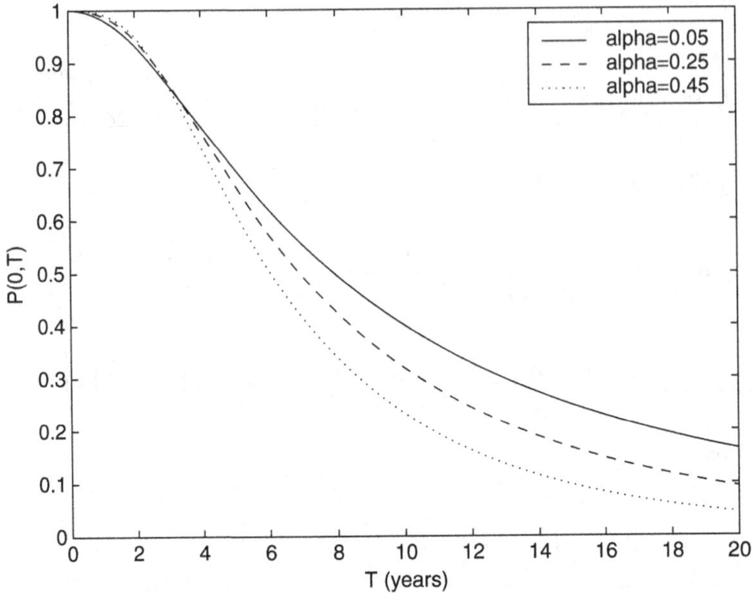

Fig. 24.3 Approximation of prices, d = 2, sigma = 0.2, alpha > 0

Figures 24.3 and 24.4 show the approximated price $P(0,T)$ under the circumstances of Proposition 6.1, for $T \in (0,20)$ in years, $d = 2$, $\sigma = 0.2$, and $\alpha \in \{-0.45, -0.25, -0.05, 0.05, 0.25, 0.45\}$.

6.3 *The Dynamics of the CIR Model*

A natural question, as we did in Sect. 4, is if the process

$$r_t = \sum_{i=1}^{d} \left(\int_0^t g(t-s)\sigma_s \mathrm{d}W_i(s) \right)^2$$

can be seen as the solution of certain SDE. Write

$$Y_i(t) := \int_0^t g(t-s)\sigma_s \mathrm{d}W_i(s),$$

then

$$r_t = \sum_{i=1}^{d} Y_i^2(t).$$

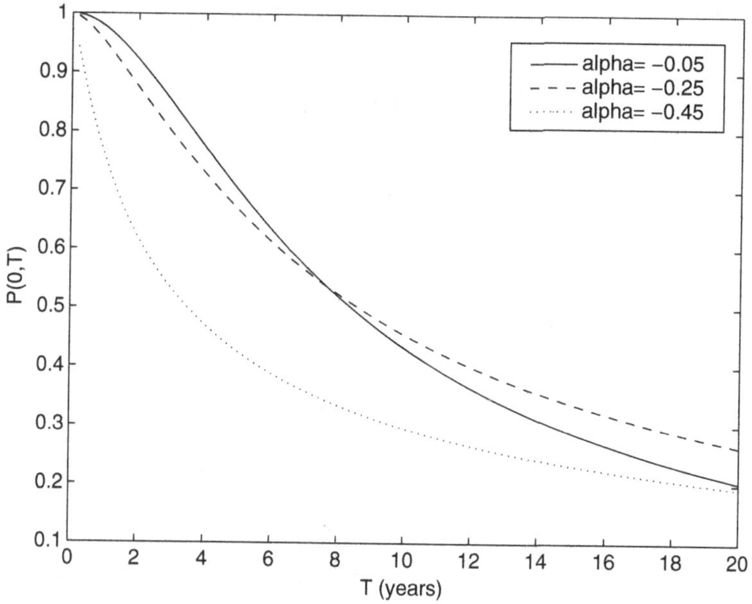

Fig. 24.4 Approximation of prices, d = 2, sigma = 0.2, alpha < 0

Assume that $g \in C^1$and it is square integrable, then Y is a semimartingale with

$$dY_i(t) = g(0)\sigma_t dW_i(t) + \left(\int_0^t g'(t-s)\sigma_s dW_i(s)\right) dt,$$

suppose $g(0) \neq 0$ as well. If we apply the Itô formula for continuous semimartingales we have

$$\begin{aligned} dr_t &= \sum_{i=1}^d 2Y_i(t)dY_i(t) + \sum_{i=1}^d d[Y_i, Y_i]_t \\ &= \sum_{i=1}^d 2g(0)\sigma_t Y_i(t)dW_i(t) + \sum_{i=1}^d 2Y_i(t)\left(\int_0^t g'(t-s)\sigma_s dW_i(s)\right) dt \\ &\quad + \sum_{i=1}^d g^2(0)\sigma_t^2 dt \\ &= 2g(0)\sigma_t\sqrt{r_t}\sum_{i=1}^d \frac{Y_i(t)}{\sqrt{r_t}} dW_i(t) \\ &\quad + \left(dg^2(0)\sigma_t^2 + \sum_{i=1}^d 2Y_i(t)\left(\int_0^t g'(t-s)\sigma_s dW_i(s)\right)\right) dt. \end{aligned}$$

Then it is easy to see, by using the Lévy characterization of the Brownian motion, that

$$\sum_{i=1}^{d} \frac{Y_i(t)}{\sqrt{r_t}} \mathrm{d}W_i(t) = \mathrm{d}B(t),$$

where B is a Brownian motion. Finally if $g'(t) = -bg(t)$, $g(0) = 1$, $\sigma_t = \sigma$, we have

$$\mathrm{d}r_t = (d\sigma^2 - 2br_t)\mathrm{d}t + 2\sigma\sqrt{r_t}\mathrm{d}B(t)$$

that is the dynamics of a CIR process. If g' is not square integrable then the process

$$Y_i(t) := \int_0^t g(t-s)\sigma_s \mathrm{d}W_i(s)$$

is not a semimartingale and we cannot apply the usual Itô formula. In the particular case that

$$g(t-s) = \mathrm{e}^{-b(t-s)} \int_0^{t-s} \mathrm{e}^{bu} \beta u^{\beta-1} \mathrm{d}u, \beta \in (-1/2, 0) \cup (0, 1/2),$$

and $\sigma_u = \sigma$

$$Y_i(t) = \int_0^t \sigma \mathrm{e}^{-b(t-s)} W_i^{\beta}(\mathrm{d}s)$$

$$W_i^{\beta}(t) := \int_0^t (t-s)^{\beta} W(\mathrm{d}s),$$

so

$$Y_i(t) = -b\int_0^t Y_i(s)\mathrm{d}s + \sigma W_i^{\beta}(t)$$

and, by the Itô formula for these processes, we have [1]

$$\mathrm{d}r_t = \sum_{i=1}^{d} 2\sigma Y_i(t) \partial W_i^{\beta}(t) - 2br(t)\mathrm{d}t + \sum_{i=1}^{d} \sigma^2 \left(\int_0^t (t-u)^{\beta} \mathrm{d}u \right) \mathrm{d}t$$

$$= \left(d\sigma^2 t^{2\beta} - 2br(t)\right) \mathrm{d}t + 2\sigma\sqrt{r_t} \sum_{i=1}^{d} \frac{Y_i(t)}{\sqrt{r_t}} \partial W_i^{\beta}(t).$$

But we do not have a characterization of the process

$$Z_t := \sum_{i=1}^{d} \int_0^t \frac{Y_i(s)}{\sqrt{r_s}} \partial W_i^{\beta}(s), t \geq 0.$$

In the case that $b = 0$,

$$Z_t := \sum_{i=1}^{d} \int_0^t \frac{W_i^\beta(s)}{\sqrt{r_s}} \partial W_i^\beta(s), t \geq 0,$$

and it can be shown that Z is 2β-self-similar [9].

Acknowledgements The work of José Manuel Corcuera and Gergely Farkas is supported by the MCI Grant No. MTM2009-08218.

References

1. Alòs, E., Mazet, O., Nualart, D.: Stochastic calculus with respect to Gaussian processes. Ann. Probab. **29**(2), 766–801 (2001)
2. Backus, D.K., Zin, S.E.: Long-memory inflation uncertainty: evidence from the term structure of interest rates. J. Money Credit Banking **25**, 681–700 (1995)
3. Biagini, F., Fink, H., Klüppelberg, C.: A fractional credit model with long range dependent default rate. LMU preprint (2011)
4. Bielecki, T.R., Rutkowski, M.: Credit Risk: Modeling, Valuation and Hedging. Springer, Berlin (2002)
5. Björk, T.: Arbitrage Theory in Continuous Time. Oxford University Press, New York (1998)
6. Bornemann, F.: On the numerical evaluation of the Fredholm determinants (2008). Math. Comp. **79**, 871–915 (2010)
7. Comte, F., Renault, E.: Long memory continuous time models. J. Econom. **73**(1), 101–149 (1996)
8. Fink, H.: Prediction of fractional convoluted Lévy processes with application to credit risk. (2010) submitted for publication
9. Guerra, J.M.E., Nualart, D.: The 1/H-variation of the divergence integral with respect to the fractional Brownian motion for H>1/2 and fractional Bessel processes. Stochastic Process. Appl. **115**(1), 91–115 (2005)
10. Jeanblanc, M., Yor, M., Chesney, M.: Mathematical Methods for Financial Markets. Springer Finance, London (2009)
11. Kailath, T.: Fredholm resolvents, Wiener-Hopf equations, and Riccati differential equations. IEEE Trans. Inf. Theory **IT-15**(6), 665–672 (1969)
12. Lamberton, D., Lapeyre, B.: Introduction to Stochastic Calculus Applied to Finance, 2nd edn. Chapman & Hall, London (2008)
13. Mocioalca, O., Viens, F.: Skorohod integration and stochastic calculus beyond the fractional Brownian scale. J. Funct. Anal. **222**(2), 385–434 (2005)
14. Revuz, D., Yor, M.: Continuous Martingales and Brownian Motion. Springer, Berlin (1999)
15. Varberg, D.E.: Convergence of quadratic forms in independent random variables. Ann. Math. Statist. **37**, 567–576 (1966)

Chapter 25
Parametric Regularity of the Conditional Expectations via the Malliavin Calculus and Applications

A.S. Üstünel

Abstract Let (W, H, μ) be the classical Wiener space and assume that $U_\lambda = I_W + u_\lambda$ is an adapted perturbation of identity where the perturbation u_λ is an H-valued map, defined up to μ-equivalence classes, such that its Lebesgue density $s \to \dot{u}_\lambda(s)$ is almost surely adapted to the canonical filtration of the Wiener space and depending measurably on a real parameter λ. Assuming some regularity for u_λ, its Sobolev derivative and integrability of the divergence of the resolvent operator of its Sobolev derivative, we prove the almost sure and L^p-regularity w.r. to λ of the estimation $E[\dot{u}_\lambda(s)|\mathcal{U}_\lambda(s)]$ and more generally of the conditional expectations of the type $E[F \mid \mathcal{U}_\lambda(s)]$ for nice Wiener functionals, where $(\mathcal{U}_\lambda(s), s \in [0,1])$ is the filtration which is generated by U_λ. These results are applied to prove the invertibility of the adapted perturbations of identity, hence to prove the strong existence and uniqueness of functional SDE, convexity of the entropy and the quadratic estimation error, and finally to the information theory.

Keywords Entropy • Adapted perturbation of identity • Wiener measure • Invertibility

Received 3/28/2011; Accepted 11/22/2011; Final 2/11/2012

A.S. Üstünel (✉)
LTCI CNRS Dépt. Infres, Institut Telecom, Telecom ParisTech,
46, rue Barrault, 75013, Paris, France
e-mail: ustunel@enst.fr

F. Viens et al. (eds.), *Malliavin Calculus and Stochastic Analysis: A Festschrift in Honor of David Nualart*, Springer Proceedings in Mathematics & Statistics 34,
DOI 10.1007/978-1-4614-5906-4_25,

1 Introduction

The Malliavin calculus studies the regularity of the laws of the random variables (functionals) defined on a Wiener space (abstract or classical) with values in finite-dimensional Euclidean spaces (more generally manifolds) using a variational calculus in the direction of the underlying quasi-invariance space, called the Cameron–Martin space. Although its efficiency is globally recognized by now, for the maps taking values in the infinite-dimensional spaces the Malliavin calculus does not apply as easily as in the finite-dimensional case due to the absence of the Lebesgue measure and even the problem itself needs to be defined. For instance, there is a notion called signal-to-noise ratio which finds its roots in engineering which requires regularity of infinite-dimensional objects with respect to finite-dimensional parameters (cf.[1, 7–11]). Let us explain the problem along its general lines briefly: imagine a communication channel of the form $y = \sqrt{\lambda}x + w$, where x denotes the emitted signal and w is a noise which corrupts the communications. The problem of estimation of the signal x from the data generated y is studied since the early beginnings of the electrical engineering. One of the main problems dealt with is the behavior of the L^2-error of the estimation w.r. to the signal-to-noise ratio λ. This requires elementary probability when x and w are independent finite-dimensional variables, though it gives important results for engineers. In particular, it has been recently realized that (cf. [7, 20]), in this linear model with w being Gaussian, the derivative of the mutual information between x and y w.r. to λ equals to the half of the mean quadratic error of estimation. The infinite-dimensional case is more tricky and requires already the techniques of Wiener space analysis and the Malliavin calculus (cf. [20]). The situation is much more complicated in the case where the signal is correlated to the noise; in fact we need the λ-regularity of the conditional expectations w.r. to the filtration generated by y, which is, at first sight, clearly outside the scope of the Malliavin calculus.

In this paper we study the generalization of the problem mentioned above. Namely assume that we are given, in the setting of a classical Wiener space, denoted as (W, H, μ), a signal which is of the form of an adapted perturbation of identity (API):

$$U_\lambda(t, w) = W_t(w) + \int_0^t \dot{u}_\lambda(s, w)ds\,,$$

where $(W_t, t \in [0, 1])$ is the canonical Wiener process, $\dot{u}_\lambda$ is an element of $L^2(ds \times d\mu)$ which is adapted to the Brownian filtration ds-almost surely, and λ is a real parameter. Let $\mathcal{U}_\lambda(t)$ be the sigma algebra generated by $(U_\lambda(s), s \le t)$. What can we say about the regularity, i.e., continuity and/or differentiability w.r.t. λ, of the functionals of the form $\lambda \to E[F \mid \mathcal{U}_\lambda(t)]$ and $\lambda \to E[F \mid U_\lambda = w]$ (the latter denotes the disintegration) given various regularity assumptions about the map $\lambda \to \dot{u}_\lambda$, like differentiability of it or its H-Sobolev derivatives w.r. to λ? We prove that the answer to these questions depend essentially on the behavior of the random resolvent operator $(I_H + \nabla u_\lambda)^{-1}$, where ∇u_λ denotes the Sobolev derivative of u_λ, which is a quasi-nilpotent Hilbert–Schmidt operator; hence its resolvent exists always. More precisely we prove that if the functional

$$(1 + \rho(-\delta u_\lambda)\delta\left((I_H + \nabla u_\lambda)^{-1}\frac{\mathrm{d}}{\mathrm{d}\lambda}u_\lambda\right) \tag{25.1}$$

is in $L^1(d\lambda \times d\mu, [0, M] \times W)$ for some $M > 0$, where δ denotes the Gaussian divergence and $\rho(-\delta u)$ is the Girsanov–Wick exponential corresponding to the stochastic integral $\delta u = \int_0^1 \dot{u}_s \mathrm{d}W_s$, i.e.,

$$\rho(-\delta u) = \exp\left(-\delta u - \frac{1}{2}|u|_H^2\right),$$

then the map $\lambda \to L_\lambda$ is absolutely continuous almost surely where L_λ is the Radon–Nikodym derivative of $U_\lambda\mu$ w.r. to μ and we can calculate its derivative explicitly. This observation follows from some variational calculus and from the Malliavin calculus. The iteration of the hypothesis (25.1) by replacing $\delta((I_H + \nabla u_\lambda)^{-1}\frac{\mathrm{d}}{\mathrm{d}\lambda}u_\lambda)$ with its λ-derivatives permits us to prove the higher-order differentiability of the above conditional expectations w.r. to λ and these results are exposed in Sect. 3. In Sect. 4, we give applications of these results to show the almost sure invertibility of the adapted perturbations of the identity, which is equivalent to the strong existence and uniqueness results of the (functional) stochastic differential equations of the following type:

$$X_t(w) = W_t(w) - \int_0^t \dot{u}(s, (X_r(w), r \le s))\mathrm{d}s,$$

written in a popular manner, where $U = I_W + u$, u is an H-valued functional of the form $u(w)(t) = \int_0^t \dot{u}_s(w)\mathrm{d}s$, and $\dot{u}$ is adapted to the Brownian filtration.

In Sect. 5, we apply the results of Sect. 3 to calculate the derivatives of the relative entropy of $U_\lambda\mu$ w.r. to μ in the general case, i.e., we do not suppose the a.s. invertibility of U_λ, which demands the calculation of the derivatives of the nontrivial conditional expectations. Some results are also given for the derivative of the quadratic error in the case of anticipative estimation as well as the relations to the Monge–Kantorovich measure transportation theory and the Monge–Ampère equation. In Sect. 6, we generalize the celebrated result about the relation between the mutual information and the mean quadratic error (cf. [1,8,9]) in the following way: we suppress the hypothesis of independence between the signal and the noise as well as the almost sure invertibility of the observation for fixed exterior parameter of the signal. With the help of the results of Sect. 3, the calculations of the first- and second-order derivatives of the mutual information w.r. to the ratio parameter λ are also given.

2 Preliminaries and Notation

Let W be the classical Wiener space $C([0, T], \mathbb{R}^n)$ with the Wiener measure μ. The corresponding Cameron–Martin space is denoted by H. Recall that the injection $H \hookrightarrow W$ is compact and its adjoint is the natural injection $W^\star \hookrightarrow H^\star \subset L^2(\mu)$.

Since the image of μ under the mappings $w \to w + h$, $h \in H$ is equivalent to μ, the Gâteaux derivative in the H direction of the random variables is a closable operator on $L^p(\mu)$-spaces and this closure is denoted by ∇ and called the Sobolev derivative (on the Wiener space) (cf., e.g., [12, 13]). The corresponding Sobolev spaces consisting of (the equivalence classes) of real-valued random variables will be denoted as $\mathbb{D}_{p,k}$, where $k \in \mathbb{N}$ is the order of differentiability and $p > 1$ is the order of integrability. If the random variables are with values in some separable Hilbert space, say cc, then we shall define similarly the corresponding Sobolev spaces and they are denoted as $\mathbb{D}_{p,k}(\Phi)$, $p > 1$, $k \in \mathbb{N}$. Since $\nabla : \mathbb{D}_{p,k} \to \mathbb{D}_{p,k-1}(H)$ is a continuous and linear operator its adjoint is a well-defined operator which we represent by δ. A very important feature in the theory is that δ coincides with the Itô integral of the Lebesgue density of the adapted elements of $\mathbb{D}_{p,k}(H)$ (cf.[12, 13]).

For any $t \geq 0$ and measurable $f : W \to \mathbb{R}_+$, we note by

$$P_t f(x) = \int_W f\left(e^{-t}x + \sqrt{1 - e^{-2t}}\,y\right) \mu(dy),$$

it is well-known that $(P_t, t \in \mathbb{R}_+)$ is a hypercontractive semigroup on $L^p(\mu)$, $p>1$, which is called the Ornstein–Uhlenbeck semigroup (cf.[12, 13]). Its infinitesimal generator is denoted by $-\mathcal{L}$ and we call $\mathcal{L}$ the Ornstein–Uhlenbeck operator (sometimes called the number operator by the physicists). The norms defined by

$$\|\phi\|_{p,k} = \|(I + \mathcal{L})^{k/2}\phi\|_{L^p(\mu)} \tag{25.2}$$

are equivalent to the norms defined by the iterates of the Sobolev derivative ∇. This observation permits us to identify the duals of the space $\mathbb{D}_{p,k}(\Phi) : p > 1, k \in \mathbb{N}$ by $\mathbb{D}_{q,-k}(\Phi')$, with $q^{-1} = 1 - p^{-1}$, where the latter space is defined by replacing k in Eq. (25.2) by $-k$, this gives us the distribution spaces on the Wiener space W (in fact we can take as k any real number). An easy calculation shows that, formally, $\delta \circ \nabla = \mathcal{L}$, and this permits us to extend the divergence and the derivative operators to the distributions as linear, continuous operators. In fact $\delta : \mathbb{D}_{q,k}(H \otimes \Phi) \to \mathbb{D}_{q,k-1}(\Phi)$ and $\nabla : \mathbb{D}_{q,k}(\Phi) \to \mathbb{D}_{q,k-1}(H \otimes \Phi)$ continuously, for any $q > 1$ and $k \in \mathbb{R}$, where $H \otimes \Phi$ denotes the completed Hilbert–Schmidt tensor product (cf., for instance, [12,13,17]). We shall denote by $\mathbb{D}(\Phi)$ and $\mathbb{D}'(\Phi)$, respectively, the sets

$$\mathbb{D}(\Phi) = \bigcap_{p>1,k\in\mathbb{N}} \mathbb{D}_{p,k}(\Phi)$$

and

$$\mathbb{D}'(\Phi) = \bigcup_{p>1,k\in\mathbb{N}} \mathbb{D}_{p,-k}(\Phi),$$

where the former is equipped with the projective and the latter is equipped with the inductive limit topologies.

Let us denote by $(W_t, t \in [0,1])$ the coordinate map on W which is the canonical Brownian motion (or Wiener process) under the Wiener measure and let $(\mathcal{F}_t, t \in [0,1])$ be its completed filtration. The elements of $L^2(\mu, H) = \mathbb{D}_{2,0}(H)$ such that $w \to \dot{u}(s,w)$ are ds-a.s. $\mathcal{F}_S$ measurable will be noted as $L^2_a(\mu, H)$ or $\mathbb{D}^a_{2,0}(H)$. $L^0_a(\mu, H)$ is defined similarly (under the convergence in probability). Let $U : W \to W$ be defined as $U = I_W + u$ with some $u \in L^0_a(\mu, H)$, we say that U is μ-almost surely invertible if there exists some $V : W \to W$ such that $V\mu \ll \mu$ and that

$$\mu\{w : U \circ V(w) = V \circ U(w) = w\} = 1.$$

The following results are proved with various extensions in [14–16]:

Theorem 2.1. *Assume that $u \in L^0_a(\mu, H)$ and let L be the Radon–Nikodym density of $U\mu = (I_W + u)\mu$ w.r. to μ, where $U\mu$ denotes the image (push forward) of μ under the map U. Then we have*

$$E[L \log L] \le \frac{1}{2}\|u\|^2_{L^2(\mu,H)} = \frac{1}{2}E\int_0^1 |\dot{u}_s|^2 ds\,.$$

Assume moreover that $E[\rho(-\delta u)] = 1$, then the equality

$$E[L \log L] = \frac{1}{2}\|u\|^2_{L^2(\mu,H)} \tag{25.3}$$

holds if and only if U is almost surely invertible and its inverse can be written as $V = I_W + v$, with $v \in L^0_a(\mu, H)$. In particular, this is equivalent to the fact that the following functional stochastic differential equation:

$$dX_t = -\dot{u}_t \circ X dt + dW_t,\ X_0 = 0$$

has unique strong solution and the solution X is equal to the process defined as $(t,w) \to V(w)(t)$, where V is the almost sure inverse of U.

Finally the condition $E[L \log L - \log L] < \infty$ (without the condition $E[\rho(-\delta u)] = 1$) and the equality (25.3) imply again that U is almost surely invertible and its inverse can be written as $V = I_W + v$, with $v \in L^0_a(\mu, H)$ and hence we also have $E[\rho(-\delta u)] = 1$.

The following result, which is key for the proof of Theorem 2.1, gives the relation between the entropy and the estimation (cf. [14] for the proof):

Theorem 2.2. *Assume that $u \in L^2_a(\mu, H)$ and let L be the Radon–Nikodym density of $U\mu = (I_W + u)\mu$ w.r. to μ, where $U\mu$ denotes the image (push forward) of μ under the map U and let $(\mathcal{U}_t, t \in [0,1])$ be the filtration generated by $(t,w) \to U(t,w)$. Assume that $E[\rho(-\delta u)] = 1$. Then the following relations hold true:*

(i)

$$E[L\log L]=\frac{1}{2}E\int_0^1|E[\dot{u}_s\mid \mathcal{U}_s]|^2 ds\,.$$

(ii)

$$L\circ U\;E[\rho(-\delta u)|U]=1$$

μ-almost surely.

3 Basic Results

Let (W,H,μ) be the classical Wiener space, i.e., $W=C_0([0,1],\mathbb{R}^d)$, $H=H^1([0,1],\mathbb{R}^d)$, and μ is the Wiener measure under which the evaluation map at $t\in[0,1]$ is a Brownian motion. Assume that $U_\lambda:W\to W$ is defined as

$$U_\lambda(t,w)=W_t(w)+\int_0^t\dot{u}_\lambda(s,w)ds\,,$$

with $\lambda\in\mathbb{R}$ being a parameter. We assume that $\dot{u}_\lambda\in L^2_a([0,1]\times W,dt\times d\mu)$, where the subscript "$_a$" means that it is adapted to the canonical filtration for almost all $s\in[0,1]$. We denote the primitive of $\dot{u}_\lambda$ by u_λ and assume that $E[\rho(-\delta u_\lambda)]=1$, where ρ denotes the Girsanov exponential:

$$\rho(-\delta u_\lambda)=\exp\left(-\int_0^1\dot{u}_\lambda(s)dW_s-\frac{1}{2}\int_0^1|\dot{u}_\lambda(s)|^2ds\right).$$

We shall assume that the map $\lambda\to\dot{u}_\lambda$ is differentiable as a map in $L^2_a([0,1]\times W,dt\times d\mu)$; we denote its derivative w.r. to λ by $\dot{u}'_\lambda(s)$ or by $\dot{u}'(\lambda,s)$ and its primitive w.r. to s is denoted as $u'_\lambda(t)$.

Theorem 3.1. *Suppose that $\lambda\to u_\lambda\in L^p_{loc}(\mathbb{R},d\lambda;\mathbb{D}_{p,1}(H))$ for some $p\geq 1$, with $E[\rho(-\delta u_\lambda)]=1$ for any $\lambda\geq 0$ and also that*

$$\textbf{(H)}\quad E\int_0^\lambda(1+\rho(-\delta u_\alpha))\left|E[\delta(K_\alpha u'_\alpha)|U_\alpha]\right|^p d\alpha<\infty\,, \tag{25.4}$$

where $K_\alpha=(I_H+\nabla u_\alpha)^{-1}$. Then the map

$$\lambda\to L_\lambda=\frac{dU_\lambda\mu}{d\mu}$$

is absolutely continuous and we have

$$L_\lambda(w)=L_0\exp\int_0^\lambda E\Big[\delta(K_\alpha u'_\alpha)|U_\alpha=w\Big]d\alpha.$$

Proof. Let us note first that the map $(\lambda, w) \to L_\lambda(w)$ is measurable thanks to the Radon–Nikodym theorem. Besides, for any (smooth) cylindrical function f, we have

$$\begin{aligned}\frac{\mathrm{d}}{\mathrm{d}\lambda}E[f\circ U_\lambda] &= E[(\nabla f\circ U_\lambda, u'_\lambda)_H]\\ &= E[((I_H+\nabla u_\lambda)^{-1\star}\nabla(f\circ U_\lambda), u'_\lambda)_H]\\ &= E[(\nabla(f\circ U_\lambda), (I_H+\nabla u_\lambda)^{-1}u'_\lambda)_H]\\ &= E[f\circ U_\lambda\,\delta\{(I_H+\nabla u_\lambda)^{-1}u'_\lambda\}]\\ &= E[f\circ U_\lambda\,E[\delta(K_\lambda u'_\lambda)|U_\lambda]]\\ &= E[f\,E[\delta(K_\lambda u'_\lambda)|U_\lambda = w]L_\lambda],\end{aligned}$$

where $(\cdot,\cdot)_H$ refers to the scalar product in the Cameron–Martin space H. Hence, for any fixed f, we get

$$\frac{\mathrm{d}}{\mathrm{d}\lambda}\langle f, L_\lambda\rangle = \langle f, L_\lambda E[\delta(K_\lambda u'_\lambda)|U_\lambda = w]\rangle,$$

both sides of the above equality are continuous w.r. to λ; hence we get

$$< f, L_\lambda > - < f, L_0 >= \int_0^\lambda < f, L_\alpha E\left[\delta(K_\alpha u'_\alpha)|U_\alpha = w\right] > \mathrm{d}\alpha\,.$$

From the hypothesis, we have

$$E\int_0^\lambda L_\alpha|E[\delta(K_\alpha u'_\alpha)|U_\alpha = w]|\mathrm{d}\alpha = E\int_0^\lambda |E[\delta(K_\alpha u'_\alpha)|U_\alpha]|\,\mathrm{d}\alpha < \infty\,.$$

By the measurability of the disintegrations, the mapping $(\alpha, w) \to E[\delta(K_\alpha u'_\alpha)|U_\alpha = w]$ has a measurable modification; hence the following integral equation holds in the ordinary sense for almost all $w \in W$:

$$L_\lambda = L_0 + \int_0^\lambda L_\alpha E[\delta(K_\alpha u'_\alpha)|U_\alpha = w]\mathrm{d}\alpha,$$

for $\lambda > 0$. Therefore the map $\lambda \to L_\lambda$ is almost surely absolutely continuous w.r. to the Lebesgue measure. To show its representation as an exponential, we need to show that the map $\alpha \to E[\delta(K_\alpha u'_\alpha)|U_\alpha = w]$ is almost surely locally integrable. To achieve this it suffices to observe that

$$\begin{aligned}E\int_0^\lambda |E[\delta(K_\alpha u'_\alpha)|U_\alpha = w]|\mathrm{d}\alpha &= E\int_0^\lambda |E[\delta(K_\alpha u'_\alpha)|U_\alpha = w]|\frac{L_\alpha}{L_\alpha}\mathrm{d}\alpha\\ &= E\int_0^\lambda |E[\delta(K_\alpha u'_\alpha)|U_\alpha]|\frac{1}{L_\alpha\circ U_\alpha}\mathrm{d}\alpha\\ &= E\int_0^\lambda |E[\delta(K_\alpha u'_\alpha)|U_\alpha]|E[\rho(-\delta u_\alpha)|U_\alpha]\mathrm{d}\alpha < \infty\end{aligned}$$

by hypothesis and by Theorem 2.2. Consequently we have the explicit expression for L_λ given as

$$L_\lambda(w) = L_0 \exp \int_0^\lambda E[\delta(K_\alpha u'_\alpha)|U_\alpha = w]\mathrm{d}\alpha. \qquad \square$$

Remark 3.1. An important tool to control the hypothesis of Theorem 3.1 is the inequality of T. Carleman which says that (cf. [2], Corollary XI.6.28)

$$\|\det_2(I_H + A)(I_H + A)^{-1}\| \le \exp \frac{1}{2}\left(\|A\|_2^2 + 1\right),$$

for any Hilbert–Schmidt operator A, where the left-hand side is the operator norm, $\det_2(I_H + A)$ denotes the modified Carleman–Fredholm determinant, and $\|\cdot\|_2$ denotes the Hilbert–Schmidt norm. Let us remark that if A is a quasi-nilpotent operator, i.e., if the spectrum of A consists of zero only, then $\det_2(I_H + A) = 1$; hence in this case the Carleman inequality reads

$$\|(I_H + A)^{-1}\| \le \exp \frac{1}{2}\left(\|A\|_2^2 + 1\right).$$

This case happens when A is equal to the Sobolev derivative of some $u \in \mathbb{D}_{p,1}(H)$ whose drift $\dot{u}$ is adapted to the filtration $(\mathcal{F}_t, t \in [0,1])$.

From now on, for the sake of technical simplicity, we shall assume that u_λ *is essentially bounded uniformly w.r. to* λ.

Proposition 3.1. *Let* $F \in L^p(\mu)$*, then the map* $\lambda \to E[F|U_\lambda = w]$ *is weakly continuous with values in* $L^{p-}(\mu)$.[1]

Proof. First we have

$$\begin{aligned}
\int_W |E[F|U_\lambda = w]|^p \mathrm{d}\mu &= \int_W |E[F|U_\lambda = w]|^p \frac{L_\lambda}{L_\lambda} \mathrm{d}\mu \\
&= \int_W |E[F|U_\lambda]|^p \frac{1}{L_\lambda \circ U_\lambda} \mathrm{d}\mu \\
&= \int_W |E[F|U_\lambda]|^p E[\rho(-\delta u_\lambda)|U_\lambda] \mathrm{d}\mu < \infty,
\end{aligned}$$

hence $E[F|U_\lambda = w] \in L^{p-}(\mu)$ for any $F \in L^p(\mu)$. Besides, for any $f \in C_b(W)$,

$$E[f \circ U_\lambda\, F] = E[f E[F|U_\lambda = w]\, L_\lambda]$$

therefore

[1] $p-$ denotes any $p' < p$ and $q+$ any $q' > q$.

$$|E[f \circ U_\lambda \, F]| \leq \|F\|_p \|f \circ U_\lambda\|_q \leq C_q \|F\|_p \| \|f\|_{q+} \, .$$

This relation, combined with the continuity of $\lambda \to f \circ U_\lambda$, due to the Lusin theorem, in L^q for any $f \in L^{q+}$, implies the weak continuity of the map $\lambda \to [F|U_\lambda = w]\, L_\lambda$ with values in $L^{p-}(\mu)$; since $\lambda \to L_\lambda$ and $\lambda \to (L_\lambda)^{-1}$ are almost surely and strongly continuous in $L^p(\mu)$, the claim follows. □

Theorem 3.2. *Assume that $F \in \mathbb{D}_{p,1}$ for some $p > 1$ and that*

$$E \int_0^\lambda |\delta(FK_\alpha u'_\alpha)| \mathrm{d}\alpha < \infty$$

for any $\lambda > 0$, then $\lambda \to E[F|U_\lambda = w]$ is μ-a.s. absolutely continuous w.r. to the Lebesgue measure $d\lambda$, and the map $\lambda \to E[F|U_\lambda]$ is almost surely and hence L^p-continuous.

Proof. Using the same method as in the proof of Theorem 3.1, we obtain

$$\begin{aligned}\frac{\mathrm{d}}{\mathrm{d}\lambda} E[\theta \circ U_\lambda \, F] &= \frac{\mathrm{d}}{\mathrm{d}\lambda} E[\theta \, E[F|U_\lambda = w]\, L_\lambda] \\ &= E[\theta \, L_\lambda \, E[\delta(F\, K_\lambda u'_\lambda)|U_\lambda = w]]\end{aligned}$$

for any cylindrical function θ. By continuity w.r. to λ, we get

$$\begin{aligned}&E\left[\theta\Big(L_\lambda E[F|U_\lambda = w] - L_0 E[F|U_0 = w]\Big)\right] \\ &\quad = \int_0^\lambda E\left[\theta L_\alpha E[\delta(FK_\alpha u'_\alpha)|U_\alpha = w]\right] \mathrm{d}\alpha \, .\end{aligned}$$

By the hypothesis

$$E \int_0^\lambda |L_\alpha E[\delta(FK_\alpha u'_\alpha)|U_\alpha = w]| \mathrm{d}\alpha < \infty$$

and since θ is an arbitrary cylindrical function, we obtain the identity

$$L_\lambda E[F|U_\lambda = w] - L_0 E[F|U_0 = w] = \int_0^\lambda L_\alpha \, E[\delta(FK_\alpha u'_\alpha)|U_\alpha = w] \mathrm{d}\alpha$$

almost surely and this proves the first part of the theorem since $\lambda \to L_\lambda$ is already absolutely continuous and strictly positive. For the second part, we denote $E[F|U_\lambda]$ by $\hat{F}(\lambda)$ and we assume that $(\lambda_n, n \geq 1)$ tends to some, λ, then there exists a subsequence $(\hat{F}(\lambda_{k_l}), l \geq 1)$ which converges weakly to some limit; but, from the first part of the proof, we know that $(E[F|U_{\lambda_{k_l}} = w], l \geq 1)$ converges almost surely to $E[F|U_\lambda = w]$ and by the uniform integrability, there is also strong convergence in $L^{p-}(\mu)$. Hence, for any cylindrical function G, we have

$$\begin{aligned} E[\hat{F}(\lambda_{k_l})\,G] &= E[E[F|U_{\lambda_{k_l}} = w]E[G|U_{\lambda_{k_l}} = w]L_{\lambda_{k_l}}] \\ &\to E[E[F|U_\lambda = w]E[G|U_\lambda = w]L_\lambda] \\ &= E[\hat{F}(\lambda)\,G]\,. \end{aligned}$$

Consequently, the map $\lambda \to \hat{F}(\lambda)$ is weakly continuous in L^p; therefore it is also strongly continuous. □

Remark. Another proof consists of remarking that

$$E[F|U_\lambda = w]|_{w=U_\lambda} = E[F|U_\lambda]$$

μ-a.s. and that $\lambda \to E[F|U_\lambda = w]$ is continuous a.s. and in L^{p-} from the first part of the proof and that $(L_\lambda, \lambda \in [a,b])$ is uniformly integrable. These observations, combined with the Lusin theorem, imply the continuity in $L^0(\mu)$ (i.e., in probability) of $\lambda \to E[F|U_\lambda]$ and the L^p-continuity follows.

We shall need some technical results; to begin with, let U_λ^τ denote the shift defined on W by

$$U_\lambda^\tau(w) = w + \int_0^{\cdot\wedge\tau} \dot{u}_\lambda(s)\mathrm{d}s,$$

for $\tau \in [0, 1]$. We shall denote by $L_\lambda(\tau)$ the Radon–Nikodym density

$$\frac{\mathrm{d}U_\lambda^\tau\mu}{\mathrm{d}\mu} = L_\lambda(\tau).$$

Lemma 3.1. *We have the relation*

$$L_\lambda(\tau) = E[L_\lambda|\mathcal{F}_\tau]$$

almost surely.

Proof. Let f be an $\mathcal{F}_\tau$-measurable, positive, cylindrical function; then it is straightforward to see that $f \circ U_\lambda = f \circ U_\lambda^\tau$; hence

$$E[f\,L_\lambda] = E[f \circ U_\lambda] = E[f \circ U_\lambda^\tau] = E[f\,L_\lambda(\tau)].$$

□

Lemma 3.2. *Let $\mathcal{U}_\lambda^\tau(t)$ be the sigma algebra generated by $\{U_\lambda^\tau(s);\ s \le t\}$. Then, we have*

$$E[f|\mathcal{U}_\lambda^\tau(1)] = E[f|U_\lambda^\tau]$$

for any positive, measurable function on W.

Proof. Here, of course the second conditional expectation is to be understood w.r. to the sigma algebra generated by the mapping U_λ^τ and once this point is fixed the claim is trivial. □

Proposition 3.2. *With the notations explained above, we have*

$$L_\lambda(\tau) = L_0(\tau) \exp \int_0^\lambda E[\delta\{(I_H + \nabla u_\alpha^\tau)^{-1} u'^\tau_\alpha\} | U_\alpha^\tau = w] d\alpha.$$

Moreover, the map $(\lambda, \tau) \to L_\lambda(\tau)$ is continuous on $\mathbb{R} \times [0, 1]$ with values in $L^p(\mu)$ for any $p \geq 1$.

Proof. The first claim can be proved as we have done in the first part of the proof of Theorem 3.1. For the second part, let f be a positive, measurable function on W; we have

$$E[f \circ U_\lambda^\tau] = E[f \, L_\lambda(\tau)].$$

If $(\tau_n, \lambda_n) \to (\tau, \lambda)$, from the Lusin theorem and the uniform integrability of the densities $(L_{\lambda_n}(\tau_n), n \geq 1)$, the sequence $(f \circ U_{\lambda_n}^{\tau_n}, \, n \geq 1)$ converges in probability to $f \circ U_\lambda^\tau$; hence, again by the uniform integrability, for any $q > 1$ and $f \in L^q(\mu)$,

$$\lim_n E[f \, L_{\lambda_n}(\tau_n)] = E[f \, L_\lambda(\tau)].$$

From Lemma 3.1, we have

$$\begin{aligned} E[L_{\lambda_n}(\tau_n)^2] &= E[L_{\lambda_n}(\tau_n) \, E[L_{\lambda_n} | \mathcal{F}_{\tau_n}]] \\ &= E[L_{\lambda_n}(\tau_n) \, L_{\lambda_n}], \end{aligned}$$

since, from Theorem 3.1, $L_{\lambda_n} \to L_\lambda$ strongly in all L^p-spaces, it follows that $(\lambda, \tau) \to L_\lambda(\tau)$ is L^2-continuous, hence also L^p-continuous for any $p > 1$. □

Proposition 3.3. *The mapping $(\lambda, \tau) \to L_\lambda(\tau)$ is a.s. continuous; moreover the map*

$$(\tau, w) \to (\lambda \to L_\lambda(\tau, w))$$

is a $C(\mathbb{R})$-valued continuous martingale and its restriction to compact intervals (of λ) is uniformly integrable.

Proof. Let us take the interval $\lambda \in [0, T]$, from Lemma 3.1 we have $L_\lambda(\tau) = E[L_\lambda | \mathcal{F} \circ \tau]$, since $C([0, T])$ is a separable Banach space and since we are working with the completed Brownian filtration, the latter equality implies an a.s. continuous, $C([0, T])$-valued uniformly integrable martingale. □

Theorem 3.3. *Assume that*

$$E \int_0^\lambda \int_0^1 \left(|\delta(\dot{u}_\alpha(s) K_\alpha u'_\alpha)| + |\dot{u}'_\alpha(s)|^2 \right) ds < \infty$$

for any $\lambda \geq 0$, *then the map*

$$\lambda \to E[\dot{u}_\lambda(t)|\mathcal{U}_\lambda(t)]$$

is continuous with values in $L_a^p(\mu, L^2([0,1], \mathbb{R}^d))$, $p \geq 1$.

Proof. Let $\xi \in L_a^\infty(\mu, H)$ be smooth and cylindrical, then, by similar calculations as in the proof of Theorem 3.2, we get

$$\begin{aligned}\frac{\mathrm{d}}{\mathrm{d}\lambda} E[(\xi \circ U_\lambda, u_\lambda)_H] &= \frac{\mathrm{d}}{\mathrm{d}\lambda} < \xi \circ U_\lambda, u_\lambda >= \frac{\mathrm{d}}{\mathrm{d}\lambda} < \xi \circ U_\lambda, \hat{u}_\lambda > \\ &= E \int_0^1 \dot{\xi}_s L_\lambda(s) E\left[\delta(\dot{u}_\lambda(s) K_\lambda u'_\lambda) + \dot{u}'_\lambda(s) | U_\lambda^s = w\right] \mathrm{d}s\,,\end{aligned}$$

but the l.h.s. is equal to

$$E[(\nabla \xi \circ U_\lambda[u'_\lambda], u_\lambda)_H + (\xi \circ U_\lambda, u'_\lambda)_H],$$

which is continuous w.r. to λ provided that ξ is smooth and that $\lambda \to (u'_\lambda, u_\lambda)$ is continuous in L^p for $p \geq 2$. Consequently, we have the relation

$$\begin{aligned}&< \xi \circ U_\lambda, u_\lambda > - < \xi \circ U_0, u_0 > \\ &= E \int_0^\lambda \int_0^1 \dot{\xi}_s L_\alpha(s) E\left[\delta(\dot{u}_\alpha(s) K_\alpha u'_\alpha) + \dot{u}'_\alpha(s) | U_\alpha^s = w\right] \mathrm{d}s \mathrm{d}\alpha\end{aligned}$$

and the hypothesis implies that $\lambda \to L_\lambda(s) E[\dot{u}_\lambda(s)|U_\lambda^s = w]$ is μ-a.s. absolutely continuous w.r. to the Lebesgue measure $d\lambda$. Since $\lambda \to L_\lambda(s)$ is also a.s. absolutely continuous, it follows that $\lambda \to E[\dot{u}_\lambda(s)|U_\lambda^s = w]$ is a.s. absolutely continuous. Let us denote this disintegration as the kernel $N_\lambda(w, \dot{u}_\lambda(s))$, then

$$N_\lambda(U_\lambda^s(w), \dot{u}_\lambda(s)) = E[\dot{u}_\lambda(s)|U_\lambda^s]$$

a.s. From the Lusin theorem, it follows that the map $\lambda \to N_\lambda(U_\lambda^s, \dot{u}_\lambda(s))$ is continuous with values in $L_a^0(\mu, L^2([0,1], \mathbb{R}^d))$ and the L^p-continuity follows from the dominated convergence theorem. □

Remark 3.2. In the proof above we have the following result: assume that $\lambda \to f_\lambda$ is continuous in $L^0(\mu)$, then $\lambda \to f_\lambda \circ U_\lambda$ is also continuous in $L^0(\mu)$ provided that the family

$$\left\{\frac{\mathrm{d}U_\lambda \mu}{\mathrm{d}\mu}, \lambda \in [a,b]\right\}$$

is uniformly integrable for any compact interval $[a,b]$. To see this, it suffices to verify the sequential continuity; hence assume that $\lambda_n \to \lambda$, then we have

$$\mu\{|f_{\lambda_n} \circ U_{\lambda_n} - f_\lambda \circ U_\lambda| > c\} \le \mu\{|f_{\lambda_n} \circ U_{\lambda_n} - f_\lambda \circ U_{\lambda_n}| > c/2\} + \mu\{|f_\lambda \circ U_{\lambda_n} - f_\lambda \circ U_{\lambda_n}| > c/2\},$$

but

$$\mu\{|f_{\lambda_n} \circ U_{\lambda_n} - f_\lambda \circ U_{\lambda_n}| > c/2\} = E[L_{\lambda_n} 1_{\{|f_{\lambda_n} - f_\lambda| > c/2\}}] \to 0$$

by the uniform integrability of $(L_{\lambda_n}, n \ge 1)$ and the continuity of $\lambda \to f_\lambda$. The second term tends also to zero by the standard use of Lusin theorem and again by the uniform integrability of $(L_{\lambda_n}, n \ge 1)$.

Corollary 3.1. *The map $\lambda \to E[\rho(-\delta u_\lambda)|U_\lambda]$ is continuous as an $L^p(\mu)$-valued map for any $p \ge 1$.*

Proof. We know that

$$E[[\rho(-\delta u_\lambda)|U_\lambda] = \frac{1}{L_\lambda \circ U_\lambda}.$$

□

Corollary 3.2. *Let $Z_\lambda(t)$ be the innovation process associated to U_λ, then*

$$\lambda \to \int_0^1 E[\dot{u}_\lambda(s)|\mathcal{U}_\lambda(s)]\mathrm{d}Z_\lambda(s)$$

is continuous as an $L^p(\mu)$-valued map for any $p \ge 1$.

Proof. We have

$$\log L_\lambda \circ U_\lambda = \int_0^1 E[\dot{u}_\lambda(s)|\mathcal{U}_\lambda(s)]\mathrm{d}Z_\lambda(s) + \frac{1}{2}\int_0^1 |E[\dot{u}_\lambda(s)|\mathcal{U}_\lambda(s)]|^2 \mathrm{d}s,$$

since the l.h.s. of this equality and the second term at the right are continuous, the first term at the right should be also continuous. □

Theorem 3.4. *Assume that*

$$E\int_0^\lambda |\delta\{\delta(K_\alpha u'_\alpha)K_\alpha u'_\alpha - K_\alpha \nabla u'_\alpha K_\alpha u'_\alpha + K_\alpha u''_\alpha\}|\mathrm{d}\alpha < \infty$$

for any $\lambda \ge 0$. Then the map

$$\lambda \to \frac{\mathrm{d}}{\mathrm{d}\lambda} L_\lambda$$

is a.s. absolutely continuous w.r. to the Lebesgue measure $d\lambda$ and we have

$$\frac{\mathrm{d}^2}{\mathrm{d}\lambda^2} L_\lambda(w) = L_\lambda E[\delta D_\lambda | U_\lambda = w],$$

where

$$D_\lambda = \delta(K_\lambda u'_\lambda) K_\lambda u'_\lambda - K_\lambda \nabla u'_\lambda K_\lambda u'_\lambda + K_\lambda u''_\lambda \,.$$

Proof. Let f be a smooth function on W; using the integration by parts formula as before, we get

$$\begin{aligned}
\frac{\mathrm{d}^2}{\mathrm{d}\lambda^2} E[f \circ U_\lambda] &= \frac{\mathrm{d}}{\mathrm{d}\lambda} E[f \circ U_\lambda \,\delta(K_\lambda u'_\lambda)] \\
&= E[(\nabla f \circ U_\lambda, u'_\lambda)_H \delta(K_\lambda u'_\lambda)] \\
&= E[(K_\lambda^\star \nabla (f \circ U_\lambda), u'_\lambda)_H \delta(K_\lambda u'_\lambda) \\
&\quad + f \circ U_\lambda \delta(-K_\lambda \nabla u'_\lambda K_\lambda u'_\lambda + K_\lambda u''_\lambda)] \\
&= E\left[f \circ U_\lambda \left\{\delta(\delta(K_\lambda u'_\lambda) K_\lambda u'_\lambda) - \delta(K_\lambda \nabla u'_\lambda K_\lambda u'_\lambda) + \delta(K_\lambda u''_\lambda)\right\}\right] .
\end{aligned}$$

Let us define the map D_λ as

$$D_\lambda = \delta(K_\lambda u'_\lambda) K_\lambda u'_\lambda - K_\lambda \nabla u'_\lambda K_\lambda u'_\lambda + K_\lambda u''_\lambda \,.$$

We have obtained then the following relation:

$$\frac{\mathrm{d}^2}{\mathrm{d}\lambda^2} E[f \circ U_\lambda] = E[f \, L_\lambda \, E[\delta D_\lambda | U_\lambda = w]],$$

hence

$$< \frac{\mathrm{d}}{\mathrm{d}\lambda} L_\lambda, f > - < \frac{\mathrm{d}}{\mathrm{d}\lambda} L_\lambda, f > |_{\lambda=0} = \int_0^\lambda E[f \, L_\alpha \, E[\delta D_\alpha | U_\alpha = w]] \mathrm{d}\alpha \,.$$

The hypothesis implies the existence of the strong (Bochner) integral and we conclude that

$$L'_\lambda - L'_0 = \int_0^\lambda L_\alpha E[\delta D_\alpha | U_\alpha = w] \mathrm{d}\alpha$$

a.s. for any λ, where L'_λ denotes the derivative of L_λ w.r.t. λ. □

Theorem 3.5. *Define the sequence of functionals inductively as*

$$\begin{aligned}
D_\lambda^{(1)} &= D_\lambda \\
D_\lambda^{(2)} &= (\delta D_\lambda^{(1)}) K_\lambda u'_\lambda + \frac{\mathrm{d}}{\mathrm{d}\lambda} D_\lambda^{(1)} \\
&\dots \\
D_\lambda^{(n)} &= (\delta D_\lambda^{(n-1)}) K_\lambda u'_\lambda + \frac{\mathrm{d}}{\mathrm{d}\lambda} D_\lambda^{(n-1)} \,.
\end{aligned}$$

Assume that

$$E\int_0^\lambda |\delta D_\alpha^{(n)}|\mathrm{d}\alpha < \infty$$

for any $n \geq 1$ and $\lambda \in \mathbb{R}$, then $\lambda \to L_\lambda$ is almost surely a C^∞-map and denoting by $L_\lambda^{(n)}$ its derivative of order $n \geq 1$, we have

$$L_\lambda^{(n+1)}(w) - L_0^{(n+1)}(w) = \int_0^\lambda L_\alpha E[\delta D_\alpha^{(n)}|U_\alpha = w]\mathrm{d}\alpha\,.$$

4 Applications to the Invertibility of Adapted Perturbations of Identity

Let $u \in L_a^2(\mu, H)$, i.e., the space of square integrable, H-valued functionals whose Lebesgue density, denoted as $\dot{u}(t)$, is adapted to the filtration $(\mathcal{F}_t, t \in [0,1])$ dt-almost surely. A frequently asked question ire the conditions which imply the almost sure invertibility of the API $w \to U(w) = w + u(w)$. The next theorem gives such a condition:

Theorem 4.1. *Assume that $u \in L_a^2(\mu, H)$ with $E[\rho(-\delta u)] = 1$, let u_α be defined as $P_\alpha u$, where $P_\alpha = \mathrm{e}^{-\alpha\mathcal{L}}$ denotes the Ornstein–Uhlenbeck semigroup on the Wiener space. If there exists a λ_0 such that*

$$\begin{aligned} &E\int_0^\lambda E[\rho(-\delta u_\alpha)|U_\lambda]\Big|E[\delta(K_\alpha u'_\alpha)|U_\alpha]\Big|\mathrm{d}\alpha \\ &= E\int_0^\lambda E[\rho(-\delta u_\alpha)|U_\lambda]\Big|E[\delta((I_H + \nabla u_\alpha)^{-1}\mathcal{L}u_\alpha)|U_\alpha]\Big|\mathrm{d}\alpha < \infty \end{aligned}$$

for $\lambda \leq \lambda_0$, then U is almost surely invertible. In particular the functional stochastic differential equation

$$\begin{aligned} \mathrm{d}V_t(w) &= -\dot{u}(V_s(w), s \leq t)\mathrm{d}t + \mathrm{d}W_t \\ V_0 &= 0 \end{aligned}$$

has a unique strong solution.

Proof. Since u_α is an $H - C^\infty$-function (cf. [17]) the API $U_\alpha = I_W + u_\alpha$ is a.s. invertible (cf.[18], Corollary 1). By the hypothesis and from Lemma 2 of [18], $(\rho(-\delta u_\alpha), \alpha \leq \lambda_0)$ is uniformly integrable. Let L_α and L be, respectively, the Radon–Nikodym derivatives of $U_\alpha\mu$ and $U\mu$ w.r. to μ. From Theorem 3.1,

$$L_\lambda(w) = L(w)\exp\int_0^\lambda E[\delta(K_\alpha u'_\alpha)|U_\alpha = w]\mathrm{d}\alpha$$

for any $\lambda \le \lambda_0$ and also that $\int_0^\lambda |E[\delta(K_\alpha u'_\alpha)|U_\alpha = w]|d\alpha < \infty$ almost surely. Consequently

$$L_\lambda - L = \left(\exp \int_0^\lambda E[\delta(K_\alpha u'_\alpha|U_\alpha = w]d\alpha - 1\right) L \to 0$$

as $\lambda \to 0$, in probability (even in L^1). We claim that the set $(L_\alpha \log L_\alpha, \alpha \le \lambda_0)$ is uniformly integrable. To see this let $A \in \mathcal{F}$, then

$$\begin{aligned} E[1_A L_\alpha \log L_\alpha] &= E[1_A \circ U_\alpha \, \log L \circ U_\alpha] \\ &= -E[1_A \circ U_\alpha \, \log E[\rho(-\delta u_\alpha)|U_\alpha]] \\ &\le -E[1_A \circ U_\alpha \, \log \rho(-\delta u_\alpha)] \\ &= E\left[1_A \circ U_\alpha \left(\delta u_\alpha + \frac{1}{2}|u_\alpha|_H^2\right)\right]. \end{aligned}$$

Since $(|u_\alpha|^2, \alpha \le \lambda_0)$ is uniformly integrable, for any given $\varepsilon > 0$, there exists some $\gamma > 0$, such that $\sup_\alpha E[1_B|u_\alpha|^2] \le \varepsilon$ as soon as $\mu(B) \le \gamma$ and this happens uniformly w.r. to B, but as $(L_\alpha, \alpha \le \lambda_0)$ is uniformly integrable, there exists a $\gamma_1 > 0$ such that, for any $A \in \mathcal{F}$, with $\mu(A) \le \gamma_1$, we have $\mu(U_\alpha^{-1}(A)) \le \gamma$ uniformly in α and we obtain $E[1_A \circ U_\alpha |u_\alpha|_H^2] \le \varepsilon$ with such a choice of A. For the first term above we have

$$E[1_A \circ U_\alpha \delta u_\alpha] \le E[1_A L_\alpha]^{1/2} \|u_\alpha\|_{L^2(\mu,H)} \le \varepsilon$$

again by the same reasons. Hence we can conclude that

$$\lim_{\alpha \to 0} E[L_\alpha \log L_\alpha] = E[L \log L].$$

Moreover, as shown in [14, 15], the invertibility of U_α is equivalent to

$$E[L_\alpha \log L_\alpha] = \frac{1}{2} E[|u_\alpha|_H^2] \to \frac{1}{2} E[|u|_H^2],$$

therefore

$$E[L \log L] = \frac{1}{2} E[|u|_H^2]$$

which is a necessary and sufficient condition for the invertibility of U. □

In several applications we encounter a situation as follows: assume that $u : W \to H$ is a measurable map with the following property:

$$|u(w+h) - u(w)|_H \le c|h|_H$$

a.s., for any $h \in H$, where $0 < c < 1$ is a fixed constant, or equivalently an upper bound like $\|\nabla u\|_{op} \leq c$ where $\|\cdot\|_{op}$ denotes the operator norm. Combined with some exponential integrability of the Hilbert-Schmidt norm ∇u, one can prove the invertibility of $U = I_W + u$ (cf. Chap. 3 of [17]). Note that the hypothesis $c < 1$ is indispensable because of the fixed-point techniques used to construct the inverse of U. However, using the techniques developed in this paper we can relax this rigidity of the theory:

Theorem 4.2. *Let $U_\lambda = I_W + \lambda u$ be an API with $u \in \mathbb{D}_{p,1}(H) \cap L^2(\mu, H)$, such that, for any $\lambda < 1$, U_λ is a.s. invertible. Assume that*

$$E \int_0^1 \rho(-\delta(\alpha u)) |E[\delta((I_H + \alpha \nabla u)^{-1} u) | U_\alpha]| \mathrm{d}\alpha < \infty . \tag{25.5}$$

Then $U = U_1$ is also a.s. invertible.

Proof. Let $L = L_1$ be the Radon–Nikodym derivative of $U_1 \mu$ w.r. to μ. It suffices to show that

$$E[L \log L] = \frac{1}{2} E[|u|_H^2]$$

which is an equivalent condition to the a.s. invertibility of U, cf. [15]. For this it suffices to show first that $(L_\lambda, \lambda < 1)$ converges in $L^0(\mu)$ to L then that $(L_\lambda \log L_\lambda, \lambda < 1)$ is uniformly integrable. The first claim follows from the hypothesis (25.5) and the second claim can be proved exactly as in the proof of Theorem 4.1. □

5 Variational Applications to Entropy and Estimation

In the estimation and information theories, one often encounters the problem of estimating the signal u_λ from the observation data generated by U_λ and then verifies the various properties of the mean square error w.r. to the signal-to-noise ratio, which is represented in our case with the parameter λ. Since we know that [15]

$$E[L_\lambda \log L_\lambda] = \frac{1}{2} E \int_0^1 |E[\dot{u}_\lambda(s) | \mathcal{U}_\lambda(s)]|^2 \mathrm{d}s ,$$

the behavior of the mean square error is completely characterized by that of the relative entropy. Let θ denote the entropy of L_λ as a function of λ:

$$\theta(\lambda) = E[L_\lambda \log L_\lambda] .$$

From our results, it comes immediately that

$$\begin{aligned}\frac{\mathrm{d}\theta(\lambda)}{\mathrm{d}\lambda} &= E[L'_\lambda \log L_\lambda] \\ &= E[L_\lambda\, E[\delta(K_\lambda u'_\lambda)|U_\lambda = w] \log L_\lambda] \\ &= E[E[\delta(K_\lambda u'_\lambda)|U_\lambda] \log L_\lambda \circ U_\lambda] \\ &= -E[\delta(K_\lambda u'_\lambda) \log E[\rho(-\delta u_\lambda)|U_\lambda]]\,.\end{aligned}$$

Similarly

$$\begin{aligned}\frac{\mathrm{d}^2\theta(\lambda)}{\mathrm{d}\lambda^2} &= E\left[L''_\lambda \log L_\lambda + (L'_\lambda)^2 \frac{1}{L_\lambda}\right] \\ &= E[L''_\lambda \log L_\lambda + L_\lambda\, E[\delta(K_\lambda u'_\lambda)|U_\lambda = w]^2] \\ &= E[E[\delta D_\lambda|U_\lambda = w] L_\lambda \log L_\lambda + +L_\lambda\, E[\delta(K_\lambda u'_\lambda)|U_\lambda = w]^2] \\ &= E[E[\delta D_\lambda|U_\lambda] \log L_\lambda \circ U_\lambda + E[\delta(K_\lambda u'_\lambda)|U_\lambda]^2]\,.\end{aligned}$$

In particular we have

Theorem 5.1. *Assume that*

$$E\left[E[\delta D_\lambda|U_\lambda]\left(\int_0^1 E[\dot{u}_\lambda(s)|\mathcal{U}_\lambda(s)]\mathrm{d}Z_\lambda(s) + \frac{1}{2}\int_0^1 |E[\dot{u}_\lambda(s)|\mathcal{U}_\lambda(s)]|^2 \mathrm{d}s\right)\right]$$
$$< E\left[E[\delta(K_\lambda u'_\lambda)|U_\lambda]^2\right]$$

for some $\lambda = \lambda_0 > 0$, *then there exists an* $\varepsilon > 0$ *such that the* **entropy is convex** *as a function of* λ *on the interval* $(\lambda_0 - \varepsilon, \lambda_0 + \varepsilon)$. *In particular, if* $u_0 = 0$, *then the same conclusion holds true on some* $(0, \varepsilon)$.

5.1 Applications to the Anticipative Estimation

In this section we study briefly the estimation of $\dot{u}_\lambda(t)$ with respect to the final filtration $\mathcal{U}_\lambda(1) = \sigma(U_\lambda)$.

Theorem 5.2. *Assume that*

$$E\int_0^\lambda L_\alpha |E[\dot{u}'_\alpha(s) + \delta(\dot{u}_\alpha(s) K_\alpha u'_\alpha)|U_\alpha]|^p \mathrm{d}\alpha < \infty\,,$$

for a $p \geq 1$, *then,* dt*-a.s., the map* $\lambda \to L_\lambda E[\dot{u}_\lambda(t)|U_\lambda = x]$ *and hence the map* $\lambda \to E[\dot{u}_\lambda(t)|U_\lambda = x]$ *are strongly differentiable in* $L^p(\mu)$ *for any* $p \geq 1$ *and we have*

$$\begin{aligned}\frac{\mathrm{d}}{\mathrm{d}\lambda} E[\dot{u}_\lambda(t)|U_\lambda = x] &= E[\dot{u}'_\lambda(t) + \delta(\dot{u}_\lambda(t) K_\lambda u'_\lambda)|U_\lambda = x] \\ &\quad - E[\dot{u}_\lambda(t)|U_\lambda = x] E[\delta(K_\lambda u'_\lambda)|U_\lambda = x]\end{aligned}$$

$d\mu \times dt$*-a.s.*

Proof. For a smooth function h on W, we have

$$\begin{aligned}
&\frac{\mathrm{d}}{\mathrm{d}\lambda} < E[\dot{u}_\lambda(t)|U_\lambda = x], h\, L_\lambda > \\
&= \frac{\mathrm{d}}{\mathrm{d}\lambda} < E[\dot{u}_\lambda(t)|U_\lambda], h \circ U_\lambda > \\
&= E[\dot{u}'_\lambda(t) h \circ U_\lambda + \dot{u}_\lambda(t)(\nabla g \circ U_\lambda, u'_\lambda)_H] \\
&= E[E[\dot{u}'_\lambda(t)|U_\lambda] h \circ U_\lambda + h \circ U_\lambda \delta(\dot{u}_\lambda(t) K_\lambda u'_\lambda)] \\
&= E\left[h L_\lambda(x) \left(E[\dot{u}'_\lambda(t)|U_\lambda = x] + E[\delta(\dot{u}_\lambda(t) K_\lambda u'_\lambda)|U_\lambda = x]\right)\right].
\end{aligned}$$

The hypothesis implies that this weak derivative is in fact a strong one in $L^p(\mu)$; the formula follows by dividing both sides by L_λ and by the explicit form of L_λ given in Theorem 3.1. □

Using the formula of Theorem 5.2, we can study the behavior of the error of noncausal estimation of u_λ (denoted as NCE in the sequel) defined as

$$\begin{aligned}
NCE &= E \int_0^1 |\dot{u}_\lambda(s) - E[\dot{u}_\lambda(s)|\mathcal{U}_\lambda(1)]|^2 \mathrm{d}s \\
&= E \int_0^1 |\dot{u}_\lambda(s) - E[\dot{u}_\lambda(s)|U_\lambda]|^2 \mathrm{d}s.
\end{aligned}$$

To do this we prove some technical results:

Lemma 5.1. *Assume that*

$$E \int_0^\lambda \int_0^1 |\dot{u}''_\alpha(s) + \delta(\dot{u}'_\alpha(s) K_\alpha u'_\alpha)|^p \mathrm{d}s\mathrm{d}\alpha < \infty \tag{25.6}$$

for some $p > 1$, for any $\lambda > 0$, then the map

$$\lambda \to L_\lambda E[\dot{u}'_\lambda(s)|U_\lambda = x]$$

is strongly differentiable in $L^p_a(d\mu, L^2([0,1]))$, and its derivative is equal to

$$L_\lambda E[\dot{u}''_\lambda(s) + \delta(\dot{u}'_\lambda(s) K_\lambda u'_\lambda)|U_\lambda = x]$$

$ds \times d\mu$-a.s.

Proof. Let h be a cylindrical function on W, then, using, as before, the integration by parts formula, we get

$$\frac{d}{d\lambda}E[L_\lambda E[\dot{u}'_\lambda(s)|U_\lambda = x]\,h] = \frac{d}{d\lambda}E[\dot{u}'_\lambda(s)\,h\circ U_\lambda]$$
$$= E[\dot{u}''_\lambda(s)\,h\circ U_\lambda + h\circ U_\lambda\,\delta(\dot{u}'_\lambda(s)K_\lambda u'_\lambda)]$$
$$= E\left[h\,L_\lambda\left(E[\dot{u}''_\lambda(s) + \delta(\dot{u}'_\lambda(s)K_\lambda u'_\lambda)|U_\lambda = x]\right)\right].$$

This proves that the weak derivative satisfies the claim, the fact that it coincides with the strong derivative follows from the hypothesis (25.6). □

Let us define the variance of the estimation as

$$\beta(\lambda, s) = E\left[|E[\dot{u}_\lambda(s)|\mathcal{U}_\lambda(1)]|^2\right].$$

We shall calculate the first two derivatives of $\lambda \to \beta(\lambda, s)$ w.r.t. λ in order to observe its variations. Using Lemma 5.1, we have immediately the first derivative as

$$\frac{d}{d\lambda}\beta(\lambda, s) = E\left[E[\dot{u}_\lambda(s)|U_\lambda = x]L_\lambda\left(E[\dot{u}'_\lambda(s) + \delta(\dot{u}_\lambda(s)K_\lambda u'_\lambda)|U_\lambda = x]\right.\right.$$
$$\left.\left.-\frac{1}{2}E[\dot{u}_\lambda(s)|U_\lambda = x]E[\delta(K_\lambda u'_\lambda)|U_\lambda = x]\right)\right]. \tag{25.7}$$

The proof of the following lemma can be done exactly in the same manner as before, namely, by verifying first the weak differentiability using cylindrical functions and then assuring that the hypothesis implies the existence of the strong derivative and it is left to the reader:

Lemma 5.2. *Assume that*

$$E\int_0^\lambda |\delta(\delta(K_\alpha u'_\alpha)K_\alpha u'_\alpha) + \delta(K_\alpha u''_\alpha - K_\alpha \nabla u'_\alpha K_\alpha u'_\alpha)|^p d\alpha < \infty,$$

for some $p \geq 1$. Then the map

$$\lambda \to L_\lambda E[\delta(K_\lambda u'_\lambda)|U_\lambda = x]$$

is strongly differentiable in $L^p(\mu)$ and we have

$$\frac{d}{d\lambda}(L_\lambda E[\delta(K_\lambda u'_\lambda)|U_\lambda = x]) = L_\lambda E\left[\delta(\delta(K_\lambda u'_\lambda)K_\lambda u'_\lambda)|U_\lambda = x\right]$$
$$+L_\lambda E\left[\delta(K_\lambda u''_\lambda - K_\lambda \nabla u'_\lambda K_\lambda u'_\lambda)|U_\lambda = x\right].$$

Combining Lemma 5.1 and Lemma 5.2 and including the action of L_λ, we conclude that

$$\beta''(\lambda) = E\Big[E[\dot{u}''_\lambda + \delta(\dot{u}'_\lambda K_\lambda u'_\lambda)|U_\lambda]E[\dot{u}_\lambda(s)|U_\lambda]\Big]$$
$$+E\Big[E[\dot{u}'_\lambda(s)|U_\lambda]\Big(E[\dot{u}'_\lambda(s) + \delta(\dot{u}_\lambda(s)K_\lambda u'_\lambda)|U_\lambda]$$
$$-E[\dot{u}_\lambda(s)|U_\lambda]E[\delta(K_\lambda u'_\lambda)|U_\lambda]\Big)\Big]$$
$$+E\Big[E[\delta\left\{\dot{u}''_\lambda(s)K_\lambda u'_\lambda - \dot{u}'_\lambda(s)K_\lambda \nabla u'_\lambda K_\lambda u'_\lambda\right\}$$
$$+\delta\left\{\dot{u}_\lambda(s)K_\lambda u''_\lambda + \delta(\dot{u}_\lambda(s)K_\lambda u'\lambda)K_\lambda u'_\lambda\right\}|U_\lambda]E[\dot{u}_\lambda(s)|U_\lambda]\Big]$$
$$+E\Big[E[\delta(\dot{u}_\lambda(s)K_\lambda u'_\lambda)|U_\lambda]\Big(E[\dot{u}'_\lambda(s) + \delta(\dot{u}_\lambda(s)K_\lambda u'_\lambda)|U_\lambda]$$
$$-E[\dot{u}_\lambda(s)|U_\lambda]E[\delta(K_\lambda u'_\lambda]|U_\lambda]\Big)\Big]$$
$$-E\Big[E[\dot{u}_\lambda(s)|U_\lambda]\Big(E[\dot{u}'_\lambda(s) + \delta(\dot{u}_\lambda(s)K_\lambda u'_\lambda)|U_\lambda]$$
$$-E[\dot{u}_\lambda(s)|U_\lambda]E[\delta(K_\lambda u'_\lambda)|U_\lambda]\Big)E[\delta(K_\lambda u'_\lambda)|U_\lambda]\Big]$$
$$-\frac{1}{2}E\Big[E[E[\dot{u}_\lambda(s)U_\lambda]^2\Big\{E[\delta(\delta(K_\lambda u'_\lambda)K_\lambda u'_\lambda + K_\lambda u''_\lambda$$
$$-K_\lambda \nabla u'_\lambda K_\lambda u'_\lambda)|U_\lambda]\Big\}\Big].$$

Assume now that $\lambda \to u_\lambda$ is linear, then a simple calculation shows that

$$\beta''(0) = E[|\dot{u}(s)|^2].$$

Hence the quadratic norm of the noncausal estimation of u, i.e., the function

$$\lambda \to E\int_0^1 |E[\dot{u}_\lambda(s)|\mathcal{U}_\lambda(1)]|^2 ds$$

is convex at some vicinity of $\lambda = 0$.

5.2 *Relations with Monge–Kantorovich Measure Transportation*

Since $L_\lambda \log L_\lambda \in L^1(\mu)$, it follows the existence of $\phi_\lambda \in \mathbb{D}_{2,1}$, which is 1-convex (cf. [3]) such that $(I_W + \nabla\phi_\lambda)\mu = L_\lambda \cdot \mu$ (i.e., the measure with density L_λ) (cf. [4]). From the L^p-continuity of the map $\lambda \to L_\lambda$ and from the dual characterization of the Monge–Kantorovich problem, [19], we deduce the measurability of the transport potential ϕ_λ as a mapping of λ. Moreover there exists a noncausal Girsanov-like density Λ_λ such that

$$\Lambda_\lambda L_\lambda \circ T_\lambda = 1 \tag{25.8}$$

μ-a.s., where Λ_λ can be expressed as

$$\Lambda_\lambda = J(T_\lambda)\exp\left(-\frac{1}{2}|\nabla\phi_\lambda|_H^2\right),$$

where $T_\lambda \to J(T_\lambda)$ is a log-concave, normalized determinant (cf.[5]) with values in $[0, 1]$. Using the relation (25.8), we obtain another expression for the entropy:

$$\begin{aligned} E[L_\lambda \log L_\lambda] &= E[\log L_\lambda \circ T_\lambda] \\ &= -E[\log \Lambda_\lambda] \\ &= E\left[-\log J(T_\lambda) + \frac{1}{2}|\nabla\phi_\lambda|_H^2\right]. \end{aligned}$$

Consequently, we have

$$\begin{aligned} \frac{1}{2}E\int_0^1 |E[\dot{u}_\lambda(s) \mid \mathcal{U}_\lambda(s)]|^2 \mathrm{d}s &= E\left[-\log J(T_\lambda) + \frac{1}{2}|\nabla\phi_\lambda|_H^2\right] \\ &= E\left[-\log J(T_\lambda)\right] + \frac{1}{2}d_H^2(\mu, L_\lambda\cdot\mu), \end{aligned}$$

where $d_H(\mu, L_\lambda \cdot \mu)$ denotes the Wasserstein distance along the Cameron–Martin space between the probability measures μ and $L_\lambda \cdot \mu$. This result gives another explanation for the property remarked in [10] about the independence of the quadratic norm of the estimation from the filtrations with respect to which the causality notion is defined. Let us remark finally that if

$$d_H(\mu, L_\lambda \cdot \mu) = 0,$$

then $L_\lambda = 1$ μ-almost surely; hence $E[\dot{u}_\lambda(s) \mid \mathcal{U}_\lambda(s)] = 0$ $ds \times d\mu$-a.s. Let us note that such a case may happen without having $u_\lambda = 0$ μ-a.s. As an example let us choose an API, say $K_\lambda = I_W + k_\lambda$ which is not almost surely invertible for any $\lambda \in (0, 1]$. Assume that $E[\rho(-\delta k_\lambda)] = 1$ for any λ. We have

$$\frac{\mathrm{d}K_\lambda\mu}{\mathrm{d}\mu} = \rho(-\delta m_\lambda)$$

for some $m_\lambda \in L_a^0(\mu, H)$, define $M_\lambda = I_W + m_\lambda$, then $U_\lambda = M_\lambda \circ K_\lambda$ is a Brownian motion and an API; hence (cf. [16]) it should be equal to its own innovation process and this is equivalent to say that $E[\dot{u}_\lambda(s) \mid \mathcal{U}_\lambda(s)] = 0$ $ds \times d\mu$-a.s.

6 Applications to Information Theory

In this section we give first an extension of the results about the quadratic error in the additive nonlinear Gaussian model which extends the results of [1, 8–10] in the sense that we drop a basic assumption made implicitly or explicitly in these works, namely the conditional form of the signal is not an invertible perturbation of identity. Afterwards we study the variation of this quadratic error with respect to a parameter on whose depends the information channel in a reasonably smooth manner.

Throughout this section we shall suppose the existence of the signal in the following form:

$$U(w,m) = w + u(w,m)$$

where m runs in a measurable space $(M, \mathcal{M})$ governed with a measure ν and independent of the Wiener path w; later on we shall assume that the above signal is also parametrized with a scalar $\lambda \in \mathbb{R}$. We suppose also that, for each fixed m, $w \to U(w,m)$ is an API with $E_\mu[\rho(-\delta u(\cdot,m))] = 1$ and that

$$\int_0^1 \int_{W\times M} |\dot{u}_s(w,m)|^2 \mathrm{d}s\mathrm{d}\nu\mathrm{d}\mu < \infty .$$

In the sequel we shall denote the product measure $\mu \otimes \nu$ by γ and P will represent the image of γ under the map $(w,m) \to (U(w,m),m)$; moreover we shall denote by P_U the first marginal of P.

The following result is known in several different cases (cf. [1, 8–10]), and we give its proof in the most general case:

Theorem 6.1. *Under the assumptions explained above the following relation between the mutual information $I(U,m)$ and the quadratic estimation error holds true:*

$$\begin{aligned} I(U,m) &= \int_{W\times M} \log \frac{\mathrm{d}P}{\mathrm{d}P_U \otimes \mathrm{d}\nu} \mathrm{d}P \\ &= \frac{1}{2} E_\gamma \int_0^1 \left(|E_\mu[\dot{u}_s(w,m)|\mathcal{U}_s(m)]|^2 - |E_\gamma[\dot{u}_s|\mathcal{U}_s]|^2 \right) \mathrm{d}s, \end{aligned}$$

where $(\mathcal{U}_s(m),\ s \in [0,1])$ is the filtration generated by the partial map $w \to U(w,m)$.

Proof. Let us note that the map $(s,w,m) \to E_\mu[f_s|\mathcal{U}_s(m)]$ is measurable for any positive, optional f. To proceed to the proof, remark first that

$$\frac{\mathrm{d}P}{\mathrm{d}P_U \otimes \mathrm{d}\nu} = \frac{\mathrm{d}P}{\mathrm{d}\gamma} \frac{\mathrm{d}\gamma}{\mathrm{d}P_U \otimes \mathrm{d}\nu} \tag{25.9}$$

$$\frac{\mathrm{d}\gamma}{\mathrm{d}P_U \otimes \mathrm{d}\nu} = \frac{\mathrm{d}\mu \otimes \mathrm{d}\nu}{\mathrm{d}P_U \otimes \mathrm{d}\nu} = \left(\frac{\mathrm{d}P_U}{\mathrm{d}\mu} \right)^{-1} \tag{25.10}$$

since $P_U \sim \mu$. Think of $w \to U(w,m)$ as an API on the Wiener space for each fixed $m \in M$. The image of the Wiener measure μ under this map is absolutely continuous w.r. to μ; denote the corresponding density as $L(w,m)$. We have for any positive, measurable function f on $W \times M$

$$\begin{aligned} E_P[f] &= E_\gamma[f \circ U] \\ &= \int_{W\times M} f(U(w,m),m)\mathrm{d}\nu(m)\mathrm{d}\mu(w) \\ &= \int_M E_\mu\left[f \frac{\mathrm{d}U(\cdot,m)\mu}{\mathrm{d}\mu}\right]\mathrm{d}\nu(m) \\ &= E_\gamma[fL]. \end{aligned}$$

Hence $(w,m) \to L(w,m)$ is the Radon–Nikodym density of P w.r. to γ. From [15] we have at once

$$E_\mu[L(\cdot,m)\log L(\cdot,m)] = \frac{1}{2}E_\mu \int_0^1 |E_\mu[\dot{u}_s(\cdot,m)|\mathcal{U}_s(m)]|^2 \mathrm{d}s.$$

Calculation of $\mathrm{d}P_U/\mathrm{d}\mu$ is immediate:

$$\hat{L} = \frac{\mathrm{d}P_U}{\mathrm{d}\mu}(w) = \int_M L(w,m)\mathrm{d}\nu(m).$$

Moreover, from the Girsanov theorem, we have

$$E_\gamma[f \circ U\, \rho(-\delta u(\cdot,m))] = E_\gamma[f]$$

for any $f \in C_b(W)$. Denote by $\mathcal{U}_t$ the sigma algebra generated by $(U_s : s \le t)$ on $W \times M$. It is easy to see that the process $Z = (Z_t, t \in [0,1])$, defined by

$$Z_t = U_t(w,m) - \int_0^t E_\gamma[\dot{u}_s|\mathcal{U}_s]\mathrm{d}s,$$

is a γ-Brownian motion and any $(\mathcal{U}_t,\ t \in [0,1])$- local martingale w.r. to γ can be represented as a stochastic integral w.r. to the innovation process Z (cf. [6]). Let $\hat{\rho}$ denote

$$\hat{\rho} = \exp\left(-\int_0^1 E_\gamma[\dot{u}_s|\mathcal{U}_s]\mathrm{d}Z_s - \frac{1}{2}\int_0^1 |E_\gamma[\dot{u}_s|\mathcal{U}_s]|^2\mathrm{d}s\right). \tag{25.11}$$

Using again the Girsanov theorem we obtain the following equality:

$$E_\gamma\left[f \circ U\hat{\rho}\right] = E_\gamma\left[f \circ U\rho(-\delta u(w,m))\right]$$

for any nice f. This result implies that

$$E_\gamma[\rho(-\delta u)|U] = \hat{\rho}$$

γ-almost surely. Besides, for nice f on W,

$$\begin{aligned} E_{P_U}[f] &= E_\gamma[f\circ U] = E_\gamma[fL] = E_\gamma[f\hat{L}] \\ &= E_\gamma[f\circ U\hat{L}\circ U\,\rho(-\delta u)] \\ &= E_\gamma[f\circ U\hat{L}\circ U\,\hat{\rho}], \end{aligned}$$

which implies that

$$\hat{L}\circ U\,\hat{\rho} = 1$$

γ-almost surely. We have calculated all the necessary ingredients to prove the claimed representation of the mutual information $I(U,m)$:

$$\begin{aligned} I(U,m) &= E_P\left[\log\left(\frac{dP}{d\gamma}\cdot\frac{d\gamma}{dP_U\otimes d\nu}\right)\right] \\ &= E_P\left[\log\frac{dP}{d\gamma}+\log\frac{d\gamma}{dP_U\otimes d\nu}\right] \\ &= E_\gamma\left[\frac{dP}{d\gamma}\log\frac{dP}{d\gamma}\right]-E_P\left[\log\frac{dP_U}{d\mu}\right] \\ &= E_\gamma\left[\frac{dP}{d\gamma}\log\frac{dP}{d\gamma}\right]-E_{P_U}\left[\log\frac{dP_U}{d\mu}\right] \\ &= E_\gamma\left[\frac{dP}{d\gamma}\log\frac{dP}{d\gamma}\right]-E_\mu\left[\frac{dP_U}{d\mu}\log\frac{dP_U}{d\mu}\right] \\ &= E_\gamma[L\log L]-E\gamma\left[\log\frac{dP_U}{d\mu}\circ U\right] \\ &= \frac{1}{2}E_\gamma\int_0^1 |E_\mu[\dot{u}_s(w,m)|\mathcal{U}_s(m)]|^2 ds - E_\gamma[-\log\hat{\rho}] \end{aligned}$$

and inserting the value of $\hat{\rho}$ given by the relation (25.11) completes the proof. □

Remark. The similar results (cf. [1, 9, 10]) in the literature concern the case where the observation $w\to U(w,m)$ is invertible γ-almost surely; consequently the first term is reduced just to the half of the $L^2(\mu,H)$-norm of u (cf. [15]).

The following is a consequence of Bayes' lemma:

Lemma 6.1. *For any positive, measurable function g on $W\times M$, we have*

$$E_\gamma[g|U] = \frac{1}{\hat{L}\circ U}\left(\int_M L(x,m)E_\mu\Big[g \mid U(\cdot,m)=x\Big]d\nu(m)\right)_{x=U}$$

γ-almost surely. In particular

$$E_\gamma[g|U=x] = \frac{1}{\hat{L}(x)}\int_M L(x,m)E_\mu\Big[g \mid U(\cdot,m)=x\Big]d\nu(m)$$

P_U and μ-almost surely.

Proof. Let $f \in C_b(W)$ and let g be a positive, measurable function on $W \times M$. We have

$$\begin{aligned}
E_\gamma[g\, f\circ U] &= \int_M E_\mu[E_\mu[g \mid U(\cdot,m)]\, f\circ U(\cdot,m)]d\nu(m)\\
&= \int_M\int_W L(w,m)\, E_\mu[g \mid U(\cdot,m)=w]\, f(w)d\mu(w)d\nu(m)\\
&= \int_W f(w)\left(\int_M L(w,m)E_\mu[g \mid U(\cdot,m)=w]\,d\nu(m)\right)d\mu\\
&= \int_W \frac{\hat{L}(w)}{\hat{L}(w)} f(w)\left(\int_M L(w,m)E_\mu[g \mid U(\cdot,m)=w]\,d\nu(m)\right)d\mu\\
&= E_\gamma\left[\frac{1}{\hat{L}\circ U} f\circ U\left(\int_M L(w,m)E_\mu[g \mid U(\cdot,m)=w]\,d\nu(m)\right)_{w=U}\right]
\end{aligned}$$

□

From now on we return to the model U_λ parametrized with $\lambda \in \mathbb{R}$ and defined on the product space $W \times M$, namely we assume that

$$U_\lambda(w,m) = w + u_\lambda(w,m)$$

with the same independence hypothesis and the same regularity hypothesis of $\lambda \to u_\lambda$ where the only difference consists of replacement of the measure μ with the measure γ while defining the spaces $\mathbb{D}_{p,k}$.

Lemma 6.2. *Let $\hat{L}_\lambda(w)$ denote the Radon–Nikodym derivative of P_{U_λ} w.r. to μ. We have*

$$\hat{L}_\lambda(w) = \hat{L}_0(w)\exp\int_0^\lambda E_\gamma\Big[\delta(K_\alpha u'_\alpha)|U_\alpha = w\Big]d\alpha$$

μ-almost surely.

Proof. For any nice function f on W, we have

$$\frac{d}{d\lambda}E_\gamma[f\circ U_\lambda] = \frac{d}{d\lambda}E_\gamma[f\, L_\lambda] = \frac{d}{d\lambda}E_\mu[f\,\hat{L}_\lambda].$$

On the other hand

$$
\begin{aligned}
\frac{\mathrm{d}}{\mathrm{d}\lambda} E_\gamma[f \circ U_\lambda] &= E_\gamma[f \circ U_\lambda \delta(K_\lambda u'_\lambda)] \\
&= E_\gamma[f \circ U_\lambda E_\gamma[\delta(K_\lambda u'_\lambda)|U_\lambda]] \\
&= E_\gamma[f L_\lambda(x,m) E_\gamma[\delta(K_\lambda u'_\lambda)|U_\lambda = x]] \\
&= E_\mu[f \hat{L}_\lambda E_\gamma[\delta(K_\lambda u'_\lambda)|U_\lambda = x]].
\end{aligned}
$$

□

Remark. Note that we also have the following representation for $L_\lambda(w,m)$:

$$
L_\lambda(w,m) = L_0(w,m) \exp \int_0^\lambda E_\mu\Big[\delta(K_\alpha u'_\alpha(\cdot,m))|U_\alpha(\cdot,m) = w\Big]\mathrm{d}\alpha
$$

μ-a.s.

Lemma 6.3. *Let $\lambda \to \tau(\lambda)$ be defined as*

$$
\tau(\lambda) = E_\gamma[\hat{L}_\lambda \log \hat{L}_\lambda],
$$

where $\hat{L}_\lambda(w) = \int_M L_\lambda(w,m)\mathrm{d}\nu(m)$ as before. We have

$$
\begin{aligned}
\frac{\mathrm{d}\tau(\lambda)}{\mathrm{d}\lambda} &= E_\gamma\left[E_\gamma[\delta(K_\lambda u'_\lambda)|U_\lambda] \log \hat{L}_\lambda \circ U_\lambda\right] \\
&= E_\gamma\left[E_\gamma[\delta(K_\lambda u'_\lambda)|U_\lambda](-\log \hat{\rho}_\lambda)\right]
\end{aligned}
$$

where $\hat{\rho}_\lambda$ is given by Eq. (25.11) as

$$
\hat{\rho}_\lambda = \exp\left(-\int_0^1 E_\gamma[\dot{u}_\lambda(s)|\mathcal{U}_\lambda(s)]\mathrm{d}Z_\lambda(s) - \frac{1}{2}\int_0^1 |E_\gamma[\dot{u}_\lambda(s)|\mathcal{U}_\lambda(s)]|^2 \mathrm{d}s\right).
$$

Besides, we also have

$$
\frac{\mathrm{d}^2\tau(\lambda)}{\mathrm{d}\lambda^2} = E_\gamma\left[E_\gamma[\delta D_\lambda|U_\lambda](-\log \hat{\rho}_\lambda) + E_\gamma[\delta(K_\lambda u'_\lambda)|U_\lambda]^2\right],
$$

where

$$
D_\lambda = \delta(K_\lambda u'_\lambda) K_\lambda u'_\lambda + \frac{\mathrm{d}}{\mathrm{d}\lambda} K_\lambda u'_\lambda .
$$

Proof. The only thing that we need is the calculation of the second derivative of $\hat{L}_\lambda$: let f be a smooth function on W, then, from Lemma 6.1,

$$\begin{aligned}\frac{d^2}{d\lambda^2}E_\gamma[f\circ U_\lambda] &= \frac{d}{d\lambda}E_\gamma[f\circ U_\lambda\,\delta(K_\lambda u'_\lambda)]\\ &= E_\gamma\left[f\circ U_\gamma\delta\left(\delta(K_\lambda u'_\lambda)K_\lambda u'_\lambda+\frac{d}{d\lambda}(K_\lambda u'_\lambda)\right)\right]\\ &= E_\gamma[f\circ U_\gamma\,\delta D_\lambda]\\ &= E_\gamma[f(x)\,E_\gamma[\delta D_\lambda|U_\lambda=x]\,\hat{L}_\lambda(x)].\end{aligned}$$

□

As an immediate consequence we get

Corollary 6.1. *We have the following relation:*

$$\begin{aligned}\frac{d^2}{d\lambda^2}I(U_\lambda,m) = E_\gamma\Big[&E_\mu[\delta(D_\lambda(\cdot,m))|U_\lambda(m)](-\log E_\mu[\rho(-\delta u_\lambda(\cdot,m))|U_\lambda(m)])\\ &+E_\mu[\delta(K_\lambda u'_\lambda(\cdot,m))|U_\lambda(m)]^2\Big]\\ &-E_\gamma\Big[E_\gamma[\delta(D_\lambda)|U_\lambda](-\log\hat{\rho}_\lambda)+E_\gamma[\delta(K_\lambda u'_\lambda)|U_\lambda]^2\Big].\end{aligned}$$

References

1. Duncan, T.: On the calculation of mutual information. SIAM J. Appl. Math. **19**, 215–220 (1970)
2. Dunford, N., Schwartz, J.T.: Linear Operators, vol. 2. Interscience, New York (1967)
3. Feyel, D., Üstünel, A.S.: The notion of convexity and concavity on Wiener space. J. Funct. Anal. **176**, 400–428 (2000)
4. Feyel, D., Üstünel, A.S.: Monge-Kantorovitch measure transportation and Monge-Ampère equation on Wiener space. Probab. Theor. Relat. Fields **128**(3), 347–385 (2004)
5. Feyel, D., Üstünel, A.S.: Log-concave measures. TWMS J. Pure Appl. Math. **1**(1), 92–105 (2010)
6. Fujisaki, M., Kallianpur, G., Kunita, H.: Stochastic differential equations for the non linear filtering problem. Osaka J. Math. **9**, 19–40 (1972)
7. Guo, D., Shamai, S., Verdú, S.: Mutual information and minimum mean-square error in Gaussian channels. IEEE Trans. Inf. Theory **51**(4), 1261–1282 (2005)
8. Gelfand, I.M., Yaglom, A.M.: Calculation of the amount of information about a random function contained in another such function. Usp. Mat. Nauk **12**, 3–52 (1957) (transl. in Amer. Math. Soc. Transl. **12**, 199–246, 1959)
9. Kadota, T.T., Zakai, M., Ziv, J.: Mutual information of the white Gaussian channel with and without feedback. IEEE Trans. Inf. Theory, **IT-17**(4), 368–371 (1971)
10. Mayer-Wolf, E., Zakai, M.: Some relations between mutual information and estimation error in Wiener space. Ann. Appl. Probab. **7**(3), 1102–1116 (2007)

11. Pinsker, M.S.: Information and Information Stability of Random Variables and Processes. Holden-Day, San Francisco (1964)
12. Üstünel, A.S.: Introduction to analysis on Wiener space. In: Lecture Notes in Math, vol. 1610. Springer, Berlin (1995)
13. Üstünel, A.S.: Analysis on Wiener Space and Applications. http://arxiv.org/abs/1003.1649 (2010)
14. Üstünel, A.S.: A necessary and sufficient condition for the invertibility of adapted perturbations of identity on the Wiener space. C.R. Acad. Sci. Paris, Ser. I **346**, 897–900 (2008)
15. Üstünel, A.S.: Entropy, invertibility and variational calculus of adapted shifts on Wiener space. J. Funct. Anal. **257**(11), 3655–3689 (2009)
16. Üstünel, A.S.: Persistence of invertibility on the Wiener space. COSA, **4**(2), 201–213 (2010)
17. Üstünel, A.S., Zakai, M.: Transformation of Measure on Wiener Space. Springer, Berlin (1999)
18. Üstünel, A.S., Zakai, M.: Sufficient conditions for the invertibility of adapted perturbations of identity on the Wiener space. Probab. Theory Relat. Fields **139**, 207–234 (2007)
19. Villani, C.: Topics in optimal transportation. In: Graduate Series in Math., vol. 58. Amer. Math. Soc., Providence (2003)
20. Zakai, M.: On mutual information, likelihood ratios and estimation error for the additive Gaussian channel. IEEE Trans. Inform. Theory **51**, 3017–3024 (2005)